NEW TRENDS IN SYSTEM RELIABILITY EVALUATION

FUNDAMENTAL STUDIES IN ENGINEERING

Vol. 1 Coelho/Physics of Dielectrics for the Engineer
Vol. 2 Szücs/Similitude and Modelling
Vol. 3 Scharf, Lisieski/Amplitude Distribution Spectrometers
Vol. 4 Mohilla, Ferencz/Chemical Process Dynamics
Vol. 5 Baker et al./Explosion Hazards and Evaluation
Vol. 6 Sobczyk/Stochastic Wave Propagation
Vol. 7 König/Shakedown of Elastic-Plastic Structures
Vol. 8 Čačko, Bílý, Bukoveczky/Random Processes
Vol. 9 Kubáček/Foundations of Estimation Theory
Vol. 10 Catuneanu, Mihalache/Reliability Fundamentals
Vol. 11 Boros/Measurement Evaluation
Vol. 12 Baker et al./Similarity Methods in Engineering Dynamics, Revised Edition
Vol. 13 Baker, Tang/Gas, Dust and Hybrid Explosions
Vol. 14 Piotrowski/Theory of Physical and Technical Measurement
Vol. 15 Misra/Reliability Analysis and Prediction
Vol. 16 Misra/New Trends in System Reliability Evaluation

FUNDAMENTAL STUDIES IN ENGINEERING 16

NEW TRENDS IN SYSTEM RELIABILITY EVALUATION

edited by

Krishna B. MISRA

Reliability Engineering Centre, Indian Institute of Technology, Kharagpur, India

ELSEVIER
Amsterdam - London - New York - Tokyo 1993

ELSEVIER SCIENCE PUBLISHERS B.V.
Sara Burgerhartstraat 25
P.O. Box 211, 1000 AE Amsterdam, The Netherlands

ISBN: 0-444-81660-7

This book is printed on acid-free paper.

Printed in The Netherlands

Acknowledgments

Let me at the outset express my sincere thanks to the contributors of the chapters, who did an excellent job of putting their ideas together in a logical and informative sequence in the form of chapters on a theme devised by me. The contributors deserve a great deal of appreciation to whom I might have appeared unreasonable at times and more demanding but this was done only to get the work expedited. I have no hesitation in saying that the contributors have set an example of an excellent level of cooperation to bring the book in the present form in a record time of less than a year. I consider it my privilege to work with such a cooperative team of contributors who without bothering about limitations of time and resources were determined to achieve the set objectives.

I would like to put on record my sincere thanks for the facilities made available by the Ministry of Human Resources Development, New Delhi and Reliability Engineering Centre, Indian Institute of Technology, Kharagpur. The facilities made available to me by the National Environmental Engineering Research Institute (NEERI), Nagpur during my stay at the institute on deputation, are gratefully acknowledged and appreciated.

I would like to acknowledge the valuable assistance rendered by my wife, Veena Misra and my two daughters Vinita and Kavita during the preparation of my two books with Elsevier. Vinita (who is presently at Tampa, U.S.A.) , especially, assisted me in handling the work connected with the manuscripts of the contributors from USA, for the present book and I would like to affectionately acknowledge her help. I would also like to acknowledge the secretarial support provided by Mr. Dipankar Chakraborti and Mr. Subinoy Biswas of IIT-Kharagpur and subsequently by Mrs. Devaki P. Nayar and Mrs. Bhanumati Swamy of NEERI, Nagpur.

Lastly, I would like to thank the publishers, particularly, Drs. Eefke Smit, Mr. Erik Oosterwijk and their publication department for all the understanding, cooperation in bringing out timely publication of this book.

K.B. Misra

who acted like a mother

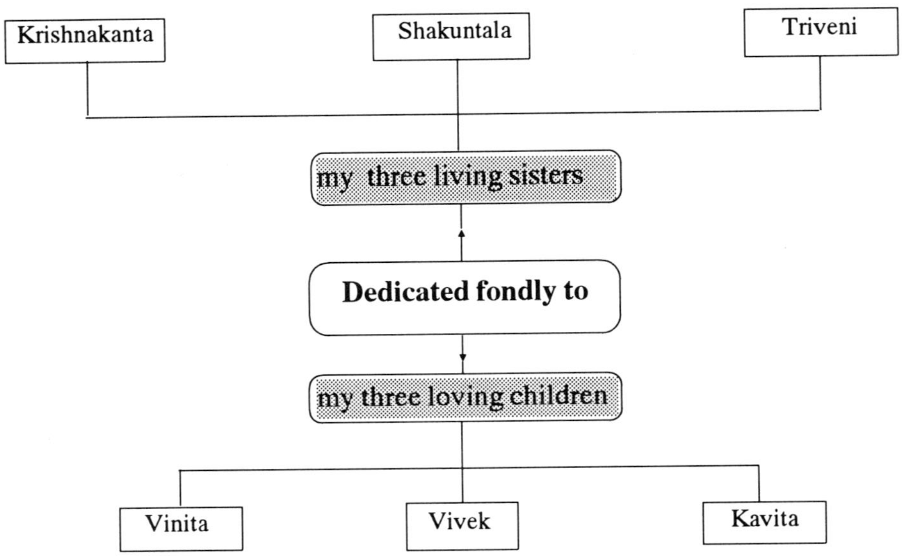

They set *new trends* in our family

FOREWORDS

Last year, Professor Misra, authored a book titled **Reliability Analysis and Prediction**. This text was an outstanding contribution to the field. One reviewer, who himself had written three books on reliability, gave the opinion that this text *"contains the most complete coverage of any book on reliability yet written"*.

This new text from Professor Misra **New Trends in System Reliability Evaluation** extends his earlier work. Professor Misra is the editor of this new text and also an author of four of its seventeen chapters. The remaining authors are well known experts from all around the world.

There are questions raised periodically at reliability symposia and in reliability media as to the pending death of the reliability engineer. CAD/CAE Tools would automate a lot of the routine task now being performed in reliability; the fear is this automation will take away work from reliability engineers.

New Trends in System Reliability Evaluation is poised for the future. It discusses and applies new tools and technology to the reliability profession. It addresses : multiple variable inversion techniques, decomposition and reduction, digraphs and causal trees, Petri nets, fuzzy sets theory, Dempster-Shafer theory, and expert systems to name some. It also brings one up-to-date on the latest advances in some of the key focus areas such as mechanical, human, and software reliability. The oft overlooked area of common cause and dependent failure modeling is elaborated upon. This text is exciting in helping to expand the application domain of the reliability engineer. It helps you feel the opportunity and the analysis power that lies ahead in the field of reliability.

Professor Misra has also provided an extensive and categorized listing of references. This text truly builds on a comprehensive and scholarly basis. At the same time, the topics are developed in a manner such that they can be readily applied by a reliability practitioner.

This text opens up new concepts and frontiers to explore in reliability and will be a catalyst in raising the profession to a new level. This book should be seen as a trend setter. It will be a most useful reference and guide to a practitioner. It intertwines the disciplines of decision sciences with the practice of reliability. It is expected that this book, along with Professor Misra's state-of-art text published last year, (**Reliability Analysis and Prediction** : *A Methodology Oriented Treatment*, published by Elsevier Science Publishers) will form a complimentary set of reliability texts that every practitioner should have in his/her library.

Samuel J. Keene, Ph.D
President, IEEE Reliability Society 1991-92
IBM Corporation
Boulder, CO 80301, USA

May 13, 1993
U.S.A.

There has been a phenomenal growth of literature in the field of **Reliability Engineering** during the post war period and the old methodologies are rapidly being replaced by newer and the efficient ones. Therefore, it is necessary to take stock, occasionally, of this changing scenario and assign priorities for the future. Professor K.B. Misra, National Environmental Engineering Research Institute (NEERI), Nagpur, has brought together a number of contributions from well known workers in this field in the form of this volume, which is suitably titled **New Trends in System Reliability Evaluation**. The aim of this book is to identify directions in which efforts are being made or are likely to be made and those that are expected to revolutionize the art and science of system reliability assessment. The present volume presents an insight into the futuristic development and provides an authoritative coverage of a number of techniques, models and alternative approaches.

Prof. Misra deserves our appreciation for having brought out a book, which is bound to be a useful guide to all those interested in the science of system reliability evaluation. I am sure this volume will provide students, researchers and engineers interested in system reliability with a mine of useful information and knowledge.

S.K. JOSHI

Director-General, Council of Scientific & Industrial Research,
Secretary, Department of Scientific & Industrial Research *and*
President, Indian National Science Academy.

New Delhi
April 30, 1993

PREFACE

As a student of reliability for nearly three decades, the editor has witnessed reliability literature grow both-in quantity and in quality. In 1960, there were hardly any books on reliability and the only journals available on reliability were IEEE Transactions on Reliability and Microelectronics & Reliability, which were started during the sixties. Very soon, by 1970, the number of pages of quarterly IEEE Transactions on Reliability rose to 202 in addition to 500 pages of Microelectronics and Reliability. Today (in 1992) there are more than 200 books on reliability and IEEE Transactions on Reliability has over 650 pages earmarked annually besides the voluminous (505 pages) proceedings of Reliability and Maintainability Annual symposium. In addition to this, we have 1800 pages of Microelectronics & Reliability and there exist at least three more international journals on reliability engineering, which were all started during 1980s.

All these help generate and disseminate knowledge at a very fast rate, today. There has been a sea change in the quality of state-of-art in the subject since 1970 and naturally in this plethora of information, it becomes difficult to see distinctly the direction for future growth, particularly looking to the size of the literature available today. Therefore, it becomes necessary for us to identify distinct directions for our efforts in search of better and efficient methodologies. The present book is an attempt to accomplish this task. The book besides identifying the shortcomings of the current state-of-art in the subject, outlines the trends, which will lay the foundation of the future of reliability engineering. Naturally, it has not been an easy task for the editor to select the topics for the book, however the dialogue with several experts in the area and the session on *Reliability Engineering in Future* held during the European Conference on Safety and Reliability, in Denmark in 1992 helped crystallize the thinking of the editor on the subject and the results are presented in the form of this book.

Apart from the introductory section, which outlines the future trends in system reliability evaluation, the material has been broadly organized in six distinct sections. In the first section, the editor has tried to confine the discussion to the sphere of improvement of existing techniques of system reliability evaluation, in terms of computational time and efficiency. The recent concept of using multiple variable inversions in preference to single variable inversion, has been included as the first contribution under this section. Several new algorithms, proposed recently, have been included in this chapter along with their comparisons. The decomposition and transformation techniques have been found to be very helpful and economical in handling large and complex systems. Therefore, they find an appropriate emphasis in this book. Another area which is likely to be proved advantageous in reducing the computation time is the feasibility of parallel processing on the computers. Obviously all these concepts together are likely to improve substantially the art and science of system reliability evaluation.

Since mid seventies, there has been tremendous improvement in the analysis of k-out-of-n systems, which are considered useful for a variety of system models and applications. The same is true about the fault tree analysis, which has become quite indispensable for system reliability studies, today. Consecutive k-out-of-n systems, which came into existence only in 1980s, have also been analysed extensively. It was, therefore, felt necessary to provide a coverage of these improvements under the section on *Towards Improved Analyses.*

The application of graph theory to reliability was introduced during late sixties and early seventies. This provided an analyst an opportunity to employ better tools for reliability analysis.

Recent developments in this area consist of the use of digraphs, causal trees, Petri nets and neural nets and these are likely to prove useful in reliability studies. Therefore, a section on these new tools was considered relevant for this book. A discussion on neural nets could not, however, be included due to certain reasons.

Inspite of considerable progress in reliability evaluation methodologies, there are still areas in a man-machine system performance, which have been always considered difficult if not impossible for modeling. Failure dependency and simultaneous catastrophic failures of subsystems are difficult to model. There are other areas which have come into relevance on account of an increased use of computers and microprocessors, which often execute several tasks based on the software developed. Therefore, not only hardware reliability but software reliability is also important in system reliability evaluation. There are other areas like mechanical systems reliability modeling where not much progress has been made. Therefore, a section which provides a discussion of the difficulties and the newer concepts in modeling in the important and critical areas like human reliability, common cause and dependency failures, software reliability and mechanical systems reliability, has been included in this book. Each of these problems poses a challenge to a reliability analyst and it was felt necessary to discuss this at some length so that future trends could be identified and a concerted effort could be made in future in these directions.

The next section provides a thorough discussion of the new approaches that are being considered as alternatives to the probabilistic approach. The probabilistic approach has not been able to provide a satisfactory answer to the problem of handling uncertainty as there is a considerable amount of uncertainty present in the probabilistic approach which mainly stems from the imprecise data besides the problem of subjectivity and imprecision that is inherent in system reliability evaluation. Fuzzy Sets Theory and Dempster-Shafer Theory have generally been considered appropriate to handle these problems. Therefore, it was considered necessary to include the discussion of these approaches.

Lastly, the task of system reliability evaluation need not be confined to specialists or experts and a non-specialist user should be able to carry out the reliability analysis. This is a new direction, which is likely to further develop into a fully grown area and consists of use of expert systems and the use of special purpose hardware gadget, like a reliability analyzer. Therefore, it was considered appropriate by the editor to include a chapter each on the expert system and the reliability analyser to evoke interest in readers in these promising new areas.

With the discussion of all these important aspects of system reliability evaluation, it is expected that the art and science of system reliability evaluation will get the necessary fillip and will achieve new heights, in the foreseeable future. Even after this, if any important topic is left out, the editor takes the entire blame for the same.

There are three chapters in this book, which have two parts. To distinguish the two parts of the same chapter, the first part has a different script from the other part. Also, at places, different size of fonts, have been used to distinguish a particular section of the book but the same style has been maintained throughout the book. Also the same font size has been maintained for a given section. This has been done primarily to provide clarity and a distinctiveness to a particular aspect or section.

K.B. Misra

CONTENTS

Acknowledgements vi
Forewords vii
Preface ix

I : Future Trends

1 Part I : Reliability Engineering for the Future **1**
Gerard Collas
1.1 Introduction 1
1.2 Product and Process Assurance Management 2
1.3 Areas for Improving the State-of-Art 3
1.4 Real Causes of Unreliability 4
1.5 System Description and Modeling 6
1.6 Conclusions 7
Appendix 9

1 Part II : New Trends in System Reliability Evaluation **11**
Krishna B. Misra
1.1 Introduction 11
1.2 Organization 12
1.3 Towards Improved Techniques 13
1.4 Towards Improved Analyses 16
1.5 Towards Better Tools 18
1.6 Towards Better Models 20
1.7 Towards Alternative Approaches 23
1.8 Towards Development of Non-specialist's Aid 25
1.9 Conclusions 27
References 28

II : Towards Improved Techniques

2 Multiple Variable Inversion Techniques **39**
Malathi Veeraraghavan and Kishor S. Trivedi
2.1 Introduction 39
2.2 Preliminaries 43
2.3 Algorithms 48
2.4 Comparisons 68
2.5 Conclusions 72
References 73

3 Part I : Decomposition and Reduction Techniques **75**
Frank Beichelt
3.1 Introduction 75
3.2 Stochastic Networks 76
3.3 Network Decomposition 78
3.4 Network Reduction 95
3.5 Conclusions 113
References 114

3 Part II: Probabilistic Graph-Reduction Techniques **117**
Andrew M.Shooman
3.1 Introduction 117
3.2 Previous Contributions 121
3.3 Basic Graph Reductions 123
3.4 Advanced Graph Reductions 129
3.5 Approximate Graph Reductions 138
3.6 Graph Reduction Algorithms 142
3.7 Test Problems 153
3.8 Results 158
3.9 Future Developments 162
References 163

4 Parallel Algorithms and Implementations **165**
Narsingh Deo and Muralidhar Medidi
4.1 Introduction 165
4.2 The Butterfly Parallel Machine 166
4.3 Sequential Algorithms : Review 167
4.4 Parallel Algorithms 174
4.5 Performance Comparison 177
4.6 Conclusions 181
References 182

III: Towards Improved Analyses

5 Reliability of k-out-of-n Systems **185**
Ali M. Rushdi
5.1 Introduction 185
5.2 Assumptions, Notátion, and Nomenclature 188
5.3 Systems with Components of Equal Reliabilities 191
5.4 Approximate Methods 193
5.5 Conventional Methods : Inclusion-Exclusion & Disjoint Products 194

5.6	Recursive Techniques	200
5.7	Iterative Algorithms Based on a Generating Function : The BH Algorithms	206
5.8	An Iterative Algorithm Based on Binary Recursion : The AR Algorithm	210
5.9	Comparison of Various Algorithms	216
5.10	New Trends and Developments	220
References		221
Appendix 5.1 to 5.6		221

6 Consecutive-k-out-of-n Systems **228**

S.G. Papastavridis and Markos V. Koutras

6.1	Introduction	228
6.2	Recursive Formulae	234
6.3	Exact Formulae	235
6.4	Limits and Bounds	237
6.5	Optimization Problems	241
6.6	Conclusions	241
References		242

7 Fault Tree Analysis **249**

Hiromitsu Kumamoto

7.1	Introduction	249
7.2	Risk and Decision	250
7.3	Safety Goals	259
7.4	Accident Causation Mechanisms	266
7.5	Risk Management	279
7.6	Automated Fault Tree Synthesis	286
References		310

IV: Towards Better Tools

8 Digraphs and Causal Trees **313**

Takehisa Kohda and Koichi Inoue

8.1	Introduction	313
8.2	Digraph Approach	314
8.3	Causal Tree Approach	318
8.4	Interactive Method for Obtaining Minimal Cut Sets	323
8.5	Conclusions	335
References		335
Appendix		336

9 Use of Petri Nets for System Reliability Evaluation **339**

Gurdeep S. Hura

9.1 Introduction 339
9.2 Modeling 345
9.3 Extension of Petri Nets 347
9.4 Analysis of Petri Nets 352
9.5 Applications in System Reliability Evaluation 354
9.6 Future Trends 364
References 364

V: Towards Better Models

10 Human Reliability Modeling **369**

Joseph Sharit

10.1 Introduction 369
10.2 Quantitative Modeling of Human Error 370
10.3 Qualitative Modeling of Human Reliability 388
10.4 Critical Analysis 401
10.5 Synthesis 403
10.6 Concluding Remarks 407
References 408

11 Common Cause and Dependent Failure Modelling **411**

Per Hokstad

11.1 Introduction 411
11.2 Classification of Dependent Failures 414
11.3 Defences against CCFs 416
11.4 Incorporation of CCFs into Reliability Model 421
11.5 Parametric Models 424
11.6 Parameter Estimation 434
11.7 Failure Events Reports 437
11.8 Conclusions 441
References 441

12 Software Reliability Modelling **445**

Tom Downs

12.1 Introduction 445
12.2 Software Reliability Models 447
12.3 Assessing the Performance of Software Reliability Models 462
12.4 Applications and Performance of Models that Incorporate Testing 468
12.5 Conclusions 477
References 477

13 Mechanical Systems Reliability Modeling **479**

S.S. Rao

13.1 Introduction 479
13.2 Role of Reliability in Mechanical Systems Design 481
13.3 Differences between Reliability Modeling of Electronic and Mechanical Systems 482
13.4 Failure Modes of Mechanical Systems 483
13.5 Modeling of Strength, Load and Geometry 484
13.6 Probability of Failure of Weakest-Link Systems 486
13.7 Probability of Failure of Fail-Safe Systems 490
13.8 Illustrative Examples 492
13.9 Conclusions 501
References 501

VI : Towards Alternative Approaches

14 Use of Fuzzy Sets Theory (Part-I : Theory) **503**

Krishna B. Misra

14.1 Introduction 503
14.2 Set-theoretic Operations on Fuzzy Sets 509
14.3 Fuzzy Sets and Multi-valued Logic 511
14.4 Algebraic Properties of Fuzzy Sets 514
14.5 Extended Operations 517
14.6 Fuzzy Relations 519
14.7 Fuzzy Sets of Type 2 521
14.8 Fuzzy Numbers 525
14.9 Representations of a Fuzzy Number 532
14.10 Fuzzy Operators 537
14.11 Fuzzy Measures 540
14.12 Classification of Events and Probabilities 544
14.13 Possibility Theory 547

14 Use of Fuzzy Sets Theory (Part - II : Applications) **551**

Takehisa Onisawa and Krishna B. Misra

14.14 FST in Performance Evaluation 551
14.15 Reliability Analysis of General System 556
14.16 Fuzzy Probability 559
14.17 Fuzzy Fault Tree Analysis 563
14.18 Dependency Considerations 569
14.19 Environmental Considerations 575
14.20 Common Cause Failures 579
14.21 Conclusions 580
References 583

15 Dempster-Shafer Theory and its Applications **587**
Toshiyuki Inagaki
15.1 Introduction 587
15.2 Dempster-Shafer Theory 588
15.3 Identification and Location of Faults 607
15.4 Safety-Control and Management 613
15.5 Concluding Remarks 622
References 623

VII : Towards Development of Non-Specialist's Aid

16 Expert Systems for Reliability Evaluation **625**
D. J. Russomanno, R. D. Bonnell and J. B. Bowles
16.1 Introduction 625
16.2 Expert System Architecture 626
16.3 Characterization of Knowledge and Reasoning in Expert Systems for Failure Modes and Effects Analysis (FMEA) 629
16.4 Prototype Computer-Aided Reliability Analysis 631
16.5 Second Generation Expert Systems for Reliability Evaluation 636
16.6 A Second Generation Expert System for FMEA 640
16.7 Conclusions 650
References 651

17 Reliability Analyzer **653**
Krishna B. Misra
17.1 Introduction 653
17.2 Basic Model of Reliability Analyzer 656
17.3 Generation of Spanning Trees 664
17.4 Algorithms for Path Sets and Cut Sets 670
17.5 Microprocessor Based Reliability Analyzer 677
17.6 Development of ESCAF : A New Reliability Analyzer 679
17.7 S.ESCAF : A Modified Version 690
17.8 ESCAF Capabilities 692
17.9 Conclusions 700
References 700

About the Editor **701**

About the Contributors **702**

Subject Index **709**

New Trends in System Reliability Evaluation
K.B. Misra (Editor)

CHAPTER 1
(Part-I)

RELIABILITY ENGINEERING FOR THE FUTURE

Gerard Collas
System Reliability Engineering Manager
Bull
Rue Jean Jaures, P.O. Box 53
78340, Les Clayes sous Bois, France

ABSTRACT

At the European Safety and Reliability Conference 1992, a panel discussion was held on *Reliability Engineering for the Future*, where, the word reliability was intended to cover all the aspects of dependability activities. During this session, various aspects of the current state-of-art and the new developments for the future in dependability engineering were thoroughly discussed among the panellists and the attendees. This chapter is based on the summary of discussions held during the session and highlights different aspects of dependability engineering activities, viz., product and process assurance management, problems of unreliability in the field, system description and modeling and finally future areas of possible improvements. The conclusion provides a brief description of different domains of activities of the dependability engineering which must be developed in the coming years.

1.1 Introduction

For past several years, during the international reliability and maintainability conferences held in Europe and in the United States, various aspects of dependability engineering which integrate reliability, maintainability and safety analyses better in the product life cycle have been often discussed. The word *dependability* covers not only the estimation and evaluation of system reliability, availability and maintainability but also all the factors and means which permit a complete system to deliver its service to the users for which it is designed. Today, the system integrates not only hardware (as electronic equipment, electromechanical, and mechanical devices) and software (as firmware and operating system) but also its applications, communication links, environmental (logical and physical) and other activities (design, support, maintenance, customer). All these elements are significant to determine the real dependability of the system offered to the end user.

During the European Safety and Reliability Conference, 1992 held in Copenhagen, the author was requested to conduct a panel discussion on :*Reliability Engineering for the Future*. The list of panellists included specialists representing various aspects of dependability and their addresses and biographies are given in the Appendix of this Chapter. Jane Rise of Ericsson Quality Institute (Sweden) presented the Ericsson Process approach to quality, dependability and safety; Joel Nachlas from Virginia Tech (USA) addressed some theoretical aspects of the reliability engineering and some aspects of practical implementation; Finn Jensen, a reliability consultant (Denmark) made presentation on *Components versus Systems*, an analysis of real causes of unreliability in the field and on reliability efforts to be undertaken in the future; Emmanuel Arbaretier of Sofreten (France), emphasized that a description and simulation of the system in its functional and failure aspects can be the best recourse to incorporate reliability during the early design phase. However, the objective of this panel discussion was to try to find answer to the following question : What will be the focus and trends for reliability activities for the next decade or so? As mentioned earlier, the word reliability here implies all the aspects of dependability engineering.

On the basis of these presentations, subsequent comments, discussions and exchanges of opinions between the conference attendees and panellists and from my own personal opinion, I have tried to present the state-of-art in this chapter and it is envisaged that areas of research in dependability engineering in the future would comprise of : *product and process assurance management*, *improvement of existing knowledge*, *identification of real problems of unreliability* and *advancement of state-of-art in system description and modeling*.

Consequently, the conclusion of this chapter provides an enumeration of various activities concerning dependability engineering which must be strengthened and developed in the years to come.

1.2 Product and Process Assurance Management

We observe that in the U.S. and Western Europe, quality assurance has always been important to, and has been enforced by the government and military organisations. During past 20 years, the economic competition has stimulated broad private sector commitments to quality assurance initiatives. It is also true that in the U.S. and Western Europe, the primary support for reliability engineering has been from the government and the military organisations. Today, we are caught up in a transition in which these organisations are likely to reduce their level of support for the evolution of this discipline in future. Therefore, the question which concerns us all is whether industry will recognise the importance of the discipline and step in to support its continued evolution? In Europe, there is already an evidence that industry will do so. In the U.S., this does not appear to be the case.

The first important point in implementing a dependability programme is the process to develop the product. The process management must be considered as an integral part of the plan to achieve quality and dependability of products and systems. This is very important for the system hardware but for the software, it is the only method to achieve a reliable product. This aspect of management addresses all the phases of product life cycle, in particular, the concept and definition phase and the design and development phase. This type of approach is the one that has been proposed in the standards IEC 300 which covers dependability programmes including reliability, maintainability and maintenance support performance.

Some of the useful activities which improve the intrinsic dependability, during the development phases are: the implementation of technological and manufacturing rules, the use of quality methods as *Quality Function Deployment*, reliability analyses by way of FMECA, determination of enlarged margins as six sigmas method (Motorola) to improve product robustness, design of experiment to verify the parameters sensitivity, technical real time reviews and inspections for specifications and software code, the use of robustness tests as STRIFE, BEST (Hewlett Packard), GUARDBAND (IBM).

An important recent phenomenon is the interest in RAMCAD (as a combination of Reliability, Availability and Maintainability analyses and Computer Aided Design) and in the *Concurrent Engineering* (an integrated design approach, considering all the characteristics and parameters of a product during the whole life cycle from conception to disposal). The idea behind these efforts is to move reliability engineering effort to support equipment design into an automated design process. This is especially a good idea but the implementation unfortunately has not been perfect, in that the reliability analysis is still performed subsequent to prospective design selection. The analysis is rapid because it is automated but it is still not incorporated in the design activity, with the result, it is not really concurrent. To arrive at the really integrated parameters of the product such as performances, design quality and dependability measures, produceability and life cycle costs, it is necessary that the members of product team change their culture so that a programme of information, formation and training can be implemented effectively.

1.3 Areas for Improving the State-of-Art

Undoubtedly the modern reliability engineering is a matured field of activity that is supported by a foundation of solid scientific and technological knowledge. By and large, the techniques currently in use effectively support the design of safe and reliable products and systems. However, there are several important problems that have not yet been successfully resolved.

At the component level, we do not have *accurate models* of transient and intermittent failures. This is mostly due to the fact that we do not have a satisfactory understanding of the physical causes of these failures. These are especially important because they usually occur when an equipment is in use by the customer.

Another significant gap in our understanding is our inability to define *proper general models for component dependencies*. Nearly all reliability analyses assume independence of component failure processes. In case of electronic devices, this is sometimes inappropriate and in the case of mechanical components, the independence assumption is almost always incorrect. As a consequence, many of our reliability models overstate equipment reliability.

Our knowledge of the *influence of environment on aging of components* is much more limited than we generally imply. The Arrhenius reaction rate equation, the inverse power law model, and the Eyring model are all reasonable but do not form a comprehensive body of knowledge. Consequently, component deterioration during idle periods or while in storage is not well understood and is thus not usually represented accurately in reliability models. Further, we often do not evaluate all environmental factors. We do not treat differential (mechanism specific) age acceleration. We have no established models for the effect of low temperature on microelectronic devices, and we have no general theory at all on the *age acceleration of mechanical components*.

Another aspect that limits our understanding is the fact that extensive data bases have been created for electrical and electronic components but these tend to contain many errors. The data bases carry formally defined algorithms for using the data to obtain reliability predictions. The algorithms are often out- of-date and they have never included a scientifically defensible method for extrapolating the tabulated information to new technologies or even new designs for existing technologies.

At the system level, we have not yet developed satisfactory techniques for integrating consideration of software faults into system analyses. Proper design rule have been defined for software and some clever methods for evaluating and reducing software defect levels have been created but none of these have been effectively incorporated into a unified system reliability modeling method.

In the same manner, reasonable warranty analyses have been defined but have not been integrated into reliability analysis routines or into equipment design practices. There is a reasonably widespread recognition of the importance of manufacturing and assembly quality for improving system reliability. Nevertheless, formal methods for reflecting process quality metrics in reliability models and predictions have not been developed. Even in case of integrated circuits yield models, the probability models for yield are not incorporated in reliability analyses.

Many large complex systems are intended to perform several distinct tasks using a subset of their components for each task. Frequently, such systems can be operational for one function while they undergo maintenance on the components that support the other function that do not employ them. The models for reliability and more important for the availability of these systems have not been fully developed.

Our studies of preventive maintenance planning tend to consider a single component or a single system without reflecting the more realistic situation in which the maintenance tasks are organised to include several sets of components that are serviced simultaneously. More important, we tend to consider only age and block replacement policies while actual practice often includes opportunistic repair strategies.

Equipment design people have become more sensitive to reliability issues or at least to the consequences of failure. Consequently, they are creating designs that fail *gracefully*. New types of reliability models are needed to represent this type of equipment behaviour.

1.4 Real Causes of Unreliability

We now insist on the verification and measures of product reliability which is an important area of dependability activity and particularly on field data.

If we look into the real causes of unreliability in the field, particularly, of the electronic and electromechanical systems, we find that a majority of the failures are caused by :

- human factors including operator error and abuse,
- misapplication of the equipment,
- environmental interactions,
- software problems,
- inadequate test coverage,
- poor quality assurance during manufacture,
- faulty installation,
- inadequate maintenance,
- unclear specifications.

Very few reported studies of field failures in electronic systems actually identify component intrinsic reliability as a problem. The majority of the failures could be traced to handling problems or design misapplication of the components.

If the above statements are reasonably representative of the real world situations, does it then mean that we don't have to bother about component reliability? No, far from it! But we have to understand better the real causes of component failures, including especially early, infant mortality failures and long term wearout failures.

The infant mortality failures are always caused by deficiencies during manufacturing, and new steps in process control should be considered to resolve such problems. The term *proactive process control* covers the new approaches in this area. The long-term wearout in typical electronic components is not normally believed to be a problem. However, for systems with a useful life expectancy in the vicinity of 15-20 years, the component wearout can become a problem, especially, with the sub-micron technologies and III-V based products now being installed. Therefore for components, the focus in the future should be on eliminating early failures and studying the long-term wearout. The middle *useful-life* period will probably have a very low hazard rate, often considered as zero (at least as far as the *pure* or *intrinsic* components reliability is concerned).

However, many systems problems listed before will probably dominate the field failure pictures for many years to come. Reliability engineers should, therefore, turn their talents to a system or top-down approach, sadly absent in practically all reliability studies being performed today.

How do we start getting a handle on the real causes of system unreliability? How do we quantify the human factors and all the other problems areas listed earlier? There can of course be several approaches. Most of them would be fairly unsophisticated, but all will require hard work!

If the management agrees that failure of a product in the hands of its customers is a high-priority issue to resolve, it should be able to convince the company that getting to the root causes of the failures is a must. This is much easier said than done. Failure reports from the field tend to be obscure, lacking in data, and often misleading. Detailed failure analyses are seldom performed. We must change the existing attitudes towards field failure reporting. Only in this way - and this is a major step - can we finally hope to identify the causes of unreliability, i.e., are they design related, software related, process related, or possibly even caused by maintenance person dropping a screwdriver into a complex electronic equipment? - a situation that will remain unreported in the official file!

Even such situations can, however, be quantified, of course with a great deal of uncertainty. But this uncertainty can be taken into account in the analysis of system reliability. The Bayesian approach is one way of going about this. There are other approaches as well, like - fuzzy sets theory or Dempster-Shafer's theory, that are being talked about to handle this situation.

The major issue is that professional engineers need to emphasize ever more strongly the need for a system approach to reliability, to accept the absence of the real world of failure causes, and to finally establish a methodology for quantifying these situations and assessing reliability. The tools are there, we need to go out and use them.

1.5 System Description and Modeling

Another important area of dependability activity is the description and simulation of a system by the integration of functional analysis, malfunction or failure analysis, maintenance analysis and logistic support analysis.

For functional analysis, many methodologies and languages can be used (APTE, FAST, RELIASEP, SADT, MERISE) to study for what purpose a system is designed, what is expected of a system and how it normally operates. In a functional simulator, all equations of the components of the system consider them in a nominal state, dynamic simulation does not take into account the effects of degradation of the functional flows and the model is specified for a given set of environmental conditions.

Since the functional description is insufficient, we need a new description language, a malfunction model which takes into account quantitative physics, the simulation of failures, a representation of degraded states of all functions, a description of the logical dependence between these states. In a malfunction simulator, there is a possibility of degrading the mission requirements of a system (level of performance), the degradation of the external environment can be simulated, the failure states of all the components are represented and their effects on the equations of the system are taken into account.

The *malfunctional knowledge* considers two types of failure states:

- hardware failure or inherent failure state for which a repair action is necessary to remove it and is generally given by reliability libraries or data base,
- functional failure or temporary unsatisfactory physical state of a function which can be removed with better inputs to the system or by the introduction of another operation mode; this statement is relative and depends on the definition of the mission. The functional failure states can be described, for qualitative physics, by logical assertion about behaviour of the function such as : *very high*, *stable*, *unpure*, and, for

quantitative description, by discrete or continuous spectrum of possible values for the attributes of the functions such as : X = 5 V, 2 < X < 5.

Functional or malfunctional analyses, referring to the basic knowledge of systems, can be considered as input models. As a complementary of knowledge models, output models permit a system evaluation. These are dependability and mathematical tools such as graphs of states (Markov chains, Petri nets, Queue networks), FMECA, reliability diagrams, digraphs, fault trees and event trees, probability networks. The functional and malfunctional knowledge is necessary for an automated generation of evaluation models : propagation of failures for FMECA, logical dependency between failure states and failure models for fault trees, state function of a system for nominal-degraded missions for reliability diagrams, strategic measurements to obtain a good coverage rate for testability graphs.

As an example, *SOFIA* - a software tool (developed by SOFRETEN) for analysis of system dependability employs this approach. This functional/malfsimulator permits to describe a system in blocks, functions, states of blocks, states of functions, logical dependency rules between functions and logical dependency rules between logical states of blocks and functions. The application modules are *FMECA*, *SAFETY* for generation and evaluation of fault trees, *RELDIAG* for generation and evaluation of reliability diagrams, *SIMTEST* for design and evaluation of test devices, *FAILDIAG* for generation of expert system of failure diagnosis.

1.6 Conclusions

To summarize, in order that a substantial improvement takes place in the state-of-art, based on the panel discussion, the author feels that the following aspects of dependability engineering must be developed in the near future

1.6.1 New Mathematical Models

For system analyses and modeling, a hard work is needed to be done for logical coherency, logical traceability, resolution of logical equations, completeness of the knowledge and an equivalence of models.

The need is the same if we consider a design quality metric to represent the status of a design in a concurrent engineering approach.

The graphs of states, such as Petri nets, need a continuous improvement of their capabilities, efficiency and analytical power.

At the level of failures modeling, the easy problems have been solved and the methods required now are the ones that will make reliability models more realistic. A major constituent of the scientific extensions is the need of a unified theory of failure that captures the behaviour of mechanical as well as electronic devices and reflects the influence of environment on both.

1.6.2 New Methodologies

The system analyst will have to devote all his efforts to the functional and malfunctional description of his system if dependability tools and models are automatically deduced from a central knowledge data base. The architecture work will change to more interactive and iterative in simultaneous evaluations of models. New analytical and evaluation techniques need to be developed to make highly dependable systems and products. Another major point is that both existing and new techniques must permit computer based implementation and must clearly add value to the equipment designs evaluated.

1.6.3 New Approaches

Real efforts must be continued to arrive at a complete integration of dependability analyses in *Concurrent Engineering* approach. If some important steps are taken in electronic hardware and software, it is not so true in mechanical system activities. There is a need of technical enhancement to CAD to make dependability and life cycle costs analyses fully integrated in future design efforts and of integration of reliability analyses methods to include common techniques for hardware, software and systems.

The constitution of real team work, which is completely involved in terms of the technical performance characteristics, production, and support features, can also be an important factor to improve dependability.

Dependability engineering requires both internal and external integration. Analyses such as those for warranty commitments, maintenance planning, and software interactions that were previously treated as ancillary must be folded into coherent reliability evaluation algorithms. At the same time, reliability analysis activity itself must be integrated into the equipment design process so that the reliability is one of the component of a vector of design effectiveness measures.

1.6.4 General Problem of Cost

Even considering all the improvements we need for reliability engineering, we should not forget that it is always very important to explain to management that dependability improvements are not free but that the return on the investment is important (dependability analyses and studies have an impact on the initial cost but the final global cost is lower).

Therefore, a book highlighting the new trends in system reliability evaluation providing the state-of-art and highlighting the new developments in reliability techniques and analyses, some critical modeling areas and new approaches can be a very useful source of information for any person who intends to evaluate system dependability.

Acknowledgements

First of all, I would like to express my gratitude to Kurt Petersen who accepted the proposal to organize this panel discussion during ESRC'92, then to Emmanuel Arbaretier, Finn Jensen, Joel Nachlas, Jane Rise and the attendees of the panel discussion in Copenhagen for their presentations and participation which helped define the areas for future developments in dependability engineering.

Appendix

Panel Chairman

Gerard Collas
Bull
Rue Jean Jaures, P.O. Box 53
78340, Les Clayes sous Bois, FRANCE

Gerard Collas is an Engineering System Reliability Manager in R & D Department of Bull, France. Prior to joining Bull at the end of 1984, he had over twenty years experience in Quality Assurance, Reliability and Statistics for component industries, computer activities and aerospace and military product companies. He holds an Engineer Diploma from Ecole Superieure d'Electricite. He teaches some courses in Reliability Engineering at University of Angers, at Quality Master of complex systems in Grenoble and at "Movement Francais pour la Qualite" in Paris. He is a member of I.E.E.E. and S.R.E. He received the annual Stan Ofsthun Award of the Society of Reliability Engineers in 1991.

Panellists

Jan Rise
Ericsson Quality Institute
HF/LME/Q
S-126 25 Stockholm, SWEDEN

Jan Rise is a senior Quality Management consultant at the Ericson Quality Institute. He is the Convenor of the working group (WG4) on Verification and Evaluation procedures within the Technical Committee (TC56), Dependability of the International Electrotechnical Commission (IEC). Jan Rise has extensive management and consulting experience within Quality, Reliability and Maintainability, gained from earlier positions as a consultant (10 years) serving the defence forces and from work with the electric power and aircraft industry. He received the degree of Teknisk Licentiat in Mathematical Statistics from the Lund Institute of Technology in 1983, a Master of Science in Operations Research from Stanford University 1971 and the degree of Civilingenor (#M.Sc) in Electrical Engineering from the Lund Institute of Technology in 1968. He has a long record of activity from the Society of Reliability Engineers as President of the Scandinavian Chapter (1982-87), International vice President (1987-89) and International President (1990-91).

Joel A. Nachlas
Department of Industrial Engineering and Operations Research
Virginia Polytechnic Institute and State University
Blacksburg, Virginia 24061, U.S.A.

Joel A. Nachlas is an Associate Professor of Industrial Engineering and Operations Research at Virginia Tech., where he has served on the faculty since 1974. He received his BES from the John Hopkins University in 1970, M.S. in 1972 and Ph.D in 1975 both from the University of Pittsburgh. His research interests are in the applications of probability and statistics to problems in reliability and quality control. His recent work addressed problems in acceptance sampling, multivariate quality control, microelectronics reliability, and composite materials testing and reliability. He is the winner of the 1990 P.K. McElroy Award and is currently an associate editor of the IEEE Transactions on Reliability. He is a senior member of the ASQC and IIE and a member of SRE, ORSA, and TIMS.

Finn Jensen
Reliability Consultant
Pile Alle 11, DK-2840 Holte
DENMARK

Finn Jensen is a private reliability consultant and a senior research scientist at the Engineering Academy of Denmark, with whom he recently completed a study of field failure data analysis and a separate study on component life time modeling. Dr. Jensen has taught reliability engineering for industries and universities in Europe, the United States, and the Far East. His clients have included IBM, Apple Computer, Digital Equipment, AT & T, Ericsson, Danfoss, and Motorola. He is the Chief Editor of the International Journal on Quality and Reliability Engineering and is author and co-author of two books on reliability engineering. He received the first annual award of the Society of Reliability Engineers in 1989.

Emmanuel Arbaretier
SOFRETEN
41 rue des trois Fontanot
BP 312-92003 Nanterre Cedex, FRANCE

Emmanuel Arbaretier worked for four years in Thomson-CSF-SDC Radar Company, where he was responsible for the use of American Standards in Dependability and Integrated Logistics Support studies. Then he joined SOFRETEN as an ILS project manager. Now he is in charge of the development of simulation and optimisation software (SOFIA, DIANA and DSELSYS) in the field of Dependability Evaluation and Logistic Support Analysis.

New Trends in System Reliability Evaluation
K.B. Misra (Editor)

CHAPTER 1

(Part-II)

NEW TRENDS IN SYSTEM RELIABILITY EVALUATION

Krishna B. Misra

Director-grade-Scientist,
Disaster Prevention & Impact Minimization Division,
National Environmental Engineering Research Institute,
Nagpur 440 020, India

ABSTRACT

There was a serious discussion on *Reliability Engineering for the Future* during the European Conference on Safety and Reliability held at Copenhagen, in June 1992. The panellists discussed the current state-of-art and critically examined the shortcomings of present models, practices, methodologies and approaches in the context of dependability engineering. The editor's motivation to put these thoughts into a detailed logical and meaningful presentation of involved problems in the form of a book came as a consequence of this event. This stocktaking will guide and channelize the on-going efforts and the future developments in the area of system reliability evaluation.

1.1 Introduction

Although there have been tremendous advances in the art and science of system reliability evaluation during the post war period and new techniques have replaced the old that were once used in 1950s, 1960s or 1970s, yet it is very difficult to assess reliability with a very high accuracy or precision. Basically, this inaccuracy in assessment stems mainly from the inaccuracy of data, lack of exactness of the models and even from the limitations of the current methods themselves. These are some of the problems that are still difficult, if not impossible, to overcome. Besides, the current techniques are not free from certain pitfalls, which unfortunately are inherent on account of the process of system reliability evaluation itself. We must not forget that an accurate prediction or assessment helps in understanding of causes of failures that can eventually be eliminated or minimized.

Notwithstanding the importance of the subject, it was rather surprising that, except for a very recent and comprehensive book [1], there did not appear to be any serious contribution, that highlighted the shortcomings of the current state-of-art and focussed on the on-going efforts aimed at seeking better models, improved solutions and even in the direction of alternative approaches to the problem of system reliability evaluation. These improvements could flow from our capability in handling of data inaccuracies or from building proper models or even by way of improving the current practices.

Although reliability analysts were aware of some of these problems and of their possible solutions but as the methodologies were fast improving, these considerations did not receive adequate attention which otherwise would have helped mature the current practices. In order to make fast and firm progress towards more reliable and durable solutions in a period of a decade or two, it is time, we present a thorough discussion of these aspects and provide them due consideration, they deserve. Notwithstanding the above, reliability analysts are having a relook at the adequacies of the present approaches to system performance evaluation and it would only be proper to have an appraisal of the on-going efforts in this direction. The present book aims to fill this gap in the literature and provides an adequate coverage of the issues that are being resolved in the areas of system reliability evaluation and highlights the directions in which the future improvements in the current state-of-art are envisaged.

1.2 Organization

The book has been organised in six sections and each section consists of the chapters, which highlight certain aspects of system reliability evaluation in which the new trends in the domain of its consideration are foreseen. Therefore, the foremost objective of present book is to provide an insight or rather a foresight into these futuristic developments that are likely to revolutionize the state-of-art in near future. At the same time, it would help serve as a benchmark for a reader not only to understand and appreciate the newer developments but to profitably guide him to reorient his efforts.

The following is the organization of the book :

(i) Towards Improved Techniques
- Ch. 2 : Multiple Inversion Techniques
- Ch. 3 : Decomposition and Reduction Techniques
  Probabilistic Graph-Reduction Techniques
- Ch. 4 : Parallel Algorithms and Implementations

(ii) Towards Improved Analyses
- Ch. 5 : k-out-of-n Systems
- Ch. 6 : Consecutive k-out-of-n Systems
- Ch. 7 : Fault Tree Analysis

(iii) Towards Better Tools
- Ch. 8 : Digraphs and Causal Trees
- Ch. 9 : Use of Petri Nets

(iv) Towards Better Models
Ch. 10 : Human Reliability Modeling
Ch. 11 : Common Cause and Dependent Failures Modeling
Ch. 12 : Software Reliability Modeling
Ch. 13 : Mechanical Systems Reliability Modeling

(v) Towards Alternative Approaches
Ch. 14 : Use of Fuzzy Sets Theory
Pt. I : Theory and Pt. II : Applications
Ch. 15 : Dempster-Shafer Theory and its Applications

(vi) Towards Development of Non-Specialist's Aid
Ch. 16 : Expert Systems for Reliability Evaluation
Ch. 17 : Reliability Analyzer

1.3 Towards Improved Techniques

This section describes some of the developments that have helped improve the existing techniques. Several new ideas, like multiple inversion technique, decomposition and reduction techniques, and parallel processing, have been introduced to produce highly efficient and computationally economical algorithms for system reliability evaluation.

In order to understand these developments, let us first of all define the system modelling criterion. Each component of a system can be represented by a two terminal graph. If this component is good, a signal can pass from its one terminal to the other indicating that it performs its intended function adequately and the probability of this successful accomplishment of the task by a component is known as the reliability of the component or element. Let us also understand that we are talking about a coherent system [1]. Then the system forms a graph like structure and is better known as a *probabilistic graph*. This structure can also be designated as a *network*.

Generally speaking, system reliability can be defined as the probability that at least one minimal set of system components (or elements), known as a *path set* in the parlance of graph theory [2] is good. Since a path set is generally used in the sense of a chain of elements required to establish continuity between two specified terminals, the system reliability is called *terminal-pair reliability*. However, it is possible that a path set may consist of an acyclic graph in case of flow networks [10] instead of just being a simple chain structure. If this set of elements happens to form a tree like structure with specified k-nodes out of the total of n nodes of the system, the system reliability is known as *k-terminal reliability*. This implies that k-nodes of the network are able to communicate with each other. Lastly, if all n nodes are able to communicate with each other while the path set has a tree like structure (known as a spanning tree), the system reliability is known as *all-terminal reliability*. The last two definitions are usually employed in the context of communication system reliability. It is, sometimes, advantageous to use duality in graphs and we may define a *cut set* as a minimal set of elements whose failure results in a system failure. Therefore, instead of computing system reliability, we may go for computing system unreliability and thus indirectly calculate system reliability.

Basically, the survey of literature [1] indicates that the system reliability algorithms can be broadly classified into two categories, viz.,

- those, which use a path set/cut set approach
- those, which do not use path sets (or cut sets) but employ either reduction or transformation or a combination of these two approaches.

An extensive and up-to-date review of the techniques that have been in vogue since 1960s is provided in [1]. In fact, the presentation in [1] does not assume any prior knowledge of the mathematical tools that are required to evaluate system reliability.

Ever since the application of graph theory to the problem of system reliability evaluation was suggested [2,3], several hundreds of papers have now been published, each claiming to improve upon the art and science of system reliability evaluation. A vast majority of them use path set approach and the problem of computing system reliability, basically, consists of a two step procedure [2], viz.,

(i) determining minimal path sets/cut sets of the system

and (ii) combining the path sets/cut sets so determined in (i) logically, using the law of Union and ensuring that the law of idempotence also holds good.

Both these steps are not as simple to execute as they appear to be, particularly, in case of a large system. For example, a 33-element system [1] generated 1681 path sets. To enumerate all these path sets is in itself a problem and then to combine them in step (ii) will still be more complicated than imagined. Therefore, while on one hand the efforts have been directed in developing computationally efficient procedure for steps (i) and (ii), on the other hand methods have been developed which obtain system reliability without recourse to path sets or cut sets.

There have been other attempts [4-17] to improve upon reliability evaluation procedure. Among these, topological method originally proposed in [2] was presented in a formalized way in [7]. This had an advantage of generating all the noncancelling terms of a system reliability expression directly. However this approach cannot be used economically for large networks. For that matter, it is worth mentioning here that the computation time increases exponentially with the network complexity. Basically the problem of computing system reliability [11-14, 16, 17] is NP-hard. Therefore, it is essential that we develop practical methods which allow computation of system reliability in reasonable time. Among the approaches falling in this category, we have matrix method [3], decomposition [37-39] and transformation techniques [36]. In fact, best results are obtained by combining decomposition and transformation methods [35].

There have been several landmarks in the literature that have significantly improved upon the practice of system reliability evaluation since 1970s. However, a substantial improvement in the path set approach was proposed in [18], realizing the fact, the system reliability has to be numerically evaluated repetitively during the design phase of a system and it was felt necessary to obtain the system reliability expression in a compact form. This was also necessary from another consideration, i.e., to minimize round off errors in multiplicative computations involving uncomplemented variables in reliability expressions that were in vogue at that time [6]. Therefore, the first ever attempt [18] was made in 1975 to obtain a compact expression involving complemented and uncomplemented variables both. This was popularly known as *AMG-Algorithm* in the literature. This was soon followed by others, notably by Abraham [19] in 1979. Both [18] and [19] were based on the enumeration of *path sets* or *cut sets* of a network or a system.

1.3.1 Multiple Variable Inversion

Ever since minimizing system reliability expression [18], using tie sets or cut sets, became the practice for handling large and complex systems and to reduce computational time, very often one resorted to several single variable inversions [19,20]. Considerable work [21-29] since then has been done in the area of Boolean minimization and inversion for obtaining disjoint system reliability expression in order to reduce computing time and data processing effort. This includes preprocessing of path sets [29] and ordering them in an optimal way. Some of these problems and algorithms have been discussed in detail in [1]. However, multiple variable inversion [30-31] has been found to be more efficient and economical and is expected to become quite popular. This has not only helped considerably in reducing the expression but also improved upon the accuracy and time by providing a more compact expression than was possible with [19]. It is, therefore, considered as one of the major directions of thrust in the area of system reliability evaluation and Chapter 2 aptly discusses this problem and also projects the developments that are likely to take place in future in this area.

1.3.2 Decomposition and Transformation Techniques

Decomposition techniques for system reliability evaluation have been in vogue for a long time. For series parallel network, decomposition [3] has been straightforward. But for reliability evaluation of non-series parallel networks by making use of *factoring theorem*, or what is also known as Bayes theorem, was suggested in [3]. In doing so, a non-series parallel network is decomposed to a series parallel category, for which several efficient techniques are now available.

In fact, factoring theorem has been successfully used [37-39] in system reliability evaluation. Another approach that has been successfully used for system reliability evaluation is the transformation technique such as described in [33, 34]. This technique has been widely used in system reliability evaluation for obtaining reasonably accurate results. However, combination of decomposition and transformation techniques provide an efficient solution to the problem. Decomposition and transformation techniques [35, 36, 40, 41], of recent, have revolutionized the whole concept of system reliability evaluation and reduced the time and effort in obtaining a system reliability expression. These have not only been applied to two-terminal reliability problem but has been extended to *k-terminal* and *all-terminal* reliability problems as well. Therefore, it was considered appropriate to include a chapter on Decomposition and Transformation Techniques. This is done in Chapter 3 in Part I and II, which provide an excellent coverage and documentation of various aspects of this approach.

1.3.3 Parallel Processing Algorithms

Yet another significant recent development has been the use of parallel processing algorithms to reduce computational time in case of very large and complex systems. This concept also has been introduced in [32] and is presented in this book in Chapter 4. In less than a decade, this approach is likely to mature and the concept will be extended to *k-terminal* and *all-terminal* reliability evaluation problems as well. At present it is in an infancy stage.

Thus in the opinion of the editor, it is expected that the material presented in Chapters 2,3 and 4 would lead to substantial improvements that will eventually, in a decade or so, revolutionise the practice of system reliability evaluation.

1.4 Towards Improved Analyses

In this section, some selected topics such as *k-out-of-n System analysis, analysis of Consecutive k-out-of-n System* and *Fault Tree Analysis* have been included. There has been considerable improvement in these analyses, since these subjects were formally introduced and even as of today, there is a considerable potential for improvement in these analyses. For example, the concept of automatic generation of fault tree and modularization have helped improve the fault tree analysis and thereby the art of system reliability evaluation.

1.4.1 k-out-of-n Systems

The most general model in the reliability studies of series parallel systems is a k-out-of-n model. All other models can be derived as special cases of this model. Therefore, it was natural that the interest in this model originated in early 1970s and several papers [42-44] evoked further interest in this model. A detailed survey of this category of models is provided in [1]. Chapter 5 discusses among other things the approximate and recursive techniques that have been used for system reliability evaluation using this model. Besides providing a comparison of existing techniques [45], future trends have also been indicated in the chapter.

1.4.2 Consecutive k-out-of-n Systems

The recent developments in the area of k-out-of-n include consecutive k-out-of-n models. The first paper [46] in this area was published in 1980 and since then, several articles have been published [47- 53]. Since 1985, practically every issue of the IEEE Transactions on Reliability contains few papers devoted to this model. A system is known as consecutive k-out-of-n : F system, if for the failure of any consecutive k components ($k \leq n$), the system fails. In fact [50] provided a relationship between consecutive k-out-of-n : G and consecutive k-out-of-n : F systems. Further, if these n components are arranged in a line, the system is *linear.* However, if they form a circle, the system is *circular.* This definition has been further generalized [54] to define *linear and circular (m,n) lattice systems.*

All these foregoing developments have provided an extensive insight in the system modeling process and have proved their efficacy in handling various practical and realistic situations. In fact, these models have found applications in the reliability assessment of of complex and highly connected communication networks, supervision or monitoring systems, pipeline or fluid transmission systems, nuclear systems and space relay systems, a bus-structured multiprocessor system, fault tolerant and hypercube multiprocessor systems, etc. Chapter 7 provides a brief summary of all these developments.

1.4.3 Fault Tree Analysis

Since its inception in 1962, fault tree analysis (FTA) has now acquired a status of a well-established technique for assessing the safety and reliability of engineering systems [55]. Fault tree analysis has been extensively and very successfully applied in nuclear, chemical, aerospace industries and in other high risk or consequences situations. FTA has become all pervasive in the context of probabilistic risk analyses of high risk systems as are encountered in the chemical and nuclear industries. It has both qualitative as well as quantitative aspects of the analyses while performing reliability and risk studies and several software packages and algorithms are available today to perform these tasks. An exhaustive, informative and state-of-art review of the fault tree analysis is available in [1].

The fault tree analysis is a well-documented methodology [59,61,62]. However, even after such an extensive consideration and amount of effort [55-78], the current FTA is wrought with certain shortcomings. One such area is the fault tree construction itself. Even today, the fault tree structure depends on the subjective judgement and experience of an analyst. To overcome this difficulty, several attempts [56-58,60,75] in the direction of automatic generation of fault trees, have been made in the past. Truncation [64], modularization [66,71,72] and faster computations [68,70,73] have been suggested to handle large fault trees. Since FTA has been effectively and widely used for system design, development, modification and verification processes [76] and is the most popular technique for the probabilistic risk assessment studies [78], it was considered appropriate to provide a thorough and state-of-art discussion on this subject. Therefore, the objective of Chapter 7 is to realize this objective while highlighting various aspects of fault tree analysis and the new developments that are likely to improve upon the state-of-art. It is expected that the new trends in the area of fault tree analysis are likely to pave way for the development of efficient techniques for risk analyses.

1.5 Towards Better Tools

It has been proved beyond doubt that graph theory has been of immense benefit to system reliability modeling as well as in the analysis and evaluation procedure. Of recent, digraphs and causal trees have proved their usefulness in automatic fault tree construction. Petri nets and neural networks have also been reported to have applications in the system reliability evaluation. Therefore, in this section, these topics have been included to provide readers of the implications of their capabilities and uses. Neural Networks and their applications to system reliability evaluation have not been included in this book as the editor at this point of time could not be convinced of their efficacy and direct utility in solving system reliability evaluation problems. However, it is possible that in near future the neural networks may be proved to be useful for system reliability studies. Besides, it must be mentioned that inspite of his best efforts, the editor could not obtain the consent of any contributor to write this chapter with an emphasis on the relevance of neural networks to system reliability evaluation.

1.5.1 Digraphs and Causal Trees

As mentioned in section 1.4.3, several algorithms have been proposed in order to produce a fault tree automatically to eliminate individual subjective judgement and experience of an analyst and ensure consistency and repeatability of a fault tree even if they come from two different analysts. For example, decision table approach [58], mini fault tree approach [60], digraph approach [79-81, 83,84,86] and causal tree approach [85] all aim at providing methodology to generate fault tree automatically.

The digraph method [83] is a two-step approach to fault tree construction. The first step consists of producing a digraph from the system description and interrelationships between process variables through a multivalued logic diagram. The second step consists of constructing a conventional fault tree with the help of an algorithm.

A causal tree approach [85] employs component models that show fault propagation through the components and fault initiation by the components in both the directions for all basic events leading to any deviation in the top parameters.

Subsequently, the fault tree is constructed from a causal tree, with all feed back and feedforward loops in it.

Both, the digraphs and causal tree approaches, require a vigorous understanding of the system function, and therefore require human intervention in one form or the other to prevent generation of incomplete or inconsistent failure modes. Therefore, it becomes necessary to combine the two failures and develop an interactive methodology. Chapter 8 provides a discussion of the techniques of digraph and causal trees and introduces the concepts of interactive methodology to overcome the difficulties of digraphs and causal trees in order to help improve the fault tree analysis, eventually.

1.5.2 Petri Nets

Since the publication of [87], Petri nets have been widely used [88-97] not only for system modeling but also for verification of communication protocols, distributed databases, system reliability evaluation, fault-tree analysis, software reliability growth, software safety analysis in heterogeneous multiprocessor control systems. At present, the major strength of Petri nets lies in its ability to model concurrent events in a system and to provide a powerful formalism to describe distributed systems.

The modeling power of Petri nets have been extended to provide an enlarged family of nets such as, Timed Petri nets (TPN), Stochastic Petri nets (SPN), Generalized Stochastic Petri nets (GSPN) and Timed Petri nets (GTPN), etc. This extension facilitates the applicability of Petri Nets to a variety of system problems. For example, the work described in [96] integrates software fault tree analysis and Timed Petri nets to describe time critical events in multiprocessor control applications and to analyze if safety-critical states are reachable during the normal execution. A computerized reliability growth model generation based on Petri nets and using the notion of Extended Stochastic Petri Nets is considered in [97]. Petri nets allow modeling of the process of fault manifestation, localization and removal, in the design and software development processes, considering the reliability and complexity of the software under development.

In short, Petri nets find an extensive use in several aspects of reliability analyses. In the area of system reliability evaluation, Petri nets have been employed (although not very efficiently) to determine simple paths, k-trees and in fault tree analysis and fault diagnostics. It is expected that Petri nets, eventually, may also help solve several problems from system reliability area. It is believed that an exposure of Petri nets to reliability analysts may prove rewarding. It is with this aim in view that Chapter 9 on Petri Nets has been included in this book.

1.6 Towards Better Models

Although we have come a longway in the development of reliability engineering as a distinct discipline in its own right since 1950s, there are areas where we still do not have appropriate models to deal with, effectively. A few but important among them are :

(i) Human Reliability Modeling
(ii) Common Cause and Dependent Failure Modeling
(iii) Software Reliability Modeling
(iv) Mechanical Systems Reliability Modeling

The state-of-art in the these areas is far from being very satisfactory and we need to strengthen our research efforts to develop some realistic and appropriate models in these areas for assessing system reliability with a reasonable degree of confidence. Therefore, it is necessary that we first make an appraisal of the existing state-of-art, identify gaps and make an all out effort to overcome the existing shortcomings. Unfortunately, all the above areas are very critical and significantly influence the system reliability results.

1.6.1 Human Reliability Modeling

The influence of human behaviour, particularly, on the reliability and safety of plants and systems can't be underestimated. Human error be it in the overall management, in design, in manufacture, in commissioning, in operation, in testing or in maintenance, can certainly defeat the performance of any man-machine system, particularly when man's interface with technological systems cannot be totally eliminated. Therefore, it is a matter of concern to any system designer.

Human reliability assessment [98-107] lays emphasis on quantification of human error as well as human cognitive processing. Human reliability analysis (HRA) aims at determining the probability of human error that contributes to the system failure. It is also essential that a system designer understands the basis for human error or rather all those human activities that may lead to harmful consequences due to the situational context and human cognitive processing tendencies.

One of the widely used methods is the technique for Human Error Rate Prediction (THERP), which was developed out of the need to quantify human error in nuclear weapon system [98,99,101] by assessing Human Error Probabilities (HEP) through the use of event trees representing relevant human tasks. There was a strong criticism of this approach in [100]. The approach suffers from its inability to represent *error of commissions* [106]. A good critical review of the issues in human reliability modeling is provided in [100,104,105]. The Human Cognitive Reliability (HCR) model [103], which is based on classical engineering reliability approach and distinguishes between skill, rule and knowledge-based levels of human performance.

Of recent, some developments [106,107] have taken place in the analysis of human aspects in management and maintenance and therefore it was considered highly desirable to include human reliability modeling among the topics covered in this book and Chapter 10 is expected to provide the state-of-art information on the subject besides presenting a critical review and indications of the trends in the area.

1.6.2 Common Cause and Dependent Failures Modeling

A *dependent failure* is, generally, defined as the simultaneous failure of more than one component of a system. If all failures are due to the same shared root cause, the event is known as *common cause failure.* All identifiable causes of multiple failures which may include failure of services (electrical power, cooling, air etc.), duty/standby switching elements, common maintenance errors, external causes like earthquake, lightening etc. must be considered. It is generally thought that the use of redundancy enhances system reliability. However, very often we implicitly assume that the component failures are independent but this is not always valid. In reality, the use of redundancy and diversity makes a system more prone to the common-cause failures.

Although the importance of common-cause failures or dependent failures was recognised in the late 1950s but it is only of recent that it has become an important factor in the context of overall system reliability evaluation, particularly of control and computer systems. Today, there exist several models for quantitative assessment of the effect of common-cause failures. One of the first attempt was by Fleming [108], who introduced the highly popular *beta-factor method.* In fact, Reactor Safety Study [109] was one such comprehensive quantitative study ever made, in which the bounding or *square root method* was used to model the common cause failures. The *binomial failure rate* (BFR) method was suggested by Vesely [111]. There have been several other attempts [110-122], including what is known as *multiple dependent failure fraction method* for the analysis of common-cause failures.

The *multiple dependent failure fraction method* is a generalization of the beta factor (BF) method and produces more realistic assessment of system reliability/availability. Martin and Wright [119] combined the ease of beta-factor method with the idea of limit values to propose a method which has been found useful in chemical and nuclear plants. In the *shock model* presented in [120], two different causes of failure were assumed-one due to the natural aging and the other due to the shock caused by environmental stresses or errroneus maintenance. It is the rate of system shocks which may cause one or more components to fail simultaneously. The beta-factor and BFR were shown to be special cases of the shock model.

To discuss all these developments [108-122] and future trends, Chapter 11 on common-cause failures modeling, has been included in the book.

1.6.3 Software Reliability Modeling

With the advent of the microprocessors and ever-increasing use of computers in the management, control and safe operation of plants and systems, the problem of software reliability has come into a sharp focus. This is again not a new problem but the magnitude of software use has made it an important problem to be tackled today. Today, a failure of a computer system due to software faults is more likely than on account of a hardware failure. Like hardware area, even in the non-hardware area, modeling and the data need to be improved upon and developed. Consequently, a lot more needs to be done in the area of software reliability.

The concept of probability of failure due to an error has been used by several investigators to derive software reliability models. The probability of failure due to an error is defined as the number of failures the error causes compared to the number of failures caused by all errors, given that no errors are corrected and the work load (number of instructions executed per unit time) remains representative during the period of measurement. Generally, a software failure occurs when software errors are encountered during its execution. Encountering an error (i.e., executing an error embedded instruction) does not necessarily cause a failure, it merely provides an opportunity, the degree of which depends on the probability of simultaneously processing the failure inducing input.

Since the earliest software reliability model [123] appeared in 1972, many models have been proposed and some 41 of them existed as early as in 1982. A potential user of these models easily gets confused by this diversity. For example, linear reliability models (Jelinski-Moranda [123], Shooman [124], Musa [125]) result from the assumptions that the average size of remaining errors and work load are constant and its apparent error density (number of encountered errors per executed instruction) equals its real error density (number of initial remaining errors per software instruction). Rayleigh model (Schick-Wolverton [127]) assumes that the average size of the remaining errors increases linearly with time. Finally, the geometric reliability models (Moranda [126], Ramamoorthy-Bastani [128]) differ from linear models in that they assume that the average-error size decreases geometrically as the errors are detected.

Very recently, Trachtenberg [132] proposed a general theory on the premise that software failure rates are the product of software average error size, apparent error density and work load. A general software reliability model would permit reliability engineers to model diverse reliability factors found in a modern software. The Chapter 12 on *software reliability modeling* highlights these new developments and provides an indication of the areas where the gaps exist and to the future trends in this critical area.

1.6.4 Mechanical Systems Reliability Modeling

It is perhaps not without historical reasons that there is a great deal of information on reliability of electrical and electronic equipment but comparatively very little exists from the point of view of mechanical systems reliability.

We know that the discrete component modeling, as has been applicable to electrical or electronic components, can't be applied in case of mechanical components where the concept of a distributed system consisting of separate functional items does not work satisfactorily. A mechanical system should be treated as a closely coupled system like an engine or a structure. Here, the reliability models have to be conceived as a whole, as a closely coupled system, where the components interact with each other and are dependent upon each other. Besides the mechanical failure mechanisms like fatigue have no parallel in electronics area. A vast majority of mechanical failures are due to wear-out. Erosion, corrosion, creep, aging, fouling etc. form the part of the failure mechanism.

Early mechanical system designs were based on deterministic approach involving safety factors. Variability of stress and strength were not considered. Reference [1] provides a detailed discussion of these problems. Today, the probabilistic design of mechanical systems has become very common and therefore, it is not surprising that lot of effort is being devoted in developing practical and realistic models of mechanical components. However, the state-of-art [138-147] is far from being satisfactory in this area.

Realizing this, it was decided to devote Chapter 13 to mechanical system reliability modeling with the expectation that this chapter will help generate further interest in the subject, which is so important in the context of overall system reliability evaluation.

1.7 Towards Alternative Approaches

For a long time, the probabilistic approach to system reliability has been found adequate and highly useful in assessing the performance of engineering systems. However, the major shortcoming of probabilistic approach has been in its inadequacy in offering an effective tool to handle the problem of uncertainty. Uncertainty of results puts the entire process of system reliability assessment to question. Specifying mean and variance or confidence levels is not enough to tackle the problem of uncertainty. Although some intuitive methods of accounting, combination and propagation of the uncertainty through the use of certainty factors has been in vogue for quite some time and these have been extensively made use of, in the probabilistic risk assessment as well, but certainly these have not been found satisfactory. While Bayesian models in probability theory, have been widely used, primarily, as a numerical approach for representation and inference with uncertainty yet Bayesian approach masks the problem of uncertainty when priors are selected and ignorance remains hidden in priors. These problems have led analysts to question the basic approach of probability itself and to explore other avenues of analysis.

Amongst the recent developments that are quite encouraging and in due course can provide a sound framework to tackle the problem of uncertainty of data and models are : Zadeh's fuzzy sets theory (FST) and Shafer's evidence theory (ET). As we will see later on, both of them have been successfully applied to the problem of system reliability evaluation. It is expected that eventually, when fully developed in near future, the FST and ET will replace the three centuries-old and well-established probabilistic approach for system reliability evaluation.

1.7.1 Application of Fuzzy Sets Theory

Conventional system reliability evaluation is based on probabilistic approach which considers all failure events including the system failure as random events. This does not necessarily represent the actual situation. Some of the worst accidents in human history were evaluated to have very small probability. Yet they occured. This attracted the attention of system analysts to propose new approaches (in place of the probabilistic approach), which reflect the human perception of accidents on more pragmatic foundations. Fortunately Zadeh's fuzzy sets theory [148] and possibility theory [150] provide the basis of *possibilistic* approach to system reliability evaluation based on the premise that *small probability* does not always mean *low possibility* of an event, whereas a *low possibility* necessarily would imply *low probability*.

Fuzzy sets theory [148-168] provides an ingenious approach to cope with the cases where subjective judgement or estimation by an individual will play a central and significant role in dealing with the ambiguity or uncertainty. The imprecision of basic failure data and model uncertainties in reliability studies can be best dealt with the FST. Fuzzy sets theory has been applied in system performance (reliability) evaluation, [153,161] reliability apportionment [157], fault tree analysis [155, 156, 160, 166, 168], event tree analysis [167], risk assessment [162], human reliability assessment [158, 165], development of expert systems [178] and many other areas in system performance evaluation. Another development in system performance evaluation is the use of linguistic variables [149, 163]. This would bring evaluation process closer to human thinking.

Chapter 14 has been organized in two parts. Part I describes the basics of fuzzy sets theory and has been included to introduce a reader to the fundamentals of FST as would be required in the context of system reliability evaluation. Part

II describes the applications of FST to system performance evaluation problems including fault tree analysis, common cause failures, dependency considerations, environmental considerations etc. It also provides an indication of the direction in which future challenges in system reliability evaluation would lie.

1.7.2 Application of Dempster-Shafer Theory

Another approach to futuristic developments in the area of system reliability evaluation is the use of ET [169-175]. The advantage that ET provides, over the probabilistic framework is in its capability of explicit representation of ignorance. We are aware that various kinds of ignorance exist in our knowledge or information, particularly in dealing with large and complex systems and DS theory helps in working with this situation. There are situations when we are not able to identify a state of the system or carry out fault diagnosis based on a vague information, DS can be used effectively. While it provides a powerful methodology of representing uncertain information, very little has been done to perform reasoning or inferencing with the information represented in terms of belief functions.

There are many areas of system reliability where DS theory can be applied very effectively instead of using conventional probability theory, particularly in the field of fault location and diagnosis, safety and performance evaluation etc. To consider all these aspects, Chapter 15 on Shafer's evidence theory and its application to system performance assessment is included in this book. It is expected that in future the DS theory will be used extensively to solve many system reliability problems.

1.8 Towards Development of Non-Specialist's Aid

In addition to the developments that are taking place in the area of system reliability evaluation and have been discussed so far in this book, there are other developments which are likely to take the art of system reliability evaluation to non-experts or a non-specialist user. These developments include expert systems for system reliability evaluation. These are expected to perform all the tasks of a specialist in system reliability evaluation and eventually might replace him for carrying out these jobs. The development of expert systems for carrying out various phases of system reliability analysis is, as of today, in infancy stage but is likely to pick up very fast. We have come a long way from our initial efforts to develop computer-aided reliability analyses. Expert systems are now being developed using fuzzy set theory [178] and Dempster-Shafer theory [172] based rules, as well. Apart from these intelligent implementation in software area, there is an equally important new development in the area of the introduction of a special purpose user-friendly hardware or gadget being called as *Reliability Analyzer*. This provides another new direction in the development of a non-specialist's aid to carry out reliability analyses. Both, these software and hardware developments are likely to revolutionize the state-of- art of system reliability evaluation in near future.

1.8.1 Development of Expert Systems

Notwithstanding the development of several useful, powerful and user-friendly softwares in the realm of system reliability evaluation, some encouraging developments [176-189] have taken place recently, particularly in the direction of developing expert systems for system reliability evaluation. Efforts are continuing to develop powerful expert systems, which are likely to use expert's knowledge to overcome the present difficulties. Today we have an array of excellent and well-documented computer-aided reliability analyses with the objective of assisting a system reliability designer. Although expert systems and computer softwares have much in common as far as algorithmic tasks are concerned, expert systems are intended to capture human intution and experience and are expected to play the role of an intelligent analyst. This would particularly be more relevant in the domain of diagnostics. Decision making in fuzzy environment which is basically the power and strength of human brain is likely to be accomplished through expert systems. Knowledge about system behaviour, structure and design heuristics which is normally acquired through experience forms an essential part of expert systems' architecture. Needless to mention that we come across several problems in system reliability evaluation which necessiate this capability.

Therefore, in short, it is expected that designing of reliable systems based on pragmatic knowledge representation will become a reality in near future. In Chapter 16, we explore the feasibility of such a possibility and also indicate what can be done in future in this direction.

1.8.2 Reliability Analyser

Lastly, an entirely new and important development [190] which could not have gone unnoticed by reliability analysts and can't be underestimated, is the introduction of a special desk-top gadget (a dedicated mini or micro computer) to do the job of system reliability evaluation efficiently, possibly in a real time domain. This development promises to provide an analyst a handy tool for carrying out various reliability analyses and make system reliability assessment a routine job, which need not be confined to highly qualified analysts only.

A gadget in the form of a desk-top facility is considered preferable over the sophisticated softwares on a PC, or on a mini or microcomputer keeping in view the several iterative trials that are required during the reliability design of a system. These can best be simulated directly on a basic system configuration. Using a reliability analyzer, a reliability analyst or designer can observe the consequences of ramifications he carries out on system configuration in real-time domain. Thus this gadget can also help in the learning process of system reliability designers.

The basic idea introduced in [190-192, 194] was commercially exploited to develop a desk top aid in the form of ESCAF by Laviron [193]. Several problems [195-199] were subsequently attempted and modifications were carried out to develop a more versatile equipment-S.ESCAF. In fact, the reliability analysis of complex sequential system, which is often done by the well-known Markov technique, inherently restricts the size of a system, viz., the number of components that can be handled. However, it was demonstrated [200] that a desk top reliability analyzer such as S.ESCAF is easy to use even by a non-specialist and offers an advantage of time and enormous convenience as it is sufficient to simulate directly the functional logic of the system under study and does not require performing fault tree analysis first. Additionally, it offers a guarantee of exhaustivity of search for all failure sequences due to its automatic search procedure. It is also easy and fast to incorporate the changes in simulation logic or reliability data during the analysis.

Thus, a desk-top reliability analyzer is likely to set a new trend and prove highly successful and favourite tool for several reliability studies, particularly for system reliability evaluation. Future appears to be promising for this new and novel approach outlined in Chapter 17.

1.9 Conclusions

In short, new trends in techniques, analyses, modeling, tools and alternative approaches together with the development of user- friendly non-specialist aid are bound to take the state-of-art in system reliability evaluation to new heights in a foreseeable future. This book should be seen as a trend setter and a useful guide to all these new and promising areas, which in due course are likely to revolutionize the art and science of system reliability evaluation. It is expected that this book along with the state-of-art book published earlier [1], together can take a beginner from the fundamentals of system reliability to futuristic art and science of the subject. These books would become a necessity to any serious-minded designer of engineering systems or services.

Another development, that is taking place silently as we enter the new century, is the emphasis on the environmental friendliness of the products, systems and services. In the name of development we have not only allowed unregulated exploitation of the resources of the earth but have done considerable damage to the environment of our planet. Environmental audit, which includes proper raw material utilization, waste minimization, energy conservation, recycling and reuse, will be the order of the day. Ecology and Economy will have to go together. Therefore, the *performance* and *environment* would be the trend setting words of the coming decades of the new century for the competing nations for their techno-economic prosperity. This is the post-Rio message for the world and will naturally have influence on the design procedures in future.

REFERENCES

State-of-Art Review

[1] Misra K.B., *Reliability Analysis and Prediction (A Methodology Oriented Treatment)*, Elsevier Science Publishers BV, Amsterdam, 1992.

Algorithms on System Reliability Evaluation

[2] Misra K.B. and T.S.M. Rao, *Reliability Analysis of Redundant Networks Using Flow Graphs*, IEEE Trans. on Rel., Vol. R-19, No. 1., pp. 19-24, Feb. 1970.

[3] Misra K.B., *An Algorithm for the Reliability Evaluation of Redundant Networks*, IEEE Trans. on Rel., Vol. R-19, No. 4, pp. 146-151, Nov. 1970.

[4] Hansler E., *A Fast Recursive to Calculate the Reliability of a Communication Network*, IEEE Trans. on Com., Vol. Com-20, No. 3, pp. 637-640, June 1972.

[5] Kershenbaum A. and R.M. Van Styke, *Recursive Analysis of Network Reliability*, Networks, Vol. 3, pp. 81-94, 1973.

[6] Aggarwal K.K., J.S. Gupta and K.B. Misra, *Reliability Evaluation : A Comparative Study of Different Methods*, Microelectronics and Reliability, Vol. 14, No. 1., pp. 49-56, 1975.

[7] Satyanarayana A. and A. Prabhakar, *New Topological Formula and Rapid Algorithm for Reliability Analysis of Complex Networks*, IEEE Trans. on Rel., Vol. R-27, No. 2, pp. 82-100, June 1978.

[8] Rai S. and K.K. Aggrawal, *An Efficient Method for Reliability Evaluation of General Network*, IEEE Trans. on Rel., Vol. R-27, No. 3, pp. 206-211, Aug. 1978.

[9] Dotson W.P. and J.O. Gobian, *A New Analysis Technique for Probability Graphs*, IEEE Trans. on Circuits and Systems, Vol. 26, pp. 855-865, 1979.

[10] Misra K.B. and P. Prasad, *Comment on Reliability Evaluation of a Flow Network*, IEEE Trans. on Rel., Vol. R-31, No. 2, pp. 174-176, June 1982.

[11] Satyanarayana A. and R.K. Wood, *A Linear-Time Algorithm for Computing k-terminal Reliability in Series-Parallel Networks*, SIAM Journal of Computing, Vol. 14, pp. 818-832, 1983.

[12] Arnborg S. and A. Proskurowski, *Linear-Time Algorithms for np-hard Problems Embedded in k-tree*, Technical Report TRITA-NA-8404, Royal Institute of Technology, Stockholm, 1984.

[13] Agrawal A. and R.E. Barlow, *A Survey of Network Reliability and Dominion Theory*, Operations Research, Vol. 32, pp. 478-492, May-June, 1984.

[14] Agrawal A. and A. Satyanarayana, *An O(/E/) Time Algorithm for Computing the Reliability of a Class of Directed Networks*, Operations Research, Vol. 32, pp. 493-517, May-June 1984.

[15] Rai S. and A. Kumar, *Recursive Technique for Computing System Reliability*, IEEE Trans. on Rel., Vol. R-36, No. 1, pp. 38-44, April 1987.

[16] Yoo Y.B. and N. Deo, *A Comparison of Algorithms for Terminal- Pair Reliability*, IEEE Trans. on Rel., Vol. 37, No. 2, pp. 210-215, June 1988.

[17] Politof T. and A. Satyanarayana, *A Linear Time Algorithm to Compute the Reliability of Planer Cube-free Networks*, IEEE Trans. on Rel., Vol. R-39, No. 5., pp. 557-563, Dec. 1990.

Algorithms on Disjoint Products

[18] Aggrawal K.K., K.B. Misra and J.S. Gupta, *A Fast Algorithm for Reliability Evaluation*, IEEE Trans. on Rel., Vol. R-24, No. 1, pp. 83-85, April 1975.

[19] Abraham J.A., *An Improved Algorithm for Network Reliability*, IEEE Trans. on Rel., Vol. R-28, No. 1, pp. 58-62, April 1979.

[20] Locks M.O., *Recursive Disjoint Products : A Review of Three Algorithms*, IEEE Trans. on Rel., Vol. R-31, No. 1, pp. 33-35, 1982.

[21] Schneeweiss W.G., *Disjoint Boolean Products via Shanon's Expansion*, IEEE Trans. on Rel., Vol. R-33, No. 4, pp. 329-332, Oct. 1984.

[22] Locks M.O., *Recent Developments in Computing of System Reliability*, IEEE Trans. on Rel., Vol. R-34, No. 5, pp. 425-436, Dec. 1985.

[23] Beichelt F. and L. Spross, *An Improved Abraham Method for Generating Disjoint Sums*, IEEE Trans. on Rel., Vol. R-36, No. 1, pp. 70-74, April 1987.

[24] Locks M.O., *A Minimizing Algorithm for Sum of Disjoint Products*, IEEE Trans. on Rel., Vol. R-36, No. 4, pp. 445-453, Oct. 1987.

[25] Ball M.O. and J.S. Provan, *Disjoint Products and Efficient Computation of Reliability*, Operations Research, Vol. 36, pp. 703-715, Oct. 1988.

[26] Heidtmann K.D., *Smaller Sums of Disjoint Products by Subproduct Inversion*, IEEE Trans. on Rel., Vol. R-38, No. 3, pp. 305-311, Aug. 1989.

[27] Beichelt F. and L. Spross, *Comments on : An Improved Abraham Method for Generating Disjoint Sums*, IEEE Trans. on Rel., Vol. R-38, No. 4, pp. 422-424, Oct. 1989.

[28] Wilson J.M., *An Improved Minimizing Algorithm for Sum of Disjoint Products*, IEEE Trans. on Rel. Vol. R-39, No. 1., pp. 42-45, April 1990.

[29] Soh S. and S. Rai, *Experimental Results on Preprocessing of Paths/Cuts Terms in Some of Disjoint Products Technique*, Proceedings of the Infocom, pp. 533-542, 1991.

[30] Veeraraghavan M. and K.S. Trivedi, *An Improved Algorithm for the Symbolic Reliability Analysis of Networks*, IEEE Trans. on Rel., Vol. R-40, No. 3, pp. 347-360, Aug. 1991.

[31] Locks M.O., and J.M. Wilson, *Note on Disjoint Products Algorithms*, IEEE Trans. on Rel., Vol. R-41, No. 1, pp. 81-84, March 1992.

<u>*Algorithms on Parallel Processing*</u>

[32] Deo N. and M. Medidi, *Parallel Algorithms for Terminal-Pair Reliability*, IEEE Trans. on Rel., Vol. R-41, No. 2, pp. 201-209, June 1992.

<u>*Algorithms on Transformation and Reduction Techniques*</u>

[33] Ramamoorthy M. and Balgopal, *Block Digram Approach to Power System Reliability*, IEEE Trans. on PAS, Vol. PAS-89, pp. 802-811, May 1970.

[34] Rosenthal A. and D. Frisque, *Transformations for Simplifying Network Reliability Calculations*, Networks, Vol. 7, pp. 97-111, 1977.

[35] Gadani J.P. and K.B. Misra, *A Network Reduction and Transformation Algorithm for the Assessment of System Effectiveness Indices*, IEEE Trans. on Rel., Vol. R-30, No. 1, pp. 48-57, April 1981.

[36] Gadani, J.P. and K.B. Misra, *Quadrilateral-star Transformation : An Aid for Reliability Evaluation of Large Complex Systems*, IEEE Trans. on Rel., Vol. R-31, No. 1, pp. 49-59, April 1982.

[37] Wood R.K., *Factoring Algorithms for Computing k- terminal Network Reliability*, IEEE Trans. on Rel., Vol. R-35, No. 3, pp. 269-278, Aug. 1986.

[38] Page L.B. and J.E. Perry, *A Practical Implementation of its Factoring Theorem for Network Reliability*, IEEE Trans. on Rel., Vol. R-37, No. 3, pp. 259- 267, Aug. 1988.

[39] Page L.B. and J.E. Perry, *Reliability of Directed Networks Using the Factoring Theorem*, IEEE Trans. on Rel., Vol. R-38, No. 5, pp. 556-562, Dec. 1989.

[40] Shooman A.M. and A. Kershenbaum, *Exact Graph- Reduction Algorithms for Network Reliability Analysis*, Technical Report, IBM TJ Watson Research Center, Hawthorne, New York, 1991.

[41] Theologu O.R. and J.G. Carlier, *Factoring and Reductions for Networks with Imperfect Vertices*, IEEE Trans. on Rel., Vol. R-40, No. 2, pp. 210-217, June 1991.

Reliability k-out-of-n Systems

[42] Balagurusamy E. and K.B. Misra, *Reliability and Mean Life of a Parallel System with Non-identical Units*, IEEE Trans. on Rel., Vol. R-24, No. 5, pp. 340-341, Dec. 1975.

[43] Misra K.B. and E. Balagurusamy, *Reliability Analysis of k-out-of-n : G System with Dependent Failures*, Int. Journal System Science, Vol. 7, No. 11, pp. 1209-1215, Nov. 1976.

[44] Balagurusamy E. and K.B. Misra, *Availability and Failure Frequency of Repairable m-order Systems*, Int. Journal System Science, Vol. 7, No. 11, pp. 1209-1215, Nov. 1976.

[45] Rushdi A.M., *Comments on : An Efficient Non- Recursive Algorithm for Computing the Reliability of k-out-of-n System*, IEEE Trans. on Rel., Vol. R-40, No. 1, pp. 60-61, April 1991.

Consecutive-k-out-of-n Systems

[46] Kontoleon J.M., *Reliability Determination of a r- Successive-out-of-n : F System*, IEEE Trans, on Rel., Vol. R- 29, No. 5, p. 327, Dec. 1980.

[47] Bollinger R.C. and A.A. Salvia, *Consecutive-k-out- of-n : F Networks*, IEEE Trans. on Rel., Vol. R-31, No. 1, pp. 53-56, April 1982.

[48] Fu J.C., *Reliability of a Large Consecutive-k-out- of-n : F System*, IEEE Trans. on Rel., Vol. R-34, No. 2, pp. 127-130, June 1985.

[49] Rushdi A.M., *Efficient Computation of k-to-l-out-of- n System Reliability*, Reliability Engineering, Vol. 17, pp. 157-163, 1987.

[50] Kuo W., W. Zhang and M. Zuo, *A Consecutive k-out- of-n : G System: The Mirror Image of a Consecutive-k-out-of-n : F System*, IEEE Trans. on Rel., Vol. R-39, No. 2, pp. 244-253, June 1990.

[51] Rushdi A.M., *Some Open Questions on : Strict Consecutive-k-out-of-n : F Systems*, IEEE Trans. on Rel., Vol. R-39, No. 3, pp. 380-381, June 1990.

[52] Papastavridis S., *m-Consecutive-k-out-of-n : F System*, IEEE Trans. on Rel., Vol. R-39, No. 3, pp. 386-388, Aug. 1990.

[53] Iyer S., *Distribution of Lifetime of Consecutive k- within m-out-of-n : F Systems*, IEEE Trans. on Rel., Vol. R-41, No. 3, pp. 448-450, Sept. 1992.

[54] Boehme T.K., A Kossow and W. Preuss, *A Generalization of Consecutive k-out-of-n : F Systems*, IEEE Trans. on Rel., Vol. R-41, No. 3, pp. 451-457, Sept. 1992.

Fault Tree Analysis

[55] Lambert H.E., *Fault Tree for Decision Making in Systems Analysis*, Lawrence Livermore National Laboratory, UCRL-51829, (Available from the National Technical Information Service, Springfield, Virginia, 22151, USA), 1975

[56] Misra K.B. and R. Thakur, *Development of Fault Tree for Reliability Studies of a Data Processing System*, Int. J. of System Sciences, Vol. 8, No. 7, pp. 771-780, 1977.

[57] Willie R.R., *Computer Aided Fault Tree Analysis*, ORC 78-14, Operations Research Centre, University of California, Berkeley, Aug. 1978.

[58] Salem S.L., J.S. Wu and G. Apostolakis, *Decision Table Development and Application to the Construction of Fault Trees*, Nuclear Technology, Vol. 42, pp. 51-64, 1979.

[59] Vesely W.E., F.F. Goldberg, N.H. Roberts and D.F. Haasl, *Fault Tree Handbook*, NUREG-0492, US NRC, 1981.

[60] Taylor J.R., *Algorithm for Fault Tree Construction*, IEEE Trans. on Rel., Vol. R-31, No. 2, pp. 137- 146, June 1982.

[61] US NRC : PRA Procedures Guide - A Guide to the Performance of Probabilistic Risk Assessments for Nuclear Power Plants; NUREG/CR-2300, US NRC, 1983.

[62] Hickman J.W. et. al., *PRA Procedure Guide*, NUREC/CR-2300, U.S. Nuclear Regulatory Commission, Washington D.C., USA, Jan. 1983.

[63] Zipf G., *Computation of Minimal Cut Sets of Fault Trees : Experience with Three Different Methods*, Reliability Engineering, Vol. 7, No. 3, pp. 159-167, 1984.

[64] Modarres M. and H. Dezfuli, *A Truncation Methodology for Evaluating Large Fault Trees*, IEEE Trans. on Rel., Vol. R-33, No. 4, pp. 325-328, Oct. 1984.

[65] Lee W.E., D.L. Grosh, F.A. Tillman and C.H. Lie, *Fault Tree Analysis, Methods and Applications - A Review*, IEEE Trans. on Rel., Vol. R-24, No. 3, pp. 194-203, Aug. 1985.

[66] Wilson J.M., *Modularizing and Minimizing Fault Trees*, IEEE Trans. on Rel., Vol. R-34, No. 4, pp. 320-322, Oct. 1985.

[67] Kumamoto H. and E.J. Henley, *Automated Fault Tree Synthesis by Disturbance Analysis*, Ind. Eng. Chem. Fundam., Vol. 24, No. 2, pp. 233-239, 1986.

[68] Page L.B. and J.E. Perry, *An Algorithm for Exact Fault-Tree Probabilities Without Cut Sets*, IEEE Trans. on Rel., Vol. R-35, No. 5, pp. 544-558, Dec. 1986

[69] Spangler M.B., *Policy Issues Related to Worst Case Risk Analysis and the Establishment of Acceptable Standards of De Minimis Risk*, in Uncertainty in Risk Assessment, Risk Management, and Decision Making, Plenum Press, New York, 1987.

[70] Schneeweiss W., *Approximate Fault-Tree Analysis with Prescribed Accuracy*, IEEE Trans. on Rel., Vol. R-36, No. 2, pp. 250-254, June 1987.

[71] Han S.H., T.W. Kim and K.J. Yoo, *Development of an Integrated Fault Tree Analysis, Computer Code Module by Modularization Technique*, Rel. Engg. & Syst. Safety, Vol. 21, No. 2, pp. 145-154, 1988.

[72] Kohda T., E.J. Henley, K. Inoue, *Finding Modules in Fault Trees*, IEEE Trans. on Rel., Vol. 38, No. 2, pp. 165-176, June 1989.

[73] Patterson-Hine F.A. and B.V. Koen, *Direct Evaluation of Fault Trees Using Object-Oriented Programming Techniques*, IEEE Trans. on Rel., Vol. R-38, No. 2, pp.186-192, June 1989.

[74] Helman P. and A. Rosenthal, *A Decomposition Scheme for the Analysis of Fault Trees and Other Combinatorial Circuits*, IEEE Trans. on Rel., Vol. R-38, No. 3, pp. 312-327 and 332, Aug. 1989.

[75] De Vries R.C., *An Automated Methodology for Generating a Fault Tree*, IEEE Trans. on Rel., Vol. R-39, No. 1, pp. 76-86, April 1990.

[76] Hessian Jr. R.T., B.B. Salter and E.F. Goodwin, *Fault-Tree Analysis for System Design, Development, Modification and Verification*, IEEE Trans. on Rel., Vol. R-39, No. 1, pp.87-91, April 1990.

[77] Wei B.C., *A Unified Approach to Failure Mode, Effects and Critically Analysis (FMECA)*, Proc. Annual Reliability and Maintainability Symposium, Orlando, Florida, USA, pp. 260-271, 1991.

[78] Henley E.J. and H. Kumamoto, *Probabilistic Risk Assessment - Reliability Engineering, Design and Analysis, IEEE Press, New York, 1992.*

<u>Digraphs and Causal Trees</u>

[79] Allen D.J. and M.S. Rao, *New Algorithm for the Synthesis and Analysis of Fault Trees*, Ind. Eng. Chem. Fundam., Vol. 19, No. 1, pp. 79-85, 1980.

[80] Lambert H.E., L. Philipson and H.Roland, *The Digraph-Fault Tree Methodology and its Use in Transportation Risk Analysis*, Tera Corporation, Berkeley, California, USA, 1980.

[81] Cummings D.L., S.A. Lapp and G.J. Powers, *Fault Tree Synthesis from a Directed Graph Model for a Power Distribution Network*, IEEE Trans. on Rel., Vol. R-32, No. 1, pp. 140-149, June 1983.

[82] Alesso H.P., P. Prassinos and C.F. Smith, *Beyond Fault Trees to Fault Graphs*, Reliability Engineering, Vol. 12, No. 2, pp. 79-92, 1985.

[83] Andrews J.D.and J.M. Morgan, *Application of the Digraph Method of Fault Tree Construction to Process Plant*, Reliability Engineering, Vol. 14, No. 2, pp. 85-106, 1986.

[84] Kohda T. and E.J. Henley, *On Digraphs, Fault Trees, and Cut Sets*, Rel. Engg. and Syst. Safety, Vol. 20, No. 1, pp. 35-61, 1988.

[85] Bossche A., *Fault Tree Analysis and Synthesis*, Ph.D. Thesis, Elect. Engg. Deptt., Delft University of Tech., Delft, The Netherlands, 1988.

[86] Andrews J. and G. Brennan, *Applications of the Digraph Method of Fault Tree Construction to a Complex Control Configuration*, Rel. Engg. and Syst. Safety, Vol. 4, No. 3, pp. 357-384, 1990.

<u>Petri Nets</u>

[87] Petri C.A., *Kommunikation mit Automaten*, Schrifleir des Rheinisch-West Falischen Institutes fur Instrumentelle Mathematik an der Universitat Bonn Heftt 2, Bonn, West Germany, 1982. Translation: c.f. Greene, Supplement 1 to Tech. Rept. RADC-Tr., 65, p. 337, Vol. 1. Rome Air Development Centre, Griffiss Air Force Base, New York, pp. 89, 1965.

[88] Sifakis J., *Use of Timed Petri Nets for Performance Evaluation*, 3rd Int. Symp. Measuring, Modeling and Evaluating Computer System, Beilner and Gelenbe Eds., pp. 75-95, 1977.

[89] Khan A.A., Hura G.S. and et al., *State Equation Representation of Logic Operations Through a Petri Net*, Proc. IEEE, Vol. 69, No. 4, pp. 485-487, 1981

[90] Hura G.S., *Petri Net as a Modeling Tool*, Microelectronics and Reliability, Vol. 22, No. 3, pp. 433-439, 1982.

[91] Molloy M.K., *Performance Analysis Using Stochastic Petri Nets*, IEEE Trans. on Computers, Vol. C-31, No. 9, pp. 913-917, Sept. 1982.

[92] Reisig W., *Petri Nets*, Springer-Verlag, Heidelberg, 1985.

[93] Hura G.S. and J.W. Atwood, *The Use of Petri Nets to Analyze Coherent Fault Trees*, IEEE Trans. on Rel., Vol. R-37, No. 5, pp. 469-474, Dec. 1988.

[94] Mandaltsis D. and J. Kontoleon., *Enumeration of k-trees and Their Applications to Reliability Evaluation of Communication Networks*, Microelectronics and Reliability, Vol. 29, No. 5, pp. 733-735, 1989.

[95] Hura G.S. and Etessami, F.S., *Rule-Based Design Methodology for Solving Real-Time Control Problems*, IEEE Trans. on Software Engg., Vol. 17, No. 3, pp. 274-282, 1991.

[96] Shimeall T.J., *Software Safety Analysis in Heterogeneous Multiprocessor Control System*, Proc. Reliability and Maintainability Symposium, Orlando, Florida, USA, pp. 290-294, 1991.

[97] Shabalin A.N., *Generation of Models for Reliability Growth*, Proc. Reliability and Maintainability Symposium, Las Vegas, Nevada, USA, pp. 299-302, 1992.

<u>Human Reliability Modeling</u>

[98] Swain A.D., *Some Problems in the Measurement of Human Performance in Man-Machine Systems*, Human Factors, Vol. 6, pp. 687-700, 1964.

[99] Rasmussen J., *Human Reliability in Risk Analysis*, In A.E. Green (Ed), High Risk Safety Technology, London : John Wiley & Sons, pp. 143-170, 1982.

[100] Adams J.A., *Issues in Human Reliability*, Human Factors, Vol. 24, pp.1-10, 1982.

[101] Swain A.D. and H.E. Guttmann, *Handbook of Human Reliability Analysis with Emphasis on Nuclear Power Plant Applications* Final Report, SAND 80-0200, Sandia Laboratories, Albuquerque, New Mexico USA, 1983.

[102] Dhillon B.S., *Human Reliability with Human Factors*, Pergamon Press, New York, 1986.

[103] Hannaman G.W. and D.H. Worledge, *Some Developments in Human Reliability Analysis Approaches and Tools*, Rel. Engg. and Syst. Safety, Vol. 22, pp. 235-256, 1988.

[104] Sharit J., *A Critical Review of Approaches to Human Reliability Analysis*, International Journal of Industrial Ergonomics, Vol. 2, pp. 111-130, 1988.

[105] Apostolakis G.E., V.M. Bier and A. Mosleh, *A Crique of Recent Models for Human Error Rate Assessment*, Rel. Engg. and Syst. Safety, Vol. 22, pp. 201-217, 1988.

[106] Reason J., *Human Error*, Cambridge University Press, Cambridge, 1990.

[107] Sharit J. and D.M. Malon, *Incorporating the Effects of Time Estimation into Human-Reliability Analysis for High-Risk Situations*, IEEE Trans. on Rel., Vol. R-40, No. 2, pp. 247- 254, June 1991.

<u>Common Cause and Dependent Failure Modeling</u>

[108] Fleming K.N., *A Reliability Model for Common Mode Failures in Redundant Safety Systems*, General Atomic Report GA-13284, Dec. 1974.

[109] WASH-1400, *Reactor Safety Study, Appendix IV : Common-Mode Failures-Bounding Techniques and Special Techniques*, NUREG-75/014, US Nuclear Regulatory Commission, WA, 1975.

[110] Apostalakis G.E., *The Effect of Certain Çlass of Potential Common-Cause Failures on the Reliability of Redundant Systems*, Nucl. Engg. Design, Vol. 36, pp. 123-133, 1976.

[111] Vesely W.E., *Estimating Common-Cause Failure Probabilities in Reliability and Risk Analyses : Marshall-Olkin Specializations*, In Nuclear Systems Reliability Engineering and Risk Assessment, ed. J.B. Fussel and G.R. Burdick, Society for Industrial and Applied Mathematics, Philadelphia, PA, pp. 314-341, 1977.

[112] Edwards G.T. and A. Watson, *A Study of Common-Mode Failures*, SRD R-146, UK Atomic Energy Authority, Warrington, July 1979.

[113] Evans M.G.K., G.W. Parry and J. Wreathall, *On the treatment of Common-Cause Failures in System Analysis*, Reliability Engineering, Vol. 9, No. 2, pp. 107-115, 1984.

[114] Fleming K.N., A. Mosleh and R.I. Deremer, *A Systematic Procedure for Incorporation of Common Cause Events into Risk and Reliability Models*, Nucl. Engg. Design, Vol. 93, pp. 245-279, 1985.

[115] Waller R.A., *A Brief Survey and Comparison of Common-Cause Failure Analysis*, NUREG/CR-4314, Los Alamos National Laboratory, Los Alamos, NM, 1985.

[116] Harris B., *Stochastic Models for Common Failures*, In Reliability and Quality Control, ed. A.P. Basu, Elsevier Science Publishers B.V. (North Holland), pp. 185-200, 1986.

[117] Heising C.D. and D.M. Luciani, *Application of a Computerized Methodology for Performing Common Cause Failure Analysis : The Mocus-Bacfire Beta Factor (MOBB) Code*, Reliability Engineering, Vol. 17, No. 3, pp. 193-210, 1987.

[118] Hughes R.P., *A New Approach to Common Cause Failure*, Reliability Engineering, Vol. 17, No. 3, pp. 211-236, 1987.

[119] Martin B.R. and R.I. Wright, *A Practical Method of Common Cause Failure Modelling*, Reliability Engineering, Vol. 19, No. 3, pp. 185-199, 1987.

[120] Hokstad P., *A Shock Model for Common-Cause Failures*, Rel. Engg. & Syst. Safety, Vol. 23, No. 2, pp. 127- 145, 1988.

[121] Fullwood R.R. and R.E. Hall, *Probabilistic Risk Assesskent in the Nuclear Power Industry : Fundamentals and Applications*, Pergamon Press, Oxford, 1988.

[122] Eagle K.H., *Redundancy Design Philosophy for Catastrophic Loss Protection*, Proc. Reliability and Maintainability Symposium, pp. 1-4, 1992.

<u>Software Reliability Modeling</u>

[123] Jelinski, Z. and P.B. Moranda, *Software Reliability Research*, in Statistical Computer Performance Evaluation, W.Freiberger (ED), Academic Press, New York, pp. 465-484, 1972.

[124] Shooman M.L., *Operational Testing and Software Reliability Estimation during Program Development*, Proc. Int. Symp. Computer Software Reliability, pp. 51-57, 1973.

[125] Musa J.D., *A Theory of Software Reliability and its Applications*, IEEE Trans. Software Engineering, Vol. SE-1, pp.312-327, 1975.

[126] Moranda P.B., *Prediction of Software Reliability during Debugging*, Proc. Ann. Reliability & Maintainability Symp., Washington D.C., USA, pp. 327-332, 1975.

[127] Schick G.J. and R.W. Wolverton, *An Analysis of Competing Software Reliability Models*, IEEE Trans. on Software Engineering, Vol. SE-4, pp. 104-120, Mar. 1978.

[128] Ramamoorthy C.V. and F.B. Bastani, *Software Reliability-Status and Prospectives*, IEEE Trans. Software Engineering, Vol. SE-8, pp.354-371, 1982.

[129] Goel A.L., *Software Reliability Models : Assumptions, Limitations and Applicability*, IEEE Trans. on Software Engineering, Vol. SE-11, pp. 1411-1423, 1985.

[130] Downs T., *Reliability Problems in Software Engineering - A Review*, Computer Systems Science & Engineering, Vol. 2, pp. 131-147, 1987.

[131] Tohma Y., K. Tokunaga, S. Nagase and Y. Murata, *Structural Approach to the Estimation of the Number of Residual Software Faults Based on the Hypergeometric Distribution*, IEEE Trans. on Software Engineering, Vol. SE-15, pp. 345-355, 1989.

[132] Trachtenberg M., *A General Theory of Software- Reliability Modelling*, IEEE Trans. on Rel., Vol. R-39, No. 1, pp. 92-96, April 1990.

[133] Ohtera H. and S. Yamada, *Optimal Allocation and Control Problems for Software Testing Resources*, IEEE Trans. on Rel., Vol. R-39, No. 5, pp. 171-176, Dec. 1990.

[134] Downs T. and P. Garrone, *Some New Models of Software Testing with Performance Comparisons*, IEEE Trans. on Rel., Vol. R-40, No. 3, pp.322-328, Aug. 1991.

[135] Sofer A. and D.R. Miller, *A Nonparametric Software- Reliability Growth Model*, IEEE Trans. on Rel., Vol. R-40, No. 3, pp. 329-337, Aug. 1991.

[136] Trachtenberg M., *Why Failure Rates Observe Zipf's Law in Operational Software*, IEEE Trans. on Rel., Vol. R-41, No. 3, pp. 386-389, Sept. 1992.

[137] Khoshgoftaar T.M., B.B. Bhattacharyya and G.D. Richardson, *Predicting Software Errors, During Development, Using Nonlinear Regression Models : A Comparative Study*, IEEE Trans. on Rel., Vol. R-41, No. 3, pp. 390-395, Sept. 1992.

<u>Mechanical System Reliability Modeling</u>

[138] Rao S.S., *A Probabilistic Approach to the Design of Gear Trains*, Int. J. of Machine Tool Design and Research, Vol. 14, pp. 267-278, 1974.

[139] Rao S.S. and S.S. Gavane, *Analysis and Synthesis of Mechanical Error in Geneva Mechanisms - A Stochastic Approach : Part I - External Geneva Mechanism*, ASME Journal of Mechanical Design, Vol. 104, pp. 63-71, 1982.

[140] Dumai A. and A. Winkler, *Reliability Prediction Model for Gyroscopes*, Proc. Reliability and Maintainability Symposium, Los Angeles, California, USA, pp. 5-9, 1990.

[141] Benz G.E., I. Bazovsky Sr., *Adapting Mechanical Models to Fit Electronics*, Proc. Reliability and Maintainability Symposium, Los Angeles, California, USA, pp. 153-156, 1990.

[142] Vannoy E.H., *Improving MIL-HDBK-217 Type Models for Predicting Mechanical Reliability*, Proc. Reliability and Maintainability Symposium, Los Angeles, California, USA, pp. 341-345, 1990.

[143] Clark W.B., *Analysis of Reliability Data for Mechanical Systems*, Proc. Reliability and Maintainability Symposium, Orlando, Florida, USA, pp. 438-441, 1991.

[144] Thien-My D., Z. Lin and M. Massoud, *Mechanical Strength Reliability Evaluation Using an Iterative Approach*, Proc. Reliability and Maintainability Symposium, Orlando, Florida, USA, pp. 446-450, 1991.

[145] Barker D.B., A. Dasgupta and M. G. Pecht, *Printed-Wiring-Board Solder-Joint Fatigue-Life Calculations under Thermal and Vibration Loading*, Proc. Reliability and Maintainability Symposium, Orlando, Florida, USA, pp. 451-459, 1991.

[146] Ganguly J., *Fastener-Torque Model to Enhance Reliability under Cyclic Load*, Proc. Reliability and Maintainability Symposium, Las Vegas, Nevada, USA, pp. 196-201, 1992.

[147] English J.R. and T. Sargent, *A Simplified but Accurate Approach for Reliability Computation of a Complex System*, Proc. Reliability and Maintainability Symposium, Las Vegas, Nevada, USA, pp. 436-440, 1992.

<u>*Application of Fuzzy Set Theory*</u>

[148] Zadeh, L.A., *Fuzzy Sets*, Information and Control, Vol. 8, pp. 338-353, 1965.

[149] Zadeh, L.A., *The Concept of a Linguistic Variable and its Application to Approximate Reasoning*, Pts. 1,2 and 3, Information Science, Vol. 8, pp. 199-249, Vol. 8, pp. 301-357 and Vol. 9, pp. 43-80, 1975.

[150] Zadeh L.A., *Fuzzy Sets as a Basis for a Theory of Possibility*, Fuzzy Sets and Systems, Vol. 1, No. 1, pp. 3-28, 1978.

[151] Dubois D. and H. Prade, *Operations on Fuzzy Numbers*, Int. Journal of Syst. Sciences, Vol. 9, pp. 613-626, 1978.

[152] Dubois D. and H. Prade, *Fuzzy Sets and Systems*, Academic Press, New York, 1980.

[153] Misra K.B. and A. Sharma, *Performance Index to Quantify Reliability Using Fuzzy Subset Theory*, Microelectronics and Reliability, Vol. 21, No. 4, pp. 543-549, 1981.

[154] Mizumoto M. and K. Tanaka, *Fuzzy Sets and Their Operations*, Information and Control, Vol. 48, pp. 30-48, 1981.

[155] Tanaka H.L., T. Fan, F.S. Lai and K. Toguchi, *Fault Tree Analysis by Fuzzy Probability*, IEEE Trans. on Rel., Vol. R-32, No. 5, pp. 453-457, Dec. 1983.

[156] Sugeno M., T. Onisawa and Y. Nishiwaki, *A New Approach Based on Fuzzy Sets Concepts to Fault Tree Analysis and Diagnosis of Failures at Nuclear Power Plants*, IAEA Seminar on Diagnosis of and Response to Abnormal Occurrences at Nuclear Power Plants, Dresden, June 1984.

[157] Park K.S., *Fuzzy Apportionment of System Reliability*, IEEE Trans. on Rel., Vol. R-36, No. 1, pp. 129-132, April 1987.

[158] Onisawa T. and Y. Nishiwaki, *Fuzzy Human Reliability Analysis on the Chernobyl Accident*, Fuzzy Sets and Systems, Vol. 28, pp. 115-127, 1988.

[159] Klir G.J. and T.A. Folger, *Fuzzy Sets, Uncertainty and Information*, Prentice-Hall, Inc., Englewood Cliffs, N.J., USA, 1988.

[160] Misra K.B. and G.G. Weber, *A New Method for Fuzzy Fault Tree Analysis*, Microelectronics Reliability, Vol. 29, No. 2, pp. 195-216, 1989.

[161] Keller A.Z. and C. Kara Zaitri, *Further Applications of Fuzzy Logic to Reliability Assessment and Safety Analysis*, Microelectronics and Reliability, Vol. 29, No. 3, pp. 399-404, 1989.

[162] Misra K.B. and G.G. Weber, *Use of Fuzzy Set Theory for Level-1 Studies in Probabilistic Risk Assessment*, Fuzzy Sets and Systems, Vol. 37, pp. 139-160, 1990.

[163] Park K.S. and J.S. Kim, *Fuzzy Weighted-Checklist with Linguistic Variables*, IEEE Trans. on Rel., Vol. R-39, No. 3, pp. 389-393, 1990.

[164] Onisawa T., *An Application of Fuzzy Concepts to Modeling of Reliability Analysis*, Fuzzy Sets and Systems, Vol. 37, pp. 269-286, 1990.

[165] Onisawa T., *Fuzzy Reliability Assessment Considering the Influence of Many Factors on Reliability*, Int. J. of Approx. Reasoning, Vol. 5, pp. 265-280, 1991.

[166] Guth M.A.S., *A Probabilistic Foundation for Vagueness and Imprecision in Fault Tree Analysis*, IEEE Trans. on Rel., Vol. R-40, No. 5, pp. 563-571, Dec. 1991.

[167] Kenaranguie R., *Event-Tree Analysis by Fuzzy Probability*, IEEE Trans. on Rel., Vol. R-40, No. 1, pp. 120- 124, April 1991.

[168] Soman K.P. and K.B. Misra, *Fuzzy Fault Tree Analysis Using Resolution Identity*, Int. J. of Fuzzy Sets and Mathematics, Vol. 1, (to appear) 1993.

<u>Application of Dempster Shafer Theory</u>

[169] Dempster A.P., *Upper and Lower Probabilisties Induced by a Multi-Valued Mapping*, Ann. Math, Statist., Vol. 38, pp. 325-339, 1967.

[170] Shafer G., *Mathematical Theory of Evidence*, Princeton University Press, Princeton, N.J., 1976.

[171] Dubois D. and H. Prade,*Combination and Propagation of Uncertainty with Belief Functions*, Proc. Ninth IJCAI, Vol. 1, pp. 111-113, 1985.

[172] Gordon J. and E.H. Shortcliffe, *Dempster-Shafer Theory of Evidence and its Relevance to the Expert Systems*, In Rule-based Expert Systems - The MYCIN Experiments of the Stanford Heuristic Programming Project, Chapter 13, ed. B.G. Buchanan and E.F. Shortcliffe, Addison-Wesley, Reading, 1984.

[173] Bogler P.L., *Shafer-Dempster Reasoning with Applications to Multisensor Target Identification Systems*, IEEE Trans. Syst. Man. Cybern., Vol. 17, No. 6, pp. 968-977, 1987.

[174] Inagaki T. and Y. Ikebe, *A Mathematical Analysis of Human-Machine Interface Configurations for a Safety Monitoring Systems*, IEEE Trans. on Rel., Vol. R-37, No. 1, pp. 35-40, April 1988.

[175] Inagaki T., *Interdependence Between Safety-Control Policy and Multiple-Sensor Schemes via Dempster-Shafer Theory*, IEEE Trans. on Rel., Vol. R-40, No. 2, pp. 182-188, June 1991.

<u>Expert Systems for Reliability Evaluation</u>

[176] Reggia J.A., D.S. Nau and P.Y. Wang, *Diagnostic Expert Systems Based on a Set Covering Model*, Int. J. Man-Machine Studies, Vol. 19, pp. 437-460, 1983.

[177] Andrew P.K., *Improvement of Operator Reliability Using Expert Systems*, Reliability Engineering, Vol. 14, No. 4, pp. 309-319, 1986.

[178] Hollick A., *DIAFUZZY (Version 1) : An Inference-Engine for Approximate Reasoning*, Interatom GmbH, Berglisch Gladbach, Germany, Sept. 1987.

[179] Grant M.McD. and D.W. Harvey,*A Progress Review on the Development of FIABEX-an Intelligent Knowledge-based System for Automated Safety*, Proc. Reliability and Maintainability Analysis, Vol. 2, pp. 4C/2/1-8, UKAEA, Warrington, UK, 1989.

[180] Lehtela M., *Computer-Aided Failure Mode and Effect Analysis of Electronic Circuits*, Microelectronics and Reliability, Vol. 30, No. 4, pp. 761-773, 1990.

[181] Moureau R., *FURAX : Expert System for Automatic Generation of Reliability Models for Electrical or Fluid Networks*, Proc. 7th International Conf. on Reliability and Maintainability, Brest, France, 1990.

[182] Bowles J.B. and L.A. Klein, *Comparison of Commercial Reliability-Prediction Programs*, Proc. Reliability and Maintainability Symposium, Los Angeles, California, USA, pp. 450-455, 1990.

[183] Howell S.V., S.J. Bavuso and P.J. Haley, *A Graphical Language for Reliability Model-Generation*, Proc. Reliability and Maintainability Symposium, Los Angeles, California, USA, pp. 471-475, 1990.

[184] Elliott, M.S., *Knowledge-Based Systems for Reliability Analysis*, Proc. Reliability and Maintainability Symposium, Los Angeles, California, USA, pp. 481-489, 1990.

[185] Russomanno D.J., R.D. Bonnell and J.B. Bowles, *Computer-Aided FMEA:Toward an Artificial Intelligence Approach*, Fifth International Symposium on Artificial Intelligence, AAAI Press, pp. 103-112, 1992.

[186] Zaitri C.K., A.Z. Keller and P.V. Fleming, *A Smart FMEA (Failure Modes and Effects Analysis) Package*, Proc. Reliability and Maintainability Symposium, Las Vegas, Nevada, USA, pp. 414-421, 1992.

[187] Rawicz, A.H. and H.X. Jiang, *Diagnostic Expert- System for Mechanical Reliability in Heavy Trucks*, Proc. Reliability and Maintainability Symposium, Las Vegas, Nevada, USA, pp. 426-431, 1992.

[188] Hansen W.A., B.N. Edson and P.C. Larter, *Reliability, Availability and Maintainability Expert System (RAMES)*, Proc. Reliability and Maintainability Symposium, Las Vegas, Nevada, USA, pp. 478-482, 1992.

[189] Bavuso S.J. and J.B. Dugan, *HiRel : Reliability/Availability Integrated Work-Station Tools*, Proc. Reliability and Maintainability Symposium, Las Vegas, Nevada, USA, pp. 491-500, 1992.

<u>Reliability Analyzer</u>

[190] Misra K.B. and A.K. Raja, *A Laboratory Model of System Reliability Analyzer*, Microelectronics and Reliability, Vol. 19, No. 3, pp.259-264, 1979.

[191] Bansal V.K. and K.B. Misra, *Hardware Approach for Generating Spanning Trees in Reliability Studies*, Microelectronics and Reliability, Vol. 21, No. 2, pp. 243-253, 1981.

[192] Bansal V.K., K.B. Misra and M.P. Jain, *Minimal Pathsets and Minimal Cutsets using a Search Technique*, Microelectronics and Reliability, Vol. 22, No. 6, pp. 1067-1075, 1982.

[193] Laviron A., A. Carnino and J.C. Manaranche, *ESCAF- A New and Cheap System for Complex Reliability Analysis and Computation*, IEEE Trans. on Rel., Vol. R-31, No. 4, pp. 339-349, Oct. 1982.

[194] Bansal V.K., K.B. Misra and M.P. Jain, *Improved Implementation of a Search Technique to Find Spanning Trees*, Microelectronics and Reliability, Vol. 23, No. 1, pp. 141-147, 1983.

[195] Laviron A. and C.D. Heising, *Error Transmission in Large Complex Fault Trees Using the ESCAF Method*, Reliability Engineering, Vol. 12, No. 3, pp. 181-192, 1985.

[196] Laviron A., *Simulation and Reliability Analysis of Sequential Complex Systems with S. ESCAF*, Reliability Engineering, Vol. 12, No. 3, pp. 139-150, 1985.

[197] Laviron A., A. Blin, J.M. Lanore and C. Rivet, *S.ESCAF : Sequential Complex Systems are Analyzed with ESCAF through an Add-on-Option*, IEEE Trans. on Rel., Vol. R-34, No. 3, pp. 229-232, Aug. 1985.

[198] Laviron A., *Analysing Critical Components in Complex System Design : A Comparison of Methods with the ESCAF Approach*, Reliability Engineering, Vol. 15, No. 3, pp. 209-233, 1986.

[199] Laviron A., *ESCAF for MTBF Time Evolution, Sensitivity Coefficients, Cutset Importance and Non-coherence of Large Systems*, IEEE Trans. on Rel., Vol. R-35, No. 2, pp. 139-144, June 1986.

[200] Blot M. and A. Laviron, *Reliability Analysis with the Simulator S.ESCAF of a very Complex Sequential System : The Electrical Power Supply System of a Nuclear Reactor*, Rel. Engg. and Syst. Safety, Vol. 21, No. 2, pp. 91-106, 1988.

New Trends in System Reliability Evaluation
K.B. Misra (Editor)

CHAPTER 2

MULTIPLE VARIABLE INVERSION TECHNIQUES

Malathi Veeraraghavan
AT&T Bell Laboratories
Holmdel, NJ 07733, U.S.A.

and

Kishor S. Trivedi
Duke University
Durham, NC 27706, U.S.A.

ABSTRACT

Several algorithms have been developed to solve non-series- parallel networks combinatorially for a reliability measure called the *two-terminal network reliability*. This chapter presents a comparative review of four *sum-of-disjoint-products* (sdp) algorithms to solve the two-terminal network reliability problem. The algorithms use *multiple variable inversion* for improved efficiency. Multiple variable inversion based algorithms result in sum-of-disjoint-products expressions with fewer number of terms than algorithms that use only single variable inversion. A comparison of operations used in the different algorithms is provided along with descriptions of the algorithms. One common network example is solved for two- terminal network reliability using each of these algorithms.

2.1 Introduction

Complex networks are used to interconnect multiple processing centers in distributed systems for several applications, such as telecommunications, the airline industry, banking, etc. Distributed processing offers several advantages such as increased performance, improved availability of applications, and decreased costs through resource sharing. In designing such networks, an important consideration is often the network reliability. For the applications cited above, continued availability of communication paths between different processing centers (*network nodes*) is an important requirement. Hence, the networks are often designed with several paths between any two nodes. Ideally, if completely diverse paths were available between every pair of nodes, the probability of there existing at least one path between any two nodes at a given time would be very high. However, cost constraints in designing and maintaining such networks preclude this completely diverse solution. As a compromise, networks are designed such that between any two nodes, while there may exist a few disjoint paths, additional paths that have common links (node to node connection) are also provided. This leads to network designs that are not series-parallel.

This chapter describes several algorithms to model such non-series-parallel networks for a reliability measure called the *two-terminal network reliability*.

This is defined as follows: For any arbitrary network, the probability that a path exists for communication from a specified source node, *s*, to a given destination node, *t*, is referred to as the *two-terminal reliability* [1]. Other network reliability measures also of interest while designing networks are *all-terminal reliability* and *k-terminal reliability*. *All-terminal reliability* is the probability that for every pair of nodes *(a, b)* in the network, there is a path from node *a* to node *b*. The parameter, *k-terminal reliability* for a specified set of *k* nodes, is the probability that a path exists between every pair of nodes in this set.

This work concentrates on solution techniques for the two-terminal network reliability problem which has been researched extensively in the literature. For exact solutions, a survey of the three classes of algorithms, **inclusion-exclusion, sum-of-disjoint-products**, and **factoring**, is provided in [2].

Before we describe the basis for these algorithms at a high-level, we list a few definitions: *structure functions*, *minimal pathsets* and *minimal cutsets*. Let the state vector x be defined as $\mathbf{x} = (x_1, x_2, \cdots, x_n)$ where

$$x_i = \begin{cases} 1, & \text{if component (node/link) } i \text{ is functioning} \\ 0, & \text{if component (node/link) } i \text{ has failed.} \end{cases}$$

The *structure function* ϕ (x) is defined by:

$$\phi(\mathbf{x}) = \begin{cases} 1, & \text{if system is functioning when the state vector is } \mathbf{x} \\ 0, & \text{if system has failed when the state vector is } \mathbf{x}. \end{cases} \tag{2.1}$$

A *minimal pathset (minpath)* is vector **x** for which $\phi(\mathbf{x}) = 1$ and for all vectors $\mathbf{y} < \mathbf{x}$ [1], $\phi(\mathbf{y}) = 0$. A *minimal cutset (mincut)* is a vector **x** for which $\phi(\mathbf{x}) = 0$ and for all vectors $\mathbf{y} > \mathbf{x}$, is $\phi(\mathbf{y}) = 1$. If all the components in at least one minpath are functional, then the system is functional as well. If all the components in a mincut have failed, then the system has also failed.

There are several path enumeration algorithms, for example, Tarjan's algorithm in [3]. The dual of this method exists to find the system unreliability using minimal cutsets [4]. If the set of all the minproducts (minpaths or mincuts), $MP_i, i = 1, 2, \cdots p$, from the source node *s* to the destination node *t* for a network, *G* is already available, then the two-terminal network reliability, $R_{st}(G)$, is given by:

$$R_{st}(G) = \begin{cases} P(\cup_{i=1}^{p} MP_i), & \text{if } MP_i \text{ are minpaths} \\ 1 - P(\cup_{i=1}^{p} MP_i), & \text{if } MP_i \text{ are mincuts} \end{cases} \tag{2.2}$$

where $P(\cdot)$ stands for the probability function. A fault tree model with repeated basic events equivalent to a non-series-parallel network, can be solved for a

[1] A vector $\mathbf{y} < \mathbf{x}$ if $y_i \le x_i$, $i = 1, 2, \cdots, n$, with $y_i < x_i$ for some i.

similar union of a set of terms in its Sum-Of-Products (SOP) expression. This union would yield system unreliability if the top event of the fault tree is *SYSTEM FAILURE*. An algorithm by Bennetts [4] defines a procedure to list the mincuts of a system from its fault tree description.

The *inclusion-exclusion* formula provides a means to evaluate the union on the left-hand side of (2.2) as follows:

$$P(\bigcup_{i=1}^{p} MP_i) = \begin{cases} \sum_{i=1}^{p} P(MP_i) - \sum_{i=1}^{p}\sum_{j=1,j\neq i}^{p} P(MP_i \cap MP_j) + \cdots \\ +(-1)^{p-1}\sum_{i=1}^{p} \sum_{j=1,j\neq i}^{p} \cdots \sum_{n=1,n\neq i,j,\cdots(n-1)}^{p} \\ P(MP_i \cap MP_j \cap \cdots \cap MP_p). \end{cases} \tag{2.3}$$

Sum-of-disjoint products (sdp) algorithms are based on a different formula for evaluating the union in (2.2). This solution is:

$$P\{\bigcup_{i}^{p} MP_i\} = \begin{cases} P\{(MP_1) \cup (\overline{MP}_1 MP_2) \cup (\overline{MP}_1\overline{MP}_2 MP_3) \cup \cdots \cup \\ (\overline{MP}_1\overline{MP}_2\overline{MP}_3 \cdots \overline{MP}_{p-1} MP_p)\} \\ \\ P(MP_1) + P(\overline{MP}_1 MP_2) + P(\overline{MP}_1\overline{MP}_2 MP_3) + \cdots + \\ P(\overline{MP}_1\overline{MP}_2\overline{MP}_3 \cdots \overline{MP}_{p-1} MP_p) \end{cases} \tag{2.4}$$

Several algorithms based on the *sdp* formula have been published. A few references provided here are [5, 6, 7, 8, 9, 10, 11, 2, 12, 13, 14].

Algorithms that do no start with the set of minimal paths or minimal cuts are also used to solve the two-terminal network reliability problem. Satyanarayana and Prabhakar [15] describe an algorithm that uses a form of *inclusion-exclusion*. By using acyclic subgraphs of the network, they obtain only the non-canceling terms in the inclusion-exclusion formula. This algorithm can be used for both undirected and directed networks with both unreliable links and nodes which could be s-dependent.

The *factoring* theorem for the two-terminal reliability or s-t connectedness of graph G is given below. It is described as *pivotal decomposition* in [16]:

$$R(G) = p_e R(G|e \text{ is functional}) + (1 - p_e)R(G|e \text{ is not functional}),$$

where p_e is the probability that edge e in graph G is functional. Moskowitz [17] was the first to use this concept in reliability modeling. He introduced two new graphs G_e and $G - e$ defined as the graphs obtained from G by collapsing and removing the edge e in the graph G, respectively. The factoring theorem can then be rewritten as:

$$R(G) = p_e R(G_e) + (1 - p_e)R(G - e).$$

The main strategy provided by a factoring algorithm is a method to identify the edges of the graph (network) on which to factor. Once an edge e is found, the two graphs G_e and $G - e$ are drawn. Series-parallel reductions are applied to these two graphs. Then a subsequent factoring edge is found in each of these two graphs. This process is continued until all the graphs are series-parallel. Other network decomposition schemes have been presented in [18] and [19].

Methods based on inclusion-exclusion, *sdp* formula and factoring are combinatorial methods for obtaining two-terminal network reliability. The other approach is to use ***state enumeration*** techniques such as Markov models which can then be solved for steady-state or transient measures.

As mentioned earlier, several *sdp* algorithms exist for the two-terminal network reliability problem. This chapter provides a review of a selected set of *sdp* algorithms that use the concept of **multiple variable inversion**. The problem of computing the probability of the union of a set of events (2.4) is NP-hard [20]. Hence to compare different *sdp* algorithms, researchers have often used parameters such as execution times of programs and amount of storage space needed to solve specific example networks. A parameter used in such comparisons is the number of terms in the resulting disjoint product expression. Use of multiple variable inversion decreases this resulting number of terms. The concept is illustrated with a simple example. Typically, a Boolean product consists of single complemented or uncomplemented variables, $\bar{w}_i$ (indicating that component i has failed) and w_i (indicating that component i is functional). So, for example, a Boolean expression representing five components, with components 1 or 3 in failed stated, 4 in a functional state and components 2 and 5 in a *don't care* state, would be represented by a Boolean expression as $\bar{w}_1 w_4 + \bar{w}_3 w_4$, which consists of two Boolean products. A compressed Boolean product that is equivalent to this Boolean expression is $\overline{w_1 w_3} w_4$. This illustrates the use of multiple variable inversion within a Boolean product. Allowing such inverted subproducts within a Boolean product thus results in a significant decrease in the number of terms in the final *sdp* expression.

The first[2] multiple variable inversion technique that was applied to the two-terminal network reliability problem was proposed by Grnarov et al. [8]. They used Miller [21] cube algebra in their algorithm *PROB*. An improvement to this algorithm was published in [2, 12], algorithm *GKG-VT*. Abraham's [7] algorithm using single-variable inversion was extended using multiple variable inversion by Heidtmann in an algorithm, *KDH88* [11]. Soh and Rai [13] published an algorithm called *CAREL* using four newly defined operators, which is also based

[2] to the knowledge of the authors

on multiple variable inversion. These are the four algorithms, (*PROB, GKG-VT, KDH88, CAREL*), reviewed in this chapter.

Section 2.2 describes preliminaries common to all algorithms. A common notation has been adopted for all the algorithms. Four algorithms, *GKG-VT, PROB, KDH88,* and *CAREL* are described in Section 2.3. A comparison of these four algorithms is provided in Section 2.4. The issue of pre-ordering the starting minproducts to obtain the minimal form of the *sdp* expression is briefly addressed in Section 2.5.

2.2 Preliminaries

In this section, nomenclature, notation, assumptions, definitions of relations between Boolean products, and propositions common to all the algorithms reviewed in this chapter are described here.

2.2.1 *Nomenclature*

Nomenclature common to more than one algorithm is listed here with terms specific to an algorithm defined in its corresponding section.

Boolean Product	conjunction (AND) of at least two Boolean expressions [11].
Uniproduct	Boolean product of only uncomplemented variables [11].
Minproduct	Boolean product which represents a minpath or mincut of a coherent structure. This implies that a minproduct is a uniproduct, i.e., it contains only uncomplemented variables [11].
Inverted subproduct	part of a Boolean product which is complemented and is also not contained in another complemented or uncomplemented subproduct [11].
Cube	representation of a Boolean product of *n* variables as a set of vertices of the *n*-dimensional cube, where each vertex corresponds to a minterm of the Boolean product [2, 12].
Subcube	cube representation of a Boolean product that can be expressed without the disjunction (OR) operator[2, 12].

Coordinates of a subcube	correspond to the variables of the Boolean product equivalent to the subcube; values assumed by these coordinates are 1 (uncomplemented Boolean variables), $\bar{\nu}\bar{\nu}$ (inverted subproducts) or x (the constant Boolean True)[2, 12].

2.2.2 *Notation*

Notation common to all algorithms is described here with symbols/operators specific to each algorithm defined in its corresponding section.

n	number of components (links + nodes or basic events)
p	number of minpaths/mincuts
i, j, k, l	subscripts for components
p_i	probability of link/node (component) i is working
q_i	probability that link/node (component) i has failed
w_i	Boolean variable denoting that component i is working
$\bar{w}_i$	Boolean variable denoting that component i has failed
1	at position i stands for w_i denoting that component i is working
x	at position i stands for the constant Boolean True
$\bar{\nu}$	at positions i, j stand for the inverted subproduct $\overline{w_i w_j}$ denoting that one or more of the components i, j have failed
A, B, C, R	cubes
a_i, b_i, c_i	coordinates of cubes, $1 \leq i \leq n$
MP_i	minproduct i, $1 \leq i \leq p$
DP_i	disjoint product corresponding to MP_i, $1 \leq i \leq p$

2.2.3 *Assumptions*

The assumptions listed below apply to the algorithms reviewed in this chapter unless otherwise stated.

1. The structure function of the system is *coherent*, defined in [16], by two conditions: they are increasing functions (i.e., the system reliability increases with component reliability) and each component is relevant. In a fault tree this implies that there are no NOT gates and the minimal sum-of-products/products-of-sum expression is obtained [22]. Similarly, the set of minimal paths or cutsets are used in networks.

2. Fault-occurrences in components are s-independent events.

3. Both nodes and links are allowed to be unreliable.

4. The set of starting minpaths/mincuts is already available.

2.2.4 *Relations*

Two characteristics common to the algorithms described in this chapter are stated here. **First**, the algorithms described in this chapter are all based on the *sdp* formula. Equation (2.4) is solved to obtain the disjoint products. Probabilities of the disjoint products are then summed to obtain the two-terminal network reliability. **Secondly**, all the algorithms described in this chapter use multiple variable inversion as mentioned in Section 2.1. Hence the term *Boolean Product* and its representations *cube, subcube* are defined in Section 2.2.1 to allow inverted subproducts within a Boolean product. Relations between Boolean products that arise due to these characteristics are defined in this section [2, 12].

In (2.4), we are interested in determining sets of the form $\overline{MP}_1 MP_2$. If we let $R = \overline{MP}_1 MP_2$, $B = MP_1$ and $A = MP_2$, R has three properties:

P1: *R is disjoint from B,*
P2: *R is a subset of A, and*
P3: $R \cup B = A \cup B$.

Initially, we start with a cube listing of the minproducts. Then we apply operations on these cubes to make them mutually disjoint. All the initial cubes have only 1 or x, where 1 signifies that the component is in the minproduct represented by this cube and x signifies that the particular minproduct set does not contain the corresponding component. There are no complemented variables in the initial cubes, since by assumption, all components are functional in the initial state of the network. The inverted subproducts are created while obtaining the disjoint products by complementing cubes as per (2.4). For example, if components 1 and 2 form a minpath in a network of five components, the initial cube corresponding to this minpath is $11xxx$. Hence cubes A and B in definitions of these relations have only coordinates 1 or x.

To ensure that the algorithms generate product terms that are disjoint, we need to define the conditions required for two cubes to be disjoint. If a variable or subproduct appears complemented in one cube and uncomplemented in the

Table 2.1: Examples to test for disjointedness of cubes

Cubes	Boolean expressions	Disjoint?
$\bar{1}\bar{1}x11$ and $11xxx$	$\overline{w_1 w_2} w_4 w_5$ and $w_1 w_2$	Yes
$\bar{1}\bar{1}x1\bar{2}$ and $1x1x1$	$\overline{w_1 w_2} w_4 \overline{w}_5$and $w_1 w_3 w_5$	Yes
$\bar{1}\bar{1}x11$ and $x11xx$	$\overline{w_1 w_2} w_4 w_5$ and $w_2 w_3$	No

other, the two cubes are disjoint. Formally, the relation is stated as:

Relation 1: *Given that two cubes R and B have the following forms: $R = r_1 r_2 \cdots r_n$ and $B = b_1 b_2 \cdots b_n$ where $b_i = 1$ or x and $r_i = 1$ or x or $\bar{\nu}$, the cubes R and B are disjoint, i.e., $R \bigcap B = \Phi$, if for some integer k, $\forall\ 1 \leq i \leq n$, for which $r_i = \bar{k}$, $b_i = 1$.*

To state the condition informally, for all variables $\bar{k}$ in R, the corresponding variables in B are 1's. If this is true for at least one integer $\bar{k}$, then the cubes are disjoint. As examples, we give three sets of cubes to test for disjointedness in Table 2.1.

Relation 1 gives the test for the first property, **P1**, disjointedness. We now define the conditions when a cube is a subset of another cube (Property **P2**). Cube R is a subset of cube A if all uncomplemented variables in R correspond to uncomplemented variables or Boolean True variables in A, all Boolean True variables in R correspond to Boolean True variables in A, and an inverted subproduct in R must either correspond to Boolean True variables in A or must also be an inverted subproduct in A. The integer used to denote an inverted subproduct in R could be different from that in A. There could be more than one inverted subproduct in R and in A. Stated formally,

Relation 2: *A cube R is a subset of another cube A, iff the three conditions given below are met. Let $R = r_1 r_2 r_3 \cdots r_n$ and $A = a_1 a_2 a_3 \cdots a_n$, where $r_i = 1, x, \bar{\nu}$ and $a_i = 1, x, \bar{\nu}$.*

1. $\forall 1 \leq i \leq n$, *if* $r_i = 1$ *then* $(a_i = 1) \vee (a_i = x)$.

2. $\forall 1 \leq i \leq n$, *if* $r_i = x$ *then* $a_i = x$.

3. $\forall k, (\forall 1 \leq i \leq n \ni r_i = \bar{k}, a_i = x) \vee (\forall 1 \leq i \leq n \ni r_i = \bar{k}, a_i = \bar{j})$.

For example, $\bar{2}1\bar{2}$ is not a subset of $x1\bar{1}$. In Boolean algebra $\overline{w_1 w_3} w_2 + w_2 \bar{w}_3 \neq$

$w_2\bar{w}_3$. Also, $R = \bar{1}1\bar{1}$ is a subset of $A = \bar{2}x\bar{2}$.

Since Boolean products are allowed to have inverted subproducts in these algorithms, we need to define the conditions for two cubes to be nondisjoint due to the inverted subproduct $\bar{k}$.

Relation 3: *Two cubes are nondisjoint due to $\bar{k}$ if, given that neither cube is a subset of the other:*

1. *The condition of Relation 1 is not satisfied.*
2. *There exists at least one $\bar{k}$ in one cube which has a corresponding* 1 *in the other cube and at least one $\bar{k}$ in the first cube which does not have a corresponding* 1 *in the second cube.*

For example, $\bar{1}1x\bar{1}\bar{2}xx1\bar{2}$ and $x11xx1x11$ are not disjoint but the violation of the disjointedness property is not due to the inverted subproduct $\bar{1}$, but due to the one represented by $\bar{2}$.

2.2.5 *Propositions*

Soh and Rai classify some of the *sdp* algorithms using three propositions [13]. Operations in these algorithms are identified as belonging to one of three propositions. We use the term *Form* **(FI, FII, FIII)**, instead of *Proposition* to avoid confusion with the previously defined three Properties **(P1, P2, P3)** of the cube R (Section 2.2.4). These are described as follows:

Program State: Minproducts $MP_1, MP_2, \cdots MP_j \cdots MP_\ell$ listed.
Products $DP_1, DP_2, \cdots, DP_{j-1}$ computed.
Action: To find disjoint product corresponding to MP_j.

Form I (FI):
In order to make cube MP_j disjoint from each of disjoint products $DP_1, DP_2 \cdots DP_{j-1}$, the cube C_j is computed using:

$$C_j = \bigcup_{i=1}^{j-1} DP_i \mid \text{each literal of } MP_j \rightarrow 1. \tag{2.5}$$

The subcubes in cube C_j obtained using this process are reduced and combined with each other. The resulting cube (possibly a union of subcubes) is then combined with MP_j to obtain the disjoint product DP_j:

$$DP_j = MP_j \; OP_{\mathbf{FI}} \; C_j. \tag{2.6}$$

Form II (FII):
In order to make cube MP_j disjoint from each of the previous minproducts $MP_1, MP_2, \cdots, MP_{j-1}$, the cube C_j is computed using:

$$C_j = \bigcup_{i=1}^{j-1} MP_i \mid \text{each literal of } MP_j \rightarrow 1. \tag{2.7}$$

The subcubes in cube C_j obtained using this process are reduced and combined with each other. These cubes are then combined with MP_j to obtain the disjoint product DP_j:

$$DP_j = MP_j \ OP_{\mathbf{FII}} \ C_j. \tag{2.8}$$

Form III (FIII):
Algorithms based on this Form specify operators of the form $OP_{\mathbf{FIII}}$ that are used as shown below:

$$DP_j = (\cdots((MP_j \ OP_{\mathbf{FIII}} \ MP_1) \ OP_{\mathbf{FIII}} \ MP_2) \cdots \ OP_{\mathbf{FIII}} \ MP_{j-1}). \tag{2.9}$$

Minproduct MP_j is sequentially made disjoint with all the previous minproducts.

2.3 Algorithms

Four algorithms, *GKG-VT*, *PROB*, *KDH88*, and *CAREL* are described in Sections 2.3.1, 2.3.2, 2.3.3 and 2.3.4 respectively. In the description of each algorithm, four subsections are used to *specify additional notation, define operators, list overall procedure*, and finally, *illustrate with a common example*. The example network shown in Fig. 2.1 is repeatedly solved for its two-terminal reliability between nodes *S* and *T*, using each of these four algorithms. Intuitive explanations are provided in the description of the operators of the first algorithm *GKG-VT*. These algorithms are compared in Section 2.4. Algorithms *GKG-VT*, *PROB*, and *KDH88*, are all based on **Form III** while algorithm *CAREL* is based on **Forms I and II**.

2.3.1 *Algorithm GKG-VT*

This algorithm is based on Miller cube operations and the operations described in [8]. A need for modification of the operators defined in [8] for application to the two-terminal reliability problem is demonstrated in [2, 12].

Additional Notation

Only a few additions to the notation specified in Section 2.2 are provided here.

S, D	Cube
$\mathcal{G}, \mathcal{V}$	products of $p_i's, q_i's$

Operators

All the cube operators defined in this section are distributive over union. So we give all the definitions in terms of two cubes, A and B each of which consists of a single subcube [2, 12].

Our goal is to find an operator which evaluates $R = \overline{B} \cdot A$, where A is a cube that represents a Boolean product which may have one or more inverted subproducts and B is a minproduct with variables x or 1. Algorithm *GKG-VT* uses Form FIII defined in (2.9). The resulting cube R should satisfy properties **P1**, **P2** and **P3** defined in Section 2.2.4. A new operator, percent (%), is introduced for this purpose.

Percent (%) Operator

The percent (%) operation, $A\%B$, where $A = a_1 a_2 \cdots a_n$ and $B = b_1 b_2 \cdots b_n$, and $a_i = 1, x,$ or $\bar{\nu}$ and $b_i = 1$ or x:

$$A\%B = \begin{cases} \Phi & \text{if } \forall i, ((a_i = b_i) \vee (b_i = x)) \\ A & \text{if } \exists i \ni ((a_i = \bar{\nu}) \wedge (b_i = 1)) \\ C & \text{otherwise} \end{cases} \tag{2.10}$$

where
$C = c_1 c_2 \cdots c_i \cdots c_n$ and

$$c_i = \begin{cases} a_i & \text{if } (a_i = b_i) \vee (b_i = x) \\ \bar{\mu} & \text{if } (a_i = x) \wedge (b_i = 1) \end{cases}$$

Inverted subproducts are created in the cube C. All complemented variables created in one application of the % operation are assigned the same integer $\bar{\mu}$.

In the following paragraphs, we test to check if cube $R = A\%B$ satisfies properties **P1**, **P2**, and **P3** for each of the three cases in (2.10).

- Case 1: The condition for this case, $\forall i, ((a_i = b_i) \vee (b_i = x))$, implies that cube (set) A is a subset of cube (set) B by Relation 2 (Section 2.2.4). So the cube $A\%B$ that is disjoint from B is the empty set Φ. The empty set Φ is disjoint from B, $\Phi \subset A$ and $\Phi \cup B = A \cup B$.

Table 2.2: Equivalent Cube and Boolean Expressions

Cube	Boolean expression
$xx1$	w_3
111	$w_1w_2w_3$
$\bar{1}\bar{1}1$	$\overline{w_1w_2} \cdot w_3$

- Case 2: The resulting cube might not be disjoint from B. If for a given integer ν and all $a_i = \bar{\nu}$, $b_i = 1$, then the set A is disjoint from B. For example, $\bar{1}\bar{1}1\bar{2}\bar{2}\%x1111 = \bar{1}\bar{1}1\bar{2}\bar{2}$. However, the condition for case 2 requires that there exist at least one i, for which $a_i = \bar{\nu}$ and $b_i = 1$. So if $a_i = \bar{\nu}$ and $b_i = 1$, but $a_j = \bar{\nu}$ and $b_j = x$, and A is not disjoint from B due to some other inverted subproduct, $\bar{\mu}$, then A is not the cube R that satisfies properties **P1**, **P2** and **P3**. For example, cube $\bar{1}\bar{1}1\%1x1(= \bar{1}\bar{1}1)$ does not satisfy property **P1**. The resulting cube $\bar{1}\bar{1}1$ is not disjoint from $1x1$ (by Relation 1). In the Boolean domain, expressions, $\overline{w_1w_2}w_3$ and w_1w_3 are not disjoint.

- Case 3: The set C resulting from the "otherwise" condition satisfies the three properties **P1**, **P2**, and **P3**. For a given ν, for all coordinates $c_i = \bar{\nu}$, the corresponding coordinates in cube B are 1. So set C is disjoint from set B. Since the corresponding coordinates in cube A are x, and all other coordinates in set A and C are equal (first case in definition of set C), set C is a subset of set A. Also, using Venn diagrams, we can see that there is no part of set A that belongs to neither set B nor set C. So $C \cup B = A \cup B$. For example, $xx1\%111 = \bar{1}\bar{1}1$. Table 2.2 relates the two cubes $xx1$ and 111 and the result of the percent operation to Boolean expressions in three variables, w_1, w_2, w_3. The properties, R is disjoint from cube B (by Relation 1), R is a subset of cube A (by Relation 2), and $R \cup B = A \cup B$ ($\bar{1}\bar{1}1 \cup 111 = xx1 \cup 111$), are satisfied.

In conclusion, when the cube A has an inverted subproduct, the resulting cube, $R = A\%B$, may not be disjoint from cube B. Hence, $R = A\%B$ in the cube domain is not equivalent to the Boolean product $R = \overline{B} \cdot A$ which is needed to solve (2.4).

In order to handle inverted subproducts in input cubes, we define the modified cube operator @ [8]. There may be several different inverted subproducts; we handle each subproduct separately. If the current inverted subproduct being handled is represented by the integer $\bar{j}$, then correspondingly, the operator required is $@_j$.

The aim of the operation $A@_jB$ is to obtain a cube which is either disjoint from B or if it is not disjoint, then this is not due to an inverted subproduct represented by j per Relation 3 (Section 2.2.4).

$@_j$ Operator

The $@_j$ operation between two cubes, A and B, where A and B are defined in 2.10:

$$A@_jB = \begin{cases} A & \text{if } \exists\nu \ni \forall(a_i = \bar{\nu}), (b_i = 1) \\ A & \text{if } \forall(a_i = \bar{j}), b_i \neq 1 \\ A & \text{if } \forall i, a_i \neq \bar{j} \\ (A_1\%B) \cup A_x & \text{otherwise} \end{cases} \tag{2.11}$$

where
A_1 is a cube obtained from A by setting to 1 all $a_i = \bar{j}$, for which $b_i = 1$,
A_x is a cube obtained from A by setting to x all $a_i = \bar{j}$, for which $b_i = x$.

In the following paragraphs, we informally describe each of the cases in the definition above and provide illustrations.

- Case 1: The cubes, A and B are already disjoint. It satisfies the condition clause in Relation 1.
- Cases 2 & 3: Either the cubes are not nondisjoint due to $\bar{j}$ (as in Relation 3) or that there is no $\bar{j}$ at all in cube A.
- Case 4: We use the same example as before, $\bar{1}\bar{1}1 \quad @_1 \quad 1x1$. Cubes $\bar{1}\bar{1}1$ and $1x1$ are not disjoint (By Relation 1 or by their Boolean equivalents). The Boolean identity $\overline{w_i w_j} = \overline{w_i} w_j + \overline{w_j}$ is the basis for the A_1 and A_x cubes. A separation of the inverted subproduct identified by integer $\bar{j}$ is carried out in Case 4 of the $@_j$ operation. Cube $\bar{1}\bar{1}1$ is split into $1\bar{1}1(A_1)$ and $\bar{1}x1(A_x)$ $(\overline{w_1 w_2} w_3 = w_1 \overline{w_2} w_3 + \overline{w_1} w_3)$. Since cube B has coordinate 1 in all positions where A_x has $\bar{j}$, cubes A_x and B are disjoint. Cube A_1 has to be made disjoint from B via the percent operation. Table 2.3 shows the $\bar{1}\bar{1}1 \quad @_1 \quad 1x1$ operation with its Boolean equivalents. Cube A_1 is a subset of cube B by Relation 2. So $A_1\%B = \Phi$, the empty set. The resulting cube is A_x, $\bar{1}x1$ (Boolean $\bar{w}_1 w_3$). It is disjoint from $1x1$ and is hence also not nondisjoint due to inverted subproduct $\bar{1}$.

We can now combine the % and the $@_j$ (for all j) operations to get the new exclusive sharp ($) operator which takes care of all the cases. The cube

$R = A\$B$ satisfies properties **P1, P2, P3** for all cases of A. The equivalent Boolean expression is $R = \overline{B} \cdot A$. The \$ operator merely allows computation of R using the compact cube representations of Boolean products A and B where A may have inverted subproducts. Thus this \$ operator can be used in (2.4) written as:

$$P\{\bigcup_{i}^{p} MP_i\} = \begin{cases} P(MP_1) + P(MP_2\$MP_1) + P((MP_3\$MP_1)\$MP_2) + \\ P(\cdots((MP_p\$MP_1)\$MP_2)\$\cdots\$MP_{p-1}) \end{cases} \tag{2.12}$$

Table 2.3: Example for the "otherwise" case of the $@_j$ Operation

Cube name	Cube	Boolean term
A	$\bar{1}\bar{1}1$	$\overline{w_1 w_2} \cdot w_3$
B	$1x1$	$w_1 w_3$
$A@_jB$	$\bar{1}\bar{1}1@_j1x1$	$(\overline{w_1 w_2} \cdot w_3) \cdot (\overline{w_1 w_3})$
A_1	$1\bar{1}1$	$w_1 \overline{w_2} w_3$
$A_1\%B$	Φ	F
A_x	$\bar{1}x1$	$\overline{w_1} w_3$
$(A_1\%B) \cup A_x$	$\bar{1}x1$	$\overline{w_1} w_3$

Exclusive sharp (\$) Operator

The exclusive sharp operation between two cubes, A and B, where A and B are defined in (2.10):

$$A\$B = \begin{cases} \Phi & \text{if } \forall i, ((a_i = b_i) \vee (b_i = x)) \\ D_k & \text{if } \exists i \ni ((a_i = \bar{\nu}) \wedge (b_i = 1)) \\ C & \text{otherwise} \end{cases} \tag{2.13}$$

where Φ is the empty set, $D_j = D_{j-1}@_jB$, for $j = 2, 3, \cdots k$ (where k is the largest value of the symbol(s) $\bar{\nu}$ in cube A). $D_1 = A@_1B$ and C is the same as in (2.10).

A comparison of the definitions shows that the only difference between the exclusive \$ operator and the percent % operator is seen to be in case 2, where the \$ operation results in D_k and the % operation results in A. The percent operation does not always result in a cube R that satisfies the three properties, **P1, P2,** and **P3**, in case 2 as was shown earlier. The \$ operation does so, due to the inclusion of the $@_j$ operator to remove nondisjointedness due to any inverted subproduct $\bar{j}$. In summary, if $A = a_1 a_2 \cdots a_n$ and cube $B = b_1 b_2 \cdots b_n$, where $a_i = 1, x$, or $\bar{\nu}$ and $b_i = 1$ or x:

Operation	Function
$A\%B$	$A\%B$ satisfies **P1, P2, P3** if cubes A and B are not nondisjoint due to any $\bar{j}$
$A@_jB$	$A@_jB$ is either disjoint from B or if is not disjoint, it is not due to any $\bar{j}$
$A\$B$	$A\$B$ satisfies **P1, P2, P3** and is equivalent to $\overline{B} \cdot A$.

Procedure

The complete algorithm *GKG-VT* using the above defined operations is given below [2, 12].

1. Represent the minpaths/mincuts as cubes. The cubes have only 1 or x initially. The system has n components and there are p cubes, MP_1 to MP_p.

2. The algorithm to find the disjoint products is the same for minpaths and mincuts. The assignment of the probabilities of the disjoint products is however different. So first we set an indicator flag to distinguish between the two.
$$flag = \begin{cases} 1 & \text{if } MP_i \text{ are the minpaths} \\ 0 & \text{if } MP_i \text{ are the mincuts} \end{cases} \tag{2.14}$$

3. Sort the cubes according to increasing number of 1's. This is done because the terms which subsume other terms have fewer number of ones, i.e. they are larger sets.

4. Initialize:
$$P_{st} = P(MP_1) = \begin{cases} \prod_i p_i, \forall i, \ni a_i = 1 \text{ in } MP_1 & \text{if } flag = 1 \\ \prod_i q_i, \forall i, \ni a_i = 1 \text{ in } MP_1 & \text{if } flag = 0 \end{cases} \tag{2.15}$$
$q_i = 1 - p_i$.

5. Do loop

   for $\ell = 2, 3, \cdots, p$ do
   begin
   $S_1 = MP_\ell$ /* S is a dynamic variable */

for $j = 1, 2, \cdots \ell - 1$ do
begin
$S_{j+1} := S_j \$ MP_j$ /* use j for complemented variables */
end
$P_{st} := P_{st} + P(S_\ell)$
where if $S_\ell = a_1 a_2 \cdots a_n$ then

$$P(DP_\ell) := \mathcal{G} \cdot \prod_j (1 - \mathcal{V}_j) \tag{2.16}$$

where

$$\mathcal{G} = \begin{cases} \prod p_i \ \forall\, i \text{ such that } a_i = 1 & \text{if } flag = 1 \\ \prod q_i \ \forall\, i \text{ such that } a_i = 1 & \text{if } flag = 0, \end{cases}$$

$$\mathcal{V}_j = \begin{cases} \prod p_i \ \forall\, i \text{ such that } a_i = 0_j & \text{if } flag = 1 \\ \prod q_i \ \forall\, i \text{ such that } a_i = 0_j & \text{if } flag = 0, \end{cases}$$

PRINT $DP_\ell = S_\ell$
end

6. System reliability is given by the following equation:

$$P(\text{ System is up }) = \begin{cases} P_{st} & \text{if } flag = 1 \\ 1 - P_{st} & \text{if } flag = 0 \end{cases} \tag{2.17}$$

The probabilities associated with component failures can be replaced by the distribution functions of their times to failure. This feature is available in the program implementation incorporated in the tool SHARPE(Symbolic Hierarchical Automated Reliability and Performance Evaluator) [23].

Example

We apply Algorithm *GKG-VT* to find the reliability of the network in Fig. 2.1.

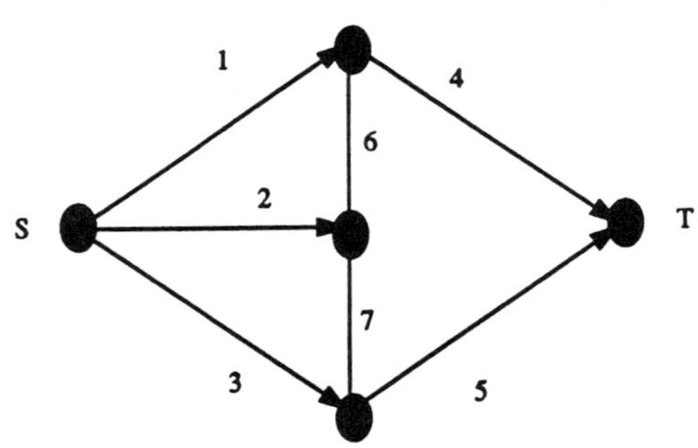

Fig. 2.1 A Sample Network

MP_1	$1xx1xxx$
MP_2	$xx1x1xx$
MP_3	$x1x1x1x$
MP_4	$x1xx1x1$
$P_{st} = MP_1 = p_1p_4$	
$\ell = 2$	
$S_1 = MP_2$	$xx1x1xx$
$j = 1$	
S_2	$S_1\$MP_1$
S_1	$xx1x1xx$
MP_1	$1xx1xxx$
S_2	$\bar{1}x1\bar{1}1xx$
DP_2	$\bar{1}x1\bar{1}1xx$
$P_{st} = P_{st} + p_3p_5(1 - p_1p_4)$	
$\ell = 3$	
$S_1 = MP_3$	$x1x1x1x$
$j = 1$	
S_2	$S_1\$MP_1$
S_1	$x1x1x1x$
MP_1	$1xx1xxx$
S_2	$\bar{1}1x1x1x$
$j = 2$	
S_3	$S_2\$MP_2$
S_2	$\bar{1}1x1x1x$
MP_2	$xx1x1xx$
S_3	$\bar{1}1\bar{2}1\bar{2}1x$
DP_3	$\bar{1}1\bar{2}1\bar{2}1x$
$P_{st} = P_{st} + p_2p_4p_6q_1(1 - p_3p_5)$	

$\ell = 4$	
$S_1 = MP_4$	$x1xx1x1$
$j = 1$	
S_2	$S_1\$MP_1$
S_1	$x1xx1x1$
MP_1	$1xx1xxx$
S_2	$\bar{1}1x\bar{1}1x1$
$j = 2$	
S_3	$S_2\$MP_2$
S_2	$\bar{1}1x\bar{1}1x1$
MP_2	$xx1x1xx$
S_3	$\bar{1}1\bar{2}\bar{1}1x1$
$j = 3$	
S_4	$S_3\$MP_3$
S_3	$\bar{1}1\bar{2}\bar{1}1x1$
MP_3	$x1x1x1x$
S_4	$\bar{1}1\bar{2}11\bar{3}1 \bigcup x1\bar{2}\bar{1}1x1$
DP_4	$\bar{1}1\bar{2}11\bar{3}1 \bigcup x1\bar{2}\bar{1}1x1$
$P_{st} = P_{st} + p_2p_4p_5p_7q_1q_3q_6 + p_2p_5p_7q_3q_4$	
Explanation of the step $\ell = 4$, j=3:	
$D_2 = D_1@_2B$, $D_1 = A@_1B$ where	
$A = (MP_4\$MP_1)\$MP_2$ and $B = MP_3$	
$D_1 = A@_1B$, (2.11) last case	
$D_1 = (A_1\%B) \bigcup A_x$	
$A_1 = \bar{1}1\bar{2}11x1$ (2.11)	
$A_1\%B = \bar{1}1\bar{2}11\bar{3}1$ (2.10)	
$A_x = x1\bar{2}\bar{1}1x1$ (2.11)	
$D_1 = \bar{1}1\bar{2}11\bar{3}1 \bigcup x1\bar{2}\bar{1}1x1$	
$DP_4 = D_2 = D_1$ (2.11 Case 2)	

2.3.2 *Algorithm PROB*

This algorithm is based on Miller cube operators. This operator on cubes A and B, evaluates $\bar{B} \cdot A$. However it generates Boolean products using single variable inversion. Grnarov et al. [8] modified this operator to generate Boolean products with multiple variable inversion.

Additional Notation

Only a few additions to the notation specified in Section 2.2 are provided here.

S_k, D_k	Cube

Operators

Miller Cube (#) Operator

This operator is defined in [21]. The notation has been changed in this section to be consistent with the notation used in this chapter.

The Miller cube # operator [21] is defined as:

$$A\#B = \begin{cases} \Phi & \text{if } a_i\#b_i = z \text{ for all } i \\ A & \text{if } a_i\#b_i = y \text{ for any } i \\ \bigcup_i(a_1 \cdots a_{i-1}\alpha_i a_{i+1} \cdots a_n) & \\ \text{where } \alpha_i = a_i\#b_i = \bar{\nu} \text{ or } 1 & \\ \text{and the union runs over all such } i & \text{otherwise} \end{cases} \tag{2.18}$$

where Φ is the empty set, and $a_i\#b_i$ is given by the coordinate # operation defined below:

b_i / a_i	$\bar{\nu}$	1	x
$\bar{\nu}$	z	y	z
1	y	z	z
x	1	$\bar{\nu}$	z

The Miller cube operation $A\#B$ is equivalent to $\bar{B}\cdot A$. However, it results in cubes that do not contain inverted subproducts. For example, $xx1\#111 = \bar{1}x1 \cup x\bar{1}1$. This is the third case in (2.18). Only single variables are inverted in each of the terms in the resulting product. Since this operator only supports single variable inversion, the number of terms in the *sdp* expression is much larger than if multiple variable inversion is used.

Exclusive Sharp ($) Operator

Grnarov et al. defined this operator as a modification to the Miller cube operator in order to allow inverted subproducts in the resulting Boolean products.

The exclusive sharp operation on cubes is defined as:

$$A\$B = \begin{cases} \Phi & \text{if } a_i\#b_i = z \text{ for all } i \\ D_k & \text{if } a_i\#b_i = y \text{ for any } i \\ C & \text{otherwise} \end{cases} \tag{2.19}$$

where Φ is the empty set, $a_i\#b_i$ is given by the coordinate # operation, $D_j = D_{j-1}@_jB$, for $j = 2,3,\cdots k$ (where k is the largest step number in which any symbol(s) $\bar{j}$ is generated). $D_1 = A@_1B$ and cube $C = c_1c_2\cdots c_i\cdots c_n$ and $c_i = a_i\#b_i$.

The $@_j$ operation is defined as:

$$A@_jB = \begin{cases} A & \text{if } a_i\#b_i = y \text{ for any } \bar{\nu} \\ A & \text{if } a_i\#b_i = y \text{ for all } a_i = \bar{\nu} \text{ for any } \nu \\ A_1\#B \cup A_x & \text{otherwise} \end{cases} \qquad (2.20)$$

where
A_1 is a cube obtained from A by setting to 1 all $a_i = \bar{j}$ for which $a_i\#b_i = y$,

A_x is a cube obtained from A by setting to x all $a_i = \bar{j}$, for which $a_i\#b_i \neq y$.

The motivation for the $@_j$ operator is explained in Section 2.3.1. It is used to separate out an inverted subproduct if this subproduct causes two cubes to be non-disjoint.

Procedure

A listing of algorithm *PROB* is given below:

step 1. Find all the simple s-t paths (minpaths).
step 2. Sort them in increasing order of the number of 1's.
step 3. Set $P_{st} = P(MP_1)$.
step 4. (loop) For $\ell = 2,3,\cdots p$ determine

$$S_\ell = (\cdots((MP_\ell\$MP_1)\$MP_2\cdots)\$MP_{\ell-1}$$
$$P_{st} = P_{st} + P(S_\ell)$$

step 5. End

Example

We apply Algorithm *PROB* to find the reliability of the network in Fig. 2.1.

Steps 1, 2	
MP_1	$1xx1xxx$
MP_2	$xx1x1xx$
MP_3	$x1x1x1x$
MP_4	$x1xx1x1$
Step 3	
$P_{st} = MP_1 = p_1p_4$	
Step 4	
$\ell = 2$	
S_2	$MP_2\$MP_1$
MP_2	$xx1x1xx$
MP_1	$1xx1xxx$
S_2	$\bar{1}x1\bar{1}1xx$
$P_{st} = P_{st} + p_3p_5(1 - p_1p_4)$	
$\ell = 3$	
S_3	$(MP_3\$MP_1)\$MP_2$
MP_3	$x1x1x1x$
MP_1	$1xx1xxx$
	$\bar{1}1x1x1x$
MP_2	$xx1x1xx$
S_3	$\bar{1}1\bar{2}1\bar{2}1x$
$P_{st} = P_{st} + p_2p_4p_6q_1(1 - p_3p_5)$	
$\ell = 4$	
S_4	$((MP_4\$MP_1)\$MP_2)\$MP_3$
MP_4	$x1xx1x1$
MP_1	$1xx1xxx$
	$\bar{1}1x\bar{1}1x1$
MP_2	$xx1x1xx$
	$\bar{1}1\bar{2}\bar{1}1x1$
MP_3	$x1x1x1x$
S_4	$\bar{1}1\bar{2}11\bar{3}1 \cup x1\bar{2}\bar{1}1x1$
$P_{st} = P_{st} + p_2p_4p_5p_7q_1q_3q_6 + p_2p_5p_7q_3q_4$	

2.3.3 *Algorithm KDH88*

This algorithm is a modification of the algorithm by [7] to use multiple variable inversion for the two-terminal network reliability problem.

Additional Notation

A change in the notation specified in Section 2.2 is provided here. The variable DP_i is redefined to represent any subcube in the set representing the disjoint product corresponding to minproduct MP_i. In the previous algorithms, DP_i represented the whole disjoint product corresponding to minproduct MP_i. To illustrate this change of notation consider the disjoint product corresponding to minproduct MP_4 for the example network of Fig. 2.1. Each subcube in DP_4 = $\bar{1}1\bar{2}11\bar{3}1 \cup x1\bar{2}\bar{1}1x1$, would be represented as DP_4 in the *KDH88* algorithm described in this section.

DP_i	any subcube in the disjoint product corresponding to MP_i

Operators

No new operators are defined for algorithm *KDH88*.

Procedure

1. List the minproducts. The system has n components and there are p minproducts, MP_1 to MP_p.

```
2.  for ℓ = 1, 2, ···, p do
    begin
3.      DPℓ = MPℓ
4.      for j = 1, 2, ··· ℓ do
        begin
5.          if j ≥ ℓ, print DPℓ
            else
            begin
6.              Mask(j,DPℓ, MPj,mask,r,Disjoint)
7.              if (Disjoint = False)
                begin
8.                  for i = 1, 2, ···, r
                    begin
9.                      DPℓ^x = DPℓ with all variables ¬mask[i] of DPℓ
                        replaced by x where MPj is x;
                        DPℓ=DPℓ^x
                        Go to step 4 for next j.
10.                     DPℓ^1 = DPℓ with all variables ¬mask[i] of DPℓ
                        replaced by 1 where MPj is 1;
                        DPℓ=DPℓ^1
                    end
11.                 DPℓ = DPℓ with all positions where MPj is 1
                    substituted in DPℓ with j. If this does not
                    happen at least once, drop from inner loop.
                end
            end
        end
    end
12. STOP
```

Procedure $Mask(j, DP_\ell, MP_j, \mathbf{mask}, r, Disjoint)$

1. $Disjoint$=False; r=0.

2. for $s = 1, 2, \cdots, j-1$ and Disjoint = False BEGIN
3. Find all positions in DP_ℓ that equal s
4. If at these positions, MP_j is 1 THEN set Disjoint = True ELSE BEGIN
5. If at these positions MP_j is equal to 1 at least once and is equal to x at least once, then increment r.
6. Set $\mathbf{mask}[r]$ to s END END

Example

We apply Algorithm *KDH88* to find the two-terminal reliability of the network in Fig. 2.1.

Step 1	MP_1	$1xx1xxx$
	MP_2	$xx1x1xx$
	MP_3	$x1x1x1x$
	MP_4	$x1xx1x1$
	n, p	7, 4
Step 2	$\ell = 1$	
Step 3	DP_ℓ	$1xx1xxx$
Step 4	$j = 1$	
Step 5	$j \geq \ell$	Print DP_1
Step 2	$\ell = 2$	
Step 3	DP_ℓ	$xx1x1xx$
Step 4	$j = 1$	
Step 5	$j < \ell$	
Step 6	*Mask*	(0, False)
Step 7	*Disjoint*	False
Step 8	r	0
Step 11	DP_2	$\bar{1}x1\bar{1}1xx$
Step 4	$j = 2$	
Step 5	$j \geq \ell$	Print DP_2
Step 2	$\ell = 3$	
Step 3	DP_ℓ	$x1x1x1x$
Step 4	$j = 1$	
Step 5	$j < \ell$	
Step 6	*Mask*	(0,False)
Step 7	*Disjoint*	False
Step 8	r	0
Step 11	DP_3	$\bar{1}1x1x1x$
Step 4	$j = 2$	
Step 5	$j < \ell$	
Step 6	*Mask*	(0,False)
Step 7	*Disjoint*	False
Step 8	r	0
Step 11	DP_3	$\bar{1}1\bar{2}1\bar{2}1x$

Step 4	$j = 3$	
Step 5	$j \geq \ell$	Print DP_3
Step 2	$\ell = 4$	
Step 3	DP_ℓ	$x1xx1x1$
Step 4	$j = 1$	
Step 5	$j < \ell$	
Step 6	*Mask*	(0,False)
Step 7	*Disjoint*	False
Step 8	r	0
Step 11	DP_4	$\bar{1}1x\bar{1}1x1$
Step 4	$j = 2$	
Step 5	$j < \ell$	
Step 6	*Mask*	(0,False)
Step 7	*Disjoint*	False
Step 8	r	0
Step 11	DP_4	$\bar{1}1\bar{2}\bar{1}1x1$
Step 4	$j = 3$	
Step 5	$j < \ell$	
Step 6	*Mask*	(1,False)
	$\mathbf{mask}[1]$	1
Step 7	*Disjoint*	False
Step 8	i, r	1
Step 9	DP_4^x	$x1\bar{2}\bar{1}1x1$
	DP_4	DP_4^x
Step 4	$j = 4$	
Step 5	$j \geq \ell$	Print DP_4
Step 10	DP_4^1	$\bar{1}1\bar{2}11x1$
	DP_4	DP_4^1
Step 11	DP_4	$\bar{1}1\bar{2}11\bar{3}1$
Step 4	$j = 5$	
Step 5	$j \geq \ell$	Print DP_4

2.3.4 *Algorithm CAREL*

CAREL uses operators that are based on the two forms, FI and FII (Section 2.2.5).

Additional Notation

A change in the notation specified in Section 2.2 is provided here. For certain operators, it is beneficial to represent the cubes using a different notation from that listed in 2.2.2. Table 2.4 shows the modified (*Mod.*) notation as well as the standard (*Std.*) notation (as per Section 2.2.2)[3]. For example, if x is replaced by 0 to represent the *don't care* state of a variable, bit manipulation (Boolean algebra) can be used. Also, a subscript is used with variables representing inverted subproducts denoting the number of variables in the inverted subproduct. For example, the set $\overline{w_1}w_2\overline{w_4w_5w_6}$ is represented as $\bar{1}_1 10\bar{2}_3\bar{2}_3\bar{2}_3$ with the modified notation instead of $\bar{1}1x\bar{2}\bar{2}\bar{2}$. Component 3 is in a *don't care* state represented by 0, and the subscript 3 is used for the inverted subproduct, 2, since there are 3 variables ($\overline{w_4w_5w_6}$) in this subproduct ($\bar{2}_3\bar{2}_3\bar{2}_3$). Similarly, the inverted subproduct 1 has only one variable ($\overline{w_1}$), and hence the subscript 1 ($\bar{1}_1$). The modified notation is **only used when specifically needed** for an operator and the standard notation is reverted to after the operation. In the description of algorithm *CAREL* usage of the term *standard* or *modified* notation can be interpreted from Table 2.4. Also, any one feature of the *modified* notation may be used with the remaining variables in the *standard* notation. For example, a cube may be represented as $011\bar{2}\bar{2}\bar{1}$, where 0 stands for the *don't care* state (i.e., *modified* notation), but the inverted subproducts ($\bar{1}$, $\bar{2}\bar{2}$) do not have a subscripts (i.e., *standard* notation).

Table 2.4: Additional Notation used in *CAREL*

Std.	*Mod.*	Represents
1	1	at position i stands for w_i, denoting that component i is working
x	0	at position i stands for the constant Boolean True
$\bar{\nu}$	$\bar{\nu}_\alpha$	the same negative integer at positions i, j, k stand for $\overline{w_i w_j w_k}$ where α is the number of variables in the inverted subproduct
	C'_j	Cube with variables in only one inverted subproduct and *don't care* variables represented by 1 and 0 respectively

Operators

[3]The modified notation introduced here is not the same as the one used in [13].

In this section, six operators are defined, *pl*, *df*, *COM*(\), *RED*(/), *CMB*(*) and *GEN*(⊕).

The *pl* and *df* operators are defined to test cubes for redundancy. The *pl* operator is similar to the Boolean OR operator and the *df* operator similar to the Boolean XOR operator.

pl Operator

This operator applies between two cubes $A = a_1, a_2, \cdots, a_n$ and $B = b_1 b_2, \cdots, b_n$, where $a_i, b_i \epsilon \{0, 1, \bar{\nu}\}$. The *modified* notation is used to allow application of Boolean like operators.

a_i	b_i	a_i *pl* b_i
0	0	0
0	1	1
1	0	1
1	1	1
0	$\bar{\nu}$	$\bar{\nu}$
$\bar{\nu}$	0	$\bar{\nu}$

Other combinations of the two input variables are not applicable for this operator.

df Operator

This operator applies between two cubes $A = a_1, a_2, \cdots, a_n$ and $B = b_1 b_2, \cdots, b_n$, where $a_i, b_i \epsilon \{0, 1\}$. The *modified* notation is used to allow application of Boolean like operators.

A *df* B is the bitwise *XOR* operation on the coordinates of A and B.

COM(\) Operator

This operator performs a COMpare on two cubes to find the variables at which the two cubes differ. It has two definitions based on the form used. Input cubes to the COM operator have the coordinates $a_i \epsilon \{1, x\}$ and $b_i \epsilon \{1, x, \bar{\nu}\}$, where a_i are the coordinates of the first operand and b_i are coordinates of the second operand. The result of the COM operation is cube, $C_j = c_1 c_2, \cdots, c_n$, where $c_i \epsilon \{x, \bar{\nu}\}$. The Form **FI** version of the COM operator uses operands MP_k and DP_j as in (2.5) to generate cubes C_j. Similarly, the Form **FII** version of the

COM operator uses operands MP_k and MP_j as in (2.7).

$COM(\backslash)$ operator: FI

$$MP_k \setminus DP_j = \begin{cases} \Phi, & \textit{if } a_i = 1 \textit{ in all those positions where } b_i = \bar{\nu} \\ C_j, & \textit{otherwise} \end{cases} \tag{2.21}$$

where C_j is $(c_1c_2\cdots c_n)$. Coordinate c_i is given by $a_i \setminus b_i$ defined as follows:

a_i \ b_i	$\bar{\nu}$	x	1
x	x	x	$\bar{\nu}$
1	x	x	x

An example of the COM operator for form **FI** is as follows: $MP_k = xx11x, DP_j = 11xx\bar{1}, MP_k \setminus DP_j = \bar{2}\bar{2}xxx$.

Modified notation is used for definition of the COM operator using Form FII. To simplify this operation, replace x by 0 as in Table 2.4. Since the input cubes are minproducts (2.7), input coordinates are either x or 1. Using 0 for x allows bit manipulation. At the end of the operations, replace variable 1 in cube C'_j by $\bar{\nu}$ to obtain cube C_j consistent with the resulting cube C_j obtained from *COM FI* operator. This cube, C_j, only consists of variables, $\bar{\nu}$ or x.

$COM(\backslash)$ operator: FII

$$C'_j = MP_k \setminus MP_j = (MP_k \;\; pl \;\; MP_j) \;\; df \;\; MP_k \tag{2.22}$$

For example, if $MP_k = x1x1x$, $MP_j = xx11x$, using the *modified* notation $MP_k = 01010$, $MP_j = 00110$, $MP_k \;\; pl \;\; MP_j = 01110$. The cube $C'_j = (MP_k \;\; pl \;\; MP_j) \;\; df \;\; MP_k = 01110 \;\; df \;\; 01010 = 00100$. Converting this back to the *standard* notation results in cube $C_j = xx\bar{1}xx$.

RED(/) Operator

Since the cubes C_j in (2.21) and (2.22) are formed independently from each set DP_i, $i = 1, 2, \cdots, j-1$ or MP_i, $i = 1, 2, \cdots, j-1$, respectively the next operation needed is to REDuce these cubes C_j by removing redundant and duplicate cubes. These cubes are listed as $C_1, C_2, \cdots C_R$. This operator also has an **FI** and **FII** version.

Modified notation is used for this operator. A subscript is added for each inverted subproduct, $\bar{\nu}$, to indicate the number of variables in each inverted subproduct.

RED (/) operator: FI

Let cube $C_j = c_1, c_2, \cdots, c_n$ *and cube* $C_k = d_1, d_2, \cdots, d_n$,

$$C_j/C_k = \begin{cases} C_j, & \text{if } c_i = \bar{\nu}_\alpha \text{ in all those } \alpha \text{ positions} \\ & \text{where } d_i = \bar{\mu}_\beta \text{ and } \beta > \alpha \\ C_k, & \text{if } d_i = \bar{\mu}_\beta \text{ in all those } \beta \text{ positions} \\ & \text{where } c_i = \bar{\nu}_\alpha \text{ and } \alpha > \beta \\ \text{Retain } C_j \text{ and } C_k, & \text{otherwise.} \end{cases} \tag{2.23}$$

For the Form **FII** definition of the operator below, *modified notation* is used in that the *don't care* variable is 0 and $\bar{\nu}$ variables are replaced by the variable 1. There is no conflict between this variable 1, and the variable 1 representing a working component, because cubes C_j have only variables $\bar{\nu}$ and x. These cubes are labeled C_j' (Table 2.4).

RED (/) operator: FII

$$C_j'/C_k' = \begin{cases} C_j' & \text{If } ((C_j' \ pl \ C_k') \ df \ C_k') = 0 \\ C_k' & \text{If } ((C_j' \ pl \ C_k') \ df \ C_j') = 0 \\ \text{Retain } C_j' \text{ and } C_k', & \text{otherwise} \end{cases} \tag{2.24}$$

Replace cubes C_j' *by* C_j *by replacing variables* 1 *by* $\bar{\nu}$.

For example, on application of the RED operator on cubes $\bar{1}_2\bar{1}_2xxx(\bar{1}_2\bar{1}_2000)$ and $\bar{2}_1xxxx(\bar{2}_10000)$, results in the elimination of cube $\bar{1}_2\bar{1}_2xxx(\bar{1}_2\bar{1}_2000)$. We assume that at the start of the RED operation, we had R cubes in set C. This is reduced to r cubes via the $RED(/)$ operator described above.

As a second step, in this operation, the set of cubes, C, are partitioned into two subsets, Independent Group, IG and Dependent Group, DG, where η cubes are in the set IG and κ cubes are in the set DG. The number of cubes $r = \eta + \kappa$, implying that a cube has to belong to one of these two sets IG or DG. Cubes in the Independent Group IG are mutually disjoint, while cubes in DG are dependent. Let cube $C_j = c_1, c_2, \cdots, c_n$ and cube $C_k = d_1, d_2, \cdots, d_n$. Coordinates $c_i, d_i \epsilon \{\bar{\nu}, 0\}$.

<u>Criterion for Partitioning Set of Cubes C</u>: For $i = 1, 2, \cdots, n$, *if there exists at least one pair* (c_i, d_i), *for* $c_i, d_i \neq 0$, *then cubes,* C_j, C_k *belong to set DG. If a cube* C_j *has no such pair with any other cube in the set of cubes C, then cube* C_j *belongs to subset IG.*

For example, if set C has cubes C_1, C_2, and C_3 defined as $C_1 = \bar{1}xx\bar{1}xxx$, $C_2 = xx\bar{2}xxxx$, and $C_3 = xxx\bar{3}x\bar{3}x$, then this set is partitioned into subset IG which consists of one entry, C_2 and subset DG which consists of two cubes, C_1 and C_3.

CMB (*) Operator

This is the combine operator. Combining cubes in the IG set is relatively straightforward since the cubes in this set are mutually disjoint. As there are no pairs $(c_i, d_i) \neq 0$ in cubes belonging to subset IG, this combining operator, creates a cube which has all the independent inverted subproducts from each of the cubes in subset IG. Thus, one cube IG_η is created by repeated application of the CMB operator. The CMB operator applied to cubes in subset DG is used to separate out inverted subproducts. The separated cubes are then recombined if possible to create a list of cubes labeled TG_j that are mutually disjoint.

Let $IG = \{IG_1, IG_2, \cdots IG_\eta\}$ *and* $DG = \{DG_1, DG_2, \cdots DG_\kappa\}$

COMBINE CUBES IN INDEPENDENT SET.

For $j = 1, 2, \cdots, \eta - 1$ *do*

$$IG_{j+1} = IG_j * IG_{j+1} = IG_j \ pl \ IG_{j+1} \tag{2.25}$$

Resulting cube is IG_η.

COMBINE CUBES IN DEPENDENT SET.

Find cubes TG_1 *and* TG_2 *from cubes* DG_1 *and* DG_2

TG_1 has $\bar{\nu}_\alpha$ in α positions where cubes DG_1 and DG_2 are both negative, while other positions are x.

TG_2 has 1 in α positions and its other positions are generated using $x \ pl \ \bar{\nu} = \bar{\nu} \ pl \ x = \bar{\nu}$ and $x \ pl \ x = x$.

Compute the subscript for the terms $\bar{\nu}$ created in TG_2. For $j = 3, 4, \cdots, \kappa$

begin

For $i = 1, 2, \cdots, s$

begin

Call **DG**(TG_i, DG_j, TG'_i) *Terminate and* TG'_i *are returned*
If (Terminate) i = i+ 1;
end
Create new set TG*, such that* $TG'_i \ \epsilon \ TG$
$TG = TG_1, TG_2, \cdots TG_s$
end
Resulting set $TG = TG_1, TG_2, \cdots TG_s$.

Details of the Procedure **DG** can be found in [13]. The cubes in set TG are then recast into *standard* notation by removing the subscripts of inverted subproducts.

For example, two cubes in set IG, $\bar{1}x\bar{1}xx$ and $x\bar{2}x\bar{2}\bar{2}$ are combined to form $IG_\eta = \bar{1}\bar{2}\bar{1}\bar{2}\bar{2}$. Two cubes in set DG, $\bar{1}x\bar{1}xx$ and $\bar{1}\bar{2}x\bar{2}\bar{2}$ are combined to generate two cubes in set TG, $TG_1 = \bar{1}xxxx$ and $TG_2 = 1\bar{2}\bar{1}\bar{2}\bar{2}$.

<u>$GEN(\oplus)$ Operator</u>

This operator is used to combine the minproduct MP_ℓ with the cubes IG_η and $TG_1, TG_2, \cdots TG_s$. These latter cubes are generated using the *COM*, *RED*, and *CMB* operators on the previous $\ell - 1$ starting minproducts. This operator is useful to the $OP_{\mathbf{FI}}$ and $OP_{\mathbf{FII}}$ operations in Section 2.2.5, where the minproduct MP_j is combined with the cubes C_j. The combine operator operates on the C_j cubes to form the IG_η and cubes in set TG.

Input cubes to the GEN operator are cube $MP_\ell = a_1a_2, \cdots, a_n$ where $a_i \epsilon \{1, x\}$, cube $IG_\eta = b_1b_2, \cdots, b_n$, where $b_i \epsilon \{\bar{\nu}, 0\}$, and cube $TG_j = c_1c_2, \cdots, c_n$, where $c_i \epsilon \{\bar{\nu}, x, 1\}$.

For each cube $TG_j, j = 1, 2, \cdots, s$*, a sub-cube contributing to the disjoint product* DP_ℓ *is GENerated using:*

$$\textit{Subcube in } DP_\ell = MP_\ell \oplus IG_\eta \oplus TG_j, \tag{2.26}$$

where $\oplus$ *is defined in Table 2.5.*

Only five of the combinations from a total of 12 are shown in Table 2.5. See [13] for a proof showing that other combinations are not possible. For example, if $MP_4 = x1xx1x1$, $IG_1 = xx\bar{2}xxxx$, and $TG_1 = xxx\bar{1}xxx$, $MP_4 \oplus IG_1 \oplus TG_1 = x1\bar{2}\bar{1}1x1$.

Procedure

1. Sort the minproducts MP according to increasing number of 1's. For cubes with the same number of 1's, use lexicographic ordering.

Table 2.5: Definition of the GEN(⊕) operator

a_i	b_i	c_i	$a_i \oplus b_i \oplus c_i$
x	x	x	x
1	x	x	1
x	$\bar{\nu}$	x	$\bar{\nu}$
x	x	1	1
x	x	$\bar{\nu}$	$\bar{\nu}$

2. $DP_1 = MP_1$

3. Do loop

For $\ell = 2, 3, \cdots, p$ do
begin

COMPARE OPERATION

For $j = 1, 2, \cdots, \ell - 1$
Generate set of R cubes $C_1, C_2, \cdots, C_R$ using
$MP_\ell \setminus MP_j$ (2.22) or $MP_\ell \setminus DP_j$ (2.21)
Inverted subproducts created at step j are represented by integer $\bar{j}$

REDUCE OPERATION

For $j = 1, 2, \cdots, R - 1$ do
For $k = j + 1, \cdots, R$ do
Generate set of r cubes $C_1, C_2, \cdots, C_r$ using
$RED(C_j, C_k)$ (2.23) or (2.24)

Split set of cubes $C = C_1, \cdots, C_r$ into two subsets IG and DG
Use *Criterion for Partitioning Set of Cubes C*
$IG = IG_1, IG_2, \cdots IG_\eta$
$DG = DG_1, DG_2, \cdots IG_\kappa$, where $\eta + \kappa = r$

COMBINE OPERATION

CUBES IN SET IG

Form one cube IG_η from cubes in set IG using (2.25).

CUBES IN SET DG

Initialize set $TG = \Phi$, $s = 1$

Obtain cubes $TG_1, TG_2, \cdots TG_s$ by combining cubes in DG.

GENERATE OPERATION

$DP_\ell = \Phi$

For $j = 1, 2, \cdots, s$

$$DP_\ell = DP_\ell \bigcup (MP_\ell \oplus IG_\eta \oplus TG_j)$$

Print DP_ℓ.

end

Example

We apply Algorithm *CAREL* to find the two-terminal reliability of the network in Fig. 2.1. The Form **FII** versions of the COM and RED operators have been assumed in this illustration.

Step 1	
MP_1	$1xx1xxx$
MP_2	$xx1x1xx$
MP_3	$x1x1x1x$
MP_4	$x1xx1x1$
Step 2	
DP_1	MP_1
Print $DP_1 = 1xx1xxx$	
Step 3	
$\ell = 2$	
COMPARE - FII	
$j = 1$	
MP_2	0010100
MP_1	1001000
$MP_2 \setminus MP_1$	1011100
MP_2	0010100
C_1'	1001000
C_1	$\bar{1}xx\bar{1}xxx$
R	1
REDUCE - FII	
r	1
$IG_1; \eta$	$C_1; 1$
COMBINE IG	
IG_1	$\bar{1}xx\bar{1}xxx$
GENERATE	
$j = 1$	
MP_2	$xx1x1xx$
IG_1	$\bar{1}xx\bar{1}xxx$
DP_2	$\bar{1}x1\bar{1}1xx$
Print $DP_2 = \bar{1}x1\bar{1}1xx$	
$\ell = 3$	
COMPARE - FII	
$j = 1$	
MP_3	0101010
MP_1	1001000
$MP_3 \setminus MP_1$	1101010
MP_3	0101010

C_1'	1000000
C_1	$\bar{1}xxxxxx$
j = 2	
MP_3	0101010
MP_2	0010100
$MP_3 \setminus MP_2$	0111110
MP_3	0101010
C_2'	0010100
C_2	$xx\bar{2}x\bar{2}xx$
R	2
REDUCE - FII	
r	2
$IG_1; IG_2; \eta$	$C_1; C_2; 2$
COMBINE IG	
IG_1	$\bar{1}xxxxxx$
IG_2	$xx\bar{2}x\bar{2}xx$
IG_2	$\bar{1}x\bar{2}x\bar{2}xx$
GENERATE	
$j = 1$	
MP_3	$x1x1x1x$
IG_2	$\bar{1}x\bar{2}x\bar{2}xx$
DP_3	$\bar{1}1\bar{2}1\bar{2}1x$
Print $DP_3 = \bar{1}1\bar{2}1\bar{2}1x$	
$\ell = 4$	
COMPARE - FII	
j = 1	
MP_4	0100101
MP_1	1001000
$MP_4 \setminus MP_1$	1101101
MP_1	0100101
C_1'	1001000
C_1	$\bar{1}xx\bar{1}xxx$
j= 2	
MP_4	0100101
MP_2	0010100
$MP_4 \setminus MP_2$	0110101

MP_4	0100101
C_2'	0010000
C_2	$xx\bar{2}xxxx$
j = 3	
MP_4	0100101
MP_3	0101010
$MP_4 \setminus MP_3$	0101111
MP_4	0100101
C_3'	0001010
C_3	$xxx\bar{3}x\bar{3}x$
R	3
REDUCE - FII	
r	3
$IG_1; \eta$	$C_2; 1$
DG_1	C_1
$DG_2; \kappa$	$C_3; 2$
COMBINE IG	
IG_1	$xx\bar{2}xxxx$
COMBINE DG	
DG_1	$\bar{1}_2 00 \bar{1}_2 000$
DG_2	$000\bar{3}_2 0\bar{3}_2 0$
TG_1	$000\bar{1}_1 000$
TG_2	$\bar{1}_1 0010\bar{3}_1 0$
s	2
GENERATE	
j = 1	
MP_4	$x1xx1x1$
IG_1	$xx\bar{2}xxxx$
TG_1	$xxx\bar{1}xxx$
DP_4	$x1\bar{2}\bar{1}1x1$
j = 2	
MP_4	$x1xx1x1$
IG_1	$xx\bar{2}xxxx$
TG_2	$\bar{1}xx1x\bar{3}x$
	$\bar{1}1\bar{2}11\bar{3}1$
DP_4	$DP_4 \bigcup \bar{1}1\bar{2}11\bar{3}1$
Print $DP_4 = x1\bar{2}\bar{1}1x1 \bigcup \bar{1}1\bar{2}11\bar{3}1$	

2.4 Comparisons

As described in Section 2.3, intuitive explanations for different aspects of multiple variable inversion algorithms is provided for algorithm *GKG-VT*. In this section, operators and procedures used in the other three algorithms, *PROB*, *KDH88*, and *CAREL* are compared to those used in algorithm *GKG-VT*. So subsections 2.4.1, 2.4.2 and 2.4.3 correspond to these three sets of comparisons, respectively. The reader can use these comparisons as a tool to gauge the differences between these four approaches.

2.4.1 *Comparison of PROB with GKG-VT*

In comparing algorithms *PROB* and *GKG-VT*, the following differences are noted:

1. Clauses 2 and 3 in (2.11) for the cases when $\forall(a_i = \bar{j}), b_i \neq 1$ and $\forall i, a_i \neq \bar{j}$ are not listed in (2.20) of algorithm *PROB*. Without these cases, two cubes that not disjoint due to a particular inverted subproduct, $\bar{j}$, are also separated into A_x and A_1. Reference [12] provides an example which shows that this omission of Case 2 could lead to incorrect results.

2. The operator % is used in the last clause in (2.11) in *GKG-VT* instead of the Miller # operator in (2.20) in algorithm *PROB*. Using the Miller cube operator results in cubes with single variable inversion and thus more terms than produced by the % operator in *GKG-VT*. This case is also illustrated with an example in [2, 12].

3. The definition of cube C in (2.10) is conditional upon the values of a_i and b_i and different from the definition in (2.19).

2.4.2 *Comparison of KDH88 with GKG-VT*

In comparing algorithms *KDH88* and *GKG-VT*, the following analogies are noted:

1. Step 11 of *KDH88* corresponds to the % operator in *GKG-VT* where the inverted subproducts are formed.

2. The procedure *Mask* is used in *KDH88* to tag the integers representing inverted subproducts that cause the cubes DP_ℓ and MP_j to be nondisjoint (Relation 3, Section 2.2.4). In *KDH88*, the number of inverted subproducts that cause the cubes to be nondisjoint, is given by r. The vector **mask** contains the integer values representing these inverted subproducts. Only these subproducts are separated out before the follow-up operation of Step 11. In *GKG-VT*, the $@_j$ operator is defined recursively; hence, clauses 2

and 3 in (2.11) are used to bypass inverted subproducts that do not cause the nondisjointedness of the cubes (Relation 3 of Section 2.2.4). The cube DP_ℓ is then separated into cubes DP_ℓ^1 and DP_ℓ^x for these inverted subproducts that make the cubes DP_ℓ and MP_j nondisjoint.

3. The product DP_ℓ^x in *KDH88* and sub-cube A_x in *GKG-VT* are disjoint from MP_j by definition. Hence, step 9 in *KDH88* results in incrementing j and continuing to the next j. In *GKG-VT*, the sub-cube A_x is used as part of the cube D_j in (2.13) and the next inverted subproduct represented by the integer, $j+1$, is handled via the operation $@_{j+1}$.

4. The product DP_ℓ^1 in *KDH88* is analogous to the cube A_1. This product needs to be made disjoint from MP_j. Hence in *KDH88*, step 10 leads into step 11, and in *GKG-VT*, the operation $A_1\%B$ is carried out.

2.4.3 *Comparison of CAREL with GKG-VT*

The COM operator of *CAREL* appears to be very similar to the % operator if *GKG-VT*. Both operators create inverted subproducts in the resulting cubes C_j and C respectively. Simplifying (2.22) leads to

$$C_j' = \begin{cases} MP_k \setminus MP_j \\ (MP_k \quad pl \quad MP_j) \quad df \quad MP_k \\ \overline{MP_k} \cdot MP_j \end{cases} \tag{2.27}$$

Similarly, using the % operator of *GKG-VT*, $MP_j\%MP_k = \overline{MP_k}MP_j$. However, there are marked differences. The interpretation of the cube $\overline{MP_k}MP_j$ is significantly different if the cubes are represented in the *modified* notation (*CAREL*) than if they were represented using the *standard* notation (*GKG-VT*). An example is shown in the tables below.

CAREL *Modified* Notation	
MP_k	01010
MP_j	00110
$\overline{MP_k}$	10101
MP_j	00110
$\overline{MP_k}MP_j$	00100
C_j'	00100
C_j	$xx1xx$

GKG-VT *Standard* Notation	
MP_k	$x1x1x$
MP_j	$xx11x$
MP_j	$xx11x$
MP_k	$x1x1x$
$\overline{MP_k}MP_j$	$x\bar{1}11x$

Boolean Expression	
MP_k	w_2w_4
MP_j	w_3w_4
$\overline{MP_k}MP_j$	$\overline{w_2w_4}w_3w_4$
	$\overline{w_2}w_3w_4$

Using bit manipulation has the advantage of easy implementation; however, since only two variables are allowed, 0 and 1, there is no way to represent the three types of variables, x, 1, and $\bar{\nu}$, for the *don't care*, uncomplemented and complemented states, respectively. Thus while the COM operator appears to be solving for the product $\overline{MP_k}MP_j$, due to the use of the *modified* notation with only two variables, its interpretation is different from the solution of $\overline{MP_k}MP_j$ in the *standard* notation of Section 2.2.2.

Comparing $\overline{MP_k}MP_j$ in *CAREL* ($MP_k \setminus MP_j$) with $\overline{MP_j}MP_k$ in *GKG-VT*($MP_k\%MP_j$), shows that the resulting cube C_j from the COM operator only stores the newly created inverted subproducts without copying the uncomplemented variables and the already existing inverted subproducts from the input cube into the result C_j while the % operator in *GKG-VT* does. From the example shown above, $\overline{MP_k}MP_j = xx1xx$ in *CAREL*, while $\overline{MP_j}MP_k = x1\bar{1}1x$ in *GKG-VT*. The resulting cube C_j in the COM operator in *CAREL* has coordinates belonging to the set $\{\bar{\nu}, x\}$. While the coordinates of cube C created in the % operator of *GKG-VT* has coordinates in set $\{\bar{\nu}, x, 1\}$.

The RED operator needed in *CAREL* has no counterpart in algorithm *GKG-VT*. This is because of the use of Forms **FI** and **FII** in *CAREL* vs. use of Form **FIII** in *GKG-VT*. By definition the cubes C are found in sequence and while they are made disjoint from previous cubes, the redundant/duplicate terms are absorbed in the operations (%, @, \$) where either Φ, the empty set or A, the first input cube is the result of the operation. With the COM operator, independent invocations of the operator are carried out for each preceding minproduct of disjoint product to find cubes in set C. Hence an explicit phase for removing redundant and duplicate cubes is needed in *CAREL*.

The next operator in *CAREL*, the CMB operator, for cubes in subset IG combine in the last cases of (2.10) and (2.13). The new cube C preserves all independent inverted subproducts and adds a new inverted subproduct for the coordinate pairs of the form $\{x, 1\}$. The CMB operator for cubes in subset DG consist of first separating out inverted subproducts and then recombining them if possible. With the $@_j$ operator of *GKG-VT*, since one of the input cubes is an intermediary in disjoint product computation while the other is a minproduct, the separation of inverted subproducts is needed for occurrences of pairs such as $\{\bar{\nu}, 1\}$ and $\{\bar{\nu}, x\}$. On the other hand, separation of inverted subproducts carried out in the CMB operation of *CAREL* is needed for occurrences of coordinate pairs such as $\{\bar{\nu}, \bar{\mu}\}$ and $\{\bar{\nu}, x\}$ between two cubes in set C that only has coordinates $\{\bar{\nu}, x\}$. Hence the definitions of TG_1, TG_2 are different from those of A_1, A_x in *GKG-VT*.

An explicit step such as GEN is not needed in *GKG-VT* since the \$ operation is applied sequentially and inverted subproducts are combined into the disjoint product at each step.

2.5 Conclusions

This chapter presented a comparative review of four *sum-of-disjoint-products* algorithms to solve the two-terminal network reliability problem. The algorithms use **multiple variable inversion technique** for improved efficiency. A comparison of operations used in the algorithms was provided in each section along with descriptions of the algorithms. The same network example was solved for two-terminal network reliability using each of these algorithms.

Due to the NP-hard nature of the two-terminal network reliability problem for general networks, an exact comparison of these algorithms from a computational view point is not done. Soh and Rai [13] provide some comparative analysis of algorithms belonging to the three Forms, **FI, FII,** and **FIII**. Issues such as ease of implementation on parallel systems are addressed in [13].

An aspect that was not pointed out in the text of the chapter has to do with minimizing the number of terms in the final resulting *sdp* expression using pre-ordering schemes to sort the minproducts before application of the algorithms. In this chapter, *GKG-VT*, *PROB* and *CAREL* point out the advantages in sorting the starting minproducts in order of increasing number of 1's. Algorithm *CAREL* further uses lexicographic ordering for minproducts with the same number of 1's. Beichelt and Spross [24] recommend ordering of variables within intermediate products formed during application of an algorithm. For example, variables in cubes C resulting from the COM operation in *CAREL*, could be reordered using their scheme. Singh [25] proposes an ordering of minproducts with the same number of 1's that could be added to any of the algorithms in this chapter to decrease the number of terms in the final *sdp* expression. A different ordering scheme is proposed by Liu, Yang and Liu [26]. Soh and Rai [27, 28] have reported performance comparisons of pre-ordering the minproducts using cardinality, lexicographic ordering, Hamming distance based ordering, and their combinations, for use in disjoint products techniques.

As cited in the introduction, there have been several single variable inversion techniques to solve the two-terminal network reliability problem using the *sdp* formula. Any of these methods can be potentially improved by using the *multiple variable inversion* technique. Recent work by Locks [29] shows extensions of algorithms in [10] and [30] using multiple variable inversion. Locks combines the ordering schemes in [25] and [26] along with the notion of multiple variable inversion applied to the algorithms of [10] and [30] to obtain an "eclectic" solution for the example network introduced in [7]. So while the first *sdp* solution for the two-terminal network reliability of this network given by Abraham [7] had 71 terms, the solution given by Locks [29] has only 35 terms.

Acknowledgment

We thank Suresh Rai and Sieteng Soh, Louisiana State University, for their helpful comments on this chapter.

REFERENCES

[1] Colbourn C.J., *The Combinatorics of Network Reliability*, Oxford University Press, New York, 1987.

[2] Veeraraghavan M., *Modeling and Evaluation of Fault-Tolerant Multiple Processor Systems*, Ph.D. thesis, Department of Electrical Engineering, Duke University, Aug. 1988.

[3] Tarjan R.E., *A Unified Approach to Path Problems*, Journal ACM, Vol. 28, pp. 577-593, 1981.

[4] Bennetts R.G., *On the Analysis of Fault Trees*, IEEE Trans. on Rel., Vol. R-24, No. 3, pp. 194-203, Aug. 1975.

[5] Lin P.M., B.J. Leon and T.C. Huang, *A New Algorithm for Symbolic Systematic Reliability Analysis*, IEEE Trans. on Rel., Vol. R- 25, No. 1, pp. 2-15, April 1976.

[6] Rai S. and K.K. Aggarwal, *An Efficient Method for Reliability Evaluation of a General Network*, IEEE Trans. on Rel., Vol. R-27, No. 3, pp. 206-211, Aug. 1978.

[7] Abraham J.A., *An Improved Algorithm for Network Reliability*, IEEE Trans. on Rel., Vol. R-28, No. 1, pp. 58-61, April 1979.

[8] Grnarov A., L. Kleinrock and M. Gerla, *A New Algorithm for Network Reliability Computation*, Proceedings of the Computer Networking Symposium, Dec. 1979.

[9] Ahmad S.H., *A Simple Technique for Computing Network Reliability*, IEEE Trans. on Rel., Vol. R-31, No. 1, pp. 41-44, April 1982.

[10] Locks M.O., *A Minimizing Algorithm for Sum of Disjoint Products*, IEEE Trans. on Rel., Vol. R-36, No. 4, pp. 445-453, Oct. 1987.

[11] Heidtmann K.D., *Smaller Sums of Disjoint Products by Subproduct Inversion*, IEEE Trans. on Rel., Vol. R-38, No. 3, pp. 305-311, Aug. 1989.

[12] Veeraraghavan M. and K.S. Trivedi, *An Improved Algorithm for the Symbolic Reliability Analysis of Networks*, IEEE Trans. on Rel., R-40, No. 3, pp. 347-358, Aug. 1991.

[13] Soh S. and S. Rai, *CAREL: Computer Aided Reliability Evaluator for Distributed Computing Networks*, IEEE Trans. on Parallel and Distributed Systems, pp. 199-213, April 1991.

[14] Rai S. and D.P. Agrawal, *Distributed Computing Network Reliability*, IEEE Computer Society Press Tutorial, 1990.

[15] Satyanarayana A. and A. Prabhakar, *New Topological Formula and Rapid Algorithm for Reliability Analysis of Complex Networks*, IEEE Trans. on Rel., Vol. R-27, No. 1, pp. 82-100, June 1978.

[16] Barlow R.E. and F. Proschan, *Statistical Theory of Reliability and Life Testing: Probability Models*, Holt, Rinehart and Winston, New York, 1975.

[17] Moskowitz F., *The Analysis of Redundancy Networks*, AIEE Transactions on Communications Electronics, pp. 627-632, 1958.

[18] Hagstrom J.N., *Using the Decomposition-Tree of a Network in Reliability Computation*, IEEE Trans. on Rel., R-32, No. 1, pp. 71-78, April 1983.

[19] Satyanarayana A., *A Unified Formula for the Analysis of Some Network Reliability Problems*, IEEE Trans. on Rel., Vol. R-31, No. 1, pp. 23-32, 1982.

[20] Ball M.O. and J.S. Provan, *Disjoint Products and Efficient Computation of Reliability*, Operations Research, Vol. 36, pp. 703-715, Oct. 1988.

[21] Miller R., *Switching Theory, Volume 1: Combinational Circuits*, Wiley, New York, 1965.

[22] Willie R.R., *Computer-Aided Fault Tree Analysis*, Tech. Rep. ORC 78-14, University of California, Berkeley, 1978.

[23] Sahner R. and K.S. Trivedi, *Reliability Modeling Using SHARPE*, IEEE Trans. on Rel., Vol. R-37, No. 2, pp. 186-193, June 1987.

[24] Beichelt F. and L. Spross, *An Improved Abraham-Method for Generating Disjoint Sums*, IEEE Trans. on Rel., Vol. R-36, No. 1, pp. 70-74, April 1987.

[25] Singh B., *A Procedure for Generating Sums of Disjoint Products*, Microelectronics and Reliability, To appear in 1993.

[26] Liu H.H., W.T. Yang and C.C. Liu, *A Minimum Algorithm for Sum of Disjoint Products by the Shannon Expansion*, Unpublished.

[27] Soh S. and S. Rai, *Experimental Results on Preprocessing of Path/Cut Terms in Sum of Disjoint Products Technique*, in the Proceedings of the Infocom, pp. 533-542, 1991.

[28] Soh S. and S. Rai, *Experimental Results on Preprocessing of Paths/Cuts Terms in Sum of Disjoint Products Technique*, IEEE Trans. on Rel., to appear in March 1993.

[29] Locks M.O. and J.M. Wilson, *On Disjoint Products and Nearly Minimal Disjoint Forms*, Unpublished.

[30] Wilson J.M., *An Improved Minimizing Algorithm for Sum of Disjoint Products*, IEEE Trans. on Rel., Vol. R-39, No. 1, pp. 42-45, April 1990.

New Trends in System Reliability Evaluation
K.B. Misra (Editor)

CHAPTER 3

(Part-I)

DECOMPOSITION AND REDUCTION TECHNIQUES

Frank Beichelt
D-0-9206 Hillbersdorf 28
Federation Republic of Germany

ABSTRACT

A stochastic network is a deterministic graph, which is assumed to be undirected and connected. This graph can be split into two edge-disjoint through subgraphs a set of separating vertices. Separate reliability analysis is carried out for the corresponding stochastic subgraphs and the results are then combined to obtain a decomposition formula for the *k*-terminal reliability of the stochastic network. This formula is used to develop a general and unifying approach to network reduction. Special cases are analyzed in detail and several numerical examples are presented.

3.1 Introduction

Network reliability analysis is often required in many important engineering areas. Some of these include, communication networks, monitoring systems as well as transportation and electrical power systems. In particular, the computer communication networks have evolved to cope with a massive demand for the transmission and processing of information. Hence, it is imperative that effective tools be developed and refined for carrying out reliability analysis of general networks.

This contribution - as most of the literature available on network reliability - refers mainly to communication networks. These arise in connection with problems when end users such as terminals, devices, and computers have to interact in solving problems by sharing information such as data files and programs. A communication network takes charge of the transportation of information between the end users. Random failures in the network can arise in different ways, e.g., 1) wearout failures of

---

The research presented here has been sponsored by Deutsche Forschungsgemeinschaft.

network components, 2) overload failures, and 3) failures due to faulty control software. The network failures may also be caused either by an act of sabotage or by natural catastrophes, like earthquakes, hurricanes. The effects of all these failures on network performance depend, of course, heavily on the underlying network topology. Hence both the reliability analysis of networks with general topological structure and the synthesis of reliable networks are of interest. This contribution deals only with network reliability analysis.

3.2 Stochastic Networks

The basic topological structure of the network under consideration is assumed to be given by an undirected and connected graph $\mathbf{G} = (\mathbf{V},\mathbf{E})$ without loops, where

$\mathbf{V} = \{1, 2, \dots, m\}$ is its *node set* and $\mathbf{E} = \{e_1, e_2, \dots, e_n\}$ is its *edge set*, $3 \le m + n < \infty$. The nodes (vertices) are interpreted as endusers and the edges as links between them. **G** becomes a *random graph* or a *stochastic network* $\tilde{\mathbf{G}}$ by assuming that its edges and nodes exist or do not exist according to a given probability distribution. As usual it is assumed that this distribution is characterized by the following two properties:

1) The states X_i of e_i , i = 1, 2, ..., n, are statistically independent random variables:

$$X_i = \begin{cases} 1, \text{ if } e_i \text{ exists,} \\ 0, \text{ otherwise.} \end{cases}$$

2) The nodes are absolutely reliable, i.e., their availabilities are equal to 1.

The stochastic network $\tilde{\mathbf{G}}$ is adopted as a mathematical model for the communication network. With regard to most applications an edge is said to be *existing (not existing)*, if it is available (failed).

To define reliability criteria for $\tilde{\mathbf{G}}$ let **K** be a distinguished subset of **V**. The nodes in **K** are called *terminals*, *terminal nodes*, *target nodes* or simply **K**-*nodes*. The **K**-*terminal reliability* of $\tilde{\mathbf{G}}$ -denoted by $R(\tilde{\mathbf{G}},\mathbf{K})$- is the probability that all nodes of **K** are connected in $\tilde{\mathbf{G}}$, i.e., there exists a path between any two nodes of **K**. Two special and most important cases are those obtained for $\mathbf{K} = \mathbf{V}$ and $\mathbf{K} = \{u,v\}$. The first is the *all-*

terminal reliability, overall reliability or *connectedness probability* $R(\tilde{G},\mathbf{V})$ and the second is the *2-terminal-reliability* or *(u,v)-reliability* $R(\tilde{G},\{u,v\}) = R_{u,v}(\tilde{G})$. The two-terminal reliability arises also in connection with computing the reliability of any coherent binary system based on its reliability block diagram. Hence it might be the most important criterion. The all-terminal reliability is particularly relevant for telephone and monitoring systems. In the sequel $R(\tilde{G},\mathbf{K})$ will be simply denoted by $R(\tilde{G})$ if a specification of the terminal node set is not necessary or evident from the context.

The main problem of network reliability analysis consists in developing effective methods for computing $R(\tilde{G})$. Despite the extensive study devoted to this problem the size of general networks that can be analyzed exactly by computerized algorithms is still rather small, certainly not exceeding 60 nodes. This unfavourable situation is due to computational difficulties inherent to the problem. For general networks up to now only such exact algorithms for obtaining $R(\tilde{G})$ are known whose computation time increases exponentially with network complexity (exponential-time algorithms). This is not surprising since it has been shown that the problem of computing $R(\tilde{G})$ is NP-hard (*Ball [1]*). Therefore, methods have great practical importance which allow the computation of $R(\tilde{G},\mathbf{K})$ even for large networks with a complicated topological structure in reasonable time. The best results have been obtained so far by applying decomposition, factorization, and reduction methods. Efficient (polynomial-time) algorithms for the exact reliability computation were found for large classes of stochastic planar networks (for surveys (see *Coulbourn [2]* and *Politof/Satyanarayana [3]*). Most of these algorithms are based on network reduction.

LIST OF BASIC NOTATION AND ABBREVIATIONS

$\mathbf{G'}$ arbitrary network (graph)

$\tilde{\mathbf{G}}'$ stochastic network with the underlying (deterministic) network $\mathbf{G'}$

$R(\tilde{G}',\mathbf{K})$ $\mathbf{K}$-terminal reliability of $\tilde{G}'$

$R(\tilde{G}',\{u,v\})$, $R_{u,v}(\tilde{G}')$ two-terminal reliability of $\tilde{G}'$

$R(\tilde{G}')$ any reliability criterion of $\tilde{G}'$ (in the sections 3.4.3 to 3.4.7 only all-terminal reliability)

e	any edge
p_e	availability of e
G'_e (G'_{-e})	graph arising from G' by contracting (removing) e
$\tilde{G}$	stochastic network to be analyzed
$\|M\|$	number of elements (cardinality) of any finite set **M**
U	separating vertex set
u	cardinality of U
S(·,·)	Stirling number of the second kind
B(·)	Bell number
A^T	transposed of a matrix **A**
∅	empty set
iff	if and only if
□	end of an example
■	end of a definition
▌	end of a proof

3.3 Network Decomposition

3.3.1 Introduction

A well-known example for decomposing the network reliability computation problem into two problems of lower complexity is the factorization formula

$$R(\tilde{G}) = p_e R(\tilde{G}_e) + (1-p_e)R(\tilde{G}_{-e}). \tag{3.1}$$

According to the "List of Basic Notation and Abbrevations", G_e arises from **G** by contracting e, i.e., the end nodes of e are fused and the loop generated in this way is deleted. This section presents a generalization of (3.1) which is based on separating node sets.

Definition 3.1 A subset **U** of **V** is called a *separating node set* of the network **G** = (**V**,**E**) iff there exist two edge-disjoint subgraphs $G^1 = (V^1,E^1)$ and $G^2 = (V^2,E^2)$ of **G** satisfying

$$G = G^1 \cup G^2, \quad G^1 \cap G^2 = (U, \emptyset).$$

■

Figure 3.1 shows an example network with 11 nodes and 24 edges which is split into two edge-disjoint subgraphs by the separating node set **U** = {1,2,3}.

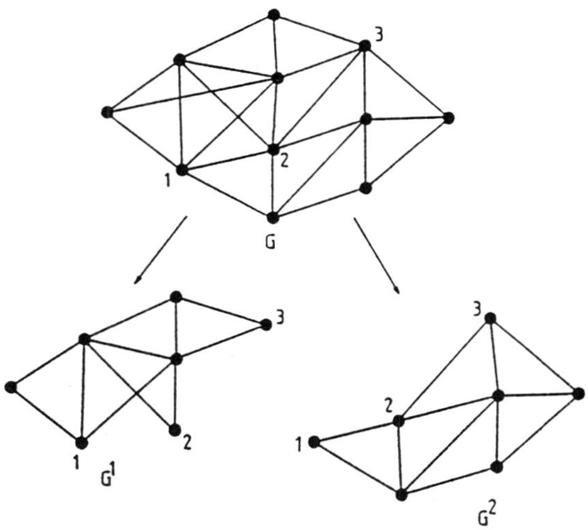

FIG. 3.1 EXAMPLE NETWORK AND ITS SPLITTING BY THE SEPARATING NODE SET $\mathbf{U}$ = {1,2,3}

It is an obvious approach to carry out reliability analysis for $\tilde{G}^1$ and $\tilde{G}^2$ separately and to combine the results to obtain R($\tilde{G}$,**K**). This approach was first applied by *Rosenthal [4]* and *Rosenthal/Frisque [5]*, however, in a somewhat heuristic way. This section is mainly based on *Beichelt [6]* and *Beichelt/Tittmann [7-10]*. *Bienstock [11, 12]* has been the first to combine combinatorial and decomposition methods to network reliability analysis, see also *Shier [13]*.

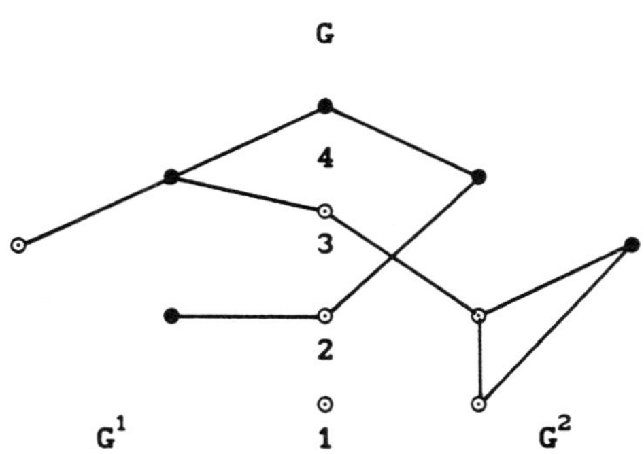

FIG. 3.2 ILLUSTRATION OF THE K-CONNECTEDNESS

3.3.2 Decomposition Formula

For any **G** and **U**, the key idea of splitting is to characterize the **K**-connectedness of **G** by certain connectedness properties of the corresponding subgraphs G^1 and G^2. Consider the example network of Fig. 3.2 with **U** = {1,2,3,4}. The **K**-nodes are the black ones. Evidently, neither G^1 nor G^2 are **K**-connected, but **G** has this property because the **K**-components of G^1 and G^2 are mutually connected via nodes of **U**. (A **K**-*component* of any graph **G'** = (**V'**,**E'**) with $K \subseteq V'$ is such a component of **G'** whose node set contains a node of **K**.)

With respect to mutually connecting **K**-components of G^2 those nodes of **U** are equivalent which belong to one and the same component of G^1. (Of course, the roles of G^1 and G^2 can be changed.) The subsets of equivalent nodes of **U** form a partition of **U**. It is called the partition *induced (generated)* by **G**.

Since the set of partitions of **U** plays an important role in deriving the decomposition formula, next some basic concepts have to be introduced. (For a more detailed treatment see, for instance, *Aigner [14].*)

Definition 3.2 A *partition* $\pi = \pi(M)$ of any set **M** is a family of disjoint subsets of **M** the union of which is **M**. The disjoint subsets constituting π are called *blocks* of π. ∎

Let $\Pi = \{\pi_1, \pi_2, \ldots, \pi_B\}$ be the partition set of **U**. The number B of partitions in Π is the *Bell-number* of **U**. It only depends on the cardinality $u = |U|$ of U: B = B(u). B(u) is the sum of the *Stirling numbers of the second kind* S(u,k):

$$B(u) = \sum_{k=1}^{u} S(u,k).$$

For instance, B(2) = 2, B(3) = 5, B(4) = 15, B(5) = 52, B(6) = 203 and B(7) = 877.

Definition 3.3 The partition π_j is said to be a *refinement* of π_i iff each block of π_j can be obtained by splitting a block π_i. (Trivial splitting is admitted.) ∎

In Π a partial ordering "≤" is introduced in the following way: There holds $\pi_i \le \pi_j$ iff π_j is a refinement of π_i. In this case, π_j is *greater* than π_i or π_i is *less* than π_j. Thus the *partition lattice* $(\Pi, \le)$ is given. Let π_1 its smallest and π_B its greatest element. Then π_1 consists only of a single block comprising all elements of **U** whereas π_B= **U**. Let

$\pi_i\pi_j$ be the greatest partition satisfying both $\pi_i\pi_j \le \pi_i$ and $\pi_i\pi_j \le \pi_j$. For instance, if $U = \{1,2,3,4\}$, $\pi_i = \{12,34,5\}$ and $\pi_j = \{12,3,45\}$, then $\pi_i\pi_j = \{12,345\}$. (The nodes not separated by a comma form a block. This notation is used throughout the paper.)

Let U_j (G_j^i) be that node set (graph) arising from U (G^i) by fusing the nodes of each block of π_j into one node, $j = 1,2,\ldots,B$; $i = 1,2$. There is, obviously, a one-to-one correspondence between the blocks of π_j and the elements of U_j. Hence, in what follows, the elements of U_j will be identified with the corresponding blocks of π_j.

Example 1 Given a separating node set $U = \{1,2,3\}$ the corresponding $B(3) = 5$ partitions are denoted by

$$\pi_1=\{123\},\ \pi_2=\{12,3\},\ \pi_3=\{13,2\},\ \pi_4=\{1,23\},\ \pi_5=\{1,2,3\}.$$

Figure 3.3 shows an example network G and the corresponding networks G_j; $j = 1,2,3,4,5$.

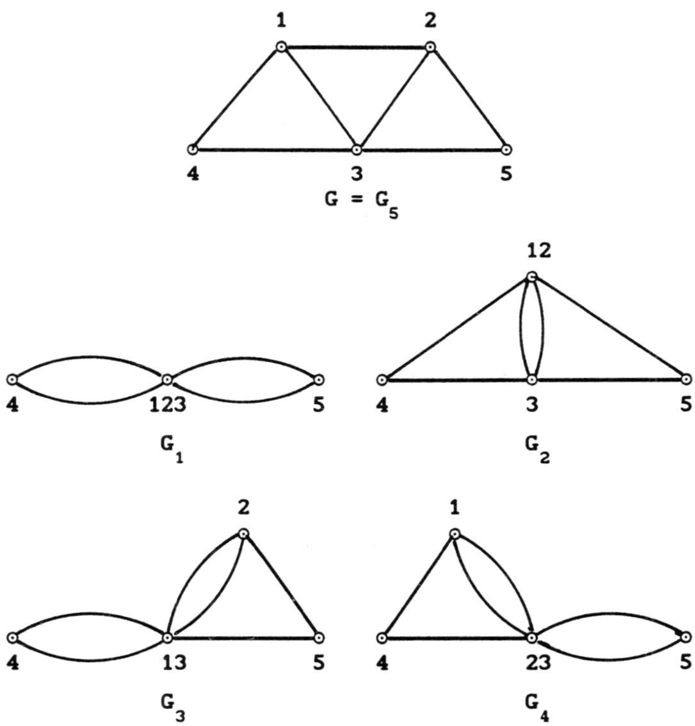

FIG. 3.3 NETWORK G AND THE CORRESPONDING NETWORKS G_j

Let L_j^i be the set of all those elements of U_j which arise from a

K-component of G^1. In order to illustrate these notations and the proof of the following theorem let us once more consider the example network of Fig. 3.2. The subgraph G^1 induces the partition $\pi_r = \{1,2,34\}$. Only the vertex 1 does not belong to a K-component of G^1. Hence, $L_r^1 = \{2, 34\}$. G^2 induces the partition $\pi_s = \{1,3,24\}$ so that $L_s^2 = \{3,24\}$. (r and s are determined by the given numbering of the partitions in Π.) The structures of L_r^1 and L_s^2 imply that all K-components of G^1 are connected by K-components in G^2 and vice versa. Hence G is K-connected. Particularly, the K-con-nectedness of G is equivalent to the $(K \cup L_r^1)$-connectedness of G_r^2, since in G^1 all K-components have a node in common with U.

The following sets are needed to introduce suitable terminal nodes for the reliability analysis of $\tilde{G}^1$ and $\tilde{G}^2$:

$$J = \begin{cases} \emptyset, & \text{if } V^2 \cap K = \emptyset, \\ U_1, & \text{otherwise.} \end{cases}$$

$$K^1 = (K \setminus V^2) \cup J, \quad K^2 = K \setminus V^1. \tag{3.3}$$

Let further be

$$A_j(G^1) = \begin{cases} 1, & \text{if } G^1 \text{ induces } \pi_j, \\ 0, & \text{otherwise,} \end{cases}$$

$$B(G_j^1, L) = \begin{cases} 1, & \text{if } L = L_j^1, \\ 0, & \text{otherwise,} \end{cases}$$

where $L \subseteq U$. For any graph $G' = (V', E')$, $K' \subseteq V'$, the indicator function of the K'-connectedness is

$$C(G', K') = \begin{cases} 1, & \text{if } G' \text{ is } K'\text{-connected,} \\ 0, & \text{otherwise.} \end{cases}$$

It is obvious to assume $C(G',K') = 1$, if $|K'| = 1$ or $|K'| = 0$.

Theorem 3.1 There holds

$$C(G,K) = \sum_{j=1}^{B} \sum_{L \subseteq U_j} A_j(G^1)\, B(G_j^1, L)\, C(G_1^1, K^1)\, C(G_j^2, K^2 \cup L).$$

Proof Let G^1 induce π_r let be and $L_r^1 = S$. Then $C(G,K)$ reduces to

$$C(G,K) = C(G_1^1, K^1)\, C(G_r^2, K^2 \cup S).$$

Case 1: $V^1 \cap K = \emptyset$

In this case there holds $K^1 = L_1^2 = U_1$ so that $C(G_1^1,K^1) = 1$. In view of $K^2 = K$ and $S = \emptyset$, it remains to show that $C(G,K) = C(G_r^2,K)$. If $C(G_r^2,K) = 1$, then the K-components of G^2 are connected by paths in G^1. Hence, $C(G,K) = 1$. The other direction is obvious.

Case 2: $V^2 \cap K = \emptyset$

Now there holds $K^2 = \emptyset$, $K^1 = K$, and $J = \emptyset$. Hence, the problem is to show that

$$C(G,K) = C(G_1^1,K)\ C(G_r^2,S). \tag{3.4}$$

If $S = \emptyset$, then $C(G_r^2,S) = 1$ and no K-component of G^1 has a node in common with U. This implies

$$C(G,K) = C(G^1,K) = C(G_1^1,K).$$

If $S \neq \emptyset$ and there is a K-component of G^1 not containing a node of U, then $C(G,K) = C(G_1^1,K) = 0$ so that (3.4) is again valid. If $S \neq \emptyset$ and all K-components of G^1 have a node in common with U, then $C(G_1^1,K) = 1$. On this condition the K-components of G^1 are connected by paths in G^2 iff $C(G_r^2,S) = 1$.

Case 3: $V^i \cap K \neq \emptyset$, $i = 1,2$

First let $C(G,K) = 0$. Then there exist at least two K-com-ponents of G, let H be one of them. If H does not contain a node of U there holds $C(G_1^1,K^1) = 0$ if $H \subseteq G^1$ and $C(G_r^2,K^2 \cup S) = 0$ if $H \subseteq G^2$. If all K-components of G contain nodes of U, then there exist at least two blocks of π_r which belong to different K-components. Thus, $C(G_r^2,K^2 \cup S) = 0$. Now let $C(G,K) = 1$. Then all K-components of G^1 contain a node of U. Therefore, $C(G_1^1,K^1) = 1$. G_r^2 connects all K-components of G^1 so that $C(G_r^2,S) = 1$. Moreover, there holds $C(G_r^2,K^2 \cup S) = 1$, since all K-vertices of G are in one and the same component. This completes the proof of the theorem. ∎

The theorem holds for any network G = G', in particular for each realization of the stochastic network $\tilde{G}$. Hence the random indicator variable of the K-conectedness of $\tilde{G}$ is

$$C(\tilde{G},K) = \sum_{j=1}^{B} \sum_{L \subseteq U_j} A_j(\tilde{G}^1)\ B(\tilde{G}_j^1,L)\ C(\tilde{G}_1^1,K^1)\ C(\tilde{G}_j^2,K^2 \cup L). \tag{3.5}$$

Note that in (3.5) the sets K^i, L and U_j are deterministic. Based on

(3.5) it is easy to derive $R(\tilde{G},\mathbf{K})$. For $\mathbf{G}^1$ and $\mathbf{G}^2$ being edge-disjoint, the product $A_j(\tilde{G}^1)B(\tilde{G}^1_j,L)C(\tilde{G}^1_1,\mathbf{K}^1)$ and $C(\tilde{G}^2_j,\mathbf{K}^2 \cup L)$ are statistically independent for all j and $\mathbf{L} \subseteq \mathbf{U}_j$. Hence, taking the mathematical expectation of $C(\tilde{G},\mathbf{K})$ yields the desired splitting formula:

$$R(\tilde{G},\mathbf{K}) = \sum_{j=1}^{B} \sum_{\mathbf{L}\subseteq \mathbf{U}_j} P_j(\tilde{G}^1,\mathbf{L})\ R(\tilde{G}^2_j,\mathbf{K}^2 \cup \mathbf{L}), \tag{3.6}$$

where

$$P_j(\tilde{G}^1,\mathbf{L}) = P\left(A_j(\tilde{G}^1)\ B(\tilde{G}^1_j,\mathbf{L})\ C(\tilde{G}^1_1,\mathbf{K}^1) = 1\right). \tag{3.7}$$

$P_j(\tilde{G}^1,\mathbf{L})$ is the probability that $\tilde{G}^1$ induces π_j and $\mathbf{L}$ and that each $\mathbf{K}$-component of $\tilde{G}^1$ contains at least one node of $\mathbf{U}$. These probabilities satisfy an inhomogeneous system of linear equations, the development of which is not difficult, but rather tedious. It will be, therefore, only deduced for the two-terminal and the all-terminal reliability. However, the probabilities (3.7) can also be computed by repeated application of the factorization formula (3.1):

$$P_j(\tilde{G}^1,\mathbf{L}) = p_e\ P_j(\tilde{G}^1_e,\mathbf{L}_e) + (1 - p_e)\ P_j(\tilde{G}^1_{-e},\mathbf{L}).$$

If the pivot-edge e has the end nodes s and t and if n_e denotes that node which arises by fusing s and t, then the set of terminal nodes belonging to $\tilde{G}^1_e$ has to be constructed in the following way:

$$\mathbf{L}_e = \begin{cases} \mathbf{K} \cap \mathbf{V}^1 & \text{if } s,t \notin \mathbf{K}, \\ \{\{\mathbf{K} \cap \mathbf{V}^1\} \setminus \{s,t\}\} \cup \{n_e\}, & \text{otherwise.} \end{cases}$$

The number $n(\mathbf{u})$ of summands in (3.6) increases exponentially with increasing $\mathbf{u}$:

$$n(\mathbf{u}) = \sum_{k=1}^{\mathbf{u}} S(\mathbf{u},k)\ 2^k.$$

For instance, $n(2) = 6$, $n(3) = 22$, $n(4) = 94$, $n(5) = 454$, and $n(6) = 2430$. However, on condition that $\mathbf{U} \subseteq \mathbf{K}$ the number $n(\mathbf{u})$ of summands in (3.6) reduces to the Bell number $B(\mathbf{u})$ since in this case each $\mathbf{L}^1_j$ equals to $\mathbf{U}$. Hence, for $\mathbf{U} \subseteq \mathbf{K}$

$$R(\tilde{G},\mathbf{K}) = \sum_{j=1}^{B} P_j(\tilde{G}^1)R(\tilde{G}^2_j,\mathbf{K}^2 \cup \mathbf{U}_j), \tag{3.8}$$

$$P_j(\tilde{G}^1) = P(A_j(\tilde{G}^1)\ C(\tilde{G}^1_1,\mathbf{K}^1) = 1). \tag{3.9}$$

Formula (3.8) can be used for the computation of the all-terminal reliability. However, if $\{u,v\} \subseteq V^1$ and $\{u,v\} \subseteq V^2$, respectively, then (3.8) is also a decomposition formula for the two-terminal reliability $R_{u,v}(\tilde{G})$.

In view of the exponential growth of $n(\mathbf{u})$ with increasing $\mathbf{u}$ and the exponential growth of computation time for $R(\tilde{\mathbf{G}},\mathbf{K})$ with increasing complexity of $\tilde{\mathbf{G}}$, the application of (3.6) is most advantegeous, if a separating node set $\mathbf{U}$ can be used with $\mathbf{u}$ small and $\mathbf{G}^1$ and $\mathbf{G}^2$ having approximately the same *size*. For a computerized generation of such "suitable" separating node sets, particularly in case of a repeated application of (3.6), the algorithm of *Lee/ Vogt/Mickle [15]* can be recommended.

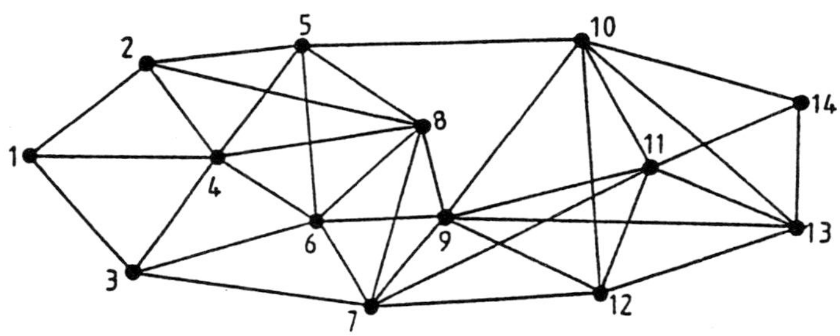

FIG. 3.4 EXAMPLE NETWORK WITH 14 NODES AND 37 EDGES

TABLE 3.1 K-TERMINAL RELIABILITY ANALYSIS (EXAMPLE 2)

K	$R(\tilde{\mathbf{G}},\mathbf{K})$			
	p 0.5	0.6	0.7	0.8
{1,2,...,14}	0.49556	0.76106	0.91461	0.97889
{1,13}	0.73580	0.89183	0.96358	0.99069
{1,5,13}	0.70869	0.88140	0.96358	0.99056
{1,3,5,7,9,11}	0.67112	0.86337	0.95487	0.98925
{1,2,7,9,10,13}	0.66141	0.86009	0.95424	0.98919
{1,3,5,6,8,11,14}	0.57545	0.80309	0.92794	0.98112
{1,2,3,7,9,10,12,13}	0.60870	0.83211	0.94473	0.98734

Example 2 Figure 3.4 shows a network with 14 nodes and 37 edges. Table 3.1 contains the results of a **K**-terminal reliability analysis on the condition that there is a common edge availability p. The results

have been obtained by applying (3.6). The **K**-terminal reliabilities $R(\tilde{G}^2_j, K^2 \cup L)$ and the probabilities $P_j(\tilde{G}^1, L)$ have been computed by factorization. In each of the cases listed in Table 3.1 the application of (3.6) proved to be 15 to 30 times faster than only applying factorization to $R(\tilde{G}, K)$. □

3.3.3 Two-Point Separating Node Sets

Let $U = \{1,2\}$. There are two partitions of **U**: $\pi_1 = \{12\} = U_1$ and $\pi_2 = \{1,2\} = U$. Let further

$$p_j(\omega) = P(\tilde{G}^1_j, \{\omega\}) \text{ and } r_j(\omega) = R(\tilde{G}^2_j, K^2 \cup \{\omega\}),$$

where $\omega \in \{\emptyset, U_1\}$ for $j = 1$ and $\omega \in \{\emptyset, 1, 2, U\}$ for $j = 2$. From (3.6),

$$R(\tilde{G}, K) = p_1(\emptyset)\, r_1(\emptyset) + p_1(U_1)\, r_1(U_1) + p_2(\emptyset)\, r_2(\emptyset) + p_2(1)\, r_2(1) + p_2(2)\, r_2(2) + p_2(U)\, r_2(U). \tag{3.10}$$

If $V^1 \cap K = \emptyset$, then (3.10) reduces to

$$R(\tilde{G}, K) = p_1(\emptyset)\, r_1(\emptyset) + p_2(\emptyset)\, r_2(\emptyset).$$

In this case, $p_j(\emptyset)$ is simply the probability that $\tilde{G}^1$ induces the partition π_j, $j = 1,2$.

If $V^1 \cap K \neq \emptyset$, then

$$p_1(\emptyset) = p_2(\emptyset) = 0. \tag{3.11}$$

If $U \subseteq K$, then there hold (3.11) and $p_2(1) = p_2(2) = 0$ so that in this case

$$R(\tilde{G}, K) = p_1(U_1)\, r_1(U_1) + p_2(U)\, r_2(U), \tag{3.12}$$

where

$$p_1(U_1) = R(\tilde{G}^1, V^1 \cap K), \qquad p_2(U) = R(\tilde{G}^1_1, K^1) - R(\tilde{G}^1, V^1 \cap K). \tag{3.13}$$

It may happen that the nodes 1 and 2 of $U = \{1,2\} \subseteq K$ are directly connected by an edge $e = (1,2)$. Of course, this case is included in (3.12), if e is attributed to G^1 or G^2, respectively. However, its separate treatment yields an interesting formula for $R(\tilde{G}, K)$. To develop this formula, instead of (3.2) now the following decomposition of **G** is taken as a basis:

$$G = F^1 \cup F^2, \quad F^1 \cap F^2 = (U,\{e\}).$$

Evidently,

$$R(\tilde{G}_e,K^1 \cup K^2) = R(\tilde{F}^1_e,K^1)\ R(\tilde{H}^2_e,K^2 \cup U_1). \tag{3.14}$$

In (3.14) the sets of terminal nodes K^i are again given by (3.3), if there G^i is exchanged by F^i, $i = 1,2$. Furthermore, assuming $0 < p_e \le 1$, one gets by applying the factorization formula (3.1) to $\tilde{G}$, $\tilde{F}^1$ and $\tilde{F}^2$

$$R(\tilde{G},K) = p_e R(\tilde{G}_e,K^1 \cup K^2) + (1-p_e)R(\tilde{G}_{-e},K), \tag{3.15}$$

$$\begin{aligned} R(\tilde{F}^1_e,K^1) &= \frac{1}{p_e}\Big[R(\tilde{F}^1,K \cap V^1)-(1-p_e)R(\tilde{F}^1_{-e},K \cap V^1)\Big], \\ R(\tilde{F}^2_e,K^2 \cup U_1) &= \frac{1}{p_e}\Big[R(\tilde{F}^2,K \cap V^2)-(1-p_e)R(\tilde{F}^2_{-e},K \cap V^2)\Big]. \end{aligned} \tag{3.16}$$

If (3.12) is applied to $R(\tilde{G}_{-e},K)$ and in the resulting formula $R(\tilde{F}^1_e,K^1)$ and $R(\tilde{F}^2_e,K^2 \cup U_1)$ are substituted by (3.16), then (3.14) and (3.15) yield

$$\begin{aligned} R(\tilde{G},K) = \frac{1}{p_e}\Big[&R(\tilde{F}^1,K \cap V^1)\ R(\tilde{F}^2,K \cap V^2) \\ &- (1 - p_e)R(\tilde{F}^1_{-e},K \cap V^1)\ R(\tilde{F}^2_{-e},K \cap V^2)\Big]. \end{aligned}$$

This formula is a generalization of a result of *Satyanarayana/Chang/Khalil [16]*.

3.3.4 Two-Terminal Reliability

Let now $K = \{u,v\}$. Then $R(\tilde{G},K) = R_{u,v}(\tilde{G})$ is the probability that there exists a path between the nodes u and v. The node u belongs exactly to one component of $\tilde{G}^1$. Hence, in the second sum of (3.6) only the terms corresponding to the 1-point subsets $\{r\}$ of U_j have to be taken into account. Further, $P_j(\tilde{G}^1,\{r\}) = P_{u,r}(\tilde{G}^1_j)$ is the probability that $\tilde{G}^1$ generates the partition π_j and that u and r belong to one and the same component of $\tilde{G}^1$ or $\tilde{G}_j$, respectively. Therefore, the two-terminal reliability $R_{u,v}(\tilde{G})$ can be written in the form

$$R_{u,v}(\tilde{G}) = \sum_{j=1}^{B} \sum_{r \in U_j} P_{u,r}(\tilde{G}^1_j)\ R_{r,v}(\tilde{G}^2_j). \tag{3.17}$$

It can be easily seen that formula (3.17) has

$$N(u) = B(u+1) - B(u)$$

terms. For instance, $N(2) = 3$, $N(3) = 10$, $N(4) = 37$, $N(5) = 151$, and

N(6) = 674.

The $P_{u,r}(\tilde{G}_j^1)$ satisfy an inhomogeneous system of linear equations. To establish this system, some further notations have to be introduced. For $r \in \mathbf{U}_j$, let $\mathbf{U}_{j,k,r}$ be the following subset of $\mathbf{U}_k$: It is $t \in \mathbf{U}_{j,k,r}$ iff there is a block in $\pi_j\pi_k$ which comprises those blocks of π_j and π_k corresponding to r and t, respectively. If $\tilde{G}^1$ induces π_k and in $\tilde{G}_k^1$ there is a path from u to t, then there is a path from u to r in $\tilde{G}_j^1$ iff $t \in \mathbf{U}_{j,k,r}$. Therefore,

$$R_{u,r}(\tilde{G}_j^1) = \sum_{k=1}^{B} \sum_{t \in \mathbf{U}_{j,k,r}} P_{u,t}(\tilde{G}_k^1), \tag{3.18}$$

where $r \in \mathbf{U}_j$, $j = 1,2,\ldots,B$.

For applying (3.17) and solving (3.18) the two-terminal reliabilities $R_{u,r}(\tilde{G}_j^1)$ and $R_{r,v}(\tilde{G}_j^2)$ have to be computed by any available method, for instance by repeated application of (3.17).

TABLE 3.2 COEFFICIENT MATRIX **C** OF (3.18) IN CASE OF **u** = 3

j	$\mathbf{U}_j$	1 {t}	2 {t,3}	{x,t}	3 {t,2}	{y,t}	4 {1,t}	{t,z}	5 {t,2,3}	{1,t,3}	{1,2,t}
1	{r}	1	1	1	1	1	1	1	1	1	1
2	{r,3}	1	1	0	1	1	1	1	1	1	0
	{x,r}	1	0	1	1	1	1	1	1	0	1
3	{r,2}	1	1	1	1	0	1	1	1	0	1
	{y,r}	1	1	1	0	1	1	1	0	1	0
4	{1,r}	1	1	1	1	1	1	0	0	1	1
	{r,z}	1	1	1	1	1	0	1	1	0	0
5	{r,2,3}	1	1	0	1	0	0	1	1	0	0
	{1,r,3}	1	1	0	0	1	1	0	0	1	0
	{1,2,r}	1	0	1	1	0	1	0	0	0	1

Example 3 Let us again consider the example network of Fig. 3.1 and its splitting by the separating node set **U** = {1,2,3}. As in Example 1 the partitions of **U** are denoted by

$$\pi_1 = \{123\},\ \pi_2 = \{12,3\},\ \pi_3 = \{13,2\},$$
$$\pi_4 = \{1,23\},\ \pi_5 = \{1,2,3\}. \tag{3.19}$$

If those elements of the U_j arising by fusing the nodes of the blocks 123, 12, 13, and 23 are denoted by w, x, y, and z, then the sets U_j are given by

$$U_1 = \{w\},\ U_2 = \{x,3\},\ U_3 = \{y,2\},\ U_4 = \{1,z\},\ U_5 = U.$$

Formula (3.17) has 10 terms so that the computation of $R_{u,v}(\tilde{G})$ is transfered to computing 20 two-terminal reliabilities of lower complexity and to solving the equation system (3.18) having 10 unknowns. Table (3.2) shows the corresponding coefficient matrix **C**. The inverse matrix of **C** is

$$C^{-1} = \begin{pmatrix} 1 & -1 & 0 & -1 & 0 & -1 & 0 & 1 & 1 & 1 \\ -1 & 0 & 0 & 1 & 0 & 1 & 0 & 0 & 0 & -1 \\ 0 & 0 & -\frac{1}{2} & 0 & \frac{1}{2} & 0 & \frac{1}{2} & -\frac{1}{2} & -\frac{1}{2} & \frac{1}{2} \\ -1 & 1 & 0 & 0 & 0 & 1 & 0 & 0 & -1 & 0 \\ 0 & 0 & \frac{1}{2} & 0 & -\frac{1}{2} & 0 & \frac{1}{2} & -\frac{1}{2} & \frac{1}{2} & -\frac{1}{2} \\ -1 & 1 & 0 & 1 & 0 & 0 & 0 & -1 & 0 & 0 \\ 0 & 0 & \frac{1}{2} & 0 & \frac{1}{2} & 0 & -\frac{1}{2} & \frac{1}{2} & -\frac{1}{2} & -\frac{1}{2} \\ 1 & 0 & -\frac{1}{2} & 0 & -\frac{1}{2} & -1 & \frac{1}{2} & -\frac{1}{2} & \frac{1}{2} & \frac{1}{2} \\ 1 & 0 & -\frac{1}{2} & -1 & \frac{1}{2} & 0 & -\frac{1}{2} & \frac{1}{2} & -\frac{1}{2} & \frac{1}{2} \\ 1 & -1 & \frac{1}{2} & 0 & -\frac{1}{2} & 0 & -\frac{1}{2} & \frac{1}{2} & \frac{1}{2} & -\frac{1}{2} \end{pmatrix}$$

It should be noted that **C** depends only on the cardinality of the separating note set.

TABLE 3.3 NUMERICAL RESULTS FOR EXAMPLE 3

j	U_j	$R_{u,r}(\tilde{G}^1_j)$	$P_{u,r}(\tilde{G}^1_j)$	$R_{r,v}(\tilde{G}^2_j)$
1	{r}	0.991915	0.898960	0.900510
2	{r,3}	0.991566	0.043999	0.985391
	{x,r}	0.945503	0.000039	0.982518
3	{r,2}	0.991519	0.037681	0.989751
	{y,r}	0.951867	0.000039	0.988359
4	{1,r}	0.988425	0.007040	0.989787
	{r,z}	0.984208	0.001434	0.949012
5	{r,2,3}	0.984130	0.002056	0.944837
	{1,r,3}	0.950395	0.000357	0.983968
	{1,2,r}	0.944031	0.000310	0.982302

Assuming a common edge availability $p = 0.8$ a reliability analysis of $\tilde{G}^1$ and G has been carried out. Table 3.3 shows the results. The scalar product of the 3rd and the 4th column gives the desired two-terminal reliability

$$R_{u,v}(\tilde{G}) = 0.981670. \qquad \square$$

Lower bounds In view of the exponential growth of the number of terms in the decomposition formula (3.17) with increasing $\mathbf{u}$ their applicability is limited, too ($\mathbf{u} \le 6$). However, very good lower bounds on $R_{u,v}(\tilde{G})$ can be derived from (3.17), if only a few partitions of Π are taken into account. This approach is now demonstrated for estimating the two-terminal reliability. (Further examples give *Tittmann/Blechschmidt [17].*)

The formula (3.17) includes the possibility that (u,v)-paths cross the set **U** several times, i.e., a (u,v)-path may contain several nodes of **U**. Such paths are comparatively long. Hence their reliability-theoretic importances are expected to be comparatively low. By neglecting these paths one gets a lower bound for $R_{u,v}(\tilde{G})$ which has proved to be rather sharp. In the sequel this bound is derived for $\mathbf{u} = 3$. Its derivation for $\mathbf{u} \ge 4$ is completely analogous.

The partitions π_j and the elements of the corresponding $\mathbf{U}_j$ are denoted as in example 3. Let further p_{uv} be the probability that in $\tilde{G}^1$ there exist (u,i)-paths for i = 1,2,3; p_{ux}; p_{uy}; p_{uz} be the probabilities that in $\tilde{G}^1$ there exist (u,i)-paths for i = 1,2; i = 1,3 and i = 2,3, respectively, but none for i = 3, i = 2 and i = 1, respectively. Finally, let p_{ui}, $i \in \mathbf{U}$, the probability that in $\tilde{G}^1$ there exists a (u,i)-path, but no (u,j)-path for $j \in \mathbf{U} \setminus \{i\}$. Then one gets from (3.17) the desired lower bound $\underline{R}_{u,v}(\tilde{G})$ for $R_{u,v}(\tilde{G})$:

$$\begin{aligned} \underline{R}_{u,v}(\tilde{G}) = \; & p_{uw}R_{w,v}(\tilde{G}_1^2) + p_{ux}R_{x,v}(\tilde{G}_2^2) \\ + \; & p_{uy}R_{y,v}(\tilde{G}_3^2) + p_{uz}R_{z,v}(\tilde{G}_4^2) \\ + \; & p_{u1}R_{1,v}(\tilde{G}_5^2) + p_{u2}R_{2,v}(\tilde{G}_5^2) + p_{u3}R_{3,v}(\tilde{G}_5^2). \end{aligned} \tag{3.20}$$

Note that $\tilde{G}_5^2 = \tilde{G}^2$! Let

$$\mathfrak{U} = \bigcup_{j=1}^{5} U_j = \{1,2,3,w,x,y,z\}$$

and R_t, $t \in \mathfrak{U}$, be the probability that u and all those nodes of **U** the fusing of which results in t are in one and the same component of $\tilde{G}^1$. Therefore, the R_t are special **K**-terminal reliabilities of $\tilde{G}^1$. The probabilities p_{ut}, $t \in \mathfrak{U}$, satisfy the linear equation system

$$\begin{aligned}
R_w &= p_{uw} \\
R_x &= p_{uw} + p_{ux} \\
R_y &= p_{uw} + p_{uy} \\
R_z &= p_{uw} + p_{uz} \\
R_1 &= p_{uw} + p_{ux} + p_{uy} + p_{u1} \\
R_2 &= p_{uw} + p_{ux} + p_{uz} + p_{u2} \\
R_3 &= p_{uw} + p_{uy} + p_{uz} + p_{u3}
\end{aligned}$$

Its unique solution is

$$\begin{aligned}
p_{uw} &= R_w & p_{u1} &= R_1 + R_w - R_x - R_y \\
p_{ux} &= R_x - R_w & p_{u2} &= R_2 + R_w - R_x - R_z \\
p_{uy} &= R_y - R_w & p_{u3} &= R_3 + R_w - R_y - R_z. \\
p_{uz} &= R_z - R_w
\end{aligned}$$

Compared to the decomposition formula (3.17), the lower bound (3.20) has fewer terms and the p_{ut} are easier to compute than the $P_{u,t}(\tilde{G}^1_j)$.

Example 4 The same situation as in example 3 is considered. Table 3.5 shows for p = 0.8 and p = 0.9 the exact two-terminal reliabilities $R_{u,v}(\tilde{G})$ and the corresponding lower bounds (3.20). The lower bounds are seen to be very close to the exact values. □

TABLE 3.4 NUMERICAL RESULTS FOR EXAMPLE 4

p	$R_{uv}(\tilde{G})$	$\underline{R}_{uv}(\tilde{G})$
0.8	0.981670	0.981666
0.9	0.997879	0.997878

Of course, the lower bound (3.20) is also applicable to directed networks. Moreover, if **G** is directed in such a way that there are no (u,v)-paths crossing **U** more than one time, then the lower bound (3.20) coincides with the exact value

3.3.5 All-Terminal Reliability

Let $R(\tilde{G})$ be the all-terminal reliability of $\tilde{G}$. Since in this case $\mathbf{K} = \mathbf{V}$, there holds always $\mathbf{U} \subseteq \mathbf{K}$ so that according to the formula (3.8)

$$R(\tilde{G},\mathbf{K}) = \sum_{j=1}^{B} P_j(\tilde{G}^1)\ R(\tilde{G}_j^2). \tag{3.21}$$

$P_j(\tilde{G}^1)$ is the probability that $\tilde{G}^1$ induces the partition π_j and that each component of $\tilde{G}^1$ has a node in common with $\mathbf{U}$.

The stochastic network $\tilde{G}_k^1$ is connected iff each component of $\tilde{G}^1$ has a node in common with $\mathbf{U}$ and $\tilde{G}^1$ induces a partition π_j of $\mathbf{U}$ satisfying $\pi_j \wedge \pi_k = \pi_1$. Hence, the $P(\tilde{G}_j^1)$ satisfy the linear equation system

$$\sum_{\{j;\ \pi_j\pi_k=\pi_1\}} P_j(\tilde{G}^1) = R(\tilde{G}_k^1). \tag{3.22}$$

To simplify the notation, let be for $i = 1,2$ and $j = 1,2,\ldots,B$

$$p_j^i = P_j(\tilde{G}^i), \quad R_j^i = R(\tilde{G}_j^i),$$

$$\mathbf{p}^i = (p_1^i,\ p_2^i,\ldots,\ p_B^i)^T, \quad \mathbf{R}^i = (R_1^i,\ R_2^i,\ldots,\ R_B^i)^T,$$

and

$$a_{jk} = \begin{cases} 1, & \text{wenn } \pi_j \wedge \pi_k = \pi_1, \\ 0, & \text{sonst.} \end{cases}$$

Evidently, $\mathbf{A} = ((a_{jk}))$ is the coefficient matrix of the equation system (3.22). Now the decomposition formula (3.21) takes the form

$$R(\tilde{G}) = \sum_{j=1}^{B} p_j^1 R_j^2 = (\mathbf{p}^1)^T \mathbf{R}^2. \tag{3.23}$$

By changing the roles of $\tilde{G}^1$ and $\tilde{G}^2$,

$$R(\tilde{G}) = \sum_{j=1}^{B} p_j^2 R_j^1 = (\mathbf{p}^2)^T \mathbf{R}^1. \tag{3.24}$$

According to (3.22), the p_j^i fulfil for $i = 1,2$ the equation system

$$\sum_{\{j;\ \pi_j\pi_k=\pi_1\}} p_j^i = R_k^i \tag{3.25}$$

or, equivalently,

$$\mathbf{A}\ \mathbf{p}^i = \mathbf{R}^i, \quad i = 1,2. \tag{3.26}$$

By combining (3.23), (3.24), and (3.26) one gets

$$R(\tilde{G}) = (\mathbf{p}^1)^T \mathbf{A} \;\; \mathbf{p}^2, \qquad R(\tilde{G}) = (\mathbf{p}^2)^T \mathbf{A} \;\; \mathbf{p}^1. \tag{3.27}$$

It is well-known that the matrix $\mathbf{A}$ is regular (*Aigner [14]*). Denoting the inverse matrix of $\mathbf{A}$ by $\mathbf{A}^{-1}$, (3.26) yields

$$\mathbf{p}^i = \mathbf{A}^{-1} \, \mathbf{R}^i, \; i = 1,2. \tag{3.28}$$

Using (3.28) the decomposition formulas (3.27) take her final shape:

$$R(\tilde{G}) = (\mathbf{R}^1)^T \mathbf{A}^{-1} \mathbf{R}^2, \qquad R(\tilde{G}) = (\mathbf{R}^2)^T \mathbf{A}^{-1} \mathbf{R}^1. \tag{3.29}$$

Note that the "coupling matrix" $\mathbf{A}$ as well as the coefficient matrix $\mathbf{C}$ is completely characterized by the partition lattice $(\Pi,\le)$ belonging to $\mathbf{U}$. The all-terminal reliability of $\tilde{G}$ is, hence, completely characterized by quantities referring only to $\tilde{G}^1$ and $\tilde{G}^2$.

In what follows three special case will be considered.

1) u = 1 In this case, the separating node set consists of a single node. Therefore, $\tilde{G}$ is connected iff both $\tilde{G}^1$ and $\tilde{G}^2$ are connected. Hence,

$$R(\tilde{G}) = R(\tilde{G}^1)\, R(\tilde{G}^2) \quad \text{or} \quad R(\tilde{G}) = R^1_1 \, R^2_1.$$

2) u = 2 Let be $\mathbf{U} = \{1,2\}$, $\pi_1 = \{12\}$, and $\pi_2 = \{1,2\}$. Then,

$$\mathbf{A} \doteq \begin{pmatrix} 1 & 1 \\ 1 & 0 \end{pmatrix}, \qquad \mathbf{A}^{-1} = \begin{pmatrix} 0 & 1 \\ 1 & -1 \end{pmatrix}.$$

It follows

$$p^1_1 = R^1_2, \quad p^2_2 = R^1_1 - R^1_2,$$

$$R(\tilde{G}) = R^1_2 R^2_1 + (R^1_1 - R^1_2)R^2_2 \, .$$

3) u = 3 Let be $\mathbf{U} = \{1,2,3\}$. The corresponding partitions are given by (3.19). Then the "coupling matrix" $\mathbf{A}$ and its inverse $\mathbf{A}^{-1}$ are given by

$$A = \begin{pmatrix} 1 & 1 & 1 & 1 & 1 \\ 1 & 0 & 1 & 1 & 0 \\ 1 & 1 & 0 & 1 & 0 \\ 1 & 1 & 1 & 0 & 0 \\ 1 & 0 & 0 & 0 & 0 \end{pmatrix}, \quad A^{-1} = \begin{pmatrix} 0 & 0 & 0 & 0 & 1 \\ 0 & -\frac{1}{2} & \frac{1}{2} & \frac{1}{2} & -\frac{1}{2} \\ 0 & \frac{1}{2} & -\frac{1}{2} & \frac{1}{2} & -\frac{1}{2} \\ 0 & \frac{1}{2} & \frac{1}{2} & -\frac{1}{2} & -\frac{1}{2} \\ 1 & -\frac{1}{2} & -\frac{1}{2} & -\frac{1}{2} & \frac{1}{2} \end{pmatrix}.$$

Formula (3.26) yields for i = 1,2,

$$p_1^i = R_5^i$$

$$p_2^i = \frac{1}{2}\left[-R_2^i + R_3^i + R_4^i - R_5^i \right]$$

$$p_3^i = \frac{1}{2}\left[R_2^i - R_3^i + R_4^i - R_5^i \right]$$

$$p_4^i = \frac{1}{2}\left[R_2^i + R_3^i - R_4^i - R_5^i \right]$$

$$p_5^i = \frac{1}{2}\left[2R_1^i - R_2^i - R_3^i - R_4^i + R_5^i \right].$$

Example 5 Let us consider again the example network of Fig. 3.1 and its splitting by the separating node set **U** = {1,2,3}. All edges are assumed to have the common edge availability p = 0.8. Table 3.5 presents numerical results of a reliability analysis of $\tilde{G}^1$ and $\tilde{G}^2$.

TABLE 3.5 NUMERICAL RESULTS FOR EXAMPLE 6

i	π_i	R_i^1	p_i^1	R_i^2
1	{123}	0.983567	0.892682	0.980261
2	{12,3}	0.938471	0.042480	0.968360
3	{13,2}	0.943555	0.037396	0.978309
4	{23,1}	0.972559	0.008393	0.939692
5	{1,2,3}	0.892682	0.002616	0.927445

The scalar product of the 4th and 5th column yields the desired connectedness probability of $\tilde{G}$:

$$R(\tilde{G}) = 0.963096.$$

This result was obtained on a PC within a few seconds.On the other hand, the underlying network G has 163160 spanning trees. This implies that the application of the "disjoint sum approach" for computing $R(\tilde{G})$ is practically impossible. On the other hand, the subgraphs G^1 and G^2 have in each case 104 and 377 spanning trees. Therefore, the R^i_j can be computed by this approach. □

3.4 Network Reduction

3.4.1 Introduction

Network reduction is a powerful tool for the reliability analysis of complex stochastic networks $\tilde{G}$. It is characterized by simplifying the topological structure of the underlying network G by substituting a subgraph of G with a *replacement graph* to obtain a stochastic network $\tilde{H}$ with a known functional relationship between $R(\tilde{G})$ and $R(\tilde{H})$. More exactly, the *reliability preserving network reduction* involves 3 steps:

1) Decomposition of G as given by (3.2):

$$G = G^1 \cup G^2, \quad G^1 \cap G^2 = (U,\emptyset).$$

2) Generation of a graph H by replacing G^2 in G with a replacement graph H^2 satisfying

$$G^1 \cap H^2 = (U,\emptyset).$$

3) The stochastic network $\tilde{H}$ belonging to $H = G^1 \cup H^2$ has to be constructed in such a way that

$$R(\tilde{G}) = h\, R(\tilde{H}). \tag{3.30}$$

The factor h is called *reduction constant*. If no other assumptions are stated, the construction of $\tilde{H}$ will be done under the assumptions of section 3.2.

Of course, the application of the reduction approach makes only sense, if the complexity of $\tilde{H}$ is lower than the one of $\tilde{G}$.

3.4.2 Elementary Reductions

To illustrate the reduction principle, next three simple, but, nevertheless, important special cases are considered. They are based in each case on 2-point separating node sets and the graph G^2 to be replaced consists of two edges e_1 and e_2 whichhave the availabilities p_1 and p_2, respectively.

Parallel reduction Let e_1 and e_2 be two parallel edges with the common end nodes i and j. A *parallel reduction* replaces these edges by a single edge ε with the end nodes i and j and the availability (Fig. 3.5) $\alpha = 1 - (1-p_1)(1-p_2)$.

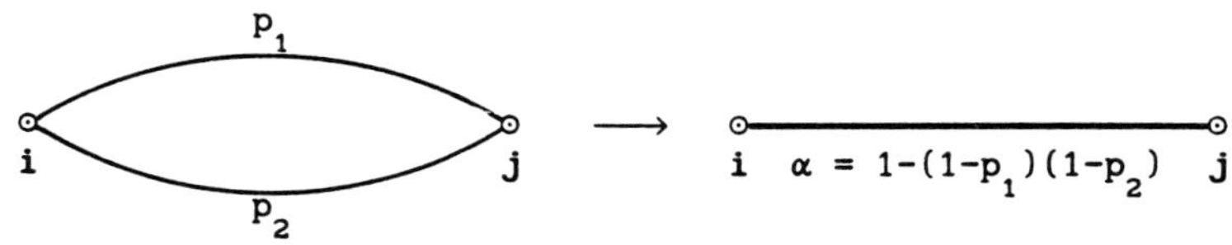

FIG. 3.5 ILLUSTRATION OF PARALLEL REDUCTION

Generally, the underlying structure **G** does not contain parallel edges. But parallel edges may arise, when the factorization formula (3.1) is used. Notwithstanding the reliability criterion to be computed, parallel reduction can always be applied.

Series reduction Let $e_1 = (i,j)$ and $e_2 = (j,k)$ be two edges in series, where the common node j is assumed to have the degree 2 and does not belong to a given set of terminal nodes **K** (with respect to computing $R(\tilde{\mathbf{G}},\mathbf{K})$. A *series reduction* replaces the edges e_1 and e_2 by the edge $\varepsilon = (i,k)$ with the availability $\alpha = p_1 p_2$ (Fig. 3.6).

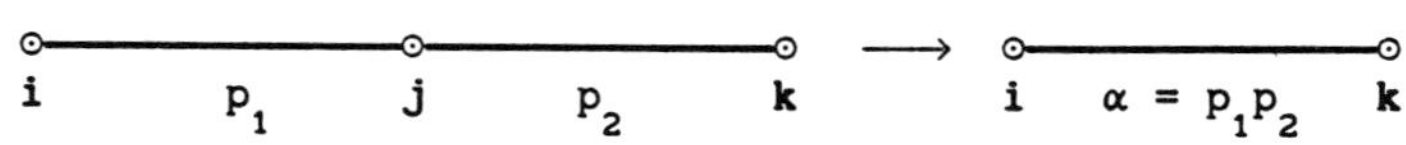

FIG. 3.6 ILLUSTRATION OF THE SERIES REDUCTION

The series reduction is applicable for computing the two-terminal reliability $R_{u,v}(\tilde{\mathbf{G}})$, if $u,v \neq j$. However, it is not applicable for computing the all-terminal reliability, since in this case there holds trivially $j \in \mathbf{K} = \mathbf{V}$.

Both in case of the parallel reduction and the series reduction the constant h is equal to 1.

Degree 2-reduction Let $e_1 = (i,j)$ and $e_2 = (j,k)$ be two edges in series, where the common node j has the degree 2 and $\{i,j,k\} \subseteq \mathbf{K}$. A *degree 2-reduction* replaces the edges e_1 and e_2 by the edge $\varepsilon = (i,k)$ with the availability

$$\alpha = p_1p_2/h, \tag{3.31}$$

where the reduction constant h is given by

$$h = 1 - (1-p_1)(1-p_2). \tag{3.32}$$

Although section 3.4.4 presents a general approach to computing the edge availabilities of the replacement graph, for a better understanding it seems to be appropriate to give a separate treatment of this special case. Let z_1, z_2, and z_ε be the indicator variables for the states of the edges e_1, e_2, and ε. Then there results from the reliability behaviour of a series connection

$$R(\tilde{G}|z_1= z_2= 0) = 0$$

$$R(\tilde{G}|z_1= 0,z_2= 1) = R(\tilde{G}|z_1= 1,z_2= 0) = R(\tilde{H}|z_\varepsilon= 0),$$

$$R(\tilde{G}|z_1= z_2= 1) = R(\tilde{H}|z_\varepsilon= 1).$$

The formula of the total probability yields

$$R(\tilde{G}) = R(\tilde{H}|z_\varepsilon= 0)[p_1(1-p_2)+(1-p_2)p_1] + R(\tilde{H}|z_\varepsilon=1)p_1p_2 \ .$$

Hence the reduction equation (3.30) is fulfilled iff α and h have the values (3.31) and (3.32), respectively. ∎

The degree 2-reduction is not applicable for computing the two-terminal reliability, but principally applicable for computing the all-terminal reliability.

In what follows, parallel-, series-, and degree 2-reductions are called *elementary reductions*.

Definition 3.4 A stochastic network $\tilde{G}$ is *sp-reducible*, if by successive application of elementary reductions the underlying network G can be reduced to a single edge. Otherwise, the stochastic network is *sp-complex*. ▪

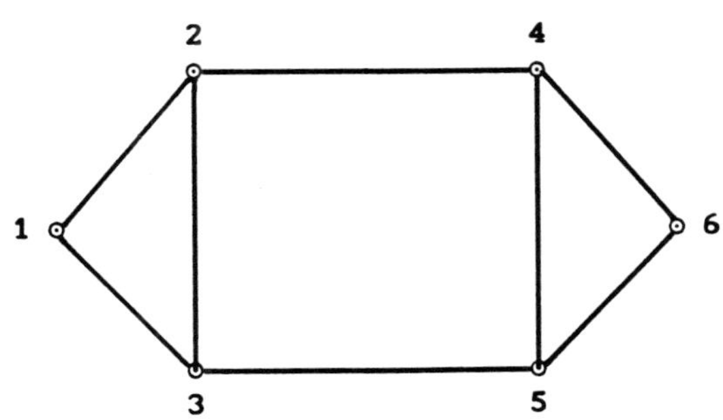

FIG. 3.7 EXAMPLE NETWORK

Fig. 3.7 shows a network **G** with the property that the corresponding stochastic network $\tilde{\mathbf{G}}$ is sp-reducible with respect to $\mathbf{K} = \{2,3\}$ and sp-complex with respect to $\mathbf{K} = \{1,6\}$.

The advantage of sp-reducible stochastic networks $\tilde{\mathbf{G}}$ is that the computation of the corresponding reliability criterion $R(\tilde{\mathbf{G}})$ can be done in polynomial time. This property have all those stochastic networks the reliability analysis of which can be completely carried out by repeated application of reduction. To this class of stochastic networks belong large classes of stochastic networks the underlying deterministic networks of which are planar graphs, for instance *k-trees* (*Colbourn [2]*) and *recurrent networks* (*Beichelt/Franken [18]*).

Polygon-to-chain-reduction There is a close connection between elementary reductions and *polygon-to-chain reductions*. Let us assume that the graph $\mathbf{G}^2$ to be replaced is a circuit connected with **G** by a two-point separating node set and consisting of at least 3 nodes where at least one of these belongs to **K**. It can be shown that by application of elementary reductions such a circuit can be transformed into one out of altogether 7 possible basic structures. These basic structures -called *polynoms*- have been found by *Wood [19]*. Furthermore, *Wood* has shown that these polygons can be transformed by special reductions to 4 types of chains. In this connection, a chain is a path between the two nodes of the separating node set. The chains differ from each other by their length and the order of the **K**-nodes. Figure 3.8 shows 2 of the 7 polygons and the corresponding chains. (The **K**-nodes are the black ones.)
A complete survey on the polygon-to-chain reduction was also given by *Kohlas [20]*.

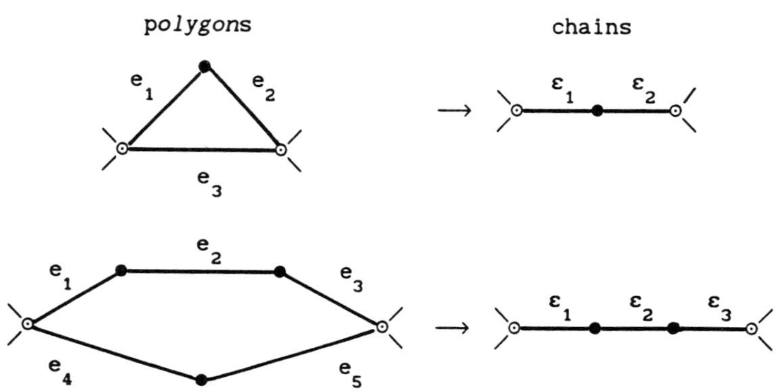

FIG. 3.8 TWO POLYGONS AND THEIR CORRESPONDING CHAINS

3.4.3 Reduction and Factorization

A simple, but rather effective method for the reliability analysis of arbitrary stochastic networks consists in the alternate application of factorization and elementary reductions. This is due to the fact that on application of the factorization formula (3.1) the resulting graphs $\mathbf{G}_e$ and $\mathbf{G}_{-e}$ may contain edges in series or in parallel, so that by application of elementary reductions a further structural simplification of the graphs $\mathbf{G}_e$ and $\mathbf{G}_{-e}$ will be possible. This approach together with the corresponding computerized algorithm has been proposed by *Wood [21]*. It is also profitable to combine the application of the decomposition formula (3.6) with the algorithm of Wood. This will be demonstrated in the following example.

Example 6 Figure 3.9 shows a network **G** with 21 nodes and 57 edges. The availability of all edges be p = 0.8. The all-terminal reliability $R(\tilde{\mathbf{G}})$ has to be computed.

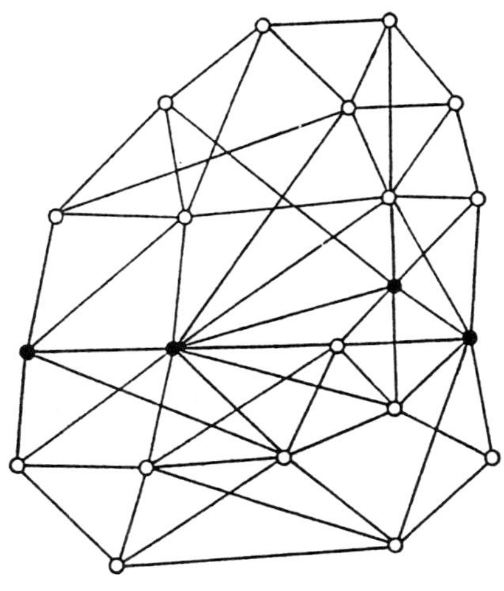

FIG. 3.9 NETWORK WITH 21 NODES AND 57 EDGES

Note that **G** has about $3.5 \cdot 10^{12}$ spanning trees and 220 000 cuts (with respect to the connectedness of **G**). Therefore, the computation of $R(\tilde{\mathbf{G}})$ is an extremely complex problem. By a direct application of the algorithm of Wood to $\tilde{\mathbf{G}}$, even the at present most fast computers would not yield $R(\tilde{\mathbf{G}})$ within reasonable time. (They would need months.) On the other hand, the application of the decomposition formula (3.29) yields $R(\tilde{G})$ within a minute, when the 4-point separating node set $U = \{1,2,3,4\}$ being black-marked in Fig. 3.9 is used and the R^i_j, i = 1, 2; j = 1, 2,..., 15, are computed by the algorithm of Wood. Table 3.6 presents the numerical results obtained. According to (3.29), the all-terminal reliability of $\tilde{G}$ is obtained by the scalar product of the 4th and 5th column and the 3th and 6th column, respectively:

$$R(\tilde{G}) = 0.97804.$$

□

TABLE 3.6 NUMERICAL RESULTS FOR EXAMPLE 6

j	π_j	R^1_j	p^1_j	R^2_j	p^2_j
1	{1234}	0.988203	0.971698	0.990075	0.907168
2	{123,4}	0.980279	0.007842	0.949216	0.039042
3	{124,3}	0.980086	0.008044	0.988481	0.001445
4	{134,2}	0.988136	0.000061	0.988402	0.001506
5	{234,1}	0.987830	0.000354	0.949414	0.038834
6	{12,34}	0.988138	0.000000	0.989866	0.000054
7	{13,24}	0.988199	0.000000	0.989934	0.000005
8	{14,23}	0.988068	0.000115	0.988330	0.000049
9	{12,3,4}	0.972228	0.000065	0.947562	0.000067
10	{13,2,4}	0.980211	0.000000	0.947549	0.000066
11	{14,2,3}	0.979893	0.000004	0.985103	0.000012
12	{13,1,4}	0.979804	0.000016	0.910178	0.001676
13	{24,1,3}	0.979716	0.000003	0.947819	0.000062
14	{34,1,2}	0.987699	0.000000	0.947709	0.000081

3.4.4 Reduction Equations

For computing the reduction constant h and the unknown availabilities α_k of the stochastic replacement structure $\tilde{H}$ according to the basic reduction equation (3.30) there is a system of reduction equations. In what follows, its derivation is demonstrated only for the all-terminal reliability $R(\tilde{G})$.

Next the decomposition formula (3.23) is applied to the stochastic replacement structure $\tilde{H} = \tilde{G}^1 \cup \tilde{H}^2$:

$$R(\tilde{H}) = \sum_{j=1}^{B} p^1_j \, R(\tilde{H}^2_j). \tag{3.33}$$

Substituting (3.33) in (3.30) yields

$$R(\tilde{G}) = \sum_{j=1}^{B} p^1_j \left[h \, R(\tilde{H}^2_j)\right] . \tag{3.34}$$

Comparing the coefficients of the p_j^1 in (3.23) and (3.34) gives

$$R_j^2 = h\, R(\tilde{H}_j^2), \quad j = 1,2,\ldots,B. \tag{3.35}$$

In this equation system, the probabilities R_j^2 are assumed to be known. They have to be computed by any available method. The $R(\tilde{H}_j^2)$ have to be expressed by the unknown availabilities α_k of the edges of $\tilde{H}_j^2$ assuming statistically independent edge states. The equation system (3.35) is generally not linear. For its unique solvability, two conditions are necessary:

1) The equation system (3.35) has exactly B unknown parameter, namely $\alpha_1, \alpha_2, \ldots, \alpha_{B-1}$, h.

2) The partition sets of **U** generated by $\tilde{G}^2$ and $\tilde{H}^2$, respectively, are identical.

Another, from the numerical point of view frequently more promising approach to the computation of the parameters α_k and h results from an application of (3.24) to $\tilde{H}$:

$$R(\tilde{H}) = \sum_{j=1}^{B} P_j(\tilde{H}^2)\, R_j^1. \tag{3.36}$$

Substituting (3.36) in (3.30) yields

$$R(\tilde{G}) = \sum_{j=1}^{B} \left[h\, P_j(\tilde{H}^2)\right] R_j^1 . \tag{3.37}$$

By comparing the coefficients of R_j^1 in (3.24) and (3.37),

$$p_j^2 = h\, P_j(\tilde{H}^2), \quad j = 1,2,\ldots,B. \tag{3.38}$$

By definition, $P_j(\tilde{H}^2)$ is the probability that $\tilde{H}^2$ induces the partition π_j and each component of $\tilde{H}^2$ contains a node out of **U**. The p_j^2 satisfy the equation system (3.26) with i = 2, whereas the probabilities $P(\tilde{H}_j^2)$ have to be expressed in dependence of the unknown edge availabilities α_k.

3.4.5 Special Cases

1) u = 1 In this case,

$$R(\tilde{G}) = R(\tilde{G}^1)\, R(\tilde{G}^2),$$

so that a reduction of $\tilde{G}$ is superfluous. (Of course, a reduction of $\tilde{G}^1$ or $\tilde{G}^2$, respectively, might be useful.)

2) u = 2 G^2 is replaced by a single edge directly connecting the two nodes of **U**. Therefore, if **U** = {1,2}, the replacement graph for G^2 is

$$\varepsilon = e_{12} = (1,2).$$

This rule does not depend on the complexity of G^2! If α denotes the availability of ε to be computed and $\pi_1 = \{12\}$ and $\pi_2 = \{1,2\}$ are the partitions of **U**, then there results from (3.35)

$$R_1^2 = h, \quad R_2^2 = h\,\alpha. \tag{3.39}$$

Example 7 As a special application of (3.39), let us consider the degree 2-reduction. In this case, G^2 consists of the two edges in series $e_1 = (1,3)$ and $e_2 = (3,2)$, where the common node 3 has the degree 2 and all the nodes 1, 2, and 3 belong to a specified set of terminal nodes **K**. Therefore,

$$R_1^2 = 1 - (1-p_1)(1-p_2), \quad R_2^2 = p_1 p_2,$$

where p_1 and p_2 are the availabilities of e_1 and e_2, respectively. Thus, (3.39) yields for α and h the very values given by (3.31) and (3.32), respectively. □

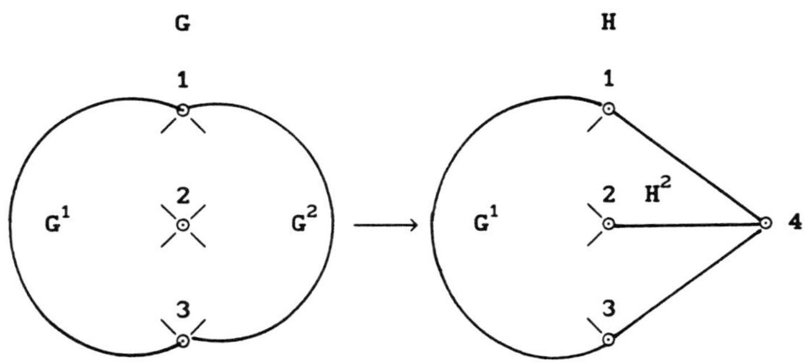

FIG. 3.10 REDUCTION IN CASE OF **u** = 3

3) u = 3 In this case, the most simple structure by which G^2 can be replaced is a "star" (Fig. 3.10), whose edge and node sets are given by {1,2,3,4} and $\{\varepsilon_1, \varepsilon_2, \varepsilon_3\}$, where $\varepsilon_1 = (1,4)$, $\varepsilon_2 = (2,4)$, and $\varepsilon_3 = (3,4)$. Let further α_k be the availability of ε_k, k = 1,2,3. The partitions of **U** are again given by (3.19):

$$\pi_1=\{123\}, \; \pi_2=\{12,3\}, \; \pi_3=\{13,2\}, \; \pi_4=\{1,23\}, \; \pi_5=\{1,2,3\}.$$

To have in the replacement structure $\tilde{H}^2$ exactly 4 free parameter, there will be now assumed that, contrary to assumption 2 of section 3.2, the node 4 may fail, too; its availability be β. Therefore, the node 4 influences also the generation of partitions of **U** by $\tilde{H}^2$. From (3.38),

$$
\begin{aligned}
p_1^2 &= h\,\beta\,\alpha_1\alpha_2\alpha_3 \\
p_2^2 &= h\,\beta\,\alpha_1\alpha_2(1-\alpha_3) \\
p_3^2 &= h\,\beta\,\alpha_1\alpha_3(1-\alpha_2) \\
p_4^2 &= h\,\beta\,\alpha_2\alpha_3(1-\alpha_1) \\
p_5^2 &= h\,[1-\beta+\beta(1-\alpha_1\alpha_2-\alpha_1\alpha_3-\alpha_2\alpha_3+2\alpha_1\alpha_2\alpha_3)]\ .
\end{aligned}
\tag{3.40}
$$

The solution of this system is

$$
\begin{aligned}
\alpha_1 &= p_1^2 \big/ (p_1^2+p_4^2) \\
\alpha_2 &= p_1^2 \big/ (p_1^2+p_3^2) \\
\alpha_3 &= p_1^2 \big/ (p_1^2+p_2^2) \\
\beta &= (p^2+p_2^2)(p_1^2+p_3^2)(p_1^2+p_4^2) \big/ \left[(p_1^2)^2 h\right] \\
h &= p_1^2+p_2^2+p_3^2+p_4^2+p_5^2 .
\end{aligned}
\tag{3.41}
$$

Delta-Star-Reduction If the graph G^2 to be replaced is a "delta" (triangle), then there arises an important special case, the *delta-star-reduction* or *delta-star-transformation* (Fig. 3.11).

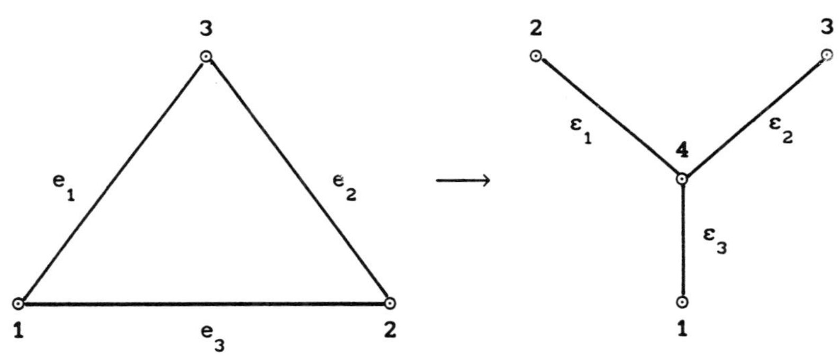

FIG. 3.11 DELTA-STAR-REDUCTION

In this case the set of the corners of the triangle {1,2,3} coincides with the separating node set **U**. Since there cannot exist a component of the triangle G^2 having no node in common with **U**, the p_j^2 satisfy the condition

$$\sum_{j=1}^{5} p_j^2 = 1.$$

Therefore, one of the 5 equations in (3.40) is superfluous and in this special case for obtaining a unique solution of (3.40), it is not necessary to assume an unreliable node 4. Thus, by dropping, for instance, the last equation of (3.40), expressing the p_j^2 by the edge availabilities p_i of the triangle, and putting $\beta = 1$ one obtains for h and the availabilities α_k of the edges of the "star" the equation system

$$\begin{aligned} p_1p_2 + p_1p_3 + p_2p_3 - 2p_1p_2p_3 &= h\ \alpha_1\alpha_2\alpha_3 \\ p_1(1-p_2)(1-p_3) &= h\ \alpha_1\alpha_2(1-\alpha_3) \\ p_3(1-p_1)(1-p_2) &= h\ \alpha_1\alpha_3(1-\alpha_2) \\ p_2(1-p_1)(1-p_3) &= h\ \alpha_2\alpha_3(1-\alpha_1)\ . \end{aligned} \tag{3.42}$$

If h is eliminated from the first equation of (3.42) and substituted in the other 3 equations, the unique solution of

$$\alpha_1 = \frac{Z}{N_1},\quad \alpha_2 = \frac{Z}{N_2},\quad \alpha_3 = \frac{Z}{N_3},\quad h = \frac{N_1N_2N_3}{Z^2}, \tag{3.43}$$

where

$$\begin{aligned} Z &= p_1p_2 + p_1p_3 + p_2p_3 - 2p_1p_2p_3, \\ N_1 &= p_2 + p_1p_3(1-p_2), \\ N_2 &= p_3 + p_1p_2(1-p_3)\ , \\ N_3 &= p_1 + p_2p_3(1-p_1). \end{aligned}$$

In particular, if $p_i = p$, $i = 1, 2, 3$, then

$$\alpha_1 = \alpha_2 = \alpha_3 = \frac{(3-2p)p}{1+p-p^2}\ . \tag{3.44}$$

Of course, instead of replacing a triangle by a star, it may also lead to a structural simplification of a given network, if a star is replaced by a triangle. Such a reduction is called *star-delta-reduction*. In this case the equation system (3.42) holds again, but the α_k are given and the p_i are unknown. However, an explicit solution of (3.42) for the p_i is not possible. Therefore, for obtaining the p_i numerical methods have to be applied.

In the following example a stochastic network is considered the all-terminal reliability of which can be computed by successive applications of delta-star- and elementary reductions. This is due to the fact that the underlying deterministic network **G** is transformed into a tree **T** so that the all-terminal reliability of the corresponding stochastic tree $\tilde{\mathbf{T}}$

is simply the product of its edge availabilities.

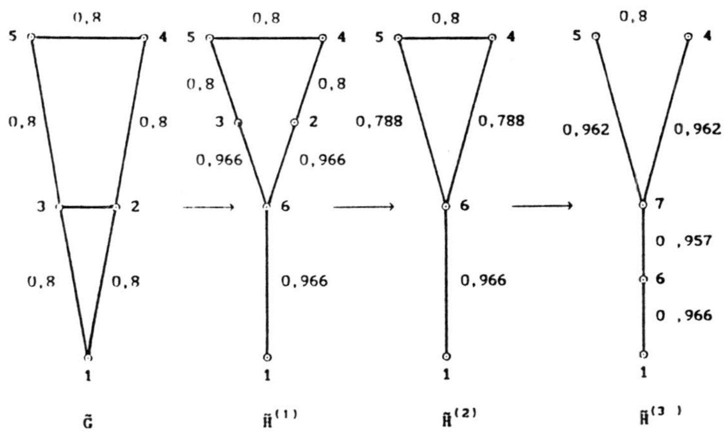

FIG. 3.12 APPLICATION OF THE DELTA-STAR-REDUCTION

Example 8 The all-terminal reliability of a stochastic network is to compute the underlying deterministic network of which is shown in Figure 3.12. For simplicity it is assumed that the edges have the common availability p = 0.8. Figure 3.12 presents also the replacement structures for $\tilde{G}$, denoted by $\tilde{H}^{(1)}$, $\tilde{H}^{(2)}$, and $\tilde{H}^{(3)}$, which arise by repeated applications of delta-star- and degree 2- reductions. The edges of this structures are weighted with their availabilities. $\tilde{H}^{(1)}$ arises from $\tilde{G}$ by application of the delta-star-reduction to the triangle given by the nodes 1, 2 and 3. The edges e_{16}, e_{26} and e_{36} of $\tilde{H}^{(1)}$ have according to (3.44) the common availability $\alpha_{16} = \alpha_{26} = \alpha_{36} = 28/29 = 0{,}996$. The corresponding h-value is $h_1 = 0{,}995$. $\tilde{H}^{(2)}$ arises from $\tilde{H}^{(1)}$ by a degree 2-reduction of the edges e_{53} and e_{36} as well as e_{42} and e_{26}. The two edges generated in this way have according to (3.31) the availabilities $\alpha_{46} = \alpha_{56} = 0{,}778$. The corresponding reduction constants are $h_{2,1} = h_{2,2} = 0{,}993$. Finally, the tree $\tilde{H}^{(3)}$ arises from $\tilde{H}^{(2)}$ by application of the delta-star-reduction to the triangle given by the nodes 4, 5 and 6. The availabilities of the edges e_{57}, e_{47} and e_{67} are according to (3.43) $\alpha_{57} = \alpha_{47} = 0{,}962$ and $\alpha_{67} = 0{,}994$. The corresponding h-value is $h_3 = 0{,}994$. Hence, the all-terminal-reliability of the tree $\tilde{H}^{(3)}$ is

$$R(\tilde{H}^{(3)}) = (0{,}962)^2 \cdot 0{,}966 = 0{,}856.$$

$R(\tilde{G})$ and $R(\tilde{H}^{(3)})$ differ from each other by the factor

$$h_1 h_{2,1} h_{2,2} h_3 = 0,976.$$

Therefore, the all-terminal-reliability of $\tilde{G}$ amounts to

$$R(\tilde{G}) = 0.976 \circ 0.856 = 0.835.$$

□

Example 9 Let us again consider the example network of Figure 3.1 with the common edge availability p =0.8 and its splitting into the subnetworks G^1 and G^2. For computing the all-terminal reliability $R(\tilde{G})$, the subnetwork G^1 is re- placed by a "star". By using the probabilities p_j^1 of Table 3.6, the solution of the equation system (3.40) is according to (3.41) (there the p_j^2 have to replaced by p_j^1)

$$\alpha_1 = 0,9907 \qquad h = 0,9836$$
$$\alpha_2 = 0,9598 \qquad \beta = 1,0024 \ .$$
$$\alpha_3 = 0,9522$$

This example yields an unexpected numerical result:

The "availability" of the node 4 is greater than 1!

Other numerical examples show that α_k and the reduction constant h may have also the property $\alpha_k \notin [0,1]$ and $h \notin [0,1]$, respectively. Therefore, the α_k are generally no probabilities. They are simply edge weights. This implies that the interpretation of $\tilde{H}$ and $\tilde{H}_j^1$ given up to now as stochastic networks has not been correct. They are as weighted graphs only auxiliary functions for computing $R(\tilde{G})$. However, from the computational point of view, these *weighted replacement graphs* can be handled as stochastic networks with independent edges.

Ramamoorty/Balgopal [22] may have been the first to propose the star-delta reduction for the reliability analysis. A treatment of both delta-star- and star-delta-reduction was given by *Banerjee/Rajamani [23,24]*. In the meantime much more work has been contributed to delta-star- and star-delta-reduction, see, for instance, *Rosenthal/Frisque [5]*, *Singh/Kankam [25]*, *Rosenthal [4]*, *Gupta/Sharma [26]*, *Gadani/Misra [27,28,29]*, *Gadani [30]*, and *Singh/Asgarpoor [31]*. The introduction of an unreliable node in the delta-star-reduction ist due to *Rosenthal/Frisque [5]*. Another, rather important type of reduction, the *quadrilateral-star-reduction*, has been introduced and analysed by *Gadani/Misra [32,33]*. A comprehensive survey on star-delta , delta-star , and quadrilateral-star -reduction has been given by *Misra [34]*.

3.4.6 A Generalized Approach to Network Reduction

In the foregoing sections, replacement structures have been proposed only in case of **u** < 4. For **u** ≥ 4 no single replacement graph is known allowing an explicit, unique solution of the equation systems (3.35) or (3.38), respectively. Therefore, *Beichelt/Tittmann [7,8]* have proposed to use more than one replacement graph in a reduction step. More exactly, given the network decomposition (3.2), the subnetwork G^2 is replaced by the replacement graphs $H^{2,1}, H^{2,2}, \ldots, H^{2,r}$, $r \geq 1$. In doing so there arise in each case the replacement graphs $H^k = G^1 \cup H^{2,k}$, $k = 1,2,\ldots,r$. The corresponding weighted re- placement graphs $\tilde{H}^k$ have to be constructed in such a way that the *generalized reduction equation*

$$R(\tilde{G}) = \sum_{k=1}^{r} h_k R(\tilde{H}^k) \tag{3.45}$$

is satisfied. The reduction constants $h_1, h_2, \ldots, h_r$ as well as the unknown edge weights ("availabilities") α_i of the $\tilde{H}^k$ will be computed under the assumptions of section 3.2. (In particular, by proper choice of the $H^{2,k}$ there will be no need to introduce unreliable nodes.) This can be done analogously to the approach exercised in section 3.4.4: Next the decomposition formula (3.23) is applied to $\tilde{H}^k$:

$$R(\tilde{H}^k) = \sum_{j=1}^{B} p_j^1 R(\tilde{H}_j^{2,k}), \quad k=1,2,\ldots,r. \tag{3.46}$$

Substitution of (3.46) in (3.45) yields

$$R(\tilde{G}) = \sum_{j=1}^{B} p_j^1 \sum_{k=1}^{r} h_k R(\tilde{H}_j^{2,k}). \tag{3.47}$$

Comparing the coefficients of the p_j^1 in (3.47) and (3.23) yields

$$R_j^2 = \sum_{k=1}^{r} h_k R(\tilde{H}_j^{2,k}), \quad j=1,2,\ldots,B. \tag{3.48}$$

If the $R(\tilde{H}_j^{2,k})$ are expressed in terms of the α_i, then (3.48) becomes a nonlinear system of equations for the α_i and h_k. Two conditions are necessary for its unique solvability:

1) The total number of edges in $H^{2,1}, H^{2,2}, \ldots, H^{2,r}$ must be B-r.
2) Each partition of **U** can be induced by at least one of the stochastic replacement structures $\tilde{H}^{2,k}$, $k = 1,2,\ldots,r$.

Analogously to (3.38), the following equation system can be established:

$$p_j^2 = \sum_{k=1}^{r} h_k \, P_j(H^{2,k}), \quad j = 1,2,\ldots,B. \tag{3.49}$$

If the probabilities $P_j(H^{2,k})$ are expressed in terms of the $\alpha_1, \alpha_2,\ldots, \alpha_{B-r}$, then (3.49) is also an equation system for the α_j and the h_k. But it has a decisive advantage compared to (3.48): replacement graphs $\tilde{H}^{2,k}$, generating disjoint sets of partitions of **U**, lead to uncoupled sets of equations of (3.49). This is due to the fact that $P_j(\tilde{H}^{2,k}) = 0$ iff π_j cannot be generated by $\tilde{H}^{2,k}$. Moreover, if the replacement graphs $H^{2,k}$ are chosen in such a way that the sets of partitions generated by different $\tilde{H}^{2,k}$ are disjoint, then the equation systems (3.48) and (3.49) yield solutions with $\alpha_j \in [0,1]$.

Figure 3.13 shows for **u** = 4 an appropriate set of r = 5 replacement graphs and Table 3.7 presents the corresponding equations (3.49). The bold (marked with little lines) edges of Figure 3.13 have the availability 1 (0). For instance, in

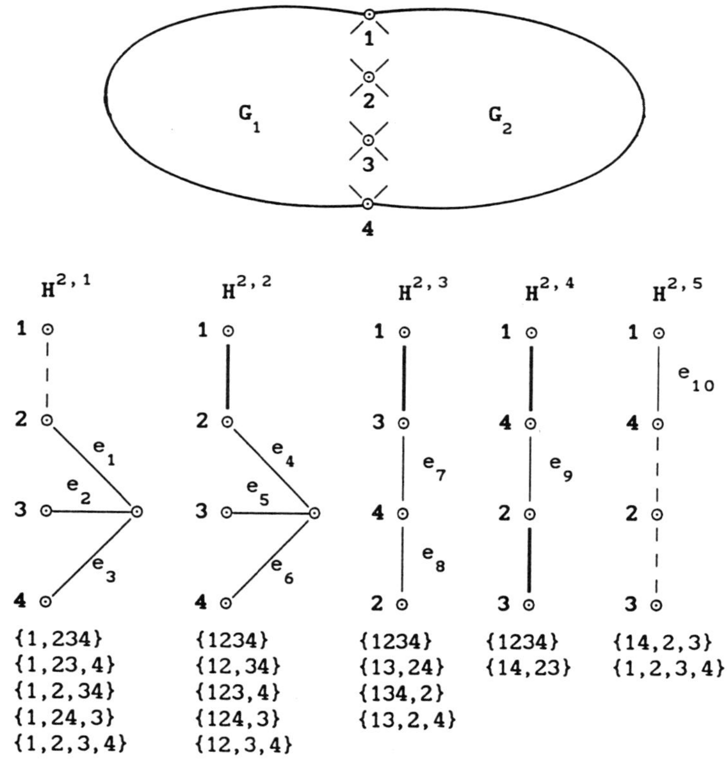

FIG. 3.13 ILLUSTRATION OF A NETWORK WITH A 4-POINT SEPARATING NODE SET; 5 REPLACEMENT GRAPHS AND THE PARTITIONS GENERATED BY THEM

$H^{2,1}$ the node 1 is isolated and in $H^{2,2}$ the nodes 1 and 2 are practically fused. Thus, $H^{2,1}$ generates only partitions of **U** containing the block {1} and there exists always a block in the partitions generated by $H^{2,2}$ containing both node 1 and node 2. Therefore, the bold and marked with little lines edges do not contribute to the total number $B - r = 15 - 5 = 10$ of weighted edges in the replacement graphs. Of course, there are other suitable sets of replacement graphs satisfying the conditions 1 and 2. But the replacement graphs proposed are trees. Hence the equation system (3.49) has a simple structure and can easily be solved. It is presented in Table 3.7. There the notation $\bar{\alpha}_j = 1 - \alpha_j$ is used and p_j^2 is reduced to p_j.

TABLE 3.7 PARTITION SET FOR **U** = {1,2,3,4} AND EQUATIONS (3.49) for α_j and h_k.

j	π_j	Equations (3.49)
1	{1234}	$p_1 = h_2\alpha_4\alpha_5\alpha_6 + h_3\alpha_7\alpha_8 + h_4\alpha_9$
2	{123,4}	$p_2 = h_2\alpha_4\alpha_5\bar{\alpha}_6$
3	{124,3}	$p_3 = h_2\alpha_4\bar{\alpha}_5\alpha_6$
4	{134,2}	$p_4 = h_3\alpha_7\bar{\alpha}_8$
5	{1,234}	$p_5 = h_1\alpha_1\alpha_2\alpha_3$
6	{12,34}	$p_6 = h_2\bar{\alpha}_4\alpha_5\alpha_6$
7	{13,24}	$p_7 = h_3\bar{\alpha}_7\alpha_8$
8	{14,23}	$p_8 = h_4\bar{\alpha}_9$
9	{12,3,4}	$p_9 = h_2(\alpha_4\bar{\alpha}_5\bar{\alpha}_6 + \bar{\alpha}_4\alpha_5\bar{\alpha}_6 + \bar{\alpha}_4\bar{\alpha}_5\alpha_6)$
10	{13,2,4}	$p_{10} = h_3\bar{\alpha}_7\bar{\alpha}_8$
11	{14,2,3}	$p_{11} = h_5\alpha_{10}$
12	{1,4,23}	$p_{12} = h_1\alpha_1\alpha_2\bar{\alpha}_3$
13	{1,3,24}	$p_{13} = h_1\alpha_1\bar{\alpha}_2\bar{\alpha}_3$
14	{1,2,34}	$p_{14} = h_1\bar{\alpha}_1\alpha_2\alpha_3$
15	{1,2,3,4}	$p_{15} = h_1(\alpha_1\bar{\alpha}_2\bar{\alpha}_3 + \bar{\alpha}_1\alpha_2\bar{\alpha}_3 + \bar{\alpha}_1\bar{\alpha}_2\alpha_3) + h_5\bar{\alpha}_{10}$

Let be

$$a = p_2p_3p_6 + (p_2 + p_3)(p_2 + p_6)(p_3 + p_6).$$

$$b = (p_2 + p_3 + p_6)p_9,$$

Then the unique solution of the nonlinear equation system for the h_k and the α_j given in Table 3.7 is

$$
\begin{aligned}
h_1 &= (p_5 + p_{12})(p_5 + p_{13})(p_5 + p_{14})/(p_5)^2, \\
h_2 &= (a + bp_2)(a + bp_3)(a + bp_6)/(a^2 b), \\
h_3 &= (p_4 + p_{11})(p_7 + p_{10})/p_{10} \\
h_4 &= p_1 + p_8 - p_4 p_7/p_{10} - a/b, \\
h_5 &= p_{11} + p_{15} - (p_{12}p_{13} + p_{12}p_{14} + p_{13}p_{14})/p_4, \\
\alpha_1 &= p_5/(p_5 + p_{14}), \quad \alpha_2 = p_5/(p_5 + p_{13}), \\
\alpha_3 &= p_5/(p_5 + p_{12}), \quad \alpha_4 = a/(a + bp_6), \\
\alpha_5 &= a/(a + bp_3), \quad \alpha_6 = a/(a + bp_2), \\
\alpha_7 &= p_4/(p_4 + p_{10}), \quad \alpha_8 = p_7/(p_7 + p_{10}), \\
\alpha_9 &= \frac{bp_1p_{10} - ap_{10} - bp_4p_7}{bp_8p_{10} + bp_1p_{10} - ap_{10} - bp_4p_7}, \\
\alpha_{10} &= \frac{p_5p_{11}}{p_5p_{11} + p_5p_{15} - p_{12}p_{13} - p_{12}p_{14}}.
\end{aligned}
\tag{3.50}
$$

For **u** = 4 it is rather tedious to construct a set of suitable replacement graphs. The main problem is to solve the corresponding equations (3.48) or (3.49). For **u** > 4, reduction seems to be no longer feasible.

3.4.7 Reduction Algorithm

The application of decomposition and reduction methods is only then useful, if the network under consideration is sufficiently complex. Let $C(\tilde{G})$ be the complexity of $\tilde{G}$. A suitable complexity measure is in case of the all-terminal reliability the number of spanning trees of **G**. If $C(\tilde{G}) \le C$, then $R(\tilde{G})$ is computed by any available method, for instance, by the algorithm of *Wood [21]*. The proper choice of the bound C depends, of course, on the computer being used and on the "available method".

The reduction algorithm proposed in the sequel can be recursively applied: It first generates the representation (3.45) of $R(\tilde{G})$, then it is applied to $R(\tilde{H}^k)$ and so on. However, a recursive application of reduction is only possible or worth while, respectively, as long as separating node sets with 1) $u \le 4$ and 2) $C(\tilde{H}^k) \ge C$ for at least one k, $k \in \{1,2,...,r\}$, can be selected. Separating node sets satisfying these two properties are called *admissible*. Admissible node sets can be found by dynamic programming (*Lee, Vogt, Mickle [15]*). A necessary condition for the existence of admissible separating node sets in each cycle of the algorithm is that the node connectivity of each subgraph of **G** does not exceed 4.

Reduction Algorithm

Input: A stochastic network $\tilde{G}$ with perfectly reliable nodes.

Output: $R(\tilde{G})$

Function: Rel($\tilde{G}$)

begin

1. RH := 0
2. If $C(\tilde{G}) \le C$,
   then compute $R(\tilde{G})$ directly by any available method. Return Rel.
3. Determine an admissible separating node set.
4. Specify G^2 and compute $R(\tilde{G}^2_j)$, j = 1,2,...,B, by any available method.
5. Solve (3.25) (i = 2) to obtain the probabilities p^2_j, j = 1,2,..., B.
6. Replace G^2 by an edge (**u** = 2), by a star (**u** = 3), or by the $H^{2,1}$, $H^{2,2}$, ..., $H^{2,5}$ of figure (3.13) (**u** = 4).
7. Solve (3.49) to obtain the h_k and the edge weights α_i of the replacement graphs.
8. Construct $\tilde{H}^k$, k = 1,2,...,5.
9. For k := 1 to *do* RH := RH + h_kRel($\tilde{H}^k$).

end;

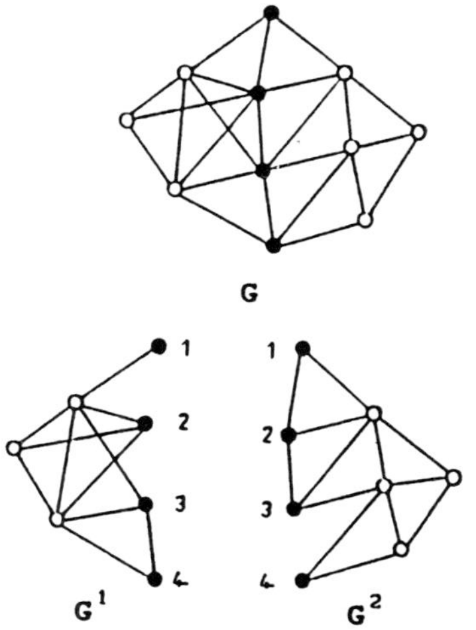

FIG. 3.14 EXAMPLE NETWORK AND ITS DECOMPOSITION

Example 10 Let us consider again the example network of figure 3.1, but this time it is split into two subgraphs G^1 and G^2 by a 4-point separating node set **U** = {1,2,3,4} (Fig. 3.14). A common edge availability p = 0.8 is assumed and the replacement graphs of figure 3.13 are applied. Tables 3.8 and 3.9 present numerical results. (The partitions of **U** are denoted as in table 3.7.) In table 3.9 "G^i repl." means that the subgraph G^i is replaced by the replacement graphs of figure 3.13 so that in the basic reduction equation (3.45) H^k is given by $H^k = G^i \cup H^{j,k}$, i = 1,2; j = 1 for i = 2 and j = 2 for i = 1; k = 1,2,...,5. Table 3.9 shows that, if G^2 is replaced, then all parameters α_j and h_k are between

TABLE 3.8 NUMERICAL RESULTS FOR THE NETWORK SPLITTING OF FIG. 3.14 (EXAMPLE 10)

j	p_j^1	R_j^1	p_j^2	R_j^2
1	0.74189	0.99127	0.87039	0.98196
2	0.03020	0.95138	0.05032	0.92759
3	0.00495	0.98319	0.00644	0.97365
4	0.00682	0.98021	0.00567	0.97405
5	0.18665	0.79183	0.03531	0.94210
6	0.00612	0.97995	0.00690	0.97215
7	0.00017	0.98901	0.00000	0.98123
8	0.00000	0.98319	0.00092	0.97831
9	0.00132	0.93552	0.00123	0.91229
10	0.00032	0.93961	0.00032	0.91995
11	0.00004	0.96503	0.00023	0.96292
12	0.00759	0.75994	0.00232	0.88939
13	0.00132	0.78503	0.00023	0.93420
14	0.00342	0.77702	0.00151	0.92807
15	0.00045	0.74189	0.00018	0.87039

TABLE 3.9 FINAL RESULTS FOR EXAMPLE 10

k	α_k: G^1 repl.	α_k: G^2 repl.	h_k: G^1 repl.	h_k: G^2 repl.
1	0.98196	0.98196	0.19921	0.03949
2	0.99297	0.99877	1.16341	0.64810
3	0.96092	0.98774	0.01083	0.00599
4	0.99457	0.98832	-0.38344	0.28810
5	0.99561	0.98909	0.00059	0.00038
6	0.97397	0.92507		
7	0.95447	0.94734		
8	0.34034	1.00000		
9	1.00001	0.99681		
10	0.15218	0.83721		

0 and 1. But if G^1 is replaced, then $\alpha_9 > 1$ and $h_4 < 0$ (If G^1 is replaced and $p = 0.9$, then $\alpha_9 < 0.$). As in example 3.5, the all-terminal reliability of $\tilde{G}$ is seen to be

$$R(\tilde{G}) = 0.963096 \qquad \square$$

3.5 Conclusions

A stochastic network $\tilde{G}$ with perfectly reliable nodes and unreliable edges is considered. By specifying a separating node set, the underlying deterministic network G is split into 2 edge-disjoint subnetworks G^1 and G^2. Reliability computations for $\tilde{G}^1$ and $\tilde{G}^2$ are done independently from each other and the results are combined to obtain a decomposition formula for the K-terminal reliability of $\tilde{G}$. Its application is most advantageous if a separating node set U can be used with $u = |U|$ being small and G^1 & G^2 having approximately the same "size". A recursive application of the decomposition formula is useful if the node connectivity of the graphs to be handled in each cycle is less than 6. But even in case of $u \geq 6$ this formula can be used for constructing bounds on the network reliability.

The decomposition formula is the base for developing a rigorous and unifying approach to network reduction. This is shown in detail for the all-terminal reliability. Up to now the applicability of network reduction was confined to the existence of separating node sets with $u < 4$. The generalized reduction approach described in this chapter allows reduction also for $u \geq 4$ by introducing more than one replacement graph for a subgraph of G. Thus, the original problem is split into several problems of lower computational complexity. In view of the generally exponential growth of computation time with increasing network complexity, the (recursive) application of the generalized reduction algorithm is expected to be profitable for $u = 4$ and $u = 5$. Many numerical examples show that the algorithm compares very well with other methods. However, for $u > 5$ the generalized reduction method does not seem to be feasible since it is very complicated to construct suitable replacement graphs.

REFERENCES

[1] Ball M.O., *Computational Complexity : An Overview*, IEEE Trans on Rel., Vol. R-35, No. 3, pp. 230-239, 1986.

[2] Colbourn Ch.J., *The Combinatorics of Network Reliability*, Oxford University Press, New York, 1987.

[3] Politof T. and Satyanarayana A., *Efficient Algorithms for Reliability Analysis of Planar Networks - A Survey*, IEEE Trans. on Rel., Vol. R-35, No. 3, pp. 252-259, 1986.

[4] Rosenthal A., *Computing the Reliability of Complex Networks*, SIAM J. Appl. Math., Vol. 35, pp. 384-393, 1977.

[5] Rosenthal A. and D. Frisque, *Transformations for Simplifying Network Reliability Calculations*, Networks, Vol. 7, pp. 97-111, 1977.

[6] Beichelt F., *Zuverlassigkeits und Instandhaltungstheorie*, Teubner-Verlag,Stuttgart, 1993.

[7] Beichelt F. and P. Tittmann, *A Combined Decomposition-Reduction Approach to the k-terminal Reliability of Stochastic Networks*, Optimization, Vol. 20, pp. 409-420, 1990.

[8] Beichelt F. and P. Tittmann, *A Generalized Reduction Method for the Connectedness Probability of Stochastic Networks*, IEEE Trans on Rel., Vol. R-39, No. 2, pp. 198-204, 1991.

[9] Beichelt F. and P. Tittmann, *Reliability Analysis of Communication Networks by Decomposition*, Microelectronics and Reliability, Vol. 31, pp. 869-872, 1991.

[10] Beichelt F. and P. Tittamann, *A Splitting Formula for the k-terminal Reliability of Stochastic Networks*, Stochastic Models Communication in Stat., Vol. 8, pp., 307-327, 1992.

[11] Bienstock D., *An Algorithm for the Reliability Analysis of Planar Graphs*, Networks, Vol. 16, pp. 411-422, 1986.

[12] Bienstock D., *Some Lattice Theoretic Tools for Network Reliability Analysis*, Math. of Oper. Res., Vol. 13, pp. 467-478, 1988.

[13] Shier D.R., *Network Reliability and Algebraic Structures*, Clarendon Press, Oxford, 1991.991.

[14] Aigner M., *Combinatorial Theory*, Springer-Verlag, New York, 1979.

[15] Lee L.G., W.G. Vogt and M.H. Mickle, *Optimal Decomposition of Large Scale Networks*, IEEE Trans. Syst., Managm. and Cybern, Vol. SMC-9, pp. 369-375, 1979.

[16] Satyanarayana A., M.K. Chang and Z.S. Khalil, *Some Results on the Overall Reliability of Stochastic Graphs*, Univ. of Calif., Berkeley, ORC 81-2, 1981.

[17] Tittmann P. and A. Blechschmidt, *Reliability Bounds Based on Networks Splitting*, J. Infor. Process. Cybern., Vol. 27, pp. 317-326, 1991.

[18] Beichelt F. and P. Franken, *Zuverlassigkeit und Instandhaltung Mathematche Methoden*, Verlag Technik, Berlin 1983; Carl Hanser-Verlag, Munchen-Wien 1984.

[19] Wood R.K., *Polygon-to-Chain-Reductions and Extensions for Reliability Evaluation of Undirected Networks*, Univ. of Calif., Berkeley, ORC 82-12, 1982.

[20] Kohlas J., *Zuverlassigkeit und Verfugbarkeit*, Teubner-Verlag, Stuttgart, 1987.

[21] Wood R.K., *Factoring Algorithms for Computing k-terminal Network Reliability*, IEEE Trans on Rel., Vol. R-35, No. 3, pp. 269-278, Aug. 1986.

[22] Ramamoorty M. and R. Balgopal, *Block Diagram Approach to Power System Reliability*, IEEE Trans. on PAS, Vol. PAS-89, pp. 802-811, 1970.

[23] Banerjee S.K. and K. Rajamani, *Parametric Representation of Probability in Two Dimensions - a New Approach in System Reliability Evaluation*, IEEE Trans. on Rel., Vol. R-21, No. 1, pp. 56-60, 1972.

[24] Banerjee S.K. and K. Rajamani, *Closed Form Solutions for Delta-Star and Star-Delta Conversions of Reliability Networks*, IEEE Trans. on Rel., Vol. R-25, No. 2, pp. 118-119, June 1976.

[25] Singh C. and M.D. Kankam, *Comments on Closed Form Solutions for Delta-Star and Star-Delta Conversions of Reliability Networks*, IEEE Trans on Rel., Vol. R-26, No. 5, pp. 336-339, 1977.

[26] Gupta H. and J. Sharma, *A Delta-Star Transformation Approach for Reliability Evaluation*, IEEE Trans. on Rel., Vol. R-27, No. 3, pp. 212-214, Aug. 1978.

[27] Gadani J.P. and K.B. Misra, *Reliability Evaluation of a System with Imperfect Nodes and Links using Network Approach*, System Science, Vol. 5, pp. 265-274, 1979.

[28] Gadani J.P. and K.B. Misra, *Delta-Star and Star-Delta Transformation for Complex Systems with Maintained Elements*, Proc. 2nd Int. Conf. on Rel. and Maintain., CNET, Lannion, France, 1980.

[29] Gadani J.P. and K.B. Misra, *Network Reliability Evaluation of Three-state Devices Using Transformation Techniques*, Microelectronics and Reliability, Vol. 21, No. 2, pp. 231-234, 1982.

[30] Gadani J.P., *System Effectiveness Evaluation Using Star and Delta Transformation*, IEEE Trans. on Rel., Vol. R-30, No. 1, pp. 43-47, April 1981.

[31] Singh C. and S. Asgarpoor, *Reliability Evaluation of a Flow Network Using Delta-Star Transformation*, IEEE Trans. on Rel., Vol. R-35, No. 4, pp. 472-477, Oct. 1986.

[32] Gadani J.P. and K.B. Misra, *Quadrilateral-Star Transformation : An Aid for Reliability Evaluation of Large Complex Systems*, IEEE Trans. on Rel., Vol. R-31, No. 1, pp. 49-59, April 1982.

[33] Gadani J.P. and K.B. Misra,

a) *A Quadrilateral-star Transformation I : Transformation Equations*, Microelectronics and Reliability, Vol. 22, No. 3, pp. 413-416, 1982.

b) *Quadrilateral-star Transformation II : Algorithmic Aspects*, Microelectronics and Reliability, Vol. 22, No. 3, pp. 417-422, 1982.

[34] Misra K.B., *Reliability Analysis and Prediction - A Methodology Oriented Treatment*, Elsevier Science Publishers, Amsterdam, 1992.

New Trends in System Reliability Evaluation
K.B. Misra (Editor)

CHAPTER 3

(Part-II)

PROBABILISTIC GRAPH-REDUCTION TECHNIQUES

Andrew M. Shooman

Networking Research Group
Motorola Codex
Mansfield, Massachusetts, U.S.A.

ABSTRACT

We consider the problem of analyzing the reliability of a network, specifically, the probability that a given set of critical nodes within the network can communicate, given the failure probabilities for component nodes and links. This is important for network management as it identifies when a network needs to be reinforced to provide adequate reliability. It is also important in planning topological modifications to a network.

This chapter presents an exact graph-reduction algorithm for solving the *k*-terminal reliability problem with node failures on an arbitrary network. *k*-terminal reliability means that a specific set of *k* target nodes must be able to communicate with one another. We model the network by an undirected probabilistic graph whose vertices represent the nodes and whose edges represent the links. A special feature of our model is that it allows nodes to be imperfect and associates a reliability measure with each node, assumed to succeed or fail independently. Therefore, the network reliability measure is based upon the reliability measures of the individual links and nodes.

The *k*-terminal reliability problem is known to be NP-complete. However, for a limited set of networks, it is possible to calculate the *k*-terminal reliability in polynomial time. We extend the Δ-Y and the Polygon-Chain graph transformations to operate on *k*-terminal reliability graphs with imperfect vertices. We present an algorithm which calculates the *k*-terminal reliability faster than Satyanarayana and Wood's algorithm and characterize the set of networks for which our computation takes polynomial time.

3.1 Introduction

In managing a network and planning topological modifications, it is important to be able to determine reliability measures quickly. In the most general case, we would like to determine a precise relationship between the failure of network components and the amount of traffic, the network can handle. Such analysis is usually quite complex and time consuming, as it involves not only a combinatorial analysis of the states arising from the failed components, but also an analysis of the routing within the network.

We are interested in obtaining an analysis quickly, in particular, quickly enough to be able to make use of this analysis in real time for network management and in the inner loop of network design procedures which generate many candidate topologies. We therefore consider here the simpler objective of finding the probability that the network (or a specified set of key nodes in the network) can all communicate. This is often meaningful in the more general, traffic related case as well because in the case of failures we may be willing to tolerate degraded performance due to congestion. Thus, as long as the network remains connected, we may say that it is functional even if it is not capable of handling the full traffic load.

This leads us to consider the k-terminal reliability problem.

3.1.1 The Problem

The problem is to compute, based on component reliabilities, the k-terminal reliability of an arbitrary network. k-terminal reliability means that a specific set of k target nodes must be able to communicate with one another, i.e. there must be a spanning tree connecting them. We model the network by an undirected probabilistic graph whose vertices represent the nodes and whose edges represent the links. We assume that each link succeeds or fails independently with an associated probability or reliability measure. A special feature of our model is that it allows nodes to be imperfect and associates a reliability measure with each node, assumed to succeed or fail independently. A perfect node is one that is operational with probability one whereas an imperfect node is working with probability less than one. Therefore, the network reliability measure is based upon the reliability measures of the individual links and nodes.

In our network model, failure means the disconnection of at least one of the k target nodes from one another. Disconnection of a single pair of target nodes from one another is sufficient to cause failure of the network, according to our problem definition. Also, failure of a single target node is sufficient to cause failure of the network. In other network models, unacceptable delay or congestion along a path connecting a pair of target nodes constitutes failure, but in our model, we consider only disconnection as failure.

We assume that all link and node failures are independent. Without this assumption, the correlation of failure events makes the network reliability problem exceedingly difficult to solve. In reality, the assumption of independent failures is not entirely correct. A lightning strike or power failure at a switching station can cause simultaneous node and multiple link failures. In addition, errors in switching software installed at multiple node sites can cause correlated node failures. The assumption of independent failures introduces less error in survivability problems when the link and node reliabilities are close to one.

3.1.2 Reliability Problems

Three popular measures of network reliability are the all-terminal, 2-terminal, and k-terminal reliabilities. All-terminal reliability requires that every node be able to communicate with every other node in the network. Thus, the all-terminal reliability problem is equivalent to the connectivity problem in graph theory. 2-terminal reliability requires that a pair of distinguished nodes, the source and the destination, be able to communicate with one another. k-terminal reliability requires that a specified set of k target nodes be able to communicate with one another. The k-terminal reliability problem is the most general and includes the other two problems. k-terminal reliability reduces to 2-terminal reliability when $k = 2$ and to all-terminal reliability when $k = n$. Fig. 3.1 illustrates the distinction among all-terminal, 2-terminal, and k-terminal reliability.

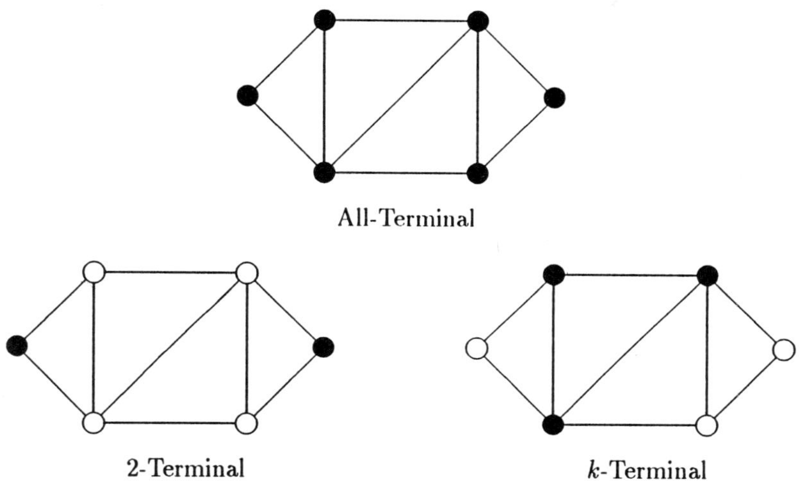

Fig. 3.1: All-Terminal, 2-Terminal, and k-Terminal Reliability

Other measures of network reliability are the expected number of nodes which can communicate with a distinguished node, the expected number of node pairs communicating, and the probability of all operative nodes (as opposed to all nodes) being connected [10].

The k-terminal reliability problem is known to be NP-complete [4]. However, for a limited set of networks, it is possible to calculate the k-terminal reliability in polynomial time. We specify an algorithm which calculates the k-terminal reliability and characterize that set of networks for which the computation takes polynomial time.

Our primary motivation for network reliability analysis is to improve the reliability of network designs. A network designer can study the reliability analysis of competing topologies to decide which is the most reliable. By studying enough problems, a network designer can learn which topological properties lead to reliable designs. In addition, the design algorithm from an automated network design tool based on numerical optimization can consider reliability as an attribute in the objective function for optimization by calling a reliability analysis algorithm as a subroutine.

3.1.3 Graph Reduction vs. Other Techniques

This chapter describes graph reduction as a method of network reliability analysis. First, we will describe two other methods briefly: state enumeration, cut-set and tie-set enumeration. For now, we assume that only edges fail.

The state enumeration method involves an evaluation of all 2^e combinations of edge successes and failures for a graph with e edges. Most practical computer networks have many links and evaluation of 2^e combinations results in an unacceptably slow algorithm.

3.1.4 Cut-Set and Tie-Set Enumeration

Four network reliability analysis methods arise from enumerating cut-sets or tie-sets while computing either the probability of success or the probability of failure. A cut-set is a set of edges whose failure disconnects the network. On the other hand, a tie-set is a set of edges whose success connects the network. In the 2-terminal reliability problem, a tie-set contains a path from source to destination. In the all-terminal reliability problem, a tie-set contains a spanning tree. We show the four cut-set and tie-set enumeration methods below:

let $G=$ network graph
$R(G)=$ reliability of graph G
$C_i=$ event that the edges in cut-set i are working
$T_i=$ event that the edges in tie-set i are working

success, cut-sets: $R(G) = P(\bigcap_{i=1}^{k} C_i)$

success, tie-sets: $R(G) = P(\bigcup_{i=1}^{k} T_i)$

failure, cut-sets: $R(G) = 1 - P(\bigcup_{i=1}^{k} \bar{C}_i)$

failure, tie-sets: $R(G) = 1 - P(\bigcap_{i=1}^{k} \bar{T}_i)$

Cut-set and tie-set enumeration is a fine method for analyzing small networks, but is not efficient for large problems because the number of combinations grows exponentially with n, the number of vertices. Identifying by inspection which combinations of failures disconnect the network is virtually impossible for non-trivial problems. Thus, we need a better system of bookkeeping for determining which combinations of failures disconnect the network.

Graph reduction has complexity related to the density of the graph. For Series-Parallel graphs and other sparse graphs, its complexity is linear in the number of edges $O(e)$. For extremely dense graphs, such as complete graphs, graph reduction also has the worst-case complexity of $O(2^e)$. However, for most exact analysis problems on realistic networks, the complexity of graph reduction is substantially better than the worst case (see Section 3.8). Our sophisticated exact graph reduction algorithm (see Section 3.6.2) is far better than the simple state enumeration or cut-set and tie-set enumeration algorithms presented in this section for large exact analysis problems. However, a sophisticated cut-set and tie-set algorithm may compete favorably with our graph reduction algorithms for large approximate analysis problems [13, 14]. For a complete comparison of graph reduction vs. state enumeration and cut-set and tie-set enumeration, see Chapter 2, "Network Reliability Analysis Methods," of Shooman's doctoral thesis [31].

3.2 Previous Contributions

It has been shown that network reliability problems are NP-complete [4]. This means that exact algorithms for computing the reliability of a general network take more than polynomial time (in the number of nodes) for the worst case. One way to achieve polynomial running time for a network reliability algorithm is to consider only a restricted class of networks. Chapter 4 of "The Combinatorics of Network Reliability" by Colbourn [4] discusses restricted classes of networks, graph transformations, reliability algorithms, and complexity results.

Among the classes of graphs considered in Chapter 4 of Colbourn are: 2-trees, 3-trees, 4-trees, Series-Parallel, normal form, Δ-Y, cube-free, and planar. A reliability-preserving graph transformation reduces a graph to a simpler one with the same reliability. The algorithms presented here use these graph transformations to reduce any graph to a single edge.

Arnborg and Proskurowski's [1] dynamic programming algorithm can be used to solve all three reliability problems (all-terminal, 2-terminal, and k-terminal) in $O(n^j)$ time, polynomial in the number of vertices but exponential in the degree of the tree (j-tree) [1]. This algorithm reduces a j-tree to a single j-clique, by recursively eliminating degree j nodes and summarizing reliability information in reliability measures associated with the j-clique adjacent to the node removed.

This method can be used on normal form graphs or 3-trees, the specific case where $j = 3$. In this algorithm, for each j-clique $C = \{x, y, z\}$ and subgraph S_c, so far reduced onto C, one calculates probability measures for each partition of the vertices x, y, and z into those which are connected to and those which are isolated from subgraph S_c. The reliability measure is extracted from the final remaining 3-clique.

Δ-Y graphs are those which can be reduced to a single edge by Series, Parallel, and Δ-Y reductions. The Δ-Y transformation converts a Δ or triangle subgraph to a Y subgraph (see Fig. 3.8). Vertices A, B, and C of the Δ subgraph are also contained in the Y subgraph. Edges of the Δ subgraph have reliabilities d, e, and f and edges of the Y subgraph have reliabilities x, y, and z. However, the Y subgraph contains a new vertex, D. The Δ-Y transformation is useful because it reduces the degree of vertices A, B, and C by one. Thus, if A, B, or C have degree 3 before applying the Δ-Y transformation, they have degree 2 after the transformation and can be further reduced by the Series transformation.

Politof and Satyanarayana [20, 21, 22] developed a linear time algorithm for all-terminal reliability on Δ-Y graphs. El Mallah and Colbourn [6] showed that every Δ-Y graph is a partial 4-tree.

[1]Throughout this chapter, we refer to k-trees and k-cliques. Here, we refer to j-trees and j-cliques so as to avoid confusion with the number of target nodes k as in the k-terminal reliability problem.

Cube-free graphs are planar and have no subgraphs homeomorphic to the 3-cube shown in Fig. 3.20. An induced subgraph H of G is 3-attached if there are three attachment nodes x, y, z in H such that all edges from H to G-H are incident with x, y, or z. All other nodes are internal. A 3-attached subgraph is a trisubgraph if it has at least two internal nodes and contains no 3-attached subgraph with more than one node. Politof and Satyanarayana [19] showed that every triconnected cube-free graph contains a trisubgraph. Furthermore, this graph remains cube-free under trisubgraph-Y replacement. Politof and Satyanarayana developed a linear-time algorithm for all-terminal reliability using Series, Parallel, Δ-Y, and trisubgraph-Y reductions.

The development of reliability algorithms for planar graphs is a current research problem. Bienstock [3] has developed a k-terminal reliability algorithm for planar graphs that runs in $2^{O(\sqrt{n})}$ time. It is known that the k-terminal reliability problem is NP-hard for planar graphs and that the 2-terminal reliability problem is NP-complete. However, the complexity of all-terminal reliability for planar graphs is still an open question.

The k-terminal reliability problem was shown to be NP-hard by Rosenthal [25] and was shown to be NP-complete even when G is planar by Valiant [37]. Two special cases of the k-terminal reliability problem are the 2-terminal reliability problem, where $k = 2$, and the all-terminal reliability problem, where $k = n$. These special cases are known to be NP-complete in general, but their complexities are unknown when G is planar.

The two transformations basic to graph-reduction methods are Series and Parallel. However, the Series and Parallel transformations are not sufficient for reducing general networks. It is necessary to add Edge Factoring to the set of graph-reductions to reduce non-Series-Parallel networks. One of the earliest references to Edge Factoring is by Bazovsky in 1961, who calls it "Bayes' Theorem in Reliability" [2].

Colbourn [4] outlines the state of the research for exact network reliability algorithms on restricted classes of networks. Satyanarayana and Wood [28] introduce a set of 7 Polygon to Chain transformations to solve the case of the irreducible Series-Parallel graph. Satyanarayana and Wood show how to construct a linear algorithm for computing the k-terminal reliability of a Series-Parallel network using the Series, Parallel, degree-2, and Polygon to Chain reductions. We compare the performance of our algorithms with that of Satyanarayana and Wood's algorithm, because it is one of the most advanced graph-reduction algorithms.

Misra [12] discusses 3 approximate transformations for reducing complex networks into series-parallel networks. These are Delta-Star, Star-Delta, and Quadrilateral-Star. Misra's Delta-Star and Star-Delta are similar to Shooman's Δ-Y (see Fig. 3.8) and Y-Δ (see Fig. 3.17) and Misra's Quadrilateral network is actually K_4, the complete graph with 4 vertices.

NP-completeness results for the all-terminal reliability of a general network show that most likely every exact algorithm has run time exponential in the number of vertices for an arbitrarily large problem. The goal is to attain polynomial run-time for partial 4-trees (Δ-Y graphs) and to reduce the constant of proportionality for the exponential run-time of large problems.

3.3 Basic Graph Reductions

We use reliability-preserving graph transformations to reduce a network to a simpler network with the same reliability. Graph-reduction algorithms apply these transformations recursively until the network has been reduced to a single edge and its two endpoint vertices, whose reliability is equal to that of the starting network.

An advantage of exact graph-reduction algorithms is that they run in polynomial time for some problems. For more complex problems which run in exponential time, exact graph-reduction algorithms can reduce the exponent of growth by performing polynomial-time reductions first, thus delaying the application of edge-factoring, an operation which contributes to exponential growth, for a smaller subproblem.

The all-terminal reliability problem requires that all nodes in the network be able to communicate with one another, i.e. all vertices in the graph be connected to one another. Four standard graph transformations are sufficient for reducing an arbitrary network to a single edge for the all-terminal reliability problem with perfect vertices. These are: Parallel, Series, Pendant, and Edge Factoring. If we allow imperfect vertices in our starting network, then we need Vertex Factoring (see Fig. 3.7) as well to solve the all-terminal reliability problem.

3.3.1 Parallel

The Parallel transformation operates on multiple edges between two vertices. Two parallel edges in the graph are removed and are replaced by a single edge with an equivalent reliability. The Parallel transformation reduces graph G to graph G' as shown in Fig. 3.2:

where

$$p = 1 - (1 - p_1)(1 - p_2)$$

Fig. 3.2: Parallel Transformation

The Parallel transformation can operate on an unlimited number of parallel edges in addition to the special case of 2 parallel edges illustrated above. In general, for e parallel edges, we compute the reliability of the new edge by:

$$p = 1 - \prod_{i=1}^{e}(1 - p_i)$$

3.3.2 Series

The Series transformation operates on vertices of degree two. The degree-two vertex is removed from the graph and its two incident edges are replaced by a single edge with an equivalent reliability. Therefore, it is necessary to condition the reliability of graph

G' on the event H that the removed vertex (in this case, vertex B) can communicate with at least one of its neighbors. This conditioning yields a global multiplicative factor which must be maintained separately and multiplied by the reliability of the single remaining edge upon termination of the subproblem recursion. The removal of vertex B affects the entire network, including the portion outside the local series subgraph, because the reliability of paths from vertex B to every other target vertex contributes to the all-terminal reliability of the graph G. The Series transformation reduces graph G to graph G' | H as shown below in Fig. 3.3. We derive the reliability of the new edge p_3 in Table 3.1.

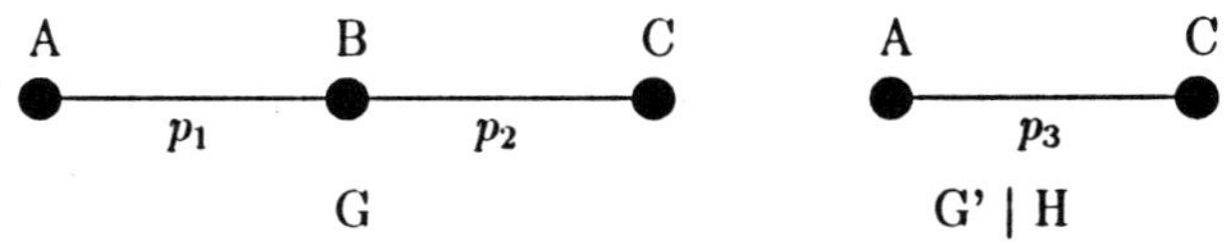

$$p_3 = \frac{p_1 p_2}{1-(1-p_1)(1-p_2)}$$

Fig. 3.3: Series Transformation

where R(G) = Ω * R(G' | H) = reliability of graph G

Ω = P(H) = multiplicative conditioning factor = $1-(1-p_1)(1-p_2)$

H = event that vertex B can communicate with at least one of its neighbors

p_1 = reliability of edge connecting vertices A and B

p_2 = reliability of edge connecting vertices B and C

p_3 = reliability of edge connecting vertices A and C

3.3.3 Pendant

The Pendant transformation operates on vertices of degree one. The degree-one vertex and its incident edge are removed from the graph. Removal of the degree-one vertex yields a multiplicative conditioning factor as in the Series transformation. The value of the multiplicative conditioning factor in this case is simply p, the reliability of the edge incident to the pendant vertex. The Pendant transformation reduces graph G to graph G' | H as shown below in Fig. 3.4:

Table 3.1: Derivation of New Edge Reliability in the Series Transformation

let H = event that vertex B is not isolated
J = event that vertices A and C can communicate

$$
\begin{aligned}
R(G) &= P(HJ) = p_1 p_2 && (3.1)\\
\Omega &\equiv P(H) = 1 - (1 - p_1)(1 - p_2) && (3.2)\\
R(G' \mid H) &= P(J \mid H) = p_3 && (3.3)\\
R(G) &= R(G' \mid H)P(H) && (3.4)\\
P(HJ) &= P(J \mid H)P(H) && (3.5)\\
p_1 p_2 &= p_3(1 - (1 - p_1)(1 - p_2)) && (3.6)\\
\Longrightarrow p_3 &= \frac{p_1 p_2}{1 - (1 - p_1)(1 - p_2)} && (3.7)
\end{aligned}
$$

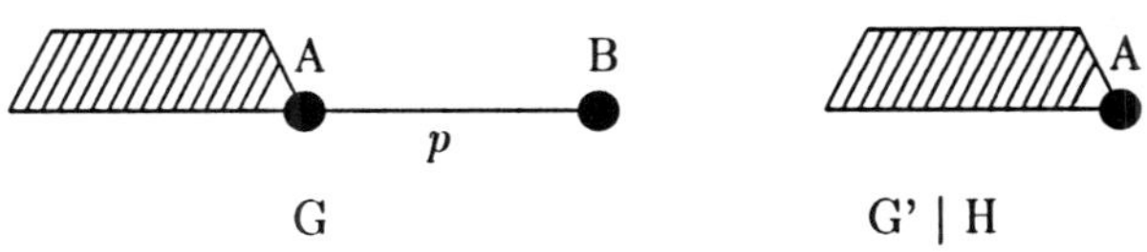

Fig. 3.4: Pendant Transformation

where R(G) = Ω * R(G' | H) = reliability of graph G
Ω = multiplicative conditioning factor = p
p = reliability of edge connecting vertices A and B
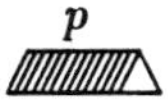 = remainder of graph G excluding Pendant subgraph

Networks which can be reduced using only the Series and Parallel transformations are said to be "Series-Parallel reducible." The reliability of a Series-Parallel graph can be computed in polynomial time, since both the Series and Parallel transformations can be performed in constant time. The Series and Parallel reductions may be performed in any order without affecting the result of the computation or the efficiency of the algorithm. A necessary condition for Series-Parallel reducibility is that a graph with n vertices has no more than $2n - 3$ edges [16, 17]. Graphs which have no multi-edges and fewer than two degree-two vertices are not Series-Parallel reducible. The simplest graph which is not Series-Parallel reducible is K_4, the complete graph with 4 vertices, as shown on the left-hand side of Fig. 3.5:

Graphs which are not Series-Parallel reducible can be reduced by Edge Factoring [4, 18, 26, 27]. This technique partitions the probability space into two sets, based on the success or failure of one particular edge. The reliability of the network before factoring is equal to the success probability of the chosen edge times the reliability of the graph (G_1), called the "short" problem, whose two vertices incident to the chosen edge are merged plus the failure probability of the chosen edge times the reliability of the graph (G_2), called the "open" problem, with the chosen edge removed. [2] We show in Fig. 3.5 Edge Factoring applied to K_4:

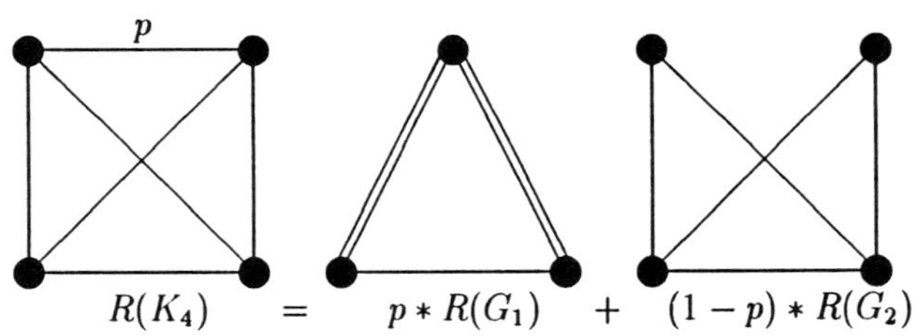

Fig. 3.5: Edge Factoring on K_4

Edge Factoring is used only as a last resort when no other transformations can be applied. Since Edge Factoring creates two new subproblems for every problem, it contributes to exponential growth of the network reliability algorithm.

Three standard graph reductions are sufficient to solve the all-terminal reliability problem for Series-Parallel graphs: Series, Parallel, and Pendant. Series-Parallel graphs are those which can be reduced to a single edge by the above transformations. The Series reduction for k-terminal reliability is more complicated than that for all-terminal reliability, since there are several cases depending on the configuration of targets and non-targets. However, the Series, Parallel, and Pendant reductions can solve the k-terminal reliability problem only for some Series-Parallel graphs, because some arrangements of targets and non-targets are unsolvable by the Series reduction.

The graph model for a k-terminal reliability problem must specify the k vertices (targets) which must be able to communicate with one another and the $n - k$ vertices (non-targets) which are not required to communicate with one another but may participate in a path from one target to another.

There exist 6 unique cases of Series reductions which can be labeled according to a binary enumeration of 3 digits. Let the symbol 'o' represent a non-target vertex and have the value of binary "0" and the symbol '•' represent a target vertex and have the value of binary "1". We now enumerate the 8 binary cases, only 6 of which are topologically unique (see Table 3.2):

Note that case 7 of the Series transformation is simply the all-terminal Series transformation described above. Cases 0, 1, 4, and 5 are the k-terminal cases which we will describe below. The remaining cases 2, 3, and 6 are unsolvable, since the conditioning event that the removed vertex can communicate with at least one of its target neighbors is ill-defined in the local series subgraph. In order to properly form the conditioning event, we must search outside the local series subgraph for target vertices, and that search may require us to examine every vertex and edge in the entire network. These

[2]We name these two problems "short" and "open" because they resemble a short circuit and an open circuit, respectively, in an electrical network.

Table 3.2: Cases of the Series Transformation

Number	*Graph*	*Comments*
0	o—o—o	solvable, k-terminal case
1	o—o—•	solvable, k-terminal case
2	o—•—o	unsolvable
3	o—•—•	unsolvable
4	•—o—o	solvable, isomorphic to (1)
5	•—o—•	solvable, k-terminal case
6	•—•—o	unsolvable, isomorphic to (3)
7	•—•—•	solvable, called "degree-2" reduction by Satyanarayana, all-terminal case

unsolvable cases have at least one non-target adjacent to the removed vertex, so the conditioning event depends on the graph structure outside the local series subgraph.

The k-terminal version of the Series reduction (cases 0, 1, 4, and 5) operates on vertices of degree two, in the case where vertex B is a non-target and vertices A and C can be either target or non-target. Vertex B can be imperfect and has an associated reliability of w. Note that, in the k-terminal Series reduction as in the all-terminal Series reduction, the degree-two vertex is removed from the graph and that the two incident edges are replaced by a single edge with an equivalent reliability. However, in the k-terminal Series reduction, it is not necessary to condition the reliability of graph G' on the event that the removed vertex B can communicate with at least one of its neighbors, because vertex B is a non-target and is not required to communicate with every other vertex in the graph. The k-terminal version of the Series reduction reduces graph G to graph G' as shown below in Fig. 3.6:

3.3.4 Edge Factoring

In order to solve the k-terminal reliability problem for "irreducible" Series-Parallel graphs as well as for non-Series-Parallel graphs, we must include Edge Factoring in our set of necessary transformations. If we allow imperfect vertices in our starting network, then we need Vertex Factoring in addition to Edge Factoring to solve the k-terminal reliability problem. Edge Factoring for the k-terminal problem with imperfect vertices is the same as Edge Factoring for the all-terminal problem, except that we must restrict Edge Factoring to those edges whose two adjacent vertices are perfectly reliable.

3.3.5 Vertex Factoring

Vertex Factoring is required to normalize imperfectly reliable vertices in irreducible configurations so that we may then perform Edge Factoring to break the irreducible configuration. Normalization means that we replace imperfectly reliable vertices by perfect ones, making the appropriate numerical adjustments. Vertex Factoring, like Edge Factoring, partitions the probability space into two sets, based on the success or

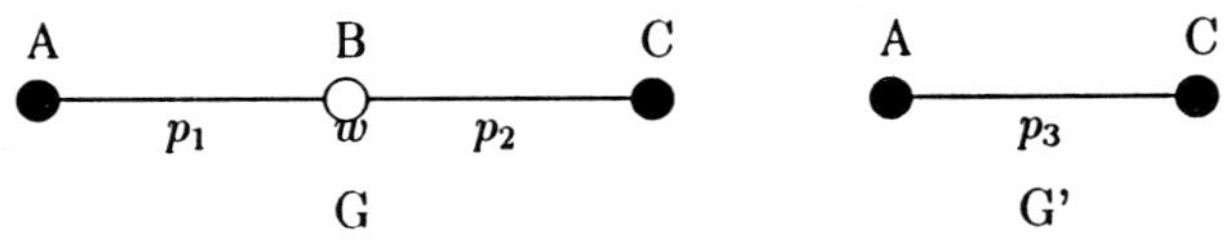

$$p_3 = wp_1p_2$$

where R(G) = R(G') = reliability of graph G
p_1 = reliability of edge connecting vertices A and B
p_2 = reliability of edge connecting vertices B and C
p_3 = reliability of edge connecting vertices A and C
w = reliability of vertex B

Fig. 3.6: k-terminal Series Transformation

failure of one particular vertex. The reliability of the network before factoring is equal to the success probability of the chosen vertex times the reliability of the graph (G_1) in which the factored vertex is up plus the failure probability of the chosen vertex times the reliability of the graph (G_2) with the chosen vertex down and its incident edges removed. We show, in Fig. 3.7, Vertex Factoring applied to vertex B of the irreducible Series-Parallel graph, G:

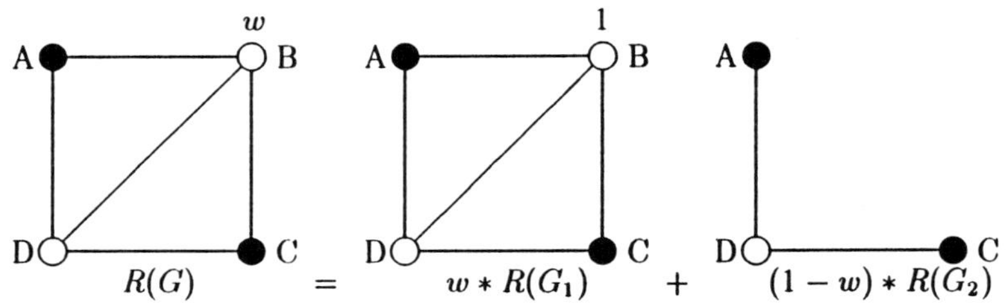

Fig. 3.7: Vertex Factoring

Vertex Factoring is also needed for the all-terminal reliability problem with imperfect vertices. However, since all vertices are required to communicate for the all-terminal reliability problem, the reliability of the case of the failed vertex is always zero, i.e. $R(G_2) = 0$. Thus, the Vertex Factoring equation for all-terminal reliability degenerates to $R(G) = w * R(G_1)$. Therefore, we replace all imperfect vertices by perfect ones as a preprocessing step and accumulate a multiplicative factor Ω, the product of the reliabilities of the imperfect vertices [36].

The set of graph transformations including Series, Parallel, Pendant, Edge Factoring, and Vertex Factoring is sufficient to reduce an arbitrary network with imperfect vertices for a k-terminal reliability problem. However, we will introduce versions of the Δ-Y and Polygon-Chain transformations which contribute to a more efficient algorithm for k-terminal reliability with imperfect vertices.

3.4 Advanced Graph Reductions

3.4.1 Δ-Y

We now introduce the Δ-Y transformation to expand the set of networks whose reliability can be computed in polynomial time. The Δ-Y transformation has been derived for resistor networks [30] and for the 2-terminal reliability of communications networks [24]. However, derivation of the Δ-Y transformation for the k-terminal reliability of communications networks is a new contribution.

The Δ-Y transformation complicates the problem of all-terminal reliability with the introduction of new vertices. These new vertices are different from the original target vertices in that they are not perfectly reliable and also are not required to communicate with the other vertices to satisfy the all-terminal reliability problem. Thus, the Δ-Y transformation changes the problem from all-terminal reliability to k-terminal reliability.

The Δ-Y transformation converts a Δ or triangle subgraph to a Y subgraph (see Fig. 3.8). Vertices A, B, and C of the Δ subgraph are also contained in the Y subgraph. Vertices A, B, and C may be either target or non-target and may be less than perfectly reliable, with associated reliabilities of a, b, and c in the Δ subgraph and reliabilities of t, u, and v in the Y subgraph. Edges of the Δ subgraph have reliabilities d, e, and f and edges of the Y subgraph have reliabilities x, y, and z. However, the Y subgraph contains a new non-target vertex, D, which must have an associated reliability, w, in order to constrain properly the transformation for k-terminal reliability. The Δ-Y transformation is useful because it reduces the degree of vertices A, B, and C by one. Thus, if A, B, or C have degree 3 before applying the Δ-Y transformation, they have degree 2 after the transformation and may be reduced further by the Series transformation.

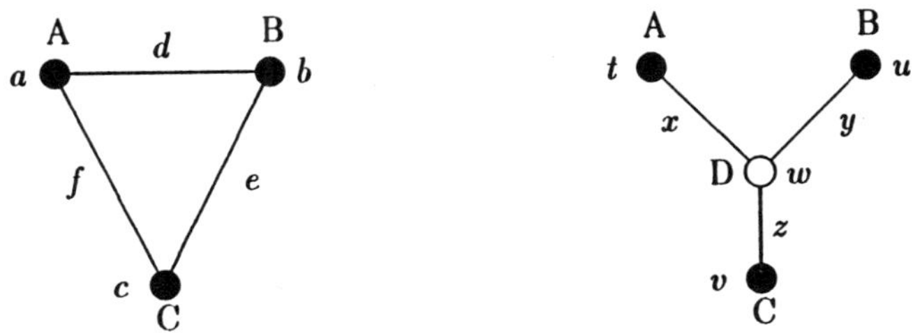

Fig. 3.8: Δ-Y Transformation

Our goal is to calculate the reliabilities t, u, v, w, x, y, and z as functions of the reliabilities a, b, c, d, e, and f. We partition the probability space into 8 events:

α=P(BC,A) =P(B and C connected, A isolated and up)
β=P(AC,B) =P(A and C connected, B isolated and up)
γ=P(AB,C) =P(A and B connected, C isolated and up)
δ=P(BC,$\bar{\text{A}}$) =P(B and C connected, A isolated and down)
ϵ=P(AC,$\bar{\text{B}}$) =P(A and C connected, B isolated and down)
ζ=P(AB,$\bar{\text{C}}$) =P(A and B connected, C isolated and down)
η=P(ABC) =P(all vertices connected)
ϕ=P(A,B,C)=P(all vertices isolated)

We now compute the probability of these events, except for ϕ, for both subgraphs, Δ and Y, to form 7 independent equations. It is not necessary to consider the ϕ equation because all sets in a partition of a probability space must sum to one. Therefore, the ϕ equation is linearly-dependent on the first 7 equations, α through η. This method of derivation follows Satyanarayana and Wood in solving the system of equations corresponding only to nonfailed induced graphs [28].

$$\alpha = P(BC, A) = abc\bar{d}e\bar{f} = tuvw\bar{x}yz \tag{3.8}$$
$$\beta = P(AC, B) = abc\bar{d}\bar{e}f = tuvwx\bar{y}z \tag{3.9}$$
$$\gamma = P(AB, C) = abcd\bar{e}\bar{f} = tuvwxy\bar{z} \tag{3.10}$$
$$\delta = P(BC, \bar{A}) = \bar{a}bce = \bar{t}uvwyz \tag{3.11}$$
$$\epsilon = P(AC, \bar{B}) = a\bar{b}cf = t\bar{u}vwxz \tag{3.12}$$
$$\zeta = P(AB, \bar{C}) = ab\bar{c}d = tu\bar{v}wxy \tag{3.13}$$
$$\eta = P(ABC) = abcdef(\bar{d}/d + \bar{e}/e + \bar{f}/f + 1) = tuvwxyz \tag{3.14}$$

These equations have a closed-form analytical solution. We solve them as follows: First, we solve for x:

$$\frac{\alpha}{\eta} = \frac{tuvw\bar{x}yz}{tuvwxyz} = \frac{\bar{x}}{x} = \frac{1-x}{x} \tag{3.15}$$
$$\alpha x = \eta(1-x) = \eta - \eta x \tag{3.16}$$
$$(\alpha + \eta)x = \eta \tag{3.17}$$
$$x = \frac{\eta}{\alpha + \eta} \tag{3.18}$$

Similarly, we solve for y and z whose derivations have the same form as that of x:

$$\frac{\beta}{\eta} = \frac{tuvwx\bar{y}z}{tuvwxyz} = \frac{\bar{y}}{y} = \frac{1-y}{y} \tag{3.19}$$
$$y = \frac{\eta}{\beta + \eta} \tag{3.20}$$
$$\frac{\gamma}{\eta} = \frac{tuvwxy\bar{z}}{tuvwxyz} = \frac{\bar{z}}{z} = \frac{1-z}{z} \tag{3.21}$$
$$z = \frac{\eta}{\gamma + \eta} \tag{3.22}$$

Now, we solve for t, u, and v, all similar in form:

$$\frac{\delta}{\eta} = \frac{\bar{t}uvwyz}{tuvwxyz} = \frac{\bar{t}}{tx} = \frac{1-t}{tx} \tag{3.23}$$
$$\delta xt = \eta(1-t) = \eta - \eta t \tag{3.24}$$
$$(\delta x + \eta)t = \eta \tag{3.25}$$

$$
\begin{aligned}
t &= \frac{\eta}{\delta x + \eta} && (3.26)\\
t &= \frac{\eta}{\delta\left(\frac{\eta}{\alpha+\eta}\right) + \eta} && (3.27)\\
t &= \frac{1}{\frac{\delta}{\alpha+\eta} + 1} && (3.28)\\
t &= \frac{\alpha + \eta}{\alpha + \delta + \eta} && (3.29)\\
\frac{\epsilon}{\eta} &= \frac{t\bar{u}vwxz}{tuvwxyz} = \frac{\bar{u}}{uy} = \frac{1-u}{uy} && (3.30)\\
u &= \frac{\beta + \eta}{\beta + \epsilon + \eta} && (3.31)\\
\frac{\zeta}{\eta} &= \frac{tu\bar{v}wxy}{tuvwxyz} = \frac{\bar{v}}{vz} = \frac{1-v}{vz} && (3.32)\\
v &= \frac{\gamma + \eta}{\gamma + \zeta + \eta} && (3.33)
\end{aligned}
$$

Finally, we solve for w:

$$
\begin{aligned}
w &= \frac{\eta}{tuvxyz} && (3.34)\\
w &= \frac{\eta}{\left(\frac{\alpha+\eta}{\alpha+\delta+\eta}\right)\left(\frac{\beta+\eta}{\beta+\epsilon+\eta}\right)\left(\frac{\gamma+\eta}{\gamma+\zeta+\eta}\right)\left(\frac{\eta}{\alpha+\eta}\right)\left(\frac{\eta}{\beta+\eta}\right)\left(\frac{\eta}{\gamma+\eta}\right)} && (3.35)\\
w &= \frac{(\alpha+\delta+\eta)(\beta+\epsilon+\eta)(\gamma+\zeta+\eta)}{\eta^2} && (3.36)
\end{aligned}
$$

We summarize the solution to the Δ-Y equations:

$$
\begin{aligned}
t &= \frac{\alpha + \eta}{\alpha + \delta + \eta} && (3.37)\\
u &= \frac{\beta + \eta}{\beta + \epsilon + \eta} && (3.38)\\
v &= \frac{\gamma + \eta}{\gamma + \zeta + \eta} && (3.39)\\
w &= \frac{(\alpha+\delta+\eta)(\beta+\epsilon+\eta)(\gamma+\zeta+\eta)}{\eta^2} && (3.40)\\
x &= \frac{\eta}{\alpha + \eta} && (3.41)\\
y &= \frac{\eta}{\beta + \eta} && (3.42)\\
z &= \frac{\eta}{\gamma + \eta} && (3.43)
\end{aligned}
$$

The Δ-Y transformation just derived is exact when the Δ vertices, A, B, and C, are perfect. However, if any of the Δ vertices is imperfect, we must impose the restriction that it be incident to only one external edge in order for the Δ-Y transformation to be exact. Rosenthal and Frisque [24] explain the correlations introduced when performing the Δ-Y transformation on imperfect Δ vertices with more than one external edge. The requirement that an imperfect Δ vertex has at most one external edge is equivalent to the requirement that an imperfect non-target Δ vertex has degree no more than 3, two edges connecting the other Δ vertices and one external edge. This restriction is discussed in Shooman's doctoral thesis [31]. If a target Δ vertex is imperfect, we can make it perfect by the all-terminal case of Vertex Factoring (see Fig. 3.7).

3.4.2 Polygon-Chain

Satyanarayana and Wood introduced a set of 7 Polygon to Chain transformations to solve the case of the irreducible Series-Parallel graph, such as graphs (b), (d), and (e) above. Their original Polygon to Chain transformations operate on perfectly reliable vertices only [28]. We have extended the Polygon to Chain transformations to operate on imperfect vertices as well [32, 33, 34].

A chain is defined as an alternating sequence of distinct vertices and edges, $V_1, (V_1, V_2), V_2, (V_2, V_3), V_3, \ldots, V_{k-1}, (V_{k-1}, V_k), V_k$. All internal vertices, $V_2, V_3, \ldots, V_{k-1}$, are of degree 2 and the end vertices, V_1 and V_k, are of degree greater than 2. The length of a chain is determined simply by the number of edges it contains.

A polygon is defined as the union of two chains with common end vertices. In other words, a polygon is a cycle with two special vertices with degree greater than 2; all other vertices are of degree 2. There exist 7 unique classes of polygons which appear in irreducible Series-Parallel graphs. Some constraints on types of polygons found in irreducible Series-Parallel graphs (see Table 3.2) help us to enumerate the 7 unique classes:

1. All internal vertices for both chains must be targets. (This corresponds to the unsolvable cases #2, 3, and 6 of the Series reduction)

2. Neither of the 2 chains composing the polygon has length greater than 3. (If so, then we could perform a number of Series reductions to reduce the chain length to exactly 3)

3. For polygons containing chains of length 3, both end vertices must be non-targets. (Unsolvable cases #3 and 6 of the Series reduction)

4. For polygons with no chains of length 3, at least one end vertex must be a non-target. (Cases #2, 3, and 6 of Series)

We enumerate the 7 unique classes of polygons in Table 3.4.2:

The Polygon to Chain reduction transforms a Polygon of one of the above types to its corresponding chain. The reliabilities of the edges belonging to the corresponding C_3 chain are derived from the reliabilities of the edges of the C_1 and C_2 chains. In addition, the reliability of the graph resulting from the Polygon to Chain reduction must be scaled by a multiplicative constant in order to preserve reliability equivalence with the graph before the reduction.

Satyanarayana and Wood derived the Polygon to chain transformations for the k-terminal reliability problem with all nodes assumed to be perfectly reliable. We extended the Polygon to Chain transformations for the k-terminal reliability problem with imperfectly reliable nodes. To accomplish this, we derived the extended Polygon to Chain

let C_1= Chain #1 component of Polygon
C_2= Chain #2 component of Polygon
C_3= new chain produced by Polygon to Chain transformation
V_1= one end vertex of Polygon
V_k= other end vertex of Polygon

Table 3.3: Classes of Polygon-Chain Transformations

Class #	C_1 *Length*	C_2 *Length*	V_1 *Type*	V_k *Type*	C_3 *Length*
1	2	1	non-target	non-target	2
2	2	1	target	non-target	2
3	2	2	non-target	non-target	2
4	2	2	target	non-target	3
5	3	1	non-target	non-target	3
6	3	2	non-target	non-target	3
7	3	3	non-target	non-target	3

transformation for all 7 classes of polygon. We show the derivation for polygon class #2 with an imperfect vertex of reliability w (see Fig. 3.9):

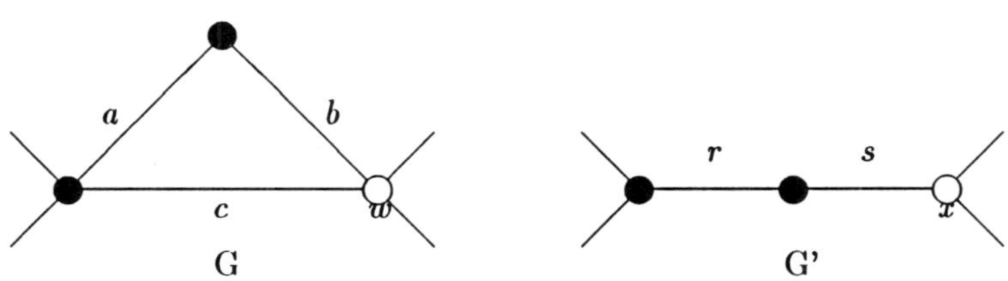

key: •= target vertex, o= non-target vertex

let R(G)= reliability of graph G
let Ω= reliability-preserving multiplicative constant
R(G) = Ω * R(G')

Fig. 3.9: Polygon-Chain #2 Transformation

The principle of the derivation is to partition the entire probability space of all combinations of link successes and failures for both the Polygon and the Chain into classes of induced subgraphs. Equating the probabilities for both Polygon and Chain for each subgraph yields a system of equations.

For polygon class #2 with imperfect vertices, there are 4 classes of non-failed subgraphs, which we name α, γ, ϵ, and ζ (see Fig. 3.10). (The other 2 possible classes of non-failed subgraphs, β and δ, appear in other classes of Polygon to chain). The failed subgraph we name ϕ. A failed subgraph is one which guarantees disconnection of the network, e.g. a subgraph containing an isolated target vertex. Note that the central vertex in the Polygon graph is of degree 2, whereas the two end vertices have degree larger than 2.

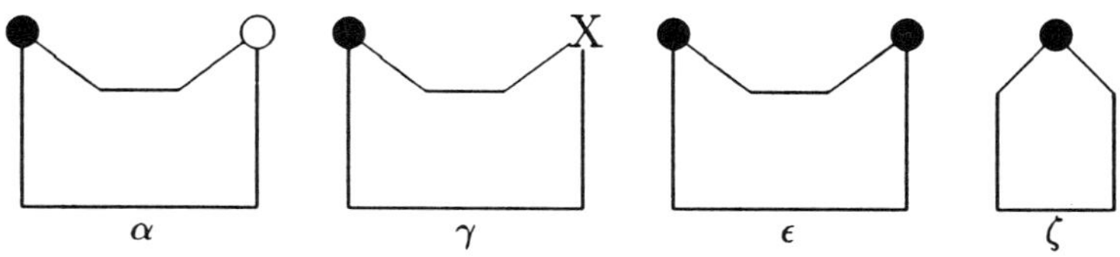

key: •= target vertex, o= non-target vertex, X= failed vertex

Fig. 3.10: Induced Subgraphs

Note that induced subgraph α has a target vertex on the left and a non-target vertex on the right. These two vertices are disconnected in the polygon, but are possibly connected outside of the Polygon subgraph. Subgraph γ differs from α only in that the non-target vertex on the right has failed. Subgraph ϵ has two target vertices, one on the left and one on the right. Subgraph ζ shows only one target vertex because the two end vertices of the polygon are connected via the polygon and can be merged into one vertex.

We now must enumerate the success and failure combinations of edges a, b, and c and vertex w in the polygon (see Table 3.4) and edges r and s and vertex x in the Chain (see Table 3.5) to obtain equations.

Polygon Equations

$$\alpha = wa\bar{b}\bar{c} \tag{3.44}$$

$$\gamma = \bar{w}a \tag{3.45}$$

$$\epsilon = w\bar{a}b\bar{c} \tag{3.46}$$

$$\begin{aligned} \zeta &= w*[(\bar{a}bc)+(a\bar{b}c)+(ab\bar{c})+(abc)] \\ &= wabc*(1+\bar{a}/a+\bar{b}/b+\bar{c}/c) \end{aligned} \tag{3.47}$$

Chain Equations

$$\alpha = \Omega xr\bar{s} \tag{3.48}$$

$$\gamma = \Omega\bar{x}r \tag{3.49}$$

$$\epsilon = \Omega x\bar{r}s \tag{3.50}$$

$$\zeta = \Omega xrs \tag{3.51}$$

Solution

$$x = \frac{\alpha+\zeta}{\alpha+\gamma+\zeta} \tag{3.52}$$

$$r = \frac{\zeta}{\epsilon+\zeta} \tag{3.53}$$

Table 3.4: Success and Failure Modes of Polygon

Polygon

n	w	a	b	c	*category*
0	0	0	0	0	ϕ
1	0	0	0	1	ϕ
2	0	0	1	0	ϕ
3	0	0	1	1	ϕ
4	0	1	0	0	γ
5	0	1	0	1	γ
6	0	1	1	0	γ
7	0	1	1	1	γ
8	1	0	0	0	ϕ
9	1	0	0	1	ϕ
10	1	0	1	0	ϵ
11	1	0	1	1	ζ
12	1	1	0	0	α
13	1	1	0	1	ζ
14	1	1	1	0	ζ
15	1	1	1	1	ζ

Table 3.5: Success and Failure Modes of Chain

Chain

n	x	r	s	*category*
0	0	0	0	ϕ
1	0	0	1	ϕ
2	0	1	0	γ
3	0	1	1	γ
4	1	0	0	ϕ
5	1	0	1	ϵ
6	1	1	0	α
7	1	1	1	ζ

note: binary "0" represents failure and binary "1" represents success

$$s = \frac{\zeta}{\alpha + \zeta} \tag{3.54}$$

$$\Omega = \frac{(\alpha + \gamma + \zeta)(\epsilon + \zeta)}{\zeta} \tag{3.55}$$

We now summarize the results of the 7 cases of the extended Polygon-Chain transformation (see Fig. 3.11 and Table 3.6):

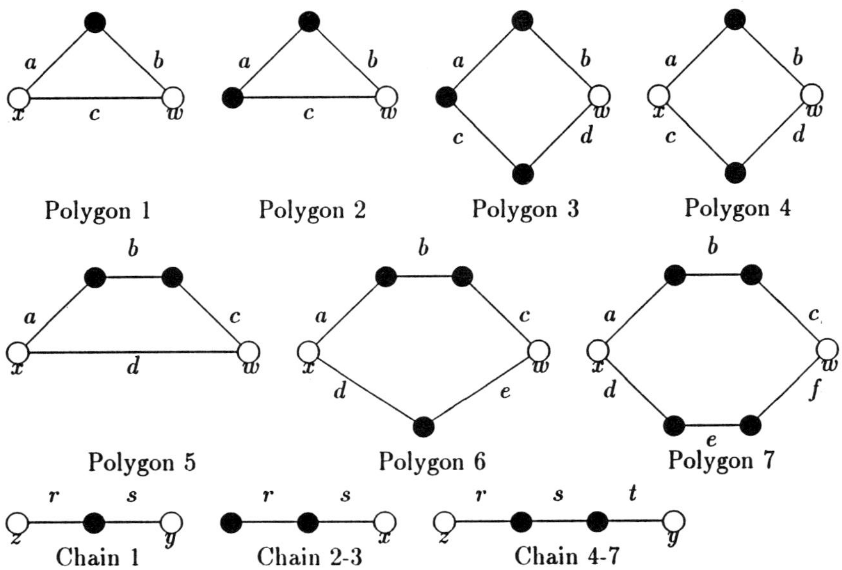

Fig. 3.11: Polygon and Chain Graphs

The extended Polygon-Chain transformations described above contribute to a k-terminal reliability algorithm that is faster than Satyanarayana and Wood's algorithm [31, 34]. These Polygon-Chain transformations also enable us to solve k-terminal reliability problems with imperfect vertices.

3.4.3 Vertex Merging

Whenever an edge is perfectly reliable (has reliability one) to machine precision, we remove it and merge its endpoint vertices into a single vertex, as shown in Fig. 3.12. The resultant vertex is a target if either or both of the original endpoint vertices were targets. Vertex merging is helpful to the graph reduction process because it creates candidates for Parallel reduction as does the "short" case of Edge Factoring. In fact, vertex merging is really a degenerate case of Edge Factoring in which the "open" case, the case in which the chosen edge fails, has probability zero and is degenerate.

Table 3.6: Polygon-Chain Transformation Equations

Case	*Polygon*	*Chain*
1	$\alpha = wxa\bar{b}\bar{c}$ $\beta = wx\bar{a}b\bar{c}$ $\gamma = w\bar{x}a$ $\delta = \bar{w}xb$ $\zeta = wxabc * (\bar{a}/a + \bar{b}/b + \bar{c}/c + 1)$	$r = \frac{\zeta}{\beta+\zeta}$ $s = \frac{\zeta}{\alpha+\zeta}$ $y = \frac{\beta+\zeta}{\beta+\delta+\zeta}$ $z = \frac{\alpha+\zeta}{\alpha+\gamma+\zeta}$ $\Omega = \frac{(\alpha+\gamma+\zeta)(\beta+\delta+\zeta)}{\zeta}$
2	$\alpha = wa\bar{b}\bar{c}$ $\gamma = \bar{w}a$ $\epsilon = w\bar{a}b\bar{c}$ $\zeta = wabc * \bar{a}/a + \bar{b}/b + \bar{c}/c + 1$	$r = \frac{\zeta}{\epsilon+\zeta}$ $s = \frac{\zeta}{\alpha+\zeta}$ $x = \frac{\alpha+\zeta}{\alpha+\gamma+\zeta}$ $\Omega = \frac{(\alpha+\gamma+\zeta)(\epsilon+\zeta)}{\zeta}$
3	$\alpha = wa\bar{b}cd$ $\gamma = \bar{w}ac$ $\epsilon = w * (\bar{a}b\bar{c}d + \bar{a}bc\bar{d} + a\bar{b}\bar{c}d)$ $\zeta = wabcd * (\bar{a}/a + \bar{b}/b + \bar{c}/c + \bar{d}/d + 1)$	< same as case 2 >
4	$\alpha = wxa\bar{b}c\bar{d}$ $\beta = wx\bar{a}b\bar{c}d$ $\gamma = w\bar{x}ac$ $\delta = \bar{w}xbd$ $\epsilon = wx * (\bar{a}bc\bar{d} + a\bar{b}\bar{c}d)$ $\zeta = wxabcd * (\bar{a}/a + \bar{b}/b + \bar{c}/c + \bar{d}/d + 1)$	$r = \frac{\zeta}{\beta+\zeta}$ $s = \frac{\zeta}{\epsilon+\zeta}$ $t = \frac{\zeta}{\alpha+\zeta}$ $y = \frac{\beta+\zeta}{\beta+\delta+\zeta}$ $z = \frac{\alpha+\zeta}{\alpha+\gamma+\zeta}$ $\Omega = \frac{(\alpha+\gamma+\zeta)(\beta+\delta+\zeta)(\epsilon+\zeta)}{\zeta^2}$
5	$\alpha = wxab\bar{c}d$ $\beta = wx\bar{a}bc\bar{d}$ $\gamma = w\bar{x}ab$ $\delta = \bar{w}xbc$ $\epsilon = wxa\bar{b}c\bar{d}$ $\zeta = wxabcd * (\bar{a}/a + \bar{b}/b + \bar{c}/c + \bar{d}/d + 1)$	< same as case 4 >
6	$\alpha = wxab\bar{c}d\bar{e}$ $\beta = wx\bar{a}bc\bar{d}e$ $\gamma = w\bar{x}abd$ $\delta = \bar{w}xbce$ $\epsilon = wx * (a\bar{d}e * (\bar{b}c + b\bar{c}) + cd\bar{e} * (\bar{a}b + a\bar{b}))$ $\zeta = wxabcde * (\bar{a}/a + \bar{b}/b + \bar{c}/c + \bar{d}/d + \bar{e}/e + 1)$	< same as case 4 >
7	$\alpha = wxab\bar{c}def$ $\beta = wx\bar{a}bc\bar{d}ef$ $\gamma = w\bar{x}abde$ $\delta = \bar{w}xbcef$ $\epsilon = wx * (\bar{a}bcd * (\bar{e}f + e\bar{f}) + a\bar{b}c * (\bar{d}ef + d\bar{e}f + de\bar{f}) + ab\bar{c}f * (\bar{d}e + d\bar{e}))$ $\zeta = wxabcdef * (\bar{a}/a + \bar{b}/b + \bar{c}/c + \bar{d}/d + \bar{e}/e + \bar{f}/f + 1)$	< same as case 4 >

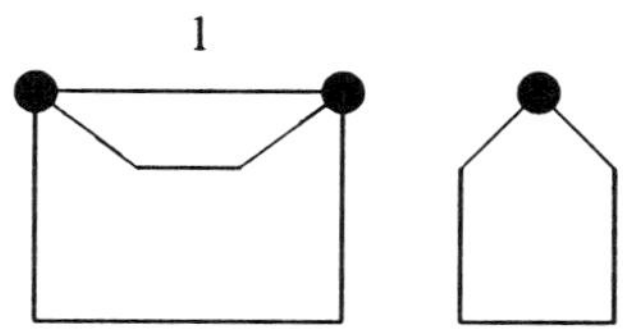

Fig. 3.12: Vertex Merging

3.5 Approximate Graph Reductions

The k-terminal reliability problem is NP-complete. Accordingly, an exact graph-reduction algorithm takes exponential time to reduce a sufficiently dense network. Therefore it is necessary to use approximate techniques, which speed-up greatly the reliability analysis algorithm while sacrificing some accuracy, in order to solve very difficult problems in a reasonable period of time. In this section, we describe 4 approximate graph reductions: K_4 Contraction (lower bound), K_4 Contraction (upper bound), Degree-3 Series (lower bound), and Y-Δ (upper bound). The numerical results of number of transformations, run-time, and accuracy for the approximate algorithms are given in Shooman's doctoral thesis [31] and in a paper [34].

3.5.1 K_4 Contraction

K_4 Contraction replaces a K_4 subgraph with a super-vertex (see Fig. 3.13) of reliability equal to that of the K_4 subgraph for a lower-bound or of reliability one for an upper-bound. The lower bound doesn't account for paths outside the subgraph contributing to the connectivity of vertices inside the subgraph, whereas the upper-bound assumes that the K_4 subgraph is perfectly reliable. Fig. 3.14 shows the reliability of K_4, $R(K_4)$, as a function of link reliability (p). Note that $R(K_4)$ is very close to one for high link reliabilities. Therefore, the upper-bound is tight for high link reliabilities and is a good approximation for reliable backbones or subnets.

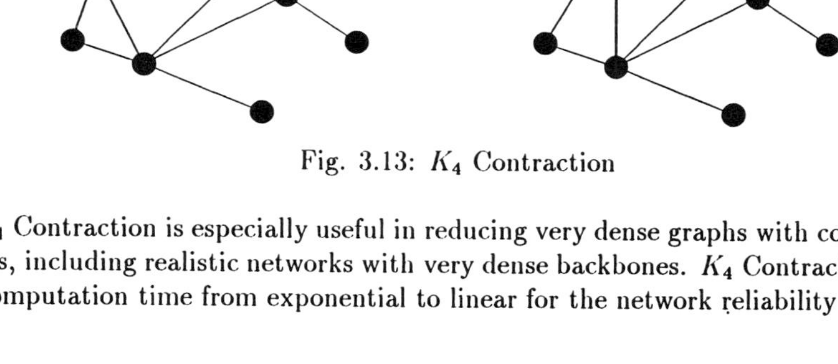

Fig. 3.13: K_4 Contraction

K_4 Contraction is especially useful in reducing very dense graphs with complete subgraphs, including realistic networks with very dense backbones. K_4 Contraction reduces the computation time from exponential to linear for the network reliability problem on

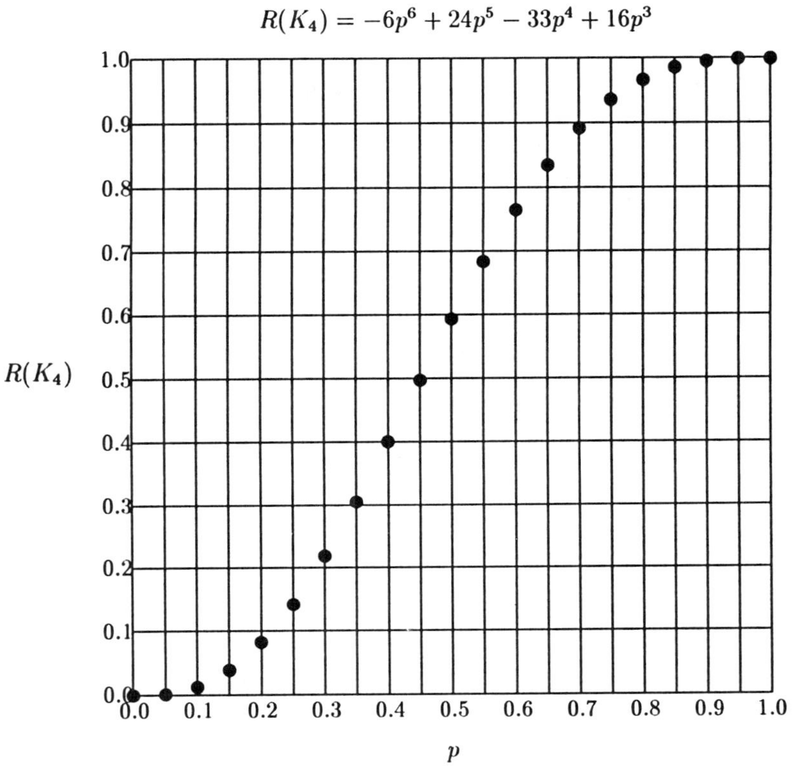

Fig. 3.14: All-Terminal Reliability of K_4 vs. Link Reliability, $R(K_4)$ vs. p

graphs with maximum degree of 4 and very dense graphs of higher degree. These include complete graphs, 3-trees, and 4-trees. K_4 Contraction can be generalized to K_n Contraction with the addition of an arbitrary size clique-finding algorithm. K_n Contraction can also be relaxed to allow complete graphs missing one edge to be contracted into a supervertex.

We could extend K_4 Contraction to K_n Contraction by generalizing the K_4 subgraph recognition algorithm to a maximal clique finding algorithm (see Table 3.11). The tightness of the upper-bound improves as n the number of vertices increases, because $\lim_{n\to\infty} R(K_n) = 1$ (see Fig. 3.15). We computed $R(K_n)$ according to the recurrence equation given by Colbourn [4], where $A_n = R(K_n)$ and $q = 1 - p$ is the probability of a single edge failure.

$$R(K_n) = A_n = 1 - \sum_{j=1}^{n-1} \binom{n-1}{j-1} A_j q^{j(n-j)}$$

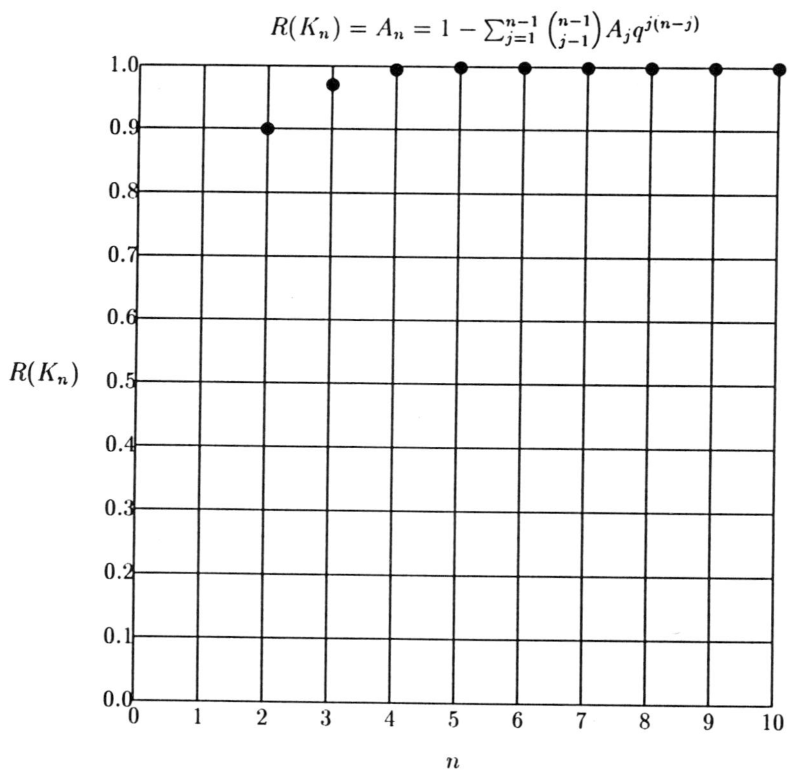

Fig. 3.15: All-Terminal Reliability of K_n vs. Number of Vertices, $R(K_n)$ vs. n for $p = 0.95$

3.5.2 Degree-3 Series

The Degree-3 Series transformation replaces two of the edges incident to a degree-3 vertex with a single edge directly connecting their endpoints. The third edge is discarded, but is accounted for partially in the global conditioning factor associated with the Degree-3 Series transformation. In Fig. 3.16, series edges AB and BC are replaced by AC and edge BD is discarded. The global conditioning factor, Ω, is the probability that vertex B can communicate with at least one of its neighbors, A, C, or D. The Degree-3 Series transformation is a lower-bound because it doesn't account for all paths involving edge BD contributing to the connectivity of the graph. The Degree-3 Series transformation expands the set of Series-Parallel graphs while sacrificing some accuracy.

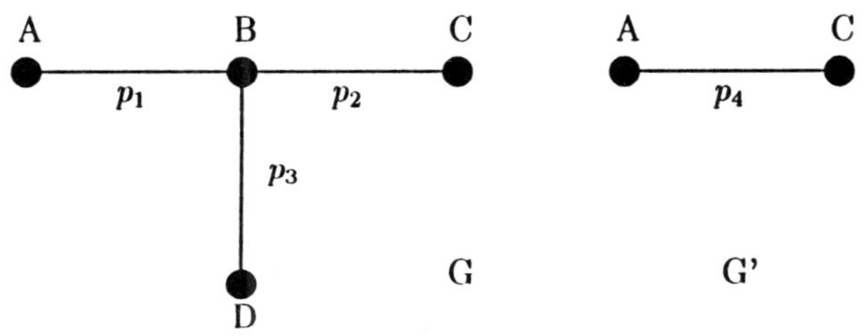

$$\begin{aligned}\Omega &= p_1||p_2||p_3 = (p_1 + p_2 - p_1p_2)||p_3 \\ &= p_1 + p_2 + p_3 - p_1p_2 - p_1p_3 - p_2p_3 + p_1p_2p_3 \\ p_4 &= \frac{p_1p_2}{\Omega}\end{aligned}$$

Fig. 3.16: Degree-3 Series Transformation

The Degree-3 Series transformation is effective in reducing moderately dense graphs of average degree 3 with few or no triangles that can be reduced by the Δ-Y transformation. These include grids, 3-trees, and realistic graphs such as models of the Arpanet and designs produced by the Mentor algorithm [9].

The Degree-3 Series transformation provides a reasonable approximation when the link reliabilities are high, i.e. greater than 0.9, but is too inaccurate for low link reliabilities. We can improve the accuracy of the Degree-3 Series transformation by choosing the edge with the lowest reliability as the one to be discarded.

3.5.3 Y-Δ

The Y-Δ transformation, like the Degree-3 Series transformation, operates on degree-3 vertices. The Y-Δ transformation removes the degree-3 vertex and its incident edges in the Y subgraph and reconnects the 3 endpoint vertices in the Δ subgraph (see Fig. 3.17). The global conditioning factor, Ω, is the probability that the central vertex D can communicate with at least one of its neighbors, A, B, or C. The Y-Δ transformation is an upper bound because the Δ graph provides extra paths that do not exist in the Y graph.

$$
\begin{aligned}
\Omega &= p_1||p_2||p_3 = (p_1 + p_2 - p_1p_2)||p_3 \\
&= p_1 + p_2 + p_3 - p_1p_2 - p_1p_3 - p_2p_3 + p_1p_2p_3 \\
p_4 &= \frac{p_1p_2}{\Omega} \\
p_5 &= \frac{p_2p_3}{\Omega} \\
p_6 &= \frac{p_1p_3}{\Omega}
\end{aligned}
$$

Fig. 3.17: Y-Δ Transformation

3.6 Graph-Reduction Algorithms

3.6.1 Heuristics

The algorithm presented in this chapter employs two heuristics to improve its performance. The first heuristic chooses the "best" edge for Edge Factoring and the second heuristic decides where to and where not to perform the Δ-Y transformation.

The Edge Factoring heuristic has two cases, the special case and the general case. In the special case, the graph to be factored has exactly 2 target vertices and they are connected by an edge. Then, the heuristic chooses that edge as the one to factor. Formally, if $|k|$=2 and $\exists(u, v)$ such that $u, v \in k$, then $(u^*, v^*) = (u, v)$ where k is set of target vertices and (u^*, v^*) is the edge chosen for Edge Factoring. In this special case of factoring the edge connecting the only 2 target vertices in the graph, the "short" case of Edge Factoring, in which we merge the endpoint vertices into a single vertex, degenerates because a graph with one or zero targets is vacuously connected and thus has reliability one. Therefore, the special case of Edge Factoring generates only one subproblem instead of two and does not contribute to exponential growth.

In the general case, the Edge Factoring heuristic chooses the edge (u^*,v^*) whose end vertices have minimum degree and a maximum cardinality intersection of their adjacent vertex sets:

$$U^+ = \min_{u \in V} deg(u) \tag{3.56}$$

$$V^+ = \min_{v \in \Gamma U^+} deg(v) \tag{3.57}$$

$$(u^*, v^*) = \max_{u^+ \in U^+, v^+ \in V^+} |\Gamma u^+ \cap \Gamma v^+| \tag{3.58}$$

Equation 3.56 states that we scan the set of vertices V for a subset of minimum degree vertices. We call this minimum degree vertex set U^+. We specify a set of vertices and not a single vertex because the minimum degree vertex is very often non-unique. Equation 3.57 states that we scan the set of vertices ΓU^+ adjacent to the minimum degree vertex U^+ for a subset of minimum degree adjacent vertices which we call V^+. Finally, Equation 3.58 states that for each pair of minimum degree endpoint vertices u^+ and v^+, we compute the intersection of their adjacent sets, $\Gamma u^+ \cap \Gamma v^+$. The pair with the maximum cardinality intersection, (u^*, v^*), is chosen as the endpoint vertices of the edge to be factored. (We choose the first such pair if the solution is non-unique).

The minimum degree criterion attempts to create candidates for Series reduction in the "open" subproblem of Edge Factoring. If either of the vertices u^* or v^* has degree 3 before Edge Factoring, then in the "open" subproblem created by Edge Factoring, it will have degree 2 and will be a candidate for the Series reduction.

The maximum cardinality intersection criterion attempts to maximize the number of candidates for Parallel reduction in the "short" subproblem of Edge Factoring. In fact, the number of candidates for Parallel reduction in the "short" subproblem created by Edge Factoring is equal to the cardinality of the intersection of the adjacent vertex sets.

The Δ-Y transformation affects the complexity of a graph in two opposite ways, one simplifying and one complicating. Therefore, it is necessary to decide for each specific application whether the benefits of the simplifying effects outweigh the costs of the complicating effects. The Δ-Y transformation reduces the degree by one of the three Δ subgraph vertices while it creates a new non-target imperfectly reliable degree-three vertex in the Y subgraph. Reducing the degree of the Δ subgraph vertices is good because it can create candidates for the Series reduction. A non-target imperfectly reliable degree-three vertex is bad because it requires extra transformations, Polygon-Chain or Vertex Factoring followed by Edge Factoring, to remove.

We tested a number of different heuristics for Edge Factoring and the Δ-Y transformation on all-terminal test problems of complete graphs, grids, and realistic networks to determine which would be the best to use for our (Shooman's) algorithm. We have determined experimentally that the combination of minimum degree and maximum common adjacent vertices heuristics for Edge Factoring is better than the minimum degree heuristic alone. Also, performing the Δ-Y transformation only on Δ subgraphs with vertex degrees 3,3,3 yields the best results for most problems (see Table 3.7). Note that all numbers reflect the total number of graph transformations needed to reduce the network, a machine-independent metric.

Run-time generally increases with the total number of transformations needed to reduce a network. However, the Polygon-Chain transformations, as implemented, take longer to perform than the other transformations. Thus, 100 total transformations to solve problem A might take longer to perform than 100 total transformations to solve problem B, because the set of transformations required to reduce problem A include more Polygon-Chain transformations than the set of transformations necessary to reduce problem B. We hope to improve the implementation of the Polygon-Chain transformations so that they take roughly the same amount of time to perform as the other transformations (see Section 3.9).

Table 3.7: Comparison of Heuristics: Heuristics Used and Number of Transformations

Edge Factoring	Δ-Y	COMPLETE GRAPHS K_4	K_5	K_6	K_7	K_8	K_9
min. deg.	none [1]	9	36	158	810	4887	34244
min. deg.	target only	5	54	772	13381	XXX	XXX
min. deg.	perfect only	5	43	545	8827	XXX	XXX
min. deg.	all vertex types	5	43	542	8708	XXX	XXX
min. deg.	3,3,3 only	5	24	110	570	3447	24164
min. deg.	3,3,3 or 3,3,4	5	24	110	570	3447	24164
max. com. adj. vert.	3,3,3 only	5	30	141	727	4330	30135
min. deg. + max. adj. [2]	3,3,3 only	5	24	110	570	3447	24164

Edge Factoring	Δ-Y	GRIDS 3x3	3x4	3x5	3x6	3x7	3x8
min. deg.	none [1]	21	66	201	606	1821	5466
min. deg.	target only	11	73	191	659	1884	5688
min. deg.	perfect only	11	76	204	656	1933	5890
min. deg.	all vertex types	11	78	204	657	1933	5892
min. deg.	3,3,3 only	17	57	159	443	1215	3327
min. deg.	3,3,3 or 3,3,4	11	73	202	640	1892	5756
max. com. adj. vert.	3,3,3 only	17	49	140	385	1083	3086
min. deg. + max. adj. [2]	3,3,3 only	17	57	159	443	1215	3327

Edge Factoring	Δ-Y	REALISTIC NETWORKS Cube	Redarpa	Arpa
min. deg.	none [1]	72	11051	178692
min. deg.	target only	48	9454	43108
min. deg.	perfect only	48	7053	40640
min. deg.	all vertex types	48	6966	40795
min. deg.	3,3,3 only	36	4584	48705
min. deg.	3,3,3 or 3,3,4	48	4809	44188
max. com. adj. vert.	3,3,3 only	36	3364	47845
min. deg. + max. adj. [2]	3,3,3 only	36	2355	46952

[1] Simple, Satyanarayana and Wood's algorithms
[2] Shooman's algorithm
XXX Excessive exponential growth; could not measure actual number of transformations

* Shooman's algorithm uses Δ-Y on 3,3,3 only and Edge Factoring on edge with minimum degree end vertices and maximum cardinality intersection of their adjacent vertex sets (see Heuristics).

* Satyanarayana and Wood has same performance as Simple algorithm for all-terminal reliability problems. Satyanarayana and Wood's algorithm was intended for k-terminal reliability analysis on Series-Parallel networks. However, Satyanarayana and Wood claim that the Polygon-Chain transformation reduces the complexity of computing the reliability of non-Series-Parallel networks, as well. We compare the performance of our algorithms with that of Satyanarayana and Wood's algorithm, because it is one of the most advanced graph-reduction algorithms.

3.6.2 Algorithms

In this section, we discuss the 8 algorithms: Simple, Δ-Y, Polygon-Chain, Shooman, K_4 Contraction Lower-Bound, K_4 Contraction Upper-Bound, Degree-3 Series, and Y-Δ. The first 4 algorithms, Simple, Δ-Y, Polygon-Chain, and Shooman, are exact, whereas the remaining 4, K_4 Contraction Lower-Bound, K_4 Contraction Upper-Bound, Degree-3 Series, and Y-Δ, are approximate. Table 3.8 shows the order of the transformations for each of the 8 algorithms.

Table 3.8: Comparison of 8 Algorithms

	Algorithm							
Transformation	1	2	3	4	5	6	7	8
Parallel	√	√	√	√	√	√	√	√
Vertex Merging	√	√	√	√	√	√	√	√
Pendant	√	√	√	√	√	√	√	√
Series	√	√	√	√	√	√	√	√
Polygon-Chain (perfect only)			√					
Polygon-Chain (extended)				√	√	√	√	√
Δ-Y		√		√	√	√	√	√
K_4 Contraction Lower-Bound					√			
K_4 Contraction Upper-Bound						√		
Degree-3 Series							√	
Y-Δ								√
Vertex Factoring	√	√	√	√	√	√	√	√
Edge Factoring	√	√	√	√	√	√	√	√

<u>Key to Algorithms</u>
1. Simple
2. Δ-Y
3. Satyanarayana and Wood
4. Shooman
5. K_4 Contraction Lower-Bound
6. K_4 Contraction Upper-Bound
7. Degree-3 Series
8. Y-Δ

The Simple algorithm consists of the Parallel, Vertex Merging, Pendant, Series, Vertex Factoring, and Edge Factoring transformations, in order of decreasing priority. This set of transformations is a simple set sufficient to reduce an arbitrary network for the k-terminal reliability problem. (The simplest set of transformations sufficient to reduce an arbitrary network for the k-terminal reliability problem is that containing only Vertex Factoring and Edge Factoring. However, that set of transformations produces an algorithm whose running time is always exponential in the number of edges and imperfect vertices). The simple algorithm has linear run-time for Series-Parallel networks, but exponential run time in the worst case for any other type of networks.

The Δ-Y algorithm adds the Δ-Y transformation to the Simple algorithm's set after Series and before Vertex Factoring in the priority order. The Δ-Y transformation attacks degree-3 vertices and reduces their degree to 2, but creates imperfect non-target vertices and "irreducible" cases of the k-terminal Series transformation which must be reduced by Vertex Factoring and Edge Factoring. Since Vertex Factoring and Edge Factoring both contribute to exponential growth and are used extensively in the Δ-Y algorithm, the Δ-Y algorithm is sometimes worse than the Simple algorithm, especially for dense networks.

Satyanarayana and Wood's algorithm adds the Polygon-Chain transformations which operate only on perfect vertices to the Simple algorithm's set after Series and before Vertex Factoring in the priority order. The Polygon-Chain transformations for perfect vertices only are not used at all in all-terminal reliability problems, since every case of the Polygon-Chain transformation involves at least one non-target vertex and all-terminal reliability problems have no non-target vertices. Therefore, Satyanarayana and Wood's algorithm reduces to the Simple algorithm for all-terminal reliability problems. Satyanarayana and Wood's algorithm is more efficient than the Simple algorithm, however, for k-terminal reliability problems, because the Polygon-Chain transformations for perfect vertices only are helpful in reducing some "irreducible" k-terminal Series subgraphs.

Our algorithm, which we called "Shooman," adds both the Δ-Y transformation and the Polygon-Chain transformations for imperfect vertices. The Polygon-Chain transformations are added to the priority list just after the Series transformation. The Δ-Y transformation is added just after the Polygon-Chain transformations and just before Vertex Factoring. Shooman's algorithm for computing the k-terminal reliability of an arbitrary network, the most advanced exact algorithm presented in this chapter, is outlined below in Table 3.9:

The four approximate algorithms, K_4 Contraction Lower-Bound, K_4 Contraction Upper-Bound, Degree-3 Series, and Y-Δ, employ approximate transformations after the Series transformation and before Vertex Factoring in the priority order.

The priority order of the graph transformations for the 4 exact algorithms described above affects the efficiency but does not affect the accuracy of solutions to network reliability analysis problems. The Parallel, Series, and Pendant transformations have the highest priority, since they are the simplest to recognize and perform. Vertex Factoring and Edge Factoring have the lowest priority since they create two subproblems and contribute to exponential growth. We perform them only when we cannot perform any other transformation. We determined empirically that the Polygon-Chain transformations should be performed before the Δ-Y transformations, because that ordering leads to a more efficient algorithm.

Shooman's algorithm is better than Satyanarayana and Wood's algorithm because it employs the Δ-Y transformation in combination with the Polygon-Chain transformations for imperfect vertices. Therefore, it is able to attack degree-2 and degree-3 vertices which Satyanarayana and Wood's algorithm could not simplify because of imperfect vertices in Polygon subgraphs or Δ subgraphs reducible only by the Δ-Y transformation. In those cases, Satyanarayana and Wood's algorithm would need Edge Factoring and Vertex Factoring to break the subgraphs unsolvable by the original Polygon-Chain for perfect vertices only. Shooman's algorithm requires the fewest total number of transformations, of the four exact algorithms presented in this section, to reduce an arbitrary

Table 3.9: Shooman's Algorithm for k-terminal Reliability

let G = graph
 G1= graph representing "short" problem after Vertex or Edge Factoring
 G2= graph representing "open" problem after Vertex or Edge Factoring

```
<perform VERTEX FACTORING on all imperfect target vertices>
compute_reliability(G)
  if G has one or zero target vertices,
    return(reliability one)
  else if G has one undirected edge,
    return(reliability of undirected edge)
  else if G has a duplicate edge,
    <perform PARALLEL transformation on duplicate edge>
    return(compute_reliability(G))
  else if G has a perfectly reliable edge,
    <perform VERTEX MERGING on the endpoints of the perfectly reliable edge>
  else if G has a degree-one vertex,
    <perform PENDANT transformation on degree-one vertex>
    return(compute_reliability(G))
  else if G has a degree-two vertex,
    <perform SERIES transformation on degree-two vertex>
    return(compute_reliability(G))
  else if G has a polygon type 1 subgraph,
    <perform POLYGON-CHAIN TYPE 1 transformation on polygon type 1 subgraph>
    return(compute_reliability(G))
  .
  . 7 categories of Polygon-Chain
  .
  else if G has a polygon type 7 subgraph,
    <perform POLYGON-CHAIN TYPE 7 transformation on polygon type 7 subgraph>
    return(compute_reliability(G))
  else if G has a delta subgraph with vertex degrees 3,3,3,
    <perform DELTA-Y transformation on triangle subgraph>
    return(compute_reliability(G))
  else if G has an imperfect vertex of reliability w,
    <perform VERTEX FACTORING on imperfect vertex>
    return(w*compute_reliability(G1)+(1-w)*compute_reliability(G2))
  else
    <perform EDGE FACTORING on edge with min degree incident vertices
       and max cardinality common adjacent vertices>
    return(p*compute_reliability(G1)+(1-p)*compute_reliability(G2))
```

network.

3.6.3 Complexity Analysis

The data structure used to store the graph representation of the network is an adjacency list. The graph is implemented as a record containing the number of vertices, the number of edges, the source and destination vertices (for the case of 2-terminal reliability), the reliability, type (target or non-target), and degree of each vertex, and a pointer to each vertex's adjacency list. The adjacency list is implemented as a linked list whose elements are records representing an edge containing the self vertex, the adjacent vertex, the multiplicity, the edge reliability, and a pointer to the next edge on the list.

In this section, we let n be the number of vertices and e be the number of edges. Scanning an adjacency list is an essential operation in the subgraph recognition and graph-reduction algorithms. The worst-case complexity of such a scanning operation is related to the length of the adjacency list. The length of an adjacency list for a single vertex is equal to the vertex degree. Thus, the longest possible adjacency list is of length $n-1$ for a complete graph and the worst-case complexity of scanning an adjacency list is $O(n)$.

We now show that the most difficult Polygon subgraph recognition algorithm (for Polygon #7) has worst-case complexity $O(n^2)$ (see Table 3.10):

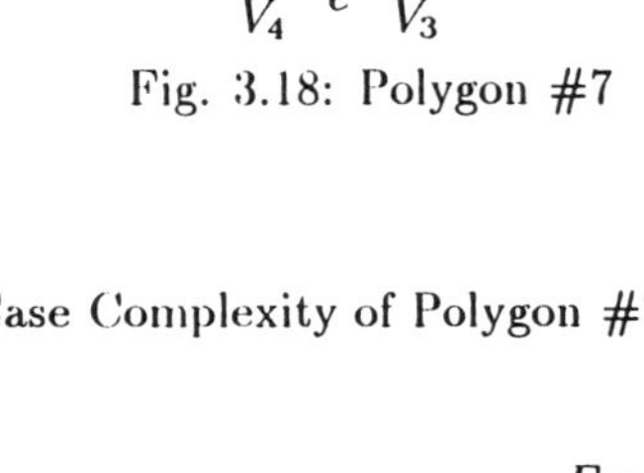

Fig. 3.18: Polygon #7

Table 3.10: Worst-Case Complexity of Polygon #7 Recognition Algorithm

Equation	Complexity	
$For\ V_2 \in V, deg(V_2) = 2, V_2 \in K$	$O(n)$	(3.59)
$For\ V_6 \in \Gamma V_2, deg(V_6) = 2, V_6 \in K, b = (V_2, V_6)$	$O(1)$	(3.60)
$For\ V_1 \in \Gamma V_2 \setminus V_6, deg(V_1) \geq 2, V_1 \notin K, c = (V_2, V_1)$	$O(1)$	(3.61)
$For\ V_5 \in \Gamma V_6, deg(V_5) \geq 2, V_5 \notin K, a = (V_6, V_5)$	$O(1)$	(3.62)
$For\ V_3 \in \Gamma V_1, deg(V_3) = 2, V_3 \in K, f = (V_1, V_3)$	$O(n)$	(3.63)
$For\ V_4 \in \Gamma V_3, deg(V_4) = 2, V_4 \in K, e = (V_3, V_4)$	$O(1)$	(3.64)
$\exists d = (V_4, V_5)$	$O(1)$	(3.65)

All seven of Equations 3.59 through 3.65 must be satisfied for a successful Polygon #7 recognition. Since the Polygon recognition algorithm attempts to satisfy the equations sequentially, the total complexity of the algorithm is the product of the complexities of each step.

The complexity of an iteration over a set is proportional to the set's cardinality. Therefore, the complexity of satisfying Equation 3.59 is $O(|V|)$ or $O(n)$. Similarly, the complexity of satisfying Equation 3.60 is $O(|\Gamma V_2|)$, where ΓV_2 is the set of all vertices adjacent to V_2. The cardinality of the adjacency set is the same as the degree. Thus, $O(|\Gamma V_2|)$ equals $O(deg(V_2))$ equals $O(1)$, a constant since a degree-2 vertex has a constant number of incident edges. We use the same reasoning for deriving the complexity of satisfying Equations 3.61 through 3.64. The complexity of checking for the existence of an edge is proportional to the minimum degree of its two endpoints. Thus, the complexity of satisfying Equation 3.65 is $O(min(deg(V_4), deg(V_5)))$ which equals $O(min(2, n))$ or $O(1)$. The total worst-case complexity of the recognition algorithm for Polygon #7 is $O(n^2)$.

We derive the worst-case complexity of a k-clique recognition algorithm in Table 3.11.

Table 3.11: Worst-Case Complexity of k-Clique Recognition Algorithm

Equation	*Complexity*	
$For\ V_1 \in V, deg(V_1) \geq k-1$	n	(3.66)
$For\ V_2 \in \Gamma V_1, deg(V_2) \geq k-1$	$n-1$	(3.67)
$For\ V_3 \in \Gamma V_1 \setminus V_2, deg(V_3) \geq k-1$	$n-2$	(3.68)
$\exists(V_2, V_3)$	$1(n-2)$	(3.69)
$For\ V_4 \in \Gamma V_1 \setminus V_2 \setminus V_3, deg(V_4) \geq k-1$	$n-3$	(3.70)
$\exists(V_2, V_4), \exists(V_3, V_4)$	$2(n-3)$	(3.71)
$\cdots$		
$For\ V_{k-1} \in \Gamma V_1 \setminus V_2 \setminus V_3 \setminus \ldots \setminus V_{k-2}, deg(V_{k-1}) \geq k-1$	$n-k+2$	(3.72)
$\exists(V_2, V_{k-1}), \exists(V_3, V_{k-1}), \exists(V_4, V_{k-1}), \ldots, \exists(V_{k-2}, V_{k-1})$	$(k-3)(n-k+2)$	(3.73)
$For\ V_k \in \Gamma V_1 \setminus V_2 \setminus V_3 \setminus \ldots \setminus V_{k-1}, deg(V_k) \geq k-1$	$n-k+1$	(3.74)
$\exists(V_2, V_k), \exists(V_3, V_k), \exists(V_4, V_k), \ldots, \exists(V_{k-1}, V_k)$	$(k-2)(n-k+1)$	(3.75)

Equation 3.66 has complexity $O(n)$ since it must scan all n vertices in the worst case to check for degree $k-1$. The loops stated in Equations 3.67, 3.68, 3.70, ..., 3.72, and 3.74 establish that the vertices of the clique are adjacent to V_1 and have degree of at most $n-1, n-2, \ldots, n-k+2, n-k+1$ since they involve scanning V_1's adjacency list excluding the vertices that have already been included in the clique. The complexity of each edge existence test in Equations 3.69, 3.71, ..., 3.73, and 3.75 is $n-i+1$ for each test following candidate vertex V_i, since each also involves scanning an adjacency list of length at most $n-i+1$.

We avoid scanning the entire adjacency list of length at most $n-1$ by starting the pointer scanning ΓV_1 for the V_i loop at the edge entry just after the one where the pointer scanning ΓV_1 for the V_{i-1} loop found candidate vertex V_{i-1}. Since V_i must be distinct from all previous candidate vertices $V_{i-1}, V_{i-2}, V_{i-3}, \ldots$, it is not necessary to examine the edge entries on adjacency list ΓV_1 corresponding to those previous candidate vertices.

$i - 2$ edge existence tests are performed after including candidate vertex V_i, so the total complexity of the edge existence tests following V_i is $(i-2)(n-i+1)$. We now calculate $T(n,k)$, the worst case total complexity of the k-clique recognition algorithm as a function of the number of vertices in the graph (n) and the clique size (k):

$$
\begin{aligned}
T(n,k) &= && (3.76)\\
&= n(n-1)(n-2)1(n-2) + \\
&\quad n(n-1)(n-2)(n-3)2(n-3) + \ldots + \\
&\quad n * \ldots * (n-k+2)(k-3)(n-k+2) + \\
&\quad n * \ldots * (n-k+1)(k-2)(n-k+1) \\
&= \sum_{i=3}^{k} \frac{n!}{(n-i)!}(i-2)(n-i+1) && (3.77)
\end{aligned}
$$

The summation in Equation 3.77 is very difficult or impossible to express in closed-form. Instead, we solve for an upper-bound by assuming that all $\binom{k-1}{2}$ edge existence tests are performed inside the k loops:

$$
\begin{aligned}
T(n,k) &= && (3.78)\\
&= \sum_{i=3}^{k} \frac{n!}{(n-i)!}\binom{k-1}{2} && (3.79)\\
&= \sum_{i=3}^{k} n(n-1)(n-2)\ldots(n-i+1)\binom{k-1}{2} && (3.80)\\
&= \sum_{i=3}^{k} O(n^i)O(k^2) && (3.81)\\
&= O(k^2 n^k) && (3.82)
\end{aligned}
$$

In any case, we expect exponential complexity because the k-Clique Recognition problem is NP-complete [8].

We now summarize the complexities of the various graph transformations (see Table 3.12):

Table 3.13 gives recurrence equations for the worst-case complexity for the class of complete graphs (K_n):

Therefore, S_n, D_n, F_n, and P_n are all of complexity class $O(n!)$. This proves that the worst-case run-time of our algorithm is exponential, which is expected for an NP-complete problem. However, maximal Series-Parallel reducible graphs, which are 2-trees [4, 16, 17], require $n-2$ Series and $n-2$ Parallel transformations and are reducible in polynomial time.

Table 3.12: Worst-Case Complexity of Graph Transformations

let T(G) = time to compute reliability of graph G
n = number of vertices
e = number of edges

Transformation	*Check Condition*	*Perform Transformation*
Insert Edge	——-	$O(1)$
Delete Edge	——-	$O(n)$
Terminate	$O(1)$	$O(n)$
Get reliability		
Parallel	$O(n^3)$	$O(n)$
Pendant	$O(n)$	$O(n)$
Vertex Merging	$O(e)$	$O(n)$
Series	$O(n)$	$O(n)$
Δ-Y	$O(n^3)$	$O(n)$
Polygon-Chain 1	$O(n^2)$	$O(n)$
Polygon-Chain 2	$O(n^2)$	$O(n)$
Polygon-Chain 3	$O(n^2)$	$O(n)$
Polygon-Chain 4	$O(n^2)$	$O(n)$
Polygon-Chain 5	$O(n^2)$	$O(n)$
Polygon-Chain 6	$O(n^2)$	$O(n)$
Polygon-Chain 7	$O(n^2)$	$O(n)$
Vertex Factoring*	$O(n)$	$O(n)$
Edge Factoring*	$O(e)$	$O(n) + O(n+e) = O(n+e)$

* creates two subproblems

Table 3.13: Worst-Case Complexity of Shooman's Algorithm on Complete Graphs

let $n =$	number of vertices
let $K_n =$	complete graph with n vertices
let $S_n =$	number of Series transformations required to reduce K_n
let $P_n =$	number of Parallel transformations required to reduce K_n
let $D_n =$	number of Δ-Y transformations required to reduce K_n
let $F_n =$	number of Edge Factoring transformations required to reduce K_n

n	S_n	P_n	D_n
2	0	0	0
3	1	1	0
4	2	1	1
5	7	9	3
6	29	46	12
7	146	245	60
8	877	1491	360
9	6140	10465	2520

$$S_n = (n-2) * S_{n-1} + 1, \quad n >= 5 \tag{3.83}$$

$$D_n = (n-2) * D_{n-1}, \quad n >= 4 \tag{3.84}$$

$$F_n = D_n \tag{3.85}$$

$$P_n > F_n, \quad n >= 5 \tag{3.86}$$

3.6.4 Criteria for Acceptable Performance

This chapter concentrates on exact algorithms, but also discusses approximate algorithms. The criteria for acceptable performance on a reliability analysis problem are different for an exact algorithm and for an approximate algorithm. For an exact algorithm, we require that the solution to a reliability analysis problem be accurate to within machine precision. For 4-byte floating point numbers in C, machine epsilon [7] is 2^{-24} or 5.96×10^{-8} but C reports 6 significant digits. We also require that the run time be reasonable. For an interactive network design program with an embedded exact reliability analysis algorithm, we could not tolerate running times longer than about ten seconds. However, for a reliability analysis algorithm running in non-interactive batch mode, we could tolerate running times up to a few hours.

For an approximate algorithm, we sacrifice some accuracy in the solution to achieve a reasonable run-time. For a network reliability analysis problem, we can tolerate a relative error in the solution of about 10^{-3}. However, for this sacrifice in accuracy, we expect a fast run-time of not more than about ten seconds.

We measure accuracy by the relative error of an approximate reliability with respect to the exact reliability. Let R_e= exact reliability, R_a= approximate reliability, and E= relative error. Then:

$$E = \frac{R_a - R_e}{R_e}$$

A positive value for the relative error means that the approximate reliability is higher than the exact reliability as would be reported by an upper-bound algorithm. Oppositely, a negative value for the relative error means that the approximate reliability is lower than the exact reliability as would be reported by an lower-bound algorithm.

3.7 Test Problems

In this section, we evaluate the performance of the 8 graph-reduction algorithms discussed in Section 3.6.2 on a set of test problems including complete graphs, grids [18], k-trees, and a cube graph of interest to graph theorists; and an Arpanet model [29], a reduced Arpanet model, and 3 realistic networks produced by a network design tool of interest to network designers. The numerical results of number of transformations, run-time, and accuracy for the approximate algorithms are given in Shooman's doctoral thesis [31] and in a paper [34].

3.7.1 Complete Graphs

Complete graphs, denoted by K_n, are the densest possible graphs for a given number of vertices, n. Complete graphs have an edge connecting each pair of vertices. Therefore, the number of edges in K_n is $\binom{n}{2}$ which grows quadratically with n. K_n is difficult to reduce with an exact algorithm because it is regular of degree $n-1$, i.e. each of its vertices has degree $n-1$.

The Δ-Y transformation is not particularly effective in reducing large complete graphs because removal of one edge from each of the Δ vertices of degree $n-1$ does not produce candidates for Series reduction nor Polygon-Chain reduction since degree $n-2$

for large n is still considerably larger than 2, the degree needed for Series or Polygon-Chain reduction. Furthermore, the creation of imperfect non-target vertices caused by the Δ-Y transformation adds complexity to large complete graphs. In this case, the costs of the Δ-Y transformation outweigh the benefits.

The 4 exact algorithms require exponential time to analyze complete graphs. Although Shooman's algorithm reduces complete graphs in fewer transformations than do the Simple, Δ-Y, or Satyanarayana & Wood algorithms, it is faster by a constant but still exponential. In fact it is hopeless to analyze complete graphs efficiently with any exact algorithm. Therefore, we must look to the approximate algorithms to reduce complete graphs efficiently. The Degree-3 Series and Y-Δ approximate algorithms are faster by a constant than any of the exact algorithms but still exponential on complete graphs. Only the K_4 Contraction Lower-Bound and Upper-Bound algorithms reduce complete graphs in polynomial time. Their accuracy is acceptable for high link reliabilities such as $p = 0.95$ but unacceptable for low link reliabilities such as $p = 0.5$. Fortunately, it is high link reliabilities that appear in realistic problems. Complete networks do not appear in practice because they are too expensive to build and because less dense networks are usually sufficiently reliable. However, the K_4 Contraction algorithms perform well in reducing complete graphs.

3.7.2 Grids

Grids are graphs whose vertices are placed at regular intervals along horizontal and vertical axes. An edge connects each pair of adjacent vertices along both the horizontal and vertical axes. An interior vertex of a grid is of degree 4 with edges emanating to the "north", "east", "south", and "west". A vertex along the perimeter of a grid has degree 3 and a corner vertex has degree 2. From this information, we can determine the number of edges in a grid as a function of its horizontal and vertical dimensions. The dimensions of a grid represent the number of vertices along the vertical and horizontal axes, respectively. Fig. 3.19 shows a 3x5 grid.

let $n =$ number of vertices
let $e =$ number of edges
let $l_1 =$ vertical dimension of grid
let $l_2 =$ horizontal dimension of grid

We compute the number of vertices in an l_1 x l_2 grid simply as $n = l_1 l_2$. We compute the number of edges as follows:

Vertex Type	*Number in Grid*	*Degree*	*Total Degree*
Corner	4	2	8
Perimeter	$2[(l_1 - 2) + (l_2 - 2)]$	3	$6[(l_1 - 2) + (l_2 - 2)]$
Interior	$(l_1 - 2)(l_2 - 2)$	4	$4(l_1 - 2)(l_2 - 2)$

Since the total degree sum in a graph is $2e$, we have:

$$2e = 8 + 6[(l_1 - 2) + (l_2 - 2)] + 4(l_1 - 2)(l_2 - 2) \quad (3.87)$$

$$e = 4 + 3[(l_1 - 2) + (l_2 - 2)] + 2(l_1 - 2)(l_2 - 2) \quad (3.88)$$

$$e = 4 + 3(l_1 + l_2 - 4) + 2(l_1 l_2 - 2l_1 - 2l_2 + 4) \quad (3.89)$$

$$e = 4 + 3l_1 + 3l_2 - 12 + 2l_1 l_2 - 4l_1 - 4l_2 + 8 \quad (3.90)$$

$$e = 2l_1 l_2 - l_1 - l_2 \quad (3.91)$$

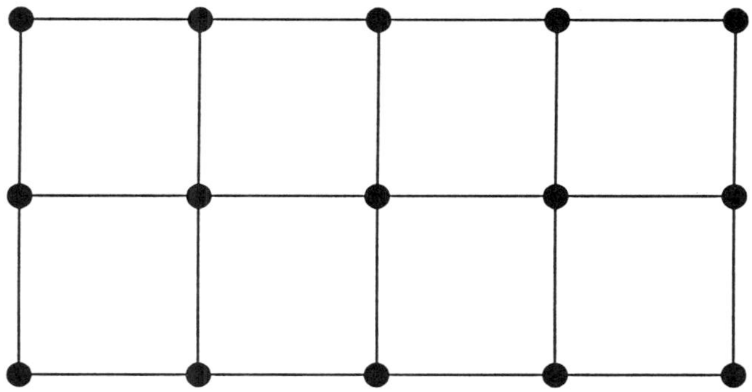

Fig. 3.19: 3x5 Grid

Grids are difficult to reduce because they have only four degree-2 vertices, one at each corner. In addition, grids contain no triangles (Δ subgraphs) so we cannot perform any Δ-Y reductions initially. Instead, we can perform a Series reduction followed by a Δ-Y reduction at each corner, but we must resort to Edge Factoring after that since the remaining Δ subgraphs have vertex degrees 3,3,5 and we perform Δ-Y reduction only on delta subgraphs with vertex degrees 3,3,3. Furthermore, grids are not k-trees (see Section 3.7.3), so we expect them to be difficult to reduce.

Grids are not k-trees, since a grid has 4 degree-2 vertices, but k-trees have vertices ranging in degree from k to $n-1$. Therefore, a degree-2 vertex could be found only in a 2-tree, but not in any k-tree with $k > 2$. But, 2-trees have triangle subgraphs and grids have no triangle subgraphs. Thus grids are not 2-trees and grids are not k-trees. Q.E.D.

The 4 exact algorithms require exponential time to analyze grids. However, Shooman's algorithm solves the reliability analysis problem on grids in approximately 25% fewer transformations than does the Simple algorithm. Of the approximate algorithms, K_4 Contraction does not help at all since grids contain no K_4 subgraphs. Both the Degree-3 Series and Y-Δ algorithms exhibit linear growth on the sequence of grids. The accuracy of the Y-Δ algorithm is better than that of the Degree-3 Series algorithm but neither meets our stated requirement of relative error less than 10^{-3} (see Section 3.6.4) for the entire sequence of grids with link reliability $p = 0.9$ or $p = 0.95$. In fact, the relative error increases with grid size. The exact algorithms perform acceptably provided that the grids do not become too large.

3.7.3 k-Trees

k-trees are graphs constructed by repeatedly inducing k-cliques on a foundation of K_k, the complete graph with k vertices. The degree of the vertices of a k-tree ranges from k to $n-1$, where n is the total number of vertices in the k-tree. All Series-Parallel graphs are partial 2-trees. Graph $G_2 = (V_2, E_2)$ is a partial of graph $G_1 = (V_1, E_1)$ if $V_2 = V_1$ and $E_2 \subset E_1$. A k-tree is defined inductively as follows [4]:

1. K_k, the complete graph on k nodes, is a k-tree
2. If G=(V,E) is a k-tree, and $K \subseteq V$ induces a k-clique in G, adding a new vertex x and edges from x to each $y \in K$, produces a k-tree

All 4 of the exact algorithms require exponential time to reduce 3-trees and 4-trees, but Shooman's algorithm performs better than the Simple, Δ-Y, and Satyanarayana & Wood's algorithms by a constant as measured by the number of transformations. However, we have found that all 4 of our approximate algorithms, K_4 Contraction (upper and lower bound), Degree-3 Series, and Y-Δ require a linear number of transformations to reduce 3-trees and that K_4 Contraction (upper and lower bound) requires a linear number of transformations to reduce 4-trees. We denote a k-tree with n vertices as ${}_kT_n$. In fact, ${}_kT_n$ has exactly $kn - \binom{k+1}{2}$ edges. All 2-trees have the same reliability [15], but this is not true for 3-trees. We claim that one K_k contraction is necessary and sufficient to reduce a k-tree in linear time. We prove that K_4 Contraction reduces 3-trees in linear time.

Reverse the 3-tree construction by induction on the number of vertices in the 3-tree. Note that each new vertex added in 3-tree construction along with its 3 edges connected to a previous 3-clique forms a 4-clique, i.e. a K_4 subgraph. Let V_i be the i-th vertex added in the 3-tree construction process. Therefore, V_n is the last vertex added to construct ${}_3T_n$. Perform K_4 Contraction on the unique 4-clique, denoted by C_n, containing V_n. If $V_{n-1} \in C_n$, then V_{n-1} has degree less than 3 and we perform a Series or Pendant transformation on it as the next step in the reduction process. Else, V_{n-1} forms a 4-clique and we perform K_4 Contraction on its 4-clique. We continue iterating on V_i for decreasing values of i ranging from n to 2. When $i = 2$, we have two vertices and one edge and we terminate the reduction. The total number of operations performed is one per iteration or $n - 2$, which is linear. Q.E.D.

3.7.4 Cube Graph

The Cube graph has 8 vertices and is regular of degree 3 (see Fig. 3.20) but has no triangles (Δ subgraphs). Therefore, the Δ-Y transformation cannot be used initially to reduce it.

The Cube graph is a small problem which is solved quickly by all methods. However, a larger graph with embedded Cube subgraphs can be a challenging problem. Nevertheless, we note that our exact algorithm (Shooman) requires the fewest number of transformations of the 4 exact algorithms to reduce the Cube graph. It requires approximately half the number of transformations as the Simple algorithm and approximately one-third the number of transformations as the Δ-Y algorithm. Among the approximate algorithms, K_4 Contraction has no embedded K_4 subgraphs upon which to operate, but Degree-3 Series and Y-Δ greatly decrease the number of transformations required. The Y-Δ algorithm meets our accuracy requirement of less than 10^{-3} relative error for approximate algorithms, but the Degree-3 Series algorithm marginally does not.

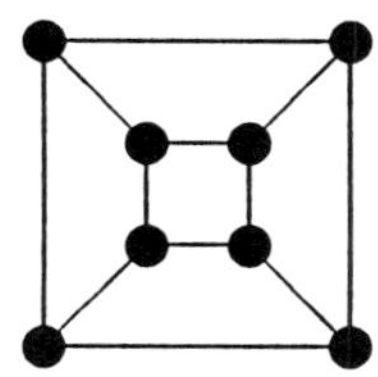

Fig. 3.20: The Cube Graph

3.7.5 Realistic Networks

We also tested our algorithms on realistic networks for both the all-terminal and the 2-terminal problems. Our test problems include an Arpanet model [29], a reduced Arpanet model, and 3 realistic networks of interest to network designers produced by a network design tool.

Real networks tend to be large and sparse. Such networks can be decomposed almost completely using the Series, Parallel, Δ-Y, and Polygon-Chain transformations without the need to resort to many Edge Factoring transformations. Our exact algorithm (Shooman) reduces the total number of transformations needed to decompose the network with respect to the Simple algorithm by a multiplicative factor ranging from 1.20 to 4.56 for the all-terminal test problems and from 1.46 to 20.30 for the 2-terminal test problems. These speedups enable us to solve exactly many realistic network problems in a reasonable length of time that would take too long to solve with the Simple algorithm.

Our algorithm (Shooman) required the fewest number of transformations of the 4 exact algorithms to reduce the each of the 10 realistic network test problems: Arpa, Redarpa, Net 1, Net 2, and Net 3, both for all-terminal and 2-terminal reliability. However, the Δ-Y algorithm usually had the shortest running time. We think that the running time of Shooman's algorithm can be shortened by improving the implementation of the Polygon recognition algorithms (see Section 3.9). Also, perhaps it is misleading to assign equal weights to each type of transformation when counting the total number of transformations, because the Polygon-Chain transformations, as presently implemented, take longer to perform than the other transformations. However, the total number of transformations is a useful machine-independent metric for evaluating the complexity of graph-reduction algorithms.

3.8 Results

We tested our network reliability algorithm against Satyanarayana and Wood's algorithm [28] and the simple one consisting only of Series, Parallel, Pendant, Vertex Factoring, and Edge Factoring. In three papers [32, 33, 34], we showed that Shooman's algorithm is faster than Satyanarayana and Wood's and the simple algorithm on all-terminal test problems consisting of complete graphs, grids [18], and miscellaneous graphs including a cube, an Arpanet model, and a reduced Arpanet model. In this section, our test problems consist of both the all-terminal and a 2-terminal case of an Arpanet model [29], a reduced Arpanet model [4], and three realistic designs produced by the Mentor network design algorithm [9].

We implemented Shooman's algorithm in approximately 2000 lines of C (including whitespace and documentation) and ran it on a number of machines, including an IBM PS/2 Model 80, a 33-MHz 386 PC, a DECstation 5000/120, and a SUN 3/60 Workstation. The numerical results of number of transformations, run-time, and accuracy for the approximate algorithms are given in Shooman's doctoral thesis [31] and in a paper [34].

3.8.1 Conclusions

We have designed and implemented a graph-reduction algorithm for computing the k-terminal reliability of an arbitrary network with possibly unreliable nodes. Our two contributions are a version of the Δ-Y transformation for k-terminal reliability and an extension of Satyanarayana and Wood's Polygon to chain transformations to handle graphs with imperfect vertices. Our exact algorithm (Shooman) is faster than or equal to that of Satyanarayana and Wood and also the simple algorithm without Δ-Y and Polygon to Chain transformations for every problem we have considered. Our exact algorithm runs in linear time on Series-Parallel graphs and faster than the above stated algorithms for large problems which run in exponential time. Our approximate algorithms reduce the computation time for the network reliability problem by 2 to 3 orders of magnitude for large problems, while providing reasonably accurate answers (relative error less than 10^{-3}) in most cases.

The difficulty of a network reliability analysis problem depends on these four factors in order of decreasing importance: The number of edges (e), the average degree ($\bar{d}$) [3], the number and type of recognizable subgraphs, and the edge reliability (p). We have plotted the test problems from Section 3.7 on a graph of Number of Edges (e) vs. Average Degree ($\bar{d}$) before (see Fig. 3.21) and after (see Fig. 3.22) Series and Parallel reduction. Our exact algorithm performs well on problems with fewer than about 50 edges after Series and Parallel reduction, but suffers from combinatorial explosion on problems with more than about 50 edges after Series and Parallel reduction (see Fig. 3.22). In the worst cases, exact network reliability analysis algorithms have a complexity of $O(2^e)$.

The approximate algorithms perform well on problems with high link reliabilities (e.g. $p > 0.9$), since the approximate transformations introduce a small relative error in those cases. We show the effect of average degree ($\bar{d}$) and link reliability (p) on the accuracy of the approximate algorithms on various test problems (see Fig. 3.23).

[3]The average degree of a graph equals $\frac{2e}{n}$.

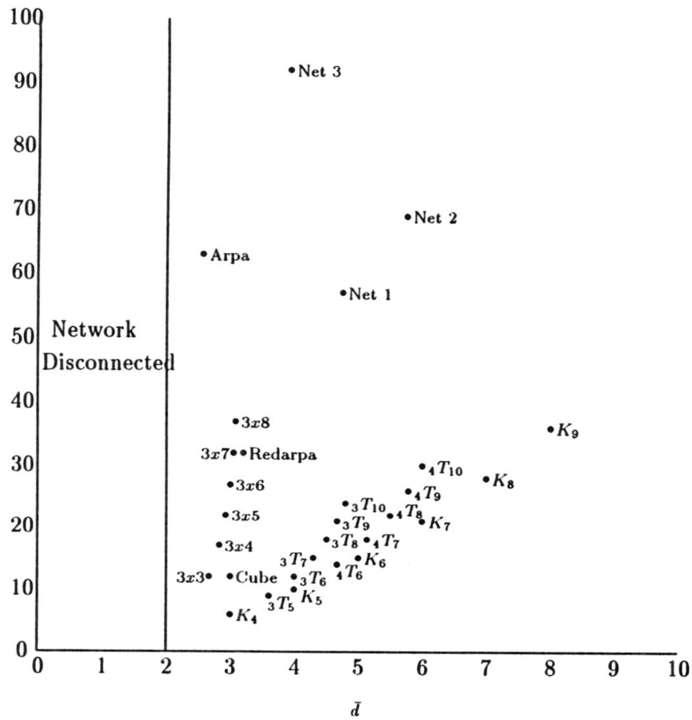

Fig. 3.21: Difficulty of Network Reliability Analysis Problems: Number of Edges (e) vs. Average Degree ($\bar{d}$) before Series and Parallel Reduction

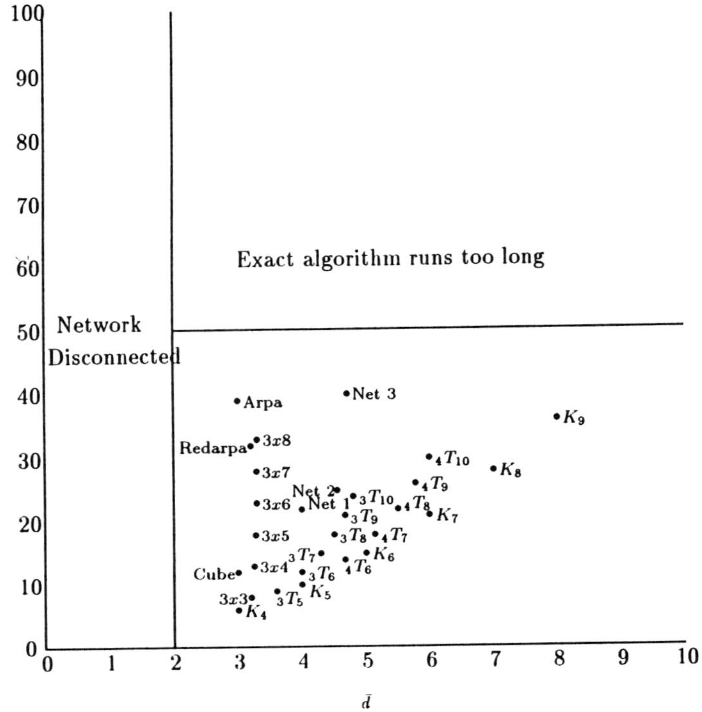

Fig. 3.22: Difficulty of Network Reliability Analysis Problems: Number of Edges (e) vs. Average Degree ($\bar{d}$) after Series and Parallel Reduction

Note that the link reliability (p) can be varied for all of our test problems and only the average degree(d) is fixed. Therefore, each test network can be either reliable or unreliable, depending on the choice of p. Note that the realistic test networks (see Section 3.7.5) lie in the lower right-hand corner of the graph corresponding to sparse designs (low $\bar{d}$) and high link reliabilities (p). Networks of very high degree are too expensive and not feasible. Also, very low link reliabilities do not occur in practice. In fact, the most practically interesting problems are also tractable. Our approximate algorithms offer the most significant improvement over the exact algorithms in the upper right-hand corner of the graph and yield the most accurate solutions on the right-hand side of the graph. The most difficult problems for the approximate algorithms to solve are those in the center of the graph, marginally reliable with degree high enough to defeat the exact algorithms and with link reliability low enough to introduce significant relative error. However, the approximate algorithms perform well on the realistic test problems.

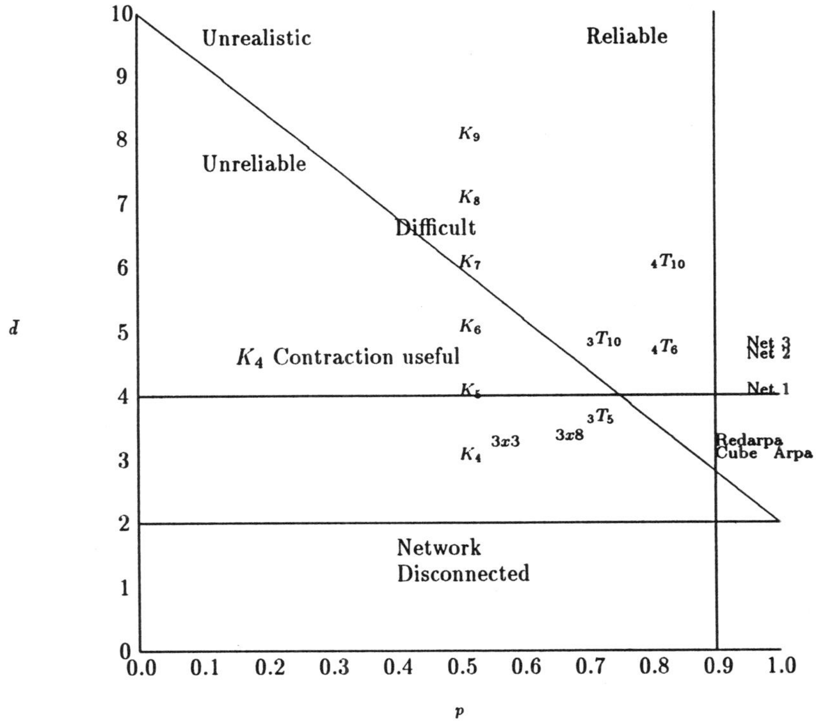

Fig. 3.23: Accuracy of Network Reliability Analysis Solutions: Average Degree ($\bar{d}$) vs. Link Reliability (p) after Series and Parallel Reduction

3.9 Future Developments

The exact and approximate algorithms presented in this chapter can handle the majority of realistic network reliability analysis problems well. However, on the most difficult problems, our exact algorithms run too long and our approximate algorithms yield upper and lower bounds that are too loose. We propose to tackle these unsolved problems in the following ways: We hope to reduce the run-time of our exact algorithm by improving the implementation of the Polygon recognizer for the Polygon-Chain transformations. Since the smaller polygons are subsets of the larger ones, we can use the same code to recognize all 7 polygons rather than 7 separate subroutines, one for each polygon. With the present implementation, some identical code is executed more than once.

In the future, we expect to see new developments in approximate algorithms, parallel algorithms, and reliability metrics.

We hope to improve our approximate algorithms by combining them into a single approximate algorithm using all of the approximate transformations employing heuristics to determine which approximate transformation to perform first on a particular graph. In addition, we would like to improve the accuracy of the Degree-3 Series transformation by adding the simple heuristic of dropping the edge with the minimum reliability rather than the last one in the data structure. This measure should reduce the relative error introduced by dropping an edge, thus improving the accuracy.

We hope to develop a new approximate algorithm called "Correlated Series" which performs Series transformations even in the ill-defined cases (#2, 3, and 6, see Table 3.2) with correlated events. We actually implemented this algorithm earlier before discovering the ill-defined Series cases and hoped it would be a valid exact algorithm. It is not valid as an exact algorithm because the correlated events introduce an error, but it is fine as an approximate algorithm because the error is small for high link reliability (p). We also discovered that this algorithm is fast. We hope to test it formally against the other approximate algorithms on our set of test problems.

Deo and Medidi [5] have implemented three graph-reduction algorithms on a shared-memory parallel computer. These algorithms are exact but operate only on 2-terminal reliability problems. In the future, we expect to see parallel graph-reduction algorithms to solve the k-terminal reliability problem.

In the future, we expect to see network reliability analysis algorithms that compute other metrics, besides connectivity, such as the expected number of nodes which can communicate with a distinguished node, the expected number of node pairs communicating, and the probability of all operative nodes (as opposed to all nodes) being connected. More complex metrics would measure a network's ability to handle its traffic requirements under partial failure. Rai and Soh [23] have developed such a metric called "capacitated reliability." Furthermore, we would expect to see call-based rather than topology-based reliability metrics which measure the ability of a call in the network to remain operational despite some component failures. These metrics would consider capacity, re-routing, priority, and preemption in addition to connectivity.

REFERENCES

[1] Arnborg S. and A. Proskurowski, *Linear Time Algorithms for np- Hard Problems Embedded in k-Trees*, Technical Report TRITA-NA- 8404, Royal Institute of Technology, Stockholm, 1984.

[2] Bazovsky I., *Reliability Theory and Practice*, Prentice-Hall, Englewood Cliffs, New Jersey, 1961.

[3] Bienstock D., *An Algorithm for Reliability Analysis of Planar Graphs*, Technical Report WP-20-85-86, GSIA, Carnegie-Mellon University, 1985.

[4] Colbourn C. J., *The Combinatorics of Network Reliability*, Oxford University Press, New York, 1987.

[5] Deo N. and Muralidhar Medidi, *Parallel Algorithms for Terminal-Pair Reliability*, IEEE Trans. on Rel., R-41, No. 2, pp. 201-209, June 1992.

[6] El-Mallah E.S. and C.J. Colbourn, *Reliability of D-Y Reducible Networks*, In Proceedings of the 16th Southeastern Conference on Combinatorics, Graph Theory and Computing, pp. 49- 54, 1985.

[7] Forsythe G. E., M.A. Malcolm and C.B. Moler, *Computer Methods for Mathematical Computations*, Prentice-Hall, Englewood Cliffs, New Jersey, 1977.

[8] Garey M.R. and D.S. Johnson, *COMPUTERS and INTRACTABILITY: A Guide to the Theory of NP-Completeness*, W.H. Freeman and Company, San Francisco, 1979.

[9] Kershenbaum A., P. Kermani and G.A. Grover, *Mentor: An Algorithm for Mesh Network Topological Optimization and Routing*, IEEE Trans. on Com., C-39, No.4, pp. 503-513, April 1991.

[10] Kershenbaum A. and R.M. Van Slyke, *Recursive Analysis of Network Reliability*, Networks, Vol. 3, pp. 81-94, 1973.

[11] Marshall C. W., *Applied Graph Theory*, Wiley-Interscience, New York, 1971.

[12] Misra K.B., *Reliability Analysis and Prediction : A Methodology Oriented Treatment*, Elsevier, Amsterdam, 1992.

[13] Murray K., *Path and Cut-Based Bounds for Network Reliability*, Ph.D Thesis, Polytechnic University, Brooklyn, New York, June 1992.

[14] Murray K., A. Kershenbaum and M.L. Shooman, *Communication Network Reliability Analysis: Approximations and Bounds*, In Proceedings of the Annual Reliability & Maintainability Symposium, Jan. 1993.

[15] Neufeld E.M., *Construction of Reliable Series-Parallel Networks: A Combinatorial Approach*, Master's Thesis, University of Saskatchewan, Department of Computational Science, Nov. 1983.

[16] Neufeld E.M. and C.J. Colbourn, *The Construction of Reliable Series-Parallel Networks*, In Congressus Numerantium, Vol. 41, pp. 21-26, May 1984.

[17] Neufeld E. M. and C.J. Colbourn, *The Most Reliable Series-Parallel Networks*, Networks, Vol. 15, 27-32, 1985.

[18] Page L. B. and J.E. Perry, *A Practical Implementation of the Factoring Theorem*, IEEE Trans. on Rel., Vol. R-37, No. 3, pp. 259-267, Aug. 1988.

[19] Politof T. and A. Satyanarayana, *A Linear Time Algorithm to Compute the Reliability of Planar Cube-Free Networks*, In TIMS/ORSA Conference, 1984.

[20] Politof T. and A. Satyanarayana, *An* $O(|v^2|)$ *Algorithm for a Class of Planar Graphs to Compute the Probability that the Graph is Connected*, Technical Report ORC 84-4, University of California at Berkeley, 1984.

[21] Politof T. and A. Satyanarayana, *Efficient Algorithms for Reliability Analysis of Planar Networks-A Survey*, IEEE Trans. on Rel., R-35, No. 3, pp. 252-259, 1986.

[22] Politof T., *A Characterization and Efficient Reliability Computation of Δ-Y Reducible Networks*, Ph.D Thesis, University of California at Berkeley, 1983.

[23] Rai S. and S. Soh, *Survivability Analysis of Complex Computer-Networks with Heterogeneous Link-Capacities*, In Proceedings of the Annual Reliability & Maintainability Symposium, pp. 374-379, Jan. 1991.

[24] Rosenthal A. and D. Frisque, *Transformations for Simplifying Network Reliability Calculations*, Networks, Vol. 7, pp. 97-111, 1977.

[25] Rosenthal A., *Computing Reliability of Complex Systems*, Ph.D Thesis, University of California at Berkeley, 1974.

[26] Rosenthal A., *Computing the Reliability of Complex Networks*, SIAM Journal of Applied Mathematics, Vol. 32, pp.384-393, March 1977.

[27] Satyanarayana A. and M.K. Chang, *Network Reliability and the Factoring Theorem*, Networks, Vol. 13, pp. 107-120, 1983.

[28] Satyanarayana A. and R.K. Wood, *A Linear-Time Algorithm for Computing k-Terminal Reliability in Series-Parallel Networks*, SIAM Journal of Computing, Vol. 14, pp. 818-832, Nov. 1985.

[29] Schwartz M., *Computer-Communication Network Design and Analysis*, Prentice-Hall, Englewood Cliffs, New Jersey, 1977.

[30] Senturia S. and B.D. Wedlock, *Electronic Circuits and Applications*, John Wiley and Sons, New York, 1975.

[31] Shooman A.M., *Exact Graph-Reduction Algorithms for Network Reliability Analysis*, Ph.D Thesis, Polytechnic University, Brooklyn, New York, June 1992.

[32] Shooman A.M. and A. Kershenbaum, *Exact Graph-Reduction Algorithms for Network Reliability Analysis*, In GLOBECOM '91 Proceedings, pp. 1412-1420, Dec. 1991.

[33] Shooman A. M. and A. Kershenbaum, *Exact Graph-Reduction Algorithms for Network Reliability Analysis*, Technical Report, IBM T.J. Watson Research Center, Hawthorne, New York, 1991.

[34] Shooman A.M. and A. Kershenbaum, *Methods for Communication-Network Reliability Analysis: Probabilistic Graph Reduction*, In Proceedings of the Annual Reliability & Maintainability Symposium, pp. 441-448, Jan. 1992.

[35] Shooman M. L., *Probabilistic Reliability: An Engineering Approach*, Krieger, Melbourne, Florida, 2nd Edition, 1990.

[36] Theologou O.R. and J.G. Carlier, *Factoring & Reduction for Networks with Imperfect Vertices*, IEEE Trans. on Rel., R-40, No. 2, pp. 210-217, June 1991.

[37] Valiant L.G., *The Complexity of Enumeration and Reliability Problems*, SIAM Journal of Computing, Vol. 8, pp. 410-421, 1979.

New Trends in System Reliability Evaluation
K.B. Misra (Editor)

CHAPTER 4

PARALLEL ALGORITHMS AND IMPLEMENTATIONS

Narsingh Deo and Muralidhar Medidi
Department of Computer Science
University of Central Florida
Orlando, Florida
U. S. A.

ABSTRACT

We describe parallelizations of three terminal-pair reliability algorithms. The BBN Butterfly GP1000, the parallel computer on which the parallel algorithms are implemented, is described first. After a review of the sequential algorithms, their parallel versions — along with implementation details like data structures — are described. The performance of the parallel algorithms is then compared with respect to their execution time and processor efficiency. We conclude this chapter with some potential research topics in parallelizing and implementing algorithms for reliability analysis.

4.1 Introduction

The problem of determining the reliability of a complex system, whose components are subject to failure, arises in a variety of applications, ranging from telecommunication, transportation, and mechanical systems, to integrated circuits and computer software. The underlying system can be naturally expressed in the form of a network, arising from the interconnections of various system elements, as exemplified by telecommunication and computer networks. For example, the vertices of the network might represent the physical locations of computers and its edges, the communication links between computer sites. Depending on whether the vertices and edges work or fail, the network itself can be considered to be either working or failed. Diverse network reliability problems, classified according to the underlying network model, capture different performance measures of the system. For an overview of these problems, see Shier [1].

The terminal-pair reliability is the probability that the source s and the sink t, two distinguished vertices in a network with known success/failure probabilities of its elements, remain connected by a path of operating vertices and edges. The terminal-pair reliability problem, the simplest of the various reliability problems, was shown to be NP-hard [2] and algorithms proposed to solve it are computationally intensive. The problem remains NP-hard even for planar networks [3]. For a review of various reliability evaluation algorithms, see Misra [4]. For some restricted networks, such as series-parallel graphs, inner-four-cycle-free graphs, and cube-free graphs, polynomial-time algorithms have been developed. For a survey of these algorithms, see Politof and Satyanarayana [5].

Our aim here was to design and implement parallel algorithms for the terminal-pair reliability problem, on general networks, and show that parallel processing pays off. We therefore looked for the fastest sequential algorithms to parallelize. In a 1988 paper, Yoo and Deo [6] compared four best-known (sequential) terminal-pair reliability algorithms reported in [7-10] and concluded that the Dotson and Gobein algorithm [7] was the fastest. Later, Page and Perry proposed a faster algorithm, PP-F2TDN [11]. A still faster algorithm, Reduce & Partition algorithm, was proposed by Deo and Medidi [12] in 1992. We will first describe the parallelizations of these three algorithms, and then examine their performance.

A directed network with unreliable vertices can be easily transformed into an equivalent network with perfectly reliable vertices for reliability computation [13], but no such transformation is known for undirected networks. Aggarwal, Gupta, and Misra [14] and, more recently, Theologou and Carlier [15] have presented techniques for computing the reliability of undirected networks with unreliable vertices. The terminal-pair reliability problem for undirected networks is a special case of the problem for directed networks. All the algorithms, in this chapter, are designed on directed networks with perfectly reliable vertices and with statistically independent edge failures.

Section 4.2 describes briefly the parallel computer on which the parallel algorithms were implemented. Section 4.3 reviews the three sequential algorithms we parallelized. Section 4.4 describes the parallel versions of these algorithms along with implementation details. Section 4.5 compares their performances. Section 4.6 gives concluding remarks and some potential research topics.

4.2 The Butterfly Parallel Machine

The BBN Butterfly GP1000 is a MIMD, shared-memory multiprocessor computer. Each processor module consists of a 16 MHz Motorala 68020 microprocessor with a MC 68881 math co-processor and has four megabytes of local memory. Each processor also has access to the memory of other processor modules. Communication and synchronization between processors are performed using the shared-memory architecture. The BBN Butterfly GP1000 at our University has 64 processors.

A local memory access requires 2 microseconds, while a non-local memory access takes, on the average, about 6 microseconds. Non-local memory accesses are achieved through a processor interconnection network. Memory contention can occur if simultaneous accesses are attempted to the same memory module. Performance deteriorates significantly if such memory contention occurs in the program frequently. The Uniform System(US) application library on the Butterfly provides routines to distribute data structures across the shared memory to reduce memory contention. The US also provides routines to generate parallel tasks. For more details of the BBN Butterfly, see Larrabee, Pennick, and Stern [16].

4.3 Sequential Algorithms: Review

Page and Perry Algorithm. Given an undirected network G, with an edge labeled e with success probability p_e, its terminal-pair reliability $Rel(G)$ can be computed, by factoring theorem, as

$$Rel(G) = (1 - p_e) \times Rel(G - e) + p_e \times Rel(G * e).$$

The networks $G - e$ and $G * e$ are obtained from G by deleting and contracting the edge e, respectively. Edge e is commonly known as the *keystone* edge. Satyanarayana and Chang [17] and Wood [18] analyzed factoring, gave its performance bounds, and presented an optimal strategy of choosing the keystone edge such that the networks remain biconnected after factoring. In directed networks, factoring requires extra caution. In a directed network, an operating edge e provides only a 1-way path while contraction of this edge means establishing a 2-way path between the end vertices. But, if a keystone edge is restricted to be either coming out of the source or going into the sink, this contraction will always work because the other directional path is irrelevant. Because of this restriction, keystone edges cannot be chosen arbitrarily.

Page & Perry algorithm computes $Rel(G)$ by simplifying network G employing series and parallel reductions, recursively computing $Rel(G - e)$ and $Rel(G * e)$, and combining these results through the factoring theorem. Recursion terminates when either there is no s-t path or when s and t are contracted into a single vertex. Since at each application of the factoring formula the number of edges in the network decreases at least by one, the recursion depth is bounded by the number of edges in G. For more details of this factoring algorithm, see [11].

Dotson and Gobein Algorithm. Dotson and Gobein [7] employ a method using boolean algebra to compute network reliability. A subgraph of G is a *success subgraph*, if it has an s-t path. The terminal-pair reliability of G is the sum of probability of occurrence of each success subgraph of G . Given a graph G and a set of edges $\{e_1, e_2, \ldots, e_l\}$ which form an s-t path in G, Dotson and Gobein used boolean algebra to partition the set P of the subgraphs of G as

$$P = [\bar{e}_1] \cup [e_1\bar{e}_2] \cup \cdots \cup [e_1e_2e_3..e_{i-1}\bar{e}_i] \cup \cdots \cup [e_1e_2e_3..e_{l-1}\bar{e}_l] \cup [e_1e_2...e_{l-1}e_l],$$

where partition $[e_1e_2e_3..e_{i-1}\bar{e}_i]$ represents the set of all subgraphs of G in which edges e_1 through e_{i-1} are functioning and edge e_i has failed. All the subgraphs in partition $P_1 = [e_1e_2...e_{l-1}e_l]$ contain the given path and are thus success subgraphs. The partition P_1 is the success partition corresponding to the given path in the set P. The remaining partitions are the complementary partitions of P_1 in P. Once the subgraphs of G are partitioned using the edges in the path, one needs to search, for success subgraphs, only in the complementary partitions. If $S = [e_1e_2..e_h\bar{e}_{h+1}..\bar{e}_{k-1}\bar{e}_k]$ is a success partition in such a partitioning scheme, then the sum of the probability of occurrence of each subgraph in the partition can be easily computed as

$$\text{probability}(S) = \prod_{i=1}^{h} p_i \times \prod_{j=h+1}^{k} (1 - p_j).$$

In Dotson & Gobein algorithm, each partition is represented by a vector of size m, called an *event*. The status of each edge is marked in the event as *present*, *absent*, or *unmarked*. As long as there are events to explore, the algorithm picks an event E and constructs the maximum subgraph G_E in E, which is derived from G by deleting edges that are marked to be absent in E. This graph G_E is tested for an s-t path. If no such path exists, there is no success subgraph in this partition. Otherwise, the subgraphs in this partition are further partitioned using the path. To maximize the number of success subgraphs collected in each event, a shortest s-t path is used in partitioning. Yoo and Deo [6] showed, empirically, that the number of events processed using this method is a small fraction of the total number of subgraphs possible. For more details, see Dotson and Gobein [7] and Yoo and Deo [6].

Reduce and Partition Algorithm. Page & Perry algorithm is fast because of the reductions performed at each step. Dotson & Gobein algorithm, on the other hand, owes its efficiency to reducing the number of states in the search tree. In Dotson & Gobein algorithm only edges labeled as absent are deleted. Similarly, edges which are labeled as present can be contracted. Edge contractions and deletions alone may not make the algorithm faster, but the ensuing reductions often decrease the problem size to make the algorithm more efficient. To achieve the effectiveness of both Dotson & Gobein, and Page & Perry approaches, network reductions should be performed at each step along with a reduction in the number of states in the search space. To use network reductions at each step, the solution should be formulated in terms of subnetworks on which reduction rules can be applied. Such a formulation can be derived as follows.

Let $Rel(G)$ be the terminal-pair reliability of network G and let $s, e_1, u_1, e_2, u_2, \ldots, u_{l-1}, e_l, t$ be a simple s-t path in G, where e_i $(1 \leq i \leq l)$ are the edges and u_i $(1 \leq i \leq l-1)$ are the internal vertices on this path. If no such path exists then $Rel(G)$ is 0. If such a path exists, we can factor on edge e_1 safely, since it emanates from s. Hence,

$$Rel(G) = (1 - p_1) \times Rel(G - e_1) + p_1 \times Rel(G * e_1).$$

In network $G * e_1$, vertices s and u_1 are contracted into s and edge e_2 is now emanating from s. Factoring on edge e_2 in network $G * e_1$, gives

$$\begin{aligned} Rel(G) &= (1 - p_1) \times Rel(G - e_1) \\ &\quad + p_1(1 - p_2) \times Rel(G * e_1 - e_2) \\ &\quad + p_1 p_2 \times Rel(G * e_1 * e_2). \end{aligned}$$

Similarly, applying factoring theorem l times on the l edges in the path and noting that $Rel(G * e_1 * e_2 * \ldots * e_l) = 1$, since s and t are contracted into a single vertex in network $G * e_1 * e_2 * .. * e_l$, we have

$$\begin{aligned} Rel(G) &= (1 - p_1) \times Rel(G - e_1) \\ &\quad + p_1(1 - p_2) \times Rel(G * e_1 - e_2) \\ &\quad + \cdots \\ &\quad + p_1 p_2 .. p_{i-1}(1 - p_i) \times Rel(G * e_1 * e_2 * .. * e_{i-1} - e_i) \\ &\quad + \cdots \\ &\quad + p_1 p_2 .... p_{l-1}(1 - p_l) \times Rel(G * e_1 * e_2 * .... * e_{l-1} - e_l) \\ &\quad + p_1 p_2 ... p_l \end{aligned} \tag{4.1}$$

Thus, to determine $Rel(G)$ of network G with a path of length l from s to t, we recursively call on l subproblems. Each of these subproblems has at least one fewer s-t path than the original problem. The equation is a simple extension of factoring, and its correctness follows from the factoring theorem. The subproblems solved by the new method precisely correspond to the complementary events generated by Dotson & Gobein method for the given s-t path in G.

Computing $Rel(G)$ based on equation (4.1) requires finding an s-t path, which requires only slightly more work than checking the existence of an s-t path (always performed in the factoring algorithm). Equation (4.1) applies factoring theorem on all the edges of this path at once, avoiding the $l-1$ intermediate recursive calls which would normally be made by the factoring algorithm.

In equation (4.1), if a longer path of length l' was to be used, then l' subproblems would have to be solved. The first l subproblems generated by these two paths are the same size, assuming that no reductions are applied. Thus factoring on a shorter path always generates fewer subproblems, and the effort spent in finding a shortest s-t path is no more than finding an arbitrary path.

The Reduce & Partition algorithm, based on (4.1), works by simplifying the input network, finding a shortest s-t path, recursively solving the subproblems generated with this path, and combining the results. Since the number of edges in the network decreases, at least by one, in each recursive call, recursion depth is limited by m, the number of edges in the network. A formal description of Reduce & Partition algorithm is given as a function REDUCE&PARTITION. The subroutine PATH3(*G, found, path, length*), used by REDUCE&PARTITION, sets variable *found* to true and returns the length of the shortest path in the variable *length*, if an s-t path exists in G. The sequence of edges in this path are returned in the linear array, *path*. PATH3 is a simple extension of the breadth-first search procedure to generate a shortest s-t path.

```
function REDUCE&PARTITION( network G with s and t )
{returns the terminal-pair reliability of the network G}
begin
multiplier := 1.0;
Reduce network G, using reduction rules, if applicable;
{multiplier may get modified if reductions are performed}
if s = t then
  return multiplier; {s and t are contracted into a single vertex}
end if
PATH3( G, found, path, length ); {returns a shortest s-t path}
if found then
  R = 0.0;
  for i := 1 to length do
      edge e := path[i];
      G := G − e;
      p_e := success probability of the edge e;
      R := R + multiplier × ( 1 − p_e ) × REDUCE&PARTITION( G );
      G := G ∗ e;
      multiplier := multiplier × p_e;
  end for
  return ( R + multiplier ); {terminal-pair reliability of the network G}
else
  return 0.0; {there is no s-t path in network G}
end if
end {REDUCE&PARTITION}
```

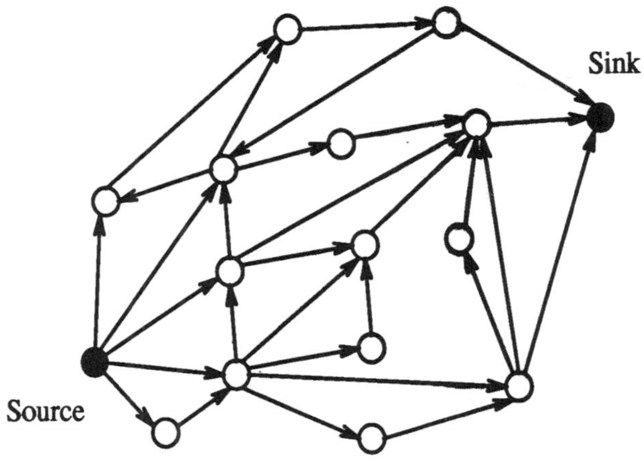

16 vertices
30 directed edges
$p_i = 0.9$, for all i
s-t reliability = 0.997186

Fig. 4.1 Benchmark Network 1

This algorithm was implemented in C on the BBN Butterfly GP1000 computer. A weighted adjacency matrix was used to represent the network, in which the entry (i, j) is the reliability of the edge from vertex i to vertex j.

Comparison. We compared the performance of the three algorithms, using two benchmark networks from the literature. As a third benchmark, we used the complete networks of different sizes, which also illustrate the inherent exponential nature of the problem. To compare with the performance of Reduce & Partition algorithm, we implemented the Page & Perry algorithm using one processor of the BBN Butterfly computer. Since our BBN Butterfly and the Macintosh II (used by Page & Perry) both employ the MC 68020 microprocessor, the two execution times are comparable.

Figure 4.1 shows an often-cited directed network in the literature, which first appeared in Fratta and Montanari [19]. Yoo and Deo [6] also used this network to compare the performance of various algorithms. This network admits substantial initial series and parallel reductions. Some people have used the pre-simplified version of this network while others have used it as-is. However, for Reduce & Partition, the pre-simplification was of no appreciable help. Table 4.1 shows the execution times reported for this network in [6, 11, 20, 21] along with execution times of our implementations.

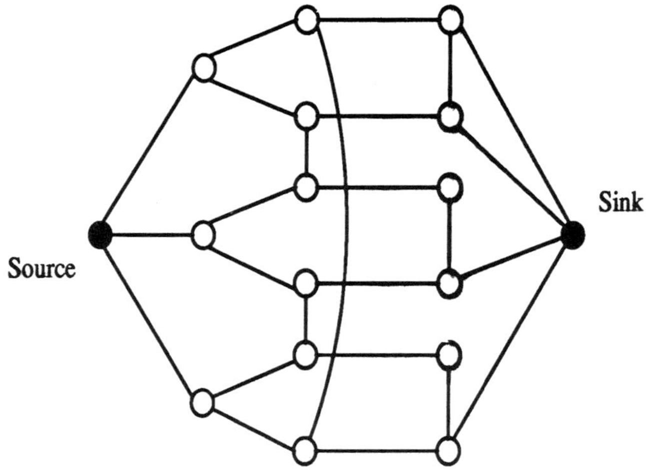

17 vertices
25 undirected edges(50 directed edges)
$p_i = 0.9$, for all i
s-t reliability = 0.998059

Fig. 4.2 Benchmark Network 2

The undirected network in Figure 4.2 was first used by Bailey and Kulkarni [22] to compare the complete enumeration algorithm of Buzacott [23], the lattice algorithm of Provan and Ball [24] and their own Markov Chain algorithm. The computation times for Page & Perry algorithms of [11, 25] are also available for this network. Table 4.2 contains these values along with the execution time of Reduce & Partition algorithm. This undirected network was transformed into an equivalent directed network by replacing each undirected edge by a pair of anti-parallel edges with same reliability and used as input. Our implementation of factoring algorithm took slightly shorter time than Page & Perry implementation because the network is not very sparse and the adjacency matrix (we used) is better suited for denser networks.

We ran the sequential Reduce & Partition algorithm on directed, complete networks of order 5 to 9. These networks are point symmetric and therefore the choice of s and t, as long as they are different, does not matter. For reliability algorithms based on factoring, a complete network represents the worst-case input for a given number of vertices. Since complete networks do not admit reductions, the results presented in this example are the limiting cases of Reduce & Partition algorithm. Table 4.3 shows the execution times of Reduce & Partition and those of Page & Perry implementation (on Macintosh II) and our implementation (BBN Butterfly — 1 processor) of the factoring algorithm.

Table 4.1 Execution Times of Various Algorithms for the Benchmark Network 1

Algorithm	Computer	Language	Time(sec.)
Nelson [9]	Honeywell DSP Level 66	FORTRAN	> 1200
Lin [8]	Honeywell DSP Level 66	FORTRAN	> 1200
Shooman [10]	Honeywell DSP Level 66	FORTRAN	> 1200
Dotson & Gobein [7]	Honeywell DSP Level 66	FORTRAN	520
Modified Dotson [6]	Honeywell DSP Level 66	FORTRAN	68.9
Fratta & Montanari [21]	IBM 360/67	Algol W	15
Dotson & Gobein [21]	CDC 6600	FORTRAN	36
Debany *et al.* [20]	Honeywell DPS 8/70M	PL/I	1.94†
Page & Perry [11]	Macintosh II	Pascal	1.22
Page & Perry [12]	BBN Butterfly(1 processor)	C	1.30
Deo & Medidi [12]	BBN Butterfly(1 processor)	C	0.55

†input network is pre-simplified using series-parallel reductions

Table 4.2 Execution Times of Various Algorithms for the Benchmark Network 2

Algorithm	Computer	Language	Time
Buzacott [22, 23]	IBM 4381-k	FORTRAN	54 min.
Provan & Ball [22, 24]	IBM 4381-k	FORTRAN	8 min.
Bailey & Kulkarni [22]	IBM 4381-k	FORTRAN	86 sec.
Page & Perry [25]	Macintosh Plus	Pascal	61 sec.
Page & Perry [11]	Macintosh II	Pascal	15 sec.
Page & Perry [12]	BBN Butterfly†	C	14.12 sec.
Deo & Medidi [12]	BBN Butterfly†	C	7.89 sec.

†with 1 processor

Table 4.3 Time required for Complete Networks of Various sizes

Vertex	Time(sec.)		
n	Reduce & Partition	Factoring	Factoring†
5	0.07	0.13	0.07
6	0.12	0.25	0.30
7	0.47	1.09	1.70
8	2.97	8.30	10.60
9	24.07	69.87	76.90

†Page & Perry [11] implementation

4.4 Parallel Algorithms

4.4.1 Factoring

The computations of the subproblems $Rel(G - e)$ and $Rel(G * e)$, to be solved recursively to compute $Rel(G)$ by factoring, are independent and can thus be performed concurrently. Before computing either of the subproblems, the parallel algorithm checks if there is a free processor. If there is, one of these two subproblems is handed to the free processor for concurrent computation. Otherwise, these two subproblems are computed one after another by the processor which generated them. The Parallel Factoring algorithm, based on this observation, is given in the form of a recursive function PARALLEL_FACTOR. This function uses a subroutine PATH1(G, *found*), which checks if there is an s-t path in G using breadth-first search and sets the variable *found* true if there is an s-t path. A formal description of the breadth-first search procedure is in [26, page 350].

```
function PARALLEL_FACTOR( network G with s and t )
{returns the terminal-pair reliability of the network G}
begin
multiplier := 1.0;
Simplify G, using reduction rules, if applicable;
{multiplier may get modified if reductions are performed}
if s = t then
  return multiplier;
else
  PATH1( G, found ); {checks if there is any s-t path in G}
  if found then
    select an edge e out of s, with success probability p_e;
    if a free processor is available then
      do in parallel
        R1 := PARALLEL_FACTOR( G-e );
        R2 := PARALLEL_FACTOR( G*e );
      {a free processor executes one of the above two tasks}
    else
      R1 := PARALLEL_FACTOR( G-e );
      R2 := PARALLEL_FACTOR( G*e );
    end if
    return ( multiplier × [(1-p_e) × R1 + p_e × R2] );
  else
    return 0.0; {there was no s-t path in G}
  end if
end if
end {PARALLEL_FACTOR}
```

In our implementation, an adjacency matrix was used to represent the network. The given success probabilities of the edges in the network are stored in a linear array. The edge out of s on a shortest s-t path, which is guaranteed not to be irrelevant, was used as the keystone edge. Page & Perry implementation, on the other hand, selects an arbitrary edge going into t as the keystone edge. Such an edge may turn out to be irrelevant, as shown in the example network in [11, Figure 4].

4.4.2 Dotson and Gobein

Dotson & Gobein algorithm essentially deletes an event from a queue, partitions the event further, and inserts the new events back into the queue. Since the events to be processed are mutually disjoint, processing on one event is independent of other available events and therefore the events can be processed simultaneously. This suggests a coarse-grain parallel algorithm with each processor grabbing an event if it can, processing the event and generating new events, if necessary. The event queue Q, that is accessible to all processors, is allocated in the shared-memory to hold the events generated but not yet processed. A formal description of the parallelized Dotson & Gobein algorithm is given as a function PARALLEL_DOTSON. The subroutine PATH2(G, *found*, *path*) used in this algorithm checks whether there is an s-t path in the network G and sets the variable *found* true if there exists a path. It also returns the sequence of edges in a shortest s-t path as the linear array, *path*. This subroutine is a simple extension of the breadth-first search procedure.

```
function PARALLEL_DOTSON( network G with s and t )
{returns the terminal-pair reliability of the network G}
begin
R := 0.0; {R, a global variable, is the terminal-pair reliability of G}
P := [ ]; {event corresponding to all the subgraphs of input network}
initialize the event queue Q with event P;
busy_procs := 0; {denotes how many processors are still busy}
while true do in parallel {all processors execute this loop simultaneously}
    lock Q;
    if Q is not empty then
        delete event E from Q;
        got_work := true;
        atomic increment busy_procs by 1;
    else
        got_work := false;
        if busy_procs = 0 then
            break; {empty queue and all processors idling denotes termination}
        end if
```

```
        end if
        unlock Q;
        if got_work then
           create network G_E from G according to E;
           PATH2( G_E, found, path);
           if found then
              E1 := SUCCESS_EVENT( E, path );
              {success partition in E due to path}
              lock R; R := R + PROBABILITY( E1 ); unlock R;
              {add the probability of occurrence of subgraphs in E1 to Rel(G)}
              COMPLEMENTARY_EVENTS( E, E1 );
              {generate all complementary events of E1 in E and insert them in Q}
              atomic decrement busy_procs by 1;
           end if
        end if
end while
return R; {the terminal-pair reliability of the network G}
end {PARALLEL_DOTSON}
```

In our implementation of this algorithm, the network was represented by an adjacency matrix along with a linear array of records about each edge. Each record contains the head vertex, tail vertex and the success probability of the edge. These records do not change during the computation and are thus static. Each event was represented by a linear array of size m, the number of edges in the network. The event queue is accessed by all the processors and is referenced only in a critical section. This bottleneck does not, however, degrade significantly the performance of Parallel Dotson & Gobein algorithm because the time spent by processors in the queue is quite small compared to the time to process an event. The *atomic* operations in the function allow only one processor at a time to access the critical variable *busy_procs*. This ensures the proper termination of the algorithm.

4.4.3 Reduce and Partition

The sequential Reduce & Partition algorithm computes $Rel(G)$ by recursively computing reliabilities of l subnetworks if there is an s-t path of length l in G. Since these l subproblems are independent of each other, they can be computed simultaneously. Before trying to solve any subproblem, the parallel algorithm checks for a free processor. If there is, the subproblem is handed to the free processor for computation. If there are no free processors, the subproblem is solved on the processor which generated it. A formal description of Parallel Reduce & Partition algorithm follows.

```
procedure PAR_REDUCE&PARTITION( network G with s and t, multiplier )
{this procedure adds (multiplier × terminal-pair reliability of G) to the
global variable R}
begin
Reduce G, using reduction rules, if applicable;
{multiplier may get modified if reductions are performed}
PATH3( G, found, path, length ); {returns the shortest s-t path in G}
{ if s = t, this returns length = 0 and found = true }
if found then
   for i := 1 to length do
        edge e := path[i];
        p_e := success probability of edge e;
        G := G − e;
        if a free processor is available then
           create task PAR_REDUCE&PARTITION( G, multiplier × (1 −p_e ));
           {the free processor executes this task concurrently}
        else
           PAR_REDUCE&PARTITION( G, multiplier × (1 − p_e ));
        end if
        G := G * e;
        multiplier := multiplier × p_e ;
   end for
   lock R; R := R + multiplier; unlock R;
   {variable R is the terminal-pair reliability of the input network}
end if
end {PAR_REDUCE&PARTITION}
```

The data structures used in the sequential Reduce & Partition algorithm are also used in implementing the Parallel Reduce & Partition algorithm. The overhead in the parallel task generation calls provided by the BBN's Uniform System is avoided by using a queue where the free processors are busy waiting for parallel tasks. Any processor generating a parallel task puts the parallel task parameters in this queue. The variable R, representing the terminal-pair reliability of the input network, is shared by all processors and is referenced only in a critical section.

4.5 Performance Comparison

All three parallel algorithms, Parallel Factoring, Parallel Dotson & Gobein and Parallel Reduce & Partition, were run on the BBN Butterfly with varying number of processors. Speedup (S_p) is the ratio of the execution time of the parallel algorithm on 1 processor to the execution time of the parallel algorithm running on p processors. Table 4.4 gives the execution times of the three parallel

Table 4.4 Execution Times of the Parallel Algorithms on Benchmark Network 1

Processor	Time (sec.)		
p	Reduce & Partition	Factoring	Dotson
1	0.55	1.37	74.01
2	0.42	1.14	37.36
3	0.34	0.64	25.03
4	0.28	0.63	18.86
5	0.26	0.67	15.15
6	0.24	0.66	12.68
7	0.26	0.65	10.88
8	0.22	0.54	9.57
9	0.22	0.49	8.54
10	0.21	0.59	7.72
11	0.21	0.49	7.04
12	0.21	0.41	6.49
13	0.21	0.43	6.01

algorithms when applied to the directed network of Figure 4.1, as the processors are varied from 1 to 13.

For this network, the Parallel Dotson & Gobein algorithm has almost ideal speedup while Parallel Factoring and Parallel Reduce & Partition algorithms do not show as good a speedup, because of the small size of the network. In Parallel Dotson & Gobein, the amount of processing is almost the same for every event in the queue, since the size of the network used for each event remains almost fixed. The processors are idle only when the queue does not contain enough events, which happens only at the very beginning and at the end of the computation. The bulk of the computation, in the middle, is evenly distributed over all the available processors, and results in excellent speedup. On the other hand, in Parallel Factoring and Reduce & Partition, processors get different subnetworks which allow reductions to different degrees. The time required to process these subnetworks varies significantly, making the load distribution across processors uneven. Moreover, the input network in Figure 4.1 is too small for these two parallel algorithms. The overhead due to the parallelization becomes appreciable compared to the sequential run time itself. However, for a given number of processors, the execution times of Parallel Factoring and Reduce & Partition are much less than the execution time of the Parallel Dotson & Gobein algorithm.

For a larger network, such as the one in Figure 4.2, the Parallel Factoring and Reduce & Partition perform better. In Parallel Factoring, processors may remain idle waiting for the parallel tasks (they generated) to terminate. In contrast, no

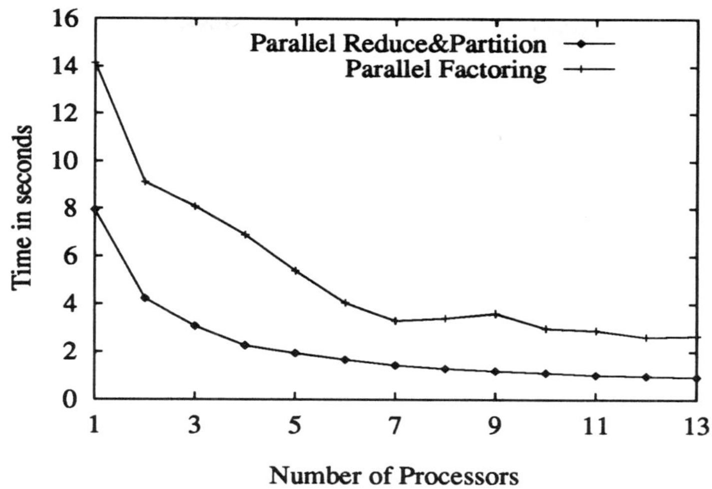

Fig. 4.3 Parallel Execution Times for the Benchmark Network 2

processor waits for any other processor to finish computation in Parallel Reduce & Partition. This improves the performance of Parallel Reduce & Partition significantly over Parallel Factoring as shown in Figure 4.3, where the execution times of these algorithms for the benchmark network in Figure 4.2 are plotted against the number of processors invoked. Parallel Factoring and Reduce & Partition algorithms took 14.14 seconds and 7.93 seconds, respectively, for this network using one processor. In contrast, the execution time of the Dotson & Gobein algorithm for the same network exceeded 10 minutes with 1 processor, and therefore we did not plot its execution times.

Figure 4.4 shows the speedup of Parallel Reduce & Partition on the complete directed networks of 8 and 9 vertices. A single processor required 3.03 seconds and 24.59 seconds, respectively to compute the terminal-pair reliabilities of these networks. This plot again illustrates that speedup of the Parallel Reduce & Partition improves rapidly as the network size increases.

Parallel processing allows computation of the terminal-pair reliability of larger networks, that cannot be handled by sequential algorithms either due to time or memory limitations. To show the relative performance of the three parallel algorithms using more processors, we used a 5×5 grid network in Figure 4.5 as input and employed up to 56 processors. Once again, this undirected network was transformed into an equivalent directed network by replacing each undirected edge by a pair of anti-parallel edges with same reliability.

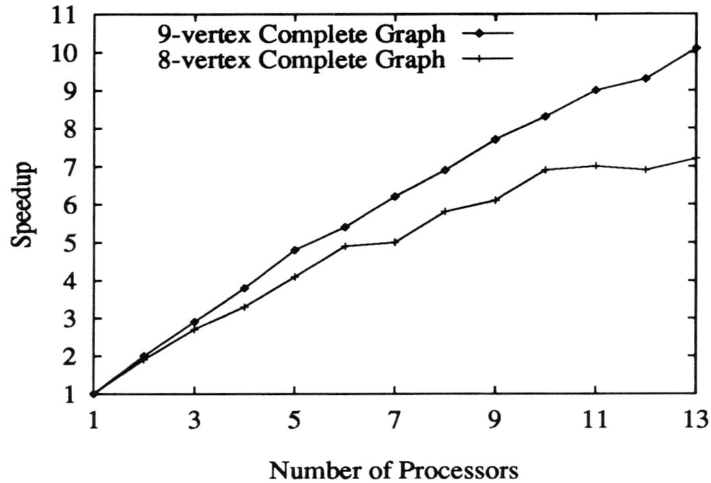

Fig. 4.4 Speedups of Parallel Reduce & Partition Algorithm

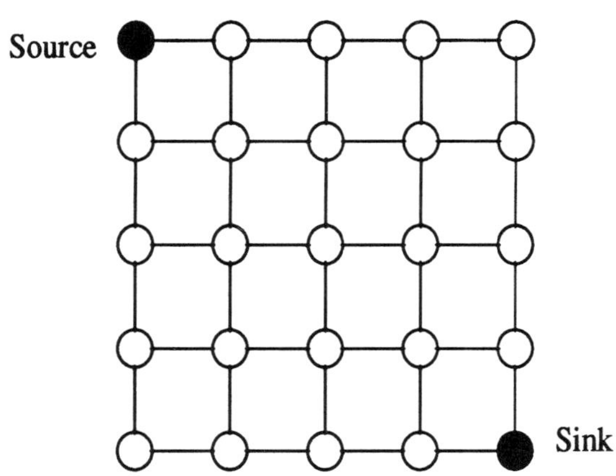

25 vertices

40 undirected edges (80 directed edges)

$$p = \begin{cases} 0.90, & \text{for all horizontal edges} \\ 0.95, & \text{for all vertical edges} \end{cases}$$

s-t reliability = 0.988341

Fig. 4.5 Grid Network

Table 4.5 Execution Times of Parallel Algorithms for the 5 × 5 Grid Network

Processor	Time (sec.)	
p	Reduce & Partition	Factoring
1	1977.93	2538.87
4	514.62	902.64
8	272.92	524.59
12	195.56	381.71
16	155.92	356.66
20	120.34	271.47
24	96.83	197.39
28	87.02	188.46
32	74.70	182.75
36	72.69	126.52
40	60.15	155.25
44	63.80	94.82
48	54.50	95.43
52	54.29	90.00
56	51.78	88.35

The execution times of the Parallel Factoring and Reduce & Partition algorithms are shown in Table 4.5 for the grid. While these recursive algorithms exhaust the search-space tree depth-first, Dotson & Gobein algorithm, which explores the tree in a breadth-first fashion, needs to store all the states in a level in the queue and hence requires a lot of memory. Dotson & Gobein algorithm, on this network, ran out of shared-memory even with the combined memory (224 Megabytes) of 56 processor modules. The speedup of the Parallel Reduce & Partition algorithm on this grid network is plotted in Figure 4.6, which illustrates that this algorithm scales well as the number of processors increase.

4.6 Conclusions

The parallel algorithms, described here, were implemented on a shared-memory parallel computer. Implementing these algorithms on distributed-memory machines like iPSC/2, where communication and synchronization between processors need to be explicitly taken care of, is a significant challenge. Further work is needed in parallelizing other reliability algorithms.

The polynomial-time reliability algorithms on special-classes of networks also need to be parallelized. The effect of network reductions on the complexity of the reliability problem has not been thoroughly studied, except for a small class of networks. Such a study might reveal the extent of parallelism that can be exploited.

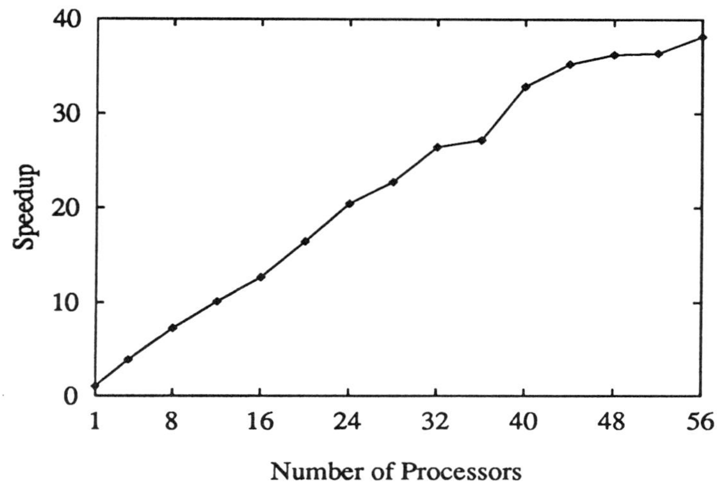

Fig. 4.6 Parallelism of Reduce & Partition, for the Grid Network

REFERENCES

[1] Shier D.G., *Network Reliability and Algebraic Structures*, Oxford University Press, New York, NY, 1991.

[2] Provan J.S. and M.O. Ball, *The Complexity of Counting Cuts and of Computing the Probability that a Graph is Connected*, SIAM J. of Comp., Vol. 12, pp. 777-788, Nov. 1983.

[3] Vertigan D., *The Computational Complexity of Tutte Invariants for Planar Graphs*, Mathematical Institute, University of Oxford, England, 1990.

[4] Misra K.B., *Reliability Analysis and Prediction (A Methodology Oriented Treatment), Elsevier Science Publishers BV, Amsterdam, 1992.*

[5] Politof T. and A. Satyanarayana, *Efficient Algorithms for Reliability Analysis of Planar Networks-a-Survey*, IEEE Trans. on Rel., Vol. R-35, No. 3, pp. 252-259, June 1986.

[6] Yoo Y.B. and N. Deo, *A Comparison of Algorithms for Terminal-pair Reliability*, IEEE Trans. on Rel., Vol. R-37, No. 2, pp. 210-215, June 1988.

[7] Dotson W.P. and J.O. Gobein, *A New Analysis Technique for Probabilistic Graphs*, IEEE Trans. on Circuits and Systems, Vol. CS-26, pp. 855-865, 1979.

[8] Lin P.M., B.J. Leon and T.Ch. Huang, *A New Algorithm for Symbolic System Reliability Analysis*, IEEE Trans. on Rel., Vol. R-25, No. 1, pp. 2-15, April 1976.

[9] Nelson A.C., J.R. Batts and R.L. Beadles, *A Computer Program for Approximating System Reliability*, IEEE Trans. on Rel., Vol. R-19, No. 2, pp. 61-65, Feb. 1970.

[10] Shooman M.L., *Probabilistic Reliability : An Engineering Approach*, McGraw-Hill, New York, NY, 1968.

[11] Page L.B. and J.E. Perry, *Reliability of Directed Networks using the Factoring Theorem*, IEEE Trans. on Rel., Vol. R-38, No. 5, pp. 556-562, Dec. 1989.

[12] Deo N. and M. Medidi, *Parallel Algorithms for Terminal-pair Reliability*, IEEE Trans. on Rel., Vol. R-41, No. 2, pp. 201- 209, June 1992.

[13] Ball M.O. and E.P. Cameron, *Experiments with Network Reliability Analysis Algorithms*, Modeling and Simulation : Proc. 17th Ann. Pittsburg Conf., Vol. 17, pp. 1799-1803, 1986.

[14] Aggarwal K., J. Gupta and K. Misra, *A Simple Method for Reliability Evaluation of a Communication Network*, IEEE Trans. on Comm., Vol. COM-23, pp. 563-566, May 1975.

[15] Theologou O.R. and J.G. Carlier, *Factoring and Reductions for Networks with Imperfect Vertices*, IEEE Trans. on Rel., Vol. R- 40, No. 2, pp. 210-217, June 1991.

[16] Larrabee A.R., K.E. Pennick and S.M. Stern, *BBN Butterfly Parallel Processor*, Programming Parallel Processors, (ed.) R.G. Babb II, Addison-Wesley, Reading, Massachusetts, pp. 43-57, 1988.

[17] Satyanarayana A. and M.K. Chang, *Network Reliability Analysis and the Factoring Theorem*, Networks, Vol. 13, pp. 107-120, 1983.

[18] Wood R.K., *Factoring Algorithms for Computing k-terminal Network Reliability*, IEEE Trans. on Rel., Vol. R-35, No. 3, pp. 269-278, Aug. 1986.

[19] Fratta L. and U.G. Montanari, *A Boolean Algebra Method for Computing the Terminal Reliability in a Communication Network*, IEEE Trans. on Circuit Theory, Vol. CT-20, pp. 203-211, 1973.

[20] Debany W.H., P.K. Varshney and C.R.P. Hartmann, *Network Reliability Evaluation using Probability Expressions*, IEEE Trans. on Rel., Vol. R-35, No. 2, pp. 161-166, June 1986.

[21] Fratta L. and U.G. Montanari, *A Recursive Method Based on Case Analysis for Computing Network Terminal Reliability*, IEEE Trans. on Comm., Vol. COM-26, pp. 1166-1177, 1978.

[22] Bailey M.P. and V.G. Kulkarni, *A Recursive Algorithm for Computing Exact Reliability Measures*, IEEE Trans. on Rel., Vol. R-35, No. 1, pp. 36-40, April 1986.

[23] Buzacott J.A., *A Recursive Algorithm for Finding Reliability Measures Related to the Connection of Nodes in a Graph*, Networks, Vol. 10, pp. 311-327, 1980.

[24] Provan J.S. and M.O. Ball, *Computing Network Reliability in Time Polynomial in the Number of Cuts*, Operations Research, Vol. 32, pp. 516-526, 1984.

[25] Page L.B. and J.E. Perry, *A Practical Implementation of the Factoring Theorem for Network Reliability*, IEEE Trans. on Rel., Vol. R-37, No. 3, pp. 259-267, Aug. 1988.

[26] Horowitz E. and S. Sahni, *Fundamentals of Data Structures in Pascal*, Computer Science Press, New York, NY, pp. 348-350, 1990.

New Trends in System Reliability Evaluation
K.B. Misra (Editor)

CHAPTER 5

RELIABILITY OF k-out-of-n SYSTEMS

Ali M. Rushdi
Department of Electrical and Computer Engineering
King Abdul Aziz University
Jeddah, Saudi Arabia

ABSTRACT

This chapter is an exposition of the recent developments in the evaluation of k-out-of-n system reliability. Careful attention is paid to ensure a clarity of presentation and a uniform treatment of the algorithms surveyed and to point out similarities, differences and interrelations among them. Moreover, all the algorithms discussed are analyzed so as to improve them whenever possible and to set measures of comparison among them. The discussion includes formulas for the case of equal-reliability components, several approximate methods as well as conventional and improved techniques based on inclusion-exclusion, disjoint products, pivoting (factoring), and generating functions. Pivoting leads to some recursive relations based on which several recursive algorithms are developed. The most efficient algorithms described are iterative algorithms that are based on generating functions or binary recursion, and that have nice interpretations in terms of very regular signal flow graphs. In the worst case, the temporal and spatial complexities of these algorithms are shown to be quadratic and linear, respectively, in the number of system components. The chapter concludes with a highlight of new trends and developments in the area of k-out-of-n system reliability and closely related areas.

5.1 Introduction

The k-out-of-n:G(F) system, introduced by Birnbaum et al. [1] in 1961, is a system of n components that functions (fails) if at least k out of its components function (fail) [2]. Situations in which this system serves as a useful model are frequently encountered in practice and include the following examples:

(a) A piece of stranded wire with n strands in which at least k are necessary to pass the required current behaves as a k-out-of-n:G system. The same concept generalizes to applications involving supply-type components with identical fixed ratings for their capacity, flow, throughput, strength or the like, such that system success is achieved when a minimum supply is met (see, e.g., [3, Chapter 3]).

(b) Reactor protection systems, sensor systems, alarm generation systems and other decision mechanisms usually employ a k-out-of-n:G voting logic [4]. Voting is also used in the realization of ultrareliable systems that are based on multichannel computations [5]. Likewise, voting is commonly used in faulty distributed computing systems to achieve mutual exclusion among groups of isolated nodes. For a uniform vote assignment in a complete graph, the system is a k-out-of-n:G system [6, Eq. (1a)] or a simple extension thereof [6, Eq. (1b)].

(c) A bus-structured multiprocessor computer system consists of n processors sharing m memory units via b common buses. If this system is required to operate in MIMD mode (i.e., with Multiple Instruction streams and Multiple Data streams), then it is logically equivalent to the series connection of a k_1-out-of-n:G system, a k_2-out-of-m:G system and a 1-out-of-b:G system, where $k_1 \geq 2$ and $k_2 \geq 1$ with the precise values of k_1 and k_2 being determined by system requirements [7].

(d) In the majority voting (MV) algorithm for managing replicated data, out of n copies of an object $\lceil (n+1)/2 \rceil$ copies must be up to form a quorum [8]. This is an $\lceil (n+1)/2 \rceil$-out-of-n:G system. A generalization of this algorithm, the hierarchical quorum consensus (HQC) algorithm is a multilevel system in which the availability of a group at level i expressed in terms of the availability of its subgroups at level (i+1) constitutes a k_i-out-of-n_i:G system. This means that HQC is nothing but an iterative composition [9, pp. 203-206] of the MV structure. Such a composition improves availability (making HQC superior to MV) provided the basic component availability is higher than a certain value [9, 10].

(e) The k-out-of-n model is useful in the study of multistage interconnection networks [11, 12]. For example, the terminal reliability of a Gamma network [13] is represented by a ladder network (of unreliable nodes and perfect links) whose behavior can be approximated by that of a k-out-of-n system. Specifically, the ladder network in [13, Fig. 5] and in [11, Fig. 1] is logically equivalent to a series connection of two components with a structure that has 8 cut sets of two components each. The reliability of a 2-out-of-6:F system is a lower bound for the reliability of this structure.

The k-out-of-n system has many attractive features. It has a symmetric structure that has many convenient mathematical descriptions such as boolean expressions, recursive equations, generating functions and so on. Nevertheless, for $1<k<n$, the k-out-of-n system lacks a graphical representation in the form of a network or a fault tree [14, 15], unless replicated network edges or fault tree inputs are allowed (see, e.g., [2, p. 333], [9, pp. 184-187] and [4, p. 70]). The k-out-of-n system plays a central role for the general class of coherent systems, as it can be used to approximate the reliability of such systems [16]. While virtually all nontrivial network reliability problems are known to be NP-hard for general networks [15], the regular structure of the k-out-of-n system allows the existence of efficient algorithms for its reliability analysis that are of quadratic-

time linear-space complexity in the worst case [17-20].

The k-out-of-n:G system covers many interesting systems as special cases. These include the perfectly reliable system (k=0), the parallel system (k=1), the voting or N-modular redundancy (NMR) system ($k=\lceil (n+1)/2 \rceil$), the fail-safe system (k=n-1), the series system (k=n), and the totally unreliable system (k=n+1). For 1<k<n, the k-out-of-n system is sometimes called a partially-redundant system [3, 21], as it lies somewhere between the extreme cases of the (non-redundant) series system and the (fully-redundant) parallel system. The k-out-of-n:G system and the k-out-of-n:F system are mirror images of each other; their successes are dual switching functions. The k-out-of-n:G system is exactly equivalent to the (n-k+1)-out-of-n:F system.

The k-out-of-n:G system is a subclass of some important systems. Beside being a coherent system for $1 \le k \le n$, it is

(a) a threshold system with a common positive weight w for each of its n components and a threshold equal to k times w [22],

(b) a special case { r = n } of the consecutive-(n-k+1)-out-of-r-from-n:F system [23],

(c) a limiting case { l = n } of the generally noncoherent k-to-l-out-of-n system [16, 24, 25].

(d) a special case $\{k_1=n-k+1,\ k_2=n+1\}$ of the $\{k_1,\ k_2\}$-out-of-n:F system (which is a system of 2 failure modes that is failed iff at least k_i of its n elements are failed in mode i, for any i).

The k-out-of-n:G(F) system and the consecutive·k-out-of-n:G(F) system [26-28] are not generally comparable in the sense that neither of them is a subclass of the other (except when they overlap at $k \le 1$ and $k \ge n$). The k-out-of-n system is a threshold system, but generally the consecutive-k-out-of-n system is not. The k-out-of-n system is structurally symmetric, i.e., the order of its components is immaterial, while the set of components in a consecutive k-out-of-n system is an ordered one. The reliability of the k-out-of-n:F system is a lower bound of that of the consecutive k-out-of-n:F system.

The k-out-of-n system is being rediscovered in the literature from time to time without being identified as such (see, e.g., [6] and [8], where that system, as well as extensions and compositions thereof appear in disguise). We hope that the present exposition may remedy this situation (at least partially), by familiarizing a large readership with the properties of the k-out-of-n system and the recent developments and future trends in its reliability analysis. Virtually all the major techniques of system reliability analysis have been applied to k-out-of-n systems. The outcome is a potpourri of algorithms that we are going to survey in the next few sections. Careful attention is paid to ensure a clarity of presentation and a uniform treatment of the various algorithms and to point out similarities, differences and interrelations among them. Moreover, all the algorithms discussed are analyzed with the intention of improving them, if possible, beside setting measures of comparison among them. Most of the algorithms discussed are not illustrated via detailed examples, since such examples are already available in the comprehensive text by Misra [2, pp. 332-353]. Though the results of this chapter are relating system reliability to component reliabilities, these results are also applicable in the context of availability.

The organization of the remainder of this chapter is as follows. Sec. 5.2 lists the assumptions, notation and nomenclature employed throughout Chapter 5. Closed-form formulas for k-out-of-n system reliability are given in Sec. 5.3 for the case of equal-reliability components. Approximate methods for the analysis of k-out-of-n systems are summarized in Sec. 5.4. The conventional techniques of inclusion-exclusion and disjoint products, together with some of their improvements or limiting cases are discussed in Sec. 5.5. Another conventional technique, namely that of pivoting or factoring leads to some recursive relations in Sec. 5.6, based on which several recursive algorithms are developed. The generating function method in Sec. 5.7 is the basis of two efficient iterative algorithms, which have an interesting interpretation in terms of recursive relations. Another efficient iterative algorithm based directly on binary recursion is developed in Sec. 5.8. This algorithm is a special case of pivoting, but two offshoots of its development lead to improved versions of the inclusion-exclusion and disjoint products methods. The algorithms surveyed are finally compared in Sec. 5.9. This is followed by a highlight of new trends and developments in Sec. 5.10.

5.2 Assumptions, Notation, and Nomenclature

5.2.1 Assumptions

1. Both the components and the system are 2-state, i.e., either good or failed.
2. Component states are statistically independent.

5.2.2 Notation

n	number of system components, $n \geq 0$.
T	a superscript indicating the transpose of a vector.
X_i, $\overline{X}_i$	indicator variables for successful and unsuccessful operation of component i. These are switching random variables that take only one of the two discrete values 0 and 1; $X_i = 1$, and hence $\overline{X}_i = 0$ iff i is good, while $X_i = 0$ and $\overline{X}_i = 1$ iff i is failed.
X	a vector of n elements representing the component successes; $\mathbf{X} = [\, X_1\ X_2 \ldots X_n\,]^T$.

$S(\mathbf{X}),\bar{S}(\mathbf{X})$	indicator variables for the successful and unsuccessful operation of the system, called system success and system failure, respectively.
$S(k,n,\mathbf{X})$	success of a k-out-of-n:G system of component successes $\mathbf{X}$, $0 \leq k \leq (n+1)$.
$Sy(A,\mathbf{X})$	a symmetric switching function of the switching variables $\mathbf{X}$, with the characteristic set A; $A = \{ a \mid 0 \leq a \leq n$, a is the number of 1's assign-ments of the X_i's for which the function is $1\}$.
Pr[.]	probability of the event [.].
E[.]	expectation of the random variable [.].
p_i,q_i	reliability and unreliability of component i; $p_i = Pr[X_i = 1] = E[X_i] = 1.0 - q_i$. Both p_i and q_i take real values in the closed interval [0.0,1.0].
d_i	the real quotient (q_i/p_i); $p_i \neq 0.0$.
g_i	the real quotient (p_i/q_i); $q_i \neq 0.0$.
$\mathbf{p}$	a vector of n elements representing the component reliabilities; $\mathbf{p} = [\, p_1\; p_2 \cdots p_{m-1}\; p_m\; p_{m+1} \cdots p_n \,]^T$.
$\mathbf{q}$	a vector of n elements representing the component unreliabilities; $\mathbf{q} = [\, q_1\; q_2 \cdots q_{m-1}\; q_m\; q_{m+1} \cdots q_n \,]^T$, $\mathbf{q} = \mathbf{1.0} - \mathbf{p}$, where **1.0** is a vector whose n elements are all ones.
$\mathbf{p}/p_m$	a vector of (n-1) elements obtained by omitting the *m*th element of vector $\mathbf{p}$; $\mathbf{p}/p_m = [\, p_1\; p_2 \cdots p_{m-1}\; p_{m+1} \cdots p_n \,]^T$, $\mathbf{X}/X_m$ is defined similarly.
$\mathbf{p}_{(i)}$	a vector of i elements ($i \leq n$) containing the last i elements of $\mathbf{p}$.
$\mathbf{p}_{<i>}$	a vector of i elements ($i \leq n$) containing the first i elements of $\mathbf{p}$.
$\mathbf{p} \mid i_m$	a special case of $\mathbf{p}$ obtained by setting its *m*th element to i, $\mathbf{p} \mid i_m = [\, p_1\; p_2 \cdots p_{m-1}\; i\; p_{m+1} \cdots p_n \,]^T$, i=0 or i=1.
$R(\mathbf{p}),U(\mathbf{p})$	reliability and unreliability of the system; $R(\mathbf{p})=Pr[S(\mathbf{X})=1]=E[S(\mathbf{X})]$, and $U(\mathbf{p})$ is the complement (to 1.0) of $R(\mathbf{p})$, i.e., $U(\mathbf{p})=1.0-R(\mathbf{p})$. $R(\mathbf{p})$ and $U(\mathbf{p})$ take real values in the closed interval [0.0,1.0].
$R(k,n,\mathbf{p})$	reliability of a k-out-of-n:G system of component reliabilities $\mathbf{p}$, $0 \leq k \leq (n+1)$.
$R(k,n,p)$	the value of $R(k,n,\mathbf{p})$ when component reliabilities are equal, i.e., when $\mathbf{p} = p * \mathbf{1.0}$. (The reader may forgive a little abuse of notation here).
$U(k,n,\mathbf{p})$	unreliability of a k-out-of-n:G system of component reliabilities $\mathbf{p}$, $0 \leq k \leq (n+1)$.

E(k,n,**p**) probability of successful operation of exactly k out of the n components of the system. This can be nonzero only for $0 \le k \le n$, and represents the probability mass function (PMF) of the total number of successes in n Bernoulli trials with variable probabilities;

$$R(k,n,\mathbf{p}) = \sum_{i=k}^{n} E(i,n,\mathbf{p}) . \tag{5.1}$$

U'(k,n,**q**) unreliability of a k-out-of-n:F system of component unreliabilities **q**, $0 \le k \le (n+1)$. This is the complemented dual of R(k,n,**p**) in the sense that the successes of the k-out-of-n:G system and the k-out-of-n:F system are dual switching functions; U(k,n,**p**) = U'(n-k+1,n,**q**).

F(k,n,**q**) probability of failure of exactly k out of the n components of the system. This is the complemented dual of E(k,n,**p**);

$$U'(k,n,\mathbf{q}) = \sum_{i=k}^{n} F(i,n,\mathbf{q}) , \tag{5.2}$$

$$F(k,n,\mathbf{q}) = E(n-k,n,\mathbf{p}) . \tag{5.3}$$

c(k,n) the binomial (combinatorial) coefficient = the number of ways of choosing k elements from a set of n objects, when repetition is not allowed and order does not matter;

binomial coefficients satisfy Pascal's identity

$$c(k,n) = c(k,n-1) + c(k-1,n-1), \quad 0 < k < n, \tag{5.4}$$

together with the boundary conditions

$$c(k,n) = 1, \quad (k=0 \text{ or } k=n) \text{ and } n \ge 0. \tag{5.5}$$

E/F the set difference of sets E and F = $\{j \mid j \in E, j \notin F\}$.

Z_n the set of the first n nonnegative integers = { 0, 1, 2, ..., n-1 }.

$|X|$ cardinality of the finite set X = the number of elements in X.

$\lceil x \rceil$ the ceiling of the real number x = the smallest integer greater than or equal to x.

5.2.3 Nomenclature

Duality: strictly speaking, the dual of a switching function is obtained by complementing the function and all its switching arguments [22]. In the reliability literature, "duality" is freely used to indicate "similarity" or "analogy".

Monotone: a monotone system is one whose reliability function is a non-decreasing function in each component reliability, i.e., $R(\mathbf{p} \mid 1_m) - R(\mathbf{p} \mid 0_m) = \delta R(\mathbf{p})/\delta p_m \ge 0.0$, $1 \le m \le n$.

Relevant: component m is relevant to the system if there exists a valid value for **p** such that $\delta R(\mathbf{p})/\delta p_m \neq 0.0$.
Coherent: a coherent system is a monotone system whose components are all relevant [29].
k-out-of-n:G system: a system that is good iff at least k out of its n components are good.
k-out-of-n:F system: a system that is failed iff at least k out of its n components are failed.
k-out-of-n system: a collective name for k-out-of-n:G and k-out-of-n:F systems; the k-out-of-n:F system is equivalent to the (n-k+1)-out-of-n:G system. The k-out-of-n system is a coherent system in the practical case of $1 \leq k \leq n$, while it is only monotone for the hypothetical limiting cases of k=0 and k=(n+1).
s-p complex: a coherent system is series-parallel (s-p) complex iff it has no components in series or in parallel [30]. A k-out-of-n system is s-p complex for $1<k<n$.
Domination: for a coherent system, domination is the absolute value of p^n in the polynomial of R(p).
Pivoting: by pivoting on component m, system reliability can be written as

$$R(\mathbf{p}) = q_m * R(\mathbf{p} \mid 0_m) + p_m * R(\mathbf{p} \mid 1_m), \tag{5.6}$$

where $R(\mathbf{p} \mid 0_m)$ and $R(\mathbf{p} \mid 1_m)$ are the reliabilities of the minors or subsystems of the original system with respect to component m.
Totally amenable: a coherent system is totally amenable if it has the property that its domination is the sum of the dominations of its minors with respect to every component m, and the same property is possessed by all its minors, minors of minors, etc. [30].

5.3 Systems with Components of Equal Reliabilities

Before tackling the general case of a k-out-of-n:G system whose component reliabilities are not necessarily equal, we digress here a little bit to consider the important special case of a system whose component reliabilities are all equal to the same value p. A closed-form expression for the reliability of such a system is obtained by using the following version of Eq. (5.1)

$$R(k,n,p) = \sum_{i=k}^{n} E(i,n,p), \tag{5.1a}$$

where E(i,n,p) is the probability of the event that exactly i components are successful. This event can occur in c(i,n) ways or sample points, each of which having a probability of $p^i(1-p)^{n-i}$ [31]. Therefore, Eq. (5.1a) reduces to

$$R(k,n,p) = \sum_{i=k}^{n} c(i,n)\, p^i\, (1-p)^{n-i}\,. \tag{5.7a}$$

Eq. (5.7a) is the classical formula for the reliability of a k-out-of-n:G system with components of equal reliabilities. For computational purposes, it can be rewritten as

$$R(k,n,p) = p^k q^{n-k} \sum_{i=k}^{n} c(i,n)\, g^{i-k}\,, \tag{5.7b}$$

where g is the quotient (p/q), $q \neq 0.0$. Alternatively, the sum in Eq. (5.7a) can be computed with the aid of the tables of the binomial distribution or those of the incomplete beta functions [6, 31, 32].

Eq.(5.7a) can be converted into an explicit polynomial in p only as follows. First, the expression

$$(1-p)^{n-i} = \sum_{j=0}^{n-i} c(j,n-i)\,(-1)^j\, p^j\,,$$

as expanded by the binomial theorem, is substituted in Eq. (5.7a) to give

$$R(k,n,p) = \sum_{i=k}^{n} \sum_{j=0}^{n-i} c(i,n)\, c(j,n-i)\,(-1)^j\, p^{(i+j)}$$

$$= \sum_{i=k}^{n} \sum_{m=i}^{n} c(i,n)\, c(m-i,n-i)\,(-1)^{m-i}\, p^m$$

$$= \sum_{i=k}^{n} \sum_{m=i}^{n} c(m,n)\, c(i,m)\,(-1)^{m-i}\, p^m\,. \tag{5.8}$$

The above expression has been simplified by using the binomial identity of Eq. (20) in [33, p. 57]. Now, we interchange the orders of the two summations in Eq. (5.8) to obtain:

$$R(k,n,p) = \sum_{m=k}^{n} c(m,n)\,(-1)^m\, p^m \sum_{i=k}^{m} c(i,m)\,(-1)^{-i}\,. \tag{5.9}$$

The second summation in Eq. (5.9) is simplified via the binomial identity of Eq.(18) in [33, p. 57] or Eq. (12.7) of [32, p. 64] to

$$\sum_{i=k}^{m} c(i,m)\,(-1)^i = (-1)^k\, c(k-1,m-1).$$

Hence, the final expression of R(k,n,p) becomes

$$R(k,n,p) = \sum_{m=k}^{n} (-1)^{m-k}\, c(k-1,m-1)\, c(m,n)\, p^m\,. \tag{5.10}$$

Eq. (5.10) is equivalent to Eq. (5.87) of [2], and is more suitable than Eq. (5.7a) for hand calculations [34].

Since the k-out-of-n:G system is coherent for $1 \le k \le n$, its domination D(k,n) is the absolute value of the coefficient of p^n in the reliability polynomial of R(k,n,p) [30], i.e.,

$$D(k,n) = c(k-1,n-1). \tag{5.11}$$

5.4 Approximate Methods

Standard approximations of the reliability of a coherent system include Bonferroni's inequalities, the min-max bounds, and the minimal cut set and the minimal tie set bounds [35]. The first set of bounds is derivable from the method of inclusion-exclusion (IE), and its application to the k-out-of-n system is outlined within the discussion of the IE method in Sec. 5.5. The other two sets of approximations use the system minsets and when applied to the k-out-of-n:G system they become

$$\max_{A_k} \prod_{a\in A_k} p_a \le R(k,n,\mathbf{p}) \le \min_{A_{n-k+1}} (1 - \prod_{a\in A_{n-k+1}} q_a) , \tag{5.12}$$

$$\prod_{A_{n-k+1}} (1 - \prod_{a\in A_{n-k+1}} q_a) \le R(k,n,\mathbf{p}) \le 1-\prod_{A_k} (1 - \prod_{a\in A_k} p_a) , \tag{5.13}$$

where A_i is a subset of { 1, 2, ..., n } with $|A_i| = i$.

We now consider some of the other approximations derived specifically for k-out-of-n systems. Mostly, these approximations rely on the use of a certain scalar value p_{mean} which represents the mean value of the elements of **p** in some sense. Selecting p_{mean} as the arithmetic mean $<p>$ of component reliabilities, Boland and Proschan [36] reported the following inequalities

$$R(k,n,\mathbf{p}) \ge R(k,n,<p>),$$

$$\text{if } \{ \sum_{i=1}^{n} p_i \ge k\} \text{ and/or } \{p_i \ge (k-1)/(n-1),\ 1\le i\le n\}, \tag{5.14}$$

$$R(k,n,\mathbf{p}) \le R(k,n,<p>),$$

$$\text{if } \{ \sum_{i=1}^{n} p_i \le (k-1)\} \text{ and/or } \{p_i \le (k-1)/(n-1),\ 1\le i\le n\}. \tag{5.15}$$

Alternatively, p_{mean} is selected in [36] as the geometric mean p_g of component reliabilities to give the lower bound

$$R(k,n,\mathbf{p}) \ge R(k,n,p_g), \tag{5.16}$$

which turns into an exact value in the case of a series system (k=n).

Serfling [37] approximated $R(k,n,\mathbf{p})$ using the cumulative distribution function of the Poisson distribution

$$Q(m,z) = e^{-z} \sum_{j=0}^{m} (z^j/j!). \tag{5.17}$$

His results for statistically independent components are

$$Q(n-k,x) - \tfrac{1}{2} \sum_{i=1}^{n} q_i^2 \le R(k,n,\mathbf{p}) \le Q(n-k,x) + \tfrac{1}{2} \sum_{i=1}^{n} q_i^2, \tag{5.18}$$

$$Q(n-k,y) \le R(k,n,\mathbf{p}) \le Q(n-k,y) + \tfrac{1}{2} \sum_{i=1}^{n} \log^2(p_i), \tag{5.19}$$

where

$$x = \sum_{i=1}^{n} q_i , \tag{5.20}$$

$$y = - \sum_{i=1}^{n} \log(p_i). \tag{5.21}$$

5.5 Conventional Methods : Inclusion-Exclusion and Disjoint Products

There are three general methods for system reliability analysis, namely; (1) inclusion-exclusion, (2) sum of disjoint products, and (3) pivoting (pivotal decomposition or factoring) [30]. These three methods, and others, are applicable to the reliability analysis of k-out-of-n systems. Such an application is discussed in the remainder of this section for the first two methods, while the case of pivoting is delegated to Sec. 5.6 since pivoting is closely related to the recursive techniques discussed therein.

5.5.1 Methods of Inclusion-Exclusion

The success of a k-out-of-n:G system is given by the minimal sum-of-products expression

$$S(k,n,\mathbf{X}) = \bigcup_{A_k} P_{A_k} = \bigcup_{j=1}^{m_T} P_j , \tag{5.22}$$

where P_{A_k} is the indicator variable for a minimal path (tie set) of the system, namely

$$P_{A_k} = \bigcap_{a \in A_k} X_a , \tag{5.23}$$

A_k is a subset of $\{1,2,...,n\}$ with $|A_k| = k$, and $m_T = c(k,n)$ is the number of tie sets for the system [2], i.e., the number of prime implicants for its success. Similarly, the system failure is

$$\overline{S}(k,n,\mathbf{X}) = \bigcup_{A_{n-k+1}} C_{A_{n-k+1}} = \bigcup_{j=1}^{m_C} C_j , \tag{5.24}$$

where $C_{A_{n-k+1}}$ is the indicator variable for a minimal cut set of the system, viz.,

$$C_{A_{n-k+1}} = \bigcap_{a \in A_{n-k+1}} \overline{X_a} , \tag{5.25}$$

and, $m_C = c(n-k+1,n)$ is the number of cut sets for the system, i.e., the number of prime implicates for its success.

The formula of inclusion-exclusion (IE) [2, 14, 32] can be used to replace the switching expressions (5.22) and (5.24) by the probabilities of their being equal to one, namely

$$R(k,n,\mathbf{p}) = \sum_{m=1}^{m_T} (-1)^{m-1} \sum_{H_m} Pr\ [(\bigcap_{h \in H_m} P_h) = 1]\ , \tag{5.26}$$

$$U(k,n,\mathbf{q}) = \sum_{m=1}^{m_C} (-1)^{m-1} \sum_{G_m} Pr[(\bigcap_{g \in G_m} C_g) = 1]\ , \tag{5.27}$$

where H_m is a subset of { 1, 2, ..., m_T } with $|H_m| = m$, and G_m is a subset of { 1, 2, ..., m_C } with $|G_m| = m$. The first IE formula (5.26) has a number of terms equal to

$$\sum_{m=1}^{m_T} c(m,m_T) = 2^{m_T} - 1\ , \tag{5.28}$$

while the "dual" IE formula (5.27) has (2^{m_C} -1) terms. These numbers of terms are dramatically large, which means that the IE formulas are extremely inefficient. The tremendous increase in the number of terms in the IE formulas is due to the existence of large numbers of identical terms with opposite/similar signs which eventually cancel/add.

Balagurusamy and Misra [38, 39], and later Heidtmann [14] proposed an improvement of the inclusion-exclusion method (henceforth referred to as improved IE or IMIE), in which the generation of cancelling terms is totally avoided, while the production of additive terms is reduced to that of a single term together with a relatively cheap computation of the multiplicity of such a term. The method is based on a classical formula in probability theory (see, e.g., [32, pp. 98,99,109]) for the probability of realization of at least k events out of n given events. If these n events are selected as

{event that component i is successful}, $1 \le i \le n$,

then the aforementioned probability is equal to R(k,n,**p**) and is given by

$$R(k,n,\mathbf{p}) = \sum_{m=k}^{n} (-1)^{m-k} c(k-1,m-1) * (\sum_{A_m} (\bigcap_{a \in A_m} p_a))\ , \tag{5.29}$$

where, again, A_m is a subset of { 1, 2, ..., n } with $|A_m| = m$. Equation (5.29) is a generalization of (5.10) that avoids the restriction of identical component reliabilities in (5.10). The "dual" of the IMIE formula (5.29) is

$$U(k,n,\mathbf{p}) = \sum_{m=n-k+1}^{n} (-1)^{m-(n-k+1)} c(n-k,m-1) * (\sum_{A_m} (\bigcap_{a \in A_m} q_a))\ . \tag{5.30}$$

The IMIE formula (5.29) produces a number of terms equal to $\sum_{m=k}^{n} c(m,n)$ only. It involves a computation of the binomial coefficients c(k-1,m-1), $k \le m \le n$. We propose herein a method of doing such a computation at the cost of (k-1)*(n-k) additions and no multiplications through an iterative utilization of Pascal's identity (5.4) and its boundary conditions (5.5). This method is similar to that of the summation coefficient matrix in [2, 39], and is achieved by constructing an array of values c(i,j), inclusively bounded in the ij-plane by the four straight lines i = 1, i = (k-1), i = (j-1), i = (j-n+k), which are the edges of a parallelogram with corners (i,j) at (1,2), (k-1,k), (k-1,n-1) and (1,n-k+1). This parallelogram is a part of the well known Pascal's triangle [33]. Fig. 5.1 shows a signal flow graph (SFG) [40, pp. 77-99] for the computations required for k=4 and n=6. In this SFG a black node is a source node of value 1, and all edge transmissions are equal to 1 indicating that there is no need for multiplications. Finally, IMIE is found to require

$$\left(\sum_{m=k}^{n} (1+c(m,n)*(m-1))\right) = \left((n-k+1)+\sum_{m=k}^{n} c(m,n)*(m-1)\right)$$

multiplications and

$$(n-k)+(k-1)*(n-k)+\sum_{m=k}^{n}(c(m,n)-1) = \left((k-1)*(n-k)\ -1+\sum_{m=k}^{n} c(m,n)\right)$$

additions. No attempt has been made to simplify the binomial sums in the above expressions, because there seems to be no simple formulas for them for k>0 [33, p. 64]. The "dual" IMIE formula requires a similar cost but with k being replaced by (n-k+1) and the number of additions being augmented by n. Heidtmann [14] also discussed how to use one or more terms of the summations over m in (5.29) or (5.30) to obtain successive upper and lower bounds for R(k,n,**p**) or U(k,n,**q**). These bounds are improved versions of those obtained from the Bonferroni's inequalities.

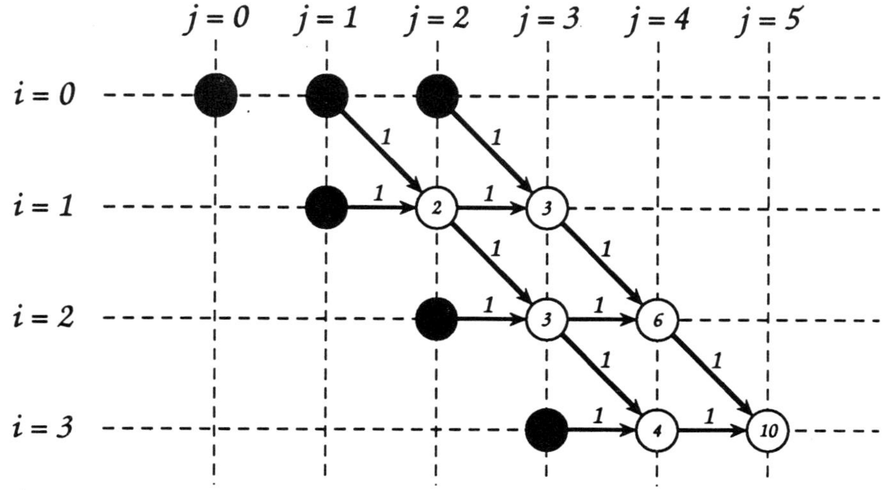

Fig. 5.1 A Signal Flow Graph to Illustrate the Computation of c(3,m-1), 4 ≤ m ≤ 6

5.5.2 Methods of Disjoint Products

The basic idea of the classical technique of disjoint products (DP) is to rewrite the minimal sum-of-products expression of system success (5.22) or failure (5.24) in the form of a disjoint sum of products, thereby allowing its direct conversion, on a one-to-one basis, into the probability domain, i.e., into a reliability or unreliability expression. The literature abounds in rules and algorithms for achieving "disjointness" (see, e.g., [41-43]). Since a k-out-of-n:G system has c(k,n) minimal paths of k components each and also has c(n-k+1,n) minimal cutsets of (n-k+1) components each, it is always possible to work with $c(k_0,n)$ minsets of k_0 components each, where k_0 is either k or (n-k+1). Contrary to a claim made by Heidtmann in [44], it is preferable to choose k_0 equal to the maximum rather than the minimum of k and (n-k+1). Such a choice minimizes $c(k_0,n)$ and hence produces a smaller probability polynomial.

Locks [44] proposed the use of disjoint products (DP) for calculating R(k,n,**p**). He concluded that a DP algorithm will result in a probability polynomial with exactly $c(k_0,n)$ terms, subject to the condition that the original minimal states be arranged in an order such that each state differs from its predecessor by just one component. Though this conclusion is right, the condition demanded for it is not really necessary with well-behaved DP algorithms, and it is not clear whether a general strategy exists for satisfying it. Moreover, the two examples given in [44] both violate that condition despite the fact that exactly $c(k_0,n)$ terms are generated in each of them. Specifically, example 1 of [44] arranges the minimal states of a 2-out-of-4:G system as 12,13,14,23,24,34 of which the two consecutive states 14 and 23 differ in two components rather than one. Likewise, example 2 of [44] lists states 125 and 134 and also states 145 and 234 as consecutive states. Later, Locks [45] rearranged the minimal states of his first example as 12,13,23,24,14,34, thereby meeting his stated condition, but without affecting the final DP formula obtained.

In the following, a new DP algorithm for computing R(k,n,**p**) or U'(n-k+1,n,**q**) is briefly outlined. Arrange the minimal states into a sequence of unordered sets of such states as follows. The first set is to contain a single arbitrarily chosen state that involves k_0 particular components. Construct the second set of the $c(k_0-1,k_0)$ states that result from replacing a single component in the state of the first set by a new component i. The next set is to consist of the $c(k_0-1,k_0+1)$ states which all share another new component j, with the other (k_0-1) components of each state being selected from the (k_0+1) previously introduced components. Continue in this manner until all components are taken into consideration and all minsets are included. Now, rewrite the switching expression of each of the sets of states beyond the first set by factoring out its common variable and replacing the rest of the expression (which turns out to be a symmetric function of a characteristic set $\{(k_0\text{-}1)\}$ by its minterm expansion. The above step does not only guarantee that terms within any set are disjoint among themselves, but will also ensure that terms belonging to different sets are disjoint. This algorithm will keep the number of terms intact, i.e., as $c(k_0,n)$.

The details of the above algorithm are better understood through a specific example. Consider the success function of a 3-out-of-5:G system (which has 10 minsets). According to our DP algorithm, this can be rewritten as

$$\begin{aligned} S(3,5,\mathbf{X}) = \; & X_3X_4X_5 \sqcup X_2(X_3X_4 \sqcup X_3X_5 \sqcup X_4X_5) \\ & \sqcup X_1(X_2X_3 \sqcup X_2X_4 \sqcup X_2X_5 \sqcup X_3X_4 \sqcup X_3X_5 \sqcup X_4X_5) \\ = \; & X_3X_4X_5 \sqcup X_2(Sy(\{2\},X_3,X_4,X_5) \sqcup X_1\, Sy(\{2\},X_2,X_3,X_4,X_5) \\ = \; & X_3X_4X_5 \sqcup X_2\, (X_3X_4\bar{X}_5 \sqcup X_3\bar{X}_4X_5 \sqcup \bar{X}_3X_4X_5) \\ & \sqcup X_1\, (X_2X_3\bar{X}_4\bar{X}_5 \sqcup X_2\bar{X}_3X_4\bar{X}_5 \sqcup X_2\bar{X}_3\bar{X}_4X_5 \sqcup \bar{X}_2X_3X_4\bar{X}_5 \\ & \sqcup \bar{X}_2X_3\bar{X}_4X_5 \sqcup \bar{X}_2\bar{X}_3X_4X_5\,). \end{aligned} \tag{5.31}$$

Now, the expression of $S(3,5,\mathbf{X})$ consists of three sets of terms, such that the terms of one set are disjoint with one another. The term of the first set is disjoint with any term of the other sets, as there exists a variable that appears uncomplemented in the former term and appears complemented in the latter term. Since any term of the second set contains the literal X_2, it is disjoint with those terms of the third set that contain the literal $\bar{X}_2$. The other terms in the third set are

$$X_1X_2\,(X_3\bar{X}_4\bar{X}_5 \sqcup \bar{X}_3X_4\bar{X}_5 \sqcup \bar{X}_3\bar{X}_4X_5) = X_1X_2\; Sy(\{1\},X_3,X_4,X_5),$$

and hence are disjoint with terms of the second set. Finally, Eq. (5.31) can be transformed to the algebraic probability domain [43, rule 1] as

$$\begin{aligned} R(3,5,\mathbf{p}) = \; & p_3p_4p_5 + p_2\,(p_3p_4q_5 + p_3q_4p_5 + q_3p_4p_5) \\ & + p_1\,(p_2p_3q_4q_5 + p_2q_3p_4q_5 + p_2q_3q_4p_5 \\ & + q_2p_3p_4q_5 + q_2p_3q_4p_5 + q_2q_3p_4p_5). \end{aligned} \tag{5.32}$$

The results of the above example are generalized in Sec. 5.8 as an offshoot of the AR algorithm of Rushdi [19]. The same results can be obtained by the manual technique of the variable-entered Karnaugh map [46].

The worst case for the DP method is that of exhaustive state enumeration (ESE) and results when the success/failure function is expressed by a minterm expansion rather than a minimal expression. Since a minterm expansion is a disjoint sum-of-product expression, it is directly convertible, on a one-to-one basis, into the reliability expression.

$$R(k,n,\mathbf{p}) = \sum_{\mathbf{a}} \left(\prod_{j=1}^{n} p_j^{a_j}\, q_j^{(1-a_j)} \right), \tag{5.33}$$

where $\mathbf{a} = [\, a_1\; a_2\; \ldots\; a_n\,]^T$ is a vector of n binary elements a_j of which at least k are 1's, while the rest are 0's. The number of terms in Eq. (5.33) is $\sum_{i=k}^{n} c(i,n)$,

and its computational cost is $((n-1) * \sum_{i=k}^{n} c(i,n))$ multiplications, plus $(n-1 + \sum_{i=k}^{n} c(i,n))$ additions. Again, whenever advantageous, k is to be replaced by (n-k+1) in the above complexity values. These values reach their maxima when k reaches the value (n+1)/2, at which exhaustive enumeration has its worst-case requirement, viz., $((n-1)*2^{n-1})$ multiplications and $(n-1+2^{n-1})$ additions.

McGrady [47] proposed a direct improvement of the exhaustive enumeration technique through a factoring out of the product of component reliabilities from the reliability expression, namely

$$R(k,n,\mathbf{p}) = (\prod_{j=1}^{n} p_j) * (\sum_{J}(\prod_{j\in J} d_j)) , \tag{5.34}$$

where J is a subset of { 1, 2, ..., n } such that $|J| \leq (n-k)$, and $d_i = q_i/p_i$, $p_i \neq 0.0$, $1 \leq i \leq n$. For example, R(3,4,**p**) expressed by exhaustive enumeration is

$$R(3,4,\mathbf{p}) = p_1p_2p_3p_4 + q_1p_2p_3p_4 + p_1q_2p_3p_4 + p_1p_2q_3p_4 + p_1p_2p_3q_4). \tag{5.35}$$

This is reduced by the method of McGrady (MG) to

$$R(3,4,\mathbf{p}) = (\prod_{i=1}^{4} p_i) * (1+d_1+d_2+d_3+d_4) . \tag{5.36}$$

The number of terms in the MG expression remains the same as that in the ESE expression. Multiplications by the p_i's in a term in the ESE expression are eliminated in the corresponding term in the MG expression, while the multiplications by the q_i's are retained in the form of multiplications by d_i's. The MG method requires $(2n+\sum_{i=2}^{n-k}((i-1)*c(i,n)))$ multiplications,

and $(n+\sum_{i=1}^{n-k} c(i,n))$ additions.

The technique proposed by McGrady can be useful with other reliability formulas as well. For example, our expression (5.32) for R(3,5,**p**) can be rewritten in the following simplified form (provided none of p_2, p_3, p_4 and p_5 is equal to 0.0)

$$\begin{aligned} R(3,5,\mathbf{p}) = p_3p_4p_5\,(1 &+ p_2((d_5+d_4+d_3) \\ &+ p_1(d_4d_5 + d_3d_5 + d_3d_4 + d_2d_5 + d_2d_4 + d_2d_3))) \end{aligned} \tag{5.37}$$

Eq. (5.37) is an improved version of a result obtained by Jain and Gopal [48] via their JG-2 algorithm.

5.6 Recursive Techniques

A Shannon's expansion [49] for a symmetric switching function of n variables **X** about one of these variables X_m, $1 \le m \le n$, can be stated as follows

$$Sy(A,\mathbf{X}) = \overline{X}_m\, Sy(B,\mathbf{X}/X_m) \cup X_m\, Sy(C,\mathbf{X}/X_m), \tag{5.38}$$

where the characteristic set A is a subset of $Z_{n+1} = \{0,1, ...,n\}$ and the sets B and C are subsets of $Z_n = \{ 0, 1, ..., n-1 \}$. Let A be given by $A = \{ a_0 , a_1 , ..., a_u \}$, $u \le n$, then, we have

$$\begin{aligned} B &= A \cap Z_n = A && \text{if } a_u \ne n \\ &= A/\{n\} && \text{if } a_u = n , \end{aligned}$$

$$A_1 = \{a_0-1,\ a_1-1,\ \ldots,\ a_u-1\},$$

$$\begin{aligned} C = A_1 \cap Z_n &= A_1 && \text{if } a_0 \ne 0 \\ &= A_1/\{-1\} && \text{if } a_0 = 0 . \end{aligned}$$

Now, the success function of a k-out-of-n:G system is the symmetric switching function [19]

$$S(k,n,\mathbf{X}) = Sy(\{k,k+1, ..., n\} ,\mathbf{X}). \tag{5.39}$$

For $1 \le k \le n$, $S(k,n,\mathbf{X})$ can be expanded via Eq. (5.38) with $A = \{k,k+1,...,n\}$, $B = \{k,k+1,...,n-1\}$ and $C = \{k-1,k,...,n-1\}$, which produces the recursive equation :

$$S(k,n,\mathbf{X}) = \overline{X}_m\, S(k,n-1,\mathbf{X}) \cup X_m\, S(k-1,n-1,\mathbf{X}), \quad 1 \le k \le n, \tag{5.40a}$$

while for the limiting cases k=0 and k=(n+1), $S(k,n,\mathbf{X})$ is given nonrecursively by

$$S(k,n,\mathbf{X}) = Sy(Z_{n+1},\mathbf{X}) = 1, \quad k=0, \quad n \ge 0, \tag{5.40b}$$

$$S(k,n,\mathbf{X}) = Sy(\phi,\mathbf{X}) = 0, \quad k=(n+1),\ n \ge 0, \tag{5.40c}$$

Since the right hand side of Eq. (5.40a) is a disjoint sum of products each term of which is an intersection of statistically independent expressions, it is readily convertible, on a one-to-one basis, into a probability expression. Hence, the following recursive relation for the reliability of a k-out-of-n:G system is obtained

$$R(k,n,\mathbf{p}) = q_m * R(k,n-1,\mathbf{p}/p_m) + p_m * R(k-1,n-1,\mathbf{p}/p_m), \quad 1 \le k \le n, \tag{5.41a}$$

$$= R(k,n-1,\mathbf{p}/p_m) + p_m * (R(k-1,n-1,\mathbf{p}/p_m) - R(k,n-1,\mathbf{p}/p_m)), \quad 1 \le k \le n. \tag{5.41b}$$

The boundary conditions (5.40b) and (5.40c) also translate from the boolean to the algebraic domain as

$$R(k,n,\mathbf{p}) = 1.0, \quad k=0, \quad n \ge 0, \tag{5.41c}$$

$$R(k,n,\mathbf{p}) = 0.0, \quad k=(n+1), \quad n \ge 0. \tag{5.41d}$$

The recursive relations (5.41) can be obtained by applying the pivoting relation (5.6) to the system reliability [30, 50], and then realizing that the two minors or subsystems of a k-out-of-n:G system are k-out-of-(n-1):G and (k-1)-

out-of-(n-1):G systems, respectively [51]. In view of Eqs. (5.4) and (5.11), the domination of the k-out-of-n:G system satisfies

$$D(k,n) = D(k,n-1) + D(k-1,n-1), \quad 1<k<n, \tag{5.42}$$

which means that it is a totally amenable system [30].

Similarly, the failure of a k-out-of-n:G system can be seen to be the symmetric switching function

$$\overline{S}(k,n,\mathbf{X}) = Sy(\{0,1,...,k-1\}, \mathbf{X}), \tag{5.43}$$

and hence can be expanded via (5.38) and then converted to the algebraic domain to yield the recursive relation

$$U(k,n,\mathbf{p})=q_m*U(k,n-1,\mathbf{p}/p_m)+p_m*U(k-1,n-1,\mathbf{p}/p_m), \quad 1\le k\le n, \tag{5.44a}$$

$$=U(k,n-1,\mathbf{p}/p_m)+p_m*(U(k-1,n-1,\mathbf{p}/p_m)-U(k,n-1,\mathbf{p}/p_m)), \quad 1\le k\le n, \tag{5.44b}$$

together with the boundary conditions

$$U(k,n,\mathbf{p}) = 0.0\ , \quad k=0\ , \quad n\ge 0, \tag{5.44c}$$

$$U(k,n,\mathbf{p}) = 1.0\ , \quad k=(n+1)\ , \quad n\ge 0. \tag{5.44d}$$

Eqs. (5.44) can also be obtained directly from Eqs. (5.41) by expressing reliabilities therein as the complements (to 1.0) of unreliabilities.

Appendix 5.1 lists a Pascal-like recursive function that computes the reliability of a k-out-of-n:G system using the relations (5.41a), (5.41c) and (5.41d). Fig. 5.2 illustrates the binary computational tree [30] for computing R(2,4,**p**) using this function. Such a tree is usually termed a recursion tree [52]. It is drawn here in the form of a (Mason) signal flow graph (SFG) [40, pp. 77-99]. In this graph, there is a direct path only from a child node to its parent node, with a path transmission of p_m (q_m) assigned to that path if the child is a right (left) one. In Fig. 5.2 and the following ones, there are two types of nodes:

(a) source nodes of known or precomputed values which are either black, white or cross hatched. A black node has a value of 1.0, a white node has a value of 0.0, while a cross hatched node has a precomputed value in the closed interval [0.0,1.0].

(b) non-source nodes drawn as shaded ones. These are of unknown values which are to be computed by the underlying algorithm. These non-source nodes include (at least) one sink node whose value is the final result sought, beside other nodes whose values are intermediate results.

In Fig. 5.2, the sink node is the root of the tree and the source nodes are the leaves of the tree. Every node has a label in the form (i,j) indicating that it stands for $R(i,j,\mathbf{p}_{<j>})$, $0\le j\le 4$, where $1\le i\le j$ for the non-source nodes, i=0 for the black nodes and i=(j+1) for the white nodes. Fig. 5.2 reveals that the computation of $R(1,1,p_1)$ is repeated three times, while that of $R(1,2,\mathbf{p}_{<2>})$ is duplicated. Unwarranted repetition of computations is one of the demerits of the present recursive algorithm. Moreover, the machine implementation of a recursive function is highly demanding in both time and space [52, 53]. The height of the recursion tree is closely related to the amount of memory that the function requires, while the number of nodes that are non leaves is equal to the number of times the recursive step is done, and hence serves as a measure of the total computation time [52].

The function in Appendix 5.1 computes the reliability of a series/parallel system recursively, i.e., by pivoting on its components, which is wasteful in time and effort [30]. The complexity of the recursive computation of R(k,n,**p**) can be decreased by limiting the use of the recursive relation (5.41a) to 1<k<n, and applying the boundary conditions of the reliabilities of the parallel (k=1) and series (k=n) systems, respectively:

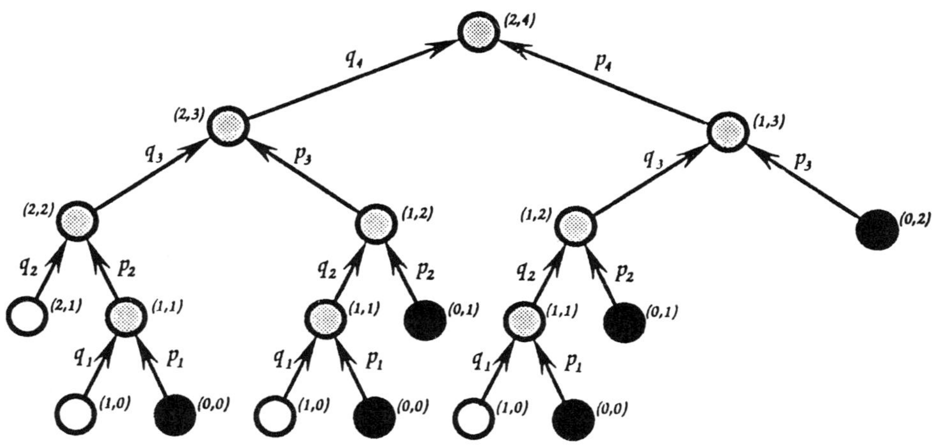

Fig. 5.2 The Binary Computation Tree for Computing R(2,4,p) Via Recursion

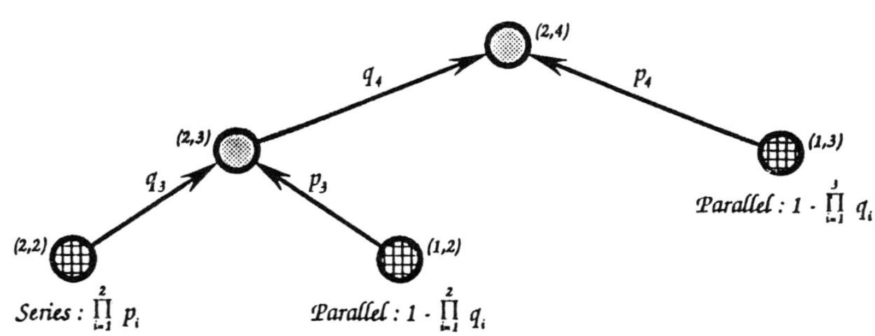

Fig. 5.3 A Variation of the Recursion Tree in Fig. 5.2

$$R(k,n,\mathbf{p}) = 1-\prod_{i=1}^{n} q_i , \qquad k=1, \quad n\geq 0 , \tag{5.45}$$

$$R(k,n,\mathbf{p}) = \prod_{i=1}^{n} p_i , \qquad k=n, \quad n\geq 0 . \tag{5.46}$$

Appendix 5.2 shows the effect of this modification on the function in Appendix 5.1. Figure 5.3 shows the corresponding effect on the tree or SFG of Fig. 5.2. The number of nodes that are non leaves, i.e., the number of shaded nodes representing s-p complex systems is two in this case. In general, the number of non leaves is equal to the number of applications of the pivoting algorithm necessary to reduce the original system to series-parallel minors or subsystems, and hence serves as a measure for the complexity of the pivoting algorithm. This number is less by one than the domination of the system, i.e., it is equal to (c(k-1,n-1) - 1). Therefore, the worst-case complexity of the conventional pivoting algorithm (i.e., direct recursion) is significantly worse than the $O(n^2/4)$ worst-case complexity of the best available algorithms. The pivoting algorithm is a very poor algorithm for computing the reliability of a k-out-of-n system, though such a system is totally amenable [30].

When either of the recursive functions in Appendices 5.1 and 5.2 is run, it sets up a stack to use while traversing the recursion tree. The intermediate results stored on the stack are discarded and a great deal of work duplication occurs when these results are required later. In this case, it is preferable to substitute another data structure for the stack, one that allows references for locations other than the top and does not discard results but retain those required for future use [52]. It will be seen in Sec. 5.8 that such a substitution is possible and leads to recursion removal in this particular case, i.e., to the construction of a purely iterative algorithm that utilizes the underlying recursive equations without any resort to recursion.

Thanks to the similarity between Eqs. (5.41) and (5.44), there exists a version of the function in Appendix 5.1 that computes unreliability instead of reliability with no change whatsoever in computational complexity. Such a version is obtained by replacing "R" wherever it appears by "U" and complementing (to one) the boundary function values for k=0 and k=(n+1). Similarly, the function in Appendix 5.2 can be appropriately modified to compute U instead of R with the same complexity. The fact that the same algorithm computes R(k,n,**p**) and U(k,n,**p**) with identical computational complexities is a characteristic of the present recursive techniques (and of some of the iterative algorithms derived therefrom).

An alternative way for computing R(k,n,**p**) is to compute its complement U'(n-k+1,n,**q**) [2, pp. 332-333]. In this case, the bodies of the functions in Appendices 5.1 and 5.2 remain intact, but the actual inputs corresponding to the dummy arguments k and **p** should have the values of (n-k+1) and **q**, respectively, and the function names may be changed. The complexity of the function in Appendix 5.2 becomes

$$c((n-k+1)-1,n-1) - 1 = c(n-k,n-1) - 1 = c(k-1,n-1) - 1,$$

i.e., it remains the same. For all techniques, there usually exist two versions of any particular method, one for the reliability $R(k,n,\mathbf{p})$ and the other for the unreliability $U'(n-k+1,n,\mathbf{q})$ which may differ (sometimes substantially) in complexity. Of course, the version with the smaller complexity has an advantage, and hence a preference, over the other.

The binary recursive relation (5.41a) can be used to derive composite recursive relations. If $R(k,n-1,\mathbf{p}/p_m)$ in the right hand side of Eq. (5.41a) is recursively expressed via Eq. (5.41a) itself, then a ternary recursive relation is obtained [20], viz.,

$$R(k,n,\mathbf{p}) = q_m *(q_r *R(k,n-2,\mathbf{p}/p_m/p_r) + p_r *R(k-1,n-2,\mathbf{p}/p_m/p_r)) + p_m *R(k-1,n-1,\mathbf{p}/p_m), \quad 1 \le k \le (n-1). \tag{5.47}$$

For the particular case k=n, the following chain recursion is applicable

$$R(k,n,\mathbf{p}) = p_n *R(k-1,n-1,\mathbf{p}/p_n), \quad k=n,\ n \ge 1. \tag{5.48}$$

Eq. (5.47) has been derived in [11] for the special case m=n and r=(n-1) with a reversal of the numbering of system components. The derivation given in [11] differs from the one given herein and is based on the conservative policy of Fratta and Montanari [54] for decomposing an event into a set of (m+1) disjoint events. The value of m=2 used in [11] has produced 3 disjoint events, and consequently led to the ternary recursive relation (5.47). The binary recursive relation (5.41a) corresponds to m=1, for which the conservative and exhaustive policies of [54] actually coincide.

Another composite recursive relation (that is derived in [11] and can be derived from (5.41a) as well) is

$$\begin{aligned} R(k,n,\mathbf{p}) = {} & \Big(\prod_{i=k+1}^{n} q_i\Big) *R(k,k,\mathbf{p}_{<k>}) \\ & + \Big(\prod_{i=k+2}^{n} q_i\Big) *p_{k+1} *R(k-1,k,\mathbf{p}_{<k>}) \\ & + \Big(\prod_{i=k+3}^{n} q_i\Big) *p_{k+2} *R(k-1,k+1,\mathbf{p}_{<k+1>}) \\ & + \ldots + p_n *R(k-1,n-1,\mathbf{p}_{<n-1>}), \quad 1 \le k \le n. \end{aligned} \tag{5.49}$$

Based on the ternary and chain recursions (5.47) and (5.48) and on the composite recursion (5.49) two recursive algorithms for computing $R(k,n,\mathbf{p})$ have been given by Rai et al. [11]. With the exception of the peripheral cases on or near the boundaries k=1 and k=n, these two algorithms cost more in time and space than the algorithms in Appendices 5.1 and 5.2 because the former algorithms traverse ternary or higher-order recursion trees while the latter traverse binary recursion trees of smaller sizes and of less duplications.

So far, we have considered direct recursive relations for R(k,n,**p**) itself or its complement. Alternatively, we may consider recursive relations for the associated probabilities E(i,n,**p**), $k \le i \le n$, in terms of which R(k,n,**p**) can be computed via (5.1). E(k,n,**p**) is the probability of the event that exactly k out of n components are successful. The indicator variable of this event for $0 \le k \le n$ is the symmetric switching function Sy({k},**X**) [19], and thanks to (5.38), it satisfies the recursive relations

$$Sy(\{k\},\mathbf{X}) = \bar{X}_m\, Sy(\{k\},\mathbf{X}/X_m) \cup X_m\, Sy(\{k-1\},\mathbf{X}/X_m), 0<k<n, \quad (5.50a)$$

$$Sy(\{k\},\mathbf{X}) = \bar{X}_m\, Sy(\{k\},\mathbf{X}/X_m), \quad k=0,\ n\ge 1, \quad (5.50b)$$

$$Sy(\{k\},\mathbf{X}) = X_m\, Sy(\{k-1\},\mathbf{X}/X_m), \quad k=n,\ n\ge 1. \quad (5.50c)$$

Therefore, E(k,n,**p**) follows the recursive relations

$$E(k,n,\mathbf{p}) = q_m{*}E(k,n-1,\mathbf{p}/p_m)+p_m{*}E(k-1,n-1,\mathbf{p}/p_m),\ 0\le k\le n,\ n\ge 1 \quad (5.51a)$$

$$= E(k,n-1,\mathbf{p}/p_m)+p_m{*}(E(k-1,n-1,\mathbf{p}/p_m)-E(k,n-1,\mathbf{p}/p_m)),$$

$$0\le k\le n,\ n\ge 1 \quad (5.51b)$$

together with the set of boundary conditions:

$$E(k,n,\mathbf{p}) = 1.0, \qquad k=n=0, \quad (5.51c)$$

$$E(k,n,\mathbf{p}) = 0.0, \qquad k=-1, n\ge 0, \quad (5.51d)$$

$$E(k,n,\mathbf{p}) = 0.0, \qquad k=(n+1), n\ge 0, \quad (5.51e)$$

or the set of boundary conditions

$$E(k,n,\mathbf{p}) = \prod_{i=1}^{n} q_i, \qquad k=0,\ n\ge 1, \quad (5.51f)$$

$$E(k,n,\mathbf{p}) = \prod_{i=1}^{n} p_i, \qquad k=n,\ n\ge 1. \quad (5.51g)$$

Note that E(k,n,**p**) obeys the same recursive relations as R(k,n,**p**) and U(k,n,**p**) but with a different region of validity and a different set of boundary conditions. The "dual" quantity F(k,n,**q**) obeys Eqs. (5.51a) to (5.51g) provided p_i and q_i are interchanged for $1 \le i \le n$. The above recursive relations are implicit in the generating function method of Barlow and Heidtmann [17], and have later appeared in Jain and Gopal [48], Rushdi [19], and Rushdi and Al-Thubaity [55]. The "dual" relations for F(k,n,**q**) have been used by Park [56] for $0<k<n$, together with appropriate boundary conditions at k=0 and k=n.

It is possible to use Eqs. (5.51) recursively to compute certain E(i,n,**p**) values, and hence compute R(k,n,**p**). We have already pointed out that such a computation is possible with the aid of Eq. (5.1). However, to eliminate some of the duplications in computation due to recursion, we replace Eq. (5.1) and partially Eq. (5.51a) by the following relations (which have been used earlier in [17,19,48] and are easily derivable from Eqs.(5.1) and (5.41a))

$$R(k,n,\mathbf{p}) = E(k,n,\mathbf{p})\ ,\quad k=n,\ n\ge 0, \quad (5.52)$$

$$R(k,n,\mathbf{p}) = R(k,n-1,\mathbf{p}/p_n)+p_n{*}E(k-1,n-1,\mathbf{p}/p_n)\ , k<n,\ n\ge 1. \quad (5.53)$$

For certain values of k, it is advantageous to compute R(k,n,**p**) as the complement (to one) of U'(n-k+1,n,**q**). In this case, the formal equations (5.51),

(5.52) and (5.53) are still to be used, but initially **p** is replaced by **1-p** and k is replaced by (n-k+1), and at the end R is replaced by its complement [48].Throughout such a computation, E(i,n,**p**) carries the "dual" of its original meaning, viz., that of F(i,n,**q**).

5.7 Iterative Algorithms Based on a Generating Function: the BH Algorithms

Barlow and Heidtmann [17] reported two programs for computing the k-out-of-n system reliability, usually referred to as the BH-1 and BH-2 algorithms, respectively. The BH-1 algorithm relies on the use of the summation relation (5.1) to express R(k,n,**p**) in terms of E(i,n,**p**), where k≤i≤n. The quantities E(i,n,**p**) are found as the coefficients of z^i in the generating function

$$G_n(z) = \prod_{i=1}^{n} (q_i + p_i z) = \sum_{i=0}^{n} E(i,n,\mathbf{p})\, z^i. \tag{5.54}$$

Appendix 5.3 lists a Pascal-like function that implements the BH-1 algorithm. It has a local memory requirement of (n+2) scalars and it requires M = ((n+1)*(n+2)/2 - 1) multiplications plus 2M+(n-k+1) additions. The BH-1 algorithm has been converted to the BH-2 algorithm by avoiding the calculation of unnecessary values B[i] for i<(j-n+k) and also bypassing the calculation of B[i] for i>k through the use of Eqs. (5.52) and (5.53). Appendix 5.4 lists a Pascal-like function that implements the BH-2 algorithm. It has a local memory requirement of (k+2) scalars. Originally, it was thought to have a computational complexity of the symmetric value k(n-k+1) [17, 19], but actually its complexity is slightly higher and lacks symmetry [20, 57]. In fact, the BH-2 function requires ((k+1)*(n-k+1)-1) multip-lications plus ((2k+1)*(n-k+1)-1) additions. The slight lacking of symmetry in the above values means that if we construct the "dual" of the BH-2 algorithm (one that computes U'(n-k+1,n,**q**) with the aid of F(i,n,**q**)), then it will have the same complexity numbers above but with k replaced by (n-k+1), and consequently it will be of almost but not exactly the same complexity as the original BH-2 algorithm. In the worst case (k ≈ (n+1)/2), the BH-2 algorithm has a linear spatial complexity of (n+5)/2 and a quadratic temporal complexity of ((n+1)*(n+3)/4-1).

Rushdi [19], and later Risse [58] have recognized that the BH algorithms do actually compute E(i,n,**p**) via the recursive equations (5.51a) and their boundary conditions (5.51c)-(5.51e). To stress this point, we give in Figs. 5.4 to 5.7 four signal flow graphs, the former two of which illustrate two versions of the computation of E(3,5,**p**) via the recursive equations (5.51a) together with either of the two sets of boundary conditions (5.51c)-(5.51e) or (5.51f)-(5.51g), while the latter two figures illustrate the computation of R(3,5,**p**) by the BH-1 and BH-2 algorithms respectively. In these four figures, and the following ones, a node of a thin circumference at coordinates (i,j) represents $E(i,j,\mathbf{p}_{<j>})$, while a node of a thick circumference at coordinates (i,j) stands for R(i,j,$\mathbf{p}_{<j>}$). In both the BH-1 and the BH-2 algorithms, nodes are visited columnwise, starting from the leftmost column (j=1) and ending at the rightmost column (j=n), with an additional summation beyond j=n in Fig. 5.6. Within each column j the bottom node is visited first and then followed by upper nodes till the top node is reached. At the end of the traversal of column j in Fig. 5.6, its nodes will contain the probability mass function of the generalized binomial distribution [59, pp. 134-135], which is based on n Bernoulli trials with variable probabilities [32, pp. 218,230,282].

For comparison purposes, we include herein a discussion of the JG-1 algorithm of Jain and Gopal [48]. This algorithm is iterative in nature though it has been called "recursive" in [48]. It has a surprising similarity with the BH-2 algorithm. It has 3 main steps, namely:

1. Compute E(k-2,j,$\mathbf{p}_{<j>}$),(k-1)≤j≤(n-2), iteratively through exhaustive enumeration (using algorithm A of [48] or the more succinct and effective algorithm TR-1 of [58]). The computational cost for evaluating E(k-2,j,$\mathbf{p}_{<j>}$) is ((j-1) * c(k-2,j)) multiplications, plus (c(k-2,j) - 1) additions.

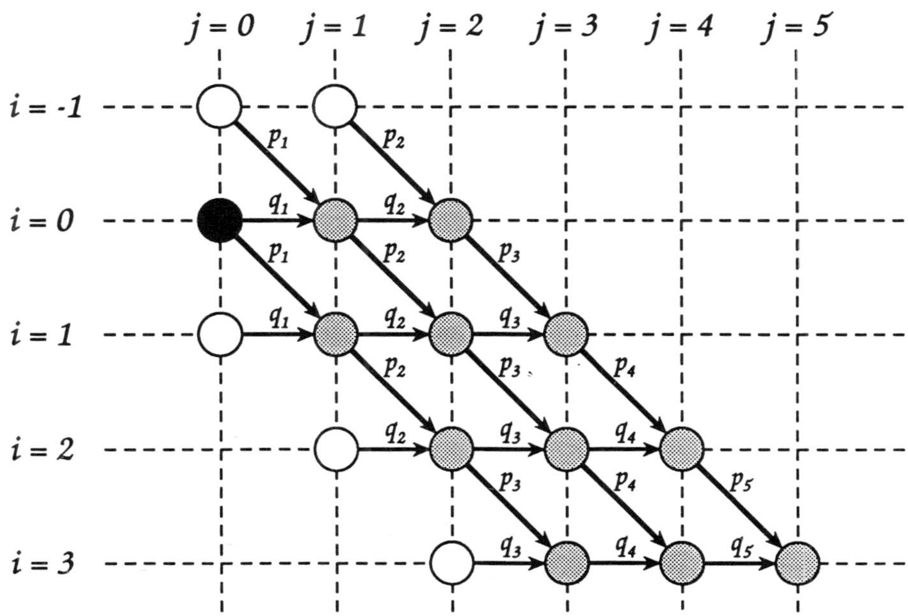

Fig. 5.4 A Signal Flow Graph to Illustrate the Computation of E(3,5,p). If p_i and q_i are interchanged for 1 ≤ i ≤ 5, the graph computes F(3,5,p)

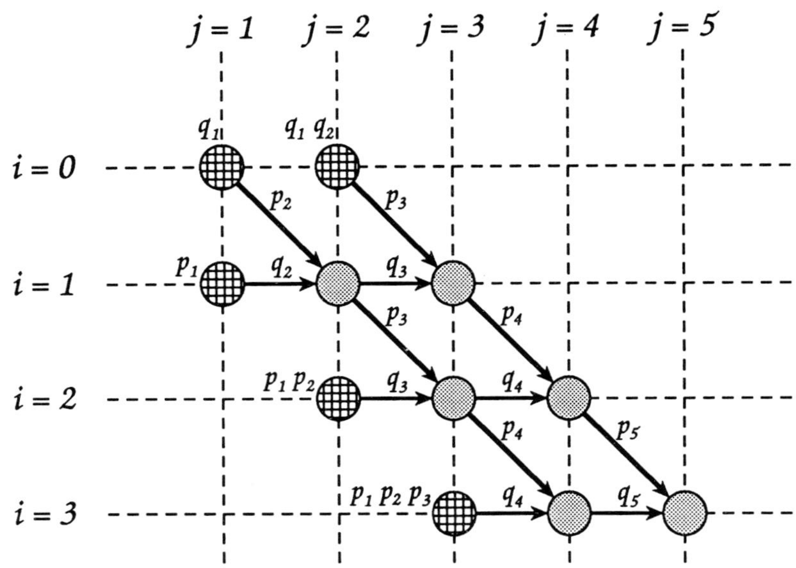

Fig. 5.5 A Variation of the Signal Flow Graph in Fig. 5.4

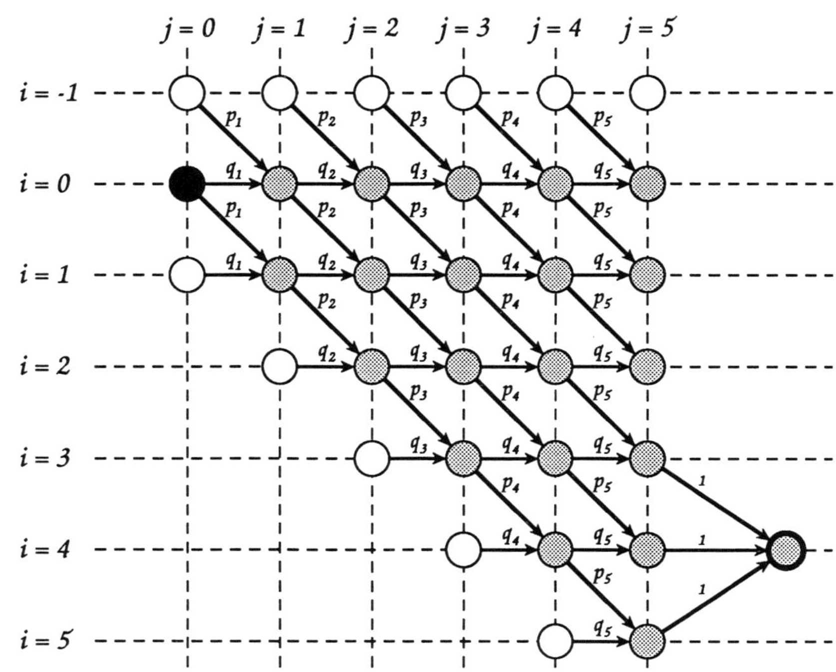

Fig. 5.6 A Signal Flow Graph to Illustrate the Computation of R(3,5,p) by the BH-1 Algorithm

2. Initialize $E(k\text{-}1,k\text{-}1,\mathbf{p}_{<k-1>})$ via (5.51g), and $R(k,k,\mathbf{p}_{<k>})$ via

$$R(k,k,\mathbf{p}_{<k>}) = p_k * E(k\text{-}1,k\text{-}1,\mathbf{p}_{<k-1>}), \quad k\geq 1, \tag{5.55}$$

This costs (k-1) multiplications.

3. Use the recursive relation (5.51a) to compute $E(k\text{-}1,j,\mathbf{p}_{<j>})$, $k\leq j\leq(n\text{-}1)$, and use (5.53) to compute $R(k,j,\mathbf{p}_{<j>})$, $(k+1)\leq j\leq n$. This costs 3(n-k) multiplications plus 2(n-k) additions.

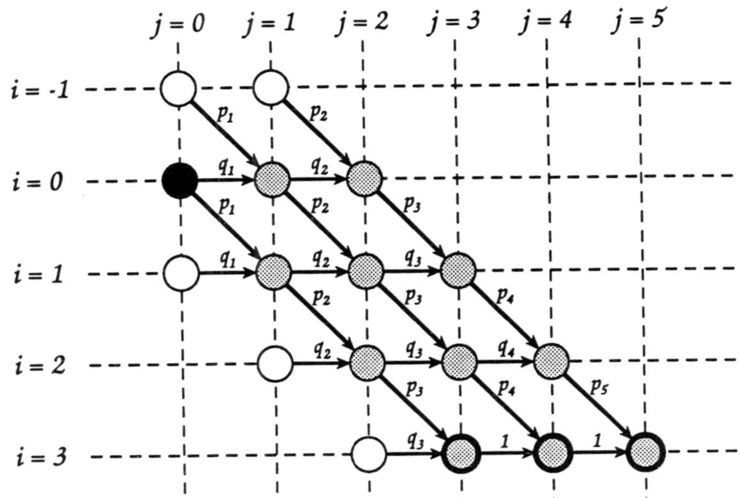

Fig. 5.7 A Signal Flow Graph to Illustrate the Computation of R(3,5,p) by the BH-2 Algorithm

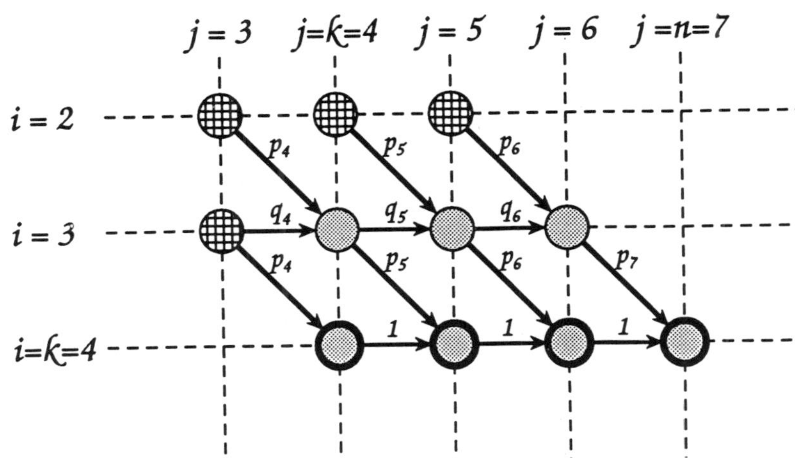

Fig. 5.8 A Signal Flow Graph to Illustrate the Computation of R(4,7,p) by the JG-1 Algorithm

Therefore, the number of multiplications required by the JG-1 algorithm is

$$\sum_{j=k-1}^{n-2} (j-1)\ c(k-2,j)+(k-1)+3(n-k),$$

which can be computed via Eq. (12.16) of [32, p. 65] as

$$(n-2)\ c(k-1,n-1)-\ c(k,n)+\ 3(n-k)+2. \tag{5.56}$$

Eq. (5.56) is equivalent to Eq. (4) of Risse [58]. The number of additions required is

$$n+\left(\sum_{j=k-1}^{n-2} (c(k-2,j)-1)\right)+2(n-k)\ =c(k-1,n-1)+2n-k-1 \tag{5.57}$$

Since algorithm JG-1 has two "dual" versions, the values in (5.56) and (5.57) should be computed with k replaced by (n-k+1) when advantageous.

The iterative structure of algorithm JG-1 is illustrated by the signal flow graph in Fig. 5.8 for the computation of R(4,7,**p**). Clearly, for k≥2 this structure is an exact replica of the bottom three rows of the corresponding graph of the BH-2 algorithm, with the non-source nodes at i=k-2, j=(k-1), k, ..., (n-2) and at i=j=k-1 in the BH-2 case being replaced by source nodes in the JG-1 case. This means that "the recursiveness of the evaluation process has not fully been exploited in JG-1". This fact has been earlier observed by Risse [58], and can be used to explain why (contrary to a claim made in [48]) the JG-1 is less efficient than the BH-2 algorithm. Risse [58], also proposed an improvement of the JG-1 algorithm by storing the results of intermediate multiplications, and thereby reduced the number of multiplications to

$$c(k-1,n-1)+6n-5k-5. \tag{5.58}$$

5.8 An Iterative Algorithm Based on Binary Recursion: The AR Algorithm

Based upon the recursive relation (5.41b) and boundary conditions (5.41c) and (5.41d) an efficient non-recursive algorithm for computing R(k,n,**p**) has been reported by Rushdi [19].For convenient and clear reference, this algorithm has been called the AR algorithm in [20], and it will be so called herein. To compute R(k,n,**p**), the AR algorithm constructs an array of values $R(i,j,\mathbf{p}_{<j>})$,inclusively bounded in the ij-plane by the four straight lines, i = 1, i = k, i = j, i = (j-n+k), which are the edges of a parallelogram with corners (i,j) at (1,1), (k,k),(k,n) and (1,n-k+1). Fig. 5.9 shows a signal flow graph (SFG) for the AR computation of R(3,5,**p**). Following the convention of the previous figures, a black node is a source node of a value 1.0, and a white node is also a source node (but of a value 0.0), while the shaded nodes are non-source nodes to be computed by the AR algorithm. The graph in Fig. 5.9 is called the block walking representation of the binary tree in Fig. 5.2 [60].

A useful feature of this graph is that it can be used to derive and illustrate the ternary, chain, and composite recursions of Eqs. (5.47), (5.48), and (5.49), respectively. Another interesting feature of this graph and also of the graph in Fig. 5.4 is that they are analogous to pass-network realizations of the symmetric switching functions whose expectations are represented by the nodes in these two graphs [61]. Fig. 5.9 may be viewed also as an inclusion-exclusion lattice in the sense described in [2, pp. 337-339], i.e., a diagonal (horizontal) movement in Fig. 5.9 may be interpreted to mean the inclusion (exclusion) of the associated component. With this interpretation, Fig. 5.9 becomes equivalent to Fig. 5.8 of [2], as both can generate all path sets of a k-out-of-n:G system, but while the latter figure has a single start point, the former has multiple start points at the black source nodes. Fig. 5.9 can also perform the "dual" job of generating all cut sets of the system by reversing the interpretations of the two types of movements and using the white source nodes as start points.

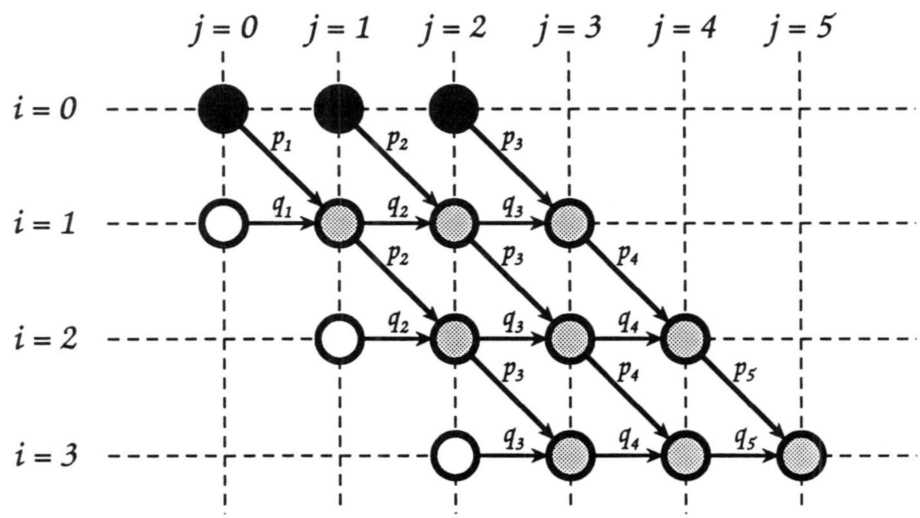

Fig. 5.9 A Signal Flow Graph to Illustrate the Computation of R(3,5,p) by the AR Algorithm

The AR algorithm has three versions depending on the order of traversing or sweeping the aforementioned parallelogram elements, namely:

1- The vertical-sweep version:

Nodes are visited columnwise, starting from the leftmost column (j = 1) and ending at the rightmost column (j = n). Within each column j, the bottom node (i = min(j,k)) is visited first, and then followed by upper nodes till the top node (i = max(1,j-n+k)) is reached.

2- The horizontal-sweep version:

Nodes are visited rowwise, starting from the topmost row (i=1) and ending at the bottom row (i=k). Within each row i, the algorithm proceeds from left (j=i) to right (j=i+n-k).

3- The diagonal-sweep version:

Nodes are visited diagonalwise, starting from the leftmost diagonal (j-i=0) and ending at the rightmost one (j-i=n-k). Within each diagonal, the algorithm proceeds downwards from the top row (i = 1) to the bottom row (i = k).

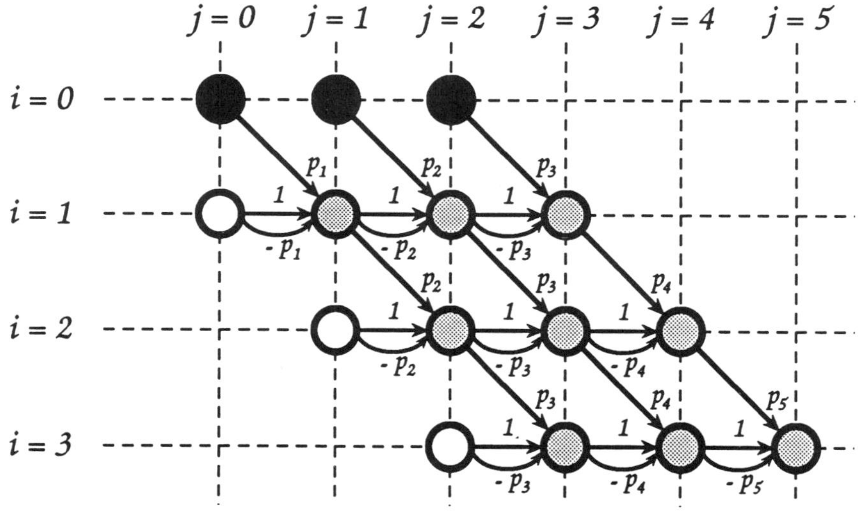

Fig. 5.10 A Variation of the Signal Flow Graph in Fig. 5.9 that can be Used to generate the IMIE Result

Appendix 5.5 lists a Pascal-like function for computing the k-out-of-n system reliability by the vertical-sweep version of the AR algorithm. This function has a local memory requirement of (k+1) scalars. Its temporal complexity is measured by N=k(n-k+1) multiplications and 2N additions. Functions implementing the horizontal-sweep or diagonal-sweep strategies have similar complexities, with the single exception that the horizontal-sweep function has a local memory requirement of (n-k+2) scalars. In the worst case ($k \approx (n+1)/2$), the AR algorithm has a linear spatial complexity of (n+3)/2 and a quadratic temporal complexity of $(n+1)^2/4$. This means that the AR algorithm has a slight advantage over the BH-2 algorithm. Both algorithms are temporally $O(n^2/4)$ and are believed to be optimal [20, 44].

Thanks to the similarity between Eqs. (5.41) and (5.44), there exist "dual" versions of the AR algorithm that compute the unreliability U(k,n,**p**) instead of the reliability R(k,n,**p**) with no change whatsoever in complexity [19]. For example, the function in Appendix 5.5 can be converted into an unreliability function by a simple modification of the initialization of array B, namely by initializing B[0] to 0.0 and initializing all other elements of B to 1.0. Alternatively, that function can be used without any change to compute U'(n-k+1,n,**q**) simply by supplying the values of (n-k+1) and **q** in place of its arguments k and **p**, respectively. In this latter case, the computational cost is the same as that of computing R(k,n,**p**) because of the symmetry of the reliability measure N=k(n-k+1), which is invariant to a replacement of k by (n-k+1). This property is unique to the AR algorithm and to its recursive predecessors (those in Appendices 5.1 and 5.2) and is not shared by other algorithms.

In the limiting cases of a parallel system (k=1) and a series system (k=n), the complexity measure N reduces to n, the parallelogram constructed in the AR algorithm degenerates to a straight line segment, the binary recursion is replaced by a chain one, and the reliability version of the AR algorithm produces Eq. (5.46) for the series system and the expression

$$R(k,n,\mathbf{p}) = p_n+q_n*(p_{n-1}+q_{n-1}*(\ldots*(p_2+q_2*p_1))), \quad k=1,\ n\geq 0, \tag{5.59}$$

for the parallel system, while the unreliability version produces Eq. (5.45) for the parallel system and the expression

$$U(k,n,\mathbf{p}) = q_n+p_n*(q_{n-1}+p_{n-1}*(\ldots*(q_2+p_2*q_1))), \quad k=n,\ n\geq 0, \tag{5.60}$$

for the series system.

Apparently unaware of the work of Rushdi [19, 24, 25], Sarge and Prasad [62] proposed an iterative algorithm for computing R(k,n,**p**) with an iterative tableau structure that duplicates the parallelogram construction of the AR algorithm with minor notational differences. However, the Sarge & Prasad (SP) algorithm has been based on the ternary recursive relation (5.47) rather than the binary recursive relation (5.41a), and hence its computational complexity in the worst case is asymptotically equal to four times that of the AR algorithm [20].

The AR algorithm is not only useful for evaluating the numerical value of R(k,n,**p**), but it can be used for obtaining a symbolic closed-from expression of it as well. To achieve this a careful study of the SFG for computing R(k,n,**p**), as exemplified by Fig. 5.9, is needed. According to Mason's rule [40], the value of the sink node R(k,n,**p**) in this loopless SFG is given by

$$R(k,n,\boldsymbol{p}) = \sum_{i,j} (N_{ij}*T_{ij}), \quad 1 \le k \le n, \tag{5.61}$$

where N_{ij} is the value of the source node (i,j) and T_{ij} is the transmission from source node (i,j) to the sink node. Of course, only the nonzero source nodes (i,j), i=0, $0 \le j \le (n-k)$, need be considered in (5.61). Since each of these nodes has a value of unity, (5.61) simplifies to

$$R(k,n,\boldsymbol{p}) = \sum_{j=0}^{(n-k)} T_{0j}, \quad 1 \le k \le n. \tag{5.62}$$

The transmission T_{0j} is the sum of c(k-1,n-j-1) path transmissions each of which is a product of p_{j+1} with (k-1) p_i's and (n-k-j) q_i's. Therefore, Eq. (5.62) reduces to

$$R(k,n,\boldsymbol{p}) = \sum_{j=0}^{n-k} (p_{j+1}*(\sum_{\boldsymbol{u}(j)} \prod_{i=j+2}^{n} p_i^{u_i}\ q_i^{(1-u_i)})), \quad 1 \le k \le n, \tag{5.63}$$

where $\mathbf{u}(j)=[\ u_{j+2}\ u_{j+3} \ldots u_n\ ]^T$ is a vector of (n-j-1) binary elements, of which (k-1) are 1's and (n-k-j) are 0's. The number of terms in (5.63) is

$$\sum_{j=0}^{n-k} c(k-1,n-j-1) = c(k,n) \tag{5.64}$$

which is the same number that can be obtained by the DP method [44, 45] or the JG-2 algorithm [48]. The "dual" of Eq. (63) has c(n-k+1,n) terms and is given by

$$U(k,n,\boldsymbol{p}) = \sum_{i=1}^{k} (q_i*(\sum_{\boldsymbol{v}(i)} \prod_{j=i+1}^{n} p_j^{(1-v_j)}\ q_j^{v_j})), \quad 1 \le k \le n, \tag{5.65}$$

where $\mathbf{v}(i)=[\ v_{i+1}\ v_{i+2} \ldots v_n\ ]^T$ is a vector of (n-i) binary elements, of which (n-k) are 1's and (k-i) are 0's. Equation (5.63) is equivalent to

$$R(k,n,\boldsymbol{p}) = \sum_{j=0}^{(n-k)} (p_{j+1}*F(n-k-j,\ n-j-1,\ \boldsymbol{p}_{(n-j-1)})), \quad 1 \le k \le n, \tag{5.66}$$

$$= (\prod_{i=n-k+1}^{n} p_i) * \sum_{j=0}^{(n-k)} ((\prod_{i=j+1}^{(n-k)} p_i) * \sum_{X} \prod_{x \in X} d_x), \quad 1 \le k \le n. \tag{5.67}$$

In (5.67) X is a subset of { j+2, j+3, ..., n } with $|X|$ = n-k-j, and $d_x = q_x / p_x$ provided $P_x \neq 0.0$.

If the numbering of system components is reversed, then Eq. (5.66) can be reduced to Eq. (3) of [48]. If fact, Eq. (5.67) is an improved version of Eq. (4) of [48], which is the basis of the JG-2 algorithm. Similarly, Eq. (5.65) can be reduced to the "duals" of Eqs. (5.66) and (5.67), viz.

$$U(k,n,\mathbf{p}) = \sum_{i=1}^{k} (q_i * E(k-i,\ n-i,\ \mathbf{q}_{(n-i)})), \quad 1 \le k \le n, \tag{5.68}$$

$$= (\prod_{j=k}^{n} q_j) * \sum_{i=1}^{k} ((\prod_{j=i+1}^{k-1} q_j) * \sum_{Y} \prod_{y \in Y} g_y), \quad 1 \le k \le n. \tag{5.69}$$

In (5.69) Y is a subset of { i+1, i+2, ..., n } with $|Y| = k-i$, and $g_y = p_y / q_y$ provided $q_y \neq 0.0$. We call Eqs. (5.67) and (5.69) improved DP formulas (IMDP), since they can result from the improved DP procedure outlined in Sec. 5.5.

If the component reliabilities are equal, both Eqs. (5.63) and (5.67) reduce to

$$R(k,n,p) = p^k * \sum_{j=0}^{n-k} (c(k-1,n-j-1) * q^{n-k-j}), \quad 1 \le k \le n, \tag{5.70}$$

while Eqs. (5.65) and (5.69) reduce to

$$U(k,n,p) = q^{n-k+1} * \sum_{i=1}^{k} (c(n-k,n-j-1) * p^{k-i}), \quad 1 \le k \le n. \tag{5.71}$$

Both (5.70) and (5.71) can be shown to be equivalent to (5.10) through the tedious use of some binomial identities.

The IMDP formula (5.67) requires a number of multiplications equal to

$$\begin{aligned} & k+(n-1)+\sum_{j=0}^{n-k-1} ((n-k-j)+(n-k-j-1)*c(n-k-j,n-j-1)) \\ &= (n-k)*c(k,n) - c(k+1,n+1) + k+n+(n-k)*(n-k+1)/2. \end{aligned} \tag{5.72}$$

In (5.72), one division is counted as one multiplication. The sum in (5.72) is evaluated by using the binomial identity (12.16) of [32, p. 65]. The number in (5.72) is remarkably less than the corresponding one for the JG-2 algorithm, as computed by Risse [58, Eq. (6)]. The number of additions required by the IMDP formula is

$$(n-k)+(n-1)+\sum_{j=0}^{n-k-1} (c(n-k-j,n-j-1)-1) = n-2+c(k,n). \tag{5.73}$$

Whenever advantageous, the "dual" IMDP formula (5.69) should replace (5.67) and (n-k+1) is to be substituted for k in the complexity expressions (5.72) and (5.73). The use of (5.67) is restricted to the case $p_i \neq 0.0$, $2 \le i \le n$. If a particular p_i is 0.0, the corresponding component should be disregarded, and (5.67) is to be used with the system considered a k-out-of-the(n-1) remaining components:G system. Similarly, if a certain q_i is 0.0, (5.69) can still be applied to a (k-1)-out-of-the(n-1) remaining components:G system.

The AR algorithm does not only encompass the IMDP technique in its SFG representation, but it also includes the improved inclusion-exclusion (IMIE) technique in that representation. To see this consider Fig. 5.10 which is the same as Fig. 5.9 with the only difference that each horizontal edge with a transmission q_i, $1 \le i \le n$, is replaced by two edges of transmissions 1.0 and $(-p_i)$, respectively. From the unity source nodes to the sink nodes there exist

(1) Paths each with a transmission equal to the product of exactly k p_i's. All possibilities of such paths are exhausted and each of them appears only once, i.e., c(k-1,k-1) times,

(2) Paths each with a transmission equal to the negative of the product of (k+1) p_i's. Again, all combinations of (k+1) p_i's appear, but now each of these is repeated c(k-1,k) = k times,

(3) Paths each with a transmission equal to the product of (k+2) p_i's. Each of these is repeated c(k-1,k+1) times,...,etc.

According to Mason's rule [40], the value of the sink node R(k,n,**p**) in the loopless SFG of Fig. 5.10 is given by the IMIE formula (5.29). Comparison of Figs. 5.9 and 5.10 clearly indicates that the IMIE formula is less efficient than the IMDP formula (5.67).

5.9 Comparison of Various Algorithms

Algorithms available for a certain problem are usually compared in terms of the criteria of correctness, simplicity, amount of work done (temporal complexity), amount of space used (spatial complexity), and optimality [63]. A comparison of the algorithms discussed in the previous sections in terms of the above criteria is now given.

An algorithm is correct if when it is given a valid input it produces the right output in a finite amount of time. Each of the algorithms surveyed is correct, since each can be shown to pass the tests in [64].

To varying degrees, the desirable feature of simplicity is possessed by all the algorithms considered. Each of these algorithms has a straightforward description in terms of either a closed-from formula (sometimes involving some sort of set enumeration) or a few lines of pseudo code. Some algorithms or parts thereof can be described or interpreted in terms of loopless signal flow graphs that are very regular in structure. The simpler algorithms are not necessarily the more efficient. The recursive algorithms are simpler than the iterative ones from the programmer's point of view but they are temporally and spatially less efficient. The simplest and most primitive algorithm of ESE is the least efficient.

To have a measure of the work done by the various algorithms, we note that the operations of multiplications and additions are basic or fundamental to these algorithms. We take the count of these basic operations as a measure of temporal complexity. Actually, the algorithms surveyed perform some other operations such as array references or jumpings to subexpressions [20, 58] in addition to their bookkeeping or overhead, but we assume that the total number of operations performed is approximately proportional to the number of basic operations. Of course, this assumption still needs some experimental verification.

The spatial complexity of an algorithm can be measured by the number of memory cells used by a program implementing the algorithm. We have not stressed the issue of spatial complexity for most of the algorithms surveyed. However, we use this issue as a criterion to exclude from further consideration any algorithms impaired by unusually high memory requirements. This applies to all recursive algorithms which are notoriously known for their extensive memory demands. It also applies to the two refinements offered by Risse [58] to the JG-1 and JG-2 algorithms of [48], as these refinements achieve better temporal complexities at the expense of the additional storage of many intermediate results. For the remaining algorithms, it seems safe to assume that spatial complexity is roughly proportional to temporal complexity as measured by the count of basic operations. Again, the validity of this latter assumption has to be checked out via experimental measurements in typical cases. A fact in support of this assumption is that the AR algorithm and the BH-2 algorithm, which are the best in temporal terms are also good space economizers.

Table 5.1 compares the numbers of multiplications (M) and the numbers of additions (A) for some of the iterative algorithms previously surveyed. The values of k and n used are those in Table 1 of [57]. The algorithms compared include the AR algorithm (Sec. 5.8), the BH-2, the BH-1 and the JG-1 algorithms (Sec. 5.7), the IMDP algorithm (Sec. 5.5 & Sec. 5.8), the IMIE, the MG and the ESE algorithms (Sec.5.5). Excluded from Table 5.1 are the JG-2 algorithm [48] (being superseded by the IMDP formula) and the SP algorithm [62] (being dominated by the AR algorithm). All the algorithms mentioned handle the general case of components with nonidentical reliabilities. For comparison purposes, Table 5.1 contains an entry for the case of independent identically distributed (IID) component successes, as handled by the formula of Eq. (5.10). If the technique of Sec. 5.5 is used for computing the binomial coefficients in Eq. (5.10), its formula has a complexity of $(3n - 2k + 1)$ multiplications, plus $(1 + (n^2 - k^2 + 3n - 3k) / 2)$ additions, while the other IID formula (Eq. (5.7b)) has the same complexity augmented by a single multiplication. Again, (n-k+1) is to replace k when advantageous. In the worst case, the IID formulas require $O(2n)$ multiplications and $O(3n^2/8)$ additions, in comparison with the $O(n^2/4)$ multiplications and $O(n^2/2)$ additions required by the AR algorithm or the BH-2 algorithm.

Apart from the AR algorithm (which has the same complexity when k is replaced by (n-k+1), and the JG-1 algorithm (which has a mixed behavior), all algorithms entered in Table 5.1 have a preference to use k if it is $\geq (n+1)/2$, and to replace k by (n-k+1) otherwise. This observation seems a little bit surprising and even counter-intuitive, but it is in line with the fact that it is preferable to analyze series systems in terms of reliability and to use unreliability in the analysis of parallel systems.

To date, the AR algorithm and the BH-2 algorithm have the least temporal complexity within the class of algorithms that basically use multiplications and additions to compute R(k,n,**p**). These two algorithms are believed [20, 44] to be optimal in the worst case, in the sense that it is unlikely that there is an algorithm in the same class that performs fewer basic operations in the worst case. Similarly, the IID formula (5.10) is believed to describe the optimal algorithm for computing R(k,n,p).

Entries in Table 5.1 indicate that the AR algorithm is suboptimal for the parallel system (k=1) and the series system (k=n). This is because we have counted multiplication by 1.0 and addition of 0.0 as valid operations. A modified version of the AR algorithm (to be called MAR) that explicitly avoids such operations (at the expense of extra overhead and the loss of symmetry) is given as a Pascal-like function in Appendix 5.6. The computational cost for that function is (k(n-k+1)-1) multiplications, and 2k(n - k) additions. The same function can be used to compute unreliability instead of reliability by supplying it with the inputs (n-k+1) and **q** in place of k and **p**, respectively. In this latter case, the computational cost of the method can be obtained from the above values by replacing k by (n-k+1). However, for fairness of comparison with other algorithms, we may augment the cost of the unreliability function by n additions to account for the computation of **q = 1.0 - p**. The reliability function is most suitable for the series system as it demands (n-1) multiplications and no additions, while the unreliability function is most suitable for the parallel system with a cost of (n-1) multiplications and n additions.

Niessen-Gillhaus and Schneeweiss [65] made a practical comparison of four algorithms for general reliability calculations, namely: ESE, IE, DP and pivoting. The results of [65] show that for k-out-of-n:G systems with large n, pivoting is the fastest method, followed closely by DP, then IE, and finally by the catastrophically inferior ESE. These results are in agreement with that part of our results in Table 5.1 that concerns the ranking of the AR, the IMDP, the IMIE and the ESE algorithms. However, it should be noted that AR, IMDP and IMIE correspond only roughly to pivoting, DP and IE, respectively. The former algorithms are derived by tailoring the latter ones specifically to k-out-of-n systems. Another difference between our results and those in [65] is that our results are in terms of estimated basic operation counts, while the results in [65] are reports of actual run times.

Table 5.1 Comparison of the Numbers of Multiplications (M) and the Numbers of Additions (A) for Some Iterative k-out-of-n Algorithms

(Values are given for the better version of the algorithm-Acronyms are given in text)

			IID		AR		BH-1		JG-1		MG	
k	n	n-k+1	M	A	M	A	M	A	M	A	M	A
2	5	4	8	7	8	16	20	42	12	9	10	10
3	5	3	10	12	9	18	20	43	16	12	20	20
6	7	2	10	9	12	24	35	72	26	17	14	14
1	15	15	16	1	15	30	135	271	14	15	30	15
5	9	5	18	35	25	50	54	113	378	82	600	264
5	11	7	20	43	35	70	77	159	1448	226	1397	572
8	15	8	30	92	64	128	135	278	38204	3453	80787	16398
4	20	17	27	62	68	136	230	464	12647	1004	2510	1370

			MAR		BH-2		IMDP		IMIE		ESE	
k	n	n-k+1	M	A	M	A	M	A	M	A	M	A
2	5	4	7	12	9	17	9	8	21	13	24	10
3	5	3	8	12	11	20	16	13	42	19	64	20
6	7	2	11	12	13	25	13	12	43	12	48	14
1	15	15	14	15	15	30	29	14	15	15	14	15
5	9	5	24	40	29	54	318	133	1216	271	2048	264
5	11	7	34	60	39	74	853	339	3689	596	5620	572
8	15	8	63	112	71	135	33656	6448	132236	16432	229376	16398
4	20	17	67	122	71	139	2133	1158	21853	1419	25669	1370

5.10 New Trends and Developments

The problem considered in this chapter is that of evaluating the reliability of a k-out-of-n system with statistically independent components whose reliabilities are not necessarily equal. Some authors believe that this problem has taken more than its share in the literature [44, 58], but this is apparently untrue. It has been only recently that this fundamental problem is satisfactorily understood and solved in terms of really efficient algorithms. What remains (and has been achieved partially), is to extend or generalize this problem in a variety of ways, e.g., by evaluating characteristics of the system other than just its reliability or availability, modifying the underlying assumptions of the model considered, or treating more general classes of systems.

Rushdi and Al-Hindi [10] utilized the AR algorithm in computing and tabulating values for the lower boundary of the region of useful redundancy for k-out-of-n systems, i.e., the region in which system reliability is better than that of a single component. If the component lifetimes are exponentially distributed, then this lower boundary can be used to determine the point that divides the time axis into intervals of short and long missions, respectively [31]. In a different direction, Rushdi and Al-Thubaity [55] modified the AR algorithm to obtain efficient algorithms for computing the first-order sensitivity of k-out-of-n system reliability. These algorithms are useful in computing several criticality or importance measures and in evaluating the instantaneous failure rate of the system. Currently, an investigation is made toward the utilization of these algorithms in assessing the uncertainty in k-out-of-n system reliability given measures of uncertainty in component reliabilities.

Another current problem of interest is to study the characteristics of k-out-of-n systems under a set of assumptions other than the one employed throughout this chapter. Examples of these include k-out-of-n systems with statistically dependent components [2, 66-68], common-cause failures [69], two failure modes [70-71], repair [2, 72-79] or spares [80-83]. It may prove fruitful to investigate the possibility of modifying some of the techniques discussed herein to suit the various modifications in the underlying assumptions.

A third major area of interest is that of reliability analysis of systems that are generalizations of the k-out-of-n system. The AR algorithm has been generalized [24, 25] to compute the reliability of the k-to-l-out-of-n system. The reliability of a threshold system has been found to satisfy a recursive relation similar to that of $R(k,n,\mathbf{p})$ [22]. Hence, it can be computed by a recursive algorithm [22] which still needs to be converted into a more efficient iterative one. The design of an efficient algorithm for computing the reliability of a consecutive-k-out-of-r-from-n system seems to be a challenging formidable task since such a system includes both the k-out-of-n and the consecutive k-out-of-n systems as special cases.

REFERENCES

[1] Birnbaum Z.W., J. D. Esary and S. Saunders, *Multi-Component Systems and Structures and their Reliability*, Technometrics, Vol. 3, No. 1, pp. 55-77, Feb. 1961.

[2] Misra K. B., *Reliability Analysis and Prediction : A Methodolgy Oriented Treatment*, Elsevier Science Publishers, Amsterdam, The Netherlands, 1992.

[3] Billinton R. and R. N. Allan, *Reliability Evaluation of Engineering Systems : Concepts and Techniques*, Plenum Press, New York,1983.

[4] Henley E.J. and H. Kumamoto, *Designing for Reliability and Safety Control.* Prentice-Hall, Englewood Cliffs, NJ,1985.

[5] Parhami B., *Voting Networks*, IEEE Trans. on Rel., Vol. 40 , No. 3, pp. 380-394, Aug. 1991.

[6] Barbara D. and H. Garcia-Molina, *The Reliability of Voting Mechanisms*, IEEE Trans. on Computers, Vol. C-36, No. 10, pp. 1197-1208, Oct. 1987.

[7] Hwang K. and T.P. Chang, *Combinatorial Reliability Analysis of Multiprocessor Computers*, IEEE Trans. on Rel., Vol. R-31, No. 5, pp. 469-473, Dec. 1982.

[8] Kumar A., *Hierarchical Quorum Consensus : A New Algorithm for Managing Replicated Data*, IEEE Trans. on Computers, Vol. C-40, No. 9, pp. 996-1004, Sep. 1991.

[9] Kaufmann A., D. Grouchko and R. Cruon, *Mathematical Methods for the Study of the Reliability of Systems,* Academic Press, New York, 1977.

[10] Rushdi A.M. and K. A. Al-Hindi, *A Table for the Lower Boundary of the Region of Useful Redundancy for k-out-of-n Systems*, Microelectronics and Reliability, Vol. 33, No.7, pp. 979-992, 1993.

[11] Rai S., A. K. Sarje, E. V. Prasad and A. Kumar, *Two Recursive Algorithms for Computing the Reliability of k-out-of-n Systems*, IEEE Trans. on Rel., Vol. R-36, No. 2, pp. 261-265, June 1987.

[12] Rai S. and D. P. Agrawal, *Distributed Computing Network Reliability*, IEEE Computer Society Press, Los Alamitos, CA, 1990.

[13] Parker D.S. and C. S. Raghavendra, *The Gamma Network*, IEEE Trans. on Computers, Vol. C-33, No. 4, pp. 367-372, Apr. 1984.

[14] Heidtmann K.D., *Improved Method of Inclusion-Exclusion Applied to k-out-of-n Systems*, IEEE Trans. on Rel., Vol. R-31, No. 1, pp. 36-40, Apr. 1982.

[15] Agrawal A. and R. E. Barlow, *A Survey of Network Reliability and Domination Theory*, Operations Research, Vol. 32, No. 3, pp. 478-492, May-June 1984.

[16] Heidtmann K.D., *Bounds on Reliability of a Noncoherent System Using its Length and Width*, IEEE Trans. on Rel., Vol. R-31, No. 5, pp. 424-427, Dec. 1982.

[17] Barlow R. E.and K. D. Heidtmann, *Computing k-out-of-n System Reliability*, IEEE Trans. on Rel., Vol. R-33, No. 4, pp. 322-323, Oct. 1984.

[18] Zhegalov S.I., *A Method of Calculating the Probability of Failure-free Operation of a System with Different Elements*, Soviet J. Computer Systems Sci. (Formerly, Engrg. Cybernetics), Vol. 24, No. 1, pp. 147-149, 1986.

[19] Rushdi A.M., *Utilization of Symmetric Switching Functions in the Computation of k-out-of-n System Reliability*, Microelectronics and Reliability, Vol. 26, No. 5, pp. 973-987, 1986

[20] Rushdi A.M., *Comment on : An Efficient Non-recursive Algorithm for Computing the Reliability of k-out-of-n Systems*, IEEE Trans. on Rel., Vol. 40, No .1, pp. 60-61, Apr. 1991.

[21] Nowicki D., *Reliability Allocation with Partial Redundancy*, Proceedings of the Annual Rel. & Maintain. Symposium, pp. 400-404, 1991.

[22] Rushdi A. M., *Threshold Systems and their Reliability*, Microelectronics and Reliability, Vol. 30, No. 2, pp. 299-312, 1990.

[23] Papastavridis S.G. and M. E. Sfakianakis, *Optimal Arrangement and Importance of th Components in a Consecutive-k-out-of-r-from-n:F System*, IEEE Trans. on Rel., Vol. R-40 No. 3, pp. 277-279, Aug. 1991.

[24] Rushdi A.M. *Efficient Computation of k-to-l-out-of-n System Reliability,* Reliabilit Engineering, Vol. 17, No. 3, pp. 157-163, 1987, Erratum : ibid, Vol. 19, No. 4, p. 321, 1987

[25] Rushdi A.M. and *F. M.-A. Dehlawi, Optimal Computation of k-to-l-out-of-n Syster Reliability,* Microelectronics and Reliability, Vol. 27, No. 5, pp. 875-896, 1987, Erratu : ibid, Vol. 28, No. 4, p. 671, 1988.

[26] Rushdi A.M., *A Switching Algebraic Analysis of Consecutive-k-out-of-n:F System* Microelectronics and Reliability, Vol. 27, No. 1, pp. 171-174, 1987.

[27] Rushdi A.M., *A Switching Algebraic Analysis of Circular Consecutive-k-out-of-n: Systems,* Rel. Eng. and Syst. Safety, Vol. 21, No. 2, pp. 119-127, 1988.

[28] Kuo W, W. Zhang and M. Zuo, *A Consecutive-k-out-of-n:G System : The Mirror Imag of a Consecutive-k-out-of-n:F System,* IEEE Trans. on Rel., Vol. R-39, No. 2, pp. 244-25 June 1990.

[29] Bergman B., *On Reliability Theory and its Applications (with discussion),* Scand. Statist., Vol. 12, No.1, pp. 1-41, 1985.

[30] Barlow R.E. and S. Iyer, *Computational Complexity of Systems and the Reliabili Polynomial,* Probability in Eng. and Informational Sci., Vol. 2, pp. 461-469, 1988.

[31] Trivedi K.S., *Probability and Statistics with Reliability, Queuing, and Computer Scien Applications,* Prentice-Hall, Englewood Cliffs, NJ, 1982.

[32] Feller W., *An Introduction to Probability Theory and its Applications, Vol. I, Thi Edition,* John Wiley & Sons, New York, 1968.

[33] Knuth D. E., *The Art of Computer Programming, Vol. 1 : Fundamental Algorithm Second Edition.* Addison-Wesley, Reading, MA, 1973.

[34] Smotherman M., R. M. Geist and K. S. Trivedi, *Provably Conservative Approximatio to Complex Reliability Models,* IEEE Trans. on Computers, Vol. C-35, No. 4, p 333-338, Apr. 1986.

[35] Barlow R.E. and F. Proschan, *Statistical Theory of Reliability and Life Testing,* Ho Rinehart and Winston, New York. 1975.

[36] Boland P.J. and F. Proschan, *Computing the Reliability of k-out-of-n Systems,* in M. Abdel-Hameed, E. Cinlar and J. Quinn (editors), Reliability Theory and Models, p 243-255, Academic Press, Orlando, FL, 1984.

[37] Serfling R.J., *The role of the Poisson Distribution in Approximating System Reliabil of k-out-of-n Structures,* in C. P. Tsokos and I. N. Shimi, The Theory and Applicatio of Reliability, Vol. 1, pp. 243-258, Academic Press, New York, 1977.

[38] Balagurusamy E. and K. B. Misra, *Reliability and Mean Life of a Parallel System w Nonidentical Units,* IEEE Trans. on Rel., Vol. R-24, No. 5, pp. 340-341, Dec. 1975.

[39] Balagurusamy E. and K. B. Misra, *Reliability Calculation of Redundant Systems w Non-identical Units,* Microelectronics and Reliability, Vol. 15, No. 2, pp. 135-138, 197

[40] Kuo B.C., *Automatic Control Systems,* Sixth Edition, Prentice-Hall, Englewood Clif NJ, 1991.

[41] Heidtmann K.D., *Smaller Sums of Disjoint Products by Subproduct Inversion,* IE Trans. on Rel., Vol. R-38, No. 3, pp. 305-311, Aug. 1989.

[42] Veeraraghavan M. and K. S. Trivedi, *An Improved Algorithm for Symbolic Reliabi Analysis,* IEEE Trans. on Rel., Vol. R-40, No. 3, pp. 347-358, Aug. 1991.

[43] Rushdi A.M. and A. A. Abdul Ghani, *A Comparison Between Reliability Analyses Bas Primarily on Disjointness or Statistical Independence : The case of the Generali INDRA Network,* Microelectronics and Reliability., Vol. 33, No.7, pp. 965-978, 1993.

44] Locks M.O. and K. D. Heidtmann, *Comments and reply on : Improved Method of Inclusion-Exclusion Applied to k-out-of-n Systems*, IEEE Trans. on Rel., Vol. R-33, No.4, pp. 321-322, Oct. 1984.

45] Locks M.O., *Recent Developments in Computing of System Reliability*, IEEE Trans. on Rel., Vol. R-34, No. 5, pp. 425-436, Dec. 1985.

46] Rushdi A.M., *Symbolic Reliability Analysis with the Aid of Variable-Entered Karnaugh Maps*, IEEE Trans. on Rel., Vol. R-32, No. 2, pp. 134-139, June 1983.

47] McGrady P.W., *The Availability of a k-out-of-n:G Network*, IEEE Trans. on Rel., Vol. R-34, No.5, pp. 451-452, Dec. 1985.

48] Jain S.P. and K. Gopal, *Recursive Algorithm for Reliability Evaluation of k-out-of-n:G System*, IEEE Trans. on Rel., Vol. R-34, No. 2, pp. 144-146, June 1985, Corrections : ibid, Vol. R-34, No. 4, p. 355, Oct. 1985.

49] Lee S.C., *Modern Switching Theory and Digital Design*, Prentice-Hall, Englewood Cliffs, NJ, 1978.

50] Misra K.B., *An algorithm for the Reliability Evaluation of Redundant Networks*, IEEE Trans. on Rel., Vol. R-19, No. 4, pp. 146-151, Nov. 1970.

51] Shen K. and M. Xie, *The Increase of Reliability of k-out-of-n Systems Through Improving a Component*, Rel. Eng. and Syst. Safety, Vol. 26, No. 3, pp. 189-195, 1989.

52] Kruse R.L., *Programming with Data Structures : Pascal Version.* Prentice-Hall, Englewood Cliffs, NJ, 1989.

53] Abbott W.R., *Comment on : Solving Nontrivial Recursive Equations on a Microcomputer*, IEEE Trans. on Rel., Vol. R-35, No. 5, pp. 616-617, Dec. 1986.

54] Fratta L. and G. Montanari, *A Recursive Method Based on Case Analysis for Computing Network Terminal Reliability*, IEEE Trans. on Com., Vol. Com-26, No. 8, pp. 1166-1177, Aug. 1978.

5] Rushdi A.M. and A. O. Al-Thubaity, *Efficient Computation of the Sensitivity of k-out-of-n System* Reliability, Microelectronics and Reliability, Vol. 33, No.15, 1993.

6] Park K.S., *Solving Nontrivial Recursive Equations on a Microcomputer*, IEEE Trans. on Rel., Vol. R-34, No. 5, pp. 561-562, Dec. 1985.

7] Pham P. and S. J. Upadhyaya, *The Efficiency of Computing the Reliability of k-out-of-n Systems*, IEEE Trans. on Rel., Vol. R- 37, No. 5, pp. 521-523, Dec. 1988.

8] Risse T., *On the Evaluation of the Reliability of k-out-of-n Systems*, IEEE Trans. on Rel., Vol. R-36, No. 4, pp. 433-435, Oct. 1987.

9] Fisz M., *Probability Theory and Mathematical Statistics*, Third Edition, John Wiley & Sons, New York, 1963.

0] Chang M.-F., W. Shi and W. K. Fuchs, *Optimal Diagnosis Procedures for k-out-of-n Structures*, IEEE Trans. on Computers, Vol. C-39, No. 4, pp. 559-564, Apr. 1990.

1] Unger S.H., *The Essence of Logic Circuits*, Prentice-Hall, Englewood Cliffs, NJ, 1989.

2] Sarje A.K. and E. V. Prasad, *An Efficient Non-recursive Algorithm for Computing the Reliability of k-out-of-n Systems*, IEEE Trans. on Rel., Vol. R-38, No. 2, pp. 234-235, June 1989.

3] Baase S., *Computer Algorithms : Introduction to Design and Analysis*, Addison-Wesley, Reading, MA, 1978.

4] Rushdi A.M., *How to Hand-check a Symbolic Reliability Expression*, IEEE Trans. on Rel., Vol. R-32, No. 5, pp. 402-408, Dec. 1983.

5] Niessen-Gillhaus U. and W. Schneeweiss, *A Practical Comparison of Several Algorithms for Reliability Calculations*, Rel. Engg. and Syst. Safety, Vol. 31, No. 3, pp. 309-319, 1991.

6] Misra K.B. and E. Balagurusamy, *Reliability Analysis of k-out-of-m:G Systems with Dependent Failures*, Int. J. Systems Sci., Vol. 7, No. 8, pp. 853-861, Aug. 1976.

[67] Scheuer E.M., *Reliability of an m-out-of-n System when Component Failure Induce Higher Failure Rates in Survivors*, IEEE Trans. on Rel., Vol. R- 37, No. 1, pp. 73-74 Apr. 1988.

[68] Schneeweiss W.G., *Usefulness of MTTF of s-independent Case in other Cases*, IEEE Trans. on Rel., Vol. R-41, No. 2, pp. 196-200, June 1992.

[69] Bai D.S., W. Y. Yun and S. W. Chung, *Redundancy Optimization of k-out-of-n System with Common-Cause failures*, IEEE Trans. on Rel., Vol. R-40, No. 1, pp. 56-59, Ap 1991.

[70] Phillips M.J., *k-out-of-n:G Systems are Preferable*, IEEE Trans. on Rel., Vol. R-29, No. 2 pp. 166-169, June 1980.

[71] Pham H. and M. Pham, *Optimal Designs of {k,n-k+1}-out-of-n:F Systems (Subject to Failure Modes)*, IEEE Trans. on Rel., Vol. R-40, No. 5, pp. 559-562, Dec. 1991.

[72] Linton D.G. and J. G. Saw, *Reliability Analysis of the k-out-of-n:F System*, IEEE Trans on Rel., Vol. R-23, No. 2, pp. 97-103, June 1974.

[73] Balagurusamy E. and K. B. Misra, *Availability and Failure Frequency of Repairab m-order Systems*, Int. J. Systems Sci., Vol. 7, No. 11, pp. 1209-1215, Nov. 1976.

[74] Gupta H. and J. Sharma, *A Method of Symbolic Steady State Availability Evaluation k-out-of-n:G System*, IEEE Trans. Reliab., Vol. R-28, No. 1, pp. 56-57, Apr. 1979.

[75] Kullstam P., *Availability, MTBF, MTTR for Repairable m-out-of-n System*, IEEE Tran on Rel., Vol. R-30, No. 4, pp. 393-394, Oct. 1981.

[76] Naik D.N., *Estimating the Parameters of a 2-out-of-3:F System*, IEEE Trans. on Rel., Vo R-30, No. 5, pp. 464-465, Dec. 1981.

[77] Kenyon R.L. and R. J. Newell, *Steady-State Availability of k-out-of-n:G System wi Single Repair*, IEEE Trans. on Rel., Vol. R-32, No. 2, pp. 188-189, June 1983.

[78] Hevesh A.H., *Comments on : Steady-State Availability of k-out-of-n:G System with Sing Repair*, IEEE Trans. on Rel., Vol. R-33, No. 4, p. 324, Oct. 1984.

[79] Angus J.E., *On Computing MTBF for a k-out-of-n:G Repairable System*, IEEE Tran on Rel., Vol. R-37, No. 3, pp. 312-313, Aug. 1988.

[80] Srivastava V.K. and A. Fahim, k-out-of-m System Availability with Minimum-Co Allocation of Spares, IEEE Trans. on Rel., Vol. R-37, No. 3, pp. 287-292, Aug. 1988.

[81] Srivastava V.K. and A. Fahim, *An Enhanced Integer Simplical Optimization Meth for Minimum Cost Spares for k-out-of-n Systems*, IEEE Trans. on Rel., Vol. R-40, No. pp. 265-270, Aug. 1991.

[82] Morrison M. and S. Munshi, *Availability of a -out-of-m+r:G System*, IEEE Trans. on Re Vol. R-30, No. 2, pp. 200-201, June 1981.

[83] Gupta H. and J. Sharma, *State Transition Matrix and Transition Diagram of k-out-of-n: System with Spares*, IEEE Trans. on Rel., Vol. R-30, No. 4, pp. 395-397, Oct. 1981.

[84] She J. and M.G. Pecht, *Reliability of a k-out-of-n Warm-standby System*, IEEE Trans. Rel., Vol. R-41, No. 1, pp. 72-75, Mar. 1992.

Appendix 5.1

A Pascal-like function that computes R(k,n,**p**)
using binary recursion

```
function R(k,n:integer;p:array [1..n] of real):real;
{validate 0≤k≤(n+1), 0.0≤p[i]≤1.0 for all i}
    function R1(k,n:integer):real;
    begin
     if k=0
         then  R1 := 1.0
```

```
            else if   k=n+1
                then  R1 := 0.0
                else  R1 := (1.0-p[n])*R1(k,n-1)+p[n]*R1(k-1,n-1)
    end;
begin
  R := R1(k,n)
end;
```

Appendix 5.2

A modification of the function in Appendix 5.1 that uses serial/parallel systems as boundary conditions

```
function R(k,n:integer;p:array [1..n] of real):real;
{validate 1≤k≤n, 0.0≤p[i]≤1.0 for all i}
    function R1(k,n:integer):real;
    var i:integer; series,parallel:real;
    begin
      if k=1
          then begin
                  parallel:= 1.0;
                  for i:=1 to n do
                      parallel:= parallel*(1.0-p[i]);
                  R1:= 1.0-parallel
                  end
          else if k=n
                  then begin
                      series:=1.0;
                      for  i:=1 to n do
                              series:= series*p[i];
                              R1:=series
                      end
                  else R1 := (1.0-p[n])*R1(k,n-1)+p[n]*R1(k-1,n-1)
    end;
begin
  R := R1(k,n)
end;
```

Appendix 5.3

A Pascal-like function for computing R(k,n,**p**) by the BH-1 algorithm

```
function R(k,n:integer;p:array [1..n] of real):real;
    {validate 0≤k≤(n+1), 0.0≤p[i]≤1.0 for all i}
    var i,j:integer;
        B: array [-1..n] of real; sum:real;
    begin
          B[-1]:=0.0;
          B[0]:=1.0;
          for i:=1 to n do
                  B[i]:=0.0;
          for j:=1 to n do
```

```
                begin {update column}
                  for i:=j downto 0 do
                      B[i]:=B[i]+(B[i-1]-B[i])*p[j]
                end; {update column}
        sum:=0.0;
        for i:=k to n do
            sum:=sum+B[i];
        R :=sum
end;
```

Appendix 5.4

A Pascal-like function for computing R(k,n,**p**) by the BH-2 algorithm

```
function R(k,n:integer;p:array [1..n] of real):real;
{validate 0≤k≤(n+1), 0.0≤p[i]≤1.0 for all i}
var i,j,nk,Top,Bot:integer;
   B: array [-1..k] of real;
begin
        nk:=n-k;
        B[-1]:=0.0;
        B[0]:=1.0;
        for i:=1 to k do
                B[i]:=0.0;
        Top:=0;
        for j:=1 to n do
                begin {update column}
                  Bot:=j;
                  if j>nk then Top:=j-nk;
                  if j>k then
                     begin
                            Bot:=k-1;
                            B[k]:=B[k]+B[k-1]*P[j];
                     end;
                  for i:=Bot downto Top do
                      B[i]:=B[i]+(B[i-1]-B[i])*p[j]
                end; {update column}
        R := B[k]
end;
```

Appendix 5.5

A Pascal-like function that computes R(k,n,**p**) by the column-sweep version of the AR algorithm

```
function R(k,n:integer; p:array[1..n] of real):real;
{validate 0≤k≤(n+1), 0.0≤p[i]≤1.0 for all i}
var i,j,nk,nk1,Top,Bot:integer;
   B: array[0..k] of real;
begin
        nk:=n-k;
```

```
        nk1:=nk+1;
        B[0]:=1.0;
        for i:=1 to k do
                B[i]:=0.0;
        Top:=1;
        for j:=1 to n do
                begin {update column}
                  Bot:=j;
                  if j>nk1 then Top:=j-nk;
                  if j>k then Bot:=k;
                  for i:=Bot downto Top do
                      B[i]:=B[i]+(B[i-1]-B[i])*p[j]
                end; {update column}
        R := B[k]
end;
```

Appendix 5.6

A Pascal-like function that computes R(k,n,**p**) by a modified column-sweep version of the AR algorithm

```
function R(k,n:integer; p:array[1..n] of real):real;
{validate 0≤k≤n, 0.0≤p[i]≤1.0 for all i}
var i,j,nk,nk1,Top,Bot:integer;
    B: array[0..k] of real;
begin
        nk:=n-k;
        nk1:=nk+1;
        B[0]:=1.0;
        B[1]:=p[1];
        Top:=1;
        for j:=2 to n do
                begin {update column}
                  if j>nk1 then Top:=j-nk;
                  if j>k then
                          begin
                            Bot:=k;
                            for i:=Bot downto Top do
                            B[i]:=B[i]+(B[i-1]-B[i])*p[j]
                          end
                  else
                          begin
                            Bot:=j-1;
                            B[j]:=B[Bot]*p[j]
                            for i:=Bot downto Top do
                                  B[i]:=B[i]+(B[i-1]-B[i])*p[j]
                          end
                end; {update column}
        R := B[k]
end;
```

New Trends in System Reliability Evaluation
K.B. Misra (Editor)
1993 Elsevier Science Publishers B.V.

CHAPTER 6

CONSECUTIVE-k-out-of-n SYSTEMS

Stavros G. Papastavridis and Markos V. Koutras
Department of Mathematics,
University of Athens,
Greece

ABSTRACT

A consecutive-k-out-of-n:F system fails if and only if at least k successive components of the system are simultaneously down. Since its introduction, a decade ago, there has been an exponential growth of the literature on this subject and several modifications, extensions and generalizations have been proposed. Due to the fact that most of them are of special interest in the study of real life engineering systems, it is necessary to take stock of the state-of-art. The present chapter conducts a survey of this rapidly expanding area, focussing mainly on the most recent results.

6.1 Introduction

The consecutive-k-out-of-n systems were introduced by Kontoleon [82] and since then the study of their reliability characteristics and all kinds of allied topics have attracted considerable interest, as the continuously expanding literature on them indicates. These systems are of special interest in the study of telecommunication and pipeline networks, vacuum systems in accelerators, computer network structures, space relay stations and several other fields such as quality control and inspection procedures, statistical run tests etc.

The basic models are described as follows: Consider a system consisting of n components ordered serially on a line, i.e., first, second, ..., i-th,...,n-th. Each component and the system can only be in one of the two states; operational (working, up, good) or failed (down). The n components of the system work independently of each other. In the classical consecutive-k-out-of-n systems and their related structures, roughly speaking, the state of the system depends on the state of certain subsets of consecutive components. In the following definition, three specific structures of that type are described.

Definition 6.1.1. A system of n components ordered on a line, is called

(a) Consecutive-k-out-of-n system, if it fails, whenever at least k consecutive components fail

(b) *m-consecutive-k-out-of-n system*, if it fails whenever at least m non-overlapping failure runs of length k (groups of k consecutive components, consisting entirely of failed components) occur.

(c) *r-within-consecutive-k-out-of-n system*, if it fails, whenever there exist k consecutive components, which include among them at least r failed ones.

It is obvious that the consecutive-1-out-of-n system is a series system while the consecutive-n-out-of-n system is a parallel system. For $1<k<n$ the reliability of a consecutive-k-out-of-n system lies between the reliabilities of the respective series and parallel structures, increasing as k increases from 1 to n. On the other hand, the m-consecutive-1-out-of-n system is the classical m-out-of-n system.

A modification of the *linear* arrangement of the n components, gives the *circular* version of the three systems defined above.

Definition 6.1.2. Assume that the first component of the system becomes next to the n-th component, so that the set of the n components forms a clockwise circular arrangement. Considering the three systems whose definitions of failure are verbatim the ones given in Definition 6.1.1 we obtain respectively the following new circular structures:

(a) *Circular consecutive-k-out-of-n system*

(b) *Circular m-consecutive-k-out-of-n system*

(c) *Circular r-within-consecutive-k-out-of-n system.*

Research has shown so far that the circular systems are considerably more difficult to study than their linear versions. Clearly, the 1-consecutive-k-out-of-n system and the k-within-consecutive-k-out-of-n system coincide with the consecutive-k-out-of-n system and the same remark applies to the circular counterparts too.

A very general way to create new coherent structures out of old ones is the consideration of the *dual coherent structure*. If $\varphi(x_1,...,x_n)$ is the structure function of a coherent structure then the structure function of the dual system is given by $1-\varphi(1-x_1,1-x_2,...,1-x_n)$ (for details see Barlow and Proschan [7]). In words, a dual system is in good state if and only if, by making its good components into failed ones and vice versa, then the corresponding original structure becomes failed. So if a system has n components with respective failure probabilities $q_1,q_2,...,q_n$ and $F(q_1,...,q_n)$ is its failure probability, then the failure probability of its dual is given by $1-F(1-q_1,1-q_2,...,1-q_n)$. This relation allows the immediate translation of most results concerning the reliability of a system to corresponding results about its dual. A notable exception to that are some questions of optimization.

Definitions 6.1.1 and 6.1.2 above introduce six coherent structures. Next we introduce their duals.

Definition 6.1.3. Correspondingly to the six systems introduced via Definitions 6.1.1 and 6.1.2 we define another six coherent systems below, namely

(a) *Consecutive-k-out-of-n: G system*
(b) *m-consecutive-k-out-of-n: G system*
(c) *r-within-consecutive-k-out-of-n: G system*
(d) *Circular consecutive-k-out-of-n: G system*
(e) *Circular m-consecutive-k-out-n: G system*
(f) *Circular r-within-consecutive-k-out-of-n: G system.*

These six systems are considered to be in their working state, if and only if, replacing good components by failed ones and failed components by good ones, then the corresponding system of Definitions 6.1.1 or 6.1.2 fails.

The verbal description given in Definition 6.1.3 above is equivalent to the algebraic one in Barlow and Proschan[7] (through structure functions or failure probabilities). To clarify matters a little bit more, with a concrete example; a consecutive-k-out-of-n: G system is good if and only if at least k consecutive components are good.

Next, let us introduce some 2-dimensional analogues of the systems mentioned above.

Consider first $n_1 x n_2$ components arranged in a rectangular grid on the Cartesian plane. This means that a component is located on the point of the plane with integer coordinates (i,j) for all $1 \leq i \leq n_1$ and all $1 \leq j \leq n_2$. The components are failure prone and work independently of each other.

Following the pattern of Definition 6.1.1 we introduce the following three systems

Definition 6.1.4. A system consisting of $n_1 x n_2$ components arranged in a rectangular grid, is called

(a) *2-dimensional consecutive-$k_1 x k_2$-out-of-$n_1 x n_2$ system,* if and only if the system fails whenever there is at least one rectangular $k_1 x k_2$ grid consisting entirely of failed components.

(b) *2-dimensional m-consecutive-$k_1 x k_2$-out-of-$n_1 x n_2$ system,* if and only if, the system fails whenever there are at least m non-overlapping rectangular $k_1 x k_2$ grids consisting entirely of failed components.

(c) *2-dimensional r-within-consecutive-$k_1 x k_2$-out-of-$n_1 x n_2$ system* if and only if, the system fails whenever there is a rectangular $k_1 x k_2$ grid, which includes at least r failed components.

Analogously with Definition 6.1.2 we may introduce the *cylindrical* version of the three systems just defined.

Definition 6.1.5. Consider $n_1 x n_2$ components arranged in a rectangular grid. We make it a cylindrical surface by considering the (1,j)-th component to be next to the (n,j)-th component, for $1 \leq j \leq n_2$. Then, the three systems described in Definition 6.1.4 give rise to three new systems whose state of failure is described verbatim like the corresponding systems of Definition 6.1.4. These systems will be called respectively

(a) *2-dimensional cylindrical consecutive-$k_1 x k_2$-out-of-$n_1 x n_2$ system*

(b) *2-dimensional cylindrical m-consecutive-$k_1 x k_2$-out-of-$n_1 x n_2$ system*

(c) *2-dimensional cylindrical r-within-consecutive-$k_1 x k_2$-out-of-$n_1 x n_2$ system*

Of course in the cylindrical case new $k_1 x k_2$ rectangular grids appear, non-existing in the rectangular case. The mathematical problems pertaining the study of 2-dimensional cylindrical structures are considerably more difficult than the corresponding rectangular cases.

So far we have six 2-dimensional systems and, needless to say, their duals create another six ones.

Definition 6.1.6. The dual coherent systems of the ones introduced in Definitions 6.1.4 and 6.1.5 above will be called

(a) *2-dimensional consecutive-$k_1 x k_2$-out-of-$n_1 x n_2$: G system*

(b) *2-dimensional m-consecutive-$k_1 x k_2$-out-of-$n_1 x n_2$: G system*

(c) *2-dimensional r-within-consecutive-$k_1 x k_2$-out-of-$n_1 x n_2$: G system*

(d) *2-dimensional cylindrical consecutive-$k_1 x k_2$-out-of-$n_1 x n_2$: G system*

(e) *2-dimensional cylindrical m-consecutive-$k_1 x k_2$-out-of-$n_1 x n_2$: G system*

(f) *2-dimensional cylindrical r-within-consecutive-$k_1 x k_2$-out-of-$n_1 x n_2$: G system*

Before closing the present paragraph, let us give some historical notes on the consecutive-k-out-of-n system and its related structures. The principle of the consecutive-k-out-of-n systems was introduced by Kontoleon [82] but their name was first used by Chiang and Niu[32]. The m-consecutive-k-out-of-n and r-within-consecutive-k-out-of-n systems were introduced by Griffith[59]. The circular consecutive-k-out-of-n systems were studied for the first time by Derman et al[37] and the circular m-consecutive-k-out-of-n systems by Alevizos et al[2]. The duals of some of them have been studied in Kuo, Zhang and Zuo[92] and Zuo and Kuo[161]. The 2-dimensional consecutive-k_1xk_2-out-of-n_1xn_2 systems, their cylindrical analogs and their duals have been introduced in Zuo[159] . Some special cases have been studied before in Salvia and Lasher[136]. The m-consecutive-k_1xk_2-out-of-n_1xn_2 systems were studied by Fu and Koutras[49].

Although the study of those systems in the reliability context originated in the early eighties, nevertheless some mathematical questions related to such structures have appeared before in various contexts. Probabilistic questions related to the consecutive-k-out-of-n systems can be found in de Moivre[100] and Feller[43]. Some ideas related to generalizations of Fibonacci Numbers pertain to the m-consecutive-k-out-of-n systems, e.g. see Philippou[123], Philippou and Makri[124, 127]. The failure distribution of an m-consecutive-k-out-of-n system is also closely related to the *distributions of order k* (see for example Hirano[62], Philippou and Makri[125], Charalambides[28], Godbole [51, 52, 53, 54] and Makri and Philippou[97]. Several problems of Quality Control and Radar Detection are connected with the r-within-consecutive-k-out-of-n systems, Greenberg[58], Saperstein [138, 139], Castella[20] and Nelson[104]. Various statistical tests based on runs are connected with those models (see Mood[101], Mosteller[103], Wolfowitz[155], Bateman[9], Burr and Cane[18], Prairie et al[128], Asano[5], Larsen et al[94], O'Brien[105], Stephens[150], Denny and Yakowitz[36], Shaughnessy[149], Woodall[156, 157], Jennen-Steinmetz and Gasser[78] and Agin and Godbole[1]). In the case of systems consisting of i.i.d. components the problem of evaluating the reliability of a consecutive-k-out-of-n system (or a related system) reduces to a combinatorial restricted occupancy problem. Among the numerous works on this and related discussions we refer to Barton and David[8], Moser and Abramson[102], Harborth[61], Konvalina[83], Prodinger[129], Bollinger[14], Hwang [65], Lascoux[95], Panaretos and Xekalaki[106, 107], Apostol[4] and Golomb[56].

For information about the applicability of those systems in reliability theory, someone can consult Blackwell[10], Kontoleon[82], Chiang and Niu[32], Kao[81], Chao and Lin[27], Chiang and Chiang[31], Hwang[66], Chao[23], Kuo et al[92] and Salvia and Lasher[136].

The twelve systems described above are the basic models that have been studied the most in the literature. There are some modifications and generalizations of these, which attracted less attention up to now, partly for reasons of limited applicability, and partly perhaps for reasons of mathematical difficulty. Some of them are

(a) *The Strict consecutive-k-out-of-n systems* which were studied by Bollinger[15], Papastavridis[109], Kossow and Preuss[84], Philippou and Makri[126], Rushdi[132, 133] etc. Their applicability was challenged by

Rushdi[134], but this criticism notwithstanding, they present challenging mathematical problems.

(b) Various coherent systems based on a graph, whose definition was inspired by the basic consecutive-k-out-of-n concept, were studied by Du and Hwang[38, 40, 41] and Santha and Zhang[137].

(c) *The consecutive-k-out-of-n systems with Markov dependence* among their components were brought up by Fu[46] and subsequently studied by Fu and Hu[48], Papastavridis and Lambiris[121] and Ge and Wang[50].

(d) Cases where various types of dependencies among failures of various components are allowed, were studied by Shanthikumar[146], Kossow and Preuss[85], Papastavridis[114], Lau[96] and Sfakianakis and Papastavridis [144].

(e) The case of *multistate consecutively connected systems* was studied by Hwang and Yao [74]

(f) A *consecutive-k-out-of-n system in a random environment* where shocks occur was studied by Chryssaphinou and Papastavridis[33]).

(g) A *consecutive-k-out-of-n system with node and link failure* was introduced and studied by Chen et al[30]

(h) The systems with *Consecutive Minimal Cut Sets*, and their special case the *Consecutively Connected Systems*, which were introduced by Shanthikumar [147, 148].

This Chapter is arranged as follows. The present and first Paragarph is of course the introduction. The second Paragraph deals with recursive formulae whereas the third Paragraph deals with exact formulae. The fourth Paragraph is devoted to bounds and limit theorems. The fifth Paragraph addresses questions of Optimality. The final sixth Paragraph discusses open problems and possible directions in future research.

Symbols and Notation

Throughout this review, we will consider a system consisting of n components, and X_i will be the time of failure of the i-th component, for $1 \leq i \leq n$, whereas T will be the time to failure of the system. Unless otherwise stated the X_i's will be stochastically independent. Let q_i be the distribution function of X_i, i.e. $q_i(t)=Pr(X_i \leq t)$. Then, q_i will be called the *failure distribution* or *failure probability* of the i-th component and $p_i=1-q_i$ its *reliability*, for $1 \leq i \leq k$. The quantity $Pr(T \leq t)$ ($Pr(T>t)$) gives the *distribution function* of T, which will be called *system failure distribution or failure probability* (*system reliability function*). The case where all n components are alike, will be refered as the i.i.d. case. In the i.i.d. case we will drop, whenever convenient, the subscript i, e.g. we will write p, p(t) instead of p_i, $p_i(t)$.

6.2 Recursive Formulae

When we study a system subjected to failures, a very important parameter we would like to know, is its failure probability (or, which amounts to the same kind of information, its reliability). Usually the best way to compute numerically the failure probability of a given system is via some kind of recursive scheme.

Theorem 6.2.1. In an m-consecutive-k-out-of-n system let F(m,n)=Pr(T≤t) be its failure distribution. The following recursive formula is satisfied, if n≥km+1

$$F(m,n) = F(m,n-1) + \sum_{s=1}^{m}\prod_{i=1}^{sk} q_{n-sk+i}(F(m-s,n-sk-1) - F(m-s+1,n-sk-1))$$

For the proof see Papastavridis[115]. This recursion enables the simultaneous computation of all F(i,j), for 1≤i≤m, 1≤j≤n, in O(nm^2) operations. It is worthmentioning that the special case m=1 (consecutive-k-out-of-n system) has been, essentially, covered by de Moivre[100].

A recursion of similar type was given for the circular m-consecutive-k-out-of-n systems, by Alevizos et al[2]. This recursion makes possible the simultaneous computation of the system failure distribution CF(i,j) of the circular structure for all 1≤i≤m, 1≤j≤n in O(nkm^3) operations. For the special case m=1 (consecutive-k-out-of-n system) several recursive algorithms have been proposed for the evaluation of system's reliability, see e.g. Hwang[64], Antonopoulou and Papastavridis[3], Rushdi[130], Wu and Chen[158], Hwang[70]. In the i.i.d. case of circular consecutive-k-out-of-n systems a qualitatively simpler recursion was given by Lambiris and Papastavridis[93], Du and Hwang[42] and Philippou and Makri[127].

A recursion for the failure probability of an i.i.d. r-within-consecutive-k-out-of-n system, is provided in an indirect way by Saperstein[138]. Certain generating functions are computed, as solutions of a system of linear equations, whose constant coefficients are polynomials. One of these generating functions provides a recursion for failure distribution of an r-within-consecutive-k-out-of-n system. Although it would be quite desirable, it seems to be not an easy task to improve upon it.

Shantikumar[147] introduced a generalization of the consecutive-k-out-of-n, the systems with consecutive minimal cutsets and provided a recursive algorithm to compute their reliability. Hwang and Yao[74] and Zuo[160] improved upon those results, they studied a circular analog too.

Going to a modification of the basic models, Chen et al[30] made the realistic assumption in an i.i.d. consecutive-k-out-of-n system, that not only nodes, but links can fail too (i.e. not only the components themselves, but the connections between them too, can fail). For the case k=2 they provided a recursive formula for the reliability of such a system, in the i.i.d. case. For bigger k, the situation looks difficult.

Some other aspects of the consecutive-k-out-of-n systems have been examined by Hwang[64, 69], Shanthikumar[145], Bollinger and Salvia[17], Chen and Hwang[29], Griffith and Govindarajula[60], Rushdi[130, 131], Chan et al[21], Satam[141].

For the study of a system useful for space exploration and communications see Hwang and Dinghua[71, 72], Hwang[67] and Chryssaphinou et al[35].

6.3 Exact Formulae

Hwang[66] once wrote that "*A recursive algorithm has the disadvantage of being a black box grinding out only numbers. The dependence of a reliability on the system parameters is hidden*". So, a deeper understanding of the reliability of a system need the added insights provided by exact formulas. *Exact formulae* are those which involve the parameters of the system, constants and a finite number of the following operations: additions, multiplications, raising to power and taking logarithms. An important subcategory of exact formulae are the *closed formulae*. Closed are the formulae which are using all the above and the factorial symbol a number of times which is finite and fixed, (i.e. independent of the parameters of the system). Thus, the formula (1+2+...+n) is exact but not closed, whereas n(n+1)/2 is closed.

Theorem 6.3.1. The failure Distribution of an i.i.d. m-consecutive-k-out-of-n system, is given by the following triple summation

$$F(m,n) = \sum_{s=m}^{[n/k]} \sum_{i=sk}^{n} \binom{s+n-i}{s} N(i-sk, n-i+1) q^i p^{n-i}$$

where

$$N(i,j) = \sum_{s=0}^{j} (-1)^s \binom{j}{s} \binom{i+j-1-sk}{i-sk}$$

for $n \geq km$.

For the proof see Godbole[51] or Papastavridis[115] or Hwang and Papastavridis[73]. Formulas expressing F(m,n), via multiple summation and multinomial coefficients were given by Philippou and Makri[125] and Hirano[62].

A similar formula for the circular m-consecutive-k-out-of-n system, is given in Hwang and Papastavridis[73].

An exact formula for the failure distribution for the i.i.d. case of an r-within-consecutive-k-out-of-n system is provided by Saperstein[139]. The formula and its proof are quite involved, signifying the more complex nature of this system vis-a-vis the previous ones. For a combinatorial approach to the same problem we refer to Kounias and Sfakianakis[87, 88]. For the mean time to failure of such a system see Huntington[63].

Up to now, we are not aware of any exact formula for the failure distribution of a non i.i.d. r-within-consecutive-k-out-of-n system.

Chao and Fu[24, 25] developed an exact formula for the reliability of a consecutive-k-out-of-n system with independent components. In their approach, the failure probability of the system appears as a specific entry of a matrix product of certain transition probability matrices. A quite different form of exact formula has also been found by Papastavridis and Hadzichristos[118]. For the evaluation of the reliability (or equivalently the failure probability) of an i.i.d. consecutive-k-out-of-n system (linear or circular) several exact formulae have appeared in the bibliography during the last 10 years. Hwang[66] derived a simple closed formula for the consecutive-2-out-of-n system while for k>2 a single summation formula was developed by Lambiris and Papastavridis[93], Hwang[66] and Goulden[57]. Some alternative methods have also been proposed by Derman et al[37], Bollinger[13], Bollinger and Salvia[16] and Iyer[77]. For the connection between the linear and circular consecutive-k-out-of-n systems see Derman et al[37], Jiongsheng[79], Hwang and Yao[76] and Koutras and Papastavridis[90]).

Some other parameters that have been studied in a consecutive-k-out-of-n system (and their related structures) are

a. The mean time to failure E[T], Kossow and Preuss[86], Papastavridis and Hadzichristos[117], Sasaki et al[140].
b. The (Birnbaum) importance of system's components, Papastavridis [111], Papastavridis and Sfakianakis[122], Sfakianakis[142].
c. The number of failed components at the time of the failure of the system, Papastavridis[113].

For the case where Markov dependence is assumed between the lifetimes of the various components of a consecutive-k-out-of-n-system, Ge and Wang[50] provided an exact formula. For the case where components' lifetimes are exchangeable, a formula for system's reliability was given by Shanthikumar[146] whereas the circular case is covered in

Papastavridis[114]. Boland et al[12] studied the reliability of a consecutive-k-out-of-n-system assuming a positive dependence between adjacent components.

For the general case of a consecutive-k-out-of-n system and its circular analogue, meaning that all kinds of dependencies are allowed between component lifetimes, formulae for their reliabilities are presented by Kossow and Preuss[85] and by Sfakianakis and Papastavridis[144]. As expected they are quite involved.

Assuming that not only nodes, but links too, can fail, in the i.i.d. consecutive-2-out-of-n model, Chen et al[30] provide a "closed" formula, involving the roots of a third degree equation.

The available exact formulae for the reliability of a consecutive-k-out-of-n system and the circular analogue, (see Lambiris and Papastavridis[93], Hwang[66], Philippou[123] and Philippou and Makri[127]), enabled Hwang and Yao[75] to deduce certain interesting results on the variation of the failure rates of such systems.

6.4 Limits and Bounds

In many cases, as the two previous chapters indicate, neither recursive nor exact formulae are available. Then, approximations become very important, particularly when both the estimate and the error of the approximation, appear in closed form. Even in cases where we have exact, but not closed formulae, good closed approximations can be important, since they give a better idea of the order of magnitude of the quantity under consideration.

Theorem 6.4.1. In an r-within-consecutive-k-out-of-n system we define the following events

a) S_i:component i fails and, for j=min(i,k), there are at least r-1 failures among components i-j+1,i-j+2,...,i-1, for i=r,r+1,...,n.
b) A_k:the r-out-of-k system consisting of components 1,2,...,k is good
c) C_i:for j=min(i-k,k-r+1), there is at least one failure among components (i-k)-j+1,(i-k)-j+2,..,i-k, for i=k+1,k+2,...,n
d) G_i:there are at most r-1 failures among components i-k+1,i-k+2,..., i-1, for i=k+1,k+2,...,n
e) E_i:component i fails and there are exactly r-1 failures among components i-k+1,i-k+2,...,i-1, for i=k+1,k+2,...,n

Then, for the lifetime T_n of the system we have ($n \geq k \geq r \geq 2$)

$$\prod_{i=k}^{n}[1-\Pr(S_i)] \le \Pr(T_n \ge t) \le \Pr(A_k) \prod_{i=k+1}^{n} [1-(1-\Pr(C_i))\frac{\Pr(E_i)}{\Pr(G_i)}]$$

For the proof see Papastavridis and Koutras[120]. Alternative approaches towards bounding the reliability of an r-within-consecutive-k-out-of-n system have been developed by Jun[80] and Sfakianakis et al[143].

For large i.i.d. r-within-consecutive-k-out-of-n systems one can easily verify that if the time t is fixed, the reliability tends to 0 as n tends to infinity. An interesting problem then is to determine proper normalizing terms $a_n>0$, b_n such that the random variable $(T_n-b_n)/a_n$ converges in law to a non-degenerate distribution. The following Theorem gives an answer to this question.

Theorem 6.4.2. In an r-within-consecutive-k-out-of-n system let $n \ge k \ge r \ge 2$ and assume that the n components have the same failure probability i.e. $q_i=q$, i=1,2,...,n. If q, as a function of time t, is of the form

$$q(t) = (\lambda t)^\alpha + o(t^\alpha)$$

and T_n denotes the lifetime of the n-th system, then

$$\lim_{n\to\infty} \Pr(n^{1/r\alpha} T_n \ge t) = \exp[-\binom{k-1}{r-1}(\lambda t)^{\alpha k}]$$

This result was first stated and proved by Papastavridis[112] via the Chen-Stein method (see also Koutras and Papastavridis[89]). A much simpler proof, based on Theorem 6.4.1, can be found in Papastavridis and Koutras[120].

It worths mentioning that in the special case of a consecutive-k-out-of-n system (r=k) the bounds of Theorem 6.4.1 take on the following more appealing form

$$\prod_{i=1}^{n-k+1} [1-a_i(t)] \le \Pr(T \ge t) \le \prod_{i=1}^{n-k+1}[1-a_i(t)p_{i-1}(t)]$$

where $p_o(t)=1$ and

$$\alpha_i(t)=q_{i+1}(t)\dots q_{i+k-1}(t), \text{ for } t \ge 0 \text{ and } i=1,2,\dots,n-k+1.$$

This last upper bound provides a slight improvement of the bound proposed by Fu[44, 45]. Several other upper and lower bounds for the reliability of a consecutive-k-out-of-n system can be found in Chiang and Niu[32], Derman et al[37], Salvia[135], Papastavridis[108], Chan et al[21], Chryssaphinou and Papastavridis[34], Barbour et al[6] etc. In the iid case of large consecutive-k-out-of-n systems with $q(t)=(\lambda t)^{\alpha}+o(t^{\alpha})$, the lifetime distribution of the system converges (after proper normalization) to the Weibull distribution. This assertion is an immediate consequence of Theorem 6.4.2 (for r=k). Several alternative approaches for the proof of this result can be found in Chao and Lin[27], Fu[44, 45], Papastavridis[110], Papastavridis and Chryssaphinou [116] and Chryssaphinou and Papastavridis[34].

Papastavridis and Koutras[119] proved that a similar limit theorem applies to consecutive-k-out-of-n systems with maintenance.

In Canfield and McCormick[19] we have "*strong*" limit theorems (i.e. convergence almost everywhere, and not just in distribution) of consecutive-k-out-of-n systems.

Fu[47] and Godbole[55] developed certain bounding techniques for the reliability of an m-consecutive-k-out-of-n system and used them to establish Poisson convergence theorems.

For a general 2-dimensional consecutive-k_1xk_2-out-of-n_1xn_2 system it is very difficult or impossible to develop simple formulae for its reliability evaluation. Certain bounds for approximating the reliability of such systems can be found in Salvia and Lasher[136], Zuo[159], Koutras et al[91] and Fu and Koutras[49]. In the last two papers, some limit theorems for the 2-dimensional consecutive-kxk-out-of-nxn system are also discussed.

6.5 Optimization Problems

The following problem, makes sense for quite a few coherent systems, beyond those we study here.

Problem. We consider a system operating with n components. Assume that we have n components in our disposal, with respective reliabilities $p_1,p_2,\ldots,p_n$ and let $p_1\leq p_2\leq\ldots\leq p_n$. We assume that each one of the n components can be placed in any one of the n positions in the system. A choice of such a placement ammounts to a permutation

$$\mathbf{a}:\{1,2,\ldots,n\}\longrightarrow\{1,2,\ldots,n\}$$

of the n numbers, (i.e. in the i-th position of the system, we will place the component of reliability $p_{a(i)}$, for i=1,2,...,n). Let $R(p_1,\ldots,p_n,\mathbf{a})$ be the reliability of the system, when **a** is the permutation of the placement of the components. The problem is to find the permutation **a** which maximizes the reliability of the system.

Clearly such an **a** always exists, since the number of all possible permutations is finite (namely n!). A particularly interesting question is whether the permutation maximizing the reliability, is independent of the particular values of $p_1,p_2,...,p_n$.

A basic result, for our models is the following

Theorem 6.5.1. The permutation which maximizes the reliability in a consecutive-k-out-of-n system is given by

$$(1,n,3,n-2,\ ...,\ n-3,4,n-1,2)$$

whereas for the circular consecutive-k-out-of-n system it is

$$(n,1,n-1,3,n-3,\ ...,\ n-4,4,n-2,2,n)$$

This result has been conjectured by Derman et al[37] and proved by Du and Hwang[39]. In this case, we observe that the maximizing permutation is independent of the particular values of the reliability of the available components, but, as research has shown, this is the exception rather than the rule. For more information on this (or similar) questions, see Wei et al[154], Malon[98, 99], Tong[151, 152], Du and Hwang[38, 39, 40], Santha and Zhang[137], Hwang[68], Boland et al[11], Zuo and Kuo[161], Kuo et al[92], and Papastavridis and Sfakianakis[122].

Another interesting optimization problem, is the following

Problem. We consider a system operating with n components. At the i-th position of the system we place a component of failure probability q_i, for i=1,2,...,n. Assuming that the product $q_1q_2...q_n=c$ is fixed, we look for q_i's which maximize the reliability of the system.

To illuminate the meaning of this condition i.e. the fixed product, Chang and Hwang[22] argue rather convincingly that the financial cost of a component of failure probability q_i, is reasonable to assume, that is given by the formula $-\ln(aq_i)/b$, where a, b are constants depending upon technology, market situation etc. So, fixed product of the q_i's, means that we have a fixed budget available. In Chang and Hwang[22] and Yao[153] a big part of this problem for consecutive-k-out-of-n systems and their circular counterpart is solved, as well as some cases of more general systems.

6.6 Conclusions

It is a quite common theme in the development of science, that a question answered creates more than one unanswered ones, and the story of the consecutive-k-out-of-n systems is not an exception. So let us see where we stand, from this point of view, going chapter by chapter.

Recursive Formulae. Good recursive schemes for the 2-dimensional systems are not available yet, and this is a missing link that should be looked upon. The same comment is applicable to the "*within*" systems. The mathematical depth and the complexity of Saperstein's[138] treatment of the original linear case indicates that improvements and extensions are not going to be easy.

Closed Formulae. We have the feeling that essentially there is not much room for non-trivial new discoveries in this area, but, of course, someone should never say never.

Limits and Bounds. Probably this is the most promising area, in the sense that important new discoveries can come by, without the need of revolutionary new techniques. We have the feeling that the ideas used in Fu[45] and a sharper use of Stein-Chen method (e.g. see Chryssaphinou and Papastavridis[34]) give some promise in that direction. The field of systems with maintenance is wide open (see Papastavridis and Koutras[119]). Perhaps the most important challenge in this direction is to make the ideas developed for consecutive-k-out-of-n systems work for more general coherent systems. It seems that it is not a futile effort (see Koutras and Papastavridis[89]).

Optimization Questions. This is a mathematically demanding area, as what is already accomplished indicates (see e.g. Du and Hwang[39]). Leaving aside the invariant permutations, it seems that every other non-trivial case, waits to be investigated.

Acknowledgements

The authors were fortunate in having the benefit of a preprint of Chao et al[26], which was instrumental in facilitating the writing of this Chapter.

REFERENCES

[1] Agin M.A. and A.P. Godbole, *A New Exact Runs Test for Randomness, Computing Science and Statistics (C. Page, R. LePage eds.)*, Proceedings of the 22nd Symposium on the Interface, Springer-Verlag, New York, pp. 281-285, 1992.

[2] Alevizos P.D., S.G. Papastavridis and P. Sypsas, *Reliability of Cyclic m-Consecutive-k-out-of-n: F Systems*, Reliability, Quality Control and Risk Assessment, Proc. IASTED Conference, 1992.

[3] Antonopoulou I. and S.G. Papastavridis, *Fast Recursive Algorithm to Evaluate the Reliability of a Circular Consecutive-k-out-of-n: F System*, IEEE Trans. on Rel., Vol. R-36, No. 1, pp. 83-84, April 1987.

[4] Apostol T.M., *Binary Vectors with Prescribed Subsets of Consecutive Ones*, J. Comb. Theory, Vol. 47A, pp. 176-190, 1988.

[5] Asano C, *Run Test for a Circular Distribution and a Table of Probabilities*, Ann. Inst. Stat. Math., Vol. 17, pp. 331-346, 1965.

[6] Barbour A.D., L. Holst and S. Janson, *Poisson Approximation*, Oxford University Press, New York, 1992.

[7] Barlow R.E. and F. Proschan, *Statistical Theory of Reliability and Life Testing*, Holt Reinhart and Winston, Inc., New York, 1975.

[8] Barton D.E. and F.N. David, *Runs in a Ring*, Biometrika, Vol. 45, pp. 572-578, 1958.

[9] Bateman G., *On the Power Function of the Longest Run as a Test for Randomness in a Sequence of Alternatives*, Biometrika, Vol. 35, pp. 97-112, 1948.

[10] Blackwell M.T.R., *The Effect of Short Production Runs on CSP-1*, Technometrics, Vol. 19, pp. 259-63, 1977.

[11] Boland P.J., F. Proschan and Y.L. Tong, *Optimal Arrangement of Components via Pairwise Rearrangements*, Nav. Research Log., Vol. 36, pp. 807-815, 1989.

[12] Boland P.J., F. Proschan and Y.L. Tong, *Linear Dependence in Consecutive-k-out-of-n: F Systems*, Prob. in Engin. and Inf. Sciences, Vol. 4, pp. 391-397, 1990.

[13] Bollinger R.C., *Direct Computation for Consecutive-k-out-of-n: F Systems*, IEEE Trans. on Rel., Vol. R-31, No. 5, pp. 444-446, 1982.

[14] Bollinger R.C., *Reliability and Runs of Ones*, Mathematics Magazine, Vol. 57, pp. 34-37, 1984.

[15] Bollinger R.C., *Strict Consecutive-k-out-of-n: F Systems*, IEEE Trans. on Rel., Vol. R-34, No. 1, pp. 50-52, 1985.

[16] Bollinger R.C. and A.A. Salvia, *Consecutive-k-out-of-n: F Networks*, IEEE Trans. on Rel., Vol. R-31, No. 1, pp. 53-56, 1982.

[17] Bollinger R.C. and A.A. Salvia, *Consecutive-k-out-of-n: F System with Sequential Failures*, IEEE Trans. on Rel., Vol. R-34, No. 1, pp.43-45, 1985.

[18] Burr E.J. and G. Cane, *Longest Run on Consecutive Observations having a Specified Attribute*, Biometrika, Vol. 48, pp. 461-465, 1961.

[19] Canfield E.R. and W.P. McCormick, *Asymptotic Reliability of Consecutive k-out-of-n Systems*, J. Appl. Prob., Vol. 29, pp. 142-155, 1992.

[20] Castella F.R., *Sliding Window Detection Probabilities*, IEEE Trans. Aerosp. Electron. Systems, Vol. AES-12, pp. 815-819, 1976.

[21] Chan F.Y., L.K. Chan and G.D. Lin, *On Consecutive-k-out-of-n: F System*, Europ. J. of Oper. Research, Vol. 36, pp. 207-216, 1988.

[22] Chang G.J. and F.K. Hwang, *Optimal Consecutive-k-out-of-n: F System under a Fixed Budget*, Prob. in the Eng. and Info. Sciences, Vol. 2, pp. 63-73, 1988.

[23] Chao M.T., *The Finite Time Behavior of CSP when Defects are Dependent*, Proc. Nat. Sci. Council, R.O.C., Vol. 13, pp. 18-22, 1989.

[24] Chao M.T. and J.C. Fu, *A Limit Theorem of Certain Repairable System*, Ann. Inst. Statist. Math., Vol. 41, pp. 809-818, 1989.

[25] Chao M.T. and J.C. Fu, *The Reliability of Large Series Systems under Markov Structure*, Adv. in Appl. Prob., Vol. 23, pp. 894- 908, 1991.

[26] Chao M.T., J.C. Fu and M.V. Koutras, *A Survey of the Reliability Studies of Consecutive-k-out-of-n: F Systems and its Related Systems*, 1993 (Preprint).

[27] Chao M.T. and G.D. Lin, *Economical Design of Large Consecutive-k-out-of-n: F Systems*, IEEE Trans. on Rel., Vol. R- 33, No. 5, pp. 411-413, 1984.

[28] Charalambides C., *Lucas Numbers and Polynomials of Order k and the Length of the Longest Circular Success Run*, Fibonacci Quarterly, Vol. 29, pp. 290-297, 1991.

[29] Chen R.W. and F.K. Hwang, *Failure Distributions of Consecutive-k-out-of-n: F Systems*, IEEE Trans. on Rel., Vol. R- 34, No. 4, pp. 338-341, 1985.

[30] Chen R.W., F.K. Hwang and W.W. Li, *Consecutive-k-out-of-n: F Systems with Node and Link Failures*, IEEE Trans. on Rel., 1993 (to appear).

[31] Chiang D.T. and R. Chiang, *Relayed Communication via Consecutive-k-out-of-n: F System*, IEEE Trans. on Rel., Vol. R-35, No. 1, pp. 65-67, 1986.

[32] Chiang D.T. and S. Niu, *Reliability of Consecutive-k-out-of-n: F System*, IEEE Trans. on Rel., Vol. R-30, No. 1, pp. 87-89, 1981.

[33] Chryssaphinou O. and S.G. Papastavridis, *Reliability of a Consecutive-k-out-of-n: F System in a Random Environment*, J. Appl. Prob., Vol. 27, pp. 452-458, 1990.

[34] Chryssaphinou O. and S.G. Papastavridis, *Limit Distribution for a Consecutive-k-out-of-n: F System*, Adv. Appl. Prob., Vol. 22, pp. 491-493, 1990.

[35] Chryssaphinou O., S.G. Papastavridis and P. Sypsas, *Relayed Communication via Parallel Redundancy*, IEEE Trans. on Rel., Vol. R-38, No. 4, pp. 454-456, 1989.

[36] Denny J.L. and S.J. Yakowitz, *Admissible Run-Contingency Type Tests for Independence and Markov Dependence*, J.A.S.A., Vol. 73, pp. 177-181, 1978.

[37] Derman C., G.J. Lieberman and S. Ross, *On the Consecutive-k-out- of-n: F System*, IEEE Trans on Rel., Vol. R-31, No. 1, pp. 57-63, 1982.

[38] Du D.Z. and F.K. Hwang, *Optimal Consecutive-2-out-of-n: F Systems of Lines and Cycles*, Networks, Vol. 15, pp. 439-447, 1985.

[39] Du D.Z. and F.K. Hwang, *Optimal Consecutive-2-out- of-n Systems*, Math. Oper. Research, pp. 187-191, 1986.

[40] Du D.Z. and F.K. Hwang, *Optimal Assignment for Consecutive 2- graphs*, SIAM Alg. Math. Disc., Vol. 8, pp. 510-518, 1987.

[41] Du D.Z. and F.K. Hwang, *Reliabilities of Consecutive 2-Graphs*, Prob. Eng. Info. Science, Vol. 1, pp. 293-298, 1987.

[42] Du D.Z. and F.K. Hwang, *A Direct Algorithm for Computing Reliability of a Consecutive-k Cycle*, IEEE Trans. on Rel., Vol. R-37, No. 1, pp. 70-72, 1988.

[43] Feller W., *An Introduction to Probability Theory and its Applications*, Vol. 1, 3rd ed., Wiley, New York, 1968.

[44] Fu J.C., *Reliability of a Large Consecutive-k-out- of-n: F System*, IEEE Trans. on Rel., Vol. R-34, No. 2, pp. 127-130, 1985.

[45] Fu J.C., *Bounds for Reliability of Large Consecutive-k-out-of-n: F Systems with Unequal Component Probabilities*, IEEE Trans. on Rel., Vol. R-35, No. 3, pp. 316-319, 1986.

[46] Fu J.C., *Reliability of Consecutive-k-out-of-n: F Systems with (k-1)-step Markov Dependence*, IEEE Trans. on Rel., Vol. R-35, No. 5, pp. 602-606, 1986.

[47] Fu J.C., *Poisson Convergence in Reliability of Large Linearly Connected Systems as Related to Coin Tossing*, Statistica Sinica, 1993 (to appear).

[48] Fu J.C. and B. Hu, *On Reliability of Large Consecutive-k-out-of- n: F System with (k-1)-step Markov Dependence*, IEEE Trans. on Rel., Vol. R-36, No. 1, pp. 75-77, 1987.

[49] Fu J.C. and M.V. Koutras, *Poisson Approximation for 2-Dimensional Patterns*, Ann. Inst. Stat. Math., 1993 (to appear).

[50] Ge G. and L. Wang, *Exact Reliability Formula for Consecutive-k- out-of-n: F Systems with Homogeneous Markov Dependence*, IEEE Trans. on Rel., Vol. R-39, No. 5, pp. 600-602, 1990.

[51] Godbole A.P., *Specific Formulae for Some Success Run Distributions*, Stat. and Prob. Letters, Vol. 10, pp. 119-124, 1990.

[52] Godbole A.P., *Degenerate and Poisson Convergence Criteria for Success Runs*, Stat. and Prob. Letters, Vol. 10, pp. 247-255, 1990.

[53] Godbole A.P., *On Hypergeometric and Related Distributions of Order k*, Commun. Statist.-Theory Meth., Vol. 19, pp. 1291-1301, 1990.

[54] Godbole A.P., *Poisson Approximations for Runs and Patterns of Rare Events*, Adv. Appl. Prob., Vol. 23, pp. 851-865, 1991.

[55] Godbole A.P., *Approximate Reliabilities of m-consecutive-k-out-of-n: Failure Systems*, Statistica Sinica, 1993 (to appear).

[56] Golomb S.W., *Distributions of Periods and Frequencies of Runs in Random Binary Sequences*, Adv. Appl. Math., Vol. 10, pp. 278-284, 1989.

[57] Goulden I.P., *Generating Functions and Reliabilities for Consecutive-k-out-of-n: F Systems*, Utilitas Mathematica, Vol. 32, pp. 141-147, 1987.

[58] Greenberg I., *The First Occurrence of n Success in N Trials*, Technometrics, Vol. 12, pp. 627-634, 1970.

[59] Griffith W.S., *On Consecutive-k-out-of-n: Failure Systems and their Generalizations*, Reliability and Quality Control (A.P. Basu ed.), Elsevier Science Publ. B.V., pp. 157-165, 1986.

[60] Griffith W.S. and Z. Govindarajula, *Consecutive-k-out-of-n Failure Systems: Reliability, Availability, Component Importance and Multistate Extensions*, Amer. J. Math. and Management Sci., Vol. 5, pp. 125-160, 1985.

[61] Harborth H., *Endliche 0-1-Folgen Mit Gleichen Teilbloechen*, J. Reine Angewandte Mathem., pp. 139-154, 1974.

[62] Hirano K., *Some Properties of the Distributions of Order k, Fibonacci Numbers and Applications* (A.N. Philippou, A.F. Horadam, G.E. Bergum, eds), D. Reidel, pp. 43-53, 1986.

[63] Huntington R., *Mean Recurrence Times for k Successes within m Trials*, J. Appl. Prob., Vol. 13, pp. 604-607, 1976.

[64] Hwang F.K., *Fast Solutions for Consecutive-k-out- of-n: F System*, IEEE Trans. on Rel., Vol. R-31, No. 5, pp. 447-448, 1982.

[65] Hwang F.K., *Selecting k Objects from a Cycle with p Pairs of Separartion s*, J. Comb. Theory, Vol. A-37, pp. 197-199, 1984.

[66] Hwang F.K., *Simplified Reliabilities for Consecutive-k-out-of-n: F Systems*, SIAM J. Alg. Disc. Meth., Vol. 7, pp. 258-264, 1986.

[67] Hwang F.K., *Relayed Consecutive-k-out-of-n: F Lines*, IEEE Trans. on Rel., Vol. R-37, No. 5, pp. 512-514, 1988.

[68] Hwang F.K., *Invariant Permutations for Consecutive-k-out-of-n Cycles*, IEEE Trans. on Rel., Vol. R-38, No. 1, pp. 65-67, 1989.

[69] Hwang F.K., *An Explicit Solution for the Number of Minimal s-Cut Sequences in a Consecutive-k-out-of-n: F System*, IEEE Trans. on Rel., Vol. R-40, No. 5, pp. 553-554, 1991.

[70] Hwang F.K., *An O(kn)-time Implementation of Antonopoulou and Papastavridis Algorithm for Computing the Reliability of a Circular Consecutive-k-out-of-n: F System*, IEEE Trans. on Rel., 1993 (to appear).

[71] Hwang F.K. and S. Dinghua, *Redundant Consecutive-k Systems*, Oper. Res. Letters, Vol. 6, pp. 293-296, 1987.

[72] Hwang F.K. and S. Dinghua, *Optimal Relayed Mobil Communication Systems*, IEEE Trans. on Rel., Vol. R-38, No. 4, pp.457-459, 1989.

[73] Hwang F.K. and S. Papastavridis, *Binary Vectors with Exactly k Non-Overlapping m-Tuples of Consecutive Ones*, Disc. Appl. Math., Vol. 30, pp. 83-86, 1991.

[74] Hwang F.K. and Y.C. Yao, *Multistate Consecutively-Connected Systems*, IEEE Trans. on Rel., Vol. 38, No. 4, pp. 472-474, 1989.

[75] Hwang F.K. and Y.C. Yao, *On Failure Rates of Consecutive-k-out-of-n: F Systems*, Prob. and Engin. and Info. Sciences, Vol. 4, pp. 57-71, 1990.

[76] Hwang F.K. and Y.C. Yao, *A Direct Argument for Kaplansky's Theorem on Cyclic Arrangement and its Generalization*, Oper. Research Letters, 1991.

[77] Iyer S., *Distribution of Time to Failure of Consecutive-k-out-of-n: F Systems*, IEEE Trans. on Rel., Vol. R- 39, No. 1, pp. 97-101, 1990.

[78] Jennen-Steinmetz C. and Gasser.Th, *The Asymptotic Power of Runs Tests*, Scand. J. Statist., Vol. 13, pp. 263-269, 1986.

[79] Jiongsheng L., *A Generalized Formula for Computing the Reliability of Consecutive-k-out-of-n: F Circular Systems*, Kexue Tongbao, Vol. 31, pp. 1533-1537, 1986.

[80] Jun C., *Reliability of a Large Consecutive-k-out-of-r-from-n: System with Unequal Component Reliability*, 1993 (Preprint).

[81] Kao S.C., *Computing Reliability from Warranty*, Proc. Amer. Stat. Assoc., Section on Stat. Comp., pp. 309-312, 1982.

[82] Kontoleon J.M., *Reliability Determination of a r-Successive-out- of-n: F System*, IEEE Trans. on Rel., Vol. R-29, No. 5, p. 437, 1980.

[83] Konvalina J., *On the Number of Combinations without Unit Separation*, J. Comb. Theory, Vol. A31, pp. 101-107, 1981.

[84] Kossow A. and W. Preuss, *Failure Probability of Strict Consecutive-k-out-of-n: Systems*, IEEE Trans. on Rel., Vol. R-36, No. 5, pp. 551-553, 1987.

[85] Kossow A. and W. Preuss, *Reliability of Consecutive-k-out-of-n: F Systems with Non-identical Component Reliabilities*, IEEE Trans. on Rel., Vol. R-38, No. 2, pp. 229-233, 1989.

[86] Kossow A. and W. Preuss, *Mean Time-to-Failure of a Linear Consecutive-k-out-of-n: F System*, IEEE Trans. on Rel., Vol. R-40, No. 3, pp. 271-272, 1991.

[87] Kounias S. and M. Sfakianakis, *The Reliability of a Linear System and its Connection with the Generalized Birthday Problem*, Statistica Applicata, Vol. 3, pp. 531-543, 1991.

[88] Kounias S. and M. Sfakianakis, *A Combinatorial Problem Associated with the Reliability of a Circular System*, Proc. Hermis 1992 Conference, 1992.

[89] Koutras M.V. and S.G. Papastavridis, *Application of the Stein- Chen Method for Bounds and Limit Theorems in the Reliability of Coherent Structures*, Nav. Research Log., 1993 (to appear).

[90] Koutras M.V. and S.G. Papastavridis, *On the Number of Runs and Related Statistics*, Statistica Sinica, 1993 (to appear).

[91] Koutras M.V., G.K. Papadopoulos and S.G. Papastavridis, *Reliabiliity of 2-Dimensional Consecutive-k-out-of-n: F Systems*, IEEE Trans. on Rel., 1993 (to appear).

[92] Kuo W., W. Zhang and M. Zuo, *A Consecutive-k-out-of-n: G System: the Mirror Image of a Consecutive-k-out-of-n: F System*, IEEE Trans. on Rel., Vol. R-39, No. 2, pp. 244-253, 1990.

[93] Lambiris M. and S.G. Papastavridis, *Exact Reliability Formulas for Linear and Circular Consecutive-k-out-of-n: F Systems*, IEEE Trans. on Rel., Vol. R-34, No. 2, pp. 124-126, 1985.

[94] Larsen R.J., C.L. Holmes and C.W. Heath, Jr., *A Statistical Test for Measuring Unimodal Clustering: A Description of the Test and its Application to Cases of Acute Leukemia in Metropolitan Atlanta*, Georgia, Biometrics, Vol. 29, pp. 301-309, 1973.

[95] Lascoux A., *Suites Recurrentes Lineaires*, Adv. Appl. Math., Vol. 7, pp. 228-235, 1986.

[96] Lau T.S., *The Reliability of Exchangeable Binary Systems*, Statist. and Prob. Letters, Vol. 13, pp. 153-158, 1992.

[97] Makri F.S. and A.N. Philippou, *Binomial Distributions of Order k on the Circle*, Runs and Patterns: Selected Papers (A. Godbole, ed.), 1993 (to appear).

[98] Malon D.M., *Optimal Consecutive-2-out-of-n: F Component Sequencing*, IEEE Trans. on Rel., Vol. R-33, No. 5, pp. 414-418, 1984.

[99] Malon D.M., *Optimal Consecutive-k-out-of-n: F Component Sequencing*, IEEE Trans. on Rel., Vol. R-34, No. 1, pp. 46-49, 1985.

[100] Moivre A.de, *The Doctrine of Chance*, Chelsea Publishing Co., 3rd Edition, pp. 254-259, 1967.

[101] Mood A.M., *The Distribution Theory of Runs*, Ann. Math. Statist., Vol. 11, pp. 367-392, 1940.

[102] Moser W.O.J. and M. Abramson, *Enumeration of Combinations with Restricted Differences and Cospan*, J. Comb. Theory, Vol. A7, pp. 162-170, 1969.

[103] Mosteller F., *Note on an Application of Runs to Quality Control Charts*, Ann. Math. Statist., Vol. 12, pp. 228-232, 1941.

[104] Nelson J.B., *Minimal-order Models for False Alarm Calculations on Sliding Windows*, IEEE Trans. on Aerospace and Electronic Systems, Vol. AES-14, pp. 351-363, 1978.

[105] O'Brien P.C., *A Test for Randomness*, Biometrics, Vol. 32, pp. 391-401, 1976.

[106] Panaretos J. and E. Xekalaki, *On Some Distributions Arising from Certain Generalized Sampling Schemes*, Commun. Statist.-Theory Meth., Vol. 15, pp. 873-891, 1986.

[107] Panaretos J. and E. Xekalaki, *A Probability Distribution Associated with Events with Multiple Occurrences*, Stat. Probab. Letters, Vol. 8, pp. 389-395, 1989.

[108] Papastavridis S.G., *Upper and Lower Bounds for the Reliability of Consecutive-k-out-of-n: F System*, IEEE Trans. on Rel., Vol. R-35, No. 5, pp. 607-610, 1986.

[109] Papastavridis S.G., *Algorithms for Strict Consecutive-k-out-of-n: F System*, IEEE Trans. on Rel., Vol. R-35, No. 5, pp. 613-615, 1986.

[110] Papastavridis S.G., *A Limit Theorem for Reliability of Consecutive-k-out-of-n: F System*, Adv. Appl. Prob., Vol. 19, pp. 746-748, 1987.

[111] Papastavridis S.G., *The Most Important Component in a Consecutive-k-out-of-n: F System*, IEEE Trans. on Rel., Vol. R- 36, No. 2,pp. 266-268, 1987.

[112] Papastavridis S.G., *A Weibull Limit for the Reliability of Consecutive-k-within-m-out-of-n System*, Adv. Appl. Prob., Vol. 20, pp. 690-692, 1988.

[113] Papastavridis S.G., *The Number of Failed Components in a Consecutive-k-out-of-n: F System*, IEEE Trans. on Rel., Vol. R-38, No. 3, pp. 338-340, 1989.

[114] Papastavridis S.G., *Lifetime Distribution of Circular Consecutive-k-out-of-n: F Systems with Exchangeable Lifetimes*, IEEE Trans. on Rel., Vol. R-38, No. 4, pp. 460-461, 1989.

[115] Papastavridis S.G., *m-Consecutive-k-out-of-n: F System*, IEEE Trans. on Rel., Vol. R-39, No. 3, pp. 386-388, 1990.

[116] Papastavridis S.G. and O. Chryssaphinou, *An Approximation for Large Consecutive-k-out-of-n: F Systems*, IEEE Trans. on Rel., Vol. R-37, No. 4, pp. 386-387, 1988.

[117] Papastavridis S.G. and I. Hadzichristos, *Mean Time to Failure for a Consecutive-k-out-of-n: F System*, IEEE Trans. on Rel., Vol. R- 36, No. 1, pp. 85-86, 1987.

[118] Papastavridis S.G. and I, Hadzichristos, *Formulas for the Reliability of a Consecutive-k-out-of-n: F System*, J. Appl. Prob., Vol. 26, pp. 772-779, 1988.

[119] Papastavridis S.G. and M.V. Koutras, *Consecutive-k-out-of-n: F Systems with Maintenance*, Ann. Inst. Statist. Math., 1992 (to appear).

[120] Papastavridis S.G. and M.V. Koutras, *Bounds for Reliability of Consecutive-k-within-m-out-of-n Systems*, IEEE Trans. on Rel., 1993 (to appear).

[121] Papastavridis S.G. and M. Lambiris, *Reliability of a Consecutive-k-out-of-n: F System for Markov-dependent Components*, IEEE Trans. on Rel., Vol. R-36, No. 1, pp. 78-79, 1987.

[122] Papastavridis S.G. and M.E. Sfakianakis, *Optimal Arrangement and Importance of the Components in a Consecutive-k-out-of-r-from-n: F System*, IEEE Trans. on Rel., Vol. R-40, No. 3, pp. 277-279, 1991.

[123] Philippou A.N., *Distributions and Fibonacci Polynomials of Order k, Longest Runs, and Reliability of Consecutive-k-out-of-n: F System,* Fibonacci Numbers and their Applications (Philippou et al, eds), D. Reidel, pp. 203-227, 1986.

[124] Philippou A.N. and F.S. Makri, *Longest Success Runs and Fibonacci-type Polynomials,* Fibonacci Quarterly, Vol. 23, pp. 338-346, 1985.

[125] Philippou A.N. and F.S. Makri, *Success, Runs and Longest Runs,* Statist. and Prob. Letters, Vol. 1, pp. 171-175, 1986.

[126] Philippou A.N. and F.S. Makri, *Closed Formulas for the Failure Probability of a Strict-Consecutive-k-out-of-n: F System,* IEEE Trans. on Rel., Vol. R-36, No. 1, pp. 80-82, 1987.

[127] Philippou A.N. and F.S. Makri, *Longest Circular Runs with an Application in Reliability, via Fibonacci-type Polynomials of Order k, Applications of Fibonacci Numbers* (G.E. Bergum et al, eds), Kluwer Academic Publishers, pp. 281-286, 1990.

[128] Prairie R.R., W.J. Zimmer and J.K. Brookhouse, *Some Acceptance Sampling Plans Based on the Theory of Runs, Technometrics,* Vol. 4, pp. 177-185, 1962.

[129] Prodinger H., *On the Number of Combinations without Fixed Distance,* J. Comb. Theory, Vol. A35, pp. 362-365, 1983.

[130] Rushdi A.M., *A Switching-algebraic Analysis of Consecutive-k-out-of-n: F Systems,* Microelectronics and Reliability, Vol. 27, pp. 171-174, 1987.

[131] Rushdi A.M., *A Switching-algebraic Analysis of Circular Consecutive-k-out-of-n: F Systems,* Reliab. Eng. and System Safety, Vol. 21, pp. 119-127, 1988.

[132] Rushdi, A.M., *On the Analysis of Ordinary and Strict Consecutive-k-out-of-n: F Systems,* IEEE Trans. on Rel., Vol. R- 37, 1988.

[133] Rushdi A.M., *Effect of Statistical Dependencies in Strict Consecutive-k-out-of-n: F Systems,* Microelectronics and Reliability, Vol. 28, pp. 309-318, 1988.

[134] Rushdi A.M., *Some Open Questions on: Strict Consecutive-k-out-of-n: F Systems,* IEEE Trans. on Rel., Vol. R- 39, No. 3, pp. 380-381, 1990.

[135] Salvia A.A., *Simple Inequalities for Consecutive-k-out-of-n: F Systems,* IEEE Trans. on Rel., Vol. R-31, No. 5, p. 450, 1982.

[136] Salvia A.A. and W.C. Lasher, *2-Dimensional Consecutive-k-out-of- n: F Models,* IEEE Trans. on Rel., Vol. R-39, No. 3, pp. 382-385, 1990.

[137] Santha M. and Y. Zhang, *Consecutive-2 Systems on Trees,* Prob. Eng. Inform. Science, Vol. 1, pp. 441-456, 1987.

[138] Saperstein B., *On the Occurrences of n Successes within N Bernoulli Trials,* Technometrics, Vol. 15, pp. 809-818, 1973.

[139] Saperstein B., *Note on a Clustering Problem,* J. Appl. Prob., Vol. 12, pp. 629-632, 1975.

[140] Sasaki M., Y. Nakai and T. Yuge, *Mean Time to Failure for a Consecutive-k-out-of-n: F System,* IEEE Trans. on Rel., Vol. R-41, No. 2, p. 306, 1992.

[141] Satam M., *Consecutive-k-out-of-n: F system: A Comment,* IEEE Trans. on Rel., Vol. R-40, No. 1, p. 62, 1991.

[142] Sfakianakis M., *Birnbaum Importance of Components in a Consecutive-k-out-of-r-from-n: F System, Reliability Quality Control and Risk Assessment,* Proc. IASTED Conference, 1992.

[143] Sfakianakis M., S. Kounias and A. Hillaris, *Reliability of a Consecutive-k-out-of-r-from-n: F System,* IEEE Trans. on Rel., Vol. R-41, No. 3, pp. 442-447, 1992.

[144] Sfakianakis M. and S.G. Papastavridis, *The Reliability of a General Consecutive-k-out-of-n: F System,* IEEE Trans. on Rel., 1993 (to appear).

[145] Shanthikumar J.G., *Recursive Algorithm to Evaluate the Reliability of a Consecutive-k-out-of-n: F System,* IEEE Trans. on Rel., Vol. R-31, No. 5, pp. 442-443, 1982.

[146] Shanthikumar J.G., *Lifetime Distribution of Consecutive-k-out-of-n: F Systems with Exchangeable Lifetimes,* IEEE Trans. on Rel., Vol. R-34, No. 5, pp. 480-483, 1985.

[147] Shanthikumar J.G., *Reliability of Systems with Consecutive Minimal Cutsets*, IEEE Trans. on Rel., Vol. R-36, No. 5, pp. 546-549, 1987.

[148] Shanthikumar J.G., *Bounding Network Reliability Using Consecutive Minimal Cutsets*, IEEE Trans. on Rel., Vol. R-37, No. 1, pp. 45-49, 1988.

[149] Shaughnessy P.W., *Multiple Runs Distributions: Recurrences and Critical Values*, J.A.S.A., Vol. 76, pp. 732-736, 1981.

[150] Stephens M.A., *Whitworth Runs on a Circle*, Ann. Inst. Statist. Math., Vol. 29, pp. 287-293, 1977.

[151] Tong Y.L., *A Rearrangement Inequality for the Longest Run, with an Application to Network Reliability*, J. Appl. Prob. Vol. 22, pp. 286-393, 1985.

[152] Tong Y.L., *Some New Results on the Reliability of Circular Consecutive-k-out-of-n: F Systems, Reliability and Quality Control* (A.P. Basu, ed.), Elsevier Science, pp. 395-399, 1986.

[153] Yao Y.C., *On Optimal Consecutive-k-out-of-n: F Systems Subject to Fixed Product of Failure Probabilities*, Prob. in Eng. and Inf. Sciences, Vol. 3, pp. 165-173, 1989.

[154] Wei V.K., F.K. Hwang and V.T. Sos, *Optimal Sequencing of Items in a Consecutive-2-out-of-n: F System*, IEEE Trans. on Rel., Vol. R- 32, No. 1, pp. 30-33, 1983.

[155] Wolfowitz J., *On the Theory of Runs with Some Applications to Quality Control*, Ann. Math. Stat., Vol. 14, pp. 280-288, 1943.

[156] Woodall W.H., *The Distribution of the Run Length of One-sided CUSUM Procedures for Continuous Random Variables*, Technometrics, Vol. 26, pp. 295-301, 1983.

[157] Woodall W.H., *On the Markov Chain Approach to the Two-sided CUSUM Procedure*, Technometrics, Vol. 26, pp. 41-46, 1984.

[158] Wu J.S. and R.J. Chen, *An O(kn) Algorithm for a Circular Consecutive-k-out-of-n: F System*, IEEE Trans. on Rel., Vol. 41, No. 2, pp. 303-305, 1992.

[159] Zuo M., *Reliability and Design of 2-Dimensional Consecutive-k-out-of-n Systems*, IEEE Trans. on Rel., 1993 (to appear).

[160] Zuo M., *Reliability of Linear and Circular Consecutively Connected Systems*, IEEE Trans. on Rel., 1993 (to appear).

[161] Zuo M. and W. Kuo, *Design and Performance Analysis of Consecutive-k-out-of-n Structure*, Nav. Research Log., Vol. 37, pp. 203-230, 1990.

New Trends in System Reliability Evaluation
K.B. Misra (Editor)
1993 Elsevier Science Publishers B.V.

CHAPTER 7

FAULT TREE ANALYSIS

Hiromitsu Kumamoto
Department of Precision Mechanics, Faculty of Engineering
Kyoto University, Kyoto 606, Japan

ABSTRACT

Logic and/or probability manipulations are only one aspect of a fault tree analysis (FTA) which should be viewed from a broader perspective including both risk assessment, risk management and system modeling. This chapter consists of two parts. The first part describes safety goals, accident causation mechanisms and systematic risk management methodologies to be reflected by FTA. The second part presents an automated synthesis algorithm because the synthesis is considered as the bottleneck of FTA.

7.1 Introduction

Accidents and failures can be reduced significantly when possible causes of abnormal events have been enumerated during a system design phase. A fault tree analysis (FTA) is a typical approach to enumeration of such causes. A fault tree (FT) is an AND/OR tree that develops a top event at its root into more basic de vents toward its leaves via intermediate events and logic gates. An AND gate requires that output event from the gate can only occur when input events to the gate occurs simultaneously, while an OR gate requires that output event can occur when one or more input events occur.

The FTA was developed by H.A. Watson of the Bell Telephone Laboratories in 1961-62 during an Air Force study contract for the Minuteman Launch Control System. The first published papers were presented at the 1965 Safety Symposium sponsored by the University of Washington and the Boeing Company, where a group including D.F. Haasl, R.J. Schroder, W.R. Jackson, and others had been applying and extending the technique. FTs were used with event trees (ETs) in the famous WASH-1400 study [3], an extensive risk assessment of nuclear power plants sponsored by the United States Atomic Energy Commission that completed in 1974.

Since the early 1970's when computer-based analysis techniques for FTs were developed, their use has become very widespread. Indeed, the use of FTA is being mandated by a number of governmental agencies responsible for worker and/or public safety. Risk assessment methodologies based on FTs and ETs are called a probabilistic risk assessment (PRA). The PRA is now widely used in various industries including nuclear, aerospace, chemical, transportation, manufacturing, etc.

Hardware failures appear to be dominant causes of various accidents such as Chernobyl, Challenger, Bhopal, Three Mile Island, etc. However, few fault tree analysis support this conjecture today. Some emphasize human errors during operation or maintenance periods, while others stress management and organizational factors over a plant life cycle. Some people emphasize lack of safety culture or ethics.

Tutorial and descriptive literature about FTA includes Lambert [27], Vesely [28], Villemeur [25], US NRC [9], and Henley & Kumamoto [24]. Recently, more emphases have been placed on management, organizational and cultural factors in performing PRA, and the FTA should be viewed from this broader perspective with which this chapter is concerned. FTs from viewpoints of probabilistic risk assessment and management (PRAM), accident causation mechanisms and accident prevention methodologies are described, rather than conventional logic/probabilistic manipulation formulas; FT synthesis method is presented in detail.

The FTA deals with risks. Subjects about risks are divided into two phases: risk assessment and risk management. This chapter first describe how FTs are related in characterizing some aspects of risks. Safety goals are then presented; their satisfactions through discoveries and corrections of design weaknesses are a major motivation of FTA. Accident causation mechanisms and systematic risk management methodologies are described; any FTA should reflect these mechanisms and methodologies. Finally, an automated synthesis algorithm of FTs are presented because the synthesis is a bottleneck of FTA

7.2 Risk and Decision

Risk is a word with various implications. Some people define risk differently from others. This disagreement causes serious confusions in the field of risk assessment and management including FTA. Webster's dictionary, for instance, defines risk as 1) the chance of loss, 2) the degree of probability of loss, 3) the amount of possible loss, 4) the type of loss that an insurance policy covers, etc. Dictionary definitions such as these are not sufficiently precise. This section gives a formal definition of risk. Some aspects of risk are investigated by FTA.

7.2.1 Outcomes and Likelihoods

Primary definition of risk: A weather forecast saying "30 percent chance of rain tomorrow" gives two outcomes together with their likelihoods; (30%, rain) and (70%, no rain). Risk is defined as a collection of such pairs of likelihoods and outcomes :

$$\{(30\%, \text{rain}), (70\%, \text{no rain})\}.$$

More generally, assume n potential outcomes in the doubtful future. Then, risk is defined as a collection of n pairs.

$$Risk \equiv \{(L_1, O_1), ..., (L_i, O_i), ..., (L_n, O_n)\}$$

where O_i and L_i denote outcome i and its likelihood, respectively. Thus, throwing a dice yields the risk,

$$Risk \equiv \{(1/6, 1), (1/6, 2), ..., (1/6, 6)\}$$

where the outcome is a particular face and the likelihood is probability 1 in 6.

In situations involving random chance, each face involves a beneficial or a harmful event as an ultimate outcome. When the faces are replaced by these outcomes, the risk of throwing the die can more explicitly be rewritten as

$$Risk \equiv \{(1/6, O_1), (1/6, O_2), ..., (1/6, O_6)\}$$

Risk profile: The distribution pattern of the likelihood-outcome pair is called a risk profile (or a risk curve); likelihoods and outcomes are displayed along vertical and horizontal axes, respectively. **Fig. 7.1** shows a simple risk profile for the weather forecast described earlier; two discrete outcomes are observed along with their likelihoods, 30% rain or 70% no rain.

In some cases, the outcomes are measured by a continuous scale, or the outcomes are so many that they may be regarded as continuous rather than discrete. Consider an investment problem where each outcome is a certain amount of monetary return (gain

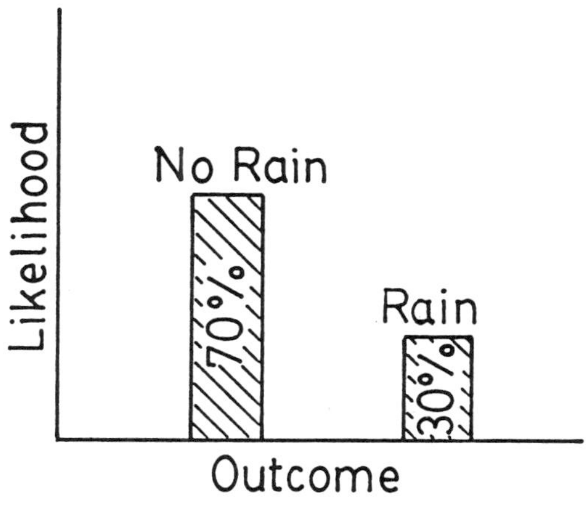

Fig. 7.1 Simple Risk Profile from Weather Forecast

or loss), and each likelihood is a density of experiencing a particular return. Potential pairs of likelihoods and outcomes then form a continuous profile. **Fig. 7.2-a** is a density profile where loosing a positive or a negative amount of money indicates loss or gain, respectively.

Objective vs subjective likelihood: In a perfect risk profile, each likelihood is expressed as an objective probability, percentage, or density per action or per unit time or during a specified time interval. Objective frequencies such as 2 occurrences per year and ratios such as 1 occurrence in one million are also used likelihoods; if the frequency is sufficiently small, it can be regarded as a probability or a ratio. Unfortunately, the likelihood is not always objective; probability, percentage, frequency, and ratio may be based on subjective evaluation. Verbal probabilities such as rare, possible, plausible and frequent are also used.

Complementary cumulative profile: The risk profile (discrete or continuous) is often displayed in terms of complementary cumulative likelihoods. For instance, the likelihood of loosing x or more money is displayed rather than the density of just loosing money x. **Fig. 7.2-b** shows a complementary cumulative risk profile obtained from the density profile in Fig. 7.2-a. Point P on the vertical axis denotes a probability of loosing zero or more money, i.e., a probability of not getting any profit. The complementary cumulative likelihood is a monotonously decreasing function of variable x, and hence has a simpler shape than the density function. The complementary representation is informative because decision makers are more interested in the likelihood of loosing x or more money than in loosing just an x amount of money; for instance, they want to know the probability of "no monetary gain" denoted by point P in Fig. 7.2-b.

Farmer curves: **Fig. 7.3** shows a famous example from the Reactor Safety Study [3] where annual frequencies of x or more early fatalities caused by 100 nuclear power plants are predicted and compared with fatal frequencies by air crashes, fires, dam failures, explosions, chlorine releases and air crashes (persons on ground). Non-nuclear frequencies are normalized by a size of population potentially affected by the 100 nuclear power plants; these are not frequencies observed on a worldwide scale. Each profile in Fig. 7.3 is called a Farmer curve; horizontal and vertical axes generally denote severity of accident and complementary cumulative frequency per unit time, respectively.

Only fatalities greater than or equal to 10 are displayed in Fig. 7.3. This is an exceptional case. Fatalities usually start with unity; in actual risk problems, a zero fatality has a far larger frequency than positive fatalities. Inclusion of a zero fatality in

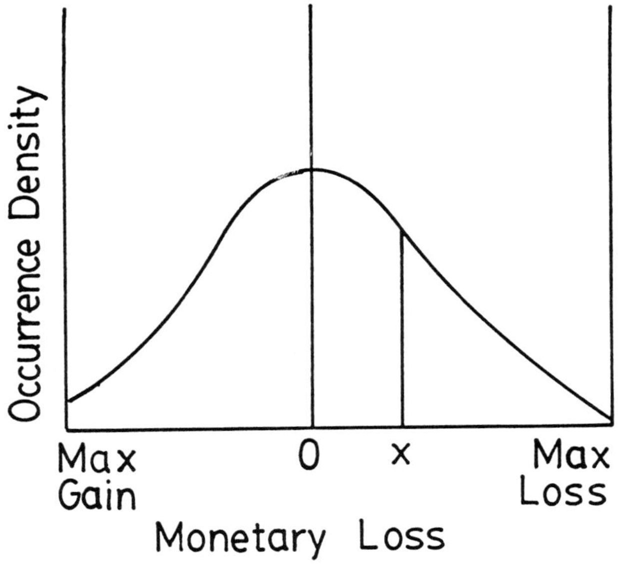

Fig. 7.2 (a) Occurrence Density Profile : Loosing Money

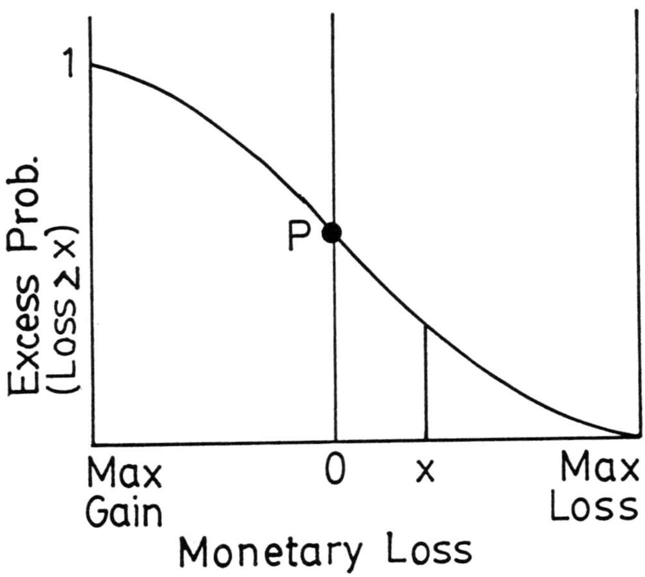

Fig. 7.2 (b) Complementary Cumulative Risk Profile from the Density Profile in Fig. 7.2 (a)

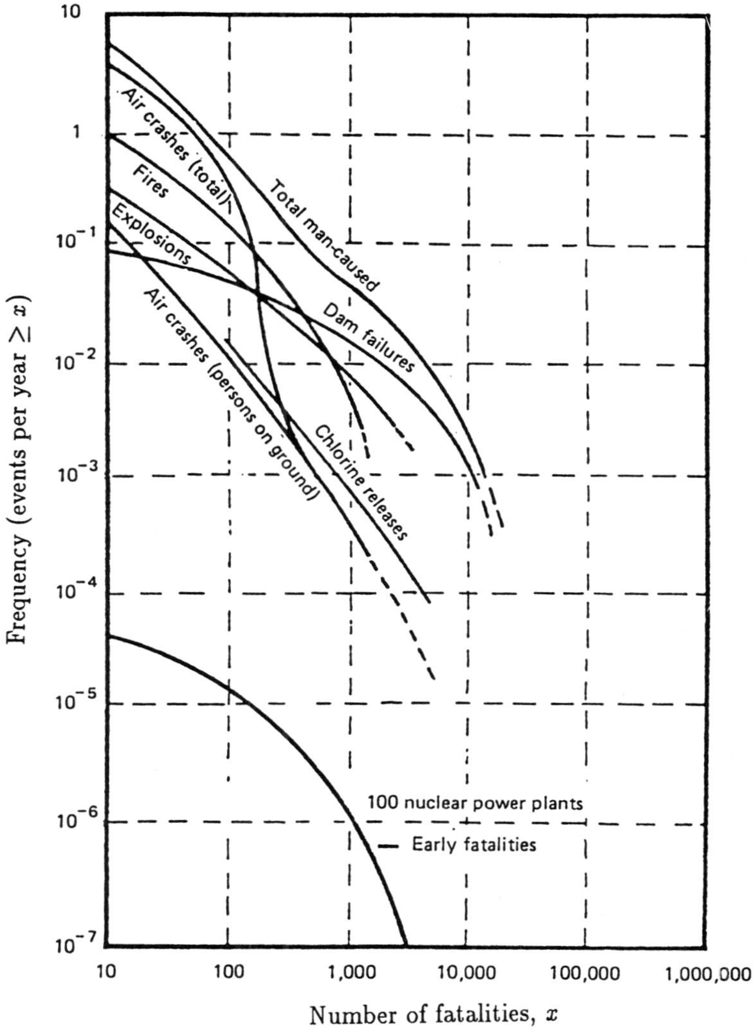

Fig. 7.3 Comparison of Annual Frequencies of x or More Fatalities

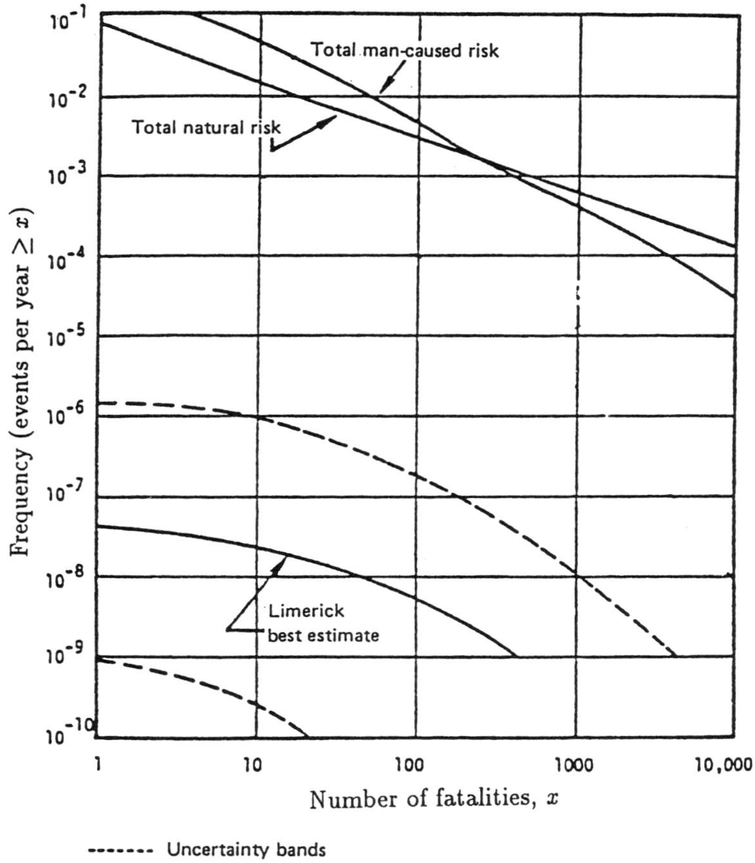

Fig. 7.4 Example of Meta-Uncertainty of Complementary Cumulative Risk Profile

the Farmer curve requires the display of an unreasonably wide range of likelihoods. On the other hand, the risk profile including the zero fatality will not necessarily be oddly shaped if the profile is conditioned by occurrence of an accident such as a core-melt, because a zero fatality now has a much smaller likelihood.

7.2.2 Meta-Uncertainty and Uncertainty about Risk

Meta-uncertainty: The risk profile itself often has associated with it uncertainties which are called meta-uncertainties. A subjective estimate of uncertainties for a complementary cumulative distribution function (CCDF) was carried out by the authors of the Limerick Study [9]. Their result is shown in **Fig. 7.4**. The subjective range of uncertainty, shown by the dotted lines, stretches over three orders of magnitude. This is a fair reflection on the present state of the art of risk assessment. The error bands are a result of two types of meta-uncertainties; uncertainty in outcome severity of an accident and uncertainty in frequency of the accident. The existence of meta-uncertainty such as this makes risk management or decision making under risk difficult and controversial.

Uncertainty: A kernel element of the risk is uncertainty represented by plural outcomes and their future likelihoods. Note that this uncertainty is a different concept from the meta-uncertainty just described.

7.2.3 *Risk Assessment and Management*

Risk assessment: A principal purpose of risk assessment is the derivation of risk profiles posed by a given situation; the weatherman performed a risk assessment when he concluded the risk profile in Fig. 7.1. The Farmer curves in Figs. 7.3 and 7.4 are final products of a methodology called a probabilistic risk assessment (PRA), which, among other things, enumerates outcomes and quantifies their likelihoods.

For nuclear power plants, the PRA proceeds as follows: 1) enumeration of sequences of untoward events that could produce a core melt, 2) clarification of containment failure modes, their probabilities and timing, 3) identification of quantity and chemical form of radioactivity released if the containment is breached, 4) modeling of dispersion of radionuclides in the atmosphere, 5) modeling of emergency response effectiveness involving sheltering, evacuation, and medical treatment, and 6) dose-response modeling in estimating health effects on a population exposed [2].

Risk management: Risk management proposes alternatives, evaluates, for each alternative, the risk profile, makes safety decisions, chooses satisfactory alternatives to control the risk, and exercises corrective actions. Terms such as risk estimation and risk evaluation only cause confusions, and should be avoided.

Assessment vs management: When the risk management is performed in relation to a PRA, the two activities are called a probabilistic risk assessment and management (PRAM).

The probabilistic risk assessment phase is more scientific, technical, formal, quantitative and objective than the management phase which involves value judgment and heuristics, and hence is more subjective, qualitative, societal and political. Ideally, the PRA is based on objective likelihoods such as mortality rates or electrical bulb failure rates inferred from statistical data and theories. However, the PRA is often compelled to use subjective likelihoods and subjective outcome severities based on intuition, expertise, and partial, defective or deceitful data, and dubious theories. These constitute the major source of meta-uncertainty in the risk profile.

Considerable research efforts are being made to establish a unified and scientific PRAM methodology where subjective assessment, value judgment, expertise and heuristics are dealt with more objectively. Nonetheless the subjective or human dimension does constitute one of the two pillars that support the entire conceptual edifice [4].

7.2.4 *Outcome Significance and Decision Making*

Significance of outcome: The significance of each outcome from each alternative must be evaluated in terms of an amount of gain or loss if an optimal, sub-optimal, feasible or satisfactory alternative is to be chosen. Significance varies directly with loss and inversely with gain. An inverse measure of the significance is called a utility, or value function. In PRA, the outcome and significance are sometimes called a consequence and a magnitude, respectively, especially when loss outcomes such as property damage and fatality are considered.

The risk profile now includes the utility U_{ij} (or significance):

$$R_j = \{(L_{ij}, O_{ij}, U_{ij}) \mid i = 1, ..., n_j\},\ j = 1, ..., m$$

This representation indicates an explicit dependence of risk profile on outcome significance: the determination of the significance is a value judgment problem and is considered mainly in the risk management phase. The significance is implicitly assumed when minor outcomes are screened out during the risk assessment phase.

7.2.5 Causal Scenario

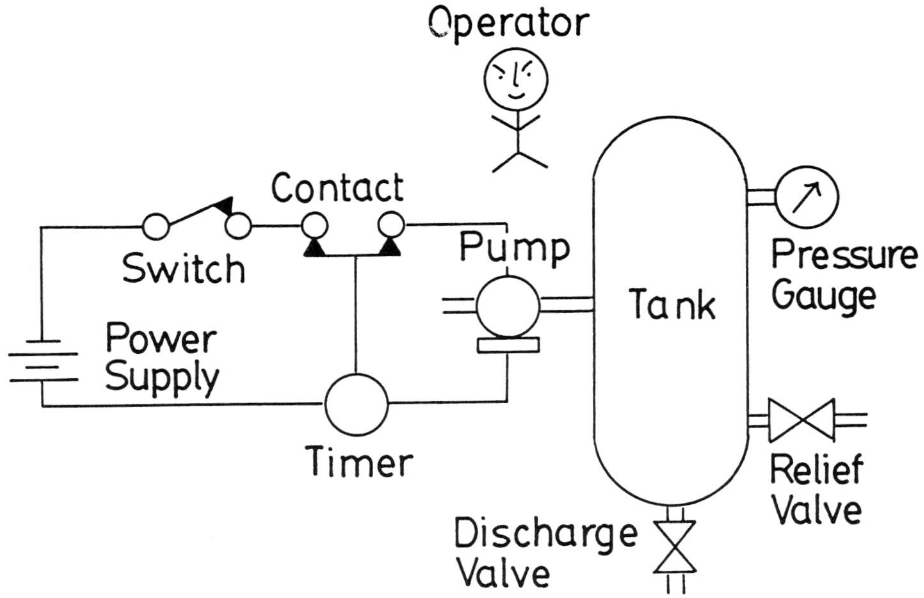

Fig. 7.5 Schematic Diagram of Pressure Tank System

The likelihood and perhaps the significance of outcome can be evaluated more easily when a causal scenario for the outcome is in place. Thus, risk R_j may be rewritten as:

$$R_j = \{(L_{ij}, O_{ij}, U_{ij}, CS_{ij}) \mid i = 1, ..., n_j\},\ j = 1, ..., m$$

where CS_{ij} denotes the causal scenario which specifies 1) causes of outcome O_{ij} and 2) event propagations for the outcome. This representation explicitly expresses a dependence of risk profile on the causal scenario identified during the risk assessment phase.

Causal scenarios and PRA: PRA uses, among other things, event tree and fault tree techniques to enumerate outcomes and causal scenarios. Such a scenario is called an accident sequence and is composed of various deleterious interactions among facilities, software, information, material, power sources, humans, and environment. In cases of nuclear power plants, accident sequences leading to a core melt are investigated thoroughly. The techniques are also used to quantify outcome likelihoods during the risk assessment phase.

Example - Pressure tank PRA [5]: The system shown in **Fig. 7.5** discharges gas from a reservoir into a pressure tank. The switch is normally closed and the pumping cycle is initiated by an operator who manually resets the timer. The timer contact closes and the pumping starts. The manual switch is normally closed.

Later (well before any over-pressure condition can exist) the timer times out and the timer contact opens. Current to the pump cuts off and pumping ceases, to prevent a tank rupture due to over-pressure. If the timer contact does not open, the operator is instructed to observe the pressure gauge and to open the manual switch, thus causing the pump to stop. Even if the timer and operator both fail, over-pressure can be relieved by the relief valve.

After each cycle, the compressed gas is discharged by opening the valve and then closing it before the next cycle begins. At the end of the operating cycle, the operator is instructed to verify the operability of the pressure gauge by observing the decrease in the tank pressure as the discharge valve is opened. To simplify the analysis, we assume that the tank is depressurized before the cycle begins. An undesired event, from a viewpoint of risk, is a pressure tank rupture by over-pressure.

The pressure gauge may fail during the new cycle even if its operability has been correctly checked by the operator at the end of the last cycle. The gauge could have failed before the new cycle if the operator committed an inspection error.

Fig. 7.6 shows the event tree for the pressure tank rupture due to over-pressure. The event tree starts with an initiating event which initiates the accident sequence. The tree describes combinations of success or failure of the system's mitigative features that lead to desired or undesired plant states. In Fig. 7.6, PO denotes the event 'pump overrun,' an initiating event which starts the potential accident scenarios. Symbol OS denotes the failure of the operator shutdown system, PP denotes failure of the pressure protection system by relief valve failure. The upper bar indicates a logic complement of the inadvertent event, i.e., successful activation of the mitigative feature. There are three sequences or scenarios displayed in Fig. 7.6. The scenario labeled $PO \cdot OS \cdot PP$ causes over-pressure and tank rupture, where symbol '$\cdot$' denotes logic intersection, (AND). Therefore, the tank rupture requires three simultaneous failures. The other two scenarios lead to safe results.

The event tree defines top events each of which can be analyzed by a fault tree which develops more basic causes such as hardware or human faults. We see, for instance, that the pump overrun is caused by 1) timer contact fails to open, or 2) voltage surge which overrides effects of the open timer contact. By linking the three fault trees (or their logic complements) along a scenario on the event tree, possible causes for each scenario can be enumerated. For instance, tank rupture occurs when the following three basic causes occur simultaneously; 1) timer contact fails to open, 2) switch contact fails to open, and 3) pressure relief valve fails to open. Probabilities for these three causes can be estimated from generic or plant specific statistical data, and eventually the probability of the tank rupture due to over-pressure can be quantified.

7.2.6 *Population Affected*

Final definition of risk: A population of a single individual is an exceptional case. Usually more than one person is affected anonymously by the risk. The population size is a factor which determines an important aspect of the risk. A comparison of risks using the Farmer curves in Figs. 7.3 and 7.4 makes no sense unless the population is specified. The risk concept includes, as a final element, the population PO_{ij} affected by outcome O_{ij}

$$R_j = \{(L_{ij}, O_{ij}, U_{ij}, CS_{ij}, PO_{ij}) \mid i = 1, ..., n_j\},\ j = 1, ..., m$$

Populations are identified during the risk assessment phase.

7.2.7 *Population vs Individual Risk*

Definitions of two types of risks: The term 'population risk' is used when a population as a whole is the risk recipient. A population risk is also called a societal risk, a collective risk, or a societally-aggregated risk. When a particular individual in the population is evaluated as the risk recipient, then the risk is called an individual risk. In other words, the population PO_{ij} in the definition of risk reduces to a single person for an individual risk.

7.2.8 *Summary*

Risk is formally defined as a collection of combinations of five primitives; 1) outcome, 2) likelihood, 3) significance, 4) causal scenario, and 5) population affected. These factors determine the risk profile. The risk assessment phase deals with primitives other than the

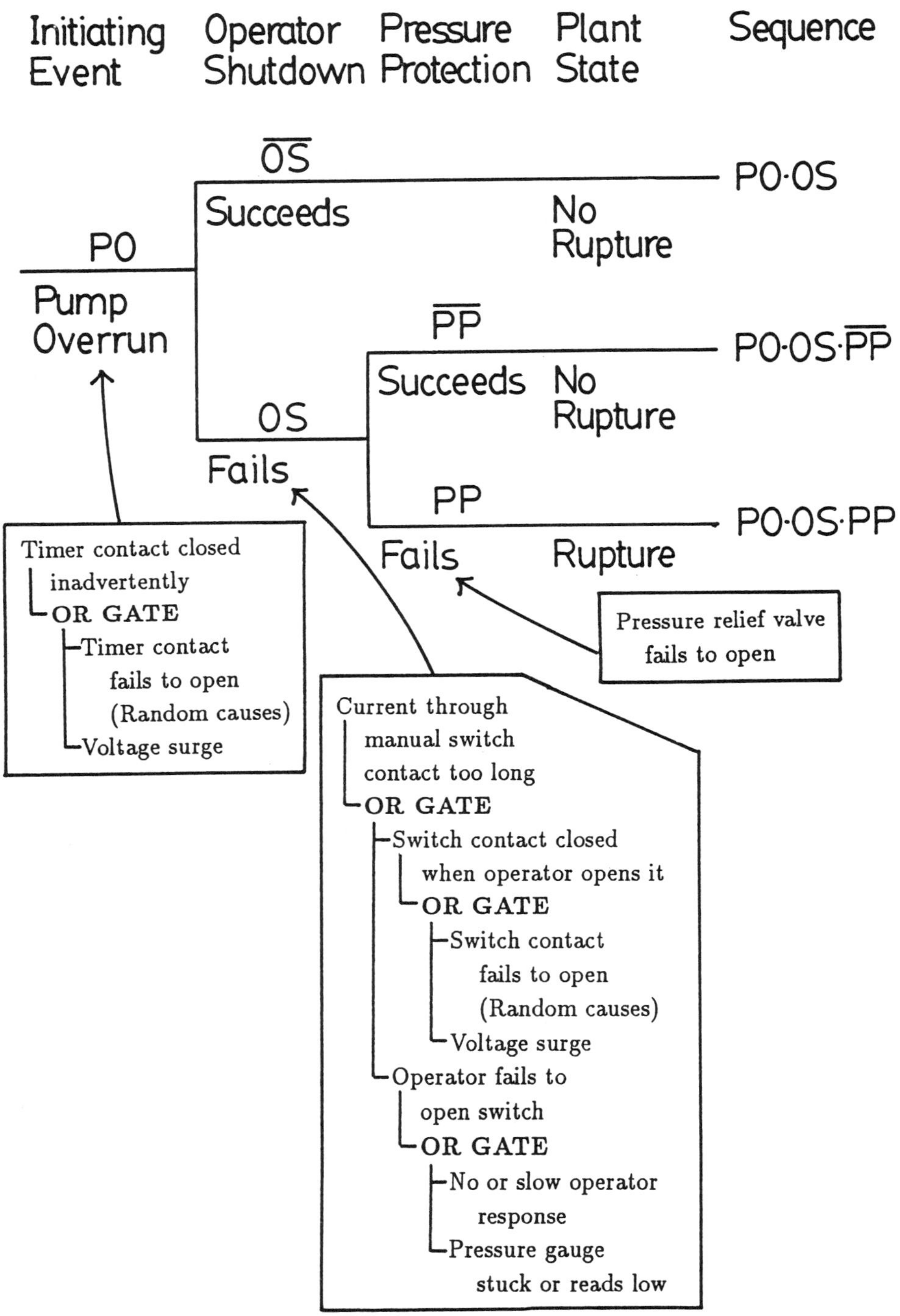

Fig. 7.6 Event Tree and Fault Tree Analyses for Pressure Tank System

outcome significance which is evaluated in the risk management phase. FTs, when used with ETs, determine outcome, likelihood, and causal scenario.

The pressure tank problem illustrates some aspects of probabilistic risk assessment. Here, the fault tree technique is used in combination with the event tree technique.

Two important types of risk are presented; individual risk and population risk. The size of the population is a crucial parameter in risk management.

7.3 Safety Goals

When a set of goals are given, the risk problem including FTA becomes more tractable; risk management tries to satisfy the goals, and the risk assessment checks the attainment of the goals. A hierarchical arrangement of the goals is given in section 7.3.1. Section 7.3.2 gives a three-layer decision structure with upper and lower bound goals. Examples of goals are given in sections 7.3.3 and 7.3.4 for normal activities and catastrophic accidents, respectively. The concept of regulatory cutoff level is introduced in section 7.3.5. The final section gives a model for varying the regulatory cutoff level as a function of population size.

7.3.1 Hierarchical Goals for PRAM

Systems subject to PRA have a hierarchical structure: components, units, subsystems, plant, and site. Safety goals also form a hierarchy. For a nuclear power plant, for instance, goals can be structured in the following way [13]:

(1) Initiating event level: occurrence frequency.
(2) Safety system level: unavailability.
(3) Containment: failure probability.
(4) Accident sequence level: sequence frequency.
(5) Plant: plant damage state frequency, core damage frequency, source term.
(6) Site: collective dose, early fatalities, latent cancer fatalities, property damage.

The safety goals at the top of the hierarchy are most important. For the nuclear power plant, the top goals are those on the site level. When the goals on the top level are given, goals on the lower levels can in theory be specified in an objective and systematic way. If a hierarchical goal system is established in advance, the PRAM process is simplified significantly; the probabilistic risk assessment phase, given alternatives, calculates performance indices for the goals on various levels, with error bands. The risk management phase proposes the alternatives, and evaluates the attainment of the goals.

To achieve goals on the various levels, a variety of techniques are proposed: (1) suitable redundancy, (2) reasonable isolation, (3) sufficient diversity, (4) sufficient independence, and (5) sufficient margin [13]. Appendix A to title 10 of the Code of Federal Regulations Part 50 (CFR Part 50) sets out 64 general design criteria for quality assurance; protection against fire, missiles, and natural phenomena; limitations on the sharing of systems; and a number of other protective safety requirements. In addition to the NRC regulations, there are numerous supporting guidelines that contribute importantly to the achievement of safety goals. These include regulatory guides (numbering in the hundreds); the Standard Review Plan for reactor license applications, NUREG-75/087 (17 chapters); and associated technical positions and appendices in the Standard Review Plan [1].

7.3.2 Upper and Lower Bound Goals

Three-layer decision structure: Cyril Comar [10] proposed the following decision structure, as cited by Spangler [2]:

(1) Eliminate any risk that carries no benefit or is easily avoided.
(2) Eliminate any large risk $\geq U$ that does not carry clearly overriding benefits.
(3) Ignore for the time being any small risk $\leq L$ that does not fall into category 1.

(4) Actively study risks falling between these limits, with the view that the risk of taking any proposed action should be weighed against the risk of not taking that action.

Of course, upper bound level U is greater than lower bound level L. In the decision structure of Comar, the term "risk" is used mainly to denote a risk level in terms of outcome likelihood, outcome size, or (outcome size $\times$ likelihood); it is not used in a strict, formal sense of risk which includes risk profiles and other risk primitives defined earlier. The term "action" means an alternative; thus, action is not identical to a risk source such as a nuclear power plant or chemical plant; elimination of a particular action does not necessarily imply elimination of the risk source; the risk source may be included by introduction of other actions or alternatives.

In Comar's second statement, the large risk $\geq U$ is reluctantly accepted if and only if it has overriding benefits; this is an exceptional case for soldiers at war (under the name of national security) or patients undergoing medical operations (rescue from a serious disease). Denote by R the risk level of an action. Then, the decision structure described above can be stated as follows:

Pre-Screening Structure

```
begin
if (risk carries no benefit)
    reduce risk below L (inclusive)
if (risk has overriding benefits)
    reluctantly accept risk
if (risk has moderate benefits)
    go to the main structure below
end
```

Main Decision Structure

```
risk R has moderate benefits
begin
if (R ≥ U)
    risk is unacceptable;
    reduce risk below U (exclusive) for justification or acceptance
if (L < R < U)
    actively study risk for justification
    begin
    if (risk is justified)
        reluctantly accept risk
    if (risk is not justified)
        reduce risk until justified or below L (inclusive)
    end
if (R ≤ L)
    accept risk
end
```

We see that only risks with moderate benefits are subject to the main decision structure which consists of three layers separated by upper and lower limits U and L, respectively: $R \geq U$, $L < R < U$ and $R \leq L$. In the top layer $R \geq U$, the risk is first reduced below U exclusive; the resultant level may locates in the middle $L < R < U$ or the bottom layer $R \leq L$. In the middle layer, the risk R is actively studied by risk-cost-benefit (RCB) analyses for justification; if it is justified, then it is reluctantly accepted; if it is not justified, then it is reduced until justification in the middle layer or inclusion in the bottom layer. In the bottom layer, the risk is automatically accepted even if it carries no benefits. The easily avoided risk in the first statement of Comar, by definition, can be reduced below L regardless of the merits.

Note that term "reduce" does not necessarily mean an immediate reduction; rather it denotes a registration into a reduction list; some risks in the top layer or some risks not justified in the middle layer are difficult to reduce immediately but can be reduced in the future; some other risks such as "background radiation", which carries no benefits, in the pre-screening structure are extremely difficult to reduce, and would remain in the reduction list forever.

The lower bound L is closely related to the de minimis risk to be described shortly, and its inclusion can be justified, for instance, for the following reasons: 1) people do not pay much attention to risks below the lower bound even if they receive no benefits, 2) it becomes extremely difficult to decrease the risk below the lower bound, 3) there are almost countless and hence intractable number of risks below the lower bound, 4) above the lower bound there are many risks in need of reduction, and 5) without such a lower bound, all company profits could be allocated for safety [14].

Upper and lower bound goals: Comer defined the upper and lower bounds by probabilities of fatality of an individual per year of exposure to the risk:

$$U = 10^{-4}/(\text{year, individual}), \quad L = 10^{-5}/(\text{year, individual})$$

Richard Wilson [11] defined the bounds for the individual fatal risk as follows:

$$U = 10^{-3}/(\text{year, individual}), \quad L = 10^{-6}/(\text{year, individual})$$

According to annual statistical data per individual, 1) "being stuck by lightening" is smaller than 10^{-6}, 2) "natural disasters" locate between 10^{-6} and 10^{-5}, 3) "all industrial work" is between 10^{-5} and 10^{-4}, and 4) "traffic accidents" and "all accidents" between 10^{-4} and 10^{-3}.

Since the upper bound suggested by Comer is $U = 10^{-4}$, the current traffic accident risk level $R \geq U$ would imply the following: 1) automobiles have overriding merits and are reluctantly accepted in the pre-screening structure, or 2) the risk level is in the reduction list of the main decision structure, i.e., the risk has moderate benefits but should at least be reduced below U.

Wilson's upper bound $U = 10^{-3}$ means that the traffic accident risk level $R \leq U$ should be subject to intensive RCB study for justification; if the risk is justified, then it is reluctantly accepted; if the risk is not justified, then it must be reduced until another justification or until it is below the lower bound L.

Note that U and L are not bounds for the population risk but for an individual risk which assumes that there is only one driver.

Wilson showed that the lower bound $L = 10^{-6}/(\text{year, individual})$ is equivalent to the risk level of any one of the following activities during a year: 1) smoking 3 US cigarettes (cancer, heart disease), 2) drinking a half liter of wine (cirrhosis of the liver), 3) visiting New York or Boston for 2 days (air pollution), and 4) spending 6 minutes in a canoe (accident). The lower bound $L = 10^{-5}$ by Comar can be interpreted in a similar way; for instance, it is comparable to drinking 5 liters of wine per year.

Spangler claims that the Wilson's annual lowèr bound $L = 10^{-6}$ is more acceptable than the Comar's bound $L = 10^{-5}$ for the following situations [2]:

(1) Whenever the risk is involuntary.

(2) Whenever there is a substantial band of uncertainty in estimating risk at such low levels.

(3) Whenever the risk has a high degree of expert and public controversy.

(4) Whenever there is a reasonable prognosis that new safety information is more likely to yield higher-than-current "best" estimates of the risk level rather than lower estimates.

Accumulation problems for lower bound risks: It should be noted that the lower bound $L = 10^{-6}/(\text{year, individual})$ would not be suitable if the risk level were measured not per year but per operation. For instance, the same operation may be performed repetitively by a dangerous forging press. The operator of this machine may think that the risk per operation is negligible because there is only one chance in one million of an accident so he removes safety interlocks to speed up the operation. However, more than ten thousands operations may be performed during a year, yielding a large annual risk level, say 10^{-2}, of injury. Another similar accumulation may be caused by multiple risk sources or by risk exposures to a large population; if enough negligible doses are added together, the result may eventually be significant [6]; if negligible individual risks of fatality are integrated over the population, a sizable number of fatalities may occur.

ALARA - As low as reasonably achievable: A decision structure similar to the ones described above is recommended by ICRP (International Commission on Radiological Protection) Report No. 26 for individual-related radiological protection [1]. Note that population-related protection is not considered.

(1) Justification of practice: No practice shall be adopted unless its introduction produces a positive net benefit.

(2) Optimization of protection: All exposures shall be kept as low as reasonably achievable (i.e., ALARA), economic and social factors being taken into account.

(3) The radiation doses to individuals shall not exceed the dose equivalent limits recommended for the appropriate circumstances by ICRP.

The third statement corresponds to upper bound U in the three-layer decision structure. However, the ICRP report lacks lower bound L, and there is a theoretical chance that the risk would be overreduced to any small number so long as it is feasible by ALARA.

The NRC adoption of ALARA radiation protection standards for the design and operation of light water reactors in May 1975 interpreted the term "as low as reasonably achievable" to mean as low as is reasonably achievable taking into account the state of technology, and the economics of improvements in relation to benefits to public health and safety, and other societal and socioeconomic considerations, and in relation to the utilization of atomic energy in the public interest [1]. Note here that the term "benefits" does not denote the benefits of atomic energy but reduction of risk levels; "utilization of atomic energy in the public interest" denotes the benefits in the usual sense for RCB analyses.

For population-related protection, USNRC proposed a conservative value of $1000 per total body person-rem (collective dose for population risk) averted for the risk/cost evaluations for ALARA [1]. The value of $1000 is roughly equal to $7.4 million per fatality averted. This ALARA value established temporarily by the commission is substantially higher than the equity value of $250,000 to $500,000 per fatality averted referenced by other agencies in risk reduction decisions. (The lower equity values apply, of course, to situations where there is no litigation.)

De minimis risk: The concept of de minimis risk is fully discussed in the book edited by Chris Whipple [8]. A purpose of de minimis risk investigation is a justification of a lower bound L below which no active study of the risk, including ALARA or RCB analyses, is required. Joyce Davis, for instance, describes in Chapter 13 of the book how the law has long recognized that there are trivial matters that need not concern it; the maxim *de minimis non curat lex*, "the law does not concern itself with trifles," expresses that principle [6]. In practice, of course, the instance of a judge actually dismissing a lawsuit on the basis of triviality is a super rare event. She suggests the following applications of de minimis risk concepts [1]:

(1) For setting regulatory priorities.

(2) As a "floor" for ALARA considerations.

(3) As a cut-off level for collective dose assessments.

(4) For setting outer boundaries of geographical zones.

(5) As a floor for definition of low level waste.

(6) As a presumption of triviality in legal proceedings.

(7) To foster administrative and regulatory efficiency.

(8) To provide perspective for public understanding, including policy judgments.

Some researchers of the de minimis say that 10^{-6}/(year, individual) risk is trivial, acceptable, or negligible, and that no more safety investment nor regulation is required at all for systems with the de minimis risk level. Two typical approaches for determining the de minimis radiation level are 1) comparison with background radiation levels, and 2) detectability of radiation [6]. The radiation is presumed to cause cancers, and the radiation level can be converted to a fatal cancer level.

ALARA vs de minimis: Cunningham [17,2] note: "We have a regulatory scheme with upper limits above which the calculated health risk is generally unacceptable. Below these upper limits are various specific provisions and exemptions involving calculated risks that are considered acceptable based on a balancing of benefits and costs, and these need not be considered further. Regulatory requirements below the upper limits are based on the ALARA principle, and any risk involved is judged acceptable given not only the magnitude of the health risk presented but also various social and economic considerations. A de minimis level, if adopted, would provide a regulatory cutoff below which any health risk, if present, could be considered negligible. Thus, the de minimis level would establish a lower limit for the ALARA range of doses."

The use of ALARA-type procedures can provide a basis for establishing an explicit standard of de minimis risk beyond which no further analysis of costs and benefits need be employed to determine the acceptability of risk [1]; in this context, the de minimis risk is a dependent variable in the ALARA procedure. An example of such a procedure is the cost benefit guideline of \$1000 per total person-rem averted. However, such a determination of the de minimis risk level by ALARA would yield different lower bounds for different risk sources; this blurs the three-layer decision structure with a universal lower bound which says that no more ALARA is required below the constant bound even if a more cost effective alternative than the \$1000 guideline is available.

7.3.3 Goals for Normal Activities

Lower bound goals: Examples of quantitative design goals for lower bound L for light-water-cooled power reactor are found in Title 10 of CFR Part 50, Appendix I [1]. They are expressed in terms of maximum permissible annual individual doses:

(1) Liquid effluent radioactivity; 3 millirems for the whole body and 10 millirems to any organ.

(2) Gaseous effluent radioactivity; 5 millirems to the whole body and 15 millirems to the skin.

(3) Radioactive iodine and other radioactivity; 15 millirems to the thyroid.

If one uses the ratio of 135 lifetime fatalities per million person-rems, then the 3 mrems whole body dose for liquid effluent radioactivity computes to a probability of 4 premature fatalities in 10 million. Similarly, 5 mrems of whole body dose for gaseous effluent radioactivity yields a probability of 6.7×10^{-7}/(lifetime, individual) fatality per year of exposure. These values comply with individual risk lower bound $L = 10^{-6}$/(year, individual) proposed by Wilson or by the de minimis risk.

Upper bound goals: According to the current radiation dose rate standard, a maximum allowable annual exposure to individuals in the general population is 500 mrems/year, excluding natural background and medical sources [6]. As a matter of fact, the average natural background in the US is 100 mrem/year, and the highest is 310 mrem/year [6]. The 500 mrems/year standard yields a probability for a premature fatality of 6.7×10^{-5}. This can be regarded as an upper bound U for an individual.

If the Wilson's bounds are used, the premature fatality likelihood lies in the middle layer, $L = 10^{-6} < 6.7 \times 10^{-5} < U = 10^{-3}$. Thus, the risk must be justified; otherwise, the risk should be reduced below the lower bound. A possibility is to reduce the risk below the maximum allowable exposure by using a safety-cost trade-off value such as the \$1000 person-rem in NRC's ALARA concept [1].

A maximum allowable annual exposure to radiological industrial workers is 5 rems/year [6] which is much less stringent than for individuals in the general population. Thus, we do have different upper bounds U's for different situations.

Having an upper bound goal as a necessary condition is better than nothing; Some unsafe alternatives are rejected as unacceptable; the chance of such a rejection is increased by gradually decreasing the upper bound level. A similar goal for the upper bound has been specified for NO_2 concentrations caused by automobiles and factories. Various upper bound goals have been proposed for airplanes, ships, automobiles, buildings, medicines, food, etc.

7.3.4 *Goals for Catastrophic Accidents*

Lower bound goals: Some lower bound goals for catastrophic accidents are stated qualitatively. A typical example is the qualitative safety goals proposed by USNRC in 1983 [12,2]. The first is related to individual risk, while the second is for population risk:
(1) **Individual risk:** Individual members of the public should be provided a level of protection from the consequences of nuclear power plant operation such that individuals bear no significant additional risk to life and health.
(2) **Population risk:** Societal risks to life and health from nuclear power plant operation should be comparable to or less than the risks of generating electricity by viable competing technologies and should not be a significant addition to other societal risks.

The NRC proposal also includes quantitative design objectives (QDOs):
(1) **Prompt fatality QDO:** The risk to an average individual in the vicinity of a nuclear power plant of prompt fatalities that might result from reactor accidents should not exceed one-tenth of one percent (0.1 percent) of the sum of prompt fatality risks resulting from other accidents to which members of the US population are generally exposed.
(2) **Cancer fatality QDO:** The risk to the population in the area near a nuclear power plant of cancer fatalities that might result from nuclear power plant operation should not exceed one-tenth of one percent (0.1 percent) of the sum of cancer fatality risks resulting from all other causes.
(3) **Plant performance objective:** The likelihood of a nuclear reactor accident that results in a large-scale core melt should normally be less than 1 in 10,000 per year of reactor operation.
(4) **Cost-benefit guideline:** The benefit of an incremental reduction of societal mortality risks should be compared with the associated costs on the basis of \$1000 per person-rem averted.

The prompt (or accident) fatality rate from all causes in the US in 1982 was 4×10^{-4} per year; 93,000 deaths in a population of 231 million. At 0.1% of this level, the prompt fatality QDO becomes 4×10^{-7} per year which is substantially below the lower bound $L = 10^{-6}$ for an individual [1]. In 1983, the rate of cancer fatalities was 1.9×10^{-3}. At 0.1% of this background rate, the second QDO is 1.9×10^{-6} which is less limiting than the lower bound.

On August 4, 1986, the NRC left unchanged the two proposed qualitative safety goals (individual and population) and the two QDOs (prompt and cancer). It deleted the plant performance objective for the large scale core melt. It also deleted the cost-benefit guideline. The following guideline was proposed for further examination:
(5) **General performance guideline:** Consistent with the traditional defense-in-depth approach and the accident mitigation philosophy requiring reliable performance of containment systems, the overall mean frequency of a large release of radioactive materials to the environment from a reactor accident should be less than 1 in 1,000,000 per year of reactor operation.

The general performance guideline is also called an FP (Fission Products) large release criteria. Off-site property damage and erosion of public confidence by accidents are considered in this criteria in addition to the prompt and cancer fatalities.

The International Atomic Energy Agency (IAEA) recommended other quantitative safety targets in 1988 [15,16]:
(1) For existing nuclear power plants, the probability of severe core damage should be below 10^{-4} per plant operating year. The probability of large off-site releases requiring short-term responses should be below 10^{-5} per plant operating year.
(2) For future plants, probabilities lower by a factor of ten should be achieved.

The future IAEA safety targets are comparable with the plant performance objective and the general performance USNRC guideline.

Risk aversion goals: Neither the USNRC QDOs nor the IAEA safety targets consider explicitly risk aversion in severe accidents; two accidents are treated equivalently if they yield the same expected numbers of fatalities, even though one accident causes more fatalities with a smaller likelihood. The Farmer curves can be used to reflect the risk aversion.

7.3.5 Regulatory Cutoff Level

The Wilson's lower bound goal $L = 10^{-6}/(\text{year, individual})$ is reasonable either from idealistic or pragmatic viewpoints when a relatively small population is affected by the risk. A typical example of such a population would be a crew of the US space shuttle. However, when a large number of people are exposed to the risk, the lower bound is not a suitable measure for the unconditional acceptance of the risk, i.e., the Wilson's lower bound is not necessarily a suitable measure for the population risk.

An individual risk of $10^{-6}/(\text{year, individual})$ is sufficiently close to the idealistic, platonic, lower bound sufficiency condition, i.e., the de minimis risk. However, such a risk level is far from idealistic for risks to large populations. The lower bound L as a de minimis level for population risks must be a sufficiently small fractional number; less than one death in the entire population per year. If some risk levels greater than this de minimis level is adopted as a lower bound, the reason must come from factors outside the risk itself. A pragmatic lower bound is called a regulatory cutoff level.

A pragmatic cutoff level is, in concept, different from the de minimis level: 1) the regulatory cutoff level is a level at or below which there are no regulatory concerns (a level by pragmatism); 2) a de minimis level is the lower bound level L at or below which the risks are accepted unconditionally; some risks below the regulatory cutoff level may not be acceptable, although the risks are not regulated: the risks are only reluctantly accepted as a necessary evil. Consequently, the de minimis level for the population risk is smaller than the regulatory cutoff level currently enforced.

Containment structures with 100-foot-thick walls, population exclusion zones of hundreds of square miles, dozens of standby diesel generators for auxiliary feedwater systems, and so on are avoided by regulatory cutoff levels implicitly involving cost considerations [7].

7.3.6 Population Size and Regulatory Cutoff Level

Paul Milvy [4] claims that a 10^{-6} lifetime risk to the U.S. population is a realistic and prudent regulatory cutoff level for the population risk. This implies 236 additional deaths over a 70-year interval (lifetime), and 3.4 deaths per year in the population of 236 million. This section briefly overviews a risk-population model as the regulatory cutoff level.

Constant likelihood model: When the regulatory cutoff level is applied to an individual, or a discrete factory, or small community population that is uniquely at risk, its consequences become extreme. A myriad of society's essential activities would have to cease. Certainly the X-ray technician and the short-order cook exposed to benzopyrene in the smoke from charcoal-broiled hamburgers are each at an individual cancer risk considerably higher than the lifetime risk of 10^{-6}. Indeed, even the farmer in an agricultural society is at a 10^{-3} to 10^{-4} lifetime risk of malignant melanoma from pursuing his trade in the sunlight. The 10^{-6} lifetime criterion may be appropriate when the whole US population is at risk, but to enforce such a regulatory cutoff level when the exposed population is small is not a realistic option. Thus, the following equation for regulatory cutoff level L_1 is too strict for a small population:

$$L_1 = 10^{-6}/\text{lifetime}$$

Constant fatality model: On the other hand, if a limit of 236 deaths is selected as the criterion, the equation for cutoff level L_2 for a lifetime is.

$$L_2 = (236/P)/\text{lifetime}, \quad P\text{: population size}$$

This cutoff level is too risky for a small population size of several hundred.

Geometric mean model: We have seen that, for small populations, L_1 from the constant likelihood model is too strict and that L_2 from the constant fatality model is too risky. On the other hand, the two models give the same result for the whole US population. Multiplying the two cutoff levels, and taking the square root yields the following equation which is based on a geometric mean of L_1 and L_2.

$$L = 0.015/\sqrt{P}$$

Using the equation with $P = 100$, the lifetime risk for the individual is 1.5×10^{-3} and the annual risk is 2.14×10^{-5}. This value is nearly equal to the lowest annual fatal occupational rate from accidents which occur in the "finance, insurance and real estate" occupational category. The geometric mean risk-population model is deemed appropriate only for populations that number 100 or more because empirical data suggest that smaller populations are not really relevant in the real world in which environmental and occupational carcinogens almost invariably expose groups of more than 100 people. The cutoff level L is consistent with the historical data of regulatory decisions by the Environmental Protection Agency.

7.3.7 Summary

Risk goals can be specified on various levels of system hierarchy in terms of a variety of measures. The safety goal on the top level is a starting point for specifying the goals on the lower levels. PRAM procedures become more useful when a hierarchical goal system is established. A typical decision procedure with the safety goals forms a three-layer structure. The ALARA principle or RCB analysis operates in the second layer. The de minimis risk gives the lower bound goal. The upper bound goal rejects risks without overriding benefits. Current upper and lower bounds goals are given for normal activities and catastrophic accidents. When a risk to large population is involved, the current lower bound goals should be considered as pragmatic goals or regulatory cutoff levels. The geometric mean model explains the behavior of the regulatory cutoff level as a function of population size.

7.4 Accident Causation Mechanisms

7.4.1 Common Features of Plants with Risks

Common aspects of plants with potentially catastrophic consequences are first clarified. These are 1) Physical containment, 2) stabilization of unstable phenomena, 3) large size, 4) new technology, 5) component variety, 6) complicated structure, 7) large inertia, 8) large consequence, 9) strict social demand for safety. The FTA frequently deals with these types of plants.

Physical containment: A plant is usually equipped with physical barriers or containments to confine hazardous materials or to shield hazardous effects. These containments are called physical barriers. For nuclear power plants, these barriers include fuel cladding, primary coolant boundary, and containment structure. For commercial airplane, various portions of airframe are physical containments. Water collecting wells and banks are simpler examples of physical containments. So long as these containments or barriers are intact, no serious accident could occur.

Stabilization of unstable phenomena: Most plants create benefits by stabilizing unstable phenomena. Together with physical containments, these plants require 1) normal control systems during routine operations, 2) safety systems during emergencies, and 3) in-site and off-site emergency countermeasures, as shown in **Fig. 7.7**. Physical barriers, normal control systems, engineering safety systems, and emergency countermeasures correspond for instance to body skin, body temperature control, immune mechanism and medical treatment, respectively.

If something goes wrong with normal control systems, incidents would occur; if emergency safety systems fail to cope with the incidents, plant accidents would occur; if in-site emergency countermeasures fail to ensure retention of accident the physical containments, the accident invades into the environment; if off-site emergency countermeasures fails to cope with the invasion, serious consequences for the public and environment would occur. FTs are typically used to analyze top events about plant, physical barriers, normal control systems, and emergency safety systems.

The stabilization of unstable phenomena is the most crucial feature of systems with catastrophic risks. For nuclear power plants, most important stabilization functions are 1) power control systems during normal operation and safety shutdown systems during emergencies, and 2) normal and emergency core cooling systems, and 3) confinement capabilities of radioactive materials during operation, maintenance, engineering modification, and accident.

Large size: Plants with catastrophic risks are frequently large in their size. Examples include commercial airplanes, space rockets, shuttles, space stations, chemical plants, metropolitan power networks, and nuclear power plants. These plants tend to become larger because of the following reasons.

1) Scale merits: Cost per product or service generally decreases as the size. This is typically observed for ethylene plants in chemical industries.

2) Demand satisfaction: Larger commercial airplanes can meet larger demands of air travel.

3) Amenity: Large airplanes allow more passenger space and require less baggage constraints.

New technology: The size increase requires new technologies. A cockpit of contemporary commercial airplane is as high as a third floor of a building. A captain at this height must be supported by new technologies such as autopilot systems and computerized displays to maneuver his airplane for landing and taking off. These technologies would reduce airplane accidents but may introduce some pitfalls during initial burn-in periods.

Component varieties: A large number of system components of various types are used. Components include not only hardware but also human beings, computer programs, procedures, instructions, specifications, drawings, charts, labels, etc. Large scale systems consist of more than one million components. Failures of some components would initiate or enable event propagations toward an accident. Human beings must perform their tasks in this jungle of hardware and software.

Complicated structure: A plant and its operating organization form a complicated structure with various lateral and vertical interactions inside. A composite hierarchy is formed in a way such as component, individual, unit, team, subsystem, department, facility, plant, corporation, and general environment.

Large inertia: An airplane can not stop suddenly; it must keep on flying to an airport anyway. A chemical plant or a nuclear power plant requires a long period to achieve a cold state after initiation of plant shutdown. A long period is also required for resuming plant operations.

Large consequence: An accident would cause large consequences; some have direct effects on individuals, societies, environment and plant; others may cause various indirect effects such as research and development schedule delays, commercial share reductions, public opposition increases, company credibility declines.

Strict social demand for safety: Society demands that airplanes and nuclear power plants should be far safer than automobiles.

7.4.2 A Taxonomy of Negative Interactions

A description of accident causation mechanisms involves a variety of negative interactions. Some of these interactions are listed here from four points of view; "why", "how", "when", and "where". Why-classification emphasizes causes of failures and errors; how-classification is based on behavioral aspects observable; when-classification is based on a time frame when a failure occurs; where-classification looks places where failures or errors

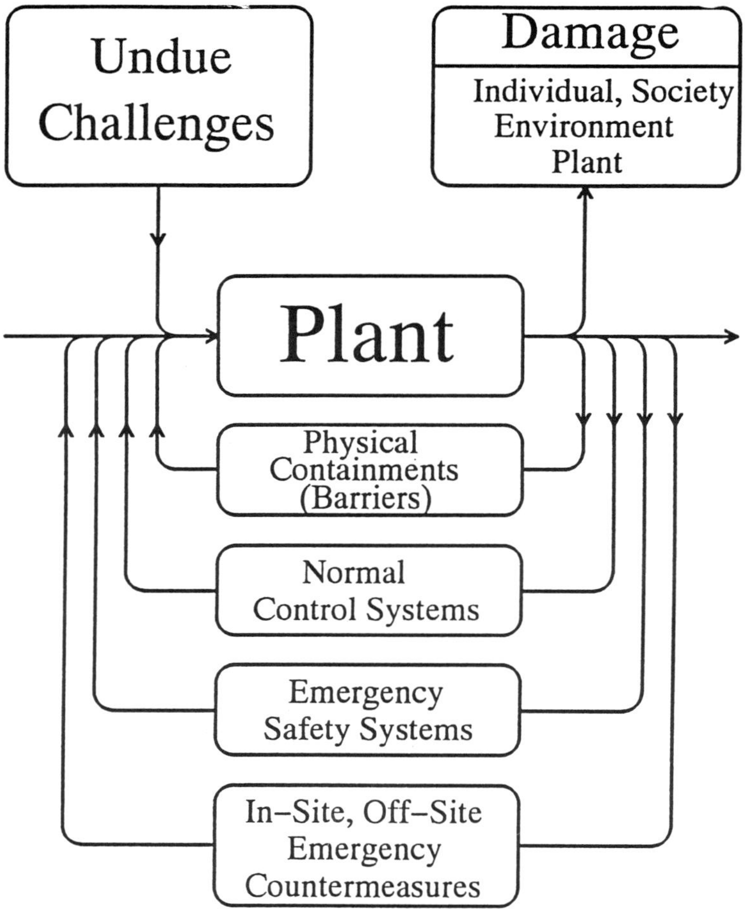

Fig. 7.7 Protection Configuration for Plant with Catastrophic Risks

occur. Most of these interactions are included in FTs.

WHY-CLASSIFICATION

Mechanical, functional and interfacing failures: For mechanical failures, a device becomes unusable and fails to perform its function because some of its components mechanically fail. For functional failures, a device is usable but fails to perform its function by causes not attributable to device or its component; a typical example is a perfect TV set when a TV station has a problem. When two devices are interfaced by a third device, a failure of this interface is called an interfacing failure; this may cause a functional failure of one or two devices interfaced; a computer printer would fail if its device driver has a bug.

Primary, secondary, and command failures: A primary failure of a device is a failure of the device itself. An example is a stuck-closed valve. The primary failure is more general than a mechanical failure because an incorrect valve operation can be regarded as a primary failure of valve operator as long as operating instructions are correct. A secondary failure is a failure where a device is forced to fail by abnormal conditions beyond design specifications. An example is a valve open failure by vibration caused by earthquakes. A command failure is a failure caused by incorrect commands to a device. An example is a valve close failure by an incorrect close command. Secondary and command failures can apply to human errors where the human is regarded as a device. Thus, if a valve operator opens a valve according to an incorrect instruction or procedure, his failure is a command failure.

Basic and intermediate failures: A basic failure is a failure with highest resolution and with available likelihood data of occurrence. An example is a mechanically stuck-closed failure of an electrical switch. Failures caused by a propagation of basic failures are called intermediate. An example is a lack of circuit electricity caused by a switch failure. A primary or mechanical failure is usually considered as a basic failure; occurrence likelihood data are often unavailable for a secondary failure; a command failure would eventually analyzed into basic failures and secondary failures by a fault tree analysis.

Parallel and cascade failures: Two or more failures may be caused by a single cause. This parallel or fanning out propagation is called a parallel failure. Two or more failures may occur sequentially starting from a cause. This sequential or consecutive propagation is called a cascade or sequential failure. An accident scenario consists of a mixture of parallel and cascade failures.

Direct, indirect, and root causes: A direct cause is a cause most adjacent in time to a device failure. A root cause is an origin of direct causes. Causes between a direct and a root cause is called indirect.

Main cause and supplemental causes: A failure may occur by simultaneous occurrence of more than one causes. One is regarded as a main cause, while others are as supplemental causes.

Inducing factors: Some causes do not necessarily yield a device failure; these only increase chances of failures. These causes are called inducing factors. A habitual smoking is an inducing factor of heart failure. Inducing factors are also called risk factors, background factors, contributing factors or shaping factors. Management and organizational deficiencies are regarded as inducing factors.

Hardware induced, human induced and system induced failures: This classification is based on what portions of a system trigger or facilitate failures. Human errors caused by an erroneous indicator is hardware induced. Hardware failures caused by incorrect operations are human induced. Human and hardware failures caused by improper management are termed system induced.

HOW-CLASSIFICATION

Random, wearout, and initial failures: A random failure occurs by a constant rate of occurrence; an example is automobile failures in its prime of life. A wearout failure occurs with an increasing rate of occurrence; an old automobile in a burn-out period suffers from wearout failures. An initial failure occurs with an decreasing rate of occurrence; an example is brand-new automobile failures in a burn-in period.

Demand and run failure: A demand failure is a failure for a device to start or stop its operation when it receives a start/stop command; this failure may be called a start or a stop failure. An example is a diesel generator failing to start upon receipt of a start signal. A run failure is the one where a device fails to continue its operation. A diesel generator failing to continue its operation is a typical example of run failures. A run failure is called a duration failure.

Persistent and intermittent failures: A persistent failure is the one where a device failure continues once its failed. For an intermittent failure, a failure only exists during sporadic time intervals. A relay may fail closed intermittently. A typical cause of an intermittent failure is an external noise to a device.

Active and latent errors [19]: In active errors, their effects are felt almost immediately, while in latent errors their adverse consequences may lie dormant for a long time, only becoming evident when they combine with other factors to breach system defenses. Active errors are typical to front-line operators. Latent errors are most likely caused by designers, high-level decision makers, construction workers, managers and maintenance personnel.

One characteristic of latent errors is that they do not immediately degrade a system, but in combination with other events, which may be active human errors or other random hardware failures, they give rise to a catastrophic failure. Two categories of latent errors can be identified; operational and organizational. Typical operational latent errors include maintenance errors, which may make a critical system unavailable or leave the system in a vulnerable state. Organizational latent errors include design errors, which yield intrinsically unsafe systems, and management or policy errors, which create conditions inducing active human errors. The latent failure concept is discussed more fully in Reason [19] and Wagenaar et al [21].

Omission and commission errors: When a necessary action is not performed at all, this failure is called an omission error. An example is an operator forgetting to read an level indicator or to manipulate a valve. A commission error is the one where a necessary step is performed but in an incorrect way.

Independent and dependent failures: Failures A and B are independent when the following product low of probability holds:

$$Pr\{A \text{ and } B\} = Pr\{A\}Pr\{B\}$$

Failures are dependent otherwise:

$$Pr\{A \text{ and } B\} \equiv Pr\{A \mid B\}Pr\{B\} \equiv Pr\{B \mid A\}Pr\{A\}$$

$$\neq Pr\{A\}Pr\{B\}$$

Independent failures are sometimes called random failures; this is misleading because failures with a constant occurrence rate are also called random.

WHEN-CLASSIFICATION

Recovery failure: A failure of recovering abnormal device state to a normal one is called a recovery failure. This is typical after maintenance and test; a valve may inadvertently left closed; a supervisor may fail to carry out specified wiring checks and wiring errors were not corrected; irregular signals being noticed by train drivers prior to the accident but not reported because of time pressure [20].

Initiating and enabling events [5]: Initiating events cause system upset conditions and triggers responses of system mitigative features. Enabling events cause failures of system mitigative features to respond to initiating events; enabling events could result in a serious accident such as a core melt and a radiological release, given an occurrence of the initiating event. In nuclear power plants, an initiating event is further classified into LOCA (Loss Of Coolant Accident) and transients.

Routine and cognitive errors: Suppose that a necessary action is known in advance as an intention. Then, a human error for performing the action is called a routine or a skill-based error. Otherwise, a series of necessary intentions for actions must be invented by cognitive efforts. A cognitive error is an incorrect generation of necessary intentions to be executed. A typical example of cognitive error is an error of diagnosis where plant state should be identified from symptoms, and appropriate actions must be planned accordingly.

Lapse, slip, and mistake: Suppose that a series of necessary actions are known, for instance, by an operation procedure. This series is nothing but a series of intentions. A lapse is a failure of recalling one of intentions, i.e., a lapse is an omission error. An example is a bypassing error of a necessary step in an operation procedure. A slip is a failure of executing particular intention when it is recalled correctly. An example is inadvertently pushing a gas pedal when a driver intends to step on a brake. Both lapse and slip are errors during an intention execution period when people try to carry out a sequence of actions according to intentions previously formed. Thus, lapse and slip are two types of routine errors. A mistake is a cognitive error, i.e., it is a failure during an intention formation period when people decide what actions are appropriate for a particular situation.

WHERE-CLASSIFICATION

Internal and external events: An internal event occurs by causes inside a system boundary, while an external event outside the boundary. Typical examples of external events include earthquakes and area power failures. Initiating events are classified into internal or external events.

Active and passive failures: A device is called active when it functions by changing its state; an example is an emergency shutdown valve which is normally closed. A device without a state change is called passive; a filter, a pipe or a containment are typical examples. An active failure is an active device failure, while a passive failure is a passive device failure.

LOCA and transient [5]: A LOCA is a breach of a pressure boundary of a Reactor Coolant System (RCS) which causes an uncontrollable loss of water inventory. There are two LOCA categories: Large LOCA and small LOCA. Transients are any abnormal condition of nuclear power plant which requires that the plant be shut down, but does not directly breach RCS integrity. A loss of offsite power is an example of transients. Another example is a loss of main feed water in a secondary loop to a steam generator.

7.4.3 Chronological Distribution of Failures

Some people might consider that accidents can only occur by failures during plant operations. This view is too narrow. Failures could occur at various chronological stages which should be considered in FTA.

(1) Siting: A site is an area within which a plant is located and which is under an effective control of an operating organization [18]. Local characteristics including natural factors and man made hazards can affect plant safety. Natural factors include geological and seismological characteristics and potential for hydrological and meteorological disturbances. Man made hazards include those arising from chemical installations, release of toxic and flammable gases, and aircraft impact. Air, food-chains and water supplies provide pathways for possible transport of hazardous materials to humans in accident situations.

(2) Design: This include 1) prototype design activities during research, development and demonstration periods, and 2) commercial design for a real use. Design errors may be committed during a series of scaling ups by insufficient budget, and by short research, development and monitoring periods. Some key technologies remain black boxes due to technology license contracts. Designers only know how to use black boxes but do not know reasons why such usages are justified. This may cause inadvertent design errors especially when black boxes are used or modified apart from original hidden specifications.

In monitoring device recalls, the Food and Drug Administration (FDA) has compiled data which show that from October 1983 to November 1988, approximately 45% of all recalls were due to preproduction related problem. These problems indicate that deficiencies had been incorporated into the device design during a preproduction phase [23].

A design deficiency can be very costly once a device design has been released to production and a device is manufactured and distributed. Costs may include not only replacement and redesign costs, with resulting modifications to manufacturing, procedures, and retraining. to enable manufacture of the modified device, but also liability costs, and loss of customer faith in the product [23].

(3) Manufacturing and construction: Defects may be introduced during the manufacturing and construction; a plant could be fabricated and constructed with deviations from original design specifications and from fabrication/construction procedures.

(3) Commissioning: Errors in design, manufacturing and construction stages may persist after commissioning which demonstrates that a completed plant is satisfactory for service before it is made operational. A simple example of commissioning failures is a software package with bugs.

(5) Operation: This is classified into 1) normal operation, 2) operation during anticipated abnormal occurrences, and 3) operation during complex events below design basis, and 4) operation during complex events beyond design basis.

4-1) Normal operation: This stage refers to a period where no challenge is posed to plant safety. The period includes start-up, stationary, or shutdown. The normal operation includes 1) daily operation, 2) maintenance, 3) testing, 4) inspection, and 5) minor engineering modifications. Tight operation schedules and unfixed instrumentation failures may induce operator errors. Safety features may inadvertently be left nullified after maintenance because some valves may incorrectly be left closed; these types of maintenance failure typically constitute latent failures. Safety systems are intentionally nullified to avoid too many false alarms. Safety systems become unavailable during a test interval. Incorrect valve operation sequences may be followed during start-up or shutdown periods.

4-2) Anticipated abnormal occurrences: These occurrences may be countered in a straightforward manner by appropriate response of normal control systems depicted in Fig. 7.7. The term "anticipated" means such an event can occur more than once in a plant life. If some of normal control systems may fail, anticipated abnormal occurrences would develop into complex events described below.

4-3) Complex events below by design basis: System designers assume that hardware, software, and human failures are possible, and can lead to abnormal occurrences, ranging from minor disturbances such as the anticipated abnormal occurrences to highly unlikely accident sequences. Additional protection is achieved by incorporation of many engineered features into the plant. These features consist of passive features such as physical barriers and active features such as emergency shutdown systems, emergency cooling systems, and filtered vent systems. Active systems in engineered features are called engineered safety systems of which performance is measured by "on demand" performance and "duration" performance. Simplest forms of safety systems include circuit breakers and rupture disks. As shown in Fig. 7.7, these safety features would be required to supplement protections afforded by normal control systems.

Design parameters of each engineered safety feature are defined by a classical, deterministic analysis which evaluates effectiveness against complex events it is intended to control. An events in a spectrum requiring the most extreme design parameters for a safety feature is termed a design basis for that feature. The engineering safety features are provided to halt progress of an event below the design basis, and, when necessary, to mitigate its consequences.

Safety features may fail to respond to complex events below design basis because something are wrong with the features themselves. High stress conditions after a plant upset may induce human errors. These failures cause events below design basis to develop complex events beyond design basis.

4-4) Complex events beyond design basis: Attention is also directed to events of very low likelihood but more severe than those considered explicitly in the design. Such an event beyond design basis is called a severe accident because the plant must have received some damage. For the nuclear power plant, for instance, some of these severe accidents could cause such deterioration in plant conditions that proper core cooling cannot be maintained, or that fuel damage occurs for other reasons. These accidents would have a potential for major radiological consequences if radioactive materials released from fuel rods were not adequately confined.

The classification of events into 1) normal operation, 2) anticipated abnormal occurrences, 3) complex events below design basis, and 4) complex events beyond design basis, is taken from No.75-INSAG-3 document from IAEA [18].

7.4.4 Event Layer and Likelihood Layer

Given event trees and fault trees, various computer codes are available to calculate probabilistic parameters of an accident. However, risk assessment is not a simple manipulation of probability formulas. Recently, more emphases are placed on how to evaluate basic occurrence probabilities to be fed into these computer codes.

EVENT LAYER

Consider the event tree and fault tree in Fig. 7.6 in section 7.2.5. We observe that the tank rupture due to over-pressure is caused if the three events occur simultaneously: 1) pump overrun, 2) operator shutdown system failure, and 3) pressure protection relief valve failure. The pump overrun event occurs if either of the two events occurs: 1) timer contact fails to open, or 2) voltage surge. These causal relations described by the event and fault trees are on an event layer level.

Event layer descriptions yield explicit causes of accident in terms of event occurrences. These causes are hardware or software failures, or human errors at a front line. Fault tree and the event tree explicitly contain these failures. Front line failures are analyzed into their ultimate resolutions by a fault tree analysis and basic events are identified. However, these basic events are not the ultimate causes of "top" event being analyzed because occurrence likelihoods of the basic events are determined by a likelihood layer described below.

LIKELIHOOD LAYER

Factors increasing likelihoods of events are true causes of accident. However, event and fault trees only describe causal relations in terms of a set of if-then statements. Occurrence probabilities of basic events, statistical dependence of event occurrences, simultaneous increase of occurrence probabilities, and occurrence probability uncertainties are greatly influenced by various shaping factors in the likelihood layer. This point is shown in **Fig. 7.8**.

The likelihood layer determines, for instance, device failure rates, statistical dependency of device failures, simultaneous increase of failure rates, and failure rate uncertainties. These shaping factors would neither appear explicitly in fault nor event trees; they can affect accident causation mechanisms by specifying occurrence probabilities of events in the trees. For instance, initiating events, operator actions, and safety system responses in event trees are affected by the likelihood layer. Similar influences exist for events in fault trees.

EVENT-LIKELIHOOD MODEL

Fig. 7.9 shows a failure type distribution in the event layer and shaping factors in the likelihood layer, which is proposed by P. Embrey [20]. When an accident such as Chernobyl, Exxon Valdez or Clapham Junction is analyzed in depth it appears at first sight to be unique. However, certain generic features of such accidents are apparent when a large number of cases are examined. Fig. 7.9 is intended to indicate, in a simplified manner, how such a generic model might be represented. The generic model is called MACHINE (Model of Accident Causation using Hierarchical Influence Network Elicitation). The direct causes in the event layer of all accidents are combinations of human errors, hardware failures and external events.

Human errors: The human errors are classified into active, latent and recovery failures. The likelihoods of these failures are influenced by factors such as training, procedures, supervision, definition of responsibilities, demand/resource matching, and produc-

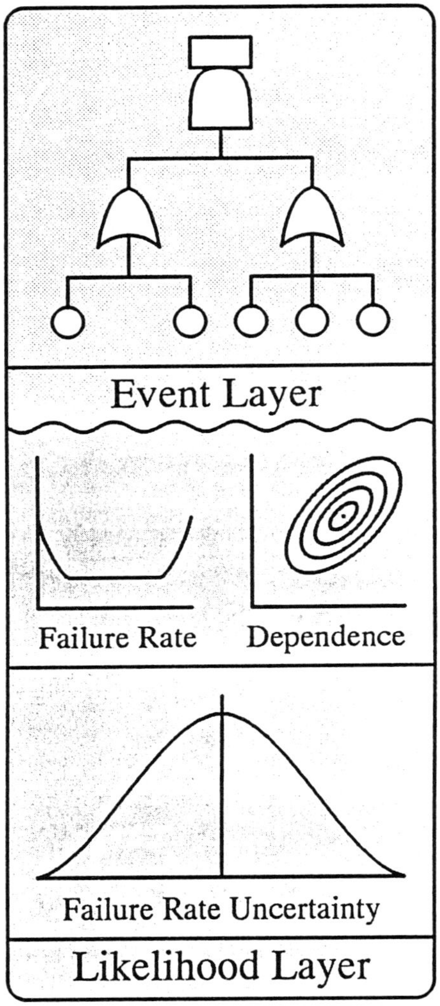

Fig. 7.8 Event Layer and Likelihood Layer for Accident Causation

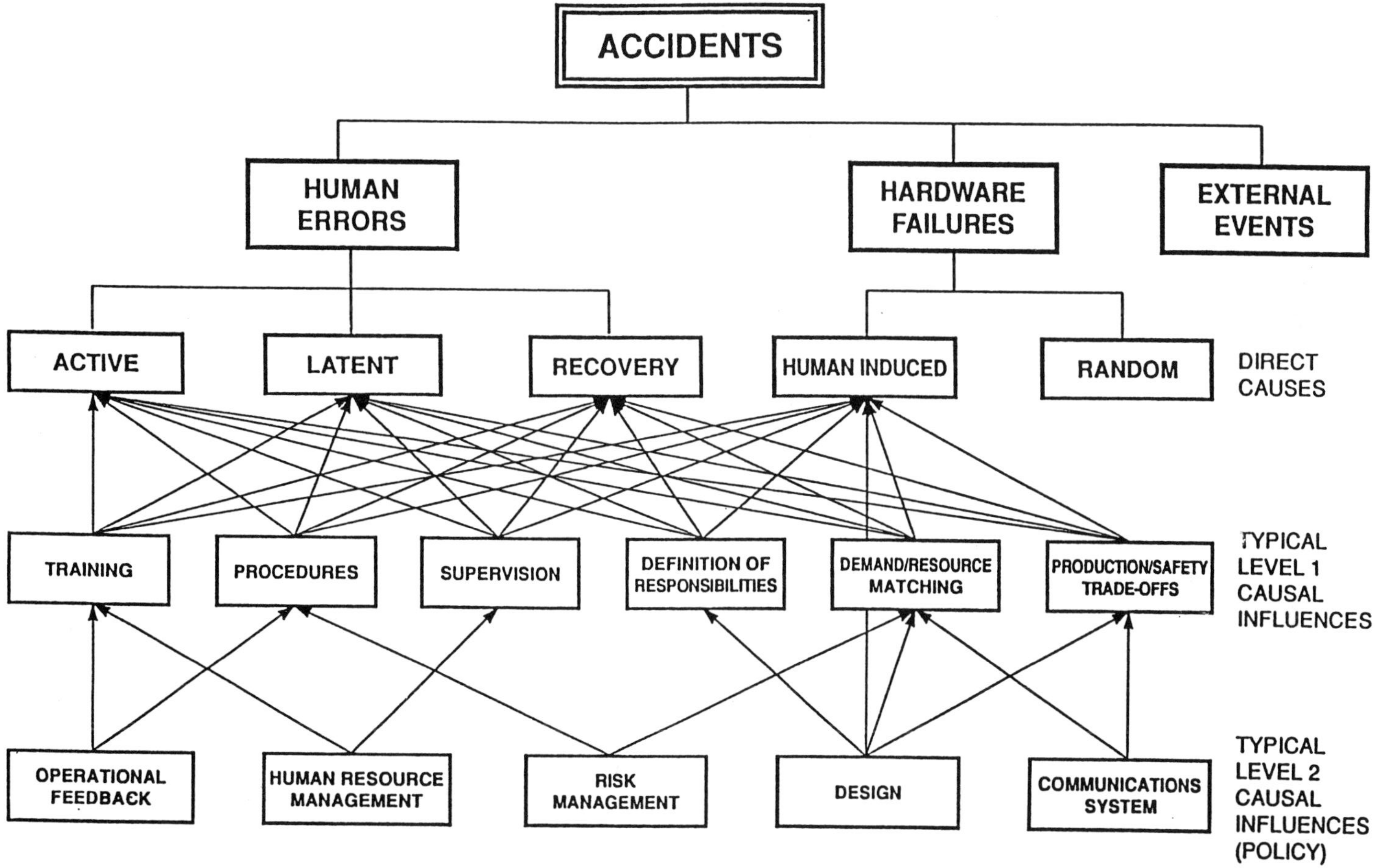

Fig. 7.9 Generic Model of Accident Causation

tion/safety trade-offs. These factors are influenced by some of the higher policy factors such as operational feedback, human resource management, risk management, design, and communications system.

Hardware failures: In the case of hardware failures, these can be categorized under two headings. "Random" failures are ordinary failures considered by reliability models, e.g. due to the anticipated processes of wear. Extensive data are available on the distribution of such failures from testing and other sources. Human induced failures comprise two subcategories, those due to human actions in areas such as assembly, testing and maintenance, and those due to inherent design errors which give rise to unpredicted failure modes or reduced life cycle.

As all reliability engineers will be aware, most failure rates for components derived from field data actually include contributions from human induced failures. To this extent, such data are not intrinsic properties of the components, but are dependent on the human influences (management, organizational) in systems where the components are employed.

External events: The third major class of direct causes are external events. These events are the characteristics of environment in which the system operates. Typical external events considered in PRA include seismic, geological, and other natural phenomena and collisions from aircraft and other objects, depending on the system. Such events are considered to be independent of any human influence within boundaries of the system being analyzed, although the risk management policy is expected to ensure that adequate defenses are available against external events which constitute significant threats to the system.

EXAMPLE: EVENT TREE ANALYSIS [22]

Simple event tree: Consider the event tree in **Fig. 7.10** which includes 1) an initiating event (IE), 2) two operator actions, and 3) two safety system responses. In this oversimplified example, damage to the reactor core can be prevented only if both operator actions are carried out correctly and both plant safety systems function accordingly. The estimated frequency of core damage (CD) for this specific initiating event is,

$$f_{CD} = f_{IE}[1 - (1 - p_1)(1 - q_{av,1})(1 - p_2)(1 - q_{av,2})]$$

where f_{CD} = frequency of core damage (caused by this particular initiating event); f_{IE} = frequency of the initiating event; p_i = probability of error of the ith operator action conditioned on its prior events; and $q_{av,i}$ = unavailability of the ith safety system conditioned on its prior events.

Safety system unavailability: Quality of organization and management should be reflected in the parameters f_{IE}, p_i, and $q_{av,i}$. Denote by q_{av} an average unavailability during an interval between periodic tests. The average unavailability is an approximation of time-dependent unavailability $q_{av,i}$, and is given by

$$q_{av} = \frac{T_0}{T} + \gamma + Q + \frac{1}{2}\lambda T$$

where

T = interval between tests,
T_0 = duration of test,
γ = probability of failure due to testing,
Q = probability of failure on demand,
λ = rate of random failure between tests.

Thus, various contributions to the average unavailability are: 1) T_0/T = test contribution while the safety system is disabled during testing; 2) γ = human error of testing; 3) Q = failure on demand; and 4) $1/2\lambda T$ = random failures between tests while the safety system is on standby.

Likelihood layer contributions: As shown in **Fig. 7.11**, these contributions are affected by maintenance activities actually performed. These activities are, in turn, affected by quality of maintenance procedures. Qualities of various procedures are determined by an overall factor such as 1) safety knowledge, 2) attitude toward plant operation and

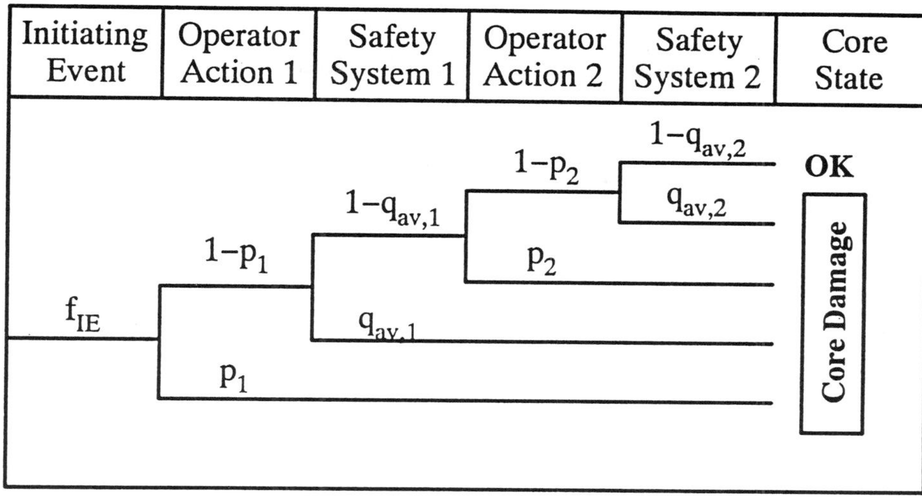

Fig. 7.10 Simple Event Tree with Two Operator Actions and Two Safety Systems

maintenance, 3) choice of plant performance goals, and 4) communication, 5) authorities, and 6) responsibilities. This figure is a simplified version of the one proposed by J.S. Wu, G.E. Apostolakis, and D. Okrent [22].

Safety knowledge: The safety knowledge refer to everyone responsible possessing deep knowledge of plant behavior, severe accident consequences, and related subjects, thus leading to an overall state of pervasive safety thinking.

Attitude: An uneventful situation of routine nuclear power plant operation often makes its working environment boring rather than challenging. People involved in plant operation are often misled by stagnant atmosphere of the plant, which can result in a slack and inattentive working attitude. This kind of attitude is often further amplified by that automatic feedback systems correct minor deviations and keep the plant functions normally without interruption. Plant personnel may misinterpret stagnation as safety. A team with a slack and inattentive attitude toward plant operation is expected to experience more difficulties in bringing the plant back to normal operation after an abnormal occurrence.

Plant performance goal: Plant performance goals are set by utility or plant managers at a higher organizational level and will influence plant personnel in making decision during plant operation. For example, if an operating team constantly receives pressure and encouragement from higher-level managers to achieve high plant availability and to increase the electricity production during daily operations, it will not be a surprise if operators weigh economic consequences higher than safety consequences during emergencies. Another extreme is a policy that plant safety will help in achieving more efficient, more economic generation of electricity.

Communication and responsibility: Some plant managers either knew operators sleeping during their shifts but took no action (lack of responsibility), or should have known about it (lack of communication). Some supervisors do not have sufficient time to be in the plant to directly observe and supervise the efforts of the work force. Some companies tend to rely solely on written communication rather than verbal face-to-face communication which is postulated be the highest approach to communication. Lessons learned by a plant in the industry are not utilized.

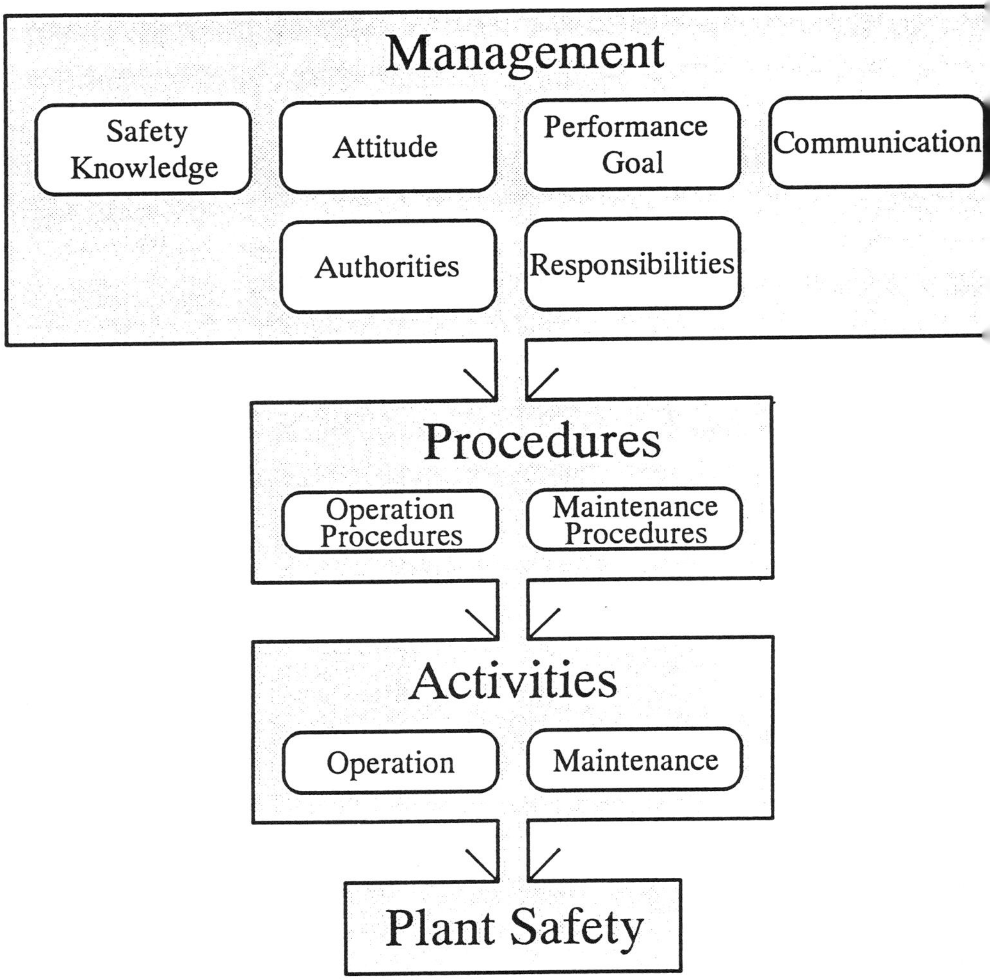

Fig. 7.11 Operation and Maintenance Affected by Management

7.4.5 Summary

Features common to plants with catastrophic risks are first presented; 1) confinement capabilities by physical containments and 2) stabilization of unstable phenomena are most important features. These plants therefore are protected by physical barriers, normal control systems, emergency safety systems, and in-site and off-site emergency countermeasures.

Various failures, errors, and events can occur for these potentially hazardous plants, and these can be regarded as series and parallel interactions between humans and plant. Some of these interactions are listed from points of view of "why", "how", "when", and "where". It is emphasized that these negative interactions occur at any time of plant life; siting, design, manufacturing, construction, commissioning, and operation. The plant operation period is divided into four phases; 1) normal operation, 2) anticipated abnormal occurrences, 3) complex events below design basis, and 4) complex events beyond design basis.

A configuration of a nuclear reactor shutdown system is presented to illustrate emergency safety systems which operate when plant states reach trip set points below safety limits but above operating ranges. Safety systems may fail in two failure modes, failed-safe and failed-dangerous, and various aspects of these failures are given through examples.

Accident causation mechanisms can be split into an event layer and an likelihood layer. Event and fault trees deal with the event layer. Recently more emphases are placed on the likelihood layer where management and organizational qualities play crucial roles for 1) occurrence probabilities, 2) dependence of event occurrences and dependent increase of probabilities, 3) uncertainties of occurrence probabilities themselves.

7.5 Risk Management

7.5.1 Risk Management Principles

Fig. 7.12 schematically shows risk management principles by IAEA document No.75-INSAG-3 [18]. Safety culture is a base of risk management. Proven engineering practice is established on safety culture. Various procedures are established on proven engineering practice. Activities are performed with strict adherence to these procedures.

The term "procedure" must be interpreted in a broader sense. It includes not only operation, maintenance, and training procedures, but also codes, standards, formulas, specifications, instructions, intentions, rules, laws etc. The activities include plant life cycle activities in section 7.4.3 ranging from siting to operation.

Adherence of a layer to a more basic one is intended but some deviations occur. Therefore, these deviations are monitored, and controlled. The term "monitor" implies verbs such as review, verify, survey, audit, test, inspect, etc. Similarly, the term "control" covers verbs such as correct, modify, repair, maintain, alarm, enforce, regulate, and so on. The multi-layer monitor/control feature in Fig. 7.12 is called a "quality assurance program."

This monitor/control program minimizes inadvertent deviations. However, some deviations may still occur. The safety assessment and verification below the safety culture layer elucidates these potential deviations to assess the monitor/control qualities for verification. In other words, layers above the safety culture try to remove and correct deviations, while the assessment layer tries to evaluate remaining deviations. The quality assurance includes the safety assessment and verification as a second element.

Safety culture The IAEA document defines the safety culture in the following way: The phrase 'safety culture' refers to a very general matter, the personal dedication and accountability of all individuals engaged in any activity which has a bearing on plant safety. The starting point for the necessary full attention to safety matters is with the senior management of all organizations concerned. Policies are established and implemented which ensure correct practices, with the recognition that their importance lies not just in the practices themselves but also in the environment of safety consciousness which they create. Clear lines of responsibility and communication are established; sound procedures

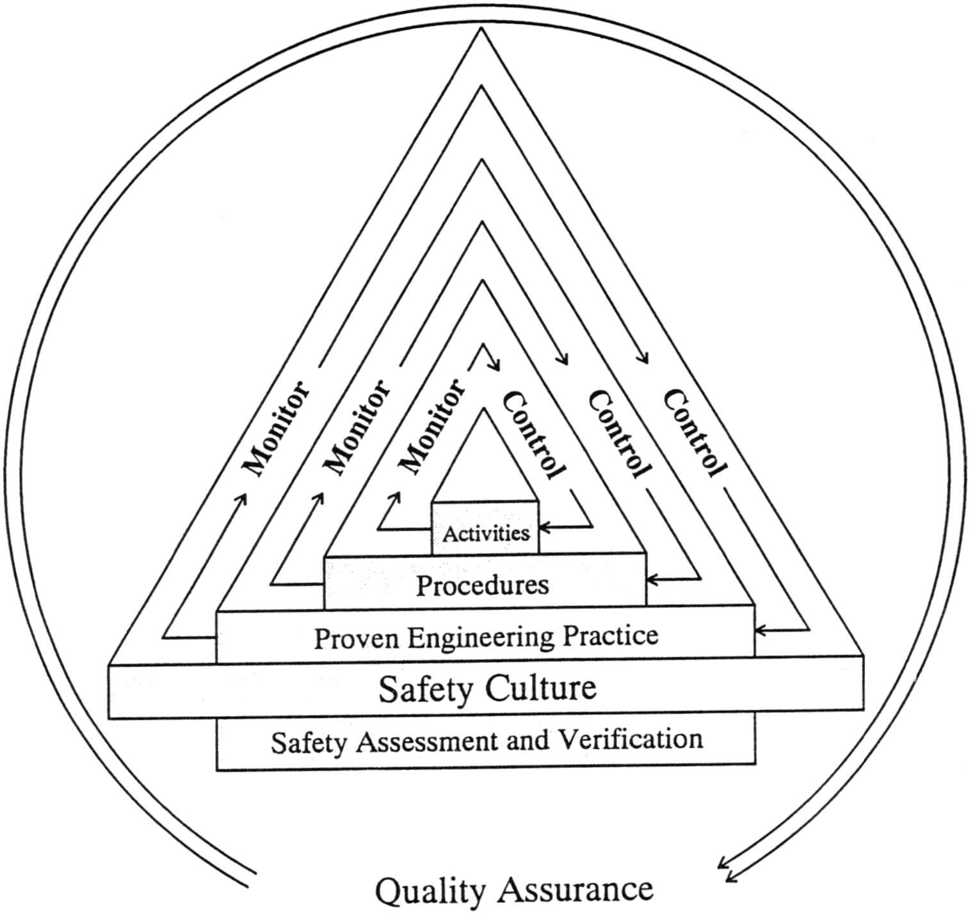

Fig. 7.12 Risk Management Principles Based on Safety Culture

are developed; strict adherence to these procedures is demanded; internal reviews are performed of safety related activities; above all, staff training and education emphasize reasons behind the safety practices established, together with the consequences for safety of shortfalls in personal performance.

These matters are especially important for operating organizations and staff directly engaged in plant operation. For the latter, at all levels, training emphasizes significance of their individual tasks from the standpoint of basic understanding and knowledge of the plant and equipment at their command, with special emphasis on reasons underlying safety limits and safety consequences of violations. Open attitudes are required in such staff to ensure that information relevant to plant safety is freely communicated; when errors are committed, their admission is particularly encouraged. By these means, an all pervading safety thinking is achieved, allowing an inherently questioning attitude, prevention of complacency, commitment to excellence, and fostering of both personal accountability and corporate self-regulation in safety matters.

Proven engineering practices: Devices are designed, manufactured and constructed by technologies which are proven by testing and experience, which are reflected in approved codes and standards and other appropriately documented statements, and which are ensured by proper selection, training and qualification of workers. The use of proven engineering continues throughout the plant's life [18].

Quality assurance: The quality assurance activities are a component of good management. This is applied throughout activities to ensure with high confidence that all items delivered and services and tasks performed meet specified requirements. Organizational arrangements should provide a clear definition of the responsibilities and channels of communication and co-ordination for the quality assurance. These arrangements are founded on the principle that the responsibility for achieving quality in a task rests with those performing it, others verify that the task has been properly performed, and yet others audit the entire process, i.e., a multiple monitor/control configuration together with safety assessment. The authority of the quality assurance staff is established firmly within the organization [18].

When repairs and modifications are made, analysis is conducted and review is made to ensure that the system is returned to a configuration covered in the safety analysis and technical specifications. If opportunities for advancement or improvement over existing practices are available and seem appropriate, such changes are applied cautiously only after demonstration that the alternatives meet the requirements

Quality assurance practices thus cover 1) validation of designs, 2) supply and use of materials, 3) manufacturing, inspection and testing methods, and 4) operational and other procedures to ensure that specifications are met. The associated documents are subject to strict procedures for verification, issue, amendment and withdrawal [18].

Safety assessment and verification: Safety assessment is made before construction and operation of a plant begin. The assessment is well documented and independently reviewed. It is subsequently updated in the light of significant new safety information.

Safety assessment includes systematic critical review of the ways in which structures, systems and components might fail, and identifies the consequences of such failures. The assessment is undertaken expressly to reveal any underlying design weaknesses. The results are documented in detail to allow independent audit of scope, depth and conclusions of the critical review [18]. The FTA is a typical approach for the safety assessment.

7.5.2 Accident Prevention and Consequence Mitigation

Fig. 7.13 shows a key flow of risk management consisting of accident prevention and accident management. Accident prevention (left-upper box) is divided into failure prevention and progression prevention, while accident management (left-lower box) into in-site consequence mitigation and off-site consequence mitigation. The FTA primarily deals with the accident prevention phase.

In medical terms, failure prevention corresponds to infection prevention, progression prevention to outbreak prevention, and consequence mitigation to treatment and recovery after outbreak.

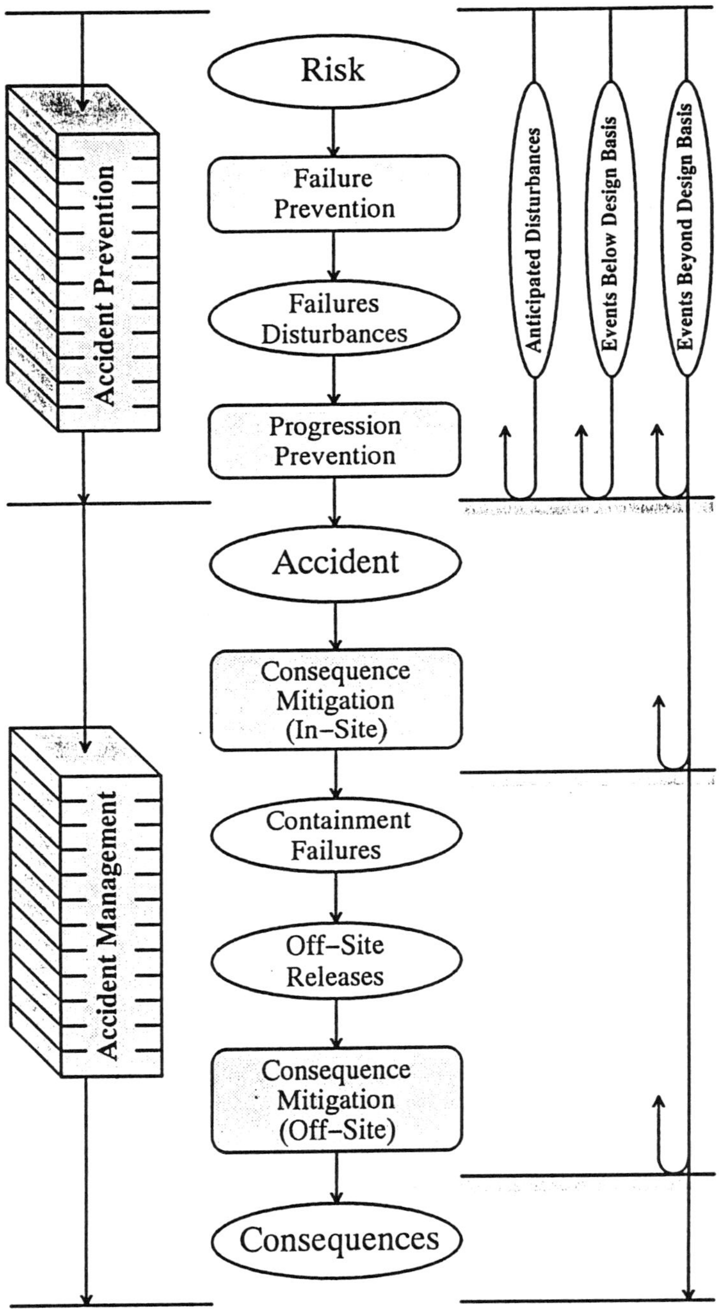

Fig. 7.13 Risk Management Process Consisting of Four Phases

Risk management scheme in Fig. 7.13 requires definitions of accident and consequences. For nuclear power plant, these are defined as [18]:

Accident: Core damage ranging over fuel failure, severe fuel damage, fuel melt, and uncontrolled fuel melt.

Consequences: Adverse health effects of radiation on plant workers and the public, and radioactive contamination of land, air, water or food products.

As shown in a right-upper portion of Fig. 7.13, neither anticipated disturbances nor events below design basis yield accidents if the progression prevention works successfully. On the other hand, if something is wrong with the progression prevention or if extreme initiating events are involved, these disturbances or events would develop to events beyond design basis with four possibilities: 1) the progression prevention still works and prevents accidents, 2) the in-site consequence mitigation works and prevents containment failures and hence off-site releases, 3) the off-site consequence mitigation works and minimizes off-site consequences, and 4) all features fail and large consequences occur.

7.5.3 Failure Prevention

The first means of preventing accidents is to strive for such high quality in design, manufacture, construction and operation of the plant that deviations from normal operational states are infrequent [18]. Such a deviation may occur from two sources: 1) facility and 2) human. Facility related deviations include ones not only for plant itself but also physical barriers, normal control systems, and emergency safety systems (See Fig. 7.7); some deviations become initiating events while others are enabling events. Human related deviations are further classified into those from individual, team, and organization.

FAILURE PREVENTION FROM FACILITY

Facility failures are prevented, among other things, by proven engineering practice and quality assurance shown in Fig. 7.12. Some examples are listed below.

Safety margins: For instance, a metal bolt with a larger diameter than predicted by theoretical calculation is used. Devices are designed by conservative rules and criteria according proven engineering practice.

Standardization: Functions, materials, specifications are standardized 1) to decrease device failures, 2) to facilitate device inspection, and 3) to facilitate prediction of a remaining device lifetime.

Maintenance: A device is periodically inspected and replaced/renewed before its failure. This is a periodic preventive maintenance. Devices are continuously monitored, and replaced/renewed before failures. This is called a condition- based maintenance. This type of monitor and control activities are typical elements of the quality assurance program.

Change control: Formal arrangements for handling of variations and deviations are an important aspect of quality assurance programs [18]. Failures occur frequently due to insufficient review of a system modification. A famous Flixborough accident in UK occurred in 1974 when a pipeline was temporarily installed to bypass a one of six reactors which was under maintenance. Twenty eight people died of explosion caused by ignition of flammable material that flew out of the defective bypass line. Change controls also avoid insufficient temporary standards to behave as a formal one.

Hidden processes review: Typical processes frequently oversighted include 1) inspection and maintenance activities, 2) industrial robots in a hot standby, etc. These processes become targets of safety assessment and verification programs in Fig. 7.12. A pipe downstream of a solenoid exhaust valve of a reactor was blocked and maintenance was taken place with the valve closed; a plant explosion occurred in Texas, US (1989) when the valve failed open, and a large amount of ethylene gas leaked; twenty three people died.

Other approaches for failure prevention include 1) use of proven device and procedures, 2) review of new technology to avoid too much emphases on merits, 3) evaluation of fatigue due to long- term repetitive use, 4) improvement of operation and maintenance procedures, and 5) application of FMEA, quality engineering, etc.

HUMAN ERRORS

One of the most important lessons of abnormal events, ranging from minor incidents to serious accidents, is that they have so often been results of incorrect human actions. Frequently such events have occurred 1) when plant personnel did not recognize the safety significance of their actions, 2) when they violated procedures, 3) when they were unaware of conditions in the plant, 4) when they were misled by incomplete data or incorrect mindset, or 5) when they did not fully understand the plant in their charge. The operating organization must ensure that its staff is able to manage the plant satisfactory according to the risk management principles illustrated in Fig. 7.12. For instance, approved procedures should be followed by operation, maintenance and inspection staff [18].

The human error component of events and accidents has been too great in the past. The remedy is a twofold attack, through design, including automation, and through optimal use of human ingenuity in unusual circumstances. Human errors occur from individuals, teams, or organizations.

7.5.4 Progression Prevention

The second step for accident prevention is to ensure that a perturbation or incipient failure would not develop into a more serious situation. In no human endeavor can one ever guarantee that the failure prevention would be totally successful. Plant designers therefore assume that component, system and human failures are possible, and can lead to abnormal occurrences, ranging from minor disturbances to highly unlikely accident sequences [18]. These occurrences will not cause accidents if physical barriers, normal control systems, and engineered safety features remain healthy and operate correctly, which is ensured by failure preventions for these barriers, systems and operators , i.e., prevention of enabling events.

Physical barriers: Each physical barrier is designed conservatively, its quality is checked to ensure that margins against failure are retained, its status is monitored, and all plant processes capable of affecting it are controlled and monitored in operation. Safety system designers ensure to the extent practicable that the different safety systems protecting the physical barriers are functionally independent under accident conditions.

For instance, multiple inspections are conducted during and after fabrication and installation of the primary system boundary for a nuclear reactor. They use ultrasonic, radiographic and surface methods. Hydraulic overpressure testing to pressures well above those expected in operation confirms the strength of the system before it is made radioactive [18].

This barrier itself may be protected by special measures; for instance, a containment structure of nuclear power plants is equipped with devices that controls pressure and temperature due to accident conditions; such devices include hydrogen ignitors, filtered vent systems, area spray systems, etc. [18].

Normal control systems: Minor disturbances (usual disturbances and anticipated abnormal occurrences) for the plant or the barrier are dealt with normal feedback control systems to provide a tolerance of any failures which might otherwise allow faults or abnormal conditions to develop into accidents. This reduces the frequency of demands on emergency safety systems. These controls protect the physical barriers by keeping the plant in a well defined region of operating parameters where barriers will not be jeopardized. Care in system design prevents cliff edge effects which might permit small deviations to precipitate grossly abnormal plant behavior and cause damage [18]

Engineered safety features and systems: Initiation and operation of the engineered safety features should be highly reliable. This reliability is achieved 1) by appropriate use of fail-safe design, 2) by protection against common cause failures, 3) by independence between safety systems (inter-independence) and between safety systems and normal control systems (outer- independence), and 4) by monitor and recovery provisions. Designs of these systems ensure that failure of a single component would not cause loss of function served by a safety system (a single failure criterion) [18].

Inter-Independence: Safety systems often make use of redundancy and diversity of design and physical separations of parallel components, where appropriate to reduce likelihood of loss of vital safety function. For instance, diesel-driven and steam-driven generators are installed for emergency power supply; different algorithms are used to calculate the same quantity. The independence between systems in a redundant configuration is achieved by 1) diversity, or 2) physical separation by distance or barrier.

The conditions under which equipment is required to perform a safety function may differ from those to which it is normally exposed, and its performance may be affected by aging or by maintenance conditions as plant operation goes on. The environmental conditions under which equipment is required to function are identified as part of a design process. Among these are conditions expected in a wide rage of accidents, including extremes of temperature, pressure, radiation, vibration, humidity and jet impingement. Effects of external events such as earthquakes are also considered.

Because of the importance of fire as a source of possible simultaneous damage to several components, design provisions to prevent and combat fires in the plant are given special attention. Fire resistant materials are used to the extent possible. Fire-fighting capability is included in the design specifications. Lubrication systems use non-flammable lubricants or are protected against initiation and effects of fires. The design takes advantage of methods identified for preventing common cause failures.

Outer-Independence: Engineered safety systems should be independent from normal process control systems. For instance, safety shutdown systems for a nuclear power plant are independent in function from reactivity control systems used for normal operation. Common sensors or devices may only be used if reliability analysis indicates that this is acceptable. Electrical busses and logic circuits of the shutdown system are separate from instruments used for normal control so that no interference is possible between demands of normal control and demands of safe shutdown.

Recovery: Not only plant itself but also barriers, normal control systems, and engineered safety systems are inspected and tested regularly to reveal any degradation which might lead to abnormal operating conditions or inadequate safety performance. Operators are trained to recognize readily an onset of an accident and to respond properly and in a timely manner to such abnormal conditions. They have also been well trained in appropriate operating procedures, with which they have become familiarized.

Automatic actuation: The plant operating staff are provided with appropriate safety equipment, instrumentation and operating procedures for response to and control of accidents below design basis. Design is such that abnormal developments are first met automatically by restoration of normal conditions by means of feedback characteristics of neutronic and process controls. These are backed up by normal capability for shutdown, continued cooling and protection against release of radioactive materials.

Further protection is available through automatic actuation of engineered safety systems. By means of such measures, any onset of abnormal behavior would be dealt with automatically by appropriately designed systems for at least a predetermined period of time, during which the operating staff could assess systems, review possibilities and decide on a subsequent course of action for conditions not adequately responded to by the automatic functioning of plant systems. The design makes provision for diagnostic aids and symptom based emergency procedures for use in these circumstances. Typical decision intervals for operator action range from 10 to 30 minutes or longer depending on a situation [18].

Symptom based procedures: Plant operation procedures generally describe responses based on a diagnosis of the event occurring (event based procedures). If the event cannot be diagnosed in time, or if further evaluation of the event causes the initial diagnosis to be discarded, a symptom based procedures define responses to symptoms observed rather than plant conditions as deduced from these symptoms [18].

7.5.5 Consequence Mitigation

Consequence mitigation covers a period after occurrence of accident, for instance, a core damage for nuclear reactors. The occurrence of accident means that events occurred are beyond a design basis; events below the design basis, by definition, could never developed into the accident because normal control systems or engineered safety features are assumed to operate as intended. It should be noted that plant controls could be restored after the accident.

Since accidents could nonetheless occur, other procedural measures are provided for managing their course and mitigating their consequences. These additional measures are defined on the basis of operating experience, safety analysis and results of safety research. Attention is given in design, siting, procedures and training to controlling progressions and consequences of accidents. Limitation of accident consequences are based on safe shutdown, continued core cooling, adequate confinement integrity and off-site emergency preparedness. High consequence severe accidents are therefore extremely unlikely if they are effectively prevented or mitigated by defense in depth.

As shown in Fig. 7.13, the consequence mitigation consists of 1) in-site consequence mitigation, and 2) off-site consequence mitigation. These mitigations are described below for nuclear power plants.

In-Site consequence mitigation: This includes preplanned and ad hoc operational practices which, in circumstances in which plant design specifications are exceeded, would make optimum use of existing plant equipment in normal and unusual ways to restore control. This phase would have an objective of restoring the plant to a safe state with the reactor shut down, continued fuel cooling assured, radioactive material confined and the confinement function protected. In such circumstances, engineered safety features would act to confine any radioactive material released from the core so that discharge to the environment would be minimal. These engineered safety features include physical barriers, some of which have the single purpose of confining radioactive material [18].

Off-Site consequence mitigation: Off-site countermeasures are available, going beyond the level of protection provided in most human endeavors, to compensate for a remote possibility that safety measures at the plant might fail. In such a case, effects on the surrounding population or the environment would be mitigated by protective actions, such as sheltering or evacuation of the population, and by prevention of a radioactive material transfer to people by food-chains and other pathways.

7.5.6 Summary

Risk management is an opposite side of accident causation mechanisms and consists of four phases: 1) failure prevention, 2) progression prevention, 3) in-site consequence mitigation, and 4) off-site consequence mitigation. The first two are called an accident prevention, while the second two an accident management. Risk management principles are proven engineering practice and quality assurance built on a basis of safety culture. The quality assurance consists of 1) multi-layer monitor/control provisions which remove and correct deviations, and 2) safety assessment and verification provisions which evaluate remaining deviations.

Failure prevention applies not only to failures of inanimate devices but also errors of human beings; errors from individual, team, and organization. It strives for such high quality in design, manufacture, construction, and operation of a plant that deviations from normal operational states are infrequent. Progression prevention ensures that a perturbation or incipient failure would not develop into a more serious situation such as accident. Consequence mitigations cover a period after occurrence of accident, and manage course of accident and mitigate consequences.

7.6 Automated Fault Tree Synthesis

7.6.1 Introduction

An FT is a graphical representation of an accident causation mechanism in an event layer. A manual generation of the FT for a large complicated system, an activity of the FTA, is however a tedious, time-consuming and error-prone task. Digraphs, truth tables, and mini FTs have been proposed so far for automated and/or semi-automated FT generations. This section describes a new automated generation method which is based on 1) a semantic network representation of the system to be analyzed, 2) a rule-based event development to search for more basic causes, and 3) a recursive algorithm to generate FTs by utilizing the semantic network and the rulebase. See Andrews and Brennan [26] for previous and recent research activities for automated FT generations.

It is obvious that the system must first be modeled by a suitable representation method, because the representation is a place where aspects specific to the system are included. An FT generation method can extract no more information than that contained in the system model, while the generation method itself and its output become too complicated if unsuitable system models are used.

The digraph can represent only fractions of system specific aspects, that is, monotone increasing or decreasing or cut-off relations among process variables defined at digraph nodes. These relations are expressed by gains (10, 1, 0, -1, -10) labeled on arrows among the nodes. In particular, it is not clear which one would dominate when two arrows with opposite gains flow into the same node. The generation algorithm based on the digraph becomes too complicated and difficult to understand.

The truth table expresses a system component in terms of tabular relations among output, state, and input in table columns. A value such as high, low, or zero is usually assumed in each cell. The resultant generation algorithm is in nature an algebraic manipulation, and becomes too complicated when negations or loops are encountered. FTs thus generated often have little resemblance with those generated manually.

The mini FTs are a library of small FTs for a set of typical components or subsystems. However, these mini FTs must be integrated manually when a system topology is given as a interconnected networks of these system elements.

Suppose now that a suitable system model is given. For a top event to be analyzed, an algorithm uses the model to generate an FT. Most of the FT generation methods proposed so far have only used a procedural algorithm for this purpose. It seems, however, that an FTA expert heavily uses heuristics to successively analyze the top event into more basic events. At the same time, she also seems to have a procedural framework for a somewhat systematic application of her heuristics to the system model. Thus, a practical FT generation approach requires three elements: 1) the system representation, 2) the expert heuristics, and 3) the procedural framework guiding the generation. This section clarifies these three elements. The system to be analyzed is modeled by a semantic network, the expert heuristics are expressed as a set of if-then rules, and the procedural framework is specified as a recursive function.

In section 7.6.2, an important type of events is first introduced because the FT is nothing but a graphical representation of event causal relations. The remaining two types of events will be defined later in section 7.6.5. A concept of 'flow' is introduced, and the event is defined as a triple (flow, attribute, value). In section 7.6.3, basic components or pieces of equipment are defined from viewpoints of controlling the attributes of flows. Some components controls an aperture through which the flow propagates, and others controls a generation of the flow. The semantic network representation as a system model is presented in section 7.6.4. The network consists of two types of nodes: 1) equipment nodes and 2) flow nodes. An arrow between these two nodes denotes particular relationship between the equipment and the flow. The event development rules are classified and presented in section 7.6.5. The recursive algorithm is listed and described in section 7.6.6. Section 7.6.7 applies the proposed FT generation method to two systems: 1) a relay circuit with a self-holding mechanism, and 2) a hypothetical nuclear reactor with various protection functions.

7.6.2 Events and Flows

Consider a water tap in a kitchen. The water flows out when the tap is opened, provided that a pump supplies the water to the tab. Four concepts are involved in this simple story. First, the water is a subject which has a potential to move through a flow path; this provides a concept of a flow. The water tap opens or closes the flow path through which the water may flow; this leads to a concept of aperture of the flow. Third, the water is pumped to the tap; this yields a concept of generation rate of the flow. Forth, when the water flows out of the tap, it has a positive flow rate; this leads to a concept of flow rate of the flow. The system is viewed as a pipeline structure of such flows, taps and pumps. These system elements can be interpreted differently depending on the particular system to be analyzed. A variety of flows go through the flow paths and cause events specific to the system.

FLOW AND EVENT PROPAGATION

The top event at the root of an FT occurs when a set of basic events at the leaves propagate to the top by various causal relations. The propagation is a result of mutual influences of flows via system components. This point will become clearer when an equipment library is described in section 7.6.3. A flow is defined as any material, information, energy, event or phenomenon which can move or propagate. Typical examples are:

(1) **Material flow;** liquid, gas, steam.
(2) **Information flow;** signal, datum, command, alarm.
(3) **Energy flow;** light, heat, sound, vibration.
(4) **Others;** fire, spark, pressure high event.

For instance, the light as a flow is generated when an electrical flow is supplied to a bulb. A fire alarm as a flow is issued when a fire event exists. The existence of the fire will be denoted by the positive flow rate of the fire just like the existence of the light.

FLOW ATTRIBUTE

As will be described in section 7.6.3, a piece of equipment in a system controls or determines an attribute of a flow. For instance, an electrical switch controls an aperture of electricity, and a battery determines a generation rate of the electricity. The flow rate of the electricity, on the other hand, is determined when both aperture and generation rate are specified. Typical attributes of the flow thus include flow rate, generation rate, aperture, temperature, pressure, concentration, and so on.

The synthesis approach mainly focuses on the first three attributes. An event such as fire is expressed as a flow in an exceptional case. Then, a convention is that the positive flow rate denotes an occurrence or an existence of the event, while the zero flow rate a non-occurrence or a non-existence. The aperture of a flow is defined similarly to that of a valve; an 'open' valve corresponds to an 'on' switch, whereas a 'close' valve corresponds to an 'off' switch. An anti-fire door acts as a valve, and forces the fire flow rate through the door to be zero. The flow aperture is closed if, for instance, one of apertures of valves in series are closed, or one of switches in series are 'off.' The flow rate of a flow becomes zero if 1) its aperture is closed or 2) its generation rate is zero. For instance, a flow rate of electricity through a bulb is zero if the bulb has a cut-off failure or a battery is dead. The generation rate may often be viewed as a potential such as a voltage difference. The potential can move the flow such as electric current when the aperture is open.

ATTRIBUTE VALUES

Aperture: Attribute values of aperture are Fully_Closed(F_Cl), Increase(Inc), Constant(Cons), Decrease(Dec), Fully_Open(F_Op), Open, and Not_Fully_Open(Not_F_Op). Values Inc, Cons, and Dec respectively mean that the aperture increases, remains constant, and decreases between F_Cl (excluded) and F_Op (excluded), as shown in **Fig. 7.14**. In a digital representation, only two attribute values are considered, i.e., F_Cl and F_Op, while additional values Inc, Cons, Dec are considered in an analogue representation.

Value Open means that the aperture is not F_Cl, i.e., that the aperture is either Inc, Cons, Dec, or F_Op. Value Not_F_Open is defined as the aperture being F_Cl, Inc, Cons, or Dec. Thus, attribute values Open and N_F_Op are composite, while values F_Cl, Inc, Cons, Dec, and F_Cl are atomic.

Generation rate: The generation rate has attribute values Zero, Inc, Cons, Dec, Max, Positive, and Not_Max. The first five values are atomic, while the last two are composite. Values Inc and Cons respectively means that the generation rate increases,

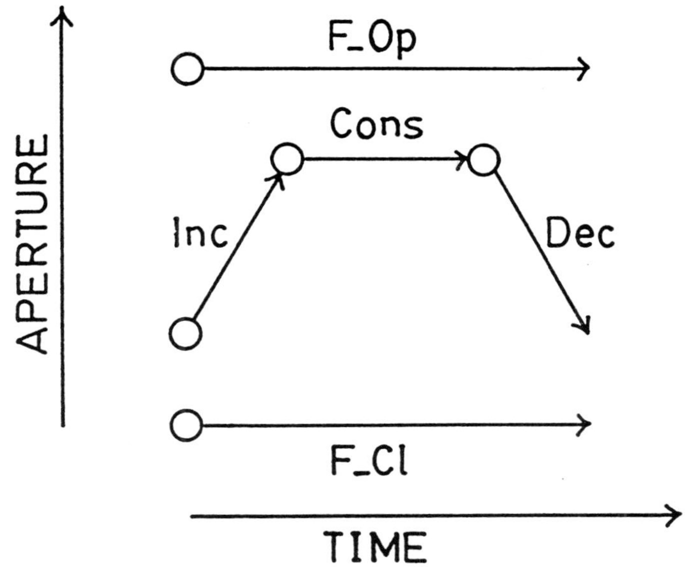

Fig. 7.14 Atomic Values of Aperture

remains constant, and decreases between Zero (excluded) and Max (excluded). Composite value Positive implies that the rate is either Inc, Cons, Dec, or Max. Another composite value Not_Max means that the rate is Zero, Inc, Cons, or Dec.

Flow rate and others: The flow rate has the same attribute values as the generation rate. The temperature or pressure may have a similar set of values to the generation or the flow rate, except that the value 'Zero' is replaced by 'Min.'

RELATIONS OF APERTURE, GENERATION RATE AND FLOW RATE

The three typical attributes mainly considered in this synthesis method, i.e., the aperture, the generation rate, and the flow rate, are not independent of each other. The flow rate is determined when the aperture and the generation rate are specified. **Table 7.1** shows the relationship. Each row has a fixed aperture value denoted in the first column, while each column a fixed generation rate value denoted in the first row. Each cell denotes a flow rate value. It is obvious that the flow rate becomes Zero when either the aperture is F_Cl or the generation rate is Zero. The flow rate is not determined uniquely when the aperture is Inc and the generation rate is Dec. This is also true when the aperture is Dec and the generation rate is Inc. In these two cases, the flow rate is assumed to be either Inc or Dec, excluding a rare chance of the flow rate becoming Cons. In other words, these two opposite combinations of the aperture and the generation rate belong to causes of the flow rate being Inc (or Dec). The flow rate values at the other cells can be understood easily.

FLOW TRIPLE AS EVENT OCCURRENCE

A flow triple is defined as a particular combination (flow, attribute, value). This can be interpreted as an occurrence or existence of an event. For example, (command, flow rate, Zero) means that the command does not exist, while (command, flow rate, Positive) an existence of the command. The triple (cooling water, flow rate, Dec) means that the

Table 7.1 Determination of Flow Rate in Terms of Flow Aperture and Generation Rate

Positive (Inc, Cons, Dec, Max)

	Zero	Inc	Cons	Dec	Max
F_Cl	Zero	Zero	Zero	Zero	Zero
Inc	Zero	Inc	Inc	Inc,Dec	Inc
Cons	Zero	Inc	Cons	Dec	Cons
Dec	Zero	Inc,Dec	Dec	Dec	Dec
F_Op	Zero	Inc	Cons	Dec	Max

Not_F_OP (F_Cl, Inc, Cons, Dec); Open (Inc, Cons, Dec, F_Op)

Not_Max (Zero, Inc, Cons, Dec)

flow rate of the cooling water decreases, while (cooling water, flow rate, Inc) that the flow rate increases.

Suppose an event that temperature of a cooling water increases. A relevant flow is a cooling water, and its attribute is the temperature. Thus the existence of the event can be expressed as (cooling water, temperature, Inc). The 'cooling water temperature increase' can in turn be considered as a flow, and the existence of the same event can be expressed as (cooling water temperature increase, flow rate, Positive). Although it is more natural to consider that the temperature itself is an attribute of the cooling water, the latter expression may sometimes be used for convenience.

7.6.3 Basic Equipment Library

In section 7.6.2 it is suggested that the triple (flow, flow rate, value) can express a variety of events about a flow. The flow rate is dependent on the other two attributes, i.e., aperture and generation rate, as shown in Table 7.1. Typical pieces of basic equipment are related to the aperture or the generation rate. In this section each piece of equipment is defined as a controller of an attribute of a flow which, in turn, may control other pieces of equipment as a command flow. Basic pieces of equipment controlling aperture or generation rate are catalogued in **Fig. 7.15** where a fragment of a semantic network is associated with each piece in the second column. A circle represents a flow node denoting a particular flow, while a box or hexagon an equipment node denoting a particular equipment. A labeled arrow between flow and equipment nodes represents a particular relation between the flow and the equipment.

APERTURE CONTROLLER

There are two types of basic equipment for the aperture control; one without and the other with a command flow. Equipment 1 through 5 in Fig. 7.15 are digital devices with only two aperture values F_Cl or F_Op, while equipment 6 is an analog device with five values F_Cl, Inc, Cons, Dec, and F_Op. An anti-fire door can be a digital or an analog device with or without a command.

Table 7.2 shows relationships between apertures of the flow and apertures of two pieces of equipment, where the two aperture controllers are in series along the flow. Each row has a fixed aperture value of equipment 1, while each column a fixed aperture value of equipment 2. Each cell denotes a flow aperture. The aperture of the flow becomes F_Cl

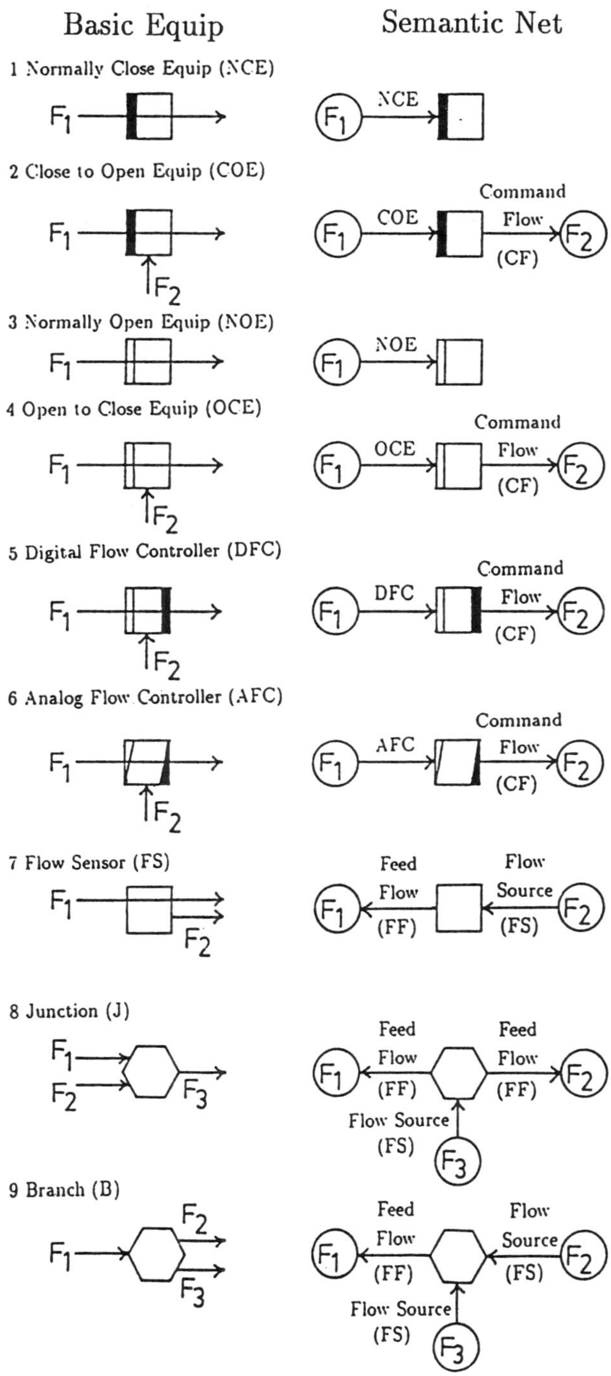

Fig. 7.15 Library of Basic Pieces of Equipment and Semantic Networks

Table 7.2 Determination of Flow Aperture in Terms of Equipment Apertures in Series

(Upper brace over columns Inc, Cons, Dec, F_Op: Open; left brace over rows Inc, Cons, Dec: Not_F_Op; right brace over rows Inc, Cons, Dec, F_Op: Open; lower brace over columns Inc, Cons, Dec: Not_F_Op)

	F_Cl	Inc	Cons	Dec	F_Op
F_Cl	F_Cl	F_Cl	F_Cl	F_Cl	F_Cl
Inc	F_Cl	Inc	Inc	Inc,Dec	Inc
Cons	F_Cl	Inc	Cons	Dec	Cons
Dec	F_Cl	Inc,Dec	Dec	Dec	Dec
F_Op	F_Cl	Inc	Cons	Dec	F_Op

when either equipment 1 or 2 has aperture F_Cl. The flow aperture is not determined uniquely when the aperture of one equipment is Inc and the other is Dec. In such a case, the flow aperture is either Inc or Dec. Aperture values of the flow at other cells in Table 7.2 can be understood easily.

Table 7.1 in section 7.6.2 and Table 7.2 here will be used to derive a set of event development rules which search for causes of events related to the flow rate: The flow rate of a flow is reduced to the aperture and the generation rate of the flow by Table 7.1. The aperture of the flow is then reduced to the apertures of pieces of equipment by Table 7.2. The generation rate is also reduced to other events by definitions of the generation rate controllers described later.

Aperture controller without commands: (1) Normally Close Equipment (NCE): This is a digital equipment normally closed (F_Cl). Its F_Op state is caused only by a failure of the NCE itself. Examples of the NCE include normally closed valve, normally 'off' switch, plug, insulator, oil fence, etc. The oil fence can be regarded as an NCE because it stops a movement of oil as a flow. The F_Op state corresponds to an oil leak failure from the fence.

In the semantic network for NCE in the second column of Fig. 7.15, symbol F_1 denotes the flow of which aperture is forced to be normally closed by the box representing the NCE. The arrow labeled as NCE points to the equipment that closes the flow aperture of F_1. Suppose an event that the flow aperture of F_1 is Open. This can be traced back to a cause that the NCE pointed by the arrow has an Open failure. Generally, causes of an event can be searched for in labeled arrow directions in a semantic network system representation.

(2) Normally Open Equipment (NOE): This is a digital equipment normally open (F_Op). Its F_Cl state is caused only by a failure of the NOE itself. Examples include normally open valve, normally 'on' switch, pipe, electric wire etc. Since a flow such as liquid and air can flow through a pipe, it can be regarded as an NOE. The F_Cl state corresponds to a blockage of the pipe.

In the semantic network for NOE in the second column of Fig. 7.15, symbol F_1 denotes the flow of which aperture is forced to be normally open by the box representing the NOE. The arrow labeled as NOE points to the equipment that opens the flow aperture of F_1. Suppose an event that the flow aperture of F_1 is closed. This can be traced back to a cause that the NOE pointed by the arrow has a close failure.

Aperture controller with commands: In the NCE or NOE, the transition of the equipment aperture is only caused by the equipment failure. Some equipment has a command to change its aperture.

(1) Close to Open Equipment (COE): This equipment is normally at F_Cl state, and a transition from F_Cl to F_Op state occurs by a command as a flow. Note that such a transition may also occurs when the equipment spuriously changes its state when the command remains constant. It is assumed that the command can not cause the reverse transition, and that the transition from F_Op to F_Cl state occurs only by the failure of the COE possibly after the command changes the equipment aperture into F_Op.

An emergency button normally at 'off' state is an example of the COE. An oil fence can be regarded as a COE when a human action removing the fence is considered as a command to the COE. This suggests that the same equipment can be with or without a command depending upon a degree of details of system representation.

Symbol F_1 for the COE in Fig. 7.15 denotes a flow of which aperture is controlled by the COE. Symbol F_2 represents the command flow. The arrow labeled as the COE points to the COE itself, while the arrow labeled as CF points to the command flow. Suppose an event that the flow aperture of F_1 is F_Op. A cause of the event can be traced back to the flow rate change of F_2. The direction of changes depends on gain types of COE described below.

Two types of COE exists; a positive gain and a negative gain type. Note in the following definitions that the 'positive gain', in general, means that flow rate of the aperture controlled flow F_1 has the same tendency as the flow rate of command flow F_2.

1) Positive gain: The equipment becomes F_Op only when flow rate of command F_2 becomes Positive.

2) Negative gain: The equipment becomes F_Op only when flow rate of command F_2 becomes Zero. The negative gain nature of COE will be denoted by a small circle at the head of the arrow pointing to the command flow. Such a small circle, by convention, means a logic negation.

(2) Open to Close Equipment (OCE): This equipment is normally at F_Op state, and a transition from F_Op to F_Cl state occurs by a command. Such a transition may also occurs when the equipment spuriously changes its state when the command remains constant. It is assumed that the command can not cause the reverse transition. The reverse transition from F_Cl to F_Op occurs only by a failure of the OCE possibly after the command changes the equipment aperture into F_Cl.

Symbol F_1 for the OCE in Fig. 7.15 denotes a flow of which aperture is controlled by the OCE. Symbol F_2 represents the command flow. The arrow labeled as the OCE points to the OCE itself, while the arrow labeled as CF points to the command flow. An emergency shutdown valve is an example of OCE, where flow F_1 may be a feed flow to a chemical reactor.

Similarly to the COE, two types of OCE exists; a positive gain and a negative gain type.

1) Positive gain: The equipment becomes F_Cl only when flow rate of command F_2 becomes Zero. The flow rate of the aperture controlled flow F_1 has the same tendency as the flow rate of command flow F_2.

2) Negative gain: The equipment becomes F_Cl only when flow rate of command F_2 becomes Positive.

(3) Digital Flow Controller (DFC): This is a generalization of COE or OCE, and a transition from F_Cl to F_Op and its reverse are both permitted. An example is a pressure switch which opens or closes when a measured pressure is high or not high, respectively. In this case a pressure high event may artificially be regarded as a command flow, and its existence is expressed as (pressure high event, flow rate, Positive). A positive or a negative gain type exists. Symbols F_1 and F_2 for DFC in Fig. 7.15 denote the aperture controlled flow and the command flow, respectively. The arrow labeled as DFC points to the DFC itself, and the arrow labeled as CF points to the command flow. Two gain types exist.

1) Positive gain: The equipment becomes F_Op or F_Cl when command flow rate becomes Positive or Zero, respectively. Again the flow rate of aperture controlled flow has the same tendency as the command flow F_2.

2) Negative gain: The equipment becomes F_Op or F_Cl when command flow rate becomes Zero or Positive, respectively.

(4) Analog Flow Controller (AFC): This is a generalization of the DFC and an example includes a flow control valve. Symbols F_1 and F_2 in Fig. 7.15 respectively denotes the aperture controlled flow and the command flow. The labeled arrows can be interpreted similarly to the case of DFC. The aperture of F_1 can assume either F_Cl, Inc, Cons, Dec, and F_Op depending on the gain type of AFC.

1) Positive gain: The aperture of F_1 becomes F_Cl, Inc, Cons, Dec, and F_Op when the flow rate of command F_2 becomes Zero, Inc, Cons, Dec, and Max, respectively. An example is an air-to-open control valve, where the air is a command flow.

2) Negative gain: The aperture of F_1 becomes F_Cl, Inc, Cons, Dec, and F_Op when the flow rate of command F_2 becomes Max, Dec, Cons, Inc, and Zero, respectively. The flow rate of F_1 has the opposite tendency to the flow rate of F_2. An example is an air-to-close control valve.

GENERATION RATE CONTROLLER

This type of equipment generates one or more flows depending on values of particular attributes of flows fed to the equipment. Some dependencies on the flow rate attribute of the feed flows are described first.

Dependence on flow rate: (1) Junction: A new flow is generated by a simple sum of feed flows. An example is a fanning-in node of electrical wires.

(2) Branch: More than one flows are generated from a single feed flow. An example is a fanning-out node of electrical wires.

(3) Flow Sensor: A new flow such as a signal is generated from a single feed flow. Examples of the Flow Sensor include leakage detector, alarm bell, light bulb, power converter, etc. The light bulb is a Flow Sensor because it generates light from the electricity supplied. The power source such as diesel generator may sometimes be considered as the Flow Sensor because it generates electricity from fuel as a feed flow.

(4) Logic Gates: Other pieces of equipment controlling the generation rate include logic gates such as NOT, OR, AND, and NAND. For instance, the OR gate generates a new flow with positive flow rate when one or more feed flows have a positive flow rate.

Dependence on attributes: (1) Temperature Sensor: This generates a new flow such as an alarm signal in response to the temperature attribute of the feed flow.

(2) Pressure Sensor: This generates a new flow such as an alarm signal in response to the pressure attribute of the feed flow. Just a brief note here about the pressure switch. When the switch is regarded as controlling an aperture of flow F_1, then this is modeled as a DFC. On the other hand, when the switch and aperture controlled flow F_1 are viewed as a single unit, then flow F_1 can be regarded as being generated from the pressure switch, given that the generation rate of F_1 is always Positive. The mapping between the basic equipment and a real world component is dependent on the context.

7.6.4 Semantic Network as System Representation

The structural aspects of the system to be analyzed by the FT is represented by a semantic network. Different FTs may be generated depending on the boundary conditions for the network even if the top event is fixed.

SEMANTIC NETWORK REPRESENTATION

The semantic network is easily obtained by using the basic equipment library like Fig. 7.15 when a system schematic is given: First, a mapping between the basic equipment and the system component is established; Second, the fragment of the semantic networks representing the basic pieces of equipment are integrated, yielding the system semantic network.

As a demonstration, consider a schematic of **Fig. 7.16**. This system is a simplified portion of an ECCS (Emergency Core Cooling System) of a nuclear reactor. Lines A and AA are electric wires with a DC voltage difference. It is assumed that these lines

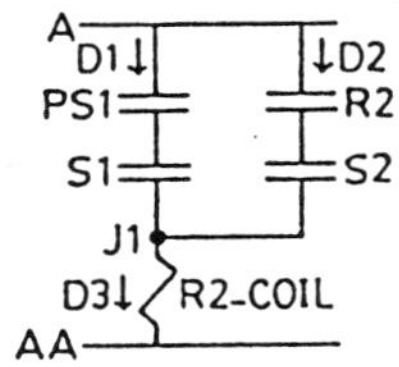

Fig. 7.16 A Relay Circuit with a Self-Holding Capability to Detect Drywell Pressure High Event

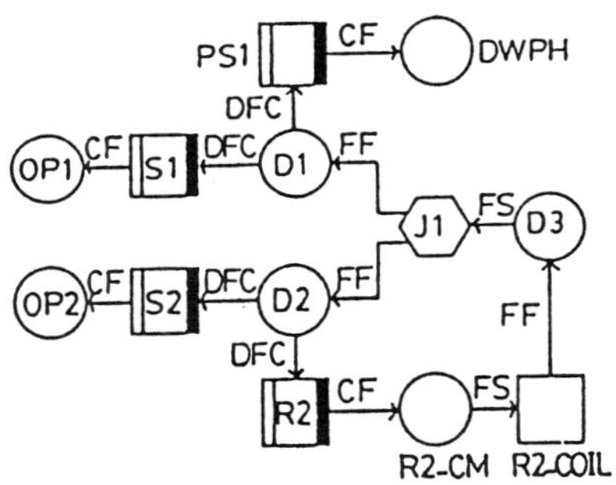

Fig. 7.17 A Semantic Network Representation of Relay Circuit

are always functioning. PS1 is a pressure switch; S1 and S2 are manual switches; R2 is a relay contact. PS1 becomes 'on' when a drywell pressure becomes high. As a result, DC current D1 flows through relay coil R2_COIL as current D3 if S1 is 'on.' This, in turn, energizes R2_COIL. As a result, relay contact R2 becomes 'on' by EMG (Electro-Magnetic) command from R2_COIL, and DC current D2 starts flowing if S2 is 'on.' Currents D1 and D2 are joined by Junction J1, and D3 can flow through R2_COIL even if current D1 stops flowing by some reasons such as PS1 incorrectly becoming 'off.' In the real ECCS, the EMG signal energized by R2_COIL is further used as a signal to operate an emergency core cooling pump when the drywell pressure is high. A semantic network is shown in **Fig. 7.17**.

The following pieces of equipment, flows and labeled arrows are identified in the network.

(1) Equipment: PS1(pressure switch), S1 and S2 (manual switches), R2 (relay contact), R2_COIL (relay coil for contact R2), J1 (funning-in junction).

(2) Flows: D1 through D3 (DC currents), R2_CM (EMG command to relay contact R2), OP1 and OP2 (manual operations for switches S1 and S2), DWPH (drywell pressure high event). Note that the manual operations and the pressure high event are modeled as the flows. This is not so unnatural because they as command flows propagate through the system.

(3) Labeled arrows:

1) From flow to equipment; FS (points to equipment generating the flow), DFC (points to DFC equipment).

2) From equipment to flow; CF (points to a command flow),FF (points to feed flow from the flow generator)

Note that a cause to effect (forward) direction is inverse of the arrow directions. In other words, the effect to cause (backward) direction essential in FT is expressed by the arrow directions.

The semantic network contains a loop consisting of R2_CM, R2_COIL, D3, J1, D2, R2 and R2_CM. This loop represents the self-holding capability of the relay system of Fig. 7.16. In order to show the capability, the loop is traced in an opposite direction of the arrows. Assume that the R2_COIL is energized and that manual switch S2 is 'on.' Then flow R2_CM has a positive flow rate because the R2_COIL is an FS (flow source) of R2_CM. Flow R2_CM is a command flow to relay R2 and this relay is in an 'on' state because it has a positive gain. Then DC current D2 has a positive flow rate because the generation rate of D2 is always positive by assumption. Flow D2 is a feed flow to Junction J1, and DC current D3 also has a positive flow rate, resulting again in an energized R2_COIL.

BOUNDARY CONDITIONS

Fixed and/or free boundary conditions can be specified explicitly for a subset of flow or equipment nodes.

Conditions at flow nodes: Boundary conditions at the flow nodes are in terms of flow triple. Consider again the relay system and the network in Figs. 7.16 and 7.17. It was assumed that the power lines A and AA are perfect, and have a voltage difference. Thus, the generation rates (or potentials) of D1 and D2 are always positive. This type of fixed boundary conditions are expressed as (D1, generation rate, Positive) and (D2, generation rate, Positive). The drywell pressure high event may or may not occur. This can be represented as (DWPH, flow rate, ?). Similarly, (OP1, generation rate, ?) and (OP2, generation rate, ?) hold. Note that fixed or free boundary conditions are required for the terminal flow nodes such as DWPH, OP1, and OP2.

Conditions at equipment nodes: Some boundary conditions on the equipment failure are implicit and are determined by event development rules. For instance, consider developing the flow rate Zero event of F_1 for NCE in Fig. 7.15. Assume that NCE is being suspected for the Zero flow rate. At this situation, an event development rule yields an event that NCE being F_Cl, and at the same time registers the possibility of F_Cl event as surely_occurring by definition of NCE. Possibilities such as this will be propagated upward toward the top event by a 3-valued logic.

Other conditions on the equipment failure modes are explicitly stated for the given semantic network. Consider the relay coil R2_COIL in Fig. 7.17. This is modeled as a Flow Sensor because the relay coil generates the EMG signal R2_CM when DC current D3 is fed. In general, such a Flow Sensor may generate a spurious flow even if the flow rate of the feed flow is Zero. However, for the relay coil such a spurious failure is improbable. Thus a relay coil failure that R2_COIL remains energized without current D3 is prohibited. This special treatment is registered as a boundary condition at the R2_COIL equipment node.

7.6.5 Event Development Rules

TYPES OF EVENTS

Two more types of events are considered in addition to the flow triple. These are equipment_suspected event and equipment failure. The three types of events are related each other through event development rules, as shown in **Fig. 7.18**. The equipment_suspected event indicates that a specific piece of equipment is suspected as a cause of an event. This event is developed into equipment failure, i.e., a failure of the equipment itself, and a command failure to the equipment. The command failure can be stated as a flow triple event such as (command, flow rate, Zero). The equipment failure usually becomes a basic event of the FT when its occurrence possibility is uncertain.

POSSIBILITIES OF EVENTS

Each event will be assigned a truth value of a 3-valued logic; surely_occurring (*yes*), surely_not_occurring (*no*), and uncertain (*unknown*). Only *unknown* events will be retained in the FT. A typical example of surely_occurring event is that the NCE is F_Cl because the NCE is by definition is normally closed. An example of surely_not_occurring event is that the generation rate of D1 in Fig. 7.17 is Zero because this contradicts with the fixed boundary condition that the rate is always Positive. On the other hand, (OP1, flow rate, Zero) is an uncertain event because flow OP1 has a free boundary condition (OP1, flow rate, ?). The truth values will be propagated toward the top event through AND and OR gates in the FT, and the values of other events are determined. **Table 7.3** shows the 3-valued logic for the gates.

TYPES OF DEVELOPMENT RULES

Four types of event development rules are included in a rulebase.

(1) Flow triple to flow triple: Consider first a flow triple (D1, flow rate, Zero) in the semantic network of Fig. 7.17. There exist two aperture controllers around flow D1. The existence of such controller can easily be identified by inspecting arrows around the flow node labeled as NCE, NOE, COE, OCE, DFC, or AFC. The flow triple event can be developed into an OR combination of 1) (D1, generation rate, Zero), or 2) (D1, aperture, F_Cl); the second event is included because of the existence of the aperture controllers for D1. This development is justified by Table 7.1. In a similar way, other attribute values of the flow rate other than Zero can be developed based on Table 7.1.

Consider second a flow triple (OP1, flow rate, Zero). There are no aperture controllers around flow OP1 and the flow aperture is assumed to be always open by convention. The flow triple event is developed into a single event (OP1, generation rate, Zero). An event development rule can be written for a general flow for each case with or without the aperture controllers. Note that the triple (D1, generation rate, Zero) thus developed turns out to be false by the boundary condition at node D1. On the other hand, the triple (OP1, generation rate, Zero) is uncertain and becomes a basic event of the FT because OP1 has a free boundary condition.

(2) Flow triple to equipment_suspected: Consider again the triple (D1, aperture,

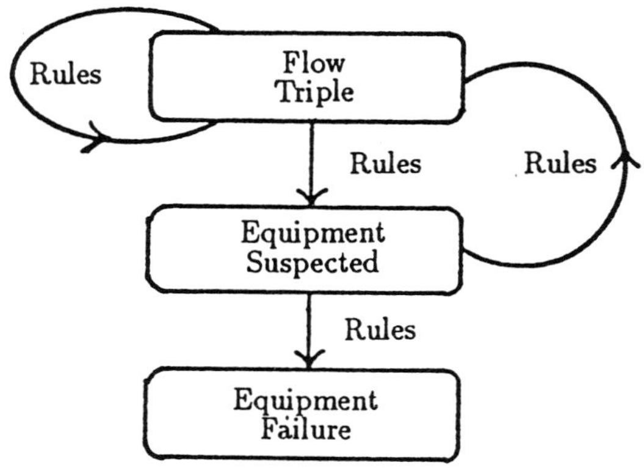

Fig. 7.18 Classification of Events and Rules

Table 7.3 Three-Valued Logic for Reflecting Free and Fixed Boundary Conditions for Fault Tree Generation

A	B	A AND B	A OR B
yes	*yes*	*yes*	*yes*
yes	*no*	*no*	*yes*
yes	*unknown*	*unknown*	*yes*
no	*yes*	*no*	*yes*
no	*no*	*no*	*no*
no	*unknown*	*no*	*unknown*
unknown	*yes*	*unknown*	*yes*
unknown	*no*	*no*	*unknown*
unknown	*unknown*	*unknown*	*unknown*

F_Cl) in Fig. 7.17. All the aperture controllers around flow D1 are suspected, thus yielding an OR combination of two equipment suspected events: 1) PS1 is suspected to be F_Cl, or 2) S1 is suspected to be F_Cl. If there were an NCE around D1, then this would also be suspected. However, the NCE is normally closed, and hence yielding a surely_occurring event.

For the triple (D3, generation rate, Zero), Junction J1 is suspected for the zero generation rate event.

(3) Equipment_suspected to flow triple: Consider the equipment_suspected event that PS1 is F_Cl. The PS1 is a DFC with a positive gain, and has command DWPH, as shown in Fig. 7.17. Thus, an event development rule for DFC yields a triple (DWPH, flow rate, Zero) for the command flow.

Consider second another equipment_suspected event about Junction J1. This Junction was suspected as a cause of (D3, generation rate, Zero). Thus, the equipment being suspected is developed into an AND combination of the two triples of the feed flow (FF): 1) (D1, flow rate, Zero), and 2) (D2, flow rate, Zero).

(4) Equipment_suspected to equipment failure: Consider the suspected event that PS1 is F_Cl. This is developed into the equipment failure that the pressure switch is inadvertently stuck at an 'off' state. For the equipment_suspected of Junction J1, no equipment failure is generated because J1 is assumed to be failure free.

The same equipment_suspected event of PS1 is developed by the two rules in (3) and (4) described above. In general, more than one rules applicable to an event represent OR cases of event development. Further, whether an event should be developed into an AND or an OR combination of causes is specified in the then-part of each rule.

CLASSIFICATION RULES

Some exceptional rules may be used, for convenience, to divide or classify statically a flow triple into its subclasses, not into causes preceding in time. For instance, (flow, aperture, F_Op) is a subclass of (flow, aperture, Open). A difference between the causes and the subclasses appear in dealing with event recurrences, as described in step 2-1 of gate inclusion and exclusion below event.

ACQUISITION OF RULES THROUGH TABLES AND EQUIPMENT DEFINITIONS

For the three attributes of aperture, flow rate, and generation rate, event development rules can be obtained systematically. Table 7.1 is used for specifying rules from a flow triple to flow triples, i.e., the flow rate is developed into the generation rate and the aperture. Table 7.2 is used to develop flow apertures into apertures of pieces of equipment suspected. The definitions of basic equipment in Fig. 7.15 yields equipment failures and command failures.

Examples of the rules thus obtained are listed below: The first two rules develops the Zero flow rate event into the F_Cl aperture or the Zero generation rate event. The third rule develops the F_Cl aperture event into equipment_suspected events. An equipment_suspected event is reduced to a command failure by the forth rule. The fifth rule yields a surely_occurring event. The Zero generation rate event are processed by the sixth and the seventh rules.

(1) if ((flow, flow rate, Zero) **and** (there exists equipment controlling aperture of the flow)) **then** ((flow, aperture, F_Cl) **or** (flow, generation rate, Zero)).

(2) if ((flow, flow rate, Zero) **and** (no equipment exists to control aperture of the flow)) **then** (flow, generation rate, Zero).

(3) if (flow, aperture, F_Cl) **then** ((suspect aperture controllers around the flow as OR causes of the F_Cl aperture).

(4) if ((equipment is suspected as a cause of flow aperture being F_Cl) **and** (the equipment is OCE of positive gain)) **then** (flow rate of command to the equipment is Zero).

(5) if ((equipment is suspected as a cause of flow aperture being F_Cl) **and** (the equipment is NCE)) **then** ((the equipment F_Cl mode is surely_occurring).

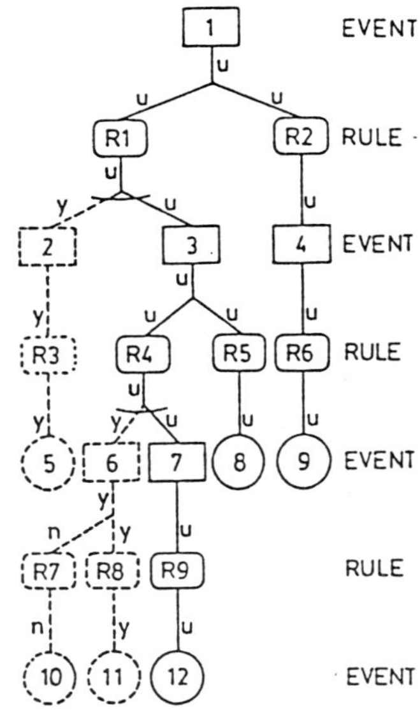

Fig. 7.19 A Schematic Diagram for Fault Tree Generation Mechanism with 3-Valued Logic

(6) if (flow, generation rate, Zero) **then** (suspect equipment pointed by flow source arrow).

(7) if ((equipment pointed by FS arrow is suspected as a cause of Zero generation rate) **and** (the equipment is a Branch)) **then** (flow rate of feed flow to the equipment is Zero).

7.6.6 Recursive Algorithm for FT Generation

An FT generation process is illustrated in **Fig. 7.19** where events and rules appear alternately. An event development rule is included as an AND or an OR gate in FT by retrieving relevant rules to an event. A rule yielding an AND or OR combinations of causes becomes the AND or OR gates, respectively. A new set of events as causes or subclasses are included by executing the development or the classification rules, respectively. The 3-valued logic in Table 7.3 is used to propagate the truth values towards the top via intermediate events and gates. Only the uncertain events or gates denoted by solid lines in Fig. 7.19 are retained in FT, solely enumerating non-trivial events.

The process in Fig. 7.19 can be generalized into a rigid algorithm as a recursive function *and_or(dlist)* which returns a value. This function consists of gate inclusion and exclusion part and the event inclusion and exclusion part.

OUTLINE OF RECURSIVE FUNCTION

Structure: The function and_or has the following structure. Comments are enclosed by /* and */.

function and_or(dlist)
argument dlist
/* *dlist*: A path from top event to current gate or event */

local data, rdata, rdlist, rules, events, poss, rposs
/* *data*: Gate or event currently investigated */
/* *rdata*: Gate or event investigated next */
/* *rdlist*: A path from top event to the next gate or event */
/* *rules*: List of rules applicable to current event */
/* *events*: List of events generated by a rule in *rules* */
/* *poss*: Truth value for current gate or event */
/* *rposs*: Truth value for next gate or event */
Step 1) Bringing current object: Current object as a gate or an event is picked up.
Step 2) Gate inclusion and exclusion below event: This is executed if an event is picked up as a current object.
Step 3) Event inclusion and exclusion below gate: This is executed if a gate is picked up as a current object.

Downward development and upward propagation: At some instance of time, the recursive function processes the current object which is either a gate or an event. The downward FT development is terminated at the object when the function returns a value which is either *yes*, *no*, or *unknown*. For instance, the function processing event 6 in Fig. 7.19 returns *yes*, while the function with event 7 returns *unknown*. These two functions are called from the function processing gate R4, which receives the two returned values and calculates their AND combination, yielding *unknown*. Then, the function with gate R4 returns the value *unknown* to the function dealing with event 3. The recursive call corresponds to a downward development of FT, while returning a value represents an upward propagation of truth values.

Event recurrence and options: The FT is generated by tracing the semantic network along directions of labeled arrows by function *and_or*. Consider the event that the flow rate of R2_CM is Zero in the network in Fig. 7.17. Because the flow node is on the loop, the same flow-attribute pair (R2_CM, flow rate) may be encountered in a process of FT development. In order to prevent an infinite iteration, a truth value must be returned when such a recurrence is encountered. There are three options to process the recurrence of the flow-attribute pair at each flow node. The first option is the most general one. The second and the third options are used as on-off switches for house events.

(1) Returning value *unknown*: This means that a one-step earlier event with the recurring flow-attribute pair is included as a cause in FT. This is a most general option, yielding all the one-step earlier time series conditions for the recurrent pair. For instance, (R2_CM, flow rate, Zero) is included as a cause of the same event, which indicates that the event continues to exists. On the other hand, the inclusion of (R2_CM, flow rate, Positive) indicates that the flow rate changes from Positive to Zero.

(2) Returning value *no*: This assumes that the event with the recurrent pair is not occurring earlier in time. As a result, the event does not appear in FT. The truth value is propagated upward accordingly.

(3) Returning value *yes*: This assumes that the event with the recurrent pair is occurring earlier in time. As a result, the event does not appear in FT. However, the truth value propagation upward differs from the one in option (2).

Other options: There are two cases where other options are used. The one is the case where no rules are applicable to the current event, and the other is the case where no events are generated after firing a rule because such an event has no counterpart in the semantic network currently investigated.

(1) No rules applicable: Usually value *no* is returned because the lack of rules means lack of causes. However, value *unknown* may be returned to alarm an incompleteness of the rulebase, including the current event in FT.

(2) No event generated: Usually value *no* is returned. However, value *unknown* may be returned to alarm an incompleteness of the semantic network, including the current gate in FT.

BRINGING CURRENT OBJECT

This can be stated as follows:

1) *data* ⇐ *first*(*dlist*) /* Object *data* is on top of dlist and is an event or a rule currently investigated. */

GATE INCLUSION AND EXCLUSION BELOW EVENT

This phase is performed when *data* is an event. This event has been temporary included in the FT by a parent *and_or* which calls the current *and_or*(*dlist*). In step 2-1, 2-2 or 2-3, development of the current event into gates is terminated because the event truth value is known without further recursive calls. When the truth value is either *yes* or *no*, event *data* is excluded from FT by the parent *and_or*. The event remains in the FT when the truth value is *unknown*. The first step processes event recurrence, the second a flow triple based on its boundary condition, and the third an equipment failure based on its implicit and/or explicit boundary conditions. Applicable rules to the current event is obtained in step 2-4. It is important to note that each rule can be interpreted as an AND or an OR gate. Hence, either 'gate' or 'rule' is used interchangeably according to a context. Event development is again terminated in step 2-5 when the applicable rule set is empty. Step 2-6 initializes the truth value for the current event. The value becomes *yes* when at least one output from the gate in the applicable rule set becomes *yes*, i.e., a logic OR combination among the gates are calculated. The truth value is returned to the parent *and_or* in steps 2-7-8 or 2-8.

2-1) Return *yes*, *no*, or *unknown* according the FT generation option if *data* is a flow triple obtained by an event development rule and if its (flow, attribute) pairs recurs for the first time in *dlist*. Note that this step is not applicable when *data* is not a cause but a subclass of an event generated by an event classification rule.
2-2) Return *yes*, *no*, or *unknown* if *data* is a flow triple and if its (flow, attribute) pair is subject to a fixed or a free boundary condition.
2-3) Return *yes*, *no*, or *unknown* according to an explicit or an implicit boundary condition of the failure mode if *data* is an equipment failure event. /* Note that *yes* or *no* are returned if the failure mode is surely_occurring or surely_not_occurring, respectively. Value *unknown* is returned otherwise, and the equipment failure is contained as a basic event of FT. */

2-4) The *data* is now a flow triple or an equipment_suspected because the equipment failure always returns a value in step 2-3. Retrieve relevant rules to *data*, and sort them according to priorities. Denote by list *rules* the resulting rule set. /* Applicable rule set is determined*/

2-5) Return *no* or *unknown* according to the FT generation option if *rules* is empty. /* Event development into gates is terminated*/

2-6) *poss* ⇐ *no*. /* Initial setting of occurrence possibility (truth value) of *data* currently investigated. Logic variable *poss* becomes *yes* if at least one *yes* is returned from the child *and_or* in the following iteration in step 2-7. */

2-7) Iterate the following steps while *rules* is not empty.

- 2-7-1) *rdata* ⇐ *first*(*rules*). /* *rdata* is a rule with the highest priority among the remaining rules */
- 2-7-2) Remove *rdata* from *rules*. /* Avoid duplicate execution of *rdata*. */
- 2-7-3) Include gate *rdata* in FT at a place immediately below event *data*. /* An AND or an OR gate is included here */
- 2-7-4) *rdlist* ⇐ *cons*(*rdata*, *dlist*). /* Revise the path from the top event to include gate *rdata*. */
- 2-7-5) *rposs* ⇐ *and_or*(*rdlist*). /* *rposs* is a possibility value of gate *rdata* returned from the recursive call *and_or*(*rdlist*) */

2-7-6) Exclude gate *rdata* from FT if *rposs* is *yes* or *no*. /* The AND or OR gate is excluded if its output has a truth value *yes* or *no* */

2-7-7) *poss* ⇐ *poss OR rposs*. /* Revise the possibility of event *data* by the 3-valued OR logic because more than one rules in the rule set indicate a logic OR combination of gates. */

2-7-8) Immediately return *yes* when *poss* becomes *yes*. /* Exit immediately from the iteration because *poss* remains *yes* once it becomes *yes*. */

2-8) Return *poss*. /* Value *no* or *unknown* is returned from current *and_or*(*dlist*). */

EVENT INCLUSION AND EXCLUSION BELOW GATE

In this phase the current object *data* is a rule (or a gate). This rule is first fired in step 3-1. The development of the gate into events is terminated in step 3-2 when no event is generated. Otherwise, an AND or an OR combination of events were generated when the rule was fired in step 3-1. Step 3-3 initializes the truth value for the combination. Step 3-4 is an iteration where each event in the AND or the OR combination is processed. Note that the event is an input to the current gate. The event is picked up in step 3-4-1 and included in the FT. A recursive function call is performed in step 3-4-5. The truth value for the event within the combination is determined by the returned value from the recursive call. The event with *yes* or *no* truth value is excluded from the FT. On the other hand, for the AND combination, its truth value initialized in step 3-3 becomes *no* in step 3-4-7 when some event turns out to have value *no*; for the OR combination, the truth value becomes *yes* in step 3-4-8 when some event turns out to have value *yes*. The iteration is terminated and a value is returned to the parent *and_or* as soon as the truth value change occurs for each type of combination. When the whole iteration is completed, a truth value for the combination (or the current gate) is returned in step 3-5 from the current *and_or*(*dlist*) to the parent *and_or*.

3-1) Fire rule *data* to generate a set of AND_causes or a set of OR_causes. Sort the causes according to priorities. Denote by list *events* the resulting event set.

3-2) Return *no* or *unknown* according to the FT generation option if *events* is empty.

3-3) *poss* ⇐ *yes* if *events* is a list of AND_causes. *poss* ⇐ *no* if events is a list of OR_causes.

3-4) Iterate the following steps while *events* is not empty.

3-4-1) *rdata* ⇐ *first*(*events*). /* *rdata* is an event with the highest priority */

3-4-2) Remove *rdata* from *events*. /* Avoid duplicate development of *rdata*. */

3-4-3) Include event *rdata* in FT at a place immediately below gate *data*.

3-4-4) *rdlist* ⇐ *cons*(*rdata*, *dlist*). /* Revise the path from the top event to include event *rdata*. */

3-4-5) *rposs* ⇐ *and_or*(*rdlist*). /* *rposs* is a possibility value of event *rdata* returned from the recursive call */

3-4-6) Exclude *rdata* from FT if *rposs* is *yes* or *no*.

3-4-7) *poss* ⇐ *poss AND rposs* if *events* was a list of AND_causes. Immediately return *no* when *poss* becomes *no*.

3-4-8) *poss* ⇐ *poss OR rposs* if *events* was a list of OR_causes. Immediately return *yes* when *poss* becomes *yes*.

3-5) Return *poss*. /* Value *yes*, *no*, or *unknown* is returned from current *and_or*(*dlist*) */

7.6.7 Practical Applications

A RELAY SYSTEM

Consider again the relay circuit in Figs. 7.16 and 7.17. Suppose the top event 'Flow rate of drywell pressure high signal R2_CM is zero.' A fault tree generated is shown in **Fig. 7.20**.

- 0) Flow rate of R2_CM is zero (Event 1)
  - 5) OR GATE
  - 6) Flow rate of D3 is zero (Event 7)
    - 11) AND GATE
    - 12) Flow rate of D2 is zero (Event 14)
      - 15) OR GATE
      - 16) Equipment S2 is suspected (Event 20)
        - 17) OR GATE
        - 18) Fully_Closed failure of S2 (Event 22)
        - 18) Flow rate of OP2 is zero (Event 24)
      - 16) Equipment R2 is suspected (Event 19)
        - 17) OR GATE
        - 18) Fully_Closed failure of R2 (Event 26)
        - 18) Flow rate of R2_CM is zero (Event 28)
    - 12) Flow rate of D1 is zero (Event 13)
      - 15) OR GATE
      - 16) Equipment S1 is suspected (Event 34)
        - 17) OR GATE
        - 18) Fully_Closed failure of S1 (Event 36)
        - 18) Flow rate of OP1 is zero (Event 38)
      - 16) Equipment PS1 is suspected (Event 33)
        - 17) OR GATE
        - 18) Fully_Closed failure of PS1 (Event 40)
        - 18) Flow rate of DWPH is zero (Event 42)
  - 6) Zero output failure of R2_COIL (Event 44)

Fig. 7.20 A Fault Tree for the Relay System in Fig. 7.16

An event recurrence was encountered at Event 28 where value unknown was returned in step 2-1 of function *and_or*. Event 1, the top event, can be expressed as:

$$1 = (22 + 24 + 26 + 28)(36 + 38 + 40 + 42)$$

Event 28, the recurrent event is actually a house event, and two cases are available: (1) If Event 28 is true, then the top event becomes

$$1 = 36 + 38 + 40 + 42$$

This corresponds to a case where drywell pressure high signal R2_CM continues to remain 'off', thus causing the top event to occur. Note that the same result is obtained when value yes is returned in step 2-1 of function *and_or*.

(2) If Event 28 is false, then the top event becomes

$$1 = (22 + 24 + 26)(36 + 38 + 40 + 42)$$

This corresponds to a case where the high pressure signal ceased to be 'on', thus causing the top event. The same result is obtained when value *no* is returned in step 2-1. Note that the same semantic network as Fig. 7.17 can be used to generate an FT with top event 'R2_CM is on' under boundary condition that DWPH event does not exist. Such an FT shows possible causes of spurious activation of the relay circuit. Similar FT to Fig. 7.20 has been successfully generated for an ECCS model of a practical size.

HYPOTHETICAL SWIMMING POOL REACTOR

Consider a hypothetical swimming pool reactor in **Fig. 7.21**. System components are listed in **Table 7.4** with their correspondence to the basic equipment library.

During a normal period, pressurized air (AIR) flows through solenoid valve C9 and mechanical valve C10 in series. As a result, inlet and outlet actuators C3 and C4 respectively force inlet and outlet valves C1 and C2 to be open. The coolant enters the pool via inlet valve C1, and exists the pool via outlet valve C2. Switches C5 through C8, C12, and C14 are at a plus side, and hence the input signals to NAND gate C15 are all 'on', suppressing the trip signal to be issued from the NAND gate.

Suppose that a water level low event occurs by some reasons such as piping failure outside the pool. In this case the following protection mechanisms are activated, preventing the reactor from overheating.

(1) Reactor Trip: Trip signal is issued from the NAND gate, thus stopping the nuclear reaction.

(2) Pool Isolation: Valves C1 and C2 are closed to prevent the leakage of the coolant from the pool.

Electrode C11 and float C13 detect the coolant level low event. C11 forces the solenoid switch C12 to be an 'off' state. Consequently, solenoid valve C9 closes, and trip suppressing signal T12 from C12 to the NAND gate becomes 'off.' On the other hand, C13 closes mechanical valve C10, forces mechanical switch C14 to be an 'off' state, thus changing trip suppressing signal T14 into an 'off' state. By these nullification of the trip suppressing signals, the trip signal from the NAND gate becomes 'on.'

Since the pressurized air is blocked by valve C9 or C10, the pistons in the actuators C3 and C4 fall, and valve C1 and C2 are closed, thus isolating the coolant in the pool. Redundant trip suppressing signals T5 through T8 from magnetic switches C5 through C8 also become an 'off' state.

A semantic network representation is shown in **Fig. 7.22**, where flow nodes are listed in **Table 7.5**. Equipment nodes have been listed in Table 7.4. The aperture controllers are C1, C2, C9, and C10. Others are generation rate controllers. Assume the existence of LOW LEVEL SIGNAL 13. Then, by convention, the flow rate of this flow is Positive. Signal T14 in Fig. 7.21 becomes 'off', i.e., the flow rate of T14 becomes Zero. Therefore, mechanical switch C14 is modeled as a NOT in Fig. 7.22 or in Table 7.4.

Mechanical valve C10 changes from an open to a closed state by LOW LEVEL SIGNAL 13, and hence C10 is modeled as an OCE. The gain of C10 is negative since the valve

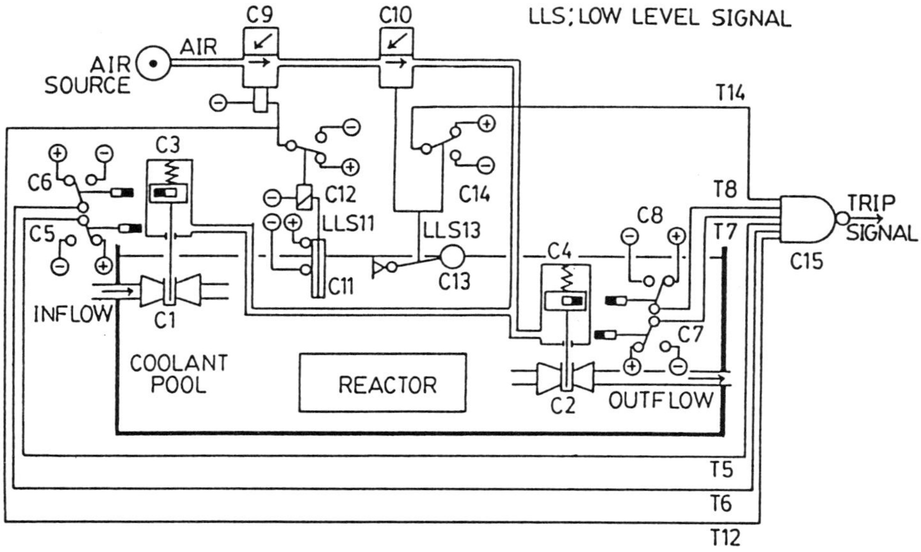

Fig. 7.21 A Schematic Diagram of a Hypothetical Swimming Pool Reactor

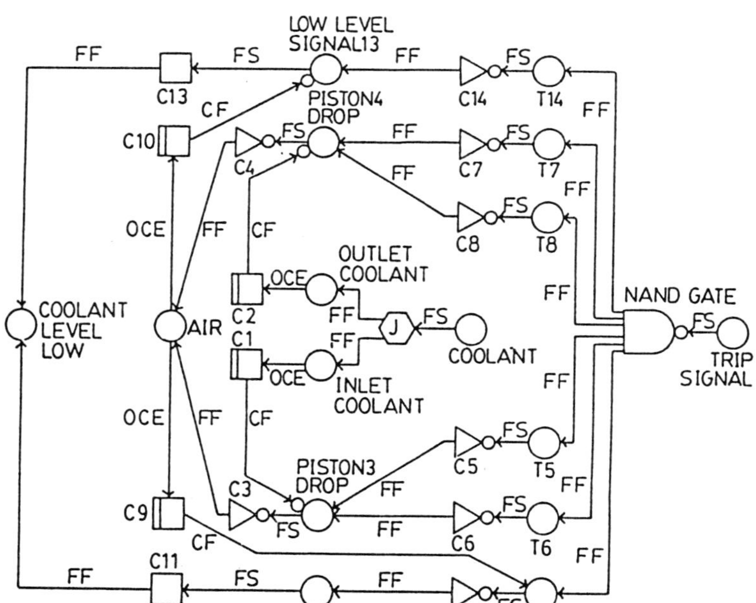

Fig. 7.22 A Semantic Network Representation of the Reactor

Table 7.4 Set of Component and Their Library Types for Swimming Pool Reactor

Symbol	Description	Library
C1	Inlet valve	OCE
C2	Outlet valve	OCE
C3	Inlet actuator	NOT
C4	Outlet actuator	NOT
C5	Magnet switch 5	NOT
C6	Magnet switch 6	NOT
C7	Magnet switch 7	NOT
C8	Magnet switch 8	NOT
C9	Solenoid valve	OCE
C10	Mechanical valve	OCE
C11	Electrode bar	Flow Sensor
C12	Solenoid switch	NOT
C13	Float	Flow Sensor
C14	Mechanical switch	NOT
C15	NAND gate	NAND
J	Junction node	Junction

Table 7.5 Flow Nodes in Semantic Network in Fig. 7.22

Flow Name	Description
AIR	Actuator air
COOLANT	Inlet or outlet coolant
COOLANT LEVEL LOW	Coolant level low event
INLET COOLANT	Inlet coolant
LOW LEVEL SIGNAL 11	Low level signal from electrode
LOW LEVEL SIGNAL 13	Low level signal from float
PISTON 3 DROP	C3 piston drop event
PISTON 4 DROP	C4 piston drop event
OUTLET COOLANT	Outlet coolant
Ti	Trip suppressing signal from switch Ci
TRIP SIGNAL	Trip Signal from NAND gate

Table 7.6 Module Flow Nodes in Semantic Network in Fig. 7.22

Module Flow Node
AIR
COOLANT LEVEL LOW
LOW LEVEL SIGNAL 13
PISTON 3 DROP
PISTON 4 DROP
T12

```
0) Event 1: Flow rate of TRIP SIGNAL is zero
    5) OR GATE
      5) AND GATE
      6) Event 12: Flow rate of T14 is positive
            11) OR GATE
            12) Module event 18: Flow rate of LOW LEVEL SIGNAL 13 is zero
            12) Basic event 28: [Positive output failure of C14]
      6) Event 11: Flow rate of T7 is positive
            11) OR GATE
            12) Module event 34: Flow rate of PISTON 4 DROP is zero
            12) Basic event 78: [Positive output failure of C7]
      6) Event 10: Flow rate of T8 is positive
            11) OR GATE
            12) Module event 34: Flow rate of PISTON 4 DROP is zero
            12) Basic event 86: [Positive output failure of C8]
      6) Event 9: Flow rate of T5 is positive
            11) OR GATE
            12) Module event 92: Flow rate of PISTON 3 DROP is zero
            12) Basic event 102: [Positive output failure of C5]
      6) Event 8: Flow rate of T6 is positive
            11) OR GATE
            12) Module event 92: Flow rate of PISTON 3 DROP is zero
            12) Basic event 110: [Positive output failure of C6]
      6) Module event 54: Flow rate of T12 is positive
    6) Basic event 112: [Zero output failure of NAND gate]
```

Fig. 7.23 Tree Top of FT with Top Event No TRIP SIGNAL under LOW LEVEL COOLANT

closes when LOW LEVEL SIGNAL 13 exists. The negative gain nature is denoted by a small circle at the head of the arrow labeled as CF from C10 to LOW LEVEL SIGNAL 13.Mechanical valve C10 in turn controls the aperture of AIR. The aperture is also controlled by solenoid valve C9 modeled as OCE with command flow T12. The gain of this OCE is positive because C9 closes when T12 becomes 'off.' The two OCEs are observed around AIR in Fig. 7.22

The aperture of the outlet coolant is controlled by valve C2 as an OCE with command flow as an event 'PISTON 4 DROP.' The aperture of the inlet coolant is controlled valve C1 as an another OCE. Flow COOLANT in Table 7.5 denotes either the inflowing or the outflowing movement of the coolant, and has Junction J as its generation rate controller with feed flows of INLET COOLANT and OUTLET COOLANT. In other words, the flow rate of COOLANT becomes Zero when the flow rates of INLET COOLANT and OUTLET COOLANT become both Zero. This indicates a success of the pool isolation.

Assume the following boundary conditions for generation of FT:

(1) The flow rate of the COOLANT LOW LEVEL is a positive constant (Cons). This means the occurrence of the COOLANT LOW LEVEL event

(2) Generation rates of AIR, OUTLET COOLANT, and INLET COOLANT are positive constant (Cons). This means that the flow rates of these flows remain positive unless their apertures become F_Cl.

Consider 'Flow rate of trip signal is Zero' as a top event. The generation algorithm traces the semantic network, for instance, in the following order: 1) NAND gate as a flow source (FS) of the trip signal; 2) trip suppressing signal T14 as a feed flow (FF) to the NAND gate, 3) mechanical switch C14 as a flow source of T14, 4) LOW LEVEL SIGNAL 13 as a feed flow to switch C14, 5) float C13 as a flow source of LOW LEVEL SIGNAL 13, 6) COOLANT LEVEL LOW as a feed flow to float C13, and so on.

When the labeled arrows are traced in similar ways, the same flow node may be encountered more than once from different accesses. A subtree as a module of the FT is generated when the same flow triple is being developed at this multiple-visit node. **Table 7.6** shows a set of such module nodes for the semantic network in Fig. 7.22 for the top event about the trip signal. For instance, flow node AIR may be encountered via the following paths of FT development: 1) TRIP SIGNAL, T7, PISTON 4 DROP, AIR, 2) TRIP SIGNAL, T8, PISTON 4 DROP, AIR, 3) TRIP SIGNAL, T5, PISTON 3 DROP, AIR, 4) TRIP SIGNAL, T6, PISTON 3 DROP, AIR.

The above list of paths also indicates that PISTON 4 DROP and PISTON 3 DROP are also module flow nodes. Further, the module FT starting with the PISTON 3 drop includes the module FT starting with the AIR. The module FT starting with LOW LEVEL SIGNAL 13 appears both in the FT with TRIP SIGNAL and in the module FT with AIR. T12 is a similar module node to LOW LEVEL SIGNAL 13. The COOLANT LEVEL LOW is also a module flow node but the module FT there is reduced to a surely_occurring event because of the boundary condition there. The module FTs thus identified form a nested structure. A tree top of the FT generated by the proposed method is shown in **Fig. 7.23**. Other details of the tree is not shown here by a lack of space.

7.6.8 Summary

An automated fault tree generation method is presented. This is based on 1) concept of flow, attribute, and value, 2) an equipment library, 3) semantic network representation of the target system, 4) event development rules, and 5) a recursive function as an algorithmic framework. Boundary conditions can be specified for the network. The event development rules can be obtained systematically. The recursive function is based on a 3- valued logic to propagate the implicit and explicit boundary conditions. Events and gates are temporary included in FT are excluded when the propagated value becomes *yes* or *no*, and only *unknown* events or gates remain in FT. Module FTs can be identified. The generation method is demonstrated for a relay system and a hypothetical swimming pool reactor.

REFERENCES

[1] Spangler M.B., *A Summary Perspective on NRC's Implicit and Explicit Use of De Minimis Risk Concepts in Regulating for Radiological Protection in the Nuclear Fuel Cycle*, Chapter 12 of De Minimis Risk ed. by Whipple C., Plenum Press, New York, 1987.

[2] Spangler M.B., *Policy Issues Related to Worst Case Risk Analysis and the Establishment of Acceptable Standards of De Minimis Risk: in Uncertainty in Risk Assessment, Risk Management, and Decision Making*, Plenum Press, New York, 1987.

[3] USNRC: Reactor Safety Study, *An Assessment of Accident Risk in U.S. Commercial Nuclear Power Plants*, WASH-1400, NUREG-75/014, 1975.

[4] Milvy P., *De Minimis Risk and the Integration of Actual and Perceived Risks from Chemical Carcinogens*, Chapter 7 of De Minimis Risk ed. by Whipple C., Plenum Press, New York, 1987.

[5] Lambert H.E., *Case Study on the Use of PSA Methods: Determining Safety Importance of Systems and Components at Nuclear Power Plants*, IAEA-TECDOC-590, International Atomic Energy Agency, 1991.

[6] Davis J.P., *The Feasibility of Establishing a De Minimis Level of Radiation Dose and a Regulatory Cutoff Policy for Nuclear Regulation*, Chapter 13 of De Minimis Risk ed. by Whipple C., Plenum Press, New York, 1987.

[7] Pitz G.F., *Risk Taking, Design, and Training*, Chapter 10 of Risk-Taking Behavior ed. by J. Frank Yates, John Wiley & Sons, New York, 1992.

[8] Whipple C. ed., *De Minimis Risk*, Plenum Press, New York, 1987.

[9] US NRC: *PRA Procedures Guide - A Guide to the Performance of Probabilistic Risk Assessments for Nuclear Power Plants*, NUREC/CR-2300, US NRC, 1983.

[10] Comar C., *Risk - A Pragmatic De Minimis Approach*, Science 203, 4378, Jan. 1979 (See also introductory chapter of De Minimis Risk ed. by Whipple C., Plenum Press, New York, 1987).

[11] Wilson R., *Commentary - Risks and their Acceptability*, Science, Technology, and Human Values, 9-2, pp. 11-22, Spring, 1984.

[12] US NRC: *Safety Goals for Nuclear Power Plant Operations*, NUREG- 0880, Rev. 1, May 1983.

[13] IAEA: *Case Study on tbe Use of PSA Methods - Backfitting Decisions*, IAEA-TECDOC-591, IAEA, April 1991.

[14] Byrd III D. and L. Lave, *Significant Risk is not the Antonym of De Minimis Risk*, Chapter 5 of De Minimis Risk ed. by Whipple C., Plenum Press, New York, 1987.

[15] Hirsch H., T. Einfalt et al., *IAEA Safety Targets and Probabilistic Risk Assessment*, Report prepared for Greenpeace International, Aug. 1989.

[16] IAEA: *Basic Safety Principles for Nuclear Power Plants*, Safety Series No. 75-IMSAG-3.

[17] Cunninggham III G., *The De Minimis Concept in Radiation Protection*, Annual Meeting of Nuclear Safety Research Association of Japan, Tokyo, June 1983.

[18] International Nuclear Safety Advisory Group: *Basic Safety Principles for Nuclear Power Plants*, Safety Series, No. 75-INSAG-3, IAEA, 1988.

[19] Reason J., *Human Error*, Cambridge University Press, 1990.

[20] Embrey P., *Incorporating Management and Organizational Factors into Probabilistic Safety Assessment*, Personal Communication.

[21] Wagenaar W.A., P.T. Hudson and J.T. Reason, *Cognitive Failures and Accidents*, Applied Cognitive Psychology, Vol. 4, pp. 273-294, 1990.

[22] Wu J.S., G.E. Apostolakis and D. Okrent, *On the Inclusion of Organizational and Managerial Influences in Probabilistic Safety Assessments of Nuclear Power Plants*, The Analysis, Communication, and Perception of Risk, ed. by B.J. Garrick and W.C. Gekler, Plenum Press, New York, pp. 429-439, 1991.

[23] FDA, Center for Devices and Radiological Health: *Preproduction Quality Assurance Planning - Recommendations for Medical Device Manufacturers,* The Food and Drug Administration, Rockville, Maryland, Sept. 1989.

[24] Henley E.J. and H. Kumamoto, *Probabilistic Risk Assessment - Reliability Engineering, Design, and Analysis,* IEEE Press, New York, 1992.

[25] Villemeur A., *Reliability, Availability, Maintainability and Safety Assessment,* Vol. 1 and 2, John Wiley & Sons, New York, 1991 (Translated into English by A. Cartier and M.C. Lartisien).

[26] Andrews J. and G. Brennan, *Applications of the Digraph Method of Fault Tree Construction to a Complex Control Configuration,* Reliab. Engg and System Safety, 4-3, 357/384, 1990.

[27] Lambert H.E., *Fault Tree for Decision Making in Systems Analysis,* Lawrence Livermore National Laboratory, UCRL-51829, 1975 (Available from the National Technical Information Service, Springfield, Virginia, 22151, USA).

[28] Vesely W.E., F.F. Goldberg, N.H. Roberts and D.F. Haasl, *Fault Tree Handbook,* NUREG-0492, US NRC, 1981.

New Trends in System Reliability Evaluation
K.B. Misra (Editor)

CHAPTER 8

DIGRAPHS AND CAUSAL TREES

Takehisa Kohda and Koichi Inoue
Department of Aeronautical Engineering
Faculty of Engineering
Kyoto University, Japan

ABSTRACT

In this chapter, the fault tree synthesis by digraphs and causal trees is discussed. Both of these tools are considered as an intermediate step for constructing a fault tree and can deal with complex systems with control loops. However, the complete automatic construction is still difficult and human intervention is necessary to assure the correct result. Following the basic concept of both methods, an interactive method is described for obtaining minimal cut sets, which utilizes the human judgement in an interactive way to supplement the missing knowledge and to obtain the consistent results.

8.1 Introduction

Fault Tree Analysis (FTA) [1,2] plays a very important role as qualitative and quantitative method in probabilistic risk assessment (PRA) as well as reliability analysis. As its merit, 1) FTA has a well established, standard procedure to obtain the causes of a system failure and their occurrence probability, 2) many algorithms and software packages are available for quantitative and qualitative analysis, and 3) the construction process of a fault tree, a basic logic relation between system failure and component failures, gives a new insight into the system analysis. However, the development of a fault tree needs a lot of time and cost. Further the fault tree structure varies depending on the subjective judgement or experience of system analysts and designers, leading to an erroneous analysis. To overcome these difficulties, several automatic construction methods have been proposed such as decision table approach [3], mini fault tree approach [4], digraph approach [5-13] and causal tree approach[14-16].

The decision table approach assumes that components are the minimal units to be modeled and the system state can be uniquely specified by combining component models. A decision table, an extension of truth table, represents a component as a logical relation between output and input states. The system failure conditions, or fault tree can be obtained by combining component decision tables. Since mini fault trees are considered as a database of fault tree analysis of subsystems and components, a system fault tree can be obtained by logically combining mini fault trees. However, a mini fault tree is required for each deviation of each component variable, and a lot of input is necessary to create one component model. These two methods also need special construction procedures required to take account of feedback and feedforward loops. The digraph approach develops component models by using general cause-consequence relations among process variables and abnormal events, which represent the dynamic behavior qualitatively. A system model can be obtained by combining these component models based on its flow sheet or map. Disturbance propagation paths leading to the system failure can be obtained as links of cause-consequence relations, which correspond to cut sets in FTA. Like a digraph, a causal tree is also an intermediate step for constructing fault tree. A causal tree is very similar to a fault tree and shows the conditional and unconditional propagation paths for all basic events to leading to the top event. Both digraphs and causal trees can deal with complex systems including control loops. However, in all methods, human intervention and judgement are needed to prevent the generation of incomplete, inconsistent and incorrect failure modes.

Firstly, we briefly review the digraph approach and causal tree approach to fault tree construction. Secondly, an interactive method is given for obtaining minimal cut sets. Though the method follows the basic concept of digraphs and causal trees, it utilize the interaction with the system analyst to obtain the consistent result.

8.2 Digraph Approach

Fault tree construction methods based on digraphs have been developed [5-13]. A digraph provides an intermediate step which explicitly describes the relationships between process variables. These studies first develop fault trees by enumerating node states and disturbances, step by step, and considering the effects of protective actions of control loops on them at the same time. If a node is related to control loops, then a special logical operator is applied to the node in the development of a fault tree. This section presents the basic digraph method based on [13].

8.2.1 Digraphs

Nodes and Edges

A digraph is composed of nodes and directed edges. The former represents process variables such as flow, temperature, or pressure, and occurrences of certain types of abnormal events, while the latter indicates relationships between the connected variable deviations under specified conditions. When one process variable X can affect another Y, a directed edge is used to connect X and Y as shown in Fig. 8.1. The direction of the edge is form the cause variable (X) to the effect variable (Y). Each edge has a gain representing the relationship between the node deviations. Edge gains are represented as +10, +1, 0, -1, and -10. Values of 1 and 10 indicate moderate and strong relationships, respectively. If an edge links variable X to variable Y, then its gain can be interpreted as dY/dX. Therefore the sign indicates the direction in which the effect variable changes when the cause variable is increased. A zero gain indicates a nullification of the relationship between two variables. Edge gains can be conditional or normally true.

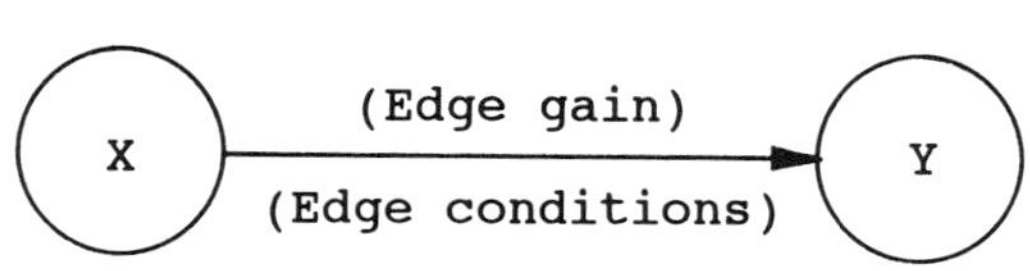

Fig. 8.1 Basic Digraph Element

The deviations in nodes are also expressed as +10, +1, 0, -1, and -10. Similarly, the sign denotes the direction of a deviation compared with the normal value, (positive or negative) and values of 0, 1, and 10 indicate that the magnitude of a deviation is none, moderate, or very large, respectively. For a node representing a component failure or occurrence of an abnormal event, the logic values 1 and 0 indicate that the failure state exists or does not exist, respectively. To obtain the deviation in the dependent process variable caused by a deviation in the independent variable. the general rule is to multiply the value of the independent variable deviation by the gain. For example, if nodes X and Y are connected by an edge with a gain of +1 and the deviation in X is -1, then Y has a -1 deviation. Table 8.1 shows the consistent relationship of node variables under each gain edge. The range of a node value may be restricted due to its physical conditions.

Table 8.1 Consistent Relationships between Node Variables

Edge gain	Source node variable	Sink node variable
+10(-10)	+10	+10(-10)
	+1	+10(-10)
	-1	-10(+10)
	-10	-10(+10)
+1(-1)	+10	+10(-10)
	+1	+1(-1)
	-1	-1(+1)
	-10	-10(-10)
0	+10	*
	+1	*
	-1	*
	-10	*

Note: The symbol '*' indicates that any value among {-10,-1,0,+1,+10} is acceptable.

System Digraphs

To construct a digraph which can be used to develop a fault tree for a specific undesired event, the first step is to identify the process variable deviation which represents the top event. The digraph is then developed back from the top event variable by considering both the actions of components which could cause a deviation in this variable and the laws of physics. New, local variables which are added to the digraph and connected to the top event variable by the appropriate edges are in turn further developed in the same manner. The construction terminates when all undeveloped variables on the diagram are those corresponding to inputs at the boundary of the system. Permissible deviations in these boundary variables should be also stated and may appear in the fault tree as basic events, whose causes are outside the scope of the analysis.

For a systematic construction of a system digraph, the database of unit model digraphs should be developed such as a table of input-output relationships of each equipment with process variables (or input-output variables). A unit model digraph is derived by considering the way in which the component functions in terms of the basic laws governing energy, mass and momentum. In addition, to identify the effects that the component failure modes produce, the unit model digraph performs the same function as mini fault trees or decision tables in other fault tree construction methods.

8.2.2 Fault Tree Construction

For simple systems which do not have any control loops, their fault tree can be developed very easily by tracing the potential causes of a deviation in the top event variable back through the system digraph. In this situation the fault tree structure will be very simple with process deviations being conditioned by OR gates unless conditional edges are encountered. For a conditional edge, the output variable deviation only occurs when the consistent input deviation occurs AND the stated condition exists. However, the main advantage of the digraph method is that it can be applied to analyze complex systems with control loops. For complex systems, the fault tree construction can only proceed when all the control loops in the system have been identified and classified as either negative feedback or negative feedforward.

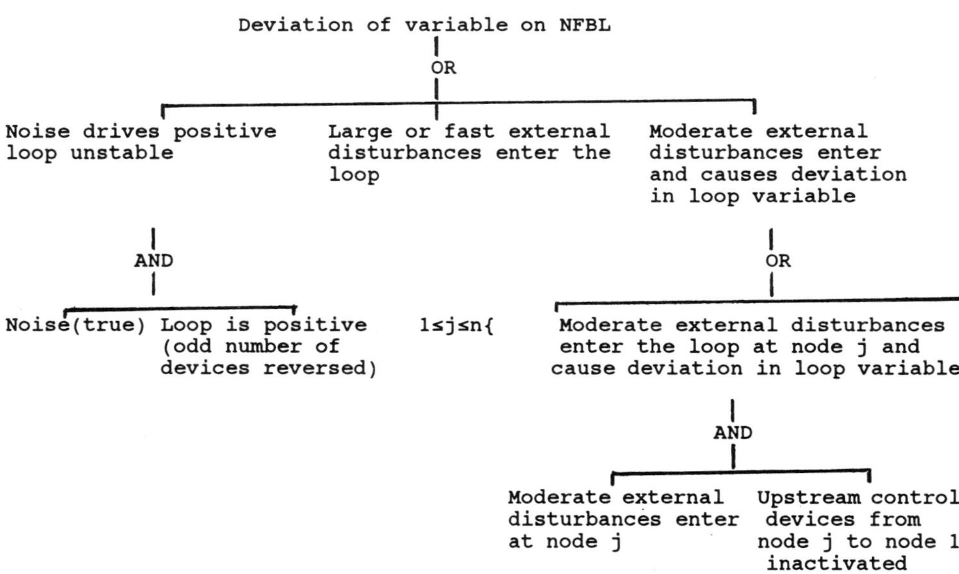

Note: n = number of nodes on the system digraph for NFBL.
Node 1 is the original point of entry on NFBL.

Fig. 8.2 NFBL Operator

Negative Feedback Loops

A negative feedback loop (NFBL) has the ability to correct moderate disturbances in the sensed process variable. On a digraph, an NFBL can be identified as a path which starts and ends at the same node and for which the product of the normal gains is negative. When tracing causes of the top event back through the digraph during fault tree construction, on encountering the first node on an NFBL a general NFBL operator [8] in Fig. 8.2 is applied. The NFBL operator essentially indicates a list of all the ways in which NFBLs can generate or pass a disturbance. These are summarized as:

1. An NFBL causes a disturbance when the net gain is positive, i.e., an odd number of devices in the loop are reversed. In practice. as one reversed device is sufficient to produce the minimal cut sets, only single reversals are considered. When the loop gain is positive, it act as an amplifier and will be driven unstable by noise.
2. An NFBL can not rectify a disturbance if it is too large or too fast for the control loop to correct.
3. An NFBL will pass a correctable disturbance if one or more control devices on the loop are inactive, provided they do not halt the propagation of the disturbance.

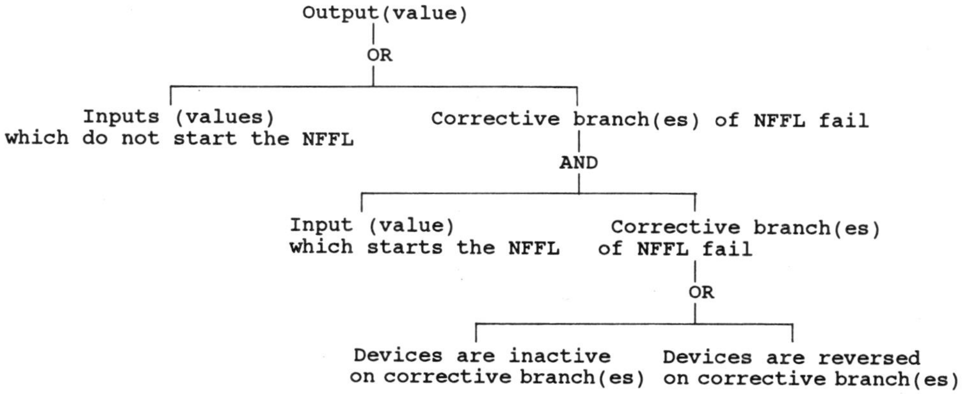

Fig. 8.3 NFFL Operator

Negative Feedforward Loops

Whereas feedback loops correct disturbances which currently exist in the system, a negative feedforward loop (NFFL) can prevent the disturbance propagating through the system. It achieves this by sensing an upstream variable and manipulating a downstream variable. On a digraph, an NFFL can be identified as follows: 1) it has two or more paths from one node to another node. and 2) the sign of the product of the normal gains on one path is different from that of the others. An NFFL can only cancel disturbances which enter at the node which starts the loop, i.e., the sensed node. Any disturbance encountering elsewhere on the loop will propagate through it. A general NFFL operator shown in Fig. 8.3 is only applied when the node which terminates the NFFL is encountered. The NFFL operator combines all the ways in which an NFFL will fail to correct a disturbance. These situations are:

1. A source disturbance enters the loop at some point other than the sensed node.

2. A correctable source disturbance enters the loop at the sensed node but the NFFL fails to correct them due to either inactive or reversed devices.

The fault tree construction also requires the consistency of deviation. The range of deviations at nodes must be checked to meet their definitions.

8.3 Causal Tree Approach

Causal tree approach [14-16] is based on component models that show all fault propagation through the components and fault initiation by the components in both directions (upstream and downstream). A causal tree constructed from component models shows the propagation paths for all basic events leading to any deviation in the top parameter. The fault tree is constructed from a causal tree with considering all control loops (feedback and feedforward loops) in it. In this sense, like a digraph, a causal tree serves as an intermediate step, from which a fault tree can be constructed. This section explains the causal tree approach based on [14,15].

8.3.1 Component Models

Variables and Relations

In a physical system, failures can propagate only by causing deviations in some variables. Although most variables have a continuous range of possible deviations, it is sufficient for fault tree synthesis to distinguish a limited number of deviation levels. Similar to the digraph approach, the deviation is classified into 1) very high (+10), 2) high (+1), 3) normal (0), 4) low (-1) and 5) very low (-10). Further, the following the range indicators are given: w(whole range), δ_x(single range) δ_xu(up range), δ_xd(down range) and δ_x,δ_y(moderate range). For example, X(w) denotes {X(-10), X(-1), X(0), X(+1), X(+10)}. Variables are classified into three types: 1) component variables (all the variables used in component models), 2) environmental variables indicating environmental effects on the system and 3) system variables for interconnecting of components and environmental variables.

Fault propagation is described by relations between variables or by relations between variables and component states. Three types of relations can be distinguished:

1) a relation indicating deviation propagation from one variable to another variable within the same component,

2) a relation indicating a deviation introduced by a component state (non-normal state),

3) a relation indicating a state transition originating from a deviation in a variable.

A relation might be attached with certain conditions such as a component's state or a variable. Thus, the general form of a relation is represented as shown in Fig. 8.4. Depending on the type of relation, 'cause' and 'effect' parameters can be both a variable with a range indicator and a component state. If the relation holds in the normal component state, then 'component state' is omitted since the occurrence probability of the normal state is assumed to be close to 1. The propagation factor may have (+1) for a positive relation, (-1) for a negative relation, or (0) to indicate that under the given conditions the normal relation is broken.

In some cases where energy or mass flow is concerned, failures may propagate both upstream and downstream. Two variables such as flow and pressure or current and voltage have a special relation with each other, which depends on the direction of the failure propagation with respect to the direction of transport (current or mass flow). When a failure propagates downstream, the relation is positive: an increase in pressure will cause an increase in flow. On the other hand, when a failure propagates upstream the relation is negative: a decrease in pressure will cause an increase flow. These two related variables are combined into one so-called 'twin variable'. A subscript 'p' denotes the potential variable (voltage, pressure), while subscript 't' is reserved for the transport variable (current, flow). The propagation factor for such a special relation has two values: the first one holds for the deviation arriving from other variable inside the component and the second one holds for deviations entering the component at the twin variable in question. In this manner the flow direction is included in the component model since the factors for the upstream twin variable and the downstream twin variable have opposite values as shown in Fig. 8.5.

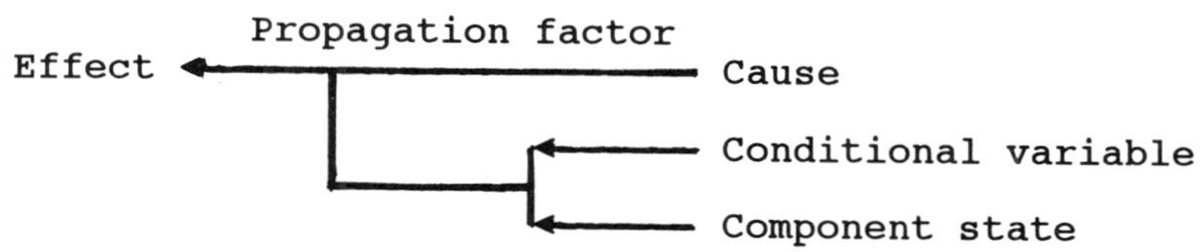

Fig. 8.4 General Form of a Relation

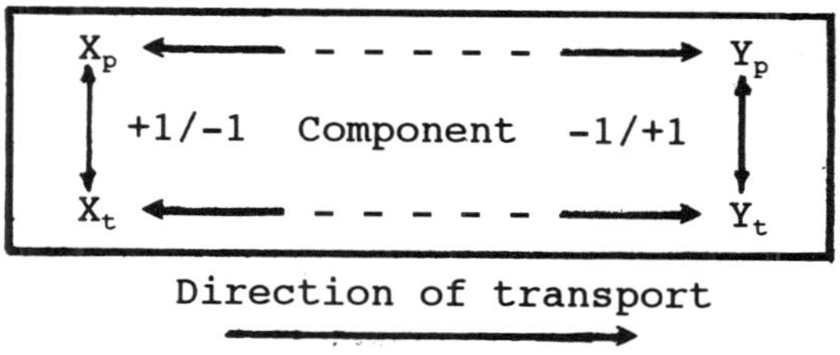

Fig. 8.5 Special Relation of Twin Variable in a Component

Component Models

A component is assumed to have one normal state and one or more degraded or failed states. Each state is represented by a relation pattern describing fault propagation through the component and fault initiation by the component in that particular state. A component model contains a relation pattern for each state. Since a considerable part of the relations holds for the more than one state, it is not necessary to specify all relations for each state. It suffices to record all the relations for the normal state, while only those relations that do not conform to the normal state are specified for all other states. Since the relation between the potential part and the transport part of a twin variable depends on the direction of transport only and is identical for all twin variables, these twin relations can be omitted when the direction of transport is evident from the component model. Each component model should contain a minimum and sufficient set of relations necessary to constitute paths between variables and events that best describe the physical fault propagation though the component.

All physical systems have three essential elements in common: s set of components, an environment and a configuration. The system is represented by three tables: a table of all components used to establish the system function, a table of environmental variables with range indicators and a configuration table showing the connection of system variables to component variables and environmental variables. The system model together with the required component models provides a complete description of the system.

8.3.2 *Causal Trees*

Often systems have control loops (feedback and feedforward loops) to improve system performance. These loops have the ability to compensate or overcompensate for the effects of certain failures. Control loops require special procedures in fault tree synthesis, and should be found before the fault tree is constructed. As a intermediate step, a causal tree is constructed, from which feedback and feedforward loops can be easily found. The causal tree has a structure very similar to that of a fault tree and shows the conditional and unconditional propagation paths for all basic events leading to deviations in the top parameter (top event).

Nodes and Branches

Causal trees are constructed by composing the appropriate propagation paths using the relation available from component and system models. Each node corresponds to a node parameter which is either a variable with a range indicator indicating one or more deviations or a component state indicating a component failure. Two different types of nodes are used in causal trees: 1) an unconditional node showing that each incoming branch is able to cause one of the events indicated by the node parameter and 2) a conditional node showing that all incoming branches are required to cause the event(s) indicated by the node parameter. As a third type of node, a negation node or NOT node represents the absence of an event. Furthermore, each branch is accompanied by two figures representing, from left to right, either the relation's propagation factor and range or the relation's range and impact. Ranges can be identified from the corresponding range indicators. The r_i's indicate unspecified ranges. Although the range attached to node parameters are restricted to those deviations related to the top parameter, each node variable is always developed for its whole range of deviations, even when only a subrange is of interest. This is because deviation outside the range of interest, which might seem to be irrelevant, still may represent corrective actions of feedforward loops and cannot be ignored without further investigation.

Casual Tree Construction

Starting at the node representing the top variable, a causal tree is constructed as follows:

1) A node representing a system variable (SY) is always an unconditional node. New nodes are added, representing the component variable (CO) and the environmental variable (EN) which are connected to the system variable in question. Each branch is assigned a propagation factor +1 and the appropriate range for the parameter below (see Fig. 8.6).
2) A node representing a component variable is developed as an unconditional node as well. New nodes are added for all unconditional relations affecting this variable and for the system variable connected to the relevant component variable (see Fig. 8.7). Conditional nodes with unconditional subnodes are added for each conditional relation. A conditional relation having a zero propagation factor indicates a disconnection in the normal propagation path and, therefore, is drawn parallel to the unconditional (normal) relation.

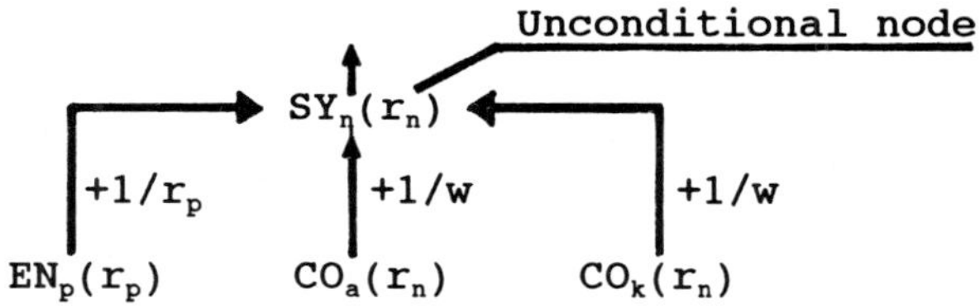

Fig. 8.6 System Variable Node

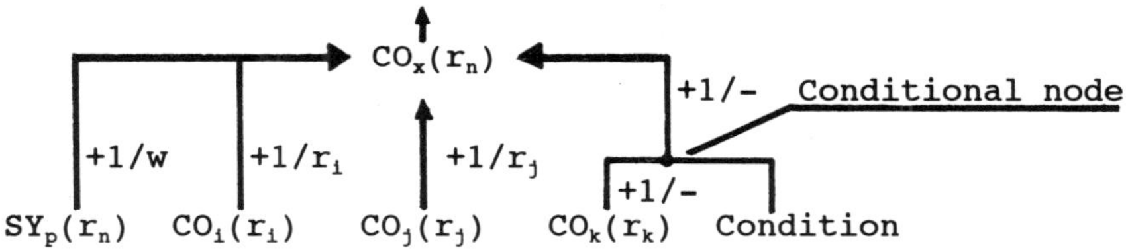

Fig. 8.7 Component Variable Node

3) A node corresponding to an environmental variable is not developed any further.
4) When a node represents a component state which is able to occur due to spontaneous failures only, it represents a basic event and a further development is not required. However, when the component state may result from an induced failure as well, the node is developed as an unconditional node, as shown in Fig. 8.8. One node representing a spontaneous failure is added, and conditional nodes are added when relations representing induced component failures are found.

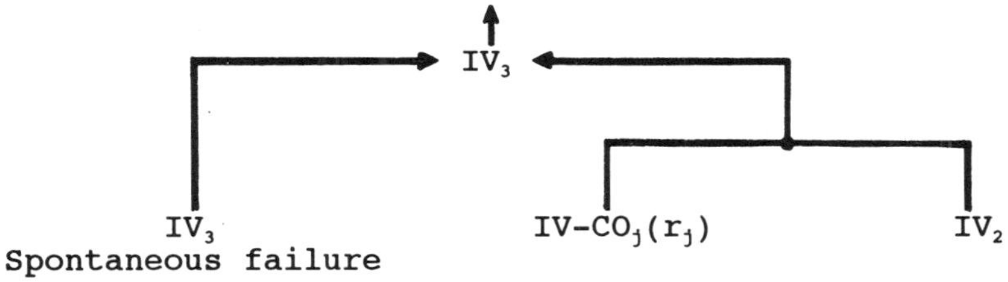

Fig. 8.8 Component State Node with Spontaneous and Induced Failures

5) The range of a node variable is restricted to those values that are relevant to the node variable above. However. in the case of environmental variables the irrelevant range must be added as well since ignoring this range could mask a corrective action of feedforward loop.

Since the component models allow fault propagation in both directions, loops exist inside components. These loops correspond to a sudden reversal in fault propagation direction and are inconsistent. Other problems during causal tree construction include inconsistent conditions. To avoid these inconsistencies, the following rules must be obeyed:

1) In considering a path from top to bottom, we note that repeated variables are not allowed unless the corresponding loop is a system feedback loop or the loop contains a node representing a component state. A loop is a feedback loop when there is an unconditional path with a propagation factor of -1 and the range of the last variable includes 0. Inconsistent loops must be deleted up to (not including) the first unconditional node.
2) The range of a variable must, after being adjusted for the propagation factor, fit in the range of the variable above it and must be consistent with conditions (variable and states) higher up the branch. The resulting range of an environmental variable must be a subrange of the possible deviations stated in the system model.
3) A component state is not allowed to appear twice in the same path and must be consistent with conditions higher up the branch.

For compactness, several additional rules are available:

1) It is not allowed to leave a component twice through the same variable under identical conditions.
2) If the node parameter corresponding to the present node is also an input to a conditional node higher up the branch, the corresponding path can be deleted until the first conditional node.
3) If the range of the node parameter fits in the range of a node parameter affecting an unconditional node higher up the branch, the corresponding path can be deleted until the first unconditional node. Otherwise only the disjoint range should be retained.

8.3.3 Fault Tree Construction

When constructing a fault tree, special attention must be paid to feedback and feedforward loops. The nodes of the causal tree are divided into three groups: non-loop nodes, nodes that belong to one or more feedback loops, and nodes at which two branches of a feedforward loop meet.

Non-loop Nodes

For non-loop nodes the process is very simple. An unconditional node of the causal tree is translated into an 'OR'-gate in the fault tree. As a matter of course the range indicator must be restricted to those deviations/events that are capable of causing the top event. Similarly, conditional nodes allow fault propagation only when all their input events apply and can be translated into 'AND' gate. Finally, the negation can be translated into an inverting gate.

Feedback Loop Nodes

A feedback loop is represented by an unconditional path in the causal tree, so only unconditional nodes appear in feedback loops. Nodes that are part of the feedback path of a loop are developed as non-loop nodes. However, nodes in the feedforward path require a special attention. The development of such a node for extreme deviation (+10/-10) is similar to that of a non-loop node, except when the node is the first node of a loop. In this case an OR-gate representing all events that change the loop in a positive feedback loop is added. For moderate deviations, a loop variable is developed as follows: deviations that propagate through the loop will directly affect the loop variable in question while deviations entering the loop at this node can represent only when the loop is incapable of reacting. The development of a node, which is part of more loops, depends on the number of descending loop variables. If there is only one descending variable shared by all loops, the development is similar to that of a single loop node. However, when more descending loop variables exist, a moderate deviation in one of the loops will be compensated by the other loops. This means that a moderate deviation can propagate through this node only when all (other) loops fail to react.

Feedforward Loop Nodes

A node at which two branches of a feedforward loop meet is developed as follows: Starting with an OR-gate, an AND-gate is added for each basic event that propagates thorough both branches with opposite propagation factors. One input of the AND-gate is used for the basic event while the second input is connected to an OR-gate representing all events that nullify or inverse the propagation factor through the compensating branch. Afterwards the node is developed normally except for the basic events used before, which are neglected.

8.4 Interactive Method for Obtaining Minimal Cut Sets

Automatic construction of fault tree using such as digraphs does not always generate consistent results, and a human intervention is necessary (for the details, see [12,13,17]). Since 1) component models such as in the digraph or causal tree approach can be expressed in the form of if-then rules, and 2) construction of fault tree by digraphs and causal trees is equivalent to identifying all minimal combination of conditions that cause a specified deviation at the specified variable (in general, these combinations corresponds to minimal cut sets), an interactive method [18] using if-then rules is given for obtaining all minimal cut sets. This method is a revised method of [17], where the system analyst's judgements are used to obtain the consistent result.

Firstly, the meaning of minimal cut sets is considered to develop a basic procedure. Secondly, the necessary knowledge for the method to obtain minimal cut sets is classified into three types, and how to use human judgement and intervention is considered. The method guarantees the consistency of minimal cut sets as long as knowledge of system representation, generally-accepted knowledge, and system analysts' judgements are correct. An illustrative example of a nitric-acid cooler system shows the method in detail.

8.4.1 Minimal Cut Sets

A minimal cut set is defined as a minimal combination of abnormal conditions causing system failure. Since system failure is usually defined as a deviation of the system function or system output from the specified one (i.e., normal condition), the minimal cut set must cause the deviation. In this sense, the minimal cut set must include a basic deviation propagation sequence ('disturbance situation') causing system failure. However, the system is usually designed to protect against hazards; the disturbance situation may be protected by control or protective systems. The minimal cut set must also prevent related control and protective systems from nullifying the effect of the disturbance situation. Whether a minimal cut set must include this condition or not depends on the effect of a disturbance situation on the system. If the disturbance situation condition can be prevented by a control loop, its malfunction is necessary for the disturbance situation to cause the system failure.

Thus, minimal cut sets are classified into two types:

(1) a disturbance situation causing system failure by itself, because 1) its effect is too large for the protective and control systems to nullify OR 2) it has no related control loops.

(2) a logical AND combination of a disturbance situation and failure conditions of the related control and protective systems.

This classification shows that the first step to obtain minimal cut sets is to identify a disturbance situation causing system failure. Depending on the effect of the related protective and control systems on it, type-(1) or type-(2) minimal cut sets are analytically obtained.

8.4.2 Three Types of Knowledge

Knowledge necessary to obtain minimal cut sets is composed of three types: (1) knowledge representing system such as system structure, component function and failure, (2) knowledge necessary to determine whether system failure occurs and whether the control system functions or not and so on, and (3) evaluation of failure and disturbance effects on control systems and the system function.

System Knowledge

The first knowledge is usually obtained from characteristic functions and design documentation of the subject system. Firstly, consider a component function. A characteristic function of a component is usually given as:

$$Y_i = f(X_1, X_2, \cdots, X_n) \quad \text{or} \quad \frac{dY_i}{dt} = g(X_1, X_2, \cdots, X_n) \tag{8.1}$$

where Y_i is output variable of the component function, X_j denotes input variable or control variable of the component which affects output Y_i, f and g are characteristic functions. From the viewpoint of reliability analysis, we take much attention to a deviation from a normal or stationary condition as well as cause-effect logical relations between these deviations. Thus, from Eq. (8.1), the cause-effect relation for each pair of input variable X_j and output variable Y_i can be obtained as

$$(r, X_j, Y_i)$$

where r denotes the correlation between variable deviations dX_j and dY_i, and is expressed as {'+1','-1','0'}. Value of r is determined as:1) r=1 if $df/dX_j>0$ or $dg/dX_j>0$, 2) r=-1 if $df/dX_j<0$ or $dg/dX_j<0$, and 3) r=0 if $df/dX_j=0$ or $dg/dX_j=0$. Since we also consider the direction of a deviation, a variable also takes one of {'+1','-1','0'}, where '+1', '-1', and '0' denote positive deviation, negative deviation from the normal or stationary range, and no deviation being in the normal range, respectively. Table 8.2 shows the consistent relations. For each cause-consequence relation, the conditions where it holds must be attached. If the characteristic function holds for a normal component, the normal condition must be attached to it Normal conditions are omitted in obtaining the minimal combinations of occurrence conditions.

Table 8.2 Consistent Relations in (r, X, Y)

1. Input-output relation '+1':	(1, -1, -1)	(1, 0, 0)	(1, 1, 1)
2. Input-output relation '-1':	(-1, -1, 1)	(-1, 0, 0)	(-1, 1, -1)
3. Input-output relation '0':	(0, -1, *)	(0, 0, *)	(0, 1, *)

Note: '*' denotes any value of { -1, 0, 1}.

Even if a design documentation does not give the quantitative information, it usually specify the relations of output variables to input and control variables qualitatively like decision tables or FMEA formats. In this case, the relations can be defined qualitatively as positive correlation, negative correlation, and no correlation (i.e., cut of information flow). Expert knowledge can be also represented qualitatively in such a way that an increase in pressure causes an increase in flow under a given condition. This kind of knowledge is also expressed in the same way.

Consider the effect of a component failure or an abnormal event. A component failure may cause 1) a different input-output variable relation from one under the normal condition or 2) a deviation of output or process variable directly. For type-1), the input-output relation can be defined similarly using (r, X_j, Y_i). For example, a temperature sensor stuck failure can be represented as (0, T, OT) under the sensor stuck failure condition, where T denotes monitored temperature and OT denotes output signal. For type-2), by introducing a dummy variable V which denotes a component failure or occurrence of an abnormal event, the effect can be also represented by (r, X_j, Y_i). For example, if a component failure decreases Y_i, then it is represented as (-1, V, Y_i).

Since the system function is usually defined as a combination or flow of component functions, the system can be represented as the connection of component functions, i.e., connection of variables X_j and Y_i in (r, X_j, Y_i) representations. Thus, type-1 knowledge, knowledge related to the system, can be represented in the form of

$$\text{If (occurrence conditions), then } (r, X_i, Y_i). \tag{8.2}$$

<u>Generally-accepted Knowledge</u>

The second knowledge is composed of the following propositions generally accepted by the system analysis [5-16]:

(1) System failure must be caused by an abnormal event such as a component failure or source disturbance (disturbance causing a deviation of variable).

(2) A negative feed back loop (NFBL) can cause a source disturbance when the loop gain is positive, i.e., an odd number of devices in the loop are reversed, because noise always exists in a feedback loop, and positive gain can amplify its effect.

(3) A zero-gain event, such as valve stuck, can prevent a deviation from propagating across its connection. This means the effect of upstream variables is halted by the zero-gain event. In the valve stuck example, the command to close the valve is not

executed. The disturbance of protective actions cannot be passed through the connection conditioned by a zero-gain event.

(4) A feedback loop (FBL) has a loop of variables obtained by backward search in cause-effect relations from the controlled variable to itself. A feedforward loop (FFL) has two sets of cause-effect sequences, each of which goes from the sensed variable to the controlled variable.

(5) From the viewpoint of the magnitude of their effect on related control loops, source disturbances are classified as uncorrectable or correctable.

(6) A normal NFBL can rectify a source disturbance entering the NFBL at any component. On the other hand, a negative feed forward loop (NFFL) can rectify nothing but the deviation caused by a source disturbance entering the NFFL at the sensed variable. The disturbances entering the NFFL at other places than the sensed variable are not sensed by the NFFL, because the NFFL monitors an upstream variable and changes a downstream variable.

(7) A related NFBL fails to rectify a deviation caused by a correctable source disturbance if (a) at least one control device not related to the source disturbance propagation is inactivated in the remainder of the loop, or (b) a reverse action of the normal corrective procedure occurs. Condition (a) prevents the corrective action of the NFBL based on proposition (3).

(8) A related NFFL fails to amend a correctable source disturbance when (a) at least one device on the corrective branch is inactive, causing the NFFL to be inactive. or (b) the loop gain is positive because an odd number of devices are reversed. Situation (a) corresponds to the case where a corrective action is halted and situation (b) corresponds to the case where a corrective action is reversed to nullify its effect. Compared with NFBLs, NFFLs cannot amplify the effect of noise by the positive loop gain, because the effect of noise is temporary, and thus not amplified.

Propositions (1) and (2) show that a disturbance situation causing a specified system failure is obtained either if 1) the occurrence condition in Eq. (8.2) is failure or abnormal event which causes a deviation on output or process variable, or 2) the controlled variable of NFBL gets the same value that it initially took when the backward search started.

System such as chemical and nuclear plants have control and protective systems which can prevent system failure from generating. Thus, we must identify protective and control systems relating to a disturbance situation in order to evaluate their effect on it. Proposition (4) identifies all control loops in the system represented by a set of if-then rules. Backward search in if-then rules is sequentially performed until 1) the same variable appears twice or 2) no backward search is possible. In the former case, a FBL candidate is found. The remaining connections of variables in case 2) are examined whether a pair of these connections have two variables in common, and neither of them is not a part of NFBLs. Proposition (6) identifies all related control loops and protective actions that can nullify the effect of a disturbance situation. Failure conditions of related correctable protective and control systems are obtained based on propositions (7) and (8).

<u>Evaluation of Disturbances</u>

The last knowledge about the effect of control loops on the disturbance situation is not easily obtained in advance of the analysis, because their effect varies depending on the situation. Consider a stuck-failure of one valve in two series connecting flow control valve system. If the control system should decreases the flow, a stuck failure is not critical because the other valve can decrease it, while it is critical for the system to increase the flow because both of them must open to increase the flow. The number of situations to be considered becomes too large. Thus, it is difficult and may be impossible or impractical to investigate and specify the effect of component failure on the control loops in all possible cases beforehand. Therefore, in the proposed method, when a disturbance situation causing a specified system failure is obtained, the effect of related control loops on it is evaluated by the system analyst. The consistency of stored knowledge on the system representation must be also guaranteed by the system analyst examining the consistency of identified loops and assumed component failures.

8.4.3 Interactive Method

The method uses three types knowledge as follows. Based on type-(2) knowledge, a disturbance situation causing a specified system failure is obtained by backward search through the input-output relations in if-then rules representing type-(1) knowledge. Then type-(3) knowledge is obtained by interaction with the system analyst to evaluate the effect of protective and control actions. Based on the judgement, the method analytically obtains minimal cut sets as logical AND combination of a disturbance situation and the corrective protective and control system failure conditions.

The procedure to obtain minimal cut sets are described as:

Step 1: Check of consistency of if-then rules

Based on system representations by Eq. (8.2), all control loops obtained by proposition (4) and component failures or source disturbances assumed in the analysis are to be examined by the system analyst.

- 1-1: If an unexpected loop exists or a failure to be considered is missing, then system representations in Eq. (8.2) should be corrected by the system analyst.
- 1-2: Otherwise, both the effective control loops for every process variable and failure conditions for each control loop are derived based on proposition (6).

Step 2: Derivation of disturbance situations causing system failure

From the process variable deviation representing the system failure, all possible disturbance situations that can cause it are obtained by backward search in if-then rules based on proposition (1).

Step 3: Evaluation of control and protection systems

For each disturbance situation obtained in step 2, the related correctable control loops are identified based on the relation between process variables and control loops at step 1. The method shows a disturbance situation, its occurrence conditions, and the related control loops to the system analyst, who evaluates the situation and the effect of the related control loops to give an appropriate combination of protective loops which can rectify the effect of the disturbance situation.

- 3-1: If no appropriate combination exists, a disturbance situation by itself constitutes a system failure condition.
- 3-2: Otherwise, a system failure condition is a logical AND combination of the disturbance situation and the negation of the appropriate protective system functions.

Omitting normal conditions from system failure conditions and minimizing them, we can obtain minimal cut sets for the disturbance situation.

Step 4: Derivation of minimal cut sets

Omitting redundant minimal cut sets obtained at step 3, we can obtain the final minimal cut sets for the specified system failure.

Step 3 produces redundant conditions for disturbance situations, because unnecessary positive gain conditions for corrective-related NFBLs exist. Since the interaction of the control loops is complicated, it may be difficult to assess the effect of a disturbance situation. So, this method removes redundant conditions after obtaining logical AND combinations. Step 4 collects minimal combinations of abnormal conditions from all minimal conditions for each disturbance situation. Minimization operations at steps 3 and 4 correspond to the application of Haasl's rule [19] in fault tree synthesis algorithms. For example, an odd number of (3,5, or ...) reverse gain events are excluded and only one reverse gain event is selected for the NFBL malfunction.

8.4.4 Application to Nitric-acid Cooler System

Consider the nitric-acid cooler problem described in Lapp and Powers [5] and modified by Allen and Rao [9], and Kumamoto and Henley [11]. Fig. 8.9 shows the system. The cooler system keeps the output nitric-acid temperature constant by controlling the input flow of cooling water based on the output temperature sensed. A temperature sensor, a temperature controller, and a control valve constitute a feedback loop. Cooling water is supplied by a pump from the tank into the heat exchanger through lines 9 and 8. If the pump gets failed, shutdown valve 1 can be automatically closed to shut down the input nitric-acid flow. If the temperature control system fails, the operator finding an abnormal condition of the heat exchanger can override the control system to manually handle valves 5 and 1.

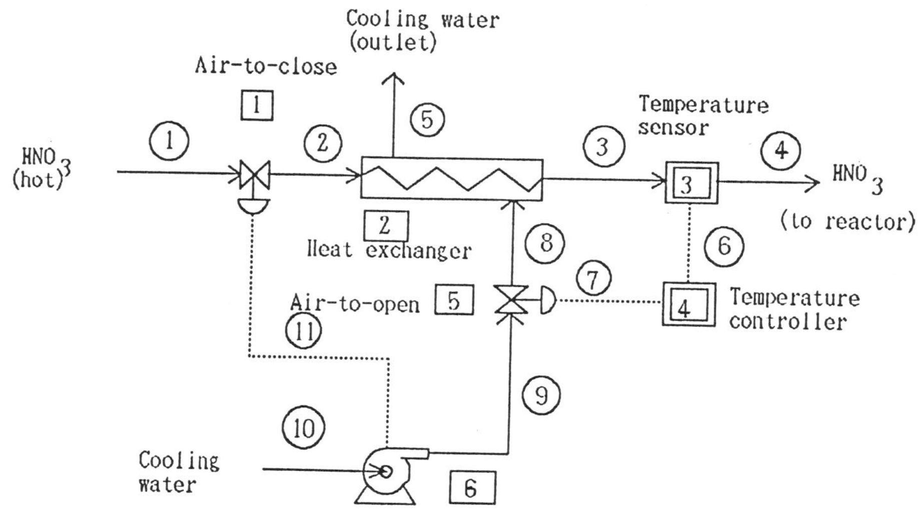

Fig. 8.9 Nitric-acid Cooler System

The same conditions in [11] are also assumed regarding disturbances and control loops as:

(1) 7 source disturbances are considered: 1) an input flow temperature deviation TIN(-), 2) loss of main pressure (LMP), 3) coolant pump failure (PF), 4) temperature indicator fails low (TIFL), 5) temperature sensor fails low (TSFL), 6) temperature

Table 8.3 System Representation by If-then Rules

IF (HEN),	THEN (+1, T2, T3)
IF (HEN),	THEN (+1, M2, T3)
IF (HEN),	THEN (-1, M8, T3)
IF (TSN),	THEN (+1, T3, P6)
IF (TCN),	THEN (+1, P6, P7)
IF (CV5N),	THEN (+1, P7, M8)
IF (CV5N),	THEN (+1, P9, M8)
IF (SV1N),	THEN (-1, P11, M2)
IF (PN),	THEN (+1, P10, P9)
IF (N),	THEN (+1, T1, T2)
IF (N),	THEN (+1, T3, T4)
IF (N),	THEN (+1, M3, M4)
IF (TIN),	THEN (+1, P6, TI)
IF (ON),	THEN (+1, TI, P11)
IF (ON),	THEN (+1, TI, P7)
IF (L11N.AND.PF),	THEN (+1, PF(+1), P11)
IF (TIN(-)),	THEN (+1, TIN(-), T1)
IF (LMP.AND.NES),	THEN (-1, LMP(+1), P10)
IF (PF),	THEN (-1, PF(+1), P9)
IF (TIFL),	THEN (-1, TIFL(+1), TI)
IF (TSFL),	THEN (-1, TSFL(+1), P6)
IF (TCSP(-)),	THEN (-1, TCSP(-), P7)
IF (TCLAP),	THEN (-1, TCLAP(+1), P7)
IF (CV5S),	THEN (0, P7, M8)
IF (CV5R),	THEN (-1, P7, M8)
IF (SV1S),	THEN (0, P11, M2)
IF (OA),	THEN (0, TI, P11)
IF (OA),	THEN (0, TI, P7)
IF (TIS),	THEN (0, P6, TI)
IF (TSS),	THEN (0, T3, P6)
IF (TCS),	THEN (0, P6, P7)
IF (L11B.AND.PF),	THEN (0, PF(+1), P11)

Note: TI, TS, TC: temperature indicator, sensor, controller
N: normal condition indicating that the input-output relation always holds
Ti, Mi, Pi: temperature i, mass flow i, pressure i
CV, SV: control valve, shutdown valve
HE: heat exchanger
-N, -S, -R: normal, failed-stuck, failed-reversed
PF, PN: pump failure, pump normal

controller setpoint deviation (TCSP(-)), and 7) temperature controller low air pressure (TCLAP).

(2) 9 failures changing input-output relations are considered: 1) control valve 5 stuck-failure (CV5S), 2) control valve 5 reverse-failure (CV5R), 3) operator absent (OA), 4) temperature indicator stuck-failure (TIS), 5) temperature sensor stuck-failure (TSS), 6) temperature controller stuck-failure (TCS), 7) shutdown valve 1 stuck-failure (SV1S), 8) line 11 blockage (L11B), and 9) No emergency supply (NES).

(3) The control loop's functions are:

1) The automatic temperature control loop controls the flow of cooling water and opens control valve 5 in response to a high temperature.
2) The operator manually opens valve 5 in response to a high temperature indication at the temperature indicator.
3) The shutdown system negates the effect of pump failure.
4) The operator overrides the automatic control loop.

Assumptions (1) and (2) are represented as if-then rules in Table 8.3. If-them rules are divided into three parts. The first part represents normal relation, the second part represents source disturbances and the last part expresses failures changing the normal relations. In assumption (1), the values of deviations for TIN and TCSP are not specified beforehand, because they can be negative or positive, depending on the system failure condition. In assumption (2), all failures except CV5R are expressed as zero-gain events, while CV5R are represented as a reverse relation of normal operation.

Table 8.4 Summary of System Representation

1. Control loops identified

FBL1: T3 <- M8 <- P7 <- P6 <- T3
FBL2: T3 <- M2 <- P11 <- TI <- P6 <- T3
FBL3: T3 <- M8 <- P7 <- TI <- P6 <- T3
FFL1: T3 <- M2 <- P11 <- PF & T3 <- M8 <- P9 <- PF

2. Process variables and effective control loops

Variable	Control loops
T3	FBL1 FBL2 FBL3
M2	FBL2
M8	FBL1 FBL3
P6	FBL1 FBL2 FBL3
P7	FBL1 FBL3
P11	FBL2
TI	FBL2 FBL3
PF	FFL1

From these if-then rules, some analytical results are obtained as shown in Table 8.4-1. 3 feedback loops and 1 feedforward loops are identified. FBL1(T3<-M8<-P7<-P6<-T3) corresponds to the automatic control loop, FBL2 (T3<-M2<-P11<-TI<-P6<-T3) and FBL3(T3<-M8<-P7<-TI<-P6<-T3) represent the control actions by the operators, and FFL1(T3<-M2<-P11<-PF & T3<-M8<-P9<-PF) indicates the shutdown system for pump

Table 8.5 Disturbance Situations Causing System Failure

1.	DS.	TIN(1) -> T1(1) -> T2(1) -> T3(1) -> T4(1)
	OC.	TIN(1) .AND. HEN
2.	DS.	TCSP(-1) -> P7(1) -> M8(-1) -> T3(1) -> T4(1)
	OC.	TCSP(-1) .AND. CV5R .AND. HEN
3.	DS.	PF(1) -> P9(-1) -> M8(-1) -> T3(1) -> T4(1)
	OC.	PF .AND. CV5N .AND. HEN
4.	DS.	TCSP(1) -> P7(-1) -> M8(-1) -> T3(1) -> T4(1)
	OC.	TCSP(1) .AND. CV5N .AND. HEN
5.	DS.	TCLAP(1)-> P7(-1) -> M8(-1) -> T3(1) -> T4(1)
	OC.	TCLAP .AND. CV5N .AND. HEN
6.	DS.	TIFL(1) -> TI(-1) -> P7(-1) -> M8(-1) -> T3(1) -> T4(1)
	OC.	TIFL .AND. ON .AND. CV5N .AND. HEN
7.	DS.	TSFL(1) -> P6(-1) -> P7(-1) -> M8(-1) -> T3(1) -> T4(1)
	OC.	TSFL .AND. TCN .AND. CV5N .AND. HEN
8.	DS.	LMP(1) -> P10(-1) -> P9(-1) -> M8(-1) -> T3(1) -> T4(1)
	OC.	LMP .AND. NES .AND. PN .AND. CV5N .AND. HEN
9.	DS.	T3(1) -> P6(1) -> P7(1) -> M8(-1) -> T3(1) -> T4(1)
	OC.	TSN .AND. TCN .AND. CV5R .AND. HEN
10.	DS.	TIFL(1) -> TI(-1) -> P11(-1) -> M2(1) -> T3(1) -> T4(1)
	OC.	TIFL .AND. ON .AND. SV1N .AND. HEN
11.	DS.	TSFL(1) -> P6(-1) -> TI(-1) -> P11(-1) -> M2(1) -> T3(1) -> T4(1)
	OC.	TSFL .AND. TIN .AND. ON .AND. SV1N .AND. HEN
12.	DS.	T3(1) -> P6(1) -> TI(1) -> P7(1) -> M8(-1) -> T3(1) -> T4(1)
	OC.	TSN .AND. TIN .AND. ON .AND. CV5R .AND. HEN
13.	DS.	TSFL(1) -> P6(-1) -> TI(-1) -> P7(-1) -> M8(-1) -> T3(1) -> T4(1)
	OC.	TSFL .AND. TIN .AND. ON .AND. CV5N .AND. HEN

Note: DS.: disturbance situation OC.: occurrence conditions

failure. After the system representation including assumed component failures and their effects is checked and guaranteed by the system analyst, the relation between control loops and process variables are also obtained as shown in Table 8.4-2 based on proposition (6). For example, deviations at T3 can be rectified by any of control loops FBL1, FBL2, and FBL3. Using this relation, the related control loops can be identified for each disturbance situation.

Obtain minimal cut sets for high output temperature of nitric-acid, which is represented as a positive deviation of output temperature, T4(1). Step 2 obtains 13 disturbance situations causing T4(1) as shown in Table 8.5. At step 3, for each disturbance situation, the system analyst is requested to evaluate its possibility and effect on the system function. Disturbance situations 10 and 11 obtained by backward search in if-then rules are found to be impossible because the operator does not handle shutdown valve 1 to increase the output temperature in case of low output temperature. In this way, the missing knowledge which is not stored in the system can be clarified and supplemented by the interactive process with the system analyst.

A part of interactive process with the system analyst is shown in Fig. 8.10, where the underline indicates the system analyst's answer. A disturbance situation and its occurrence conditions are shown to the system analyst, who first examines the possibility of the situation occurrence. If it cannot occur, then the analysis goes to another disturbance situation. If it can occur in the system, then the analyst is requested to evaluate the effect of the related control loops on the disturbance situation and to give combinations of effective control systems. Disturbance situation 1, which shows that the output temperature can get higher as the input temperature gets higher, can be regarded as a possible situation. Since any feedback loop can rectify the disturbance caused by an input temperature deviation, the combination of correctable control systems is logical OR of NFBLs 1, 2, and 3. In the figure, three sets {NFBL1}, {NFBL2} and {NFBL3} are given as effective combinations. Thus, the minimal cut sets are obtained as logical AND combinations of the disturbance situation occurrence and failure conditions of NFBLs 1, 2, and 3 consistent with the disturbance situation occurrence condition. 9 combinations of component malfunctions cause NFBLs 1, 2, and 3 to fail in the recovery action. Temperature sensor stuck failure constitutes a one-event minimal cut set for the unavailability of all NFBLs.

Table 8.6 Effective Structure of Control Loops

Disturbance situation 1:	FBL1	OR	FBL2	OR	FBL3
Disturbance situation 2:	FBL2				
Disturbance situation 3:	FBL2	OR	FFL1		
Disturbance situation 4:	FBL2	OR	FBL3		
Disturbance situation 5:	FBL2	OR	FBL3		
Disturbance situation 6:	----				
Disturbance situation 7:	----				
Disturbance situation 8:	FBL2				
Disturbance situation 9:	FBL2				
Disturbance situation 12:	FBL2				
Disturbance situation 13:	----				

Note: '----' indicates no effective control loops.

```
***** DISTURBANCE SITUATION No.  1 *****
+++ DISTURBANCE PROPAGATION PATH +++
      1: (TIN  ,  1)    2: (T1   ,  1)    3: (T2   ,  1)
      4: (T3   ,  1)    5: (T4   ,  1)
+++ DISTURBANCE CONDITIONS +++
   TIN  :POSITIVE    HE   :NORMAL

*** CAN THIS SITUATION OCCUR ?
***  IF POSSIBLE, INPUT  1  : OTHERWISE, INPUT 0
1
+++ RELATED CONTROL LOOPS +++
   NFBL  1:  T3     M8     P7     P6     T3
   NFBL  2:  T3     M2     P11    TI     P6    T3
   NFBL  3:  T3     M8     P7     TI     P6    T3
* * * * * * * * * * * * * * * * * * * * * * * * * * * * * * *
*** INPUT EFFECTIVE COMBINATION OF PROTECTIVE SYSTEMS
*** INPUT NUMBER OF CONTROL LOOPS, OR /CR/ FOR TERMINATION
  1
*** INPUT CONTROL LOOPS: +++ I ... FOR NFBL I, -J ... FOR NFFL J
  1
*** INPUT NUMBER OF CONTROL LOOPS, OR /CR/ FOR TERMINATION
  1
*** INPUT CONTROL LOOPS: +++ I ... FOR NFBL I, -J ... FOR NFFL J
  2
*** INPUT NUMBER OF CONTROL LOOPS, OR /CR/ FOR TERMINATION
  1
*** INPUT CONTROL LOOPS: +++ I ... FOR NFBL I, -J ... FOR NFFL J
  3
*** INPUT NUMBER OF CONTROL LOOPS, OR /CR/ FOR TERMINATION

***** MINIMAL COMBINATION OF CONDITIONS *****
+++ DISTURBANCE SITUATION CONDITIONS +++

   TIN  :POSITIVE
+++ CONTROL LOOP MALFUNCTION CONDITIONS +++

  1:  CV5  :STUCK       SV1  :STUCK
  2:  CV5  :STUCK       O    :ABSENT
  3:  CV5  :STUCK       TI   :STUCK
  4:  CV5  :REVERSED    O    :ABSENT
  5:  TC   :STUCK       O    :ABSENT
  6:  TC   :STUCK       TI   :STUCK
  7:  CV5  :REVERSED    TI   :STUCK
  8:  CV5  :REVERSED    SV1  :STUCK
  9:  TS   :STUCK
```

Fig. 8.10 Part of Interactive Process

Table 8.6 shows the effective logical combination of protective systems for each disturbance situation based on the assumption (3), which is equivalent to the system analyst's judgement. Minimal cut sets for each disturbance situation can be obtained as shown in Table 8.7, where the minimal cut sets for the system failure are marked by '*'. The system has 26 minimal cut sets: 2 one-event cut sets, 11 two-event cut sets, and 13 three-event cut sets.

While the fault tree construction methods need consistency checks on minimal cut sets obtained from the fault tree, this method does not need the consistency check of minimal cut sets as long as the analysis of disturbance situations which effect control loops and control loop interactions are right. This is guaranteed in the generation of minimal combinations of conditions for each disturbance situation.

Table 8.7 Minimal Cut Sets for Disturbance Situations

1.	(TIN(1),	CV5S,	SV1S	)*
	(TIN(1),	CV5S,	OA	)*
	(TIN(1),	CV5S,	TIS	)*
	(TIN(1),	CV5R,	OA	)
	(TIN(1),	TCS,	OA	)*
	(TIN(1),	TCS,	TIS	)*
	(TIN(1),	CV5R,	TIS	)
	(TIN(1),	CV5R,	SV1S	)
	(TIN(1),	TSS)*		
2.	(CV5R,	TCSP(-1),	SV1S	)
	(CV5R,	TCSP(-1),	OA	)
	(CV5R,	TCSP(-1),	TIS	)
	(CV5R,	TCSP(-1),	TSS	)*
3.	(PF,	OA,	L11B	)*
	(PF,	TIS,	L11B	)*
	(PF,	TSS,	L11B	)*
	(PF,	SV1S)*		
4.	(TCSP(1),	OA)*		
	(TCSP(1),	TIS)*		
	(TCSP(1),	TSS)*		
5.	(TCLAP,	OA)*		
	(TCLAP,	TIS)*		
	(TCLAP,	TSS)*		
6.	(TIFL)*			
7.	(TSFL)*			
8.	(LMP,	NES,	SV1S	)*
	(LMP,	NES,	OA	)*
	(LMP,	NES,	TIS	)*
	(LMP,	NES,	TSS	)*
9.	(CV5R,	SV1S)*		
	(CV5R,	OA)*		
	(CV5R,	TIS)*		
12.	(CV5R,	SV1S)*		
13.	(TSFL)*			

NOTE: '*" denotes a minimal cut set for the system failure.

8.5 Conclusions

We review the digraph and causal tree approaches for the automatic construction of fault tree. As stated in [14,15], the main advantage of both methods is their requirement for the system analyst to have a rigorous understanding of a system function. This is partly lost in an automatic computerized approach. The approach also forces a structured approach whereby different analysts are more likely to produce fault trees of consistent logic for the same system. However, the techniques seem to need further improvement for automatic fault tree construction. For example, the five discrete states used do not adequately allow the distinction between a zero level and a very low level disturbance of a variable, and the distinction between controllable and uncontrollable disturbance is difficult to define in a consistent manner at all positions within the system. Thus, human intervention and judgement are needed to prevent the generation of incomplete, inconsistent, incorrect failure modes.

As a method using human judgement, an interactive method is given for obtaining minimal cut sets. Since the evaluation of the related control loops is difficult for all disturbance situation beforehand, the method uses the analyst's judgment on the effect of related control loops on the disturbance situation. As long as knowledge of system representation, control loop functions and failure conditions, and the system analyst's judgment are correct, the consistency of minimal cut sets are guaranteed, resulting in no necessity of human intervention to the obtained result. As shown in the example, the simple backward search in if-then rules representing cause-effect relations may obtain inconsistent disturbance situations, but human intervention can be used to check the possibility of an identified situation and eliminate them. This will occur if some knowledge regarding the system function is missing. The interactive method also tries to derive the missing knowledge by interaction with the system analyst and, in turn, give an insight into the system function which is not specified explicitly beforehand.

Though automatic construction of fault tree will be developed using a new knowledge engineering approach, some kind of interaction with the system analyst will be still necessary not only to obtain the consistent result, but also to deepen the understanding of a system function itself.

REFERENCES

[1] Lee W.E., D.L. Grosh, F.A. Tillman and C.H. Lie, *Fault Tree Analysis, Methods, and Applications - A Review*, IEEE Trans. on Rel., Vol. R-24, No. 3, pp. 194-203, Aug. 1985.

[2] Henley E.J. and H. Kumamoto, *Probabilistic Risk Assessment - Reliability Engineering, Design, and Analysis*, IEEE Press, New York, 1992.

[3] Salem S.L., J.S. Wu, G. Appostlakis, *Decision Table Development and Application to the Construction of Fault Trees*, Nuclear Technology, 42, pp. 51-64, 1979.

[4] Taylor J.R., *Algorithm for Fault-tree Construction*, IEEE Trans. on Rel., Vol. R-31, No.2, pp. 137-146, June 1982.

[5] Lapp S.A. and G.J. Powers, *Computer-aided Synthesis of Fault-trees*, IEEE Trans. on Rel., Vol. R-26, No.1, pp. 2-13, April 1977.

[6] Lambert H.E., *Comments on the Lapp-Powers' **Computer-aided Synthesis of Fault Trees***, IEEE Trans. on Rel., Vol. R-28, No. 1, pp. 6-9, April 1979.

[7] Lapp S.A. and G.J. Powers, *Up-date of Lapp-Powers Fault Tree Synthesis Algorithm*, IEEE Trans. on Rel., Vol. R-28, No. 1, pp. 12-15, April 1979.

[8] Lambert H.E., L. Philipson, and H. Roland, *The Digraph-Fault Tree Methodology and its Use in Transportation Risk Analysis*, Tera Corporation, Berkeley, California, 1980.

[9] Allen D.J. and M.S. Rao, *New Algorithm for the Synthesis and Analysis of Fault Trees*, Ind. Eng. Chem. Fundam., Vol. 19, No. 1, pp. 79-85, 1980.

[10] Cummings D.L., S.A. Lapp and G.J. Powers, *Fault Tree Synthesis from a Directed Graph Model for a Power Distribution Network*, IEEE Trans. on Rel., Vol. R-32, No. 1, pp. 140-149, June 1983.

[11] Kumamoto H. and E.J. Henley, *Automated Fault Tree Synthesis by Disturbance Analysis*, Ind. Eng. Chem. Fundam., Vol. 25, No. 2, pp. 233-239, 1986.

[12] Andrews J.D. and J.M. Morgan, *Application of the Digraph Method of Fault Tree Construction to Process Plant*, Reliability Engineering, Vol. 14, pp. 85-106, 1986.

[13] Andrews J. and G. Brennan, *Application of the Digraph Method of Fault Tree Construction to a Complex Control Configuration*, Rel. Engg. and Syst. Safety, Vol. 28, pp. 357-384, 1990.

[14] Bossche A., *Computer Aided Fault Tree Synthesis*, Microelelctronics and Reliability, Vol. 25, No. 4, pp. 773-788, 1985.

[15] Bossche A., *Computer-Aided Fault Tree Synthesis-I (System Modeling and Causal Trees)*, Rel. Engg. and Syst. Safety, Vol. 32, pp. 217-241, 1991.

[16] Bossche A., *Computer-Aided Fault Tree Synthesis-II: Fault Tree Construction*, Rel. Engg. and Syst. Safety, Vol. 33, pp. 1-21, 1991.

[17] Kohda T. and E.J. Henley, *On Digraphs, Fault Trees, and Cut Sets*, Rel. Engg. and Syst. Safety, Vol. 20, pp. 35-61, 1988.

[18] Kohda T., K. Inoue and E.J. Henley, *Interactive Method for Obtaining Minimal Cut Sets Using Knowledge Engineering Approach*, Proc. the IASTED Int. Symp. Fifth IASTED Int. Conf. Reliability and Quality Control, pp. 103-107, 1989.

[19] Hassl D.F., *Advanced Concepts in Fault Tree Analysis*, Systems Safety Symp., The Boeing Co., Seattle, Wash., 1965.

Appendix

Details of Interactive Method

Nomenclature

Condition list, condition table:	list of conditions expressed by occurrence conditions in Eq. (8.2) along a sequence of relations, a set of condition lists.
Variable list, variable table:	a sequence of variables and their deviation states expressed by a sequence of relations, a set of variable lists.
Variable states:	deviations of process variables or occurrences of certain types of failures.

Current variable (conditions):	the last variable (conditions) added to the variable list (condition list).
Backward(Forward):	in the direction from the effect (cause) variable to the cause (effect) variable.
Backward(Forward) expansion:	finds the current variable in the effect (cause) part of the relation, and adds the variable in its cause (effect) part to the variable list, and the corresponding conditions to the condition list.
Disturbances:	failures or abnormal events described by the occurrence conditions in Eq. (8.2)
Source disturbances:	disturbances which can cause a variable deviation.
Disturbance situations:	a sequence of conditions or events which cause a specified deviation at a given variable, including a source disturbance.
Downstream (Upstream):	in the (opposite) direction of the effect, Forward (Backward).
Related control loops:	Negative feedback loops and negative feedforward loops whose (controlled) variable is included in the disturbance situation.
Correctable (Uncorrectable):	means that the related control loop can (can not) rectify a deviation caused by a disturbance.
Zero-gain events:	failure such as valve stuck, whose effect is expressed by a zero-gain edge

Supplement to Steps 1, 2, & 3

Step 1 (Identification of all control loops)

1) From variables which do not appear in the cause-part of the relations, perform a variable list backward expansion. Repeat this procedure until (a) no further expansion is possible, or (b) the variable already listed appears again.
2) For case (b), collect all different sequences of variables between the same variable.
3) For case (a), collect pairs of variable lists sharing two variables in different paths, which satisfy the condition that both of them are not parts of variable sequences obtained in step 2).
4) All loops have now been identified. However they should be verified by the design engineer because (1) the if-then rules may be incorrect or (2) they may be spurious.

In finding all control loops, a case (b) list corresponds to a feedback loop, where a backward sequence of variables constitutes a loop. A feedforward loop has two paths from its sensed variable to its controlled variable. Step 3) obtains feedforward loops based on this characteristic. But when feedback control loops overlap one another as in cascade control loops, the assumption that two different backward paths constitute a feedforward loop does not always hold; parts of two feedback loops constitute two different paths between two variables. Thus, for feedforward loops, checking whether both of the two paths are related to feedback loops is necessary.

<u>Step 2 (Identification of all possible disturbance situations)</u>

1) Express the subject event by a variable state, and enter it in the variable table.
2) Obtain new lists by backward expansion.
3) Examine the current condition and variable state of each list, and go to step 4).
    a. If the current condition is a zero-gain event, then delete the corresponding list from the variable and condition tables.
    b. If the current condition is a source disturbance, then store the corresponding lists in DSS (set of disturbance situations).
    c. If the current variable state already appears in the list, and is consistent with the previous state, then store the corresponding lists in DSS.
    d. Otherwise, do not perform any operations.
4) See if any lists left in the variable and condition lists must be expanded.
    a. If such a list exits, go to step 2).
    b. Otherwise, all possible disturbance situations are obtained and we go to step 3.

To obtain disturbance situations, a deviation propagation path is traced, by the backward expansion, to the cause of the deviation. Based on propositions (1) and (2), two types of situations are assumed to cause a disturbance propagating to the given variable. One is a deviation caused by a source disturbance propagating to the given variable, and the other is a deviation caused by the reverse-gain action of an NFBL. Step 3)-b corresponds to the former situation, and step 3)-c to the latter. A propagation path, which reaches a zero-gain event before a deviation cause, cannot generate a specified deviation due to proposition (3). Thus, step 3)-a deletes this path.

<u>Step 3 (Derivation of related control loop malfunction)</u>

For each corrective related control loop, obtain malfunction conditions.

1) Set variables of the corrective control loop and assign the state given in the disturbance situation.
2) Expand variable and condition lists along the control loop.
3) Check the current condition.
    a. If the current condition is inconsistent with the condition in the disturbance situation, then delete the corresponding list.
    b. If the current condition is a zero-gain event, then store the corresponding condition list.
    c. Otherwise, do not perform any operation.
4) Check for termination of the expansion.
    a. If the expansion terminates, then check the consistency of variable states for each of the remaining lists, store the corresponding condition list and examine the next corrective control loop.
    b. Otherwise, go to step 2).

Consider the case where some control loops can correct the deviation generated by the disturbance. The disturbance can cause the object event only if all corrective control loops fail to correct it. Corrective actions are halted when some control device is inactivated (a zero-gain event) or a corrective action is reversed because of a reverse-gain failure. These two types of conditions are identified at steps 3)-b and 4)-a.

New Trends in System Reliability Evaluation
K.B. Misra (Editor)
1993 Elsevier Science Publishers B.V.

CHAPTER 9

USE OF PETRI NETS FOR SYSTEM RELIABILITY EVALUATION

Gurdeep Singh Hura
School of Applied Sciences, Nanyang Avenue
Nanyang Technological University, Singapore 2263

ABSTRACT
The chapter provides an introduction of Petri nets, their modeling capabilities, techniques of analysis and a brief discussion on several extended Petri nets. Later on, various possible uses of Petri nets for performing reliability analyses are presented. It is expected that Petri nets in future would provide a viable formalism for system reliability evaluation.

9.1 Introduction

Petri nets provide a simple and powerful method to represent and analyze the dynamic behaviour of concurrent and asynchronous systems (flow of information and control). The Petri nets are based on the notion of events and conditions. Petri nets can be viewed as an abstract formal model of information flow, as a formal automata, automata or generators of formal languages. Various properties of reachability, e.g., dead-lock free operations, safeness, etc. can be studied through the approach of Petri nets. In fact, Petri nets and related graph models have been used in a variety of areas of application, e.g., verification and design of communication protocols, event-conditions systems, multi- processor systems, data flow computations, distributed systems, computer software and hardware systems, legal systems, design and reliability analysis of LANs, fault tree analysis, etc.

9.1.1 Historical Background

The theory of Petri net had its origin in the work of Professor Carl Adam Petri for his doctoral dissertation *Kommunikation mit Automaten*[1]. Petri applied the concept of the theory of communication to asynchronous components of a computer system. This theory was subsequently modified and expanded by Petri himself using the basic concepts of information flow and the structure of concurrent systems to a generalized form commonly referred to as the *Petri net theory* [2] today. At Applied Data Research Inc., Holt [3,4] and his associates developed the notation and representation of Petri nets and investigated the applications of Petri nets to the modeling and analysis of concurrent systems. Later on, Dennis [5] defined the concept of Petri nets formally.

Various technical reports and several doctoral theses on Petri nets were produced during 1973-1979. In addition to these, two conferences were held by the Computation Structures Group, viz., the Project MAC Conference on *Concurrent Systems and Parallel Computation* at Woods Hole in 1970 and the Conference on Petri nets and Related Methods at MIT in 1975. The first IEEE (Computer Society) International Workshop on *Timed Petri Nets* was held in Italy in 1985. The second workshop on *Petri Nets and Performance Models* was held in Madison, Wisconsin, in August 1987. Subsequently, the third and fourth workshops were held in Tokyo, in December 1989 and in Melbourne in December 1991, respectively. Original European Conference on *Application and Theory of Petri nets* has now been renamed as *International Conference on Application and Theory of Petri nets* (ICATPN). The tenth ICATPN was held in Bonn, in 1989. Subsequently, in 1990, it was held in Paris and for the first time it will be held outside Europe in Chicago (USA) in 1993.

Many review papers [6-11] on Petri nets and their applications have been published. A special section on Petri nets [12] was planned by *IEEE Transactions on Software* in 1989 and *Microelectronics and Reliability* exclusively published a special issue [13] on *Petri Nets and Related Graph Models : Past, Present and Future.* Several text books have already been published, viz., Peterson's book [14] in English, Bram's book [17] is in French and Strake's [15] and Winand and Rosenstengel's [16] books are in German etc. These books concentrate on *classical Petri nets* and provide a comprehensive overview of the subject. A bibliography of Petri nets is also published every year from GMD [18]. Most recent references are regularly published in the Newsletter of the GI-Special Interest Group on *Petri Nets and Related System Models* [19]. A bibliography of Petri nets [20] and related graph models containing papers from 1979- 1985 was also published in 1986.

9.1.2 Definitions and Notation

A Petri net (place/transition; p/t net) is a bipartite directed graph with two types of nodes : places (circles), and transitions (bars) as shown in Fig. 9.1(a). These nodes are connected by sets of arcs. Each transition has a set of input places and set of output places. Furthermore, each place has a set of input and output transitions. There can be several arcs between any pair of places and transitions. If there are j arcs going from a place p_i to a transition, the *in-degree* to that transition from that place is given by j. Similarly, the *out-degree* from a transition to a place is the number of arcs from that transition to that place.

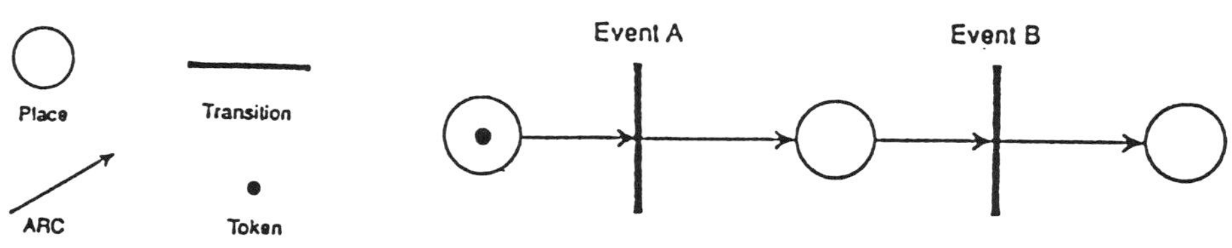

(a) Basic Components **(b) Model**

Fig. 9.1 Petri Net Modeling

Classical Petri Nets

Let the sets of positive integers, non-negative integers, and real numbers be denoted by $\mathbf{N}^+$, $\mathbf{N}$ and $\mathbf{R}$, respectively. A Petri net is a 4-tuple PN≡ (P, T, I, O), where P and T are the set of places and transitions, I and O are the transition input and output arc functions, respectively. Further, P, T, I and O are defined formally as :

$$P = \{p_i \mid p_i \text{ is a place, } 1 \le i \le I \in \mathbf{N}^+\}$$

$$T = \{t_i \mid t_i \text{ is a transition, } 1 \le i \le I \in \mathbf{N}^+, P \cap T = \phi$$

I : $P \times T \rightarrow \mathbf{N}$, where, if I(p,t) = w and w > 0, then we have an arc from p to t with a weight of w. If there is no arc from p to t, I(p,t) is defined to be zero.

O : $T \times P \rightarrow \mathbf{N}$, where, if O(t,p) = w and w > 0, then we have an arc from t to p with weight of w. If there is no arc from t to p, I(t,p) is defined to be zero.

The *structure* of a Petri net graph shows the static aspect of the system being modeled, namely connections between various components of the system and the number of constituent parts of the system.

A dual of a Petri net PN = (P,T,I,O) is defined as DPN = (T,P,I,O) where DPN is obtained by interchanging of circles and bars corresponding to places and transitions. However, the structure of dual Petri net does not change. The dynamic behaviour of a modeled system is determined by executing a Petri net model. A Petri net is executed by using tokens (see Fig. 9.1(a)), which are represented in the Petri net model by black dots inside the places. A Petri net with an initial number of tokens in places is called a *Marked Petri net.* If we specify an initial marking of a Petri net by M_o, then a marked Petri net is defined as a pair of (PN, M_o), where PN ≡ (P,T,I,O) and $M_o : P \rightarrow N$.

The tokens are non-negative integer numbers and represent data, initial conditions, or resources within a system. The number of tokens in any place at any time is not bounded but is always finite. In general, a token distribution in the places of a Petri net represents the *state* or *marking* of the system modeled by that net. The marking is defined as a column vector, where its *jth* entry represents the number of tokens in the place p_j.

The execution is achieved by means of transition firing (event A and B in Fig. 9.1(b)). A transition can fire *if and only if,* it is enabled. A transition is enabled if each of its input places contains at least as many tokens as the weight of the input arc between the place and transition. Firing of a transition t is achieved by removing w_i tokens from each input place p_i, where $I(p_i,t) = w_i$ and depositing w_j tokens into each output place p_j of the transition, where $O(t,p_j) = w_j$. This is illustrated in Fig. 9.2, which is a Petri net model for the Producer/Consumer problem, where the *arc weight* is assumed as one. An enabled transition ceases to be enabled if its enabling tokens are used by firing of some other transitions. This situation causes a conflict between two transitions sharing a common input place and is usually resolved by Petri net itself. A Petri net based model of a system can be used to prove properties such as *mutual exclusion, deadlock-free conditions, reachability,* etc. These properties along with other problems associated with concurrent systems are discussed in more details in section 9.1.4.

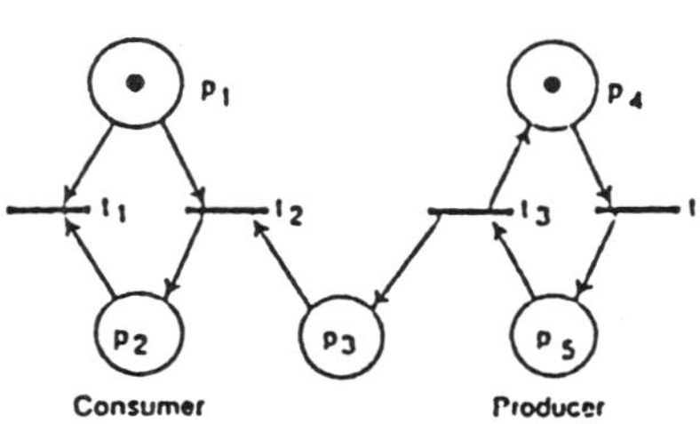

(a) Initial Petri Net

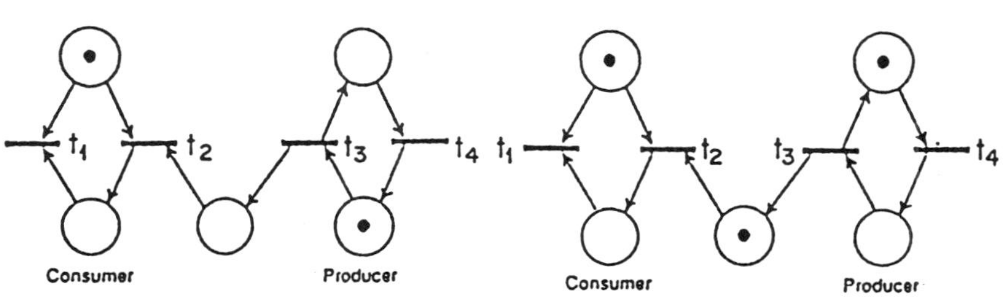

(b) After Transition t_4 is fired **(c) After Transition t_3 is fired**

Fig. 9.2 Petri Net Model for Producer/Consumer Problem

9.1.3 *Features of Petri Nets*

The interpretation of entities of a Petri net, like places, transitions, token, not only help in the representation or modeling of a system at different levels, but also in determining the dynamic properties of the system. Petri nets have the following advantages :

(i) Petri nets provide a graphical and precise representation of a system.

(ii) Pragmatic interpretation can be assigned to various entities of the Petri nets which enhances the readability of a model. Petri nets can be used to represent a system at different levels of abstractions and details, depending on the interpretation and information required.

(iii) The systems can be modeled hierarchically.

(iv) Being asynchronous in nature, Petri nets contain the necessary information required for obtaining the sequences of events of the system.

(v) The behaviour of the system can be analyzed by the movement of the tokens.

(vi) Petri nets have the ability to model the software concepts like *parallelism* or *concurrency, conflict* and *synchronization of events.*

(vii) Petri nets can be synthesized using modular approach for the design of a system whose behaviour is either known or can easily be verified.

Assumptions and Limitations

Inspite of the above features, Petri nets have the following limitations :

(i) The design correctness is not established directly.

(ii) An analysis of the Petri nets is not an easy task and depends upon the size of the system.

(iii) Although time is not explicitly represented in p/t nets; its effect, however, is represented implicitly and it is possible if it affects a partial ordering of events (occurrence of) in a system which consists of a set of dependent and independence events. This is valid, if one is only interested in the order of occurrence of events, and not in the time of occurrence of an individual event. Using this view of time, the treatment of independent events need not logically relate to the notion of time, thereby simplifying the modeling and analysis of Petri nets.

This view of time is reflected in the structure of the p/t net type representing the modeled system by means of ordering of transitions, where each transition represents a single event (primitive event). Suppose we have events A and B (Fig. 1b), and A occurs at time t_1, B at time t_2, and events A, B are dependent (it is assumed that duration of events A and B is zero, since events are primitive). Now the structure of p/t modeling this system can be made in such a way that an order is imposed on set {A, B}, where event B occurs after event A.

(iv) For the analysis, it is generally assumed that the firing of a transition (occurrence of an event) takes zero time in transferring the information from its input place to the output place, therefore the time associated with the transition simply represents the time of an operation.

(v) A major limitation of Petri net lies in its inability to determine if a place is empty. This problem can however be overcome by defining explicit places for negated conditions. This is sometimes inadequate to model situations involving counting particularly in case of unbounded place.

However, the advantages offered by Petri nets outstrip the limitations and Petri nets offer a powerful tool for the design and analysis of concurrent systems.

9.1.4 System Properties Supported by Petri Nets

As stated earlier, Petri nets have the ability to support various system properties such as :

Reachability

It defines the existence of firing sequence that transforms M_o to M_n. It is a fundamental property for studying the dynamic behaviour of a system.

Boundedness

A place is defined as *k-bounded,* if the number of tokens in that place is not more than a maximum of k, where k is an integer. A bounded Petri net can easily implement a hardware while unbounded Petri net model cannot. In other words, k-bounded Petri net defines the capacity of a net and allows a direct implementation/realization of counters, registers, buffers with appropriate values of k.

Safeness

This property is useful when Petri net is to model a hardware. A Petri net is safe, if all places in the net are safe; and a place is safe, if the number of tokens in any place is no more than a maximum of 1. In other words, any place can either have a token (k=1) or no token (k=0). It is always possible to make a place safe if (i) it is not input to more than one transition (conflict situation) and (ii) it is not output to more than one transition. *Boundedness* or *Safeness* gives a guarantee that there is no overflow.

Conservation

As pointed out earlier, tokens can represent data items, conditions and resources, an interpretation can be used for the requests, allocation and release of any resource in a computer system modeled by a Petri net. This property states that the resources (represented by tokens) can neither be created nor destroyed in a computer system. It is not always necessary to represent all resources by tokens only. There are few components, like counters, which are also represented by tokens; and also many resources have to be represented by one token. In order to handle such a situation, we define weighting of tokens and this property would then mean that the weighted sum for all reachable markings should be constant. The weights are assigned to each place of the Petri net and a weighting vector is defined for the net, where i^{th} element of the vector is a weight assigned to place p_i. Thus, the Petri is strictly conservative with respect to weighting vector.

Liveness

The problem of resource allocation in a computer system, in particular, in an operating system can be studied using conservation property. Moreover, we need to define another property - a deadlock in resource allocation problem. A deadlock in Petri net occurs when a transition or a set of transitions cannot fire. A transition is said to be *live*, if it can always fire, i.e., it is not *deadlocked.* Thus, if a transition is live, it is always possible to fire this transition by progressing through some firing sequence; and the net is live for some initial marking. Petri nets with a live marking guarantee deadlock-free operations regardless of the firing sequence chosen. Various levels of liveness can be defined in terms of dead transitions.Higher the level, stronger is the liveness of Petri net model.

Persistence

In any marking, an enabled transition can be disabled only by its own firing. A *safe persistent* Petri net can be transformed to a marked graph (structurally restricted Petri nets or a subclass of Petri nets).

Coverability

A marking M is coverable if there exists a marking M_1 (reachable from M_o) such that $M_1(P) \geq M(P)$ for each place in the net.

Reversibility

A Petri net is reversible, if for each marking M in R (M_o), M_o is in R(M). This property provides decidability in the system.

Synchronic Distance

This property provides the measure of mutual correspondence or dependency between two transitions. It is an important metric to implement fairness in resource sharing systems.

Finally, the properties discussed here depend on the structure and marking of Petri nets.

9.1.5 Structural Properties of Petri Nets

Various structural properties of Petri nets (independent of markings) are :

Controllability

A Petri net is completely *controllable*, if any marking is reachable from some given initial marking and can be verified by the following relation :

rank $\mathbf{A} = m$, where m is the number of places while $\mathbf{A}(\equiv [A])$ is a *transition-to-place incidence matrix* of the Petri net.

Boundedness

A Petri net is *bounded* for any initial marking M_o ($< \infty$) and can be provided by $\mathbf{A}\, y \geq 0$.

Conservativeness

This property holds good, if $\mathbf{A}\, y = 0$

Repetitiveness

In a Petri net, every transition fires infinitely often in the firing sequence and satisfies the following condition :

$\mathbf{A}^T x \geq 0$, where $\mathbf{A}^T$ is the transpose of matrix $\mathbf{A}$.

S- and T- Invariants

Invariants of nets provide a useful information for studying several properties like overflows, deadlocks, mutual exclusion, etc. Various formalisms and approaches exist to find the invariants of different types of extended Petri net models. There are two types of invariants, viz., *S-and T-invariants*, and both these invariants are defined as product between multisets. These invariants are defined as :

$$\mathbf{A}y = 0, \text{ S-Invariant}$$
$$\mathbf{A}^t x = 0, \text{ T-Invariants}$$

Consistency

In a Petri net, there exists M_o and a firing sequence from M_o to back to M_o such that every transition occurs (fires) at least once.

9.2 Modeling

A system in its simplest form is a set of events which occur in a particular order, where this order is defined by means of set of conditions associated with an event. For example, let us consider the job processing modeling [14] in a computer system as is given in Fig. 9.3. Petri nets provide an adequate and simple method for modeling of such systems.

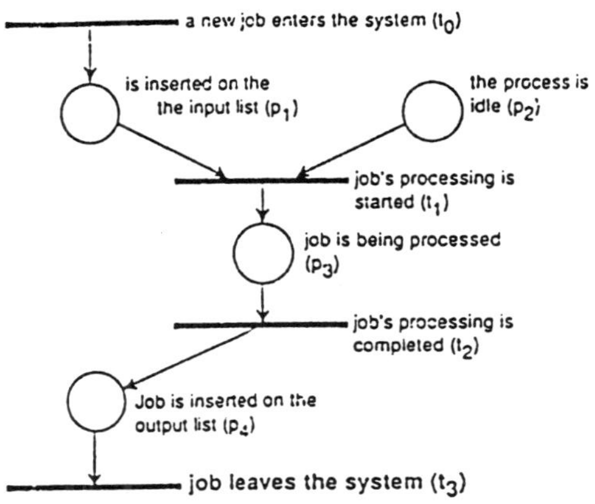

Fig. 9.3 Modeling of Job Processing in a Simple Computer System

9.2.1 Tools for Modeling

In Petri net models, transitions can represent events which occur when a set of pre-defined conditions are satisfied. The existence of these conditions are represented in the model by means of places, and their satisfaction by the availability of tokens in these places.

Pre-conditions of an event (corresponding to a transition t_i) are defined by input connected places to t_i while *post- conditions* are defined by output connected places of t_i. A Petri net model inherently provides mutual exclusion. It automatically recognizes concurrent conditions and simultaneously effects changes. This ability of Petri net avoids a major issue in programming synchronization problems, where the atomic actions are at a low comprehensive level.

The way the events are arranged in a model is defined by the *degree of concurrency* and the *level of dependence* among the events within a system, and can be easily represented, as in Fig. 9.3. Some events, when they occur, prevent occurrences of other events (mutually exclusive events), and these are easily modeled in Petri nets by means of decision structure conflict shown in Fig. 9.4 (a),(b),(c).

Conflict is a data independent method, by which decision capability is introduced into the structure of a Petri net model. However, its effect is the introduction of non-determinism due to the fact that the path taken (decision made) in presence of a number of alternatives is data dependent and can only be specified with a certain degree of interpretation. This, however, cannot be represented by simple Petri net models, since they are *uninterpreted models.* An appropriate interpretation may be assigned to the entities (places, transitions and tokens) to model any type of system. Since their components are atomic and uninterpreted, the resulting model is usually extremely large and rather complex. This problem can be handled by means of abstraction.

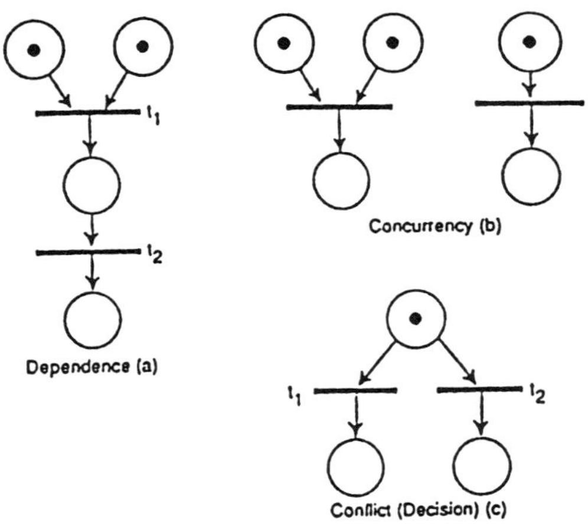

Fig. 9.4 Decision Structure Conflict Conditions

In order to appreciate the hierarchical modeling capability of Petri nets (which is achieved by means of abstraction and refinement), let us consider a model [14] presented in Fig. 9.5 (a). The model shown in Fig. 9.5 (a) represents a top level view of a system with non-essential details (at this level) abstracted away. Details present at a particular design level can be kept to a minimum by increasing the number of design levels. If more details are required, one could apply refinements to any individual transition or place of the top level model. Thus, in example of Fig.9.5 (b), place p_1 and transition t_3 are calls to the subroutines and Fig. 9.5 (a) represents the main program.

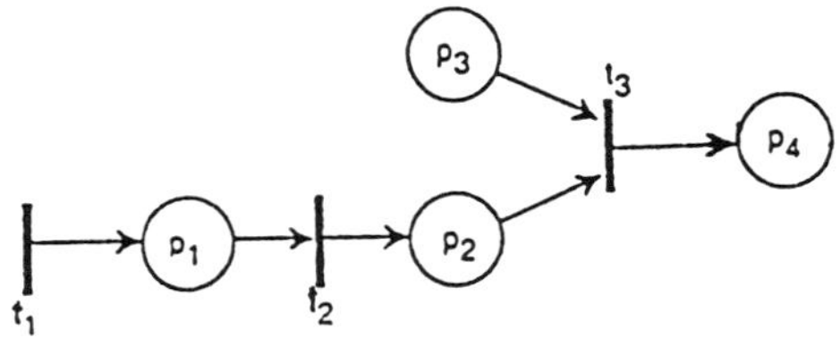

Fig. 9.5 (a) A System Modeled by a Petri Net

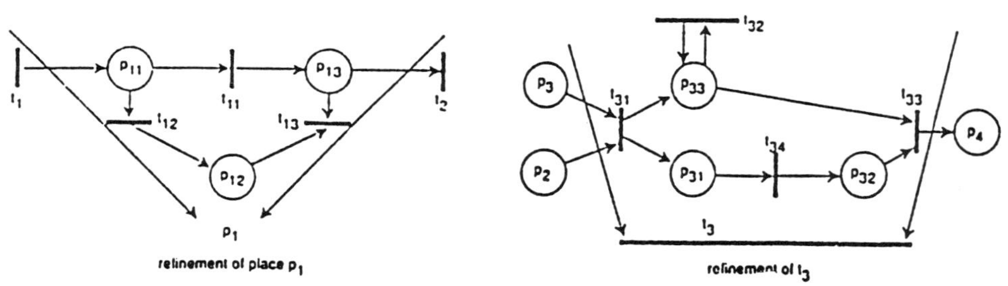

Fig. 9.5 (b) Modeling Capability of a Petri Net

9.2.2 Software Packages

Like other descriptive languages and analyses, we have graphical languages (representing various type of Petri nets) for system description and these tools help the users during construction, modification and analysis of large system description [21]. Feldbrugge and Jensen [22] provide an overview of the characteristics of all currently available Petri-net based software packages and tools, besides discussing the environment, functionality, future plans, availability, and contact persons along with their addresses. Most of the software packages are interactive and a user can analyse systems without knowing Petri net theory. These also assist a user in building net model at different level of abstractions. One of the main problems with Petri net tools has been the availability of *Graphic Petri Net Editor* and inconsistency of different tools of analysis which runs on an IBM PC. This problem seems to have been resolved in graphic editor called Design TM [23].

9.3 Extension of Petri Nets

The concept of individual tokens and the notion of time in Petri net modeling has led to the development of different types of nets. The modeling power of Petri nets can be increased by incorporating *zero testing.* This capability is achieved by introducing an inhibitor arc into Petri net models known as *extended Petri nets.* An inhibitor arc, between place p_i and transition t_j, is represented by a small circle instead of arrow on transition. A transition with inhibitor arc is enabled, when each of its connected places has at least one token and each of its inhibitor arcs contains zero tokens. After firing, each of its output-connected places, receives a token. A significant work towards increasing the modeling power of Petri nets has resulted into several extended Petri net models e.g., extended Petri nets based on individual tokens, Timed Petri nets (TPNs), Stochastic Petri nets (STNs), Generalized Stochastic Petri nets (GSPNs), Generalized TPNs, extended SPNs, Stochastic Activity Networks (SAN), Abstract Petri nets (APNs) etc. [24, 29, 33, 34, 39, 41,44, 47, 51, 64, 69].

These extended Petri net models have found their applications in a variety of areas, e.g., communication protocols, distributed data bases, formal semantics of programming language, performance evaluation of multi-processor systems, design of LANs etc. Till now, the principal application of Petri nets has been in the area of behavioural studies of various systems, or performance evaluation of computer systems but cannot be used as a practical tool for the implementation of design methodology. There is a need to develop a unified modeling tool (using layered approach) of these extended net models in the form of a common shell, which can be interfaced with any application requirements of different systems as an upper layer.

It is obvious that the structure of a system can be represented by a directed net while various aspects of the system like its function, behaviour, etc. can be described by assigning interpretations to the elements of the net, viz., places, transitions and tokens. The level of interpretation depends on the level of abstraction introduced. The interpretations may represent formal expressions, symbols or functions. For example, a net representation of the first-order predicate logic derived from the enlogic structure (one of the system invariants defined by Petri [24]; the other is synchronic structure of the system [25]), the transition nets with coloured tokens [26], transition nets with complex conditions and transmissions used by [27], TL-nets [28], predicate/transition nets [29], high-level nets [30, 31] etc., are all cases of extended Petri net based models.

9.3.1 The Predicate/Transition Nets

A designer can use this model for representing a first-order predicate logic. Here, each token has a specific colour known as token-colour and represents a specific information. The firing of transitions can be defined in several ways through different firing-colours. An expression corresponding to the relationship between *token-colour* and *firing-colour* is attached with a correspondence arc. A predicate associated with the transition controls the possible firing of that transition. Each process/task/object has its own identity within the model and their execution is controlled by means of a transition in a predefined and controlled way by the predicates [25,26].

Identifiers are assigned to both S- and T-elements. The places represent some conditions of the system and contain tokens for representing the existence of the conditions. The input (output) places of transitions define preconditions (postconditions) for the occurrence of an event represented by the transition. Some of the S-elements are marked by a token, which allows the simulation of behaviour of the system. Various features like concurrency, conflict, etc. can easily be modeled by this net. Events and facts (two classes of the enlogic structure) are represented by transitions and places.

The classical Petri nets don't allow the identification of individual conditions by tokens, whereas in *PrT-nets*, we have tokens corresponding to the predicates with their current extension, in contrast to simple tokens in a classical Petri net and marking of PrT involves the marking of individual tokens. Thus, the marking of a net is defined as a marking of predicates of S with formal sum of n-tuples of individual tokens.

This net has been used in various organization systems, e.g., distributed data bases, resource management, etc. and it has modeling flexibility, power and conciseness in modeling systems as dynamic structures. Few extensions of PrT-nets are : many-stored structures, more general places, more general arcs, multi- sets and the weak transition rules, place projections. The method of analysis of classical Petri net (place/transition) is to construct linear place-invariants through homogeneous matrix equation which is generalized to PrT-net [29,30] but it contains free variables (over the set of colours). In order to interpret place-invariant method for the analysis of PrT-net, it is neccesary to use substitution methods to bind these free variables, such that the information regarding the firing sequence leading to the marking can be used.

9.3.2 Coloured Petri Nets

The problem of dealing with free-variables during the place-invariant analysis of PrT net is overcome by *Coloured Petri nets* (CP-nets). The only difference between CP-nets and PrT-nets lies in defining the relationship between the colour-token and the firing-colour to be attached to the arcs. In case of PrT- nets, an expression corresponding to the relationship is defined and is associated with an appropriate arc, while in CP-nets, a function is associated with the arcs. Thus, place-invariant analysis method can be interpreted on CP-nets without any substitution or an addition. The functions associated with arcs are sometimes difficult to understand whereas expressions are not. However, CP-nets provide a more comprehensive modeling approach than that of PrT-nets. Further, a set of possible token-colours to each place and also a set of possible firing-colours to each transitions can be explicitly defined and associated within the net [26]. The software tool package to support hierarchical coloured Petri net can build and simulate large models of complex system, e.g., communication protocols, flexible manufacturing lines, distributed databases, etc. Other applications on CP-nets may be found in [26,28,29,32].

9.3.3 High Level Petri Nets

Jensen [31] combined both PrT-net and CP-net into one net and defined it as *High-level nets* (HL-nets). The incidence-function defined in CP-nets has been partitioned into two parts in HL- nets, viz., positive and negative. Further, the set of possible token colours at a place is explicitly defined (as in CP-nets), while the set of possible expressions and predicates are not explicitly defined in HL-nets (contrary to PrT-nets). This generalization of place-transition nets provides a controlled and manageable description of the systems and also invariants and reachability can be generalized for HL-nets. These types of net models have been used in different applications [30,31]. A detailed comparison based on various aspects like descriptive power, ease in modeling, analysis techniques, notion of invariants, etc. between PrT-nets, coloured-nets and relation-nets is given in [29].

9.3.4 *Timed and Untimed Petri Nets*

The previous classification of Petri nets was based on the concept of defining a separate identity of each of the resources/tasks/processes as tokens and did not use the notion of time and thus can be called as *Untimed Petri nets*. Obviously, untimed Petri nets cannot be used for time-related performance evaluation of systems. The extended models, based on the notion of time and have fixed time associated with either a transition or a place, are defined as *Timed Petri-Nets* (TPNs) [34, 36, 39, 40, 42-44, 46]. But the models, which have random firing time associated with a transition or a place, are defined as *Stochastic Petri-Nets* (SPNs) [51-56, 58, 61, 64, 68].]

Timed Petri Nets

Ramachandani [33] introduced time into the classical Petri net model by associating a fixed time with each transition. Further, the firing of a transition is no longer an instantaneous event as in classical Petri nets. Based on the specification of time, associated with each transition, different types of extended net models have been defined which may have *fixed firing time*, maximum and minimum firing time, and the random firing time [34, 41, 44-47, 49]. Merlin [35] defined timed Petri net in which firing delays (maximum and minimum) are associated with each transition. Transitions are enabled the same way as in classical Petri nets, but firing only occurs within the limits of time defined by static firing interval. The minimum and maximum of firing sequence can be zero and infinite, respectively. These models are very useful to model variable delays or *time-outs* and have been used in the design and implementation of communication protocols.

The TPNs have been further extended by defining a probability function distribution over the firing interval (t_{min}, t_{max}). The reachability tree analysis of Petri nets can then be used to determine the behaviour of models in terms of functionality and performance. It has been claimed that these extended TPN's provide accuracy in the specification of the systems and also don't require usable sufficient conditions for using reachability analysis in the TPN model [38-41]. Ramamoorthy and Ho [36] assumed the structure of TPN as a graph. It is assumed that the transitions have *parallelism* in their execution within a model, although this assumption is not precisely mentioned but has been used during the analysis. For a very restricted class of nets, the matrix algorithm is used to obtain the longest circuit in the graph. Coolahan and Roussopoulos [37] define the upper bound timing equations from the model and obtain the possible paths through the graph. Stotts and Pratt [38] proposed the state space of a timed Petri net. Reference [39, 40] also define models to construct state space and are analyzed as stochastic processes. By interpreting these models as stochastic processes, the rich concepts and results of stochastic process theory can be used for the performance evaluation of the system.

The state of a TPN is defined differently than in classical Petri nets as the firing of a transition is not an atomic operation. The next state is generated by a set of start firings or a set of end firing. Zuberek [39] viewed the reachability graph of TPN as a Markov chain and suggested that the performance measures should be computed using standard techniques for analyzing the Markov chain's long run behaviour. This approach of constructing reachability graph is valid for a restricted class of nets : *safe and free choice nets*. Further, the reachability graph (i.e. Markov chain) will not give correct performance measures for these restricted types of nets.

Holiday and Vernon [47] defined *Generalized Timed Petri Nets* (GTPNs) by removing all restrictions on the net except that the space be finite and compute correct performance estimate for any reachability graph. The firing duration is considered as an arbitrary real number while resources are associated with both places and transitions. Further, the durations and frequencies are state-independent as opposed to state-independent in the models of [39,40] the frequency and set of resources, transitions in GTPN have a flag associated with them which are used to compute the next state probabilities. For more details of GTPNs, their properties and applications, one can refer to technical reports [47,48,50].

9.3.5 Stochastic Petri Nets

An extended Petri net model which has random time associated with its transitions is known as a *Stochastic Petri Net* (SPN) and has been used for the performance evaluation of multiprocessor systems, distributed systems, networks, etc. The main utility of stochastic Petri net models lies in providing presentation and analysis of uncertainty and indeterminancy of the events within the systems. These nets use stochastic process theory to carry out performance evaluation of the modeled net.

In stochastic net models, a random firing time is attached with the transition and the distribution function for time defined as a continuous, real, non-negative random variable X_i is given by

$$F_i(X) = P(X_i \leq x)$$

If the reachability graph of a stochastic Petri net can be proved to be isomorphic to any kind of Markov chain process, then the solution of the system can be obtained using the properties of Markov chain. Molloy [51] considered an exponential distribution function for each transition of the net and this provided implicit necessary and sufficient condition for using the method of Markov chain.

In stochastic Petri nets, some transitions, defined as *immediate transitions,* fire instantaneously as soon as they are enabled and take zero time in depositing the tokens at the output place. Immediate transitions will always have priority over timed transitions. This does not create any conflict in the net since Dugan et al [52] have used both immediate and timed transitions in their stochastic Petri net models. In some of the stochastic Petri net models, probability arcs, random switches, inhibitor arcs, counter-alternate arcs have also been defined and used and more details about these can be found in [52,54,56,62,65]. Stochastic Petri Nets (SPNs) can be further classified as continuous time (SPNs) and discrete-time SPNs. In the former, a set of finite transition rates for exponentially distributed firing times is added to classical Petri net whereas in the latter, a set of nonzero conditional probabilities with geometrically distributed firing times are added to classical Petri nets in discrete-time SPNs.

Haas and Shedler [55] define a marking process as a generalized semi- Markov process and using *new better than used* distribution assumptions for firing times, a new concept for recurrence and regeneration in continuous time is proposed. This concept provides regenerative simulation methods for SPNs. This approach assumes that two transitions cannot fire simultaneously hence restricts the use of SPN to pure Markovian stochastic simulations.

SPNs are very effective modeling tools for the analysis and performance evaluation of computer system since the various activities within a system can be precisely described and translated by SPNs into Markovian models. Unfortunately, SPNs are suitable only for small systems and the representation becomes more difficult and unmanageable with the increase in the size and complexity of the system. Further, the reachability state space also becomes very large [51,53,54,56].

Generalized Stochastic Petri Nets

A generalized Stochastic Petri nets (GSPNs) was proposed by Marsan [57], where instead of associating an exponential distribution of firing time for each transition, we identify the activities/events within the systems which have more impact/effect on the performance analysis, we need only consider the firing times of only these transitions. This reduces the number of states and thereby a reduction in the complexity. Generalized SPNs also contain two types of transitions known as timed and immediate. Immediate transitions have higher priority than timed transitions. The probabilistic determination of, which immediate transition from many enabled transitions should fire next, is obtained through a random switch.

It has been shown that GSPNs are equivalent to Markovian models and require less computational complexity for the solutions as compared to SPNs. GSPNs are useful tools for the precise description of the operations and traffic of a system. Further, it has versatility and any change in the assumption of system operation can be represented in the original model with small modifications. A time scale decomposition algorithm of a class of GSPN models of system comprising activities whose duration differ by order of magnitude is discussed by Ammar and Islam [59]. Introduction of probabilistic branches and general distribution in the definition of SPNs is given [58,61].

Extended Stochastic Petri Nets

In extended Stochastic Petri nets [ESPNs] [52], an inhibitor arc from a place to transition is represented by an arrowhead at the transition compared to a small circle defined in a traditional method. New firing rules of transitions with inhibitor arcs have been defined. A transition in ESPNs is enabled when each of its input connected places (normal) has at least one token and no tokens are present in inhibiting input places. After firing, each of its output connected places gets a token while the inhibiting input places should contain no tokens. A probability arc from a transition to a set of output places deposits a token in one of the places in the set. The choice, which place receives the token is determined by the probability labels on each branch of the arc. Further, a counter arc from a place to a transition is also labeled with an integer value k. This changes the firing rule of transition such that a transition is enabled (as usual with normal) its counter input place contains at least k tokens. After firing, one token is removed from each of its input connected places (as usual) and k tokens are removed from the counter input place. Associated with a particular counter arc can be a counter-alternate arc, which enables an alternate transition when the count is between 1 and (k-1), inclusive [52].

The ESPNs mode was initially developed as an aid in its modeling capability in fault-tolerant computer system for the Hybrid Automated Reliability Predictor (HARP) project at Duke University. A solution package DEEP 1 (The Duke ESPN Evaluation Package) was also developed at Duke University. This package includes the works of GSPN group, the IEN group and the Duke group and contains both analytical and simulation solution methods [62, 63].

9.3.6 Stochastic Activity Networks

Stochastic Activity Networks (SANs) proposed by Movaghar and Meyer [64], a generalization of SPNs, offers various features like modeling of concurrency, timeliness, fault tolerance and degradable performance in a single model. SANs are basically suited to the performability of distributed real time systems and closely resemble the GSPNs. Two subclasses of SANs have been discussed in [64]. In one subclass, the state behaviour is described as a semi-Markov process while in the second subclass, the state behaviour is described as a Markov process. A high-level model, a stochastic activity system (SAS) has been introduced which allows the conditions for these two subclasses of SANs. Applications of these models to the construction of performability models and related software packages can be found in [65].

A different approach has been used by Molloy [66] to propose a subclass of SPN called *structurally bounded SPN* (SBSPN). A comparison of performance models with queueing network models has been discussed by Vernon et. al. [67]. Pagnow [68] compared different types of stochastic Petri net models and obtained a hierarchy chart (starting with classical Petri net model). These stochastic Petri net models have different firing rules of transitions. We still use graphical representation and reachability tree analysis of classical Petri net model in these stochastic Petri net models.

9.3.7 Abstract Petri Nets (APNs)

A new extended model called as Abstract Petri Net (APN) has been recently defined. This extended model in fact, integrates the concepts of coloured Petri nets (CPNs) and Stochastic Petri nets (SPNs) and defines input and output predicates for the transitions. It has found applications in rule-based methodology for solving real-time control problems, optimal solution for bounded buffer etc. [83]. Hura et.al. [69] provided a comparison of various extended Petri net models.

9.4 Analysis of Petri Nets

Once a system has been modeled using a Petri net graph, the system can be analyzed by using one of the three basic analytic approaches, viz., *invariant approach, matrix approach,* and the *reachability tree approach.*

9.4.1 Invariant Approach

In the invariant approach, one identifies a set of system invariants, which may be just a set of places. Let M(p) specify the number of tokens in a place p, then the net invariant I is a set of places such that $\Sigma_{p \varepsilon I} M(p)$ is constant for each reachable marking M and I does not have any proper subset that is invariant. We define an incidence matrix $\mathbf{A} \equiv (a(t,p))$ for a Petri net, where $t \varepsilon T$, $p \varepsilon P$ and

$$a(t,p) = \begin{cases} -1; & \text{if } I(p,t)=1 \text{ and } O(t,p) = 0 \\ +1; & \text{if } O(t,p)=1 \text{ and } I(p,t)= 0 \\ 0; & \text{otherwise} \end{cases}$$

It is assumed that each arc weight is unity. The arcs may have any non-negative integer value of weights. Solution of **A** Y=0 is sought such that each component of Y is zero or one. By solving the set of equations, we obtain the invariants of Petri net. Let us consider an example of Fig. 9.6 with its set of equation given by :

$$\begin{array}{c} \\ t_1 \\ t_2 \\ t_3 \\ t_4 \end{array} \begin{array}{c} \begin{array}{ccccccc} p_1 & p_2 & p_3 & p_4 & p_5 & p_6 & p_7 \end{array} \\ \left[\begin{array}{ccccccc} -1 & 1 & 0 & 0 & 0 & 0 & 0 \\ 1 & -1 & 0 & 0 & -1 & 0 & 1 \\ 0 & 0 & -1 & 1 & 1 & 0 & -1 \\ 0 & 0 & 1 & -1 & 0 & 0 & 0 \end{array}\right] \end{array} \begin{bmatrix} y_1 \\ y_2 \\ y_3 \\ y_4 \\ y_5 \\ y_6 \\ y_7 \end{bmatrix} = \begin{bmatrix} 0 \\ 0 \\ 0 \\ 0 \\ 0 \\ 0 \\ 0 \end{bmatrix}$$

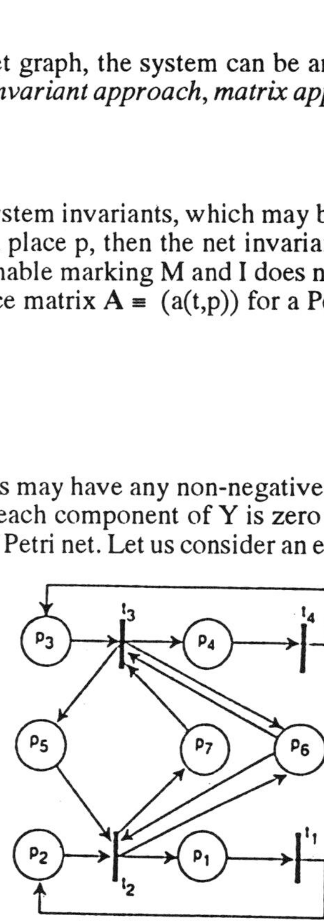

Fig. 9.6 Example of Invariant Approach

Therefore, the invariants would be :

[1 1 0 0 0 0 0] {p1,p2}, {p3,p4}, {p5,p7}, {p6}
[0 0 1 1 0 0 0]
[0 0 0 0 1 0 1]
[0 0 0 0 0 1 0]

9.4.2 Matrix Approach

In the matrix approach, we define two matrices $\mathbf{A}^-$ and $\mathbf{A}^+$ (input to and output from transitions), each matrix having m rows and n columns corresponding to m transitions and n places. Using notation and operation "#" of [14],

$A^-[j,i] = \#(p_i, I(t_j)),\ A^+[j,i] = \#(p_i, O(t_j))$

where, $\#(p_i,I(t_j))$ is the number of input places p_i for transition t_j and $\#(p_i, O(t_j))$ is the number of output places p_i for transition t_j. Suppose e[j] is a unit m-vector (1 in the jth component and ϕ, elsewhere) then transition t_j is represented by e[j]. For a given marking M, transition t_j is enabled, if $M \geq e[j].\, \mathbf{A}^-$.

The result of firing transition t_j in marking M is given by

$M - e[j].\mathbf{A}^- + e[j].\mathbf{A}^+ = M + e[j].\ \mathbf{A}$; where, $\mathbf{A}=\mathbf{A}^+ + \mathbf{A}^-$.

The matrix approach can be used for solving conservation problems as well. If we can find a (nonzero) weighting vector for which the weighted sum over all reachable marking is constant, then there exists a sequence of transition firings which takes the net from initial marking M_0 to M_1. The conservation property can be tested by finding out a positive weighting vector solving the following conditions $\mathbf{A}W = 0$.

This matrix approach for Petri net analysis also provides a useful tool for studying the reachability problem. To find if M_1 is reachable from M (where M and M_1 are the markings), we only need to find a solution for $M_1 = M + X.\ \mathbf{A}$, where X is the firing vector for the Petri net model and has non-negative integers components. Other general questions such as liveness, deadness, etc., can be answered in a similar way.

9.4.3 Reachability Approach

A finite representation of reachability set R(cn), is called a reachability tree. The reachability tree is obtained from the reachability set, which, is the set of all reachable marking from the initial marking, M_o, by employing two special steps. We should however note that, since we are making a finite representation of an infinite set, this implies many to one mapping of states (markings), and results in a loss of information. An example of a marked Petri net, and its reachability tree is given in Fig. 9.7. The construction of the reachability tree is started by labeling the root of the tree with an initial marking. For each enabled transition in this marking, an arc is drawn with that transition as its label.

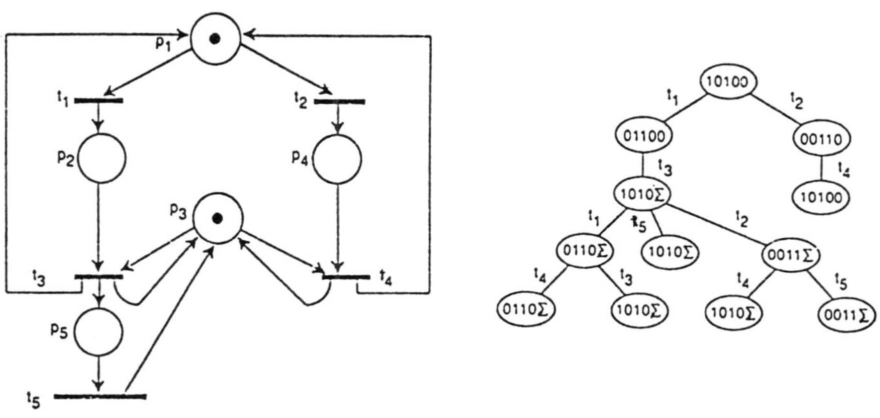

Fig. 9.7 Petri Net Model with its Reachability Tree

The label of second node of each arc is obtained by firing the associated transition (label) of that arc. This process is continued for each new marking until no transition is enabled. At that time, we get all possible states. This process results into an infinite reachability tree. The reachability set will produce all possible markings and as such any Petri net with an infinite reachability set, the corresponding reachability tree seems to become infinite. The firing sequences correspond to various reachable markings in the tree. Since many of the properties depend on reachable markings and various paths, we should develop a technique to limit the size of this tree. In order to get a finite representation and as the first special step, we get rid of the repeated transition firings. If a transition firing is repeated in a branch from the root of the tree to a marking, the marking (the current) is considered a leaf node and this branch is terminated. By utilizing ω (a very large number), the following property can be stated the second special step.

$$\omega + a = \omega$$
$$\omega - a = \omega, \quad \text{where } a < \omega$$

Given a marking x and a marking y, where y lies on the same path (branch) as x and x appears before y on the path from the root, if y > x, those components of y, which are strictly greater than corresponding components in x are replaced by ω (Fig. 9.7). If the reachability tree does not have as many ω components in any node of the tree, then we have a finite reachability set and in that case all questions about the modeled system can be answered.

The algorithm to construct the reachability tree starts by defining an initial marking as a root of the tree and processes various nodes until all of them have been considered. This method is very useful tool as a number of properties can be studied during the construction of a reachability tree. For example, both, safeness and boundedness properties can be studied using a reachability tree. A Petri net will be defined as bounded if the symbol ω does not appear in its reachability tree. Further, the position of ω (if it appears) will represent an unbounded condition of that place. Similarly, the conservation and coverability properties can also be studied from a reachability tree. Since we have obtained a finite reachability tree using the algorithm, the weighted sum of tokens can be computed for each marking. If the sums are the same for each marking, the net is defined as conservative with respect to the given weight. If ω occurs in the covering marking, then a loop is said to occur in the path between the initial marking (or root) and covering marking. Various other properties like reachability or liveness, firing sequences, etc. cannot in general be solved by reachability tree due to existence of symbol ω.

9.5 Applications in System Reliability Evaluation

The preceding sections discussed different type of Petri nets and related graph models along with their applications. This section will be devoted towards the applications of Petri nets in the area of system reliability evaluation.

9.5.1 Reliability Analysis

The reliability analysis of any system may be carried out using a graph-theoretic approach which defines a model of the system in terms of nodes and edges where edges are weighted and nodes are unweighted. The weights assigned to edges represent the probability of success or failure. For example, if we consider a computer communication network (CCN), where the network structure of the CCN is defined as host machines connected by communication subnets. The communication subnet receives the data from a host and transmits it to the requested host. The hosts are connected to switching elements known as interface message processors (IMPs) which also define the boundary of communication subnet. The IMPs are dedicated specialized computers and can be considered as nodes of the graph while the transmission lines connecting the hosts and IMPs can be represented by edges of the graph. Each edge defines a weight in terms of the number of bits per second. The communication network reliability may be defined as the probability that a communication network will function within the defined constraints and with an acceptable level of performance.

There are several reliability measures [82], which can be defined in a communication network. Some of the measures are based on the performance measurement while the others are based on reliability measurements. The most common and widely accepted measures based on reliability measurement are *terminal pair reliability* and all *terminal reliability*. The other reliability measures are *k-terminal, k- component* connectivity, etc. The network reliability of probabilistic graph can be evaluated using the path sets, cut sets, spanning trees, Hamiltonian path sets, k-trees etc. As mentioned earlier, Petri nets are bipartite directed graphs with two connected by a set of arcs. The reachability concept of Petri nets has been used for each of the above mentioned approaches. A comprehensive treatment of reliability measures and the state-of- art in system reliability evaluation can be found in [82].

A probabilistic non-oriented graph can be represented by a Petri net, by replacing the nodes by places and arcs by transitions, respectively. The transitions fire in the direction of the arc but in case of the non-oriented graph, the transition are assumed to fire in both the directions. On firing, a transition removes a token from its place(s), which did not have a token before firing in the case of an oriented graph. Thus, a transition removes a token from each of input connected places and puts a token at each of its output connected places. Tokens may be thought of representing as data items or holding some conditions represented by the places.

9.5.2 Enumeration of Simple Path Sets

In order to find the simple path sets, it is sufficient to find which of the transitions of the corresponding graph would fire such that a marking (token distributions) which has the only token at place d (destination node) is reachable from a marking which has only token at place S (source node).

This can be obtained by solving Murata's state equations [70] of Petri nets;

$$\mathbf{A}^t = \Sigma \ \Delta M \tag{9.1}$$

where $\mathbf{A}^t$ is the transpose of *transition to place incidence matrix* of the Petri net, ΔM is the change in the marking and Σ is the firing count vector.

Equation 9.1 has been used to model as a reachability problem of Petri nets in a variety of applications [72-78]. In our case, equation (9.1) reduces to

$$\mathbf{A} = \Sigma \Delta \ M \tag{9.1a}$$

where, $\mathbf{A} \equiv [\ a_{ij}\ ]$ is a connection matrix with elements $a_{ij} = 1$ if *jth* arc is incident at its node and $a_{ij} = 0$, otherwise.

Δ M ≡ is a column vector, whose elements m_i, i=1,2,...,P are given by
$m_i = 1$ for i = S = T.
= 0, otherwise

Σ ≡ is a column vector, whose element e_i, i=1,2...,T is 0, if *jth* arc is *not* included in the path.

P and T are the number of places and transitions, respectively of the Petri net representation of the system.

A success path of length L in a graph is a non- empty sequence of L edges. As the success paths are simple, all the edges are distinct. Any L column vector of $\mathbf{A}^t$ can be added (using mod-2 addition) and if addition equals to Δ M, then indices corresponding to these added columns will provide a success path of length L. The number of iterations can be further reduced by noting that three or more columns which contain values of 1 and 0 for any row cannot be added because the addition can never be equal to Δ M. Such sets of columns for every row can be identified from $\mathbf{A}^t$. Hence only those L (L > 3)columns are considered out of which any three columns do not belong to the sets as obtained above.

The following steps are required to find all success paths between a specified pair of nodes (places) i.e., $s(p_1)$ and $t(p_2)$.

(i) Find **A** and Δ M as discussed above

(ii) From **A**, find all sets S_i, for $3 \leq i \leq N$ of columns corresponding to each row which contains values of 1 and 0.

(iii) Compare all the columns **A** with Δ M. A particular column is a success path only if the addition is equal to Δ M. The indices corresponding to this particular column form a path of length one.

(iv) Add L columns (starting from L = 2) and if the addition is equal to Δ M, the indices corresponding to these added columns will give a success path of length L.

(v) Repeat step (iv) for all possible L combinations

(vi) Increment the values of L each time by 1 until L becomes equal to N-1, and go to step (iv)

Let us consider an example as shown in Fig.9.8 (a). The Petri net representation of this graph is shown in Fig.9.8 (b)

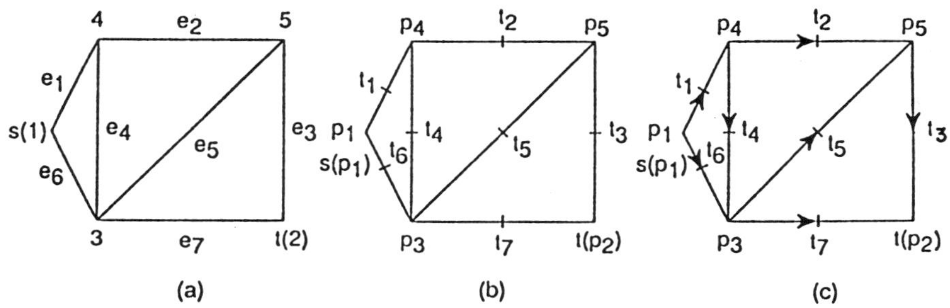

Fig 9.8 An Example of Petri Net Model

Various steps of the procedure are given as follows :

<u>Step i</u> : A and Δ M for the graph are :

$$A = \begin{array}{c} \\ p1 \\ p2 \\ p3 \\ p4 \\ p5 \end{array} \begin{array}{c} \begin{array}{ccccccc} t1 & t2 & t3 & t4 & t5 & t6 & t7 \end{array} \\ \begin{bmatrix} 1 & 0 & 0 & 0 & 0 & 1 & 0 \\ 0 & 0 & 1 & 0 & 0 & 0 & 1 \\ 0 & 0 & 0 & 1 & 1 & 1 & 1 \\ 1 & 1 & 0 & 1 & 0 & 0 & 0 \\ 0 & 1 & 1 & 0 & 1 & 0 & 0 \end{bmatrix} \end{array} ; \quad \Delta M = \begin{bmatrix} 1 \\ 1 \\ 0 \\ 0 \\ 0 \end{bmatrix}$$

<u>Step ii</u> : $S_1=(t_4,t_5,t_6,t_7), S_2=(t_1,t_2,t_4), S_3=(t_2,t_3,t_5)$.

<u>Step iii</u>: There is no success path of length 1

<u>Step iv</u>: Columns t_6 and t_7 can be added to give a success path of length 2 i.e., consisting of (e_6,e_7). There is no other success path of length 2.

<u>Steps vi and vi</u> : The other $[s(p_1),t(p_2)]$ paths, thus, obtained are; $(e_1,e_2,e_3),(e_1,e_4,e_7),(e_6,e_5,e_3),(e_1,e_4,e_5,e_3),(e_6,e_4,e_2,e_3),(e_1,e_2,e_5,e_7)$.

Alternatively, the simple paths can also be enumerated using another property of Petri nets, i.e., Conflict [73]. The conflict occurs when a place makes more than one transitions fireable simultaneously i.e., these transitions are having this place as an input place. A token in the place can either go to the output place of one transition or the transitations. Two transitions can not fire simultaneously. This situation created by a common input place is known as *conflict* and is generally resolved by Petri net itself. This property of Petri net has been used to determine all *s-t* simple paths in a directed graph. The following is a brief description of the algorithm for determining all simple *s-t* path sets, however, for details, one can refer to [73]. Consider a place as an input to two transitions where the firing of both the transitions are equiprobable. If a place is input to say m transitions, then the firing of all these m transitions are also assumed to be equiprobable.

The conflict equation (CE) of a place p_i, in general, represents the number of places which are directly connected to p_i, i.e., a token at p_i will reach any of these places after firing of any one of the transitions (connected between them) once. Similarly, the output places connected to these transitions are said to be directly connected to p_i. The CE of p_i at first level is defined by a set whose each element denotes a place directly connected to p_i and is given by

$$p_i^1 = [\, p_{i1}^{(t1)}, p_{i2}^{(t2)}, \ldots\ldots, p_{in}^{(tn)} \,] \tag{9.2}$$

$$= [\, p_{ij}^{(tj)} \,] \; ; \text{ for } j = 1, 2, \ldots, n$$

where, $p_{i1}, p_{i2}, \ldots p_{in}$ are the output places of $t_1, t_2, \ldots, t_n$, respectively and p_i is a common input place to all these transitions. p_i^1 denotes the CE of p_i at the first level.

The CE of p_i at second level can be obtained by substituting the CE (first level) of the places present in the right hand side of the above equation. This results in the term obtained by multiplying the elements of a set by their own CE (at first level),hence,the CE of p_i at second level is obtained in the form of another set as

$$p_i^2 = p_{ij} \cdot p_{im}^{(tj.tm)} \; ; \text{ for } j = 1, 2, \ldots, n \; ; m = 1, 2, \ldots, k \tag{9.3}$$

Here, each product term of the set defines an element of the set and the elements of the equation provide all possible combinations of reachable paths between p_i and the last place of each element *if only and if* the last place of the element is not p_i; as it will yield a closed path. The length of a simple path corresponding to each element of this set is two and the branches of the path correspond to the transitions associated with the elements of the equation. The path is simple if a place in an element does not occur more than once and, hence the transition will also not occur more than once in any of the elements of the equation.

Similarly, the CE of p_i at higher levels can be obtained by substituting CE (at the first level) of the place in the last position of the element of the preceding set and a new set is obtained. The expansion of the CE of p_i is continued up to the $(N\text{-}1)^{th}$ level. The elements of each of these sets (i.e., CE of p_i) at various levels will provide simple paths between p_i and the last place of each element, provided that

(i) no element contains p_i;

(ii) no place occurs more than once in any element.

In order to obtain all the simple paths between two specified places p_s and p_t (corresponding to s and t nodes), we can start with the CE of p_s at the first level and expand it up to the $(N\text{-}1)^{th}$ level and at each level the last place of each of the elements of the sets is checked for the following conditions:

(i) if the last place of an element is p_t, delete this from the set.

(ii) if a particular place occurs more than once in any of the elements, delete this element from the set.

(iii) if any element contains a place which does have any directly connected place, delete this element from the set as this element will never provide a reachable path.

(iv) if the last place of any element at any level is p_t, this gives a simple path and the corresponding element is also taken out from the set. This will not be considered for further expansion of CE at higher levels.

The CE of p_s at the $(N-2)^{th}$ level will yield all the simple paths between p_s and p_t and CE of p_s at the $(N-1)^{th}$ level will contain only the Hamiltonian paths of the graph. A Hamiltonian path from a node v_i to v_j is a simple path from v_i to v_j which passes through all nodes of the graph.

The following two properties are defined for the proposed technique and the proof and other details can be found in [73].

Property I : All simple paths and Hamiltinian paths between p_s and p_t can be obtained by expanding the CE of p_s up to the $(N-1)^{th}$ level.

Property II :The proposed technique determines all directed simple and Hamiltonian paths.

9.5.3 Enumeration of Hamiltonian Path Set

In order to obtain all Hamiltonian success paths [78] between p_1 and p_2 in the Petri net representation of Fig.9.8 (b) using (9.1a); the following steps are required:

(i) Find A and ΔM as discussed above.

(ii) Add L columns (starting from $L = N-1$) and if the addition is equal to ΔM, the indices corresponding to these added L columns will give a success path of length L (Hamiltonian path)

(iii) Repeat step (ii) for all possible combinations.

Various Hamiltonian paths between p_1 and p_2 of Fig. 9.8(b) using the technique described are given by : $t_1t_2t_5t_7$, $t_1t_4t_5t_3$, $t_6t_4t_2t_3$.

9.5.4 Enumeration of Closed Circuit

In order to find all closed circuits of various lengths [76] of Fig.9.8 (c) again using eq. (9.1a), all the steps required to determine success paths are used but for the difference in the value of ΔM. The value of ΔM is a zero column vector as in the closed circuit, it is assumed that a token starting from any place would come back to the same place after firing the transitions in a sequence leading to a closed circuit.

Since Fig. 9.8(c) represents a Petri net model of a directed graph, the matrix A will be given by :

$$A = \begin{array}{c} \\ p_1 \\ p_2 \\ p_3 \\ p_4 \\ p_5 \end{array} \begin{array}{c} \begin{array}{ccccccc} t_1 & t_2 & t_3 & t_4 & t_5 & t_6 & t_7 \end{array} \\ \begin{bmatrix} -1 & 0 & 0 & 0 & 0 & -1 & 0 \\ 0 & 0 & 1 & 0 & 0 & 0 & 1 \\ 0 & 0 & 0 & 1 & -1 & 1 & -1 \\ 1 & -1 & 0 & -1 & 0 & 0 & 0 \\ 0 & 1 & -1 & 0 & 1 & 0 & 0 \end{bmatrix} \end{array} ; \quad \Delta M = \begin{bmatrix} 0 \\ 0 \\ 0 \\ 0 \\ 0 \end{bmatrix}$$

Following the steps required to enumerate simple paths and using the following rules for addition of the entries of columns, it is observed that Fig. 9.8(c) does not contain any circuit.

$(-1) + (+1) = 0$
$(-1) + (0) = -1$
$(-1) + (0) =$ Not allowed
$(0) + (-1) = -1$

9.5.5 *Enumeration of all Trees in a Graph*

The enumeratioin of trees in a graph [77] among other areas finds its applications in computer communication system reliability evaluation. A tree of a graph is a subgraph containing all the vertices of the graph and no circuits. Using the state equation of Petri nets (9.1a), we can derive the steps for enumerating all the trees of a graph.

Here, we assume that if (p-1) distinct transitions fire in a sequence, it implies that a token has moved from a place of first transition to the place of the (p-1) edges, i.e., all the places (nodes) have been included. If the token reaches the same place after firing of these transitions, it indicates, we have determined a closed path of length (p). Thus, the enumeration of all the trees of a graph requires the determination of the firing sequence of length (p-1) which will enable the transitions in a sequence such that the token moves through all the places of the graph. Hence, it is sufficient to determine the transitions of corresponding Petri net model in such a way that the final marking and also Δ M should become unit vector. It is assumed that initial marking is a zero vector. Here, it is interpreted that the entries of Δ M simply indicates the movement of a token over all the places of the Petri net model.

A tree is a non-empty sequence of (p-1) edges. In a tree, the edges are distinct. The solutions of (9.1a) can be obtained by column operations of $\mathbf{A}^t$. Any (p-1) column vector of $\mathbf{A}^t$ can be added (using Boolean addition rules, i.e., 1+1=1, 0+1=1, 1+0=1 and 0+0=0), and if addition is equal to a unit vector, then the tree is given by the branches whose indices correspond to the added columns. All such combinations of (p-1) columns will give all the possible trees of the graph. If $\mathbf{A}^t$ contains a row with exactly one entry (corresponding to the case in a graph if only one branch is incident to this node),then the corresponding transition will be included in all trees of the graph and corresponding row and the cloumn can be deleted from $\mathbf{A}^t$ to reduce the computational effort.

In order to determine all the trees in a graph, the following steps are required to be executed:

(i) Determine $\mathbf{A}^t$

(ii) Obtain $\mathbf{A}^t_1$ by deleting the rows of $\mathbf{A}^t$ containing exactly one entry and also the corresponding column. Let p_1 be the number of rows corresponding to places in $\mathbf{A}^t$.

(iii) Add (p_1-1) columns of $\mathbf{A}^t_1$. If addition is equal to a unit vector, the indices corresponding to these added columns will give rise to a tree of the graph.

(iv) Repeat step (iii) for all possible combinations of (p_1-1) columns.

The use of Petri nets for the evaluation of 2-trees, all Tie sets and minimal cut sets between any two specified nodes can be found [76,79].

9.5.6 *Fault Tree Analysis*

The Petri net representation of fault tree and the use of certain analytical properties (through reachability concept of Petri nets) of Petri nets are shown to be useful for the fault detection also. The state equation (9.1a) provides representation of different types of logic operations as well. It gives a direct implementation of the two logic operations AND and OR gates as shown in Fig.9.9 (a) and (b).

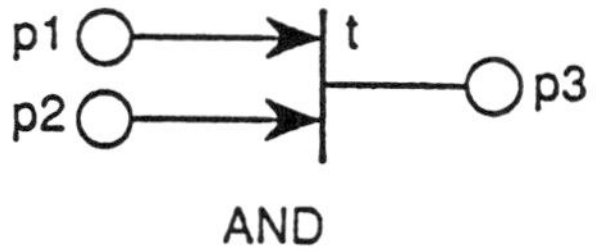

Fig. 9.9(a) Representation of AND

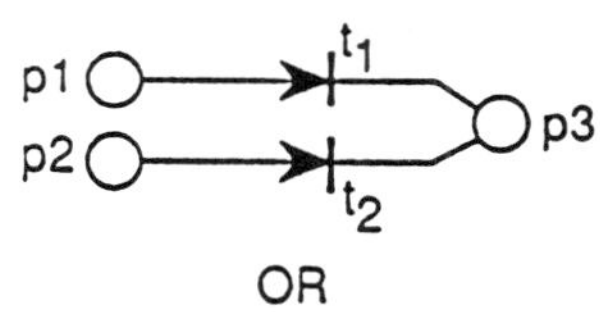

Fig. 9.9(b) Representation of OR

In Fig.9.9(a), p_1 and p_2 are the input places to t and p_3 is the output place (representing the AND operation). For the firing of t, both p_1 and p_2 should have at least one token. In other words, we should consider an initial marking M_0 as $M_0 = [1\ 1\ 0]^t$. A possible next marking for an AND operation can be given as $M_1 = [0\ 0\ 1]^t$. This marking M_1 can be obtained using (9.1) by firing t, i.e.,

$$M_1 = \begin{bmatrix} 1 \\ 1 \\ 0 \end{bmatrix} \rightarrow \begin{bmatrix} -1 \\ -1 \\ 1 \end{bmatrix} \rightarrow \begin{bmatrix} 0 \\ 0 \\ 1 \end{bmatrix}$$

In Fig. 9.9(b), p_1 and p_2 are the input places to t_1 and t_2, respectively (corresponding to an OR operation) and p_3 is the output place. The OR gate is realized through (9.1a) under any of the following three conditions :

(i) t_1 alone fires
(ii) t_2 alone fires
(iii) both t_1 and t_2 fire

In condition (iii), the p_3 can have two tokens after firing of t_1 and t_2, but for simplicity, we have assumed that the presence of token(s) at p_3 indicates the execution of the OR operation,satisfying one of the three conditions. This assumption, in fact, alters the normal computational algorithm for deciding the properties of Petri nets for representing OR logic operation.

9.5.7 Petri Net Model of Coherent Fault Trees

The Petri net model of a fault tree given in Fig. 9.10 (a) is shown in Fig.10(b). The Petri net model of fault tree is obtained by replacing the AND and OR gates by their equivalent Petri net models as discussed earlier. In the Petri net model, we replace the symbol + by OR gate while the symbol (.) is replaced by AND gate. The output of a logic gate or the last event of a tree is denoted by a rectangle, while a circle indicates a fundamental event,viz,a primary failure event of a system element. Reliability analysts are generally concerned with the evaluation of failure probability of the top event of a fault tree.

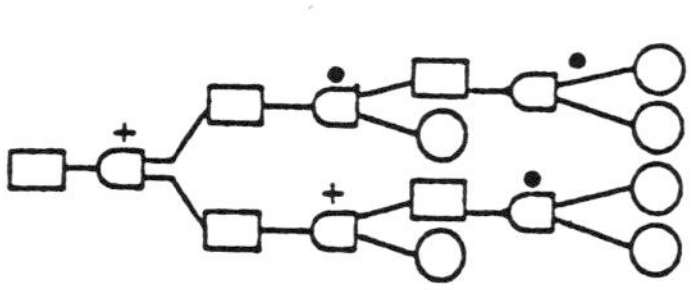

Fig. 9.10(a) An Example of Fault Tree

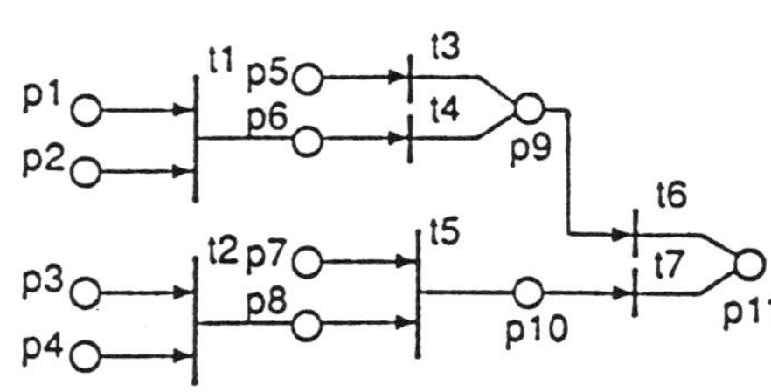

Fig. 9.10(b) Equivalent Petri-Net Representation

The Petri net model of fault tree can be interpreted as a three-level model, where the firing of fireable transitions will continue to transfer to next level until we reach the top event of the system. The firing of transitions at each level defines a new condition for the next level. In Fig 9.10 (a), the place p_{11} is the last place, where the failure probability is to be determined. From the reliability point of view, the last place p_{11} defines a condition where undesirable or unwanted event may occur.

The analysis of the modeled system can be carried out in two different ways depending upon the interpretation considered. In one approach, if the input data is given, then using this data, the function defined by the final marking (where p_{11} gets a token(s)) can be obtained through reachability concept. The failure reliability can then be obtained from this function. In the second approach, we assume some initial conditions and using these, the state of the model is analyzed at each level and the properties of interest are studied. The state of the model then becomes the initial condition for another higher level. For this condition, the states of the model at higher level is again analyzed until we reach the conditions which include final place p_{11}. The second approach seems to be advantageous as various properties can be studied at each level of the system and also at each stage, these properties can be evaluated.

Further, different types of faults, viz., stuck-at-one(s-a-1) and stuck-at-zero(s-a-0), can be detected. This provides a useful information for finding out the paths over which the faults can be propagated from the input place(s)to the last place of the fault tree which represents a combinational system. In order to explain this important feature of Petri net modeling of fault tree, we assume an initial marking as: $M_o \equiv [10111010000]^t$. As discussed above, place p_{11} is the destination place (Fig.9.10(b), therefore we have to find a marking, which has a token only at the place p_{11}, without caring for other places whether these get any token or not? In other words, our final marking should be;
$M_n = [00000000001]^t$

Here, our aim is to find all those firing sequences which will lead us from M_o to final marking M_n. The $\mathbf{A}^t$ of Fig.9.10(b) can be written as

$$
\mathbf{A}^t = \begin{array}{c} \\ p_1 \\ p_2 \\ p_3 \\ p_4 \\ p_5 \\ p_6 \\ p_7 \\ p_8 \\ p_9 \\ p_{10} \\ p_{11} \end{array}
\begin{array}{c} \begin{array}{ccccccc} t_1 & t_2 & t_3 & t_4 & t_5 & t_6 & t_7 \end{array} \\
\left[\begin{array}{ccccccc}
-1 & 0 & 0 & 0 & 0 & 1 & 0 \\
-1 & 0 & 1 & 0 & 0 & 0 & 0 \\
0 & -1 & 0 & 0 & 0 & 0 & 0 \\
0 & -1 & 0 & 0 & 0 & 0 & 0 \\
0 & 0 & -1 & 0 & 0 & 0 & 0 \\
1 & 0 & 0 & -1 & 0 & 0 & 0 \\
0 & 0 & 0 & 0 & -1 & 0 & 0 \\
0 & 1 & 0 & 0 & -1 & 0 & 0 \\
0 & 0 & 1 & 1 & 0 & -1 & 0 \\
0 & 0 & 0 & 0 & 1 & 0 & -1 \\
0 & 0 & 0 & 0 & 0 & 1 & 1
\end{array}\right] \end{array}
$$

The firing diagram indicating the firing of the transitions at different instants of time is shown in Fig.9.11.

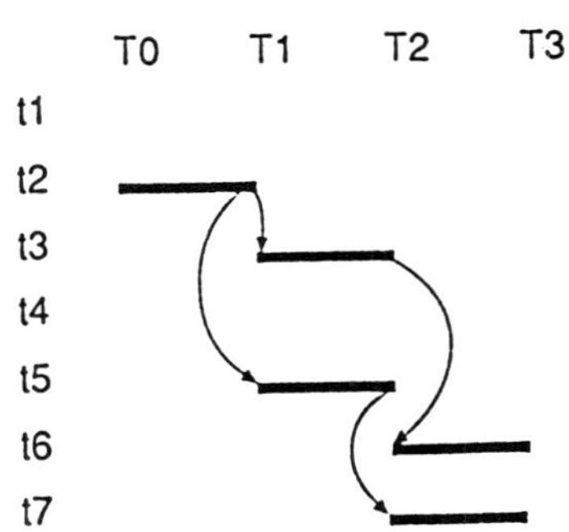

Fig. 9.11 Firing Diagram for Fault Tree

From Fig. 9.11 it is clear that the firing of t_2 at instant T_1 is directed to the firing of t_3 and t_5 at instant T_2. Different markings are obtained for the Petri net depending on whether t_3 or t_5 fires. However, after their firing, the state of the Petri net remains the same whether the firing of t_3 and t_5 occur together or separately. This diagram is very useful if there are many gates at different levels of the diagram. Using (9.1a) and the firing diagram, we get different markings starting from M_0 as

$$M_0 = \begin{bmatrix}1\\0\\1\\1\\1\\0\\1\\0\\0\\0\\0\end{bmatrix}, M_1 = \begin{bmatrix}1\\0\\0\\0\\1\\0\\1\\0\\0\\0\\0\end{bmatrix}, M_2 = \begin{bmatrix}1\\0\\0\\0\\0\\0\\1\\1\\1\\0\\0\end{bmatrix}, M_3 = \begin{bmatrix}1\\0\\0\\0\\0\\0\\0\\0\\1\\1\\0\end{bmatrix}, M_4 = \begin{bmatrix}1\\0\\0\\0\\0\\0\\0\\0\\0\\1\\1\end{bmatrix}, M_5 = \begin{bmatrix}1\\0\\0\\0\\0\\0\\0\\0\\0\\0\\1\end{bmatrix}$$

From the above markings, we can observe that M_4 or M_5 provide a token at place p_{11}. These are, therefore, the only states of interest. The firing diagram shows that the firing sequence t_2 t_5 t_7 transforms M_0 to M_5. The former deals with the transformation of a state (at first level) to the state (at the last level), while the latter deals with the transformation of a state (from any level in between) to the final state (of the last state).

9.5.8 Fault Detection

As mentioned earlier, the reachability concept derived from (9.1a) can be used for fault detection and correction. The reachability of state M_n from M_0 mainly depends on the existence of a firing sequence between these two states and if the firing sequence(s) (using (9.1a)) can be obtained between these two states, the state M_n is reachable from M_0. This condition, in fact, can be seen in equation (9.1a), where the left hand side (lhs) column vector Δ M (difference of M_n and M_0) is equal to the firing vector A^t of right hand side (rhs). The existence of firing sequence(s) between these two markings can be observed by looking at the same entries at all places of these two vectors. If we take the difference of these two vectors, we should get a zero-vector. If a zero-vector is not obtained, then non-zero entry in column vector E, where E represents the difference of lhs and rhs of equation (9.1a),has a fault at place(s) in the model. Thus, for determining the faults (s-a-1 or s- a-0), we have to look at non-zero entry of vector E and the corresponding place in the model that has a fault. Both types of faults can be considered as a similar type and as such can be detected by looking at Σ vector. After the fault is detected, it gives us the location (the place which should not have received token, but gets it). The fault detection can be obtained by considering appropriate measures to remove it from that place.

Consider a situation where we get the following marking (in Fig. 9.10(b)) after firing of t_5 as $M'_3 = [10000010110]^t$ and the actual state for fault-free condition is M_3 (as given above). The vector $E = A^t - M'_3 = [00000010000]^t$. Place p_7 has a fault, i.e.,after firing of t_5,it should be empty. Thus, in order to correct it, place should be made empty by removing the token from it. This changes the state M'_3 to M_5 after the firing of t_5.

The above was a brief discussion on how vetor E can be used to yield a fault and it can be identified after each firing of fireable transitions at different levels in the model. If at any place a fault occurs, absorb this token and change the marking by making the corresponding non-zero entry to zero. This method is applicable to detecting one fault at a time.

9.5.9 Representation of Non-Coherent Fault Tree

As shown earlier, the Petri net state (9.1a) can represent only AND and OR logic operations. This limitation was overcome by defining a generalized Petri net state equation which can not only represent AND and OR logic operations but also provides a Petri net representation of other logic operations, e.g., NOT, NAND, NOR, EX-OR and EX-NOR [71]. A brief discussion on the derivation of generalized equation will not be out of place here.

In order to derive the equation, we assume that the presence of an uncomplemented (or complemented) variable is represented by the presence (or absence) of a token. Thus, a place in the Petri net model can have at most one token. Due to some matrix operations, a place may have a token with a negative sign which means that the token can be removed physically from that place. However, a token can't be removed from a place containing no token. Thus, the following three rules in addition to normal binary addition rules must be defined

(i) $0 + (-1) = 0$
(ii) $1 + (-1) = 0$
(iii) $(-1) + (-1) = -1$

The Petri net representation of NAND logic gate contains two transitions as t' and t'' for AND and NOT operations (Fig. 9.12), respectively. The firing of these two transitions can be defined by the following generalized state equation :

$$M_{k+1} = M_k + A^{*t} D_k V_k \tag{9.4}$$

where, M_k is a pxl (p is the number of places) column vector whose element m_{jk} gives the number of tokens in place j prior to k^{th} firing. A^{*t} is the transpose of 2pxt (t is the number of transitions) modified transition-to-place incidence matrix such that $A^{*t} = [a'_{ij}\ a''_{ij}]_{px2t}$. The elements of A^{*t} are given by $a'_{ij} = -1$ if place j is an output to transition, i = 0, otherwise, $a''_{ij} = 1$, if place j is an output to transition, i = 0, otherwise. ΔK is 2tx2t diagonal matrix, called a *characteristic matrix*, whose elements d_{ik} is given by;

$$[d_{ik}] \equiv \begin{bmatrix} 1 & 0 \\ 0 & 1 - m_{oik} \end{bmatrix}$$

where, m_{oik} is the change in the number of tokens at the output place of ith transition prior to k^{th} firing, and, V_k is a 2txl column vector, whose element vector v_{ik} is given by;

$$[v_{ik}] = [v'_{ik}\ v''_{ik}]^t$$

The firing of a transition t_i is, infact, a sequence of firing of two transitions t'_i and t''_i. When $t'_i(t''_i)$ is made to fire, $v'_{ik}(v''_{ik})$ is 1, otherwise 0. Also, t'_i and t''_i are enabled to fire if

$$M_k + A^{*t} V'_k > 0$$

where, $A^{*t} = [a'_{ij}]$

$$V'_k = [v_{ik}\ v_{2k} \ldots\ v_{tk}]^t$$

$$1 - m_{oik} = 0,$$

For Fig. 9.12 (a), the order of M_k, A^{*t}, D_k and V_k are 3x1,3x2,2x2 and 2x1, respectively and m_{oik} equals $(1 - m_{Ck} = m_{Ak}.m_{Bk})$. Combining the equations for firing of the transitions, a generalized state equation (9.4) holds good with the value m_{oik} given by

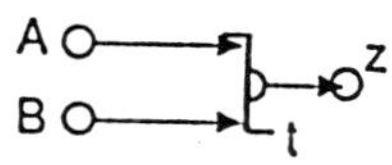

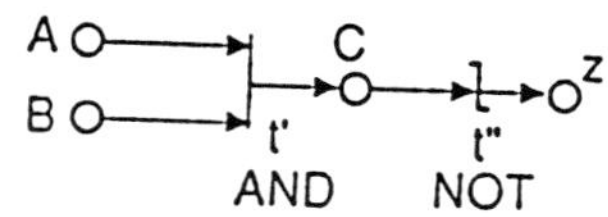

Fig. 9.12 (a) NAND Operation **Fig. 9.12 (b) Expanded Form of Fig.12(a)**

$m_{oik} = m_{oik} - \Pi\ m_{lik}$, where, $m_{oik}(m_{lik})$ is the number of tokens on output place (*lth* input place) in *ith* transition, and Π is a binary multiplication.

Similarly, (9.4) can be used to represent other logic operations e.g., NOR, EX-OR, EX-NOR, NOT. However, m_{oik} and firing rules are different for different operations and can be seen in [71].

Using the concept of fault detection as discussed earlier for coherent fault tree, we can use (9.4) for representing various logic operations of a non-coherent fault tree as well and also for the fault detection.

The problem of fault detection and propagation of different types of faults are also addressed during the modeling of fault trees. It is claimed that this type of treatment provides more insight into behaviour and is better than the conventional representation methods [80].

9.6 Future Trends

It can be seen that Petri nets and related graph models are not only useful modeling a system but also allow a designer to employ them for the verification of communication protocols, reliability evaluation, fault-tree analysis and many other issues related to system reliability evaluation. The future extensions in this area seem to be in the direction of the development of reliability software models using computer graphic facilities, in particular, the development of unified framework for solving reliability fault-tolerance problems, and other performance analysis parameters. Further, the ability of Petri nets to model concurrency may help in parallel computations.

Acknowledgement

The author would like to thank to his sons, Devendra Singh and Mandeep Singh for their patience and cooperation during the preparation of the manuscript and to the editor Prof. K.B. Misra, for his painstaking efforts in bringing the manuscript to present form.

REFERENCES

[1] Petri C.A., *Kommunikation mit Automaten*, Schrifleir des Rheinisch-West Falischen Institutes fur Instrumentelle Mathematik an der Universitat Bonn, Heft 2, Bonn, West Germany, 1962. Translation : *c.f.* Greene, Supplement 1 to Tech. Rept. RADC-Tr., Vol. 1, pp. 336, 1965, Rome Air Development Centre, Griffiss Air Force Base, New York, pp. 89, 1965.

[2] Petri C.A., *Concepts of Net Theory*, Proc. Symposium and Summer School Foundations of Computer Sc., High Tatras, Sept. 1973, Maltis, Inst. Science, pp. 137-146, 1973.

[3] Holt A., R. Shapiro and S. Warshall, *Final Report of the Information System Theory Project*, Tech. Report. RADCTR, p. 305, 1968, Rome Air Development Centre, Griffiss Air Force Base, New York, Sept., p. 352, 1968.

[4] Holt A. and F. Commoner, *Events and Conditions* (in three parts), Applied Data Research Inc., N.Y., 1970.

[5] Dennis J.B., *Concurrency in Software Systems*, Computation Structures Group Memo 65-1, Project MAC, MIT. Cambridge, p. 18, June 1972.

[6] Peterson J.L., *Petri Nets*, Computing Surveys, Vol. 9, No. 3, pp. 224-252, Sept. 1977.

[7] Agerwala T., *Putting Petri Nets to Work*, Computer (IEEE), Vol. 12, No. 12, pp. 85-94, Dec. 1979.

[8] Hura G.S., *Petri Net as a Modeling Tool*, Microelectronics and Reliability, Vol. 22, No. 3, pp. 433-439, 1982.

[9] Hura G.S., *Petri Nets*, IEEE Potentials, pp. 18-20, Oct. 1987.

[10] Hura G.S., *Petri Net Applications*, IEEE Potential, pp. 25- 27, Oct. 1987.

[11] Murata T., *Petri Nets : Properties, Analysis and Applications*, Proc. IEEE Vol. 77, No. 4, pp. 541-580, 1989.

[12] Molly M.K., T. Murata and M.K. Vernon, Guest Editors, Special Section on *Petri Net Performance Models* (PNPM), IEEE Trans. on Software Engg., Vol. 15, No. 4, pp. 365-401, April, 1989.

[13] Hura G.S., Guest Editor, Special Issue on *Petri Nets and Related Graph Models : Past, Present and Future*, Microelectronics and Reliability, Vol. 31, No. 4, pp. 559-816, 1991.

[14] Peterson J.L., *Petri Net Theory and Modeling of Systems*. Prentice-Hall, Inc., Englewood Cliffs, N.J., 1981.

[15] Starke P.H., *Petri-Netze*, VEB Deutscher Verlag der Wissenschaften, Berlin (DDR), 1981.

[16] Winand V. and B. Rosenstengel, *Petri Netze*, Eine Anwendungsorientierte Einfuhrnng, Vie weg-Verlag Braunschweig, 1981.

[17] Brams G.W. (Nom collective) : *Reseaux de Petri, Theorie et Pratique*, Masson (Two volumes). Editeur 120 Boulevard Saint-Germain, 75280 Paris Cedex 06 ISMN2-903-60712-5, 1982.

[18] *Bibliography of Net Theory*, Project PRIAMOS, Institute for Foundations of Information Technology at GMD, June 1986.

[19] *Newsletter of the Special Group, Petri Nets and Related System Models*, Gesellschaft fur Informatik (Computer Science Society in Germany), Bonn, Germany, ISSNO 173-7673.

[20] Hura G.S. and J.W. Atwood, *A Bibliiography of Petri Net and Related Graph Models : Theory and Application, 1979-1985*, Tech. Rep. WSU-CS-3, p. 37, 1986.

[21] Jensen K., P. Huber, M.N. Larsen and I.M. Martinsen, *Petri Net Package User's Manual*, Aarhus Univ., Computer Science Department, DIAMI-MD-46, Version 2.1, Sept. 1983.

[22] Feldbrugge F. and K. Jensen, *Petri Net Tool Overview*, ed. Brauer, Reisig, Rozenberg, Advances in Petri Nets, Lecture Notes in Computer Sciences, 254, 256, Springer-Verlag, Heidelberg, pp. 20-61, 1986.

[23] *Design TM Users Guide*, Meta Software Corporation, Cambridge, MA, USA

[24] Petri C.A., *Interpretations of Net Theory, Gesselschaft fur Mathematik und Datenverarbeitung, Bonn, St. Augustin, Interner Bericht ISF-75-07, 1975.*

[25] Genrich H.J. and Thieler-Mevisseng, *The Calculus of facts*, Mathematical Foundtions of Computer Science, A.Mazurkieuicz (Ed). Lecture Notes in Computer Science, 45, (Springer, Verlag, Berlin), 1976.

[26] Jensen K., *Coloured Petri Net and the Invariant Method*, Theoretical Computer Science, Vol. 14, pp. 317-336, 1981. [27]

[27] Shapiro R.M., *Towards a Design Methodology for rechnergestutzter Information Systeme*, Gessellschaft fur Mathematik and Datenverarbeitung, Bonn, St. Augustin, Bericht, Nr. 111, 1979.

[28] Nieters H., *Darstellung einiger Aspekte von Programmiersprachen mit Transitions-und Factnetzen*, Diplomarbeit, Abteilung fur Informatik, Universitat, Bonn, 1978.

[29] Reisig W., *Petri Nets*, Springer-Verlag, Heidelberg, 1985.

[30] Genrich H.J. and K. Lautenbach, *System Modeling with High Level Petri Nets*, Theoretical Computer Science, Vol. 13, pp. 109- 136, 1981.

[31] Jensen K., *High-Level Petri Nets*, Springer-Verlag, Heidelberg, pp. 166- 180, 1983.

[32] Reisig W., *Petri Nets with Individual Tokens, Application and Theory of Petri Nets*, Springer-Verlag, Heidelberg, pp. 229-249, 1983.

[33] Ramchandani C., *Analysis of Asynchronous Systems by Petri Nets*, Ph.D. Dissertation, Project MAC, MIT Rep. TR-120, 1973.

[34] Sifakis, J., *Use of Timed Petri Nets for Performance Evaluation*, 3rd Int. Symp. Measuring, Modeling and Evaluating Computer System, Beilner and Gelenbe Eds., pp. 75-95, 1977.

[35] Merlin P., *A Methodology for the Design and Implementation of Communication Protocols*, IEEE Trans. on Communication, Vol. COM-24, No. 6, pp. 614-621, 1976.

[36] Ramamoorthy C.V. and G.S. Ho, *Performance Evaluation of Asynchronous Systems Using Petri Nets*, IEEE Trans. on Software Engg., Vol. SE-6, pp. 440-449, Sept. 1980.

[37] Coolahan J.E. and N. Roussopoulos, *Timing Requirements for Time Driven Systems Using Augmented Petri Nets*, IEEE Trans. on Software Engg., Vol. SE-9, pp. 603-614, Sept. 1983.

[38] Stotts P.D. Jr. and T.W. Pratt, *Hierarchical Modeling of Software Systems with Timed Petri Nets*, Proc. Int. Workshop on Petri Nets, Torino, Italy, July 1985.

[39] Zuberek W.M., *Timed Petri Nets and Preliminary Performance Evaluation*, Proc. 7th Annual Symp. Computer Architecture, pp. 88- 96, 1980.

[40] Razouk R.R.. and C.V. Phelps, *Performance Analysis Using Timed Petri Nets*, Proc. 1984, Int. Conf. on Parallel Processing, pp. 126-129, Aug. 1984.

[41] Zuberek W.M., *Application of Timed Petri Nets to Analysis of Multi Processor Realizations of Digital Filters*, Proc. 25th Symp. on Circuits and Systems, Houston, MI., Aug. 1982.

[42] Zuberek W.M., *Performance Evaluation of Concurrent System Using Timed Petri Nets*, 13th Annual ACM Conference Science, New Orleans, LA, pp. 12-14, March 1985.

[43] Menasche M. and B. Berthomien, *Timed Petri Nets for Analyzing and Verifying Time Dependent Protocols*, 3rd Int. Workshop on Protocol, Specification Testing and Verification, Zurich, pp. 161-172, June 1983.

[44] Ajmone M. M. and V. Signore, *Timed Petri Net Performance Models of Fibre Optics LAN Architecture*, Int. Workshop on Petri Net and Performance Models, Madison, Wisconsin, pp. 66-74, Aug. 1987.

[45] Roux J.L. and B. Berthomie, *Verification of a Local Area Network with TINA.A Software Package for Time Petri Nets*, 7th European Workshop on Application and Theory of Petri Nets, Oxford, June 1986.

[46] Holliday M.A. and M.A. Vernon, *A Generalized Timed Petri Net Model for Performance Analysis*, Proc. Int. Workshop on Timed Petri Nets, Torino, Italy, pp. 181-190, 1985.

[47] Holliday M.A. and M.K. Vernon, *A Generalized Timed Petri Net Model for Performance Analysis*, Tech. Report No. 593, Computer Science Department, UW, Madison, May 1985.

[48] Holliday M.A. and M.A. Vernon, *The GTPN Analyzer : Numerical Methods and User Interfaces*, Tech. Report No. 639, Computer Science, UW, Madison, 1986.

[49] Roux J.L. and G. Juanole, *Functional and Performance Analysis Using Extended Petri Nets*, Int. Workshop on Petri Nets and Performance Models. Madison, Wisconsin, pp. 14-23, Aug. 24-26, 1987.

[50] Chen J.C. and G.S. Hura, *Design Solution for Load Balancing Problem in Distributed System Using Generalized Timed Petri Nets*, Proc. Annual Pittsburgh Conf. on Modeling and Simulation, May 1989.

[51] Molloy M.K., *Performance Analysis Using Stochastic Petri Nets*, IEEE Trans. on Computer, Vol. C-31, No. 9, pp. 913-917, 1982.

[52] Dugan J.B., K.S. Trivedi, R.M. Geist and V.F. Nicola, *Extended Stochastic Petri Nets : Application and Analysis Performance*, Paris, pp. 507-519, Dec. 1984.

[53] Natkin S.O., *Les Roseaux de Petri Stochastiques et leur Application a L'Evaluation des Systems Informatique*, Doctoral Thesis Conseratoire National des Arts et Metiers, 1980.

[54] Symons F.J.W., *Introduction to Numerical Petri Nets, A General Graphical Model of Concurrent Processing System*, Australian Telecommunication Research, Vol. 14, No. 1, pp. 28-32, 1980.

[55] Haas P. and G. Shedler, *Regenerative Simulation of Stochastic Petri Nets*, Int. Workshop on Timed Petri Nets, Torino, Italy, July 1985.

[56] Ajmone M., A. Marsan, G. Balbo and G. Conte, *A Class of Generalized Stochastic Petri Nets for the Performance Evaluation of Multiprocessor Systems*, ACM Trans. on Computer System 2, pp. 93-122, May 1984.

[57] Marsan M.A., G. Balbo, A. Bobbio, G. Chiola, G. Conte and A. Cumani, *On Petri Nets with Stochastic Timing*, Proc. Int. Workshop on Timed Petri Net IEEE-CS Press, Torino, Italy, July 1985.

[58] Marsan A., G. Balbo, G. Chiola and G. Conte, *Generalized Stochastic Petri Nets Revisited : Random Switches and Priorities*, Int. Workshop on Petri Nets and Performance Model, (IEEE Computer Society), Madison, Wisconsin, USA, pp. 44-53, 1987.

[59] Ammar H.H. and S.M.R. Islam, *Time Scale Decomposition of a Class of Generalized Stochastic Petri Net Models*, IEEE Trans. on Software Engineering, Vol. 15, No. 6, pp. 809-820, June 1989.

[60] Marsan M., G. Balbo, G. Ciardo and G. Conte, *A Software Tool for the Automatic Analysis of Generalized Stochastic Petri Net Models*, Proc. International Conference on Modeling Techniques and Tools for Performànce Analysis, Paris, May 1984.

[61] Bobbio A. and A. Cumani, *Discrete State Stochastic Systems with Phase Type Distributed Transition Time*, Proc. AMSE Int. Conf. on Modeling and Simulation, Atheuens, 1984.

[62] Dugan J.B., A. Bobbio, G. Cairdo and K. Trivedi, *The design of a Unified Package for the Solution of Stochastic Petri Net Models*, Proc. Int. Workshop on Timed Petri Nets, Torino, Italy, July 1985.

[63] *DEEP (The Duke ESPN Evaluation Package)*, Duke Univ., 1984.

[64] Meyer J.F., A. Movaghar and W.H. Sanders, *Stochastic Activity Networks : Structure Behaviour and Applications*, Int. Workshop of Timed Petri Nets, Italy, 1985.

[65] Movaghar A. and J.F. Meyer, *Performability Modeling with Stochastic Activity Networks*, Proc. 1984 Real Time System Symposium, Austin, TX., Dec. 1984.

[66] Molloy M.K., *Structurally bounded Stochastic Petri Nets*, Proc. International Workshop on Petri Nets and Performance Models, Madison, Wisconsin, pp. 156-163, Aug. 1987.

[67] Vernon M., J. Zahorjan and E.D. Lazowska, *A Comparison of Performance Petri Nets and Queueing Network Models*, Proc. Int. Workshop on Modeling Techniques and Performance Evaluation, Paris, 1987.

[68] Pagnow A., *Stochastic Nets and Performance Evaluation, LNCS, Vol. 255, Springer-Verlag, Heidelberg, 1987.*

[69] Hura G.S. and F.S. Etessami, *Petri-Nets and Various Extended Net Types : A Comparative Study*, Tech. Report, WSU-CS-88, pp. 51, 1981.

[70] Murata T., *State Equation, Controllability and Maximal Matching of Petri Nets*, IEEE Trans. Automatic Control, Vol. 22, pp. 412-416, 1977.

[71] Khan A.A., G.S. Hura and et al., *State Equation Representation of Logic Operations Through a Petri Net*, Proc. IEEE Vol. 69, No. 4, pp. 485-487, 1981.

[72] Hura G.S., *A Petri Net Approach to Enumerate all System Success Paths for Reliability Evaluation of a Complex System*, Microelectronics and Reliability, Vol. 22, No. 3, pp. 427-429, 1982.

[73] Hura G.S., *Enumeration of all Simple Paths in a Directed Graph Using Petri Net : A Systematic Approach*, Microelectronics and Reliability, Vol. 23, No. 1, pp. 157-159, 1983.

[74] Hura G.S., *Determination of all Hamiltonian Paths and Circuits and Minimum Feedback Set in a Graph Through Petri Nets*, Microelectronics and Reliability, Vol. 27, No. 2, pp. 223-227, 1987.

[75] Hura G.S., *A Petri Net Approach to Enumerating all Circuits of a Graph*, Microelectronics and Reliability, Vol. 24, No. 3, pp.387-389, 1984.

[76] Hura G.S., *On the Use of Petri Nets for the Enumeration of all Trees in a Graph*, Reliability Engineering, Vol. 7, pp. 229- 233, 1984.

[77] Hura G.S., *Enumeration of all 2-Trees in a Graph Through Petri Nets*, Microelectronics and Reliability, Vol. 23, No. 5, pp. 851-853, 1983.

[78] Hura G.S., *On the Determination of all Tie Sets and Minimal Cut set Between any Two Nodes of a Graph Through Petri Nets*, Microelectronics and Reliability, Vol. 23, No. 3, pp. 471-475, 1983.

[79] Kontoleon J. and D. Mandaltsis, *An Improved Method of Enumerating all Trees and Minimal Paths in a Graph Using Petri Nets*, Microelectronics and Reliability, Vol. 28, No. 5, pp. 783- 792, 1988.

[80] Hura G.S. and J.W. Atwood, *The Use of Petri Nets to Analyze Coherent Fault Trees*, IEEE Trans. on Rel., Vol. R-37, No. 5, pp. 469-474, Dec. 1988.

[81] Mandaltsis D. and J. Kontoleon, *Enumeration of k-trees and Their Applications to Reliability Evaluation of Communication Networks*, Microelectrics and Reliability, Vol. 29, No. 5, pp. 733- 735, 1989.

[82] Misra K.B., *Reliability Analysis and Prediction (A Methodology Oriented Treatment)*, Elsevier Science Publishers, Amsterdam, 1992.

[83] Etessami F.S. and G.S. Hura, *Rule-based Design Methodology for Solving Control Problems*, IEEE Trans. on Software Engineering, Vol. SE-17, No. 3, pp. 274-282, March 1991.

New Trends in System Reliability Evaluation
K.B. Misra (Editor)
1993 Elsevier Science Publishers B.V.

CHAPTER 10

HUMAN RELIABILITY MODELING

Joseph Sharit
Department of Industrial Engineering
State University of New York at Bufffalo, USA

ABSTRACT

This chapter discusses and critically examines a number of techniques for assessing human reliability, including methods that depend on the quantification of human error as well as approaches that emphasize human cognitive processing. The chapter concludes with an attempt at synthesis in which a number of issues potentially relevant to the problem of predicting human error are raised and briefly discussed.

10.1 Introduction

In this chapter, human reliability modeling is viewed as encompassing methods comprising what has come to be referred to as "human reliability analysis" (HRA), as well as a number of ideas and frameworks concerned with understanding human error. While it would appear that the use of HRA methods would depend on having an understanding of the underlying cognitive mechanisms that predispose the human to error, these two perspectives have been largely decoupled—a troubling circumstance in light of the difficult challenges associated with the problem of predicting human error that lie ahead.

The chapter consists of three major sections. The first section discusses some of the better known methods of HRA. These methods all have in common the goal of assessing the probability of human error, and in some cases have been motivated by regulatory agencies that demand information in this form in order to carry out probabilistic risk assessments (PRAs) that seek to determine the contribution of human error to the potential for system failure.

In the second section, the discussion reflects a shift in emphasis towards understanding the basis for human error, or more appropriately, those human activities that are capable of bringing about potentially harmful consequences due to the situational context and human cognitive processing tendencies. The approaches discussed in this section represent qualitative modeling techniques, and differ from many schemes that use cognitive processes primarily as a basis for classifying human errors.

The last section critically analyzes the techniques and approaches discussed in the previous two sections, and provides an attempt at synthesis by raising a number of possibly relevant issues we may need to turn our attention to if we intend to improve our ability to predict human error.

10.2 Quantitative Modeling of Human Error

10.2.1 THERP

The HRA technique known as THERP (Technique for Human Error Rate Prediction) has had a relatively long history and, at the time of this writing, continues to evolve. The technique grew out of a need to quantify human error in nuclear weapon systems in order to assess overall system reliability as a function of human and machine components [1]. Quantification of human reliability was viewed as a means by which human factors specialists could exercise a measure of control over a number of design-related decisions that could impact human performance. In essence, it was a conciliatory gesture to system and reliability engineers who were responsible for these decisions and who were willing to accept human factors input, but only on condition that it conform to the reliability engineering methods at their disposal. The challenge for these human factors specialists was to develop a scheme for quantifying human reliability that was sufficient in characterizing the complexity surrounding human task performance while remaining faithful to the spirit of conventional reliability methods, thereby accommodating machine as well as human components in the assessment of system reliability.

The mechanics of THERP are detailed in a work by Swain and Guttmann [2] sponsored by the U.S. Nuclear Regulatory Commission and heavily influenced by nuclear power plant (NPP) operations. In principle, however, the technique was intended to be applicable to a wide variety of industrial situations, including process control, maintenance, and manufacturing operations. As applied to NPP operations, its main functions were to use the outputs derived from HRA event trees as inputs for the analysis of system reliability (e.g., through the use of a system fault tree), and for aiding decisions involving design tradeoffs by evaluating the contribution of human error to component or subsystem reliability. The basic steps comprising THERP are outlined in Figure 10.1. In actuality, there could be any of a number of feedback loops in operation when carrying out this procedure. However, for illustrative purposes, these steps will be summarized below in the order depicted in the diagram.

The first two steps serve to identify human error-related events and the human tasks associated with these events for which HRAs will need to be performed and ultimately incorporated into PRAs. Steps 3-5 comprise the qualitative assessment stage. The analyst first performs walk-throughs and talk-throughs (interviews) to determine the "boundary conditions" under which the tasks are performed, such as time and skill requirements, alerting cues, and recovery factors. Task analysis is then performed whereby each task is decomposed into a sequence of "unit" activities. The process of walk-throughs and interviews can be repeated, although at this stage its purpose is primarily for assessing the potential for human errors associated with the individual activities. Towards this end, operators are encouraged to talk through hypothetical, yet realistic, work scenarios.

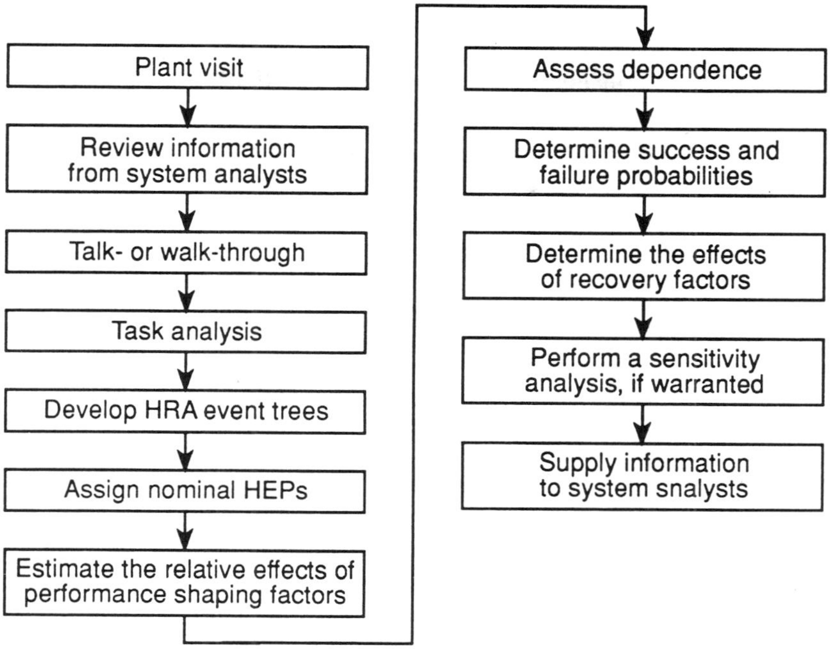

Fig. 10.1 The Steps Comprising THERP (from Ref. 1)

The relevant human task elements are then mapped into an event tree; the form of event tree preferred by Swain and Guttmann [2] is called a probability tree. Each relevant task element in this type of tree is characterized by two limbs representing either correct or incorrect performance as illustrated in Figure 10.2. In this hypothetical task, a worker checks the calibration of a series of set points consisting of three comparators that operate by OR logic to detect abnormal pressure of a plant stream. This task requires that test equipment be set up correctly; an uncorrected error would result in miscalibration of all three comparators and consequently a failure (denoted by F_1 and F_2). Note that path *a* was arbitrarily designated a success path (S_1) under the assumption that if the test equipment was set up correctly, adopting a conservative estimate of 10^{-2} for the probability of miscalibration would result in a probability of 10^{-6} of miscalibrating all three comparators, a value considered negligible in most PRAs.

The next five steps constitute the quantitative assessment stage. First, human error probabilities (HEPs) are assigned to each of the limbs of the tree corresponding to incorrect performance. These probabilities are referred to as *nominal* HEPs and represent the medians of lognormal distributions, implying that if one could observe a worker perform a particular activity a large number of times, the distribution of HEPs across many workers who perform this task would be lognormal. Associated with each nominal

HEP are upper and lower uncertainty bounds (UCBs). The square root of the ratio of the upper to the lower UCB defines the error factor, the choice of which essentially reflects the variability one associates with any given error distribution. For example, if the ratio of the *95th* to *5th* percentiles of the lognormal distribution of HEPs for a task is *4*, the standard deviation of this distribution would be approximately *0.42*. For NPP operations, larger error factors are generally assumed, reflecting variance arising from the assignment of nominal HEPs in addition to the variance associated with individual differences in worker performance.

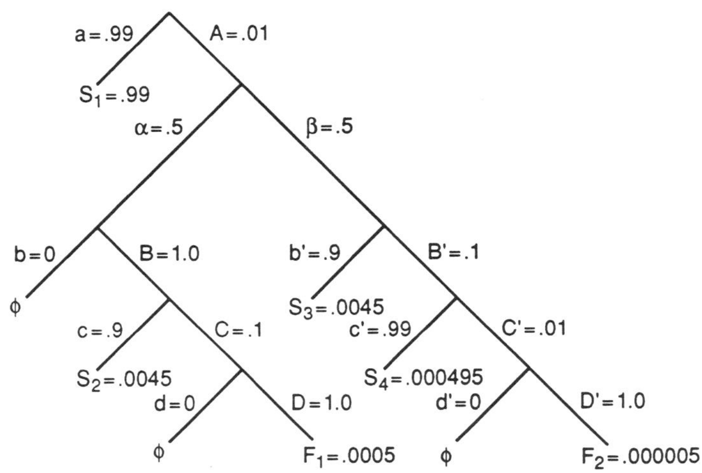

A=Failure to set up test equipment correctly

α = Small miscalibration of test equipment

B = For a small miscalibration, failure to detect miscalibration for first setpoint

C = For small miscalibration, failure to detect miscalibration for second setpoint

D = For a small miscalibration, failure to detect miscalibration for third setpoint

β = Large miscalibration of test equipment

B' = For a large miscalibration, failure to detect miscalibration for first setpoint

C' = For a large miscalibration, failure to detect miscalibration for second setpoint

D' = For a large miscalibration, failure to detect miscalibration for third setpoint

Fig. 10.2 An HRA Event Tree Representing Hypothetical Calibration Tasks (from Ref. 2)

Nominal HEPs and their associated UCBs are provided in [2] for a wide variety of NPP tasks. In principle, these estimates should be based on data derived from the actual performance of these tasks. However, the scarcity of such data requires that these estimates be derived from other sources. The most common source is expert judgment, although data from NPP simulators, process industries, or other jobs similar in psychological content to NPP operations is, in many instances, considered valid.

These nominal HEPs are then modified through a series of refinements that overall intend to capture the complexity of effects associated with human performance. First, the effects of performance shaping factors (PSFs) are estimated. These factors include the effects of work breaks, work methods, organizational structure, perceptual and decision making requirements associated with the task or equipment, team structure, psychological and physiological stressors, and psychosocial considerations—essentially, anything that could conceivably impact human performance. The modification of nominal HEPs on the basis of these considerations results in *basic* HEPs (BHEPs). For example, consider the level of stress prevalent during task performance (either very low, optimum, moderately high, or extremely high), worker experience, and task characteristics. Following the guidelines provided in [2] for modifying nominal HEPs and their associated UCBs, if moderately high stress is assumed and the task is performed in a stepwise (as opposed to continuous) manner by novices, then the nominal HEPs and UCBs are all multiplied by a factor of four; for experienced workers this factor would be two. Although stress in actuality is a continuous variable, in order to ensure the practicality of this analytical assessment technique continuous variables such as stress are typically discretized.

At this point, relationships between task elements, i.e., between adjacent limbs of the tree, are considered. This is accomplished by way of a dependency model where once again a continuous variable is discretized. In this case the variable is dependence, and the five levels considered are: zero (ZD), low (LD), medium (MD), high (HD), and complete (CD) dependence. The dependency model is nonlinear, and considers only positive dependence whereby failure on a task element increases and success on a task element decreases the probability of failure and success, respectively, associated with the subsequent task element. For example, assume the BHEP for some task element B is 10^{-2}, and assume HD between task elements A and B. The *conditional* HEP (CHEP) of B given incorrect performance on A would then be given by the equation: $CHEP = (1 + BHEP)/2 \sim 0.50$. Corresponding equations are given for computing the CHEP under LD and MD conditions; for ZD and CD, the CHEP reduces to the BHEP and 1, respectively.

Note that this dependency model considers effects spanning at most two task elements at a time. It should also be noted that instances of negative dependence can be handled, but would require the discretion of the analyst; no formal dependency model exists as for the use of positive dependence. In the hypothetical event tree presented earlier (Figure 10.2), for the case of a small set up error (α), CD was assumed between the set up task and the calibration of the first set point, as well as between the tasks of calibrating the second and third set points. The *.90* probability that the operator will be alerted by the misalignment of the second setup and therefore recheck the test setup represents an instance of negative dependence. In general, for most PRA applications the assumption of ZD as opposed to negative dependence will lead to more conservative estimates of HEPs.

Referring to Figure 10.1, success and failure probabilities for the entire task can now be computed by propagating the CHEPs associated with each limb of the event tree. Various approaches can be taken depending on the purpose of the HRA. In the simplest case, the point estimates from each limb are propagated through the event tree to arrive at the probability of failure of the entire task. In the same manner, best-case and worst-case analyses can also be performed, by propagating the lower and upper UCBs, respectively. An alternative approach for deriving upper and lower bounds is to propagate the individual point estimates (CHEPs) in order to arrive at the probability of task failure, and then assign UCBs to this total failure probability. Finally, Monte Carlo Simulation or other methods [2] can be used to evaluate the variability associated with the combinations of events comprising the HRA event tree, and thereby provide UCBs for PRAs that demand more rigorous conformance to the distributional characteristics of errors.

Following this step, the effects of recovery factors are considered. The types of recovery factors typically considered in THERP consist of: (1) alerting the operator to errors that have occurred (e.g., failure to notice a rise in temperature) through the presence of annunciators; (2) human redundancy, whereby the human's error may be caught by someone else in the work group; and (3) various types of inspections that are scheduled to occur in the workplace. In the case of recovery brought about through other workers, the dependency model could be applied to reflect the interrelationships between coworkers. As an example, consider a situation in which two operators are in a control room, with one of the operators performing a monitoring activity that is based on following a set of written procedures that is long and has no checkoff provisions. The nominal HEP for this situation is given as *0.01 (0.005-0.05)*. This value is modified by the PSF of stress to *0.02* (by virtue of it being a discrete task performed by experienced personnel). This BHEP can then be modified by the recovery factor of human redundancy, which can be assessed by assuming HD between the two operators. The new BHEP, representing the error associated with both operators, is $[0.02\ ((1 + 0.02)/2)] = 0.0102$. Recovery factors such as annunciators simply extend the relevant failure limb into two additional (success and failure) limbs, with the probability of being alerted to the annunciator feeding back into the success path of the event tree.

The final phase of the THERP procedure addresses the potential need for a sensitivity analysis, and the incorporation of the HRA assessment into systems risk assessments. One approach to sensitivity analysis is to identify the most probable errors on the event tree and determine the degree to which design modifications associated with those task elements affect the total success or failure probabilities. Sensitivity analysis can also be used to test the effects of a variety of assumptions, such as whether an assumption of MD or HD between two task elements has an appreciable effect on the total failure probability, or whether the assumption of a lognormal distribution for error probabilities need not be maintained.

In general, the tone in stepping the reader through THERP is apologetic and noncommittal. The problems encountered are consistently attributed to the lack of existing data and the absence of more rigorous performance models, the same grievances voiced by Swain [1] many years earlier. Consequently, the approaches taken to cognitive tasks such as diagnosis and decision making seem oversimplified, and the analyst is frequently advised to contribute his or her own probability assessments in cases where guidance cannot be provided with any reasonable degree of confidence. Interestingly, though methods such as THERP tend to be cast as shallow in terms of the insight they provide into the underlying causes of human error, many of the guidelines offered for carrying out THERP, such as those provided for assessing dependence, appear to be potentially more useful than the quantitative end products they were intended to produce.

10.2.2 Embellishments to THERP

The current trend among a number of human reliability analysts embracing or otherwise committed to PRAs is to accept the premise and framework of THERP but acknowledge the need to identify its most obvious shortcomings, and through various embellishments provide prescriptions for their correction. According to these investigators, the most striking deficiency has been its inability to represent what they call "errors of commission", but that are in actuality errors that Reason [3] would call mistakes or "errors in the intention." In any case, it is the more cognitively-based types of errors involving diagnosis, decision making, and intentional violations of procedures that these investigators have in mind. Three recent contributions to this problem are discussed below and address, respectfully, methods for identifying, modeling, and quantifying such errors.

One approach to the identification of errors of commission is referred to as Sneak Analysis [4]. It is based on Sneak Circuit Analysis, a method used to identify faults in electrical circuits. As applied to analysis of human error, this method is believed to be very effective in identifying latent errors that might arise from factors such as poor design, faulty maintenance, or poor management control. A prerequisite for its application is that the system's operations be flow oriented; in the case of NPP operations, this flow arises primarily from stepwise procedures.

Initially, a detailed description of the system to be analyzed needs to be provided, from which flow diagrams associated with the activities of interest are produced that guide the analyst through questions needed to identify four types of sneak conditions: (1) sneak paths, in which actions (i.e., flows) other than those desired can be taken; (2) sneak indications, in which incorrect or ambiguous output information can induce inappropriate human actions; (3) sneak labels, which are essentially the same as sneak indications but addresses control (i.e., inputs) rather than output sources; and (4) sneak timing, in which actions can be taken at inappropriate times. These analyses are carried out by a computer-interactive Sneak Analysis Tool that translates each of the flow diagrams into a series of displays containing yes/no questions posed to the analyst. Clues are provided along the way that serve as aids for thorough identification of sneak conditions. For instance, in performing a sneak path analysis, use of this tool can uncover the possibility of a sneak path in terms of whether adequate protective barriers have been provided.

The concept of sneak analysis as applied to human error in complex industrial systems has been previously addressed by Pedersen [5] within a larger systems context in which initiating events can derive from the combination of temporally staggered human and hardware precursor events. Also considered was the need for developing system design strategies that seek to minimize the consequences of such initiating events. In contrast, the Sneak Analysis Tool emphasizes identification of sneak conditions for the purpose of providing inputs to PRAs, although it can be regarded as a more general HRA tool.

As an illustration of its use in PRAs, Blackman [6] has proposed using the "errors of commission" predicted from sneak analysis as starting points (i.e., top events) in event trees referred to as COMETs (commission event trees). COMETs differ from conventional event trees in that the remainder of the sequence following the top event pertains to carrying out the error; a "success path" therefore represents a path leading to an undesirable state. Each failure event would then represent activities which intentionally or unintentionally constitute recovery from the error. These event trees, which are essentially modeling the cascading of errors, are therefore potentially well-suited for depicting the influence of well-learned behavior patterns in preventing recovery from the initiating event. If the probability of the error of commission identified by sneak analysis (or some other technique) is known, the modification of this error as provided by the COMET model would, in principle, strengthen the overall quantitative risk assessment procedure.

The remaining issue in this discussion on embellishments to THERP concerns the quantification of errors of commission. Gertman's [7] approach to this problem borrows heavily from both SLIM (discussed in the next section) and THERP. Once a set of errors of commission (or of intention) are identified, a group of experts are asked to consider situations in which all possibly relevant PSFs are either optimized or degraded. Based on these assessments, they are then asked to generate lower and upper bound HEP estimates, respectfully, for each of these errors by entering marks directly onto a logarithmic probability scale. After combining these estimates across all experts, a single pair of LB and UB HEP estimates for each error is obtained. Importance weights associated with 11 of what are considered to be the most relevant PSFs for the entire set of errors are then generated by each expert for each error, and these weights are subsequently normalized for each expert across all PSFs; for each PSF, the resulting weights are then averaged across experts. Through group consensus techniques, the experts are then asked to provide ratings of the goodness of each PSF for an assumed site specific context. Composite PSF ratings for each error are then derived by multiplying each of the PSF ratings by its corresponding importance weight and then summing. The scales associated with the weights and goodness ratings are such that the composite PSF (to be denoted C) for a given error lies between 100 and 500.

Assuming C is uniformly distributed (across various contexts), then $C_{.05}$ and $C_{.95}$ are 120 and 480, respectively. Correspondingly, the assumption that the HEP for any given error is lognormally distributed implies, by virtue of the direct estimation technique employed, that the LB and UB HEP estimates represent the 0.05 and 0.95 points on this distribution. Noting that the values of C and HEP have an inverse relationship allows the mapping equation

$$p(C \leq x) = p(HEP \geq y)$$

to be solved for y, the site specific HEP for a given error of commission, by setting

$$p(C \leq x) = (x - 100)/400$$

$$p(HEP \geq y) = 1 - \phi\,[ln(y) - \bar{x})/sd]$$

where ϕ represents the cumulative standard normal distribution, and $\bar{x}$ and sd are the mean and standard deviation of the corresponding (unstandardized) normal distribution. Given that $ln(LB) = \bar{x} - 1.645sd$ and $ln(UB) = \bar{x} + 1.645sd$, then $\bar{x} = (ln(LB) + ln(UB))/2$ and $sd = (ln(UB) - ln(LB))/3.29$. Substituting the expressions given above for $p(C \leq x)$ and $p(HEP \geq y)$ into the mapping equation and taking the inverse function results in:

$$[(ln(y) - \overline{x})/sd] = \phi^{-1}[(500 - x)/400]$$

Finally, substituting for $\overline{x}$ and *sd* and solving for *y* results in:

$$y = e\left\{\phi^{-1}\left(\frac{500-x}{400}\right)\frac{ln(LB) - ln(UB)}{-3.29} + \frac{ln(LB) + ln(UB)}{2}\right\}$$

10.2.3 SLIM-MAUD

SLIM refers to the Success Likelihood Index Methodology developed by Embrey and his colleagues [8], a procedure for deriving HEPs for specified human actions in NPPs, although the method is generally believed to be equally as applicable to other industries within which high-risk human activities take place. MAUD (Multi-Attribute Utility Decomposition) refers to a computer-interactive implementation of SLIM.

Underlying SLIM are two premises: (1) that the probability a human will carry out a particular task successfully depends on the combined effect of a number of PSFs; and (2) that these PSFs can be identified and appropriately evaluated through expert judgment. Expert judges are assumed to be capable of assessing the relative importance (or weights) of each PSF with respect to the probability for human error in the task being evaluated and, independently of this assessment, rating how good or bad each PSF actually is for each of these tasks.

On the surface, SLIM's relationship to Multi-Attribute Utility Theory (MAUT) is more than coincidental. Essentially, SLIM attempts to prioritize the attractiveness of alternatives, in this case the likelihood of success associated with various human actions under consideration, by decomposing the problem into attributes (PSFs) whose values and relative weights with respect to each "alternative" are established through expert judgment. Assuming an additive model (as SLIM does), the likelihood of success for each human action under consideration is determined by summing the product of the weights and ratings for each PSF, resulting in numbers (SLIs) that represent a scale of likelihood of success analogous to the "scaling" of alternatives achieved in MAUT through the derivation of expected utilities. The SLIs are useful in and of themselves, especially when the actions under consideration represent alternative modes of response in an emergency scenario, and the analyst is interested in determining which types of responses are least or most likely to succeed. However, for the purpose of conducting PRAs, SLIM converts the SLIs to HEPs through the use of a logarithmic relationship, although the precise form of this relationship is considered less important than the more intuitive relationship presumed to exist between the SLI values and HEPs.

The theoretical underpinnings of the SLI methodology are intended to be transparent to the user, and the actual theory is incorporated primarily in the MAUD, as opposed to the basic, implementation of SLIM. The basic procedures for implementing SLIM are as follows.

1. The first step consists of identifying an appropriate group of experts, the potential error modes associated with the human actions of interest, and the most relevant set of PSFs. Ideally, the group of experts should include individuals with operational experience in the scenarios and tasks being considered, as well as PRA, human factors,

and thermohydraulic specialists and training supervisors. The identification of all possible error modes is essential, and is generally arrived at through in-depth analysis and discussions that could include task analyses and reviews of documentation concerning emergency operating procedures. At this point, the goal is to arrive at a minimal set of PSFs that are considered to be most relevant in governing human behavior for the specific tasks and scenarios, and within the plants in which these actions take place. This is done by consensus, beginning with a pre-defined set of PSFs that the judges are encouraged to modify in light of their own experience.

2. Relative importance weights for the PSFs are derived by asking each judge to assign a weight of 100 to the most important PSF, and then assign weights to the remaining PSFs as a ratio of the one assigned the value of 100. Discussion concerning these weightings is encouraged in order to arrive at consensus weights, which can potentially minimize the variability resulting from taking the mean or other aggregation procedures. Normalized weights are then obtained by dividing each individual weight by the sum of the weights for all the PSFs.

3. Ratings from the judges are then elicited for each PSF on each task, with the lowest scale value indicating that the PSF is as poor as it is likely to be under real operating conditions, and the highest value indicating that the PSF is as good as it is likely to be in terms of promoting successful task performance. The range of possible SLI values is dictated by the ranges of values associated with the rating scale. As with the procedure for deriving weights, the individual ratings should be discussed in order to arrive at consensus ratings.

4. For each task (or action), SLIs are computed by summing the product of the normalized weights with the ratings for each PSF. An estimate of the HEP, which equals to $1 - p(success)$, can then be derived using the relationship

$$log[p(success)] = aSLI + b$$

provided that at least two tasks are available for which the probabilities of success are known so that the constants a and b can be evaluated. The "calibration" tasks selected, however, should be part of the same cluster of tasks for which the relevant set of PSFs were identified. When such reference tasks are not available, then for each task each expert can be asked to make absolute judgments of the probability of failure associated with two boundary conditions corresponding to situations where the PSFs are as good and where they are as bad as they could credibly be under real operating conditions. These judgments are facilitated through the use of a logarithmic probability scale, and are assigned SLI values of *100* and *0*, respectively. The SLI computed for a given task is then used to interpolate between these lower bound (LB) and upper bound (UB) probabilities, resulting in the following estimate of the HEP for each task:

$$HEP = [(LB)^{SLI/100}]\ [(UB)^{(1-SLI)/100}]$$

Again, consensus values for LB and UB are encouraged.

5. PRA applications typically require that measures of uncertainty bounds accompany estimates of HEPs. For HEPs derived by SLIM, one can employ the direct estimation method noted above, though care must be taken to ensure that the scale is appropriately anchored, and that the question posed to the expert concerns identifying

upper and lower bounds for the HEP such that the true HEP falls between these bounds with *95%* certainty. Alternatively, when SLI values are generated by each judge, the variance of the *log(HEPs)* can be estimated for each task; the uncertainty bounds are then set to ± 2 standard deviation units. In general, when SLIM cannot be conducted as a consensus process, aggregation of HEPs and (directly estimated) uncertainty bounds is achieved by taking the geometric mean.

The MAUD-based implementation of SLIM provides a computer-interactive environment that is highly user-friendly. MAUD is comprised of features that ensure conformance with the assumptions underlying MAUT, such as assuring that the ratings for the various PSFs by a given judge (or user) be independent of one another, and eliciting relative importance weights for the PSFs that are consistent with the judge's preferences (although in contrast to the basic implementation, judges weight the PSFs after providing the ratings). In addition, MAUD provides procedures for assisting the judge in identifying the relevant PSFs.

In eliciting ratings, MAUD defines ideal points on each scale (e.g., the ideal point for the PSF stress would be closer to the middle than to the extremes), and uses this information to rescale the PSFs in order to ensure consistency between scale direction and success likelihood. It then defines new scale values between 0 and 1 as a function of the differences between the ratings for a given PSF across the set of tasks and the ideal point for that PSF. All that is generally needed to conduct a MAUD session is knowledge of the tasks for which one wants to assess SLI values and assurance that the tasks are sufficiently homogeneous; i.e., the likelihood of success for the tasks assessed should depend on the ratings on a common set of PSFs. Up to ten such tasks can then be evaluated in a single session.

The justification for SLIM-MAUD as a HRA technique is based largely on its ability to address issues associated with three criteria: practicality, user acceptability, and usefulness. Overall, the technique is viewed as being reasonable in cost and in training requirements, applicable to a broad range of tasks (which increases its flexibility for incorporation into PRAs that are directed at either low or high system levels), and contains high face validity. In addition, the capability exists for performing sensitivity analysis that could identify the relative contributions of the PSFs to the success or failure of a task. Although it lacks the task-analytic modeling rigor associated with THERP, it is far from clear whether attempting to derive HEPs directly from PSFs as opposed to using PSFs for modifying previously assigned HEPs (as in THERP) significantly impacts, from the standpoint of its usefulness for PRAs, the assessment of HEPs. What is much more obvious is that in bypassing the construction of event trees, SLIM-MAUD is, in principle, as easily applied to a wide variety of tasks and situations, including highly infrequent events and events for which occurrences have never been documented.

10.2.4 TESEO and HEART

TESEO (Tecnica Empirica Stima Errori Operati) and HEART (Human Error Assessment and Reduction Technique) represent two relatively simplistic and straightforward techniques that, for these very reasons, can be appealing to industrial safety managers. Both techniques can be viewed as methods for guidance in assessing human reliability, and the descriptions of these techniques that follow are consistent with this view.

TESEO [9] distinguishes between *HE*, the probability of an operator error, and *HU*, human unreliability, by defining $HU = HE(1-PR)$, where PR = probability of error correction by the operator. TESEO proposes the following multiplicative model to describe a control room operator's probability of error:

$$HU = K_1 K_2 K_3 K_4 K_5$$

where K_1 represents the activity's typologic factor (e.g., $K_1 = 0.001$ for simple, routine tasks, *0.01* for routine tasks that require attention, and *0.1* for nonroutine tasks); K_2 represents a temporary stress factor associated with either routine or nonroutine tasks in terms of the time available, in seconds, to carry out the activity; K_3 refers to the human operator's characteristics, differentiated on the basis of how well-trained or knowledgeable the operator is; K_4 represents the operator's emotional state based on the extent to which the task represents an emergency situation; and K_5 refers to ergonomics considerations. The values taken on by these parameters range from *0.001* to *10*, and are assumed to be based on available data from the literature as well as a wide variety of other sources.

HEART [10] provides a number of generic task descriptions for which nominal HEPs and lower and upper bounds are assigned, identifies a variety of PSFs (referred to as error-producing conditions) that can impact these HEPs, and suggests remedial measures that can minimize the effects of each of the proposed PSFs. A number of features distinguish HEART from other techniques that share the common goal of providing estimates of HEPs. First, HEART emphasizes the practical concern of reliability assessors with the potential for changes in the probability of failure of systems by an order of magnitude of 10. This "factor of 10 criterion" is translated into a concern for identifying those PSFs that are likely to modify the probability of failure by a factor of 3. HEART also provides a much more comprehensive listing of PSFs, each accompanied by an order of magnitude corresponding to the maximum amount by which the nominal HEP might change when considering the PSF in its worst, relative to its best state. Finally, by providing a battery of remedial measures corresponding to each of the PSFs, HEART offers a form of closure, by way of design considerations, to the issue of human reliability.

The following two task descriptions, together with their respective HEPs and lower and upper bounds, represent two of the eight generic tasks alluded to above that are considered by HEART:

- Shift or restore systems to a new or original state on a single attempt with supervision or procedures (*0.26; 0.14-0.42*).

- Respond correctly to system command even when there is an augmented or automated supervisory system providing accurate interpretation of system state (*0.00002; 0.000006-0.0009*).

Examples of five of the 38 PSFs provided [10] along with their associated "orders of magnitude" are:

- A means of suppressing or overriding information or features which is too easily accessible (*x 9*).
- A need to unlearn a technique and apply one which requires the application of an opposing philosophy (*x 6*).
- No clear, direct, and timely confirmation of an intended action from the portion of the system over which control is to be exerted (*x 4*).
- A mismatch between the educational achievement level of an individual and the requirements of the task (*x 2*).
- No obvious way to keep track of progress during an activity (*x 1.6*).

Heart approaches the problem of human reliability evaluation by matching a description of the situation for which an assessment is desired with one of the eight generic tasks. It then identifies all relevant PSFs, especially those that satisfy the "factor of 3 criterion," and evaluates (presumably through subjective means) the proportion of the order of magnitude associated with each PSF for the task being considered, and the "assessed effect" of each PSF, defined as the (order of magnitude *−1*) x (assessed proportion of order of magnitude *+1*). The nominal HEP is then multiplied, in turn, by each of these assessed effects.

Although HEART appears somewhat *ad hoc* in nature, it holds up surprisingly well against more formalized HRA techniques such as THERP and SLIM-MAUD when rated on a number of criteria in a venture documented in The Human Reliability Assessor's Guide [11].

10.2.5 The HCR Model and Time Reliability Correlation

The HCR (human cognitive reliability) model [12] is essentially an application of classical engineering reliability that addresses the distinctions between skill, rule, and knowledge-based levels of human performance (these topics are discussed in later sections). This model was an outgrowth of a project entitled SHARP (Systematic Human Action Reliability Procedure) that involved the development of a framework for aiding human reliability analysts in identifying and structuring the important human interactions that: (1) occur prior to an initiating event; (2) can cause initiating events; (3) are taken to ameliorate accident scenarios; (4) have the effect of exacerbating accident scenarios; and (5) represent attempts at recovery. The model was specifically developed for analysis of control room crew responses in NPPs, and based on three requirements: it should be probabilistic, time-dependent, and capable of accounting for the variability in new responses for different control room scenarios.

The HCR model is characterized by a three-parameter Weibull function of the form:

$$p(t_n) = e\left\{-\left(\frac{t_n - C_{1i}}{C_{2i}}\right)^{C_{3i}}\right\}$$

where $p(t_n)$ = probability that a crew has not responded adequately to the problem or task by time t_n (a normalized time variable), and C_{1i}, C_{2i}, and C_{3i} are fitted parameters whose values depend upon the type of processing, either skill, rule, or knowledge-based, required of the crews in performing the task. The normalized time variable t_n controls for contributions to response times that are unrelated to human activities, and is defined as $t/T_{1/2}$, where t represents a given system time window within which the crew is required to perform the task and $T_{1/2}$ is the estimated median time taken by the crew to perform the task. The normalized curves are depicted in Figure 10.3.

The initial application of this model was based on two sources: existing simulator data that was subsequently classified into the three designated levels of human performance, and small-scale simulator experiments in which the tasks were designed to conform, as much as possible, to one of these three types of processing. PSFs such as operator crew experience and stress level were assumed to modify $T_{1/2}$ rather than affect the type of processing, so that the relationships between the three types of curves remain preserved. To account for more detailed behaviors, a Cognitive Sub-Element (CSE) model was proposed that modeled operator recovery from an initially inappropriate progression of plant state. Use of the CSE model resulted in the appearance of a "knee" in curves that turned out to be similar to the curves derived by the HCR model.

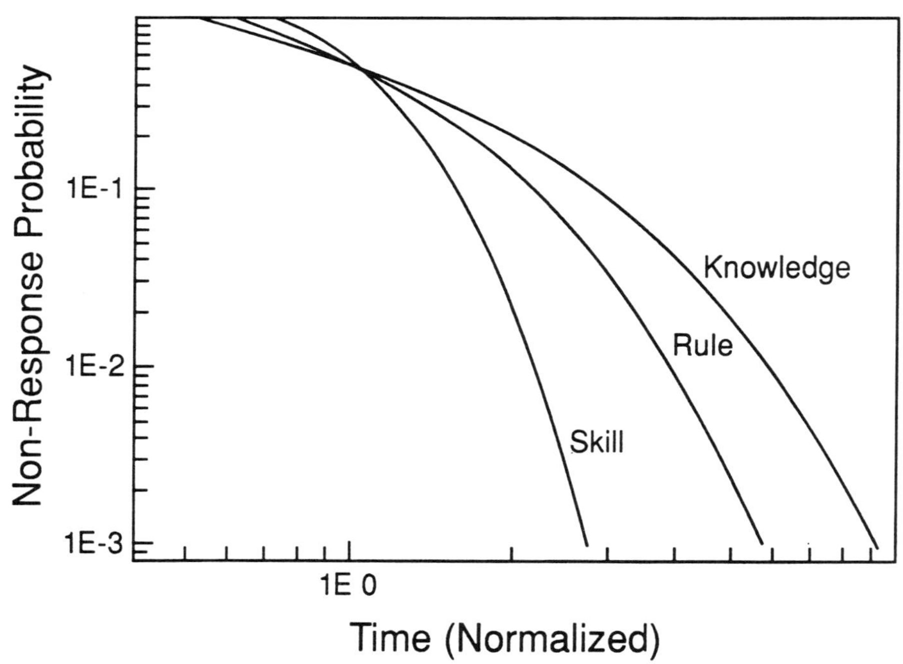

Fig. 10.3 Normalized Crew Non-Response Curves Corresponding to Skill, Rule, and Knowledge-Based Processing (from Ref. 12)

In principle, the HCR model should be able to differentiate crew responses associated with skill, rule and knowledge-based task components within a given accident scenario, while demonstrating similar response curves for these same operator crews for task components across different accident scenarios characterized by the same level of processing. However, in a study by Kantowitz, Bittner, and Fujita [13], this overall pattern did not emerge, despite good fits by three-parameter Weibull distributions. Overall, the HCR model is consistent with other classical engineering reliability approaches to human reliability assessment (e.g., [14]) in its emphasis on mathematical description through curve fitting and parameter estimation. While these quantitative descriptions are necessary for predicting human or operator crew responses, generally no claims can be made concerning the behavioral meaningfulness of the model parameters.

The HCR model is often referred to as a type of time reliability correlation (TRC). Among human reliability analysts TRCs constitute mathematical relationships between the failure for an operator or crew to appropriately deal with some compelling situation and the time t in minutes after the initiation of this situation, and are usually depicted in the form of probability versus time plots. Dougherty and Fragola [15] suggested the use of the lognormal distribution for modeling TRCs, and fitted data from a variety of simulator exercises to this family of distributions. A TRC curve is usually characterized by two parameters: the median time (m), and an error factor (f) that, as in THERP, reflects the variability of the distribution. Their findings suggest that TRCs may not be capable of distinguishing between rule and knowledge-based performance due to the relatively new symptom-based emergency procedures implemented following the Three-Mile Island incident. They also seem to have identified a construct referred to as "burden" comprised of two attributes, decisional conflict and diagnostic complexity, that are uniquely related to the TRC curves—the error factor appears to be sensitive to the decisional burden component while diagnostic burden appears related to the median.

Recognizing the need for more robust TRCs capable of reflecting a wide variety of influences, Dougherty and Fragola [15] have formulated a more general TRC model that defines a random variable T, the time needed to successfully respond to a compelling situation, as follows: $T = \tau_R \tau_U$, where τ_R and τ_U are each lognormal random variables with median and error factor (m, f_R) and $(1, f_U)$, respectively. As with the HCR model, this TRC represents the "complimentary" cumulative distribution function since at time $T = 0$ the probability of responding successfully is zero. In this formulation, f_R accounts for the uncertainty in the process while f_U accounts for the uncertainty in the model; τ_U is actually a normalized lognormal random variable since its median is 1.

The random variable τ_R can be further decomposed into the following four factors:

$$\tau_R = \tau \, K_C \, K_I \, L_R$$

where τ is the median response time, K_C adjusts τ from 1/2 to twice its value to account for taxonomic considerations (such as whether rules are available to guide crew responses), K_I adjusts τ by the same degree as does K_C but to account for influences such as PSFs, and L_R is a lognormal random variable with a median of 1 and error factor f_R (note that the median response time m is now represented in terms of $\tau K_C K_I$. Factors such as hesitancy that might arise due to decisional conflict or diagnostic complexity are assumed to increase f_R (process uncertainty); determining f_U (model uncertainty) is a more complex matter.

10.2.6 Classical Engineering Reliability Modeling

Of the variety of purposes served by models, perhaps the most fundamental is their capability for providing quantitative predictions. The appeal in adapting classical engineering reliability techniques to analysis of human reliability lies primarily in the belief that for a number of tasks human performance is sufficiently straightforward in both its observation and measurement to allow accommodation of concepts and techniques usually reserved for hardware components.

Classical engineering reliability modeling addresses both static and dynamic reliability models. In the static case, a prescribed time period of operation is implied, whereas in the dynamic case the reliability of the component is expressed as a function of time. Restricting ourselves to the dynamic case, employing a classical reliability paradigm for modeling human reliability would require deriving an instantaneous error rate, $h_e(t)$ (analogous to the hazard function), from which $R_e(t)$, the human reliability function, would be obtained.

Letting T be a random variable denoting the time at which the human commits an error, and letting $f(t)$ represent the probability density function (pdf) associated with human error, then the probability the human will commit an error by time t is:

$$F_e(t) = p\,(T \leq t), \quad t \geq 0$$

and the probability that the human will not have erred by time t, referred to as the human reliability function, is:

$$R_e(t) = 1 - \int_0^t f(t)\,dt = \int_t^\infty f(t)\,dt$$

The mean-time-to-human-error is then defined as:

$$\int_0^\infty R_e(t)\,dt = \int_0^\infty t\,f(t)\,dt$$

Note that t does not have to represent time in the conventional sense, but could imply any continuous variable—e.g., the number of turns of a device used in performing a particular task.

Deriving the instantaneous error rate involves first defining the probability of human error in the interval $[t_1, t_2]$, which is $F_e(t_2) - F_e(t_1) = R_e(t_1) - R_e(t_2)$. The error rate, which is the probability of an error, per unit time, occurring in $[t_1, t_2]$ given that an error has not occurred prior to t_1, is then:

$$\frac{\left[R_e(t_1) - R_e(t_2)\right] / R_e(t_1)}{(t_2 - t_1)}$$

Letting $[t_1, t_2] = [t, t+\Delta t]$, the hazard function for the human (i.e., the instantaneous error rate) is defined as:

$$h_e(t) = \lim_{t \to \Delta} \frac{R_e(t) - R_e(t+\Delta t)}{\Delta t R_e(t)} = \frac{1}{R_e(t)} \frac{-d[1 - F_e(t)]}{dt} = \frac{f(t)}{R_e(t)}$$

Integrating both sides of this equation results in:

$$R_e(t) = e \left\{ -\int_o^t h_e(t) dt \right\}$$

Under the assumption that the human displays a constant error rate, $h_e(t) = \lambda$ errors/(unit time), and

$$f(t) = h_e(t) e \left\{ -\int_0^t h_e(t) dt \right\} = \lambda e^{-\lambda t}$$

which is the exponential pdf. A constant error rate implies that if the human has not erred by time t, we have no more knowledge at time t about when the human might err as we did at time $t = 0$, since $R_e(t) = e^{\lambda t}$ and the probability of committing an error in $[t, t+\Delta t]$ given that no error was committed by time t is

$$\frac{e^{-\lambda t} - e^{-\lambda(t+\Delta t)}}{e^{-\lambda t}} = 1 - e^{-\lambda \Delta t}$$

which is independent of t.

This modeling approach was adopted by Askren and Regulinski [14, 16] for analysis of human error in vigilance and tracking, two time-continuous tasks that required minimal learning. Analogies were drawn between the equipment concepts of mean-time-to-first failure (MTTFF) and mean-time-between-failures (MTBF), and mean-time-to-first-human-error (MTTFHE) and mean-time-between-human-errors (MTBHE), respectively, where MTTFHE could describe errors whose effects could be corrected. In the vigilance task, the MTTFHE data was examined across a sample of subjects for both misses and false alarms, and graphical methods were used to identify the underlying probability distributions. The best fits were verified through use of the Kolmogorov-Smirnoff test. The results indicated that the (two-parameter) Weibull distribution best fit the data, implying that human reliability at time t would be best predicted by

$$R_e(t) = e^{\left\{-\left(\frac{t}{\alpha}\right)^{\beta}\right\}}$$

Both MTTFHE and MTBHE data were examined in a two-axis tracking task [16]. An attempt was also made to model human error correction, where $C_R(t)$, the instantaneous task error correction rate, is analogous to $h_e(t)$, and $C_e(t)$, the probability that a self-generated task error will be corrected by time t (the correctability function), is analogous to $R_e(t)$. Results indicated that the MTTFHE data was best fit by the Weibull distribution while the lognormal distribution best fit the MTBHE and error correction data.

An example of a situation in which both static and dynamic reliability considerations exist is the (passive) standby redundant system, where only one subsystem or component is in operation [17]. A switch is present whose purpose is to bring the next subsystem into operation if the subsystem currently in operation fails. In the case where the switch is a human, the human's task could consist of bringing the next subsystem into operation or deciding which subsystem should be activated. For this problem, imperfect switching implies either human unavailability or human error at performing the replacement activity. Assuming a two-component system and denoting $f(t_1)$ as the pdf associated with subsystem 1, then the reliability of the system at time t is:

$$R_s(t) = R_{s_1}(t) + P_{he}\int_0^t f(t_1)R_{s_2}(t-t_1)\,dt_1$$

where P_{he} = probability of human error in performing the task given the human is available, and $R_{s_i}(t)$ = reliability of subsystem $(i = 1, 2)$ at time t.

If one is willing to consider the "transition" from correct to incorrect performance as a birth-death process, then Markov modeling techniques commonly associated with the reliability analysis of systems comprised exclusively of hardware components can, in theory, be applied to modeling human reliability. Although many human reliability analysts, including those who advocate quantification of human error, may not agree with this assumption, this is precisely the approach taken by Dhillon [18], who solves a large variety of reliability problems by considering failures due to human error in addition to conventional hardware failures.

An example of this approach [18] applied strictly to human performance involves evaluating the reliability of a human during the performance of a time-continuous task in which the human is assumed to fluctuate between normal and stress states. Human errors, which are assumed to result in system failure, can occur while being in either of these states. The following four states can then be identified: 0 = human performing task correctly in the normal state; 1 = system failure due to a human error in the normal state; 2 = human performing task correctly in the stress state; and 3 = system failure due to human error in the stress state. Assume that failures are independent of one another, that the human error rates associated with the transitions from states 0 to 1 and 2 to 3 are λ_1 and λ_2, respectively, and that the transition rates from states 0 to 2 and 2 to 0 are constants denoted by α and β, respectively. Then $p_i(t)$, the probability of being in state i at time t, can be derived by solving the following system of equations:

$$\frac{dp_o(t)}{dt} + (\lambda_1 + \alpha) p_o(t) = \beta p_2(t)$$

$$\frac{dp_1(t)}{dt} = \lambda_1 p_o(t)$$

$$\frac{dp_2(t)}{dt} + (\lambda_2 + \beta) p_2(t) = \alpha p_o(t)$$

$$\frac{dp_3(t)}{dt} = \lambda_2 p_2(t)$$

Using this human modeling approach, $R_e(t) = p_o(t) + p_2(t)$ and

$$MTTFHE = \int_0^\infty R(t)\,dt = (\lambda_2 + \alpha + \beta)/b_2$$

where $b_2 = \lambda_1 (\lambda_2 + \beta) + \alpha \lambda_2$. Note that this approach can be expanded to model the transitions between any number of discrete states; of course, a much larger number of parameters would need to be specified in order to solve for the $p_i(t)$.

10.2.7 Computer Simulation

Computer simulation models are widely used for solving a variety of mostly complex problems, many of which are analytically intractable. Specialized simulation languages have been developed for accommodating human tasks; the development of SAINT (Systems Analysis of Integrated Networks of Tasks) and its microcomputer version [19] are representative of such efforts. For one of the simulation programs described in [19], mathematical relationships were developed linking human performance in sixteen categories encompassing skills such as detection, recognition, sensory short-term memory, team coordination, probability estimation, time estimation, and tracking, to a number of variables known to impact performance in these skill categories, allowing for the evaluation of human performance in complex human-machine system interactions.

A notable pioneering work in this area involved human reliability modeling of four to twenty member crews on surface ships in order to generate systems reliability and availability information associated with integrated human and equipment performance [20]. This approach demands a significant amount of groundwork in order to provide the computer model with the input (start-up) data and logic for driving the simulation. At the heart of the flow logic are human performance variables whose identification is based on one or more theories (e.g., small group theory) relevant to the principal features and objectives of the model, and whose selection is based on information available in the literature or obtained through judgment. Factors are then identified on which the variables depend, and analytical expressions, based on known or estimated qualitative relationships, are derived linking the variables to one another. To achieve model consistency, the variables and expressions are scaled throughout.

Examples of variables included in the model are an individual's physical capability (which is a function of factors such as normalized body weight and energy consumption), stress (defined as the ratio of the time required to the time available for completion of the event), and variables such as aspiration level, fatigue, and motion sickness. Standard random variable generation techniques are used to generate individual values (e.g., repair times) from distributions constructed largely on the basis of available empirical data. This data, together with values derived from fitting trend lines to known or estimated relationships that link variables to one another, represents the human reliability database which, within the context of the model's flow logic (that includes data on equipment reliability), ultimately translates into a measure of system reliability. The assessment of group performance is, in general, some aggregate function of the individual data. More recently, this approach formed the basis for the development of a computer simulation model called MAPPS (Maintenance Personnel Performance Simulation) intended for providing maintenance performance reliability information in nuclear power plants [21].

To the extent that the variables governing human performance have underlying probability distributions, computer simulation methods can provide distributional data associated with human (and group) performance, in contrast to HRA methods that make assumptions concerning such distributions. Of course, the ability to obtain probabilistic information concerning human performance comes at the expense of specifying the distributions associated with events and variables of interest, although through appropriate experimental designs simulation methods could more easily afford the opportunity for testing these and other assumptions. More problematic, however, is the potential for the logic underlying the simulation model to become confounded by the numerous assumptions that need to be made concerning human performance characteristics, since the logical flow of events will be a function of the variables affected by these assumptions. At the same time, the ability for these computer simulation models to consider the dynamic interactions that occur between a large number of variables is perhaps the most compelling feature of this methodology, and is consistent with the philosophy [22] that human performance in dynamic environments cannot be validly represented by deterministic methods.

10.3 Qualitative Modeling of Human Reliability

10.3.1 Rasmussen's Frameworks

Rasmussen's work in the areas of human performance modeling and industrial safety [23, 24, 25, 26] directed attention towards the need for a better understanding of human error and its contribution to accidents. He views many human errors as "unsuccessful experiments with unacceptable consequences," reflecting the need for humans to explore a problem space in order to become proficient at a task. The adverse consequences, which serve to characterize the human's actions as errors, typically arise as a result of unaccommodating environments such as those that do not provide the opportunity for the human to detect or reverse the error, possibly by virtue of the effects of time lags (in processes associated with human actions) interacting with other events. Under conditions of detectability and reversibility, human error is defined in terms of the outcome of the activity in relation to some operationally defined norm.

Rasmussen's designation of skill, rule, and knowledge-based levels of performance distinguishes fundamentally different categories of human information processing. Performance at the skill-based level is associated with the processing of "signals" or data from the environment that can take the form of simple feedback control, where responses are modified on the basis of the discrepancy between a desired or actual state, or feedforward control, where responses are synchronized with predictions of future states. Rule-based performance can be characterized as the sequential cuing of responses from long-term memory by "signs" that are essentially "states or situations in the environment" or goals and tasks perceived by the human. The association between signs and responses is described by rules which human workers can either invoke from memory or obtain from external sources. Knowledge-based performance typically occurs in unfamiliar situations, where rules from previous experience are not available. Performance at this level occurs when working in exceedingly complex settings or during novel situations or emergencies, where humans are likely to be reasoning with a mental model of the situation they are facing. This process involves the explicit formulation of plans directed toward attaining a goal and the manipulation of meaningful "symbols" within a problem space.

These distinctions provide a foundation for modeling human reliability. Errors at the skill-based level of performance are generally perceptual/motor in nature. Factors such as attenuated alertness can affect the perception of signals, while inadequacies in spatial-temporal coordination associated with sensory-motor patterns can result in topographic misorientation or manual variability. At the rule-based level, errors often correspond to the activation of incorrect situational cues or the omission of acts in a task. Errors at the knowledge-based level often result from incorrect mental models, and may correspond to an inability to adequately consider the side effects of actions taken earlier in the problem-solving sequence. As will become evident in the next section, Reason's [3] approach to modeling human error is built, in part, on Rasmussen's "SRK" framework, and provides a more detailed account of the failure modes that can occur at each of these levels of performance.

The distinctions between skill, rule, and knowledge-based levels of performance form the basis for inferring "mechanisms of human malfunction" in a taxonomy developed by Rasmussen (Figure 10.4) for the purpose of describing and analyzing events involving "inappropriate" human performance. In general, this perspective is viewed as more appropriate for process control tasks as opposed to the less structured work situations more commonly associated with conventional types of work-related accidents [27]. This taxonomy can be applied towards guiding data collection for causal analysis of events as well as for the prediction of human error. The latter, however, would require a detailed analysis of the work contexts in order to determine potential mismatches between task/job requirements and internal human functions. By relating these functions to internal mechanisms of malfunction, likelihood estimates for the possible causes responsible for the external modes of malfunction (i.e., errors) can, in principle, be derived. In practice,

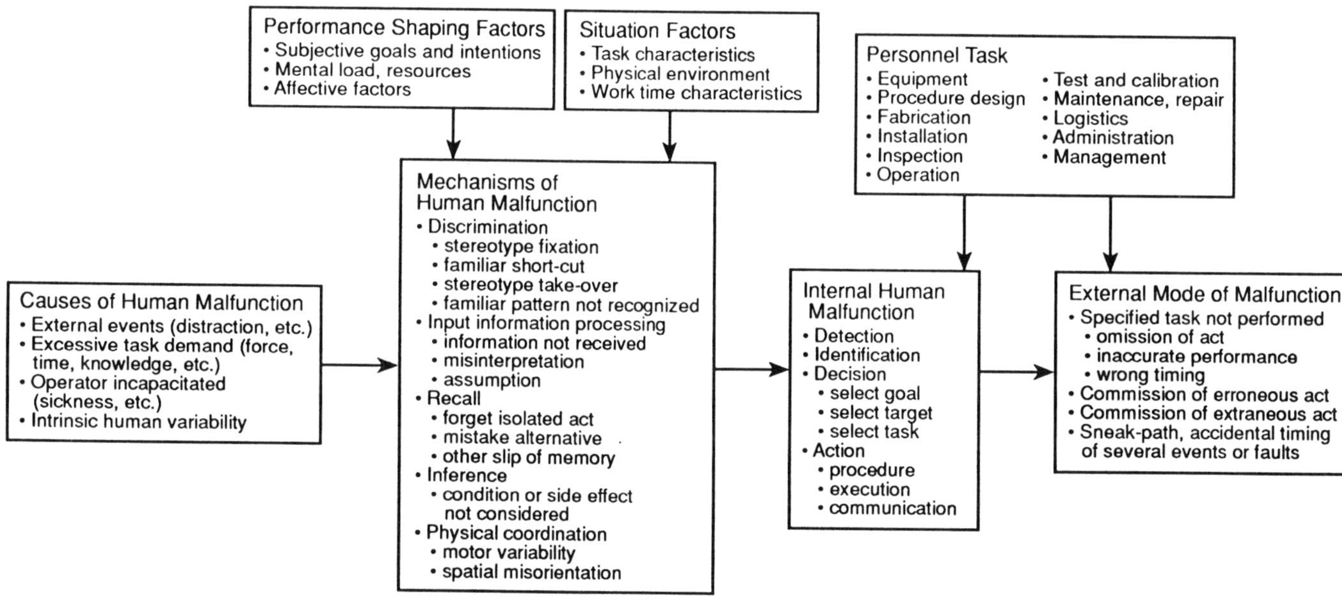

Fig. 10.4 A Taxonomy for Description and Analysis of Events Involving Human Malfunction (from Ref. 25)

especially for event analysis, the taxonomy would need to be supplemented with additional information, such as the characteristics of the system and the means by which the undesirable event was detected. The proposed taxonomy, therefore, is best viewed as a structure whose details, including the addition of categories, are determined based on the specific application.

In using this structure, guides are offered in the form of flow diagrams to aid in identifying what failed (the internal human malfunctions), how it failed (the mechanisms of human malfunction), and why it failed (the causes of human malfunction). Figure 10.5 illustrates the use of such a diagram for identifying mechanisms of human malfunction, where the implications for distinguishing between levels of human performance are clearly evident.

In Rasmussen's perspectives on human reliability design-related issues are consistently emphasized, such as the need for ensuring error detection and recoverability, predicting couplings between multiple errors (especially those derived from human and equipment malfunctions), and providing interfaces that aid in detection and problem solving.

10.3.2 Reason's Human Error Model

Reason's [3] approach to modeling human error reflects a strong cognitive psychological orientation. Behavior is categorized in terms of whether or not there was a prior intention to act. Prior intention can be further differentiated in terms of whether an unintentional action (e.g., a slip or lapse due to the action not proceeding as planned) or an intentional but mistaken action occurred. In the case where there was no prior intention to act, an action that occurs could be classified as either spontaneous (i.e., the action and the intention are one and the same) or involuntary. For the latter situation, which represents nonintentional actions, the term "error" according to Reason does not apply; in general, his model's domain comprises situations in which there was a prior intention to act.

Another distinction Reason raises concerns the classification of errors. Taxonomies of human error tend to be task or industry specific, and are usually developed with the intention of gaining insight into the causal factors and design countermeasures. Not surprisingly, these objectives are seldom fulfilled, although such error classification schemes do provide a "first-cut" that is often revealing, and can steer efforts towards minimizing unsafe behaviors in the appropriate directions. Reason distinguishes between three levels at which attempts to classify human error have been attempted: the behavioral, contextual, and conceptual levels. At the behavioral level, errors are classified according to overt behaviors. The classifications are often based on the features of the error, (e.g., whether it is an omission, commission, extraneous act, or out of sequence, or on its likelihood for being identified and corrected), or on the basis of its consequences (e.g., area of injury to worker or extent of material damage). At the contextual level, attempts are made to collect as much data as is feasible in order to hypothesize factors that could have prompted the error producing behaviors; this level, then, addresses the interplay between triggering factors and error tendencies. The conceptual level represents the greatest potential for inferring the cognitive mechanisms involved in human error and, as Reason contends, is therefore potentially the most useful because classifications at this

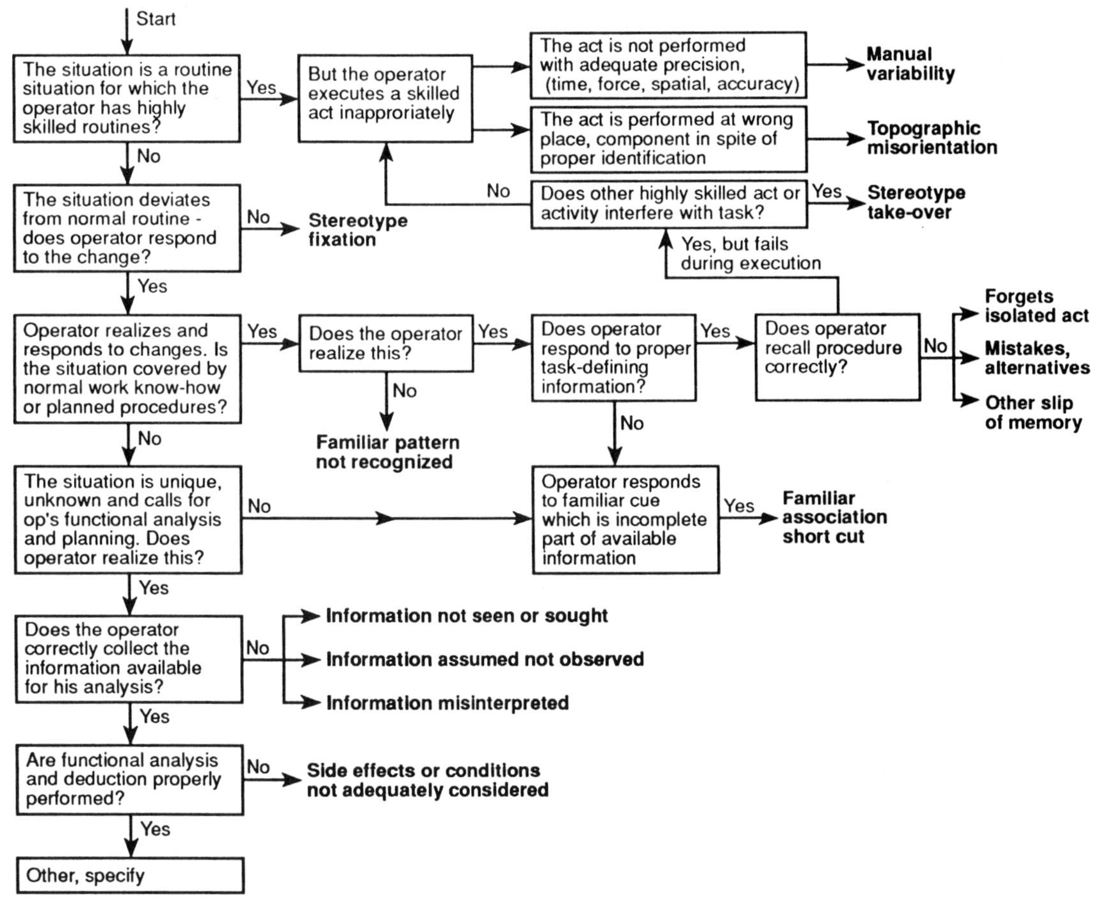

Fig. 10.5 Guide for Event Analysis to Identify the Mechanism for Human Malfunction (from Ref. 25)

level attempt to identify the causal mechanisms underlying error producing behaviors.

Reason's model, which he refers to as GEMS (Generic Error-Modelling System) is illustrated in Figure 10.6. Overall, this model is consistent with trends in cognitive theory towards framework theories that are more descriptive than predictive, and can more effectively synthesize information from natural settings, in contrast to local theories that rely on highly controlled experiments to predict results. Fundamental to the model is Rasmussen's distinction between skill, rule, and knowledge-based levels of performance. Although Rasmussen has illustrated the importance of these levels in terms of the particular forms in which information is likely to be used and the types of errors incurred, Reason's model treats the relationship between errors and level of performance in a more integrated fashion. The workings of the model are guided by an interplay between two types of control modes, the attentional and schematic modes, as well as a distinction between error types and error forms.

The attentional mode of control is sequential and laborious, and operates on a variety of sensory inputs including those derived from parallel processes associated with the schematic mode. It is responsible for the planning of activities and the monitoring that accompanies planning processes, and with detection and recovery from errors. In contrast to the resource-limited attentional mode, the schematic control mode constitutes a rapid, subconscious, and extensive network of patterns or "schemas" that reflect a number of specialized "activities." The basis for activation and coordination of these schemata are matching conditions arising from components of planning activities taking place in the restricted workspace associated with the attentional control mode, although the necessity for intentionality diminishes as a function of the prior frequency with which those schemata have been activated and the degree to which the current situational context can serve as a trigger.

The types of errors that can occur are influenced by the interplay of these control modes and, to a large extent, depend on the level of performance; specifically, error types are categorized in terms of skill-based slips and lapses, rule-based mistakes, and knowledge-based mistakes. The distinctions between these errors are summarized in Table 10.1. Unlike error types, which are linked to stages of cognitive processing, error forms are types of fallibility that are intrinsic to all types of cognitive activities. According to Reason, the two most predominant error forms are similarity and frequency biases, which were hinted at in the brief discussion above on activation of schemata.

The potential for error begins with "cognitive underspecification"; that is, somewhere in the human's cognitive processing system the specification of information is incomplete. To illustrate, assume that there is an important package that you want to take with you to work on some morning. Your strategy is to keep the package on your kitchen table to serve as a constant reminder while you are eating breakfast. However, before you leave, you go to a closet to get your coat and then proceed to take a path out of the house that bypasses the kitchen, leaving the package sitting on the kitchen table. This "error" is an example of a skill-based lapse. There was a prior intention to act but the action (forgetting, in this case) was unintentional. Due to the frequency with which the contextual cue "opening the closet to remove one's coat in the morning" is associated with taking a certain path out of the house, both frequency and similarity biases were in operation, and you would have needed to bring the attentional control mode into the control loop in order to establish whether the intended plan (of taking the package to work) was being carried out adequately. This situation is represented in the top portion

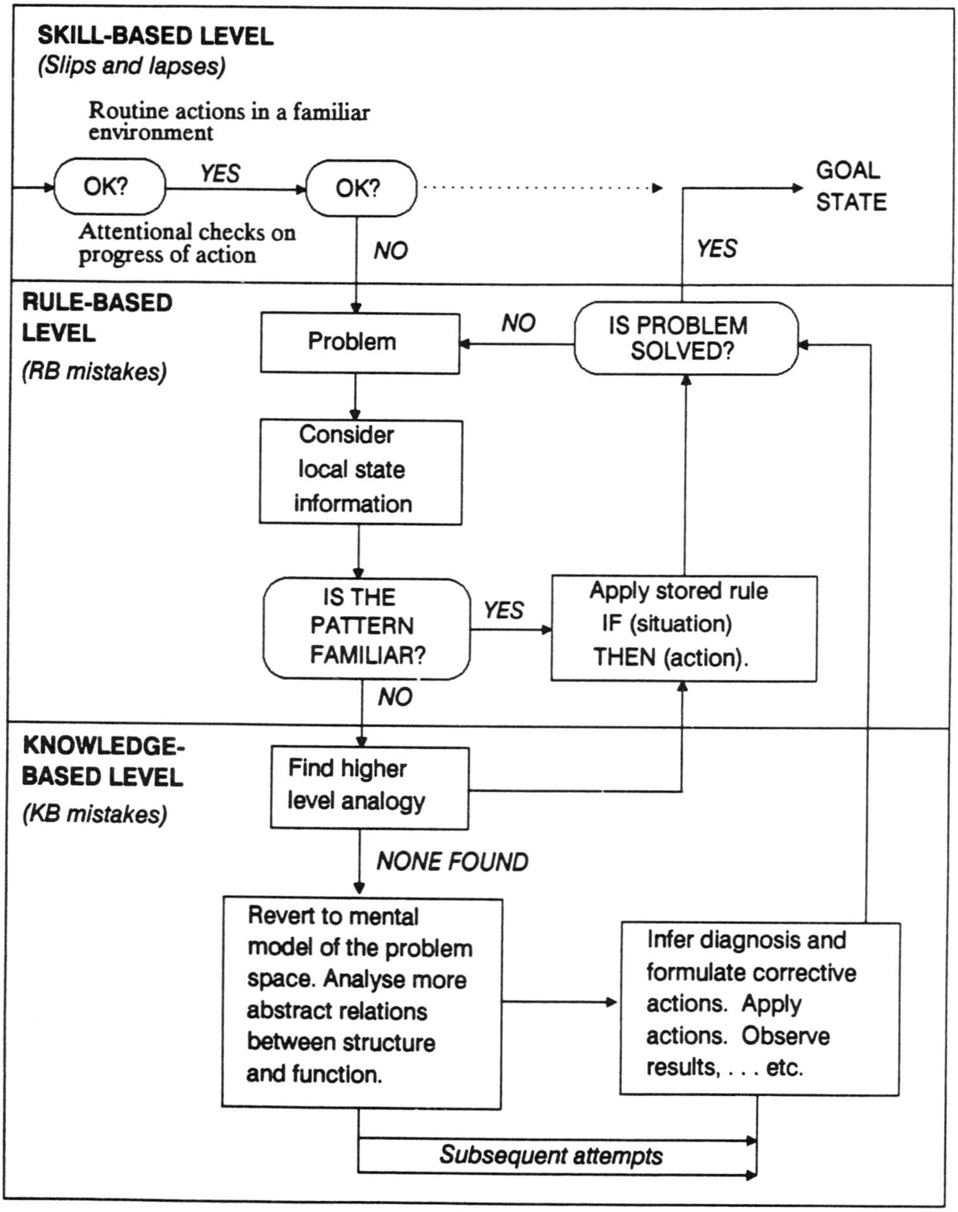

Fig. 10.6 An Outline of the Dynamics of GEMS (from Ref. 3)

Table 10.1

Summary of the Failure Models for Each of the Three Performance Levels (from Ref. 3)

Skill-based performance	
Inattention	*Overattention*
Double-caputre slips	Omissions
Omissions following interruptions	Repetitions
Reduced intentionality	Reversals
Perceptual confusions	
Interference errors	
Rule-based performance	
Misapplication of good rules	*Application of bad rules*
First exceptions	Encoding deficiencies
Countersigns and nonsigns	Action deficiencies
Informational overload	Wrong rules
Rule strength	Inelegant rules
General rules	Inadvisable rules
Redundancy	
Rigidity	
Knowledge-based performance	
Selectivity	*Problems with causality*
Workspace limitations	*Problems with complexity*
Out of sight out of mind	Problems with delayed feed-back
Confirmation bias	Insufficient consideration of processes in time
Overconfidence	Difficulties with exponential developments
Biased reviewing	Thinking in causal series not causal nets
Illusory correlation	Thematic vagabonding
Halo effects	Encysting

of Figure 10.6 confined to the skill-based level. In terms of the failure modes listed in Table 1, it would be classified as an omission following an interruption (i.e., a lapse) involving inattention.

Performance at the rule and knowledge-based levels are viewed as occurring when revisions of the current cognitive-based activities are deemed necessary, as might occur when problems with a plan are indicated. More generally, they can apply to any situation for which the schematic control mode is insufficient. As it turns out, this state of affairs normally occurs when problem-solving activities are required. At these levels of performance, frequency and similarity biases again are accountable for shaping the forms of errors. The basic premise is that the human seeks to minimize energy output and will always resort to less effortful rule-based processing, potentially employing a variety of types of pattern-matching activities that are contextually appropriate, prior to shifting to the more laborious attentional control mode in order to make available cognitive processing resources for the more analytic and conceptual analyses demanded of the problem. This state of affairs is indicated in Figure 10.6 by the path from "Find higher level analogy" back to the rule-based level, which is assumed to occur in parallel with the reasoning activities occurring at the knowledge-based level. The likelihood is that if one were to solve the problem at the knowledge-based level, the lower level "routines" activated for ultimately carrying out the solution would be less prone to error, given that they are arising from a less familiar environment; i.e., the human would be more likely to apply attentional checks to such routines due to the prior conditions that prompted their activation.

Rule-based mistakes are differentiated in terms of whether they arise from the misapplication of good rules or due to the application of bad rules. At this level of reasoning, the antecedent of a candidate rule is matched to either environmental features or internally generated information. The instantation of a rule is primarily governed by frequency and similarity biases. Consequently, the degrees of prior success of a rule and similarity to the current context will shape the form of the error.

For instance, assume cognitive underspecification is manifest in terms of the partial matching of a rule; i.e., the potential for instantiation exists despite inconsistencies between cues and the antecedent of the rule. The more successful the rule has proved in the past, the less important is the requirement for contextual appropriateness. This failure mode, referred to as rule strength (Figure 10.6), demonstrates the tradeoffs between frequency and similarity biases. In other situations, errors can arise predominantly from the influence of one of these biases, while in still other cases the role of these factors is not so clear. Moreover, there are situations where other factors such as recency and emotional value can contribute significantly to shaping the form of the error. A much more subtle example of the influence of the frequency factor relates to drivers who have experienced a number of near-miss collisions with the rear of vehicles in front of them. Although the lesson or "appropriate rule" that should emerge from these situations is that one should maintain a reasonable separation distance, the "successes" of previous accident avoidances results instead in an inadvisable rule (Table 10.1).

The two basic sources for the failures that arise at the knowledge-based level are bounded rationality and incomplete and inaccurate mental models. How these factors influence the mode of failure will depend, to some extent, on a variety of features associated with the problem, such as whether the problem space changes as a result of the human's actions and the degree to which other external factors can perturb the problem.

Although the role of frequency and similarity in shaping the error form is not always apparent, their presence can usually be implied. For example, given the demands on the attentional mode arising from problems associated with knowledge-based reasoning, humans may seek solutions based on what is available (i.e., what comes readily to mind). Use of this availability heuristic bears a close resemblance to the principle of "strength" underlying frequency gambling, and leads to the failure mode "Out of sight out of mind" (Table 10.1).

Another example concerns troubleshooting, where humans often test initial hypotheses that are based on incomplete information. The tendency is for humans to discard contradictory evidence and seek out information confirming their preliminary hypotheses (i.e., display conformation bias), which can potentially lead to failures. The basis for this "error" is that it reduces cognitive effort; it is easier to match confirmatory evidence, although unfortunately it makes it less likely that the potentially incorrect hypothesis will be abandoned. In general, however, failure modes at this level are more easily understood in terms of strategies that lead to reduced cognitive effort as a function of the characteristics of the problem space.

10.3.3 Cognitive Environment Simulation

As was the case with the computer simulation models of Seigel, Wolf, and Lautmann [20, 22] the Cognitive Environment Simulation (CES) methodology represents an attempt to capture the dynamic interaction of a large number of both task and human-related variables [27]. Unlike those models, however, CES is deterministic, and was developed specifically for modeling the processes by which humans assess situations and form intentions to act in emergency NPP operations. Human error in this context is therefore defined as a failure to form the appropriate intention for the situation under consideration. CES serves as an exploratory tool that allows the analyst to investigate situational characteristics conducive to intention failures, and the form and consequences of these failures (i.e., actions or inactions the intention failure leads to such as common mode failures). Because CES can potentially identify breakdowns in cognitive processing that lead to intention errors, it can aid in evaluating error reduction strategies as well as the effects of alternative designs involving human-machine interaction (e.g., decision aiding), group decision making, and training.

The CES software is based on an artificial intelligence problem-solving system known as EAGOL, an architecture suited for reasoning in situations that evolve over time. As a modeling tool, CES is consistent with qualitative modeling approaches to human reliability that emphasize causality of error. Driven throughout by a perspective that views the human's cognitive processing resources as limited in relation to the demands imposed by dynamic, uncertain, and complex situations, CES first attempts to characterize the overall effects of situational demands on human cognitive processing, and then considers more specialized models that address human performance as a function of the component demands. Its sensitivity is based on explorations the analyst can undertake into varieties of mismatches that can be implemented between the situational demands and the knowledge and processing resources assumed available to the human.

Inputs to CES consist of a dynamic stream of plant data that coincides with the situation of interest and that operators are assumed to be capable of observing. CES responds to these inputs through the execution of three types of activities that model the human's processing mechanisms: monitoring activities, explanation building activities, and response management activities. Monitoring activities address issues concerning what to monitor and when, and how the information being monitored should be classified (e.g., whether abnormal or expected). Activities associated with explanation building establish an ordering of possible explanations for the situation in question and determine which explanations to adopt. Response management activities dictate which responses are selected (e.g., an expected automatic system response or manual response), monitor the execution of these responses, and carry out pre-planned responses if the situation is assessed to call for such actions. Parallel execution of these three types of activities enables their intermediate results to be used by one another, reflecting a distributed as opposed to sequential processing perspective.

In addition to the plant data input, these processing mechanisms (which comprise the "inference engine") rely on a knowledge base derived from models of what humans would do in facing the circumstances that are unfolding over time in the CES model. These models are based both on general results available from the cognitive psychology literature and empirical studies of human performance in NPP emergencies. The types of knowledge the human has about the nuclear power plant and the organization of that knowledge need to be accounted for as well. Overall, the knowledge representation scheme employed within CES accommodates a wide variety of relations among concepts, including reasoning processes that can be described as rule and knowledge-based. The processing resources reflect the fact that humans are limited information processors, meaning that at each CES processing cycle in the simulation (which corresponds to "one time step of plant data") CES must specify what plant data and which (and when) relevant sources of knowledge in the knowledge base can be activated, and a number of "cognitive competencies" are available for exercising this type of control. One of these addresses the overall attentional focus at a given point in time, such as whether observed data, due to its salience, might shift the human's focus away from some processing activity, or whether the human's focus is on seeking out information in order to diagnose an abnormality. Another competency of CES, that ultimately serves to prioritize processing activities, concerns defining what would be considered "interesting findings" to the operators and the mechanisms for filtering through this information.

From an architectural standpoint, the fundamental unit in CES is represented in the form of "knowledge couplers" that express relations between items within the plant. For example, "pressurizer level" could be coupled to "steam generator tube rupture," and reasoning can occur between these two "nodes" if either of them is activated (so that activation of "decreasing pressurizer level" would suggest the possibility of a "steam generator tube rupture"). These nodes are highly flexible, capable of representing observable behaviors of the type noted above as well as plant states that are inferred from observations and corrective responses. Qualifiers in the form of contextual knowledge are capable of modifying the relations expressed by these couplers, by specifying conditions under which the relationship is relevant or that alter the relationship. The strength of the relations represented by knowledge couplers are also encoded (in terms of how strongly the activation of one node suggests the other), and these strength values could differ as a function of the direction of inference.

The qualifier and strength attributes associated with couplers govern, in part, the accessibility of knowledge. Access to relevant knowledge is also derived from contexts based on patterns formed from networks of interconnected couplers. These networks can represent almost any knowledge concerning plant structure and functions, faults, goals, and responses that operators would be expected to know. Modifications in the encoding of this knowledge constitute the adjustments the analyst may wish to explore in terms of their impact on intention failures.

Access and utilization of knowledge as encoded in knowledge couplers occurs following any single CES processing cycle, and is governed by the three types of processing mechanisms noted earlier. These mechanisms act jointly through the coordination of multiple independent "analysts" that are "spawned" when needed (i.e., when various criteria are met). Thus, corresponding to each of these three mechanisms are behavior, situation, and response plan analysts that access the knowledge base in order to: (1) acquire the criteria that establish whether new analysts should be created; (2) interrupt and discontinue analysts; and (3) provide the knowledge needed by analysts for generating expectations, drawing conclusions, and determining responses.

Behavior analysts are responsible for tracking input plant data, and could be spawned when the input data signals that some criterion has been met (e.g., a parameter out of limits), or by a situation or response plan analyst that requires information for which a behavior analyst does not yet exist. The behavior analyst is capable of determining whether plant behavior is expected, and can communicate findings concerning the observed behavior, abnormalities, and confirmation and violations of expectations to other analysts. The role of the situation analyst is to explain unaccounted for behavior, which it could do, for example, by examining other links to the unexpected data and exploring whether these potential influences are consistent with other NPP state evidence. Note that this "knowledge-driven" search process represents a cognitive competency the CES analyst could adjust for, and would entail the creation of behavior analysts for providing the relevant hypothesis-testing evidence. Response plan analysts are called upon to execute pre-planned sequences of either automatic system or human operator responses that are triggered by plant observations, accepted explanations, or anticipated effects of hypothesized influences. The actual corrective actions assumed by these analysts therefore depend on what has been observed by the behavior analysts and what explanations have been formulated by the situation analysts, as well as on the cognitive competencies encoded within the knowledge base that could pertain, for instance, to whether emergency procedures associated with certain incidents exist and how they might be utilized.

These analysts, or more generally, the three types of activities or processing mechanisms that they encompass, along with the knowledge base on which they depend, constitute the resources available within CES. Together with the "forcing function" (the input stream of plant data), they comprise the factors available to the analyst for adjustment. In making these adjustments, the analyst could track their influence on the human's intention to act. Typically, this would be accomplished by examining records of the knowledge activated and processed in response to the plant incident. A recursive appraisal of intentions to act, i.e., one which includes the execution of these intentions and their effects on plant state, is possible but would require additional mechanisms (such as input from NPP simulators) since CES is not designed to execute its intentions.

As previously noted, CES is a deterministic model—the same underlying assumptions will always lead to the same intention formations. The sources of variation in CES lie in the changes that can be made to the plant incident data and to the processing and knowledge resources. Formally, changes to CES resources and to the NPP incident data are carried out through Performance Adjustment Factors (PAFs) that enable the analyst to establish the appropriate representation (in terms of available resources) for a particular NPP situation. Initially, a relationship must be established between some factor (e.g., operator experience or stress) and the cognitive processes underlying the CES resources. This relationship is then translated into the actual adjustments (PAF settings) that need to be made to CES resources, and that ultimately determine the manifestation of intention formation failures.

Whereas CES can identify error-prone situations, CREATE (Cognitive Reliability Assessment Technique) represents a HRA method that uses this capability of CES to assess the risk associated with errors that can arise due to intention failures [28]. CREATE attempts to provide a quantitative estimate of human cognitive reliability that can be combined with hardware component reliability for the purpose of assessing the overall system reliability. To meet this objective, CES serves a similar modeling function to that of fault trees; it identifies the minimal cut sets (which in the context of CES are the minimum conjunction of conditions) that can lead to intention failures with high-risk consequences.

In applying the CREATE methodology, one begins (as with THERP and other HRAs) with a plant familiarization stage that requires an HRA analyst to collect the relevant NPP data capable of impacting human performance. In this phase, data collection (e.g., information concerning procedures available to the operator) is guided by CES's setup requirements, and for the need to derive frequency estimates for different processing demand and resource conditions. Ultimately, these estimates need to be aggregated to quantify intention errors.

CES is then run, as previously described, to identify the minimum conditions conducive to intention errors and the actions that follow these errors. Variations on NPP accident sequences are modeled through a "problem difficulty event tree" that can reveal features leading to refinement of the system event tree, from the standpoint of providing more details concerning operator problem-solving capabilities in response to event sequences underlying the accident situation. In making adjustments to CES and performing the corresponding multiple runs, the objective is to identify mismatches between the problem's demands and the resources available to the operator. Although the intention errors that come to light are, as modeled by CES, deterministic, likelihoods can be associated with the circumstances that lead to mismatches (i.e., to intention errors).

In principle, the CREATE methodology can alter the overall systems event tree (and thereby the PRA) by revealing actions, such as an error of commission, resulting from intention errors that were not previously represented, or systems that were previously considered independent but, on the basis of an error of intention that can, e.g., lead to common mode failure, are in fact interrelated. In addition, the CES modeling stage can reveal error recovery paths by the operator that were not previously represented in the PRA.

The HRA, which serves as the basis for incorporating this information into the PRA, involves determining the probability of the situation provoking the intention error (which assumes that the intention error produced by CES would, in a parallel real-world situation, be produced with a probability approaching one). This, in turn, requires assessing the likelihood of those demands actually occurring as well as the likelihood of possessing the problem-solving resources available to CES (through PAF adjustments) when confronted with those demands. As with many HRA techniques, in the absence of empirical data expert judgment is required, although these estimates are considered to be more accurate than other estimates derived from experts due to the fact that the issue of quantifying an intention error is filtered by CES into the presumably more tractable issue of estimating the probability of the conditions leading to an intention error. In general, estimates of input demands are based on systems reliability techniques while cognitive psychological expertise is used to assess likelihoods of resources available to the operator, where group consensus techniques such as SLIM-MAUD are suggested for quantifying the likelihood of resource conditions that lead to errors of intention. Finally, human reliability is assessed by aggregating the probability of error intention with the probability of execution errors (i.e., where the correct intention was formed but an error was committed in carrying out the intention). Under the assumption that these two errors are independent, the joint probability of a failure would be the sum of these two probability estimates. Overall, despite some unique features arising from the use of CES, as a tool for performing PRAs CREATE carries with it many of the problems attendant to other HRA methods.

Finally, it should be noted that other cognitive simulations within the NPP domain have been developed. In particular, Mancini [29] describes an impressive simulation model referred to as DYLAM (Dynamic Logical Analytical Simulation) in which an "operator cognitive model" is coupled to relatively elaborate machine models. For further details, the reader is advised to consult [29, 30] and the reference lists accompanying these articles.

10.4 Critical Analysis

There are seemingly an endless number of objections that can be voiced when the primary goal of a human reliability assessment technique is to derive a quantitative estimate of human error, with the more elaborate and well-documented schemes tending to prove more vulnerable. It is not surprising then that THERP has been the object of much criticism. Adams [31] essentially "pulls the rug out" from under THERP by casting many of the details that comprise this method as irrelevant. This position is based on the belief that: (1) prescriptions do not exist for defining units of behavior (such as those represented by limbs on an event tree) for which reliability can be determined; (2) human error must be defined as precisely as component failures are in order for its contribution to system failure to be meaningful; and (3) measures of human reliability cannot capture the correlational complexity associated with the behavioral sequences that underlie task performance. A number of these criticisms are dismissed by Sharit [32] on the grounds that they tend to reflect disagreements with what one considers to be the appropriate grain of analysis that should be accorded to the study of human behavior. Instead, the criticism in [32] is directed primarily at the layering of assumptions that one proceeds through in the process of deriving the HEP. Beginning with the uncertainty associated with the nominal HEP, each ensuing step of the technique contributes its own unique uncertainty, so that the probability estimate one eventually arrives at is likely to be the least useful aspect of the exercise. Moreover, this layering results in a multiplicity of assumptions that diminishes the usefulness of sensitivity analysis, the technique usually recommended for allaying any concerns regarding assumptions that are made. It is then perhaps not so surprising to find justification for short-cut techniques such as TESEO on grounds of expediency; if the process of estimating HEPs is necessarily diluted, why not simplify the process? Ironically, insights into human reliability and system safety appear to be better served by the methods and guidelines comprising THERP than by the probability estimates this method derives.

Turning to the SLIM-MAUD procedure, Apostalakis, Bier, and Mosleh [33] note that the "renormalization" of ratings of goodness, whereby the ratings closest and farthest from the ideal value for a given PSF are assigned the values one and zero, respectively, result in ratings that depend heavily on the range of ratings assigned to that PSF across the tasks under consideration. Consequently, if a new task is added to the set and its rating on that PSF falls outside the range covered by the existing tasks, all the SLIs, and therefore all the HEPs for those tasks, will be altered. The assessment of importance weights independent of ratings of goodness is also questioned, a procedure which results in the same weight for a given PSF being applied to all tasks. This leads to the possibility of an interaction between importance weights and ratings, where a rating of a PSF for a particular task could change the PSF's importance for that task relative to the other tasks. A more cynical perspective is taken by Spurgin [34] who dismisses SLIM-MAUD entirely on the basis that there is no way it can reveal how PSFs offset human performance—clearly not a perspective that embraces the subjectionist viewpoint.

Apostolakis et al. [33] also criticize a number of features associated with the human cognitive reliability model, in particular: (1) whether the normalized response times for all tasks classified into either the skill, rule, or knowledge-based types can be modeled by a Weibull (or any other single) distribution, especially for those periods beyond certain "critical times" during which the available amount of time to act is at most weakly related to the operator's or crew's probability of incorrect response; (2) the need for families of curves in place of the individual curves associated with each of these levels of performance in order to adequately account for the actual uncertainties; (3) the uncertainty deriving from identification of the correct curve due to the fact that a number of tasks cannot be characterized exclusively as either skill, rule, or knowledge-based; and (4) inaccuracies in the use of simulator data for assessing parameter values, due primarily to the inability for simulator exercises to adequately reflect the range of possible behaviors that might be encountered in actual situations.

It should be obvious that any reasonably well-documented procedure for deriving human error or nonresponse probabilities will be easily open to criticism stemming from potential inadequacies associated with its various assumptions. In contrast, qualitative modeling techniques that attempt to offer insights into causal mechanisms and thereby enhance our "understanding" of the problem, often suffer from a very different type of attack. It is not uncommon in industry to find health and safety personnel admit that these approaches are appealing, but also to confess total bewilderment as to how one might actually go about applying them to problems that concern them. While flow charts are provided in some cases (e.g., Figure 10.5) to guide the analyst, filling in the gaps is nontrivial, especially for individuals not highly trained in behavioral analysis.

Similarly, although Reason [3] presents a thoroughly convincing and entertaining excursion through the various modes of human error, industries concerned about such errors may be hard-pressed to identify the triggering factors associated with high-risk work contexts and the interplay between these factors and error tendencies. And in its deterministic path towards uncovering failures of intention, CES, with its preponderance of assumptions concerning human knowledge and processing mechanisms (a tall order in its own right), might be revealing little more than how these assumptions might interact within the context of NPP events unfolding over time. This is not meant to minimize its potential for enlightenment nor the insights that have gone into constructing this modeling tool. However, the absence of a stochastic component, especially in light of the many-to-many mappings between NPP incidents and CES PAFs, may obscure the potential for insight by exaggerating the effects of the assumptions concerning human performance. With respect to the use of CREATE for performing PRAs, despite the discipline CES brings to this undertaking, a substantial burden is still placed on the HRA analyst.

Overall, it seems that what we need is more guidance on applying qualitative approaches to industrial problems, and more behavioral insight from the probabilistic approaches along with reassurance that the quantitative risk figures they produce are meaningful.

10.5 Synthesis

With the area of human reliability assessment appearing to have reached a plateau, the notion of providing a "synthesis" in this area seems to have become in vogue, perhaps portending what Dougherty referred to as a "second generation" in HRA. What follows are some brief commentaries concerning a number of issues that could possibly affect the complexion of this generation, if and when it takes form.

10.5.1 Error Prediction

Determining the capabilities for predicting human error would seem to be the logical path for HRA to take, although Spurgin [34], who strongly advocates this view, does not believe much headway will be made on this issue in the near future. Unclear are the bounds associated with this problem. Given the variability associated with the human's information processing and motor control systems, not to mention the uncertainties arising from psychosocial influences and environmental perturbations, it is unlikely that we can predict precisely when an error will occur and what form it will take. Perfect prediction implies the possibility for eliminating human error. Acknowledging that this goal is unrealistic, what needs to be resolved is how to go about achieving convergence towards elimination of human error. Efforts in this direction will, to some extent, involve determining the appropriate bounds that can be placed on error prediction for various situations. In the ensuing sections, a number of issues are discussed that have important implications for this problem.

10.5.2 Taxonomies

The numerous taxonomies of human error that have been proposed and implemented attest to the diversity of views concerning the definition of human error and the appropriate levels of analysis for studying human error and, in part, reflect the different purposes served by taxonomies and the specific work environments they are intended for [36]. The defining features of a taxonomy usually guide the task analysis that supports its development and application; e.g., if the taxonomy emphasizes slips and lapses, the task analysis should include information on disruptions. Taxonomies of human error have been most useful for organizing accident events, which can lead to insights into design countermeasures. However, they cannot, as stated in Senders and Moray [36], be used to "predict the imminent occurrence of errors," although the position taken is that taxonomies (including Reason's taxonomy that was discussed in this chapter within the context of his model of human error) are "on the fringe of being able to predict the occurrence of errors" by virtue of their ability to identify links between task characteristics and error tendencies.

Given the relationship between human-machine design considerations and human error (Swain [37] notes that improved implementation of these considerations leads to better predictability of human error), and the potential ability for error taxonomies to provide insights into improved designs, taxonomies of human error will likely continue to be developed for addressing a number of specific problems. For example, Prabhu, Sharit, and Drury [38] discuss a taxonomy that focuses exclusively on "temporal errors" associated with humans attempting to manage the operations in computer-integrated manufacturing systems. The primary purpose of this taxonomy was to establish a framework for designing information systems that could better support (through minimization of temporal errors) human decision and control activities.

10.5.3 Databases

According to Swain [37] "the most serious problem for HRA continues to be the scarcity of data on human performance . . . in the form of time-independent and time-dependent error-relative frequencies on the tasks of interest" (p. 308). Rather than remain hostage to this demand, error prediction may ultimately benefit from more integrated and intelligent databases that cover a wide variety of incidents. An example is the Human Factors Information System [39] which has the capability for detecting trends associated with problem events, and can be applied to a variety of industries in addition to the nuclear industry for which it was developed.

10.5.4 Simulators

Although simulators provide what appears to be one of the most useful methods for exploring human performance (including the potential for committing errors and error recovery) under conditions that in principle parallel the dynamic and multivariate complexity of the real-world situations of interest, their usefulness may depend on the particular industry for which they are developed. Dougherty [35] notes a number of problems with the use of simulators in the nuclear industry, including the observation that they are limited in the scenarios they simulate and are incapable of duplicating the conditions of stress that usually accompany crisis decision making (in contrast to pilots who can acquire simulator sickness). He also notes that NPP operators have learned to play "simulator games" in the sense that "operators know that they will never fail a requalification test when they 'err toward safety'" (p. 288). NPP simulators also typically do not include personnel (such as auxiliary operators) who would contribute in the event of an emergency, nor do they include the potential for hazards (such as heat and radiation) that occur outside the control room during actual events. Swain [37] has noted that the error rates derived from simulator exercises are optimistic due to the fact that these exercises are typically performed with larger control room crews (four) than are required (two) by plant policy to be located at all times in the control room.

While some of these deficiencies can conceivably be corrected, it is unlikely that NPP simulators can become as useful for human error prediction as flight simulators [40]. One factor that undermines NPP simulators is that operators may never have actually experienced many, if any, of the situations for which exercises are being performed. In contrast, pilots can bring to simulators emotional responses that are due, if for any other reason, to the much higher frequency of close calls they are likely to have been involved in.

10.5.5 Human Performance Models

The capabilities for error prediction depend to some degree on the availability of useful models of human performance. A useful model, within the context of human reliability modeling, implies that the focus be on human tendencies under conditions of high risk. This, in turn, requires identification of the factors that comprise the predominant portion of the performance variance associated with high-risk activities. These factors are likely to include danger, time estimation, design of procedures, tradeoffs in decision making, and group decision making. Given our relatively poor understanding of the effects of many of these factors on human performance under potentially hazardous conditions, investigations of these factors in isolation would probably not seriously undermine the ultimate goal of predicting human error under conditions where these factors are likely to be interacting with one another. An example of such an effort is the conceptual analysis, based on a model of time estimation, that describes the possible dynamic effects of various assumptions (i.e., scenarios) on the propagation of discrepancies between objective time and subjective time duration experience [41].

10.5.6 Violations and Work Cultures

Reason [3] distinguishes between errors and violations, defining the latter as deliberate deviations from practices that are assumed to be necessary for maintaining plant safety. While errors are usually defined within an information processing network, he notes that violations can only be described within a relatively complex "social milieu" that encompasses procedures, rules, work practices, etc., and offers evidence that suggests that errors and violations, despite the absence of clearly defined boundaries between them, are governed by different cognitive mechanisms.

Sociotechnical analyses that can uncover details concerning management policies and the work cultures that often evolve in response to these policies represent an important and understated aspect of human reliability modeling and assessment. These types of analyses can reveal psychosocial disturbances related to the existence of inconsistencies in the design of procedures and the lack of available feedback channels from operators concerning these designs, and to perceptions by workers that management has not taken sufficient steps towards minimizing hazards that they are routinely susceptible to. Although errors and violations may be mediated by different cognitive processes, their etiologies (or what Reason refers to as "resident pathogens") may in a number of cases be identical, and are likely to be better understood through a combined cognitive and psychosocial perspective.

10.5.7 Design Considerations

The importance of incorporating human factors design considerations for reducing human error is self-evident. That the resultant errors would, as Swain [37] noted, become more predictable is less obvious but just as important for any human reliability program concerned with minimizing the potential for human error and its consequences. The problem with HRA methods that emphasize HEPs is that design tends to be motivated by differences that show up in the overall failure or success probabilities in the HRA or systems event tree. In the absence of a causal framework it becomes nearly impossible to predict the effects of design changes on human performance, potentially relegating the necessarily iterative and tightly coupled process of design into a series of independent steps.

Design considerations need to be accorded a much more important role in HRA. Methods are needed that can help better anticipate the effects of design changes, such as whether a proposed change can potentially lead to adverse consequences due to a "loss of momentum" during the performance by the operator of the newly designed task. Design methodologies are also needed to determine how to maximize error detection and error recoverability. Finally, design strategies need to be pursued that can maximize the human's performance on problem-solving and decision making tasks (such as fault diagnosis) that need to be performed under severe time constraints. Towards this end, the ecological interface design (EID) framework [42] deserves serious consideration. Human-machine interfaces based on EID principles have the potential for being much more transparent than interfaces designed based on more conventional principles, a critical feature when one considers the tendency today's highly complex systems have for appearing relatively opaque.

10.5.8 PRAs

One cannot help but wonder to what extent the requirement by PRAs to have human performance summarized in terms of an error probability has impeded insights into causality of human error. The commitment to "funnel" the complexity of human behavior into a quantitative risk figure results in what often appears as awkward attempts by various HRA methods at backing up for the purpose of mathematically describing behavioral considerations (such as PSFs) previously ignored in the "primary" analytic process. With the maturation of a variety of disciplines such as cognitive science, artificial intelligence, and fuzzy sets and systems, the earlier concern noted by Swain [1] justifying the need for deriving HEPs may no longer be valid. The possibility exists that other, perhaps "hybrid" HRA methods, can be developed that relax some of the quantitative demands, hopefully at the expense of obtaining a better understanding of human performance. The emphasis on data, which PRA promotes, may be overstated. As Rouse [43] has noted, successful design (and presumably analysis) of many complex human-machine systems is more likely to depend on obtaining answers to questions (e.g., "How would personnel be likely to respond to a particular unusual situation for which they had not been trained?") that reflect the need to gain a deeper understanding of problems and their potential alternative solutions.

10.5.9 Error Prediction and Task Analysis

As stated in [36]: "To understand and predict errors, we must understand comprehensively the properties, purpose, and operation of the human-machine system. This usually requires a detailed task analysis" (p. 60). The type of task analysis performed is critical in such an endeavor, and limiting the scope of the problem to relatively well-defined work contexts (e.g., fault diagnosis involving steam generators or transportation within the plant of hazardous materials) will help identify a suitable task-analytic methodology.

A key to prediction of human error is situation modeling, which might consist of a number of methodologies such as task analysis for the identification of "triggering factors," and Reason's model for establishing mappings between these factors and error tendencies. The situation modeling phase would most likely be characterized as recursive, allowing situational contexts to slowly take form so that the interplay between both primary and secondary triggering factors and error tendencies can become revealed more clearly. Methodologies for prioritizing the likelihood of the various potential failure modes could comprise an error-prediction stage. If previous well-documented accident cases for these work scenarios exist, they can be used to perform coarse validation studies to gauge the viability of the still tentative models of error prediction. Ultimately, design intervention strategies will need to be considered in light of the information revealed in the previous stages.

10.6 Concluding Remark

Although there are still many unanswered questions regarding the role of HRAs and the direction HRAs should take, and many gaps to fill in our understanding of human error, if there is to be a second generation in human reliability assessment the challenge of predicting human error is likely to preoccupy many of its constituents.

Acknowledgement

The author wishes to thank the CANDU Owners' Group and the Health and Safety Division of Ontario Hydro in Toronto, Canada, for support and encouragement during a project involving looking into application of human reliability techniques to radiation protection in nuclear power stations. This project provided the opportunity to test many of the concepts described in this chapter within the context of an operating station's day-to-day planning and work programs. He also wishes to thank the staff of Pickering Nuclear Generating Station for their cooperation and patience in providing details of the tasks being observed.

REFERENCES

[1] Swain A.D., *Some Problems in the Measurement of Human Performance in Man-Machine Systems*, Human Factors, Vol. 6, pp. 687-700, 1964.

[2] Swain A.D. and H.E. Guttmann, *Handbook of Human Reliability Analysis with Emphasis on Nuclear Power Plant Applications*, NUREG/CR-1278, Washington, D.C., 1983.

[3] Reason J., *Human Error*, Cambridge: Cambridge University Press, 1990.

[4] Hahn A.H. and J.A. deVries II, *Identification of Human Errors of Commission Using Sneak Analysis*, Proceedings of the Human Factors Society 35th Annual Meeting, pp. 1080-1084, 1991.

[5] Pedersen O.M., *Human Risk Contributions in Process Industry: Guides for their Pre-identification in Well-Structured Activities and for Post Incident Analysis,* Risϕ-M-2513, Risϕ National Laboratory, Denmark, 1985.

[6] Blackman H.S., *Modeling the Influence of Errors of Commission on Success Probability*, Proceedings of the Human Factors Society 35th Annual Meeting, pp. 1085-1089, 1991.

[7] Gertman D.I., *INTENT: A Method for Calculating HEP Estimates for Decision Based Errors*, Proceedings of the Human Factors 35th Annual Meeting, pp. 1090-1094, 1991.

[8] Embrey D.E., P. Humphreys, E.A. Rosa, B. Kirwan and K. Rea, *SLIM-MAUD: An Approach to Assessing Human Error Probabilities Using Structured Expert Judgement*, NUREG/CR-3518, Washington, D.C., 1984.

[9] Bello G.C. and V. Colombari, *The Human Factors in Risk Analyses of Process Plants: The Control Room Operator Model "Teseo"*, Reliability Engineering, Vol. 1, pp. 3-14, 1980.

[10] Williams J.C., *HEART - A Proposed Method for Assessing and Reducing Human Error*, 9th Advances in Reliability Technology Symposium, 1986.

[11] Humphreys P., *Human Reliability Assessors Guide*, In B.A. Sayers (Ed.), Human Factors and Decision Making: Their Influence on Safety and Reliability, Elsevier Applied Science, London, pp. 71-86, 1988.

[12] Hannaman G.W. and D.H. Worledge, *Some Developments in Human Reliability Analysis Approaches and Tools*, Reliability Engineering and System Safety, Vol. 22, pp. 235-256, 1988.

[13] Kantowitz B.H., A.C. Bittner and Y. Fujita, *Mathematical Description of Crew Response Times in Simulated Nuclear Power Plant Emergencies*, Proceedings of the Human Factors Society 34th Annual Meeting, pp. 1127-1130, 1990.

[14] Askren W.B. and T.L. Regulinski, *Mathematical Modeling of Human Performance Errors for Reliability Analysis of Systems*, Aerospace Medical Research Laboratory, AMRL-TR-68-93, 1969.

[15] Dougherty E.M. and J.R. Fragola, *Foundations for a Time Reliability Correlation System to Quantify Human Reliability*, 4th IEEE Conference on Human Factors and Power Plants, pp. 268-278, 1988.

[16] Askren B.W. and T.L. Regulinski, *Quantifying Human Performance Reliability,* Technical Report AFHRL-TR-71-72, Air Force Systems Command, Brooks Air Force Base, TX, 1971.

[17] Kapur K.C. and L.R. Lamberson, *Reliability in Engineering and Design*, John Wiley & Sons, New York, 1977.

[18] Dhillon B.S., *Human Reliability with Human Factors*, Pergamon Press, New York, 1986.

[19] Laughery K.R., *Computer Modeling of Human Operators in Systems*, Proceedings of the 1984 International Conference on Occupational Ergonomics, Vol. 2, pp. 26-34, 1984.

[20] Siegel A.I., J.J. Wolf and M.R. Lautman, *A Model for Predicting Integrated Man-Machine System Reliability: Model Logic and Description*, Naval Seas System Command, AD-A009 814, pp. 1-60, 1974.

[21] Siegel A.I., W.D. Bartter, J.J. Wolf, H.E. Knee and P.M. Haas, *Maintenance Personnel Performance Simulation (MAPPS) Model: Summary Description*, NUREG/CR-3626, Washington, D.C., 1984.

[22] Siegel A.I., J.J. Wolf and M.R. Lautman, *A Family of Models for Measuring Human Reliability*, Proceedings of the 1975 Annual Reliability and Maintainability Symposium, pp. 110-115, 1975.

[23] Rasmussen J., *Notes on Human Error Analysis and Prediction*, In G. Apostolakis and G. Volta (Eds.), Synthesis and Analysis Methods for Safety and Reliability Studies, Plenum Press, London, pp. 357-389, 1979.

[24] Rasmussen J., *Human Reliability in Risk Analysis*, In A.E. Green (Ed.), High Risk Safety Technology, John Wiley & Sons, London, pp. 143-170, 1982.

[25] Rasmussen J., *Human Errors: A Taxonomy for Describing Human Malfunction in Industrial Installations*, Journal of Occupational Accidents, Vol. 4, pp. 311-333, 1982.

[26] Rasmussen J., Information Processing and Human-Machine Interaction, North-Holland, Amsterdam, 1986.

[27] Leplat J. and J. Rasmussen, *Analysis of Human Errors in Industrial Incidents and Accidents for Improvement of Work Safety*, Accident Analysis and Prevention, Vol. 16, pp. 77-88, 1984.

[28] Woods D.D., E. Roth and H. Pople, *Cognitive Environment: System for Human Performance Assessment*, NUREG-CR-4862, Washington, D.C., 1987.

[29] Mancini G., *A Solution to Dougherty's Apprehensions?* Reliability Engineering and System Safety, Vol. 29, pp. 329-335, 1990.

[30] Amendola A., *Accident Sequence Dynamic Simulation Versus Event Trees*, Reliability Engineering and System Safety, Vol. 22, pp. 3-25, 1988.

[31] Adams J.A., *Issues in Human Reliability*, Human Factors, Vol. 24, pp. 1-10, 1982.

[32] Sharit J., *A Critical Review of Approaches to Human Reliability Analysis*, International Journal of Industrial Ergonomics, Vol. 2, pp. 111-130, 1988.

[33] Apostolakis G.E., V.M. Bier and A. Mosleh, *A Critique of Recent Models for Human Error Rate Assessment*, Reliability Engineering and System Safety, Vol. 22, pp. 201-217, 1988.

[34] Spurgin A., *Another View of the State of Human Reliability Analysis (HRA)*, Reliability Engineering and System Safety, Vol. 29, pp. 365-374, 1990.

[35] Dougherty E.M., *Human Reliability Analysis - Where Shouldst Thou Turn?*, Reliability Engineering and System Safety, Vol. 29, pp. 283-299, 1990.

[36] Senders J.W. and N.P. Moray, *Human Error: Cause, Prediction and Reduction*, Lawrence Erlbaum, Hillsdale, NJ, 1991.

[37] Swain A.D., *Human Reliability Analysis: Need, Status, Trends and Limitations*, Reliability Engineering and System Safety, Vol. 29, pp. 301-313, 1990.

[38] Prabhu P., J. Sharit and C. Drury, *Classification of Temporal Errors in CIM Systems: Development of a Framework for Deriving Human-Centered Information Requirements*, International Journal of Computer Integrated Manufacturing, Vol. 5, pp. 68-80, 1992.

[39] Goodman P.C. and C.A. DiPalo, *Human Factors Information System: A Tool to Assess Error Related to Human Performance in U.S. Nuclear Power Plants*, Proceedings of the Human Factors Society 35th Annual Meeting, pp. 662-665, 1991.

[40] Rouse W.B. and S.H. Rouse, *Analysis and Classification of Human Error*, IEEE Trans. on Systems, Man, and Cybernetics, Vol. 13, pp. 539-549, 1983.

[41] Sharit J. and D.M. Malon, *Incorporating the Effects of Time Estimation into Human Reliability Analysis for High-Risk Situations*, IEEE Trans. on Rel., Vol. R-40, No. 2, pp. 247-254, 1991.

[42] Vicente K.J. and J. Rasmussen, *Ecological Interface Design: Theoretical Foundations*, IEEE Trans. on Systems, Man, and Cybernetics, Vol. SMC-22, pp. 589-606, 1992.

[43] Rouse W.B., *Much Ado About Data*, Human Factors Society Bulletin, Vol. 30, No. 9, pp. 1-3, 1987.

New Trends in System Reliability Evaluation
K.B. Misra (Editor)

CHAPTER 11

COMMON CAUSE AND DEPENDENT FAILURE MODELING

Per Hokstad
Senior Research Scientist
Division of Safety and Reliability, SINTEF,
N-7034, Trondheim, Norway.

ABSTRACT

A review of recent developments in the modeling of dependent failures in reliability analyses is presented in this chapter. First the concept of dependent and common cause failures in reliability assessment is discussed. Later on, the classification of common cause failures, and defensive strategies against them are presented. Elements of the structured approach for qualitative and quantitative analyses of dependent failures that has been suggested and widely used e.g. by US Nuclear Regulatory Commission, is also reviewed. Some parametric models for common cause failures are discussed, including the estimation of model parameters. Finally, data sources and the analysis of common cause failure events are presented.

11.1. Introduction

In reliability studies of redundant systems, it has now become essential to consider the effect of *common cause or dependent* failures. For instance, most of the safety-related systems at nuclear power plants are designed with redundant units of equipment. Thus, the associated risk of operating these plants is dominated by accidents that cannot occur unless multiple components fail to perform their functions. These multiple component failures very often occur as a result of a common cause failure (CCF) events.

The methodology for dealing with CCFs has been developed over several years. Earlier research was carried out by Watson and Edwards and their co-workers (see [1]) at Safety and Reliability Directorate (SRD) at UK Atomic Energy Authority. More recent work [2] has been undertaken by SRD in conjunction with the US Nuclear Regulatory Commission (NRC), the Electric Power Research Institute (EPRI) and others.

A research and development effort, jointly sponsored by EPRI and NRC (see [3] and [4]) has resulted in the development of a procedural framework for CCF analysis and many methodological advancements in the field. The methodology with an example of its application is presented in [5]; also giving additional references. This frameworks and techniques have become widely used, and has contributed to the initiation of new research efforts to improve and refine methods for CCF analysis. The European CCF reliability benchmark exercise is documented in [6].

Treatment of CCFs also constitutes one of the central topics addressed in current Nordic research projects. The project ***Risk Analysis*** (RAS-470) was initiated in 1985 as part of the research programme of the Nordic Liaison Committee for Atomic Energy (NKA), e.g. see [7], [8]. Results of the Nordic CCF data benchmark exercise were reported in [9]. Further, Hirschberg, [10], provides a review of recent experience from dependent failure analysis in Nordic countries.

An extensive amount of literature exists in this field. For instance [7] provides several references in addition to those listed at the end of this chapter.

11.1.1 Dependent Failures

Several definitions of *dependent failures* have been suggested in [1]. The most appropriate definition, as given in [11], is : *The failure of a set of events, the probability of which cannot be expressed as the simple product of the unconditional failure probabilities of the individual events.*

The causes of dependent failures are many. One example is *functional dependencies* of frontline systems on support systems (such as emergency electric power or service water). *Cascading failures*, such as a pump that fails because a valve in its suction line was mistakenly closed, is another example. Further, *maintenance related errors*, such as miscalibration of sensors, could be the causes of dependency. Common *environmental stresses*, e.g. excessive humidity or corrosion is another typical example. Further, many multiple-component failure events are caused, e.g. by inadequate design, manufacturing deficiencies, installation errors or operator errors.

There are numerous ways of accounting for dependencies in the reliability model. For a non-repairable system, with times to failure, say T_A and T_B, the dependence can simply be modelled by specifying a continuous multivariate distribution for the failure times T_A and T_B (i.e. a continuous dependence model). In this context many of the concepts of positive and negative dependence are defined in terms of the conditions on covariances, e.g. see [12]. In addition to *correlation*, the most well-known condition for positive dependence is perhaps *association.* The variables T_A and T_B are associated if

$$Cov[g(T_A, T_B), h(T_A, T_B)] \geq 0,$$

where g and h are nondecreasing functions.

One way of establishing a specific multivariate distribution for dependent variables, T_A and T_B is to assume that T_A and T_B are independently and identically distributed, *given* some other variable, say W. The dependent failure model is then obtained by introducing an appropriate distribution for W, see Section 11.5.5. Apostolakis and Moieni [13] points out that in case W is actually a (constant) parameter, it must be estimated, and the same estimate will then be used in the distribution of both T_A and T_B. This introduces an *apparent* dependency for T_A and T_B, see [13], where they use the concept state-of-knowledge (s-o-k) model for this phenomena. This concept (s-o-k) applies to any parametric model (not only for the continuous ones).

A completely different approach is to perform the modeling in terms of a fault tree (logic model). Several of the above mentioned dependencies, e.g. functional dependencies, and failures caused by human errors are usually *explicitly* modeled in the logic model. Other types of dependencies, where the cause-effect is not so obvious, are often modelled in a more approximate way, introducing *common cause basic events*, see Section 11.4.1.

11.1.2 Common Cause Failures (CCF)

One way of formulating dependence stems from the idea that the components A and B are susceptible to some common stress, that can cause both components to fail *simultaneously*. The event "both A and B fail" is then referred to as a *common cause failure (CCF)*. This is due to some external *shock*. In this respect common cause failures are a special type of dependent failures.

The crucial question here is the interpretation of the term *simultaneous*. Of course there can be a strong dependence even if the two failures do not occur exactly at the same time. The concept CCF is most often used for events implying that the redundant components fail to operate on the *same demand* (e.g. a test). Typically this means that the components have failed within the same test period. However, this event could of course also be due to two independent failures occurring in the same test period. Thus, whether or not there is a true CCF should be decided on the basis of investigation of the *causes* of components that have failed.

However, due to lack of information it is often difficult to decide whether there were dependent or independent causes for the failure. Thus, in practice the term CCF is often used synonymous with *multiple* failures (as opposed to a *single* failure).

Similarly, a single failure could actually be a CCF. The CCFs are caused for instance by erroneous maintenance or environmental stresses, (humidity, high temperature, lightening, etc.). In a realistic modeling these causes could therefore lead to failures of various multiplicities. In particular, a dependent failure could also lead to just one component failing, thus a CCF of multiplicity *one* is also possible.

To conclude, the term CCF often seems to be used synonymous with multiple failures (i.e., occur close in time). However, we should neither *preclude* the possibility that a multiple failure represents the joint occurrence of two independent failures, nor that the failure of a single component is actually a dependent failure. According to this view, it is the *cause* of the failure that is true criteria for the classification.

Today the term *common cause failure (CCF)* usually means a dependent failure that results from causes not *explicitly* modeled, but are rather handled by introducing CC basic events (see Section 11.4.1). Other dependent failures, however, e.g. due to functional dependencies on support systems and some human errors are included explicitly in the logic models, and are not referred to as CCFs.

11.2. Classification of Dependent Failures

There are numerous examples of dependent failures, e.g.:

(a) Components of the same design in a similar usage fail on a common demand due to a fatal design error.
(b) Several pumps fail to run because of the moisture in more than one control cabinet. Thus, all control cabinets are susceptible to humidity intrusion, and they also have to be located in a similar environment.
(c) A basic error in the maintenance procedure would be systematic across all crews, no matter when they do the maintenance.
(d) Slips or errors made during maintenance can be the source of common-cause failures if the maintenance is performed on several redundant components during a short time interval, and the slip or error is somehow made systematically over that maintenance episode.
(e) A common, systematic, installation error leads to simultaneous high levels of degradation.

Several systems for *classification* for CCFs have been suggested, e.g. see [1] and [14]. Both the well-known classification system proposed by National Centre of Systems Reliability (NCSR) and the classification system proposed by EPRI are referred [14]. The classifications are based on CCF *causes.* The EPRI classification of dependent failures can be described in terms the following basic categories:

(i) **(Branched) Shared Root-Cause Failures** - Multiple concurrent component failures attributed to the same root cause (say lightening, fire, water intrusion, extreme vibrations or erroneous maintenance)
(ii) **Component Caused Failures** - Multiple component failures initiated by another component's unavailability that result in the classic chain reaction or "domino effect". These are also often referred to as *Cascade Failures.*
(iii) **Functional Unavailabilities** - Multiple component unavailabilities that are not failures, and could either be caused by the failure or functional unavailability of another component (i.e., power or support modules).

The PRA Procedures Guide (NRC), [15], suggests the following three main CCF categories:

1. ***Common Cause Initiating Events (External Events)***
   External and internal events that have the potential for initiating a plant transient and that increases the probability of failure in multiple systems. These events usually, but not always, result in severe environmental stresses on components and structures.
   Examples include fire, floods, earthquakes, loss of offsite power, aircraft crashes and gas clouds.
2. ***Intersystem Dependencies***
   Events or failure causes that create interdependencies among the probabilities of failure for multiple systems. Stated another way, intersystem dependencies cause the conditional probability of failure of a given system along an accident sequence to be dependent on the success or failure of systems that precede it in the sequence.
3. ***Intercomponent Dependencies***
   Events or failure causes that result in a dependence among the probabilities of failure of multiple components of subsystems. The multiple failures of interest are usually within the same system or same minimal cut set that has been identified for a system or an entire accident sequence. These dependencies should be analyzed as part of the evaluation of fault trees.

Further, both type 2 and 3 of the above categories can be split into the following 4 subtypes:

A. ***Functional Dependencies***
   Dependencies among systems/components that follow from the plant design philosophy, system capabilities and limitations and design bases. One example is a system that is not used or needed unless other systems have failed; another is a system that is designed to function only in conjunction with the successful operation of other systems.

B. ***Shared Equipment Dependencies***
   These are dependencies of multiple systems/components on the same component, subsystem or axillary equipment. Examples are (i) a collection of pumps and valves that provide both a coolant-injection and coolant-recirculation function when the functions appear as different events in the event tree, and (ii) components in different systems are fed from the same electrical bus.

C. ***Physical Interactions***
   Failure mechanisms similar to those in common cause initiators (type 1) that do not cause an initiating event but nonetheless increase the probability of multiple system failures occurring at the same time. Often they are associated with extreme environmental stresses created by the failure of one or more systems after an initiating event. For example, the

failure of a set of sensors in one system can be caused by the excessive temperature resulting from the failure of a second system to provide cooling.

D. ***Human Interaction Dependencies***
Dependencies introduced by human actions, including errors of omission and commission. The persons involved can be anyone associated with a plant life cycle activity, including designers, manufacturers, constructors, inspectors, operators and maintenance personnel. A dependent failure of this type occurs, for example, when an operator turns off a system after failing to correctly diagnose the condition of the plant - an event that happened during the Three Mile Island accident when an operator turned off the emergency core cooling system.

11.3 Defences Against CCFs

Mechanisms of failure can be grouped according to general root cause characteristics and coupling factors. Thus, a discussion of defences against CCFs leads to the consideration of the following concepts:

- *root cause:* the most basic reason or reasons why a component (or components) failed, any of which would, if corrected, prevent recurrence;
- *coupling factor:* a characteristic of a group of components or piece parts that identifies them as susceptible to the same mechanisms of failure;
- *defence:* the design and operational features that protect against causes of equipment failure (or unavailability) and/or coupling.

11.3.1 Root Causes

The *root cause* is defined as the most basic reason for an effect which, if corrected, will prevent recurrence. The concept of root cause is tied to that of defence, because there are, in many cases, several possible corrective actions (i.e. defences) that can be taken to prevent recurrence.

A number of studies have investigated the root causes of CCF events, and several classification schemes have been proposed and used to categorize these events. Tables 11.1 - 11.3 (adapted from [16]) show root causes of CCF events arranged in a hierarchy that emphasizes the time when the cause was perceived to be introduced into the system. Thus the hierarchy starts with those causes that are introduced during the design stage and proceeds through the manufacturing, construction, installation and commissioning stages (Table 11.1). Root causes introduced during plant operation (through maintenance, operation, or the environment) are given in Tables 11.2 and 11.3. An alternative classification also exists of causes, e.g. [17].

Table 11.1 Root Causes of CCF Events (Design, Manufacturing, Construction, Installation and Commissioning)

Cause type	Examples of specific cause
Design requirements and specifications inadequacy	Designer failure to predict an accident Designer failure to recognize what protective action is needed
Design error or inadequacy in design realization	Inadequate facilities for operation, maintenance, testing or calibration Inadequate components Inadequate quality assurance
Design limitations	Financial Spatial
Manufacturing error or inadequacy	Failure to follow instructions Inadequate manufacturing control Inadequate inspection Inadequate testing
Construction/installation/commissioning error or inadequacy	Failure to follow instructions Inadequate construction control Inadequate inspection Inadequate testing

Table 11.2 Root Causes of CCF Events (Operation)

Cause type	Examples of specific cause
Failure to follow procedures	Failure to follow repair procedures Failure to follow test or calibration procedures
Defective procedures	Defective repair procedures Defective test or calibration procedures
Lack of procedures	Lack of repair procedures Lack of test or calibration procedures
Supervision inadequacy	Inadequate supervisory procedures, Inadequate action or supervisory communications
Communication problems	Communication among maintenance staff
Training inadequacy	Operator training in handling emergency situations

Table 11.3 Root Causes of CCF Events (Environmental)

Cause type	Examples of specific cause
Stresses	Chemical reactions (corrosion) Electrical failure Electromagnetic interference Materials interaction (erosion) Moisture Pressure Radiation Temperature Vibration
Energetic	Earthquake Fire Flood Impact loads

11.3.2 Coupling Factors

A coupling factor is a property of components or piece parts that identifies them as being susceptible to the same mechanisms of failure. Such factors include similarity in design, location, environment, mission, and operational, maintenance, and test procedures. They identify a potential for common susceptibility, and it is preferable to think of them as factors that help to define a potential common-cause component group (see Section 11.4.1). Examples of coupling factors include:

- same design
- same hardware
- same software
- same installation staff
- same maintenance or operations staff
- same procedures
- same environment
- same location

A search for coupling factors is primarily a search for similarities in the design, manufacture, construction, installation, commissioning, maintenance, operation, environment, and location of redundant components.

11.3.3 Defences

From the previous discussion, it is seen that the strategy for defence against CCFs should be based on defences against the root causes of failure and/or defences against the couplings.

Defences that are typically employed to protect against the *root causes* of failures include design control, using environmentally qualified equipment, testing and preventive maintenance programs, review of procedures, training of personnel, quality control, and several others. Thus, some examples are

- *Personnel training.* A program to assure that the operators and maintainers are familiar with the procedures and are able to follow them correctly during all conditions of operation.
- *Quality control.* A program to assure that the product is in conformance with the documented design and that its operation and maintenance are according to approved procedures, standards, and regulatory requirements.
- *Preventive maintenance.* A program of applicable and effective preventive maintenance tasks designed to prevent premature failure or degradation of components.
- *Procedures review.* A review of operational, maintenance, and calibration/test procedures to eliminate incorrect or inappropriate actions resulting in component or system unavailability.

These defences against root causes can be aimed at reducing dependent as well as independent failures. Thus, for instance the introduction of redundancy, which is most effective for preventing independent failures of the system, can also be a means of protecting against CCFs. Proven design and standardisation, simplicity, and operational interfaces (e.g. related to by-pass) are other general factors affecting independent failures as well as CCFs.

On the other hand, a search for defences against *coupling*, is primarily a search for dissimilarities among components, including the differences in the components themselves; the differences in the way they are installed, operated, and maintained; and the differences in their environment and location. Thus, key words are separation/protection and diversity. Defences based on these principles should be applied both during design and operation.

Protection and segregation means that redundant components should be physically separated. For instance, high room temperature is a condition that could affect different equipments, and different equipments would have to be physically separated with different room cooling systems to defend against coupling from high room temperature. In general, controlling the *environment* is an important means of reducing the number of dependent failures. This concerns both the physical environment (e.g. temperature and vibration) and which the persons are allowed access.

The main type of *diversity* used as defence against CCFs are listed in Table 11.4. *Functional diversity* is a strong defence against coupled failures. Functional diversity could be functionally diverse ways of providing power, such as a diesel generator for one part of a plant, and a hydroelectric generator for the other part.

Staff diversity, e.g. having different personnel to perform testing and maintenance on functionally different systems, would be a strong defence.

Table 11.4 Types of Diversity Used as Defence Against CCFs.

Type of diversity	Description
Equipment	Mixture of interchangeable components made by different manufacturers
Functional	Totally redundant system with an entirely different principle of operation for the express purpose of reducing the likelihood of a total loss of function that might occur because all like components are vulnerable to the same cause(s) of failure.
Staff	Different teams maintain and test redundant trains

For redundant systems, the two redundant branches may be tested at different times. This is called *staggered testing*. Staggering of inspections and tests offers some protection against the coupling as compared to performing these activities simultaneously or sequentially. First, it reduces the coupling associated with human-related failures because the probability that the technician repeats an incorrect action is lower when activities are performed months, weeks, or even days apart than when these activities are performed a few minutes or a few hours apart.

A second potential advantage of staggering inspections and tests relates to the exposure time for CCF events. If multiple components have indeed failed because of internal faults, then evenly staggering inspections and tests can reduce the maximum time that the multiple failures are present, without increasing the frequency of inspection and testing of individual components.

Defences against components being unavailable can be very specific (e.g. an alarm that indicates the occurrence of a failure) or very general (e.g. a preventive maintenance program that attempts to eliminate equipment deficiencies before they develop into actual failure occurrences). Judging the adequacy of a defence is often easier for specific defences than it is for more general defences, but both types of defences can be very effective in reducing the impact of CCF occurrences.

The *cause-defence matrix* explicitly illustrates the influence of the various defences on both the root causes and the coupling factors, and is a helpful tool in evaluating the effect of various defences, see [16] and [18].

11.4 Incorporation of CCFs Into the Reliability Model

11.4.1 CC Component Groups and CC Basic Events

A framework for CCF analysis is documented in [3] and [4]. Key elements are also discussed in [5] and [17]. Reference [5] also provides a complete example of the approach. This framework is based on a four stage process for carrying out a comprehensive CCF analysis:

Stage 1: System logic model development
Stage 2: Identification of CC component groups
Stage 3: CC modeling and data analysis
Stage 4: System quantification and interpretation of results

Some elements of stages 1 and 2 will be reviewed now, in particular the identification of *common cause (CC) component groups* and the *CC basic events*, that are to be analysed by a parametric CCF model; (these models will be discussed in Section 11.5).

A CC component group will in this process be identified as a group of similar or identical components, that have a significant likelihood of experiencing a common cause event, and thus will be subject to further CCF analysis.

The common cause basic events (CCBE) represent failures of a specific set of components in a CC component group. As an example, consider a CC component group of three components, A, B and C, the Boolean representation of total failure of component A is

$$A_T = A_I + C_{AB} + C_{AC} + C_{ABC} \tag{11.1}$$

where

A_T = total failure of component A (independent or dependent)
A_I = failure of component A only (i.e. a *single* failure)
C_{AB} = failure of components A and B (and not component C) from common causes
C_{AC} = failure of components A and C (and not component B) from common causes
C_{ABC} = failure of components A, B, and C from common causes

are the common cause basic events (CCBE).

The event A_I in this expression is often referred to as "failure of component A from *independent* causes". However, using a consistent notation, *independent* should not be used as synonymous for *single* (see Section 11.2). Thus, by defining

I_A = failure of component A from independent causes
C_A = failure of component A from dependent causes

we can write $A_I = I_A + C_A$. Whether a failure event is of type I_A or C_A is in principle decided by investigating the failure cause. However, due to lack of information, it may be difficult to decide whether I_A or C_A has occurred, and for the further *parametric* analysis, one may decide to lump the two events, I_A and C_A together.

11.4.2 System Logic Model Development (Stage 1)

Now we return to the framework for CCF analyses, referred in Section 11.4.1. The analysis starts with the system familiarization and problem definition, where also system failure modes are identified.

A key step in the system analysis is the development of one or more *logic models* of the systems, e.g. in the form of fault trees or reliability block diagrams. The logic model(s) will be a basic output of stage 1 in the CCF analysis.

Regarding dependent failures, it is necessary to identify those root causes of dependency that are to be *explicitly* modeled. For instance, some external root causes as fire, flood or earthquake, can be included in the system-level analysis in terms of explicit models. Another example is *functional dependencies* on support systems (such as emergency electric power or service water). *Cascading failures*, such as a pump that fails because a valve in its suction line was mistakenly closed, is another example. Similarly, certain categories of human errors can be modeled explicitly using human reliability analysis techniques.

By identifying the dependent failures that are explicitly modeled, an analyst define the scope of the *residual* CCF analysis, i.e. those root causes of multiple failures that are *not* modeled explicitly. These residual CC events are treated using parametric CC models (see Section 11.5).

The causes of dependent failures that are not explicitly addressed are associated with many multiple-component failure events, include inadequate design, manufacturing deficiencies, installation and commissioning errors, maintenance-related errors, and environmental stresses, e.g. excessive moisture, corrosion, or contamination. The CCF models are introduced to account for these causes of dependent failures.

The CCF Reliability Benchmark Exercise [6] gives some guidelines for this process. For instance, it is stated that multiple failure events for which a clear cause-effect relationship can be identified should be explicitly modeled in the system model. The root cause events should be included in the system fault tree so that no further special dependent failure model is necessary. This applies to multiple failures caused by internal equipment failure (such as cascade failures and component-caused functional availability events) and multiple failures due to clearly identifiable human errors (e.g. in performing the steps of a prescribed procedure). Multiple failure events for which no clear root cause event can be identified can be modeled using implicit methods such as the parametric models.

11.4.3 Identification of CC Component Groups (Stage 2)

The end result of stage 2 will be to identify those component groups for which CCFs are to be included in the model. A qualitative analysis and quantitative screening is performed to arrive at this result.

Qualitative analysis

This analysis is the first step for identifying the component groups for which CCF analysis shall be performed. The analyst searches for common attributes of components and mechanisms of failure that can lead to CCFs. Key attributes for which he can assess the degree of similarity of the various components are design, location, testing/maintenance procedures, environmental conditions, etc. In addition to the search for signs of dependence, an analyst must also evaluate the effectiveness (or lack) of defensive measures against CCFs (see Section 11.3).

Guidelines to support the selection of CC groups have been provided, e.g. see [5]. Special attention should be given to evaluate the degree of diversity for redundant components. Independence can often be assumed by *complete* diversity, and passive components can often be omitted. Computer codes are available to support identification and categorization of component groups for larger and more complex problems, [19] and [20] in the CC analysis.

Quantitative screening

When the qualitative screening has been carried out and the analyst has identified component groups that are judged to be susceptible to CCFs, a *quantitative* screening for CCF candidates is then performed, by actually carrying out a complete quantitative CC analysis, except that a conservative and very simple quantification model is used.

First the fault trees are modified to explicitly include a *single* CC basic event for each component in a (potential) CC group. This means that the simple beta factor model (Section 11.5.2) is applied at this step. Thus, if components *A, B*, and *C* are identified as a CC component group, the basic events on the fault tree are expanded to include the basic event C_{ABC}, defined as the concurrent failure of A, B, and C due to a common cause. So for instance the event A_T = "A fails" is now substituted by $A_T = A_I + C_{ABC}$, where A_I denotes independent failure of component A. (Thus, in this screening process we adopt a simple view that all single failures are independent, and that all CCFs represent a simultaneous failure of all the three components.)

These substitutions are made at every point on the fault trees where the events "A fails", "B fails", or "C Fails" occur. Thus, resulting cut set involving the intersection $A_I B_I C_I$ will now also have an associated cut set involving C_{ABC}. Numerical values for the CCF basis events is estimated using the beta factor model, i.e. $P(C_{ABC}) = \beta \cdot P(A_T)$. Here $P(A_T)$ could be assessed by the total random failure frequency that would be used in the absence of any common cause considerations, and for the screening purposes a conservative value like $\beta = 0.1$ could be used.

The CC basic events that are in this way found to contribute little to the system failure/accident frequency can be dropped from further consideration. Those CC basic events that are found to be significant contributors to the system frequency are retained and can be further analyzed by a more detailed analysis of event data to support realistic estimates of model parameters.

Having finished stage 2, the CC component groups and CCBE to be analysed in stage 3 of the procedure have been identified. In this 3rd. stage the events A_T belonging to such a component group are replaced by a Boolean representation in terms of CCBE. (Equation (11.1) provides this representation for a CC group having three components.) All these expressions are now inserted in the logic model for the system, and the next question is the choice of a parametric model for the CCBE. Before turning to this topic (Section 11.5) it is stressed that all CCBE are treated as *independent* and *mutually exclusive* events. For instance C_{AC} and C_{AB} can not occur simultaneously. Thus, the minimal cut sets of a two-out-of three system has the following minimal cutsets

$$\{A_I, B_I\} \quad \{A_I, C_I\} \quad B_I, C_I\}$$
$$\{C_{AB}\} \quad \{C_{AC}\} \quad \{C_{BC}\} \quad \{C_{ABC}\}$$

Thus, for instance the cut set $\{C_{AB}, C_{AC}\}$ does not exist.

11.5 Parametric Models

11.5.1 General Considerations

The parametric models for dependent failures can be classified into :

- nonshock model, or
- shock model

In the nonshock models, we simply perform a splitting of the failure frequencies of the components according to the failure *multiplicity* of failure events. For a three component system, this corresponds to using the representation given in equation (11.1), not making any distinction between independent and dependent failures for single failure events. The shock models are defined as models where it is explicitly stated that each component experiences truly independent failures. In addition the system is subject to shocks that can cause system errors of any multiplicity (including the multiplicity of one). Thus, the shock models actually account for the fact that $A_1 = I_A + C_A$ (see Section 11.4.1).

The beta factor model, which could be classified as a special case of both these categories, will be discussed first. Next nonshock and shock models will be treated in some detail. These types of models are discussed for instance in [1] and [21-25].

Several other types of models for dependent failures are also available. The *square root bounding method* was introduced in the *Reactor Safety Study*, [26]. Further, time to failure for the component can simply be modeled as dependent stochastic variables. A few of these models will be referred to in Section 11.5.5.

All parametric models consider component groups consisting of m *identical* components, thus, assuming the following symmetry assumption.

Assumption A (symmetry)

The dependent failure of a specific set of components is independent of which components that belong to that set. Thus, for instance the failure events $\{A_T, B_T, C_T\}$ and $\{A_T, C_T, D_T\}$ have the same probability of occurring.

Next some notation is introduced. Each failure event (CCBE) can be quantified either by a *failure rate* or a *failure probability*. Here we will refer to failure *frequencies*. These can be interpreted either as rates or probabilities. In case the frequencies are interpreted as failure rates, we observe that the exponential failure model is assumed for all CCBEs.

Notation

Assuming m identical components satisfy the symmetry assumption, we can define,

Q_T = Total failure frequency of one specific component.
Q_k = Failure frequency of one specific CC basic events involving k specific components ($1 \leq k \leq m$)
Q_S = System *error* frequency

Actually, Q_k, $k = 1, 2, \ldots, m$, are the parameters of the Basic Parameter (BP)

model, see section 11.5.3. They simply represent a splitting of the *component* failure frequency, according to failure *multiplicity*.

Further, Q_S is the total frequency of an event affecting any of the *m* components, i.e. adding up the frequency of all CCBEs. Whether or not these events (of frequency Q_S) lead to a system *failure*, will depend on the system configuration (voting). The interpretation of the above concepts is illustrated in Figure 11.1. In particular we observe that for $m = 3$

$$Q_T = Q_1 + 2 \cdot Q_2 + Q_3$$

$$Q_S = 3 \cdot Q_1 + 3 \cdot Q_2 + Q_3$$

The general formulas for Q_T and Q_S are

$$Q_T = \sum_{k=1}^{m} \binom{m-1}{k-1} Q_k \tag{11.2}$$

$$Q_S = \sum_{k=1}^{m} \binom{m}{k} Q_k \tag{11.3}$$

Here the binomial coefficient $\binom{m}{k}$ represents the number of combinations of *k* components out of the total of *m*.

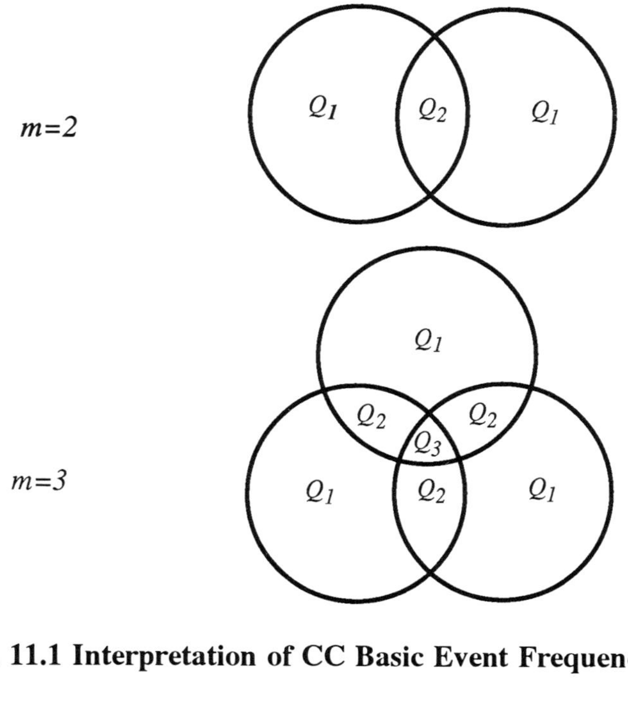

Fig. 11.1 Interpretation of CC Basic Event Frequencies (m=2 and m=3).

11.5.2 Beta Factor Model

The beta factor model was introduced by Fleming [27], and is probably the most widely used model for treating common cause failures. The main reason obviously is the extreme simplicity of this model, and its ability - at least in a rough way -to account for the occurrence of dependent failures.

There is one main assumption of this model. Given that a specific component has failed, this failure will with probability β cause **all** components (of the group) to fail, and with probability $1-\beta$ involve just the specific component in question. Thus, referring to the notation introduced in Section 11.5.1, we have

$$Q_1 = (1 - \beta)\cdot Q_T$$

$$Q_k = 0 \qquad\qquad (k = 2, 3, \dots, m-1)$$

$$Q_m = \beta\cdot Q_T$$

Further, the system error frequency equals (see equation (11.3)),

$$Q_S = [m\cdot(1 - \beta) + \beta]\cdot Q_T$$

The C - factor Model.

The *C - factor* model was discussed in [28]. This differs from the beta factor model, essentially by the use of $C = \beta/(1-\beta)$ as a parameter instead of β. The C-factor equals the ratio of the rate of dependent failures to the rate of independent failures. The use of C instead of β has some implications on the estimation of the parameter, [28].

Partial Beta Factor Model

Several approaches for trying to estimate *plant specific* beta factors have been suggested, e.g. see [29] and [30]. The Partial Beta Factor (PBF) method, described in [30], constructs the beta factor as a *product* of a number of partial betas derived from the judgement of system defences. The *reference* values are first given for factors like design control, functional diversity, operational interface, protection and segregation, testing and commissioning. These values (partial beta factors, β_i) are thought of as appropriate for systems of average quality in terms of defences. A resulting reference beta factor, β, is obtained by multiplying all these partial beta factors (i.e. $\beta=\prod\beta_i$). A structured approach is defined for making a judgement of the degree of protection against CCFs at a given plant. Then, each of the reference partial beta factors (PBF) are adjusted accordingly, and a plant specific beta factor is obtained as the product of the adjusted PBFs.

A general approach for use of failure event reports to obtain plant specific estimates for CCF analysis will be described in Section 11.7.

11.5.3 Non-Shock Models

Those multiparameter models, which are *not* categorized as shock models, include the Basic Parameter (BP) model, the Multiple Greek Letter (MGL) model and the alpha-factor model. They all allow a completely general multiplicity distributions. The number of parameters will equal the number of components (m) in the component group and there is a one-to-one relation between the parameters of these models. The choice of parametrization of course has some impact on the parameter estimation, [7] and [13].

Basic Parameter (BP) Model.

The *BP* model has been discussed for instance by Fleming et al, [21] and Apostolakis and Moieni, [13], and is simply based on the parametrization introduced in Section 11.5.1, i.e.

Q_k = failure frequency of a specific group of k components, $(k = 1, 2, \ldots, m)$.

Multiple Greek Letter (MGL) Model

The *MGL* model [13 and 21], is a direct generalization of the beta factor model allowing for CCFs of various multiplicity. With $m = 3$, the following parameters are introduced:

β = The conditional probability that the cause of a failure of specific component (say component A) will be shared by *at least one* additional component,

γ = The conditional probability that a component failure which we know is shared by at least one additional component will in fact be shared by two or more additional components.

These definitions will give the following relations between the parameters of the BP and MGL models

$$Q_1 = (1-\beta)\cdot Q_T$$

$$Q_2 = \beta(1-\gamma)\cdot Q_T / 2$$

$$Q_3 = \beta\gamma\cdot Q_T$$

For a general m - value the we have the following relation between the parameters (ρ_i) of the MGL model and the Q_k parameters of the BP model:

$$Q_k = \{\prod_{i=1}^{k} \rho_i\}(1-\rho_{k+1})Q_T \Big/ \binom{m-1}{k-1}$$

Here, $\rho_1=1$, $\rho_2=\beta$, $\rho_3=\gamma$, ... , $\rho_{m+1}=0$. Observe that the conditional probabilities in the MGL model (ρ_i), always refer to failure of *one particular component*, see the discussion in [13].

Alpha Factor Model

Definition of the *alpha factor* model [31], can be given in terms of the total frequency of system errors, Q_S. In the alpha factor model, α_k is so defined that $\alpha_k \cdot Q_S$ is the frequency of system errors of multiplicity k. Thus, given that a *system* error has occurred, α_k is the probability of having a failure of multiplicity k (with $\sum_k \alpha_k = 1$). Thus, α_k may be represented by the multiplicity distribution. In the BF model the frequency of a failure of multiplicity k equals $\binom{m}{k} \cdot Q_k = \alpha_k \cdot Q_S$. It directly follows that

$$\alpha_k = \binom{m}{k} \frac{Q_k}{Q_S} \tag{11.4}$$

Using this result in (11.2), we can easily obtain

$$Q_T = \{ \sum_{k=1}^{m} k\alpha_k \} \cdot Q_S \; / \; m \tag{11.5}$$

Now introduce the mean number of failed component at the event of a system error, i.e.

$$\mu_\alpha = \sum_{k=1}^{m} k\alpha_k \tag{11.6}$$

Thus we get the following relation between the total component frequency, Q_T and the total system frequency, Q_S:

$$Q_T = \{\mu_\alpha / m\} \cdot Q_S \tag{11.7}$$

This fundamental relation is actually quite obvious (see Appendix of [32]). Observe that if $\mu_\alpha = m$, all failures are of multiplicity m, and $Q_T = Q_S$. Further, if all failures are single ($\mu_\alpha = 1$) then $Q_T = Q_S/m$.

An alternative expression for α_k can be obtained using this relation. From (11.4), the following relation between α_k and Q_k is found

$$Q_k = \frac{m}{\binom{m}{k}} \frac{\alpha_k}{\mu_\alpha} Q_T \tag{11.8}$$

Relation of this type is often given as the defining relation for α_k. However, it is (11.4) that explains the actual interpretation of α_k.

The component group size (*m*).

All parameters, like Q's and α's will depend on the number of components, m of the relevant component group. In order to keep the notation fairly simple this dependence has been suppressed in the previous discussion. Now let us introduce $Q_k^{(m)}$ as the value of Q_k for a system consisting of m redundant components. Then it is of interest to investigate the relation between the $Q_k^{(m)}$ and $Q_k^{(m-1)}$. Assuming that the following regularity condition is to be satisfied, there is a fairly simple relation linking the distributions $\{Q_k^{(m-1)}\}$ and $\{Q_k^{(m)}\}$.

Assumptions B

- Each component has the same total frequency, Q_T, independent of m.
- Removing one of the m components will have no effect on the failure frequencies of the remaining m-1 components. Thus, a m-1 redundant system will behave exactly as a m redundant system, except that the state of component m is not observed.

Turning to Figure 11.1 it is realized that, provided both assumptions A and B are satisfied, we have the following relations between the $Q_k^{(2)}$ and the $Q_k^{(3)}$:

$$Q_2^{(2)} = Q_2^{(3)} + Q_3^{(3)}$$

$$Q_1^{(2)} = Q_1^{(3)} + Q_2^{(3)}$$

In general, the following relation shall apply

$$Q_k^{(m-1)} = Q_k^{(m)} + Q_{k+1}^{(m)}$$

This relation shows how the Q_k distribution for m-1 components can be derived, once the Q_k distribution for m components is found. Thus, given the distribution $Q_k^{(m)}$, $k = 1,2, \ldots\ m$, we can derive all distributions $Q_k^{(n)}$, where $n < m$.

Similarly, introducing $\alpha_k^{(m)}$ to represent α_k when there are m redundant components, we have

$$\alpha_k^{(m-1)} = C \cdot \{(1 - \frac{k}{m})\alpha_k^{(m)} + \frac{k+1}{m}\alpha_{k+1}^{(m)}\} \quad (k=1,2,\ldots.,m-1) \tag{11.9}$$

Here a proper normalization is provided by inserting

$$C = 1 - \frac{1}{m}\alpha_1^{(m)}$$

The need for this normalization is due to the fact that α_k is not defined for $k=0$. The relation given by equation (11.9) is referred to as a *mapping down* of the distribution.

One implication of having the relation (11.9) is rather obvious. Failure data collected from systems with different m-values can be combined to give a common set of estimates. Thus, the number of unknown parameters can thus be drastically reduced, since when the parameter estimates for the system with the highest m-value is established, the estimates corresponding to systems with smaller m-value directly follow by use of the above relation.

It is stressed that the validity of Assumptions B should not be taken for granted. However, if the parameters corresponding to the highest m-value is estimated, one should use the above formulas also to calculate the corresponding estimates for lower values of m (if relevant for the actual analysis). This mapping down will be a helpful exercise for securing that parameter estimates, which correspond to different m-values, are consistent with each other. A systematic use of mapping up/down when parameter during parameter estimation from failure event reports has also been suggested, see discussion in Section 11.7.2.

11.5.4 Shock Models

A fundamental assumption of the shock model is that there are two different sources of failures. The total failure frequency of a component can then be split into

Q_I = the frequency the component failures that are truly *independent*. Thus, only this particular component is affected. These failure events are typically the failures due to natural aging.

Q_D = the frequency of *dependent* failures that are due to certain external *system shocks*. The shocks can cause one or more component to fail simultaneously.

Thus, the total component failure frequency now equals, $Q_T = Q_I + Q_D$. The frequency, Q_D, can be split into various parts, according to the failure multiplicity. This could be done in various ways, see Section 11.5.3. It is however customary to apply a modeling close to the alpha-factor model. Let

Q_{SH}= Shock frequency (resulting in a dependent rate, Q_D, for the components involved). This corresponds exactly to the system error frequency, Q_S, introduced in Section 11.5.3. Thus, the relation between Q_{SH} and Q_D here is identical to that between Q_S and Q_T in (11.7).

Given that a shock (of rate Q_{SH}) has occurred, the multiplicity distribution is introduced as

f_k = the probability that exactly k components (among the m) fail due to a shock ($k = 0, 1, \ldots, m$). Here of course $\sum_k f_k = 1$. Observe that in contrast to the α_k- distribution, the f_k-distribution can have $f_0 > 0$.

Now the system frequency, $Q_S(k)$, for errors of multiplicity k equals

$$Q_S(1) = Q_I + Q_{SH} \cdot f_1$$

$$Q_S(k) = Q_{SH} \cdot f_k \qquad (k = 2, 3, .., m)$$

This model has also been referred to as the *Multinomial Failure Rate, (MFR)* model, see [13]. Models like the alpha-factor and beta-factor can of course be considered as special cases of the MFR model.

Binomial Failure Rate (BFR) Model

The BFR model was suggested by Vesely [33], and has been discussed for instance in [34] and [21]. The specific assumptions of the binomial failure rate model are that

- At the occurrence of a system shock - each of the components has the same probability p to fail.
- Given that a shock has occurred, the components fail independently of each other.

These assumptions lead to a binomial distribution. Thus the multiplicity distribution, f_k is given by the binomial probabilities

$$f_k = \binom{m}{k} p^k (1-p)^{m-k}$$

Two shock frequencies (rates) are often included in the BFR model; one of the shocks having $p=1$. This shock with $p=1$ is denoted as the *lethal* shock, as it causes failure of all components. Letting Q_{LS} denote the frequency of lethal shocks, we can obtain the following expressions for the basic parameters in the BFR model

$$Q_1 = Q_I + Q_{SH} \cdot p(1 - p)^{m-1}$$

$$Q_k = Q_{SH} \cdot p^k(1-p)^{m-k} \qquad (k = 2, 3, ... , m-1)$$

$$Q_m = Q_{SH} \cdot p^m + Q_{LS}$$

Observe that the frequency of single failures, Q_1, is split into the frequency of *independent* failures, Q_I and the frequency of (single) dependent failures, $Q_{SH} \cdot p(1-p)^{m-1}$. Applying equation (11.2), it is easily realized that the total failure frequency of one component equals $Q_T = Q_I + p \cdot Q_{SH} + Q_{LS}$. The contribution $p \cdot Q_{SH}$ corresponds to the fact that mean number of components that fail following a (nonlethal) shock equals $m \cdot p$. The beta factor model is a special case, letting $Q_{SH}=0$, thus having lethal shocks only.

It is observed that the shock model can be extended in a rather obvious way. In general, let us define various shock frequencies Q_{SH1}, Q_{SH2}, ..., (related to different groups of dependent failures). The total frequency of dependent failures affecting exactly k components are obtained by summing the contributions from the various shocks.

A limitation of the BFR model is the assumption that, following a shock, the components will fail independently of each other. For a more *direct* application it is also unfortunate that there is not one single parameter related to the degree of protection of the system against common cause failures. The shock rate Q_{SH} relates to factors as environment and maintenance, and is of course directly related to exposure to CCFs. Further, the parameter p essentially relates to the ability of the component to resist a shock (i.e. to component design). Consequently both these parameters are essentially independent of *system design* (and thus to common cause factors like diversity and complexity).

Random Probability Shock (RPS) Model

A generalization of the BFR model is introduced in [35], letting the failure probability, p, be a random variable, i.e. it takes different values with different shocks. This provides a more realistic model, as shocks (like vibration, temperature rise) are typically of varying strength. Letting p have a beta-distribution, it is well-known that the distribution of the number of failed components by a shock will have the so-called beta-binomial distribution, see [36]. The resulting shock model is called the *Random Probability Shock (RPS)* model, [35]. The multiplicity distribution is expressed in terms of two parameters. After a proper reparametrization, one parameter is the mean probability that a component fails due to a shock, (i.e. average p-value over all possible shocks). This parameter specifies how well the components are generally protected against the type of shock in question.

The other parameter, say D, is the correlation coefficient of the outcomes of the shock on two arbitrary components (defining the outcome of a shock as a 0, 1 variable). Therefore, this is a measure of dependence of the outcomes of the shocks. The case $D = 1$ corresponds to the case that it is impossible to get different outcomes for the components during the same shock, and the model reduces to the beta factor model. The case $D = 0$ corresponds to complete independence in the outcomes for the various components, and the model reduces to the BFR model. Thus, the dependence parameter, D could be interpreted as a measure of how well the system is protected against common cause failures.

11.5.5 Time Continuous Models for Dependent Failures

This section is terminated by referring a few alternative dependent failure models that have been suggested in the literature. Several dependence models start by assuming that the times to failure (of the m components) are identically but not independently distributed. The joint distribution can often be motivated by some physical arguments.

Common Load Model

Mankamo, [37], suggested a model where all m components are susceptible to a certain *load*, S. The components have *strengths* (protection against this load), R_1, R_2... , that are independently and identically distributed. This introduces a dependence in the failure times for the m components, and the probability of k components failing within a given time is derived from the distributions of S and R_i. A generalization of this model is introduced in [22].

The Clayton model

Clayton [38] suggested a general model, where the times to failure, T_1, T_2, ... are independent *given* the value of a stochastic variable W. A gamma distribution is assumed for the variable W, giving a joint distribution of the T_j's, depending on a single parameter ϕ. The limits $\phi \rightarrow 0$ and $\phi \rightarrow \infty$ represent independence and extreme positive dependence, respectively. This Clayton model has been treated for instance in [39] and [40] also. A possible extension of this model to cover repairable systems is indicated in [40].

11.6 Parameter Estimation

11.6.1 Estimators for CCF Model Parameters

Now we present the standard point estimators for the parameters of the most commonly used CCF models. The estimators are given for a system of m (identical) redundant components, and are based on the observation of

n_k = number of basic events involving failure of k similar components
N_D = number of system demands

Note that n_k refers to number of events. Thus, there have been $\sum_k k \cdot n_k$ individual component failures. The exposure of the events is here measured in terms of the number of demands. To obtain the time-based parameters, the number N_D should be replaced by the cumulative system exposure time. These estimators assume that all components are challenged by every system demand. Thus, system tests are assumed to be nonstaggered. More details about these estimators are found in [4]. The given estimators are based on maximum likelihood, and for instance the Bayes approach will provide alternative estimates, e,g, see [13]. Estimators for the total component failure frequency, Q_T, is also included below (when relevant).

1. Beta Factor Model

The estimator is based on an approximate method given in Appendix C of [4]

$$Q_T = \frac{1}{mN_D} \sum_{k=1}^{m} kn_k$$

$$\beta = \sum_{k=2}^{m} kn_k \Big/ \sum_{k=1}^{m} kn_k$$

2. *BP Model*
Here,

$$Q_k = \frac{n_k}{\binom{m}{k} N_D} \qquad (k=1,...,m)$$

3. *MGL Model*
Estimators are given for the case m=3. Higher order estimators are derived along the same lines. We also refer to the discussion of estimation for MGL parameters given in [13].

$$Q_T = \frac{1}{mN_D} \sum_{k=1}^{m} kn_k$$

$$\beta = \sum_{k=2}^{m} kn_k \,/\, \sum_{k=1}^{m} kn_k$$

$$\gamma = \sum_{k=3}^{m} kn_k \,/\, \sum_{k=2}^{m} kn_k$$

$$\delta = \sum_{k=4}^{m} kn_k \,/\, \sum_{k=3}^{m} kn_k$$

4. *Alpha Factor Model*
Here,

$$Q_T = \frac{1}{mN_D} \sum_{k=1}^{m} kn_k$$

$$\alpha_k = n_k \,/\, \sum_{j=1}^{m} n_j \qquad (k=1,...,m)$$

5. *BFR Model*

$$Q_I = n_I \,/\, [mN_D]$$

$$\frac{p}{1-(1-p)^m} = \sum_{k=1}^{m} kn_k \,/\, [m\sum_{k=1}^{m} n_k]$$

$$Q_{SH} = \frac{1}{1-(1-p)^m} \cdot [\sum_{k=1}^{m} n_k \,/\, N_D]$$

$$Q_{LS} = n_{LS} \,/\, N_D$$

where, n_1 is the number of single component failures due to CC shocks, n_I is the number of (single) independent failures, and n_{LS} is the number of lethal shocks. The expression $1-(-p)^m$ enters the formula because this is the fraction of *visible* CC shocks, i.e. shocks that result in at least one component failure. The parameter p is estimated by solving the second equation.

11.6.2 Example

The estimation of the beta factor is now considered in order to illustrate one type of problem that can arise while applying the above formula (essentially the same example is presented in [5]). Consider a duplicated system, say two gas detectors A and B of a safety system. The considered failure mode is *fail-to-operate* on demand, and the model of a constant failure probability on demand is assumed.

There are observed n_1 single component failures and n_2 failure events, where both detectors have failed. The total number of single-component demands is denoted by N (this is the sum of the demands for A and the demands for B). Then the relationship that exists between the component failure frequency, Q_T and the number N is $Q_T = (n_1 + 2n_2) / N$.

The number of times that there has been an effective test for the common cause failure is denoted by N_2. Then we also have the relation $Q_T \cdot \beta = n_2/N_2$ (which is the frequency of CCFs). Eliminating Q_T from these two relations, the estimator for β is seen to be of the following form

$$\beta = \frac{n_2}{n_1+2n_2} \cdot \frac{N}{N_2}$$

This estimator depends on N_2. For highly redundant systems the major number of demands are due to functional testing. Thus, N_2 and consequently also the β estimate will depend on the testing strategy. Now consider the following two test strategies, both assuming that each component is tested every second month.

1. Both detectors are tested at the same time (or at least at the same shift, every second month). In this case the number of tests against the CCFs is $N_2 = N/2$. Thus, the beta factor estimator is

$$\beta = \frac{2n_2}{n_1+2n_2}$$

This is the familiar estimator for the beta factor (see the formula for a general m given above).

2. The components are tested at *staggered* intervals, i.e. *one* of the detectors is tested each month. If there is a failure, the second detector is also tested at the same time. For this testing strategy, the number of tests against CCFs is higher than in case 1 because each successful test of a component is a confirmation of the absence of the common cause. The number of tests against the common cause failure N_2 is related to N by the relation: $N = N_2 + n_1 + n_2$. The term $n_1 + n_2$ arises because this is number of failures of the first component, leading to the test of the second component. When n_1 and n_2 are small compared to N, we approximately have $N = N_2$, and the beta factor is

$$\beta = \frac{n_2}{n_1+n_2}$$

Thus, from the same data (i.e. n_1 and n_2), we will in case 2 estimate β to be about half the value found in case 1. This illustrates the importance of recognizing that the estimators are based on specific assumptions on the testing strategy for the events in the data base, see [4]. Assuming nonstaggered testing, generally gives pessimistic estimates, and assuming staggered testing gives optimistic estimates. In practice, a parameter estimate may be most realistically assessed, assuming a strategy in between these two extremes.

11.7 Failure Event Reports

The estimation of the parameters of the chosen parametric model is an essential step in CCF analysis. A systematic approach for utilizing the information in failure events is reviewed in this section.

11.7.1 Data Sources

Plant-specific raw data records, which should be available, include

- Maintenance work orders
- Operator logs
- Work request forms
- Significant events reports

These sources must usually be supplemented by generic data. The *Licensee Event Reports (LER)* provide a substantial amount of data. The reference [5] provides the following list of sources, all giving data summaries of LERs:

- Diesel generators (NUREG/CR-1362), [41]
- Pumps (NUREG/CR-1205), [42]
- Valves (NUREG/CR-1363), [43]
- Selected instrumentation and control comp. (NUREG/CR-1740), [44]
- Primary containment penetrations (NUREG/CR-1730), [45]
- Control rods and drive mechanisms (NUREG/CR-1331), [46]
- Nuclear power experience, [47]. This reference gives a LER-based compilation of event reports supplemented by information from other sources, available on a subscription basis only.

The following are some other data sources, giving CCF parameter estimates (also essentially based on LERs):

- Pumps (NUREG/CR-2098), [48]
- Valves (NUREG/CR-2770), [49]
- Instrumentation and control assemblies (NUREG/CR-3289), [50]
- Pumps, vales, diesel generators, and breakers (EPRI NP-3967), [51]

Actual CCF estimates are also provided in reports on CCF analyses that have been carried out (e.g. Benchmark exercises). For instance Chapter 2 and 4 of [7] gives a summary of estimated CCF parameters in Scandinavian nuclear power plants. Results for various power plants are given, and comparisons of results from various parametric models and using various estimation techniques are carried out. Also see, e.g. [8] and [9]. Parameter estimation, using the BFR model, is reported in [52].

11.7.2 Impact Vector

The amount of plant-specific data for CCF analysis is usually very limited, and a variety of sources must be used to estimate the frequencies of common cause failures. Below we shortly describe the process of assessing

n_k = total number of basic events involving failure of k components within the same component group,

the estimation being based on *failure event reports*. It is referred to in [3], [4] or [5] for more details on this estimation procedure.

When the failure event reports are identified for each of the common cause component groups in the system model, the failure events should be *classified* based on the

- root cause of failure,
- coupling factor, relevant for this CC event
- component group size (m),
- failure mode, and
- observed CCBE.

Several *classification systems* for treating failure event reports have been developed, e.g. see [51]. In the first place, these systems are intended for structuring and breaking down complex event descriptions. They also provide the possibility to present the results in a systematic way, and supply a good framework for exchanging information and experience between analysts. The use of *cause-effect diagrams*, [51], is one way of systemizing the information. The special symbols used in cause-effect diagrams are also defined in [16], and examples of their use are found in [5] and [16].

Following the procedure in [3], [4], a so-called *impact vector* for each failure event is now defined. First, the indicator variables

$I_k = 1$ if there are k components that fail in an event,

$I_j = 0$ if $j \neq k$ $(0 \leq j \leq m)$

are introduced. The impact vector is now defined as, $I = \{I_0, I_1,..., I_m\}$. Thus, if k components fail, the kth element of the impact vector is 1 while all other elements are 0.

Experiences with event reports indicate that a large number of event descriptions are not clear and that the causes and the couplings are seldom defined. Therefore, an analyst may find it necessary to establish several hypotheses representing different interpretations of the event. One hypothesis may lead to an assumption that $I_2 = 1$ (and all other $I_j=0$), while another hypothesis is that the event resulted in three components to fail, i.e. $I_3=1$. The analyst then can assign probabilities to each of the hypotheses, representing his degree of confidence in that hypothesis.

Based on these considerations an *average impact vector* is formed by taking the average of the impact vectors (that correspond to the various hypotheses). Formally we write $I = \{P_0, P_1,...,P_m\}$, where P_k is the probability that the event involved failure of k components. This impact vector now describes the analyst's interpretation of the number of components that failed during a specific event.

If the event occurred at one system (the *original plant*), and the resulting estimator is to be applied for another system (the *target plant*), it may be necessary to perform a modification of this vector. The question that must be addressed is whether the same event could occur at the target plant also, i.e. is that plant more or less vulnerable to this type event than the original plant?

Further, if the target plant has a different number of redundant components, m (relevant for the given event), the impact vector must also be *mapped up* or *mapped down*, see [4]. This mapping of the impact vector is completely analogous to the mapping of the alpha factor distribution described in Section 11.5.3 (see equation (11.9)). Observe that the normalization factor required in (11.9) is not required here, as the impact vector also includes the possibility of recording 0 failures due to the event. By completing this process, we arrive at

$P_k(i)$ = the kth element of the average impact vector i $(i = 1, 2, ... , N)$

Here, N is the total number of failure events corresponding to the relevant CC component group.

Once the (average) impact vectors of all the events in the data base are assessed for the system being analyzed, the number of events in each impact category can be calculated by adding the impact vectors; that is

$$n_k = \sum_{i=1}^{N} P_k(i) \quad (k=0,1,...,m)$$

This provides the total number of basic events involving failure of k similar components of a given CC component group. The *pseudo data,* n_k, generated by the above procedure, can now be used in the estimates of the parametric CC models (see section 11.6).

11.7.3 Uncertainty of Parameter Estimation

There are several sources of uncertainty related to the use of the parametric models.

- Uncertainty with respect to the validity of the model. In particular we may question whether it is appropriate to model dependence as a set of independent and mutually exclusive CCBEs. The assumption of homogeneity (Assumption A in Section 11.5.1) and the choice of the particular parametric model can also be questioned. If *mapping up* or *mapping down* is applied, the validity of Assumptions B (Section 11.5.1) is crucial.
- Uncertainty of the parameter estimates, e.g.
  - incompleteness of failure event data sources, e.g. under-reporting of independent events and uncertainty with respect to exposure time
  - data classification and impact vector assessment
  - variation among plants
  - statistical uncertainty due to small data sample

It is commonly believed that perhaps the most significant source of uncertainty is due to the judgements required to interpret and classify the failure events. Thus, a formal procedure for specification of the uncertainty associated with the impact vector assessment has been developed.

As previously described, the analyst will set forth several hypotheses about the failure event, each hypothesis implying a specific value for the number of components which will fail. Let P_{ij} represent the analysts probability for hypothesis j about event i, and let I_{ij} be the element of the corresponding binary impact vector. Then the data can be represented as

$$E = [<P_{ij}, I_{ij}>;\ i=1, \dots, N \text{ and } j=1, \dots, M_i]$$

N represents the number of events in the data base, and M_i is the number of hypotheses about the ith event. Note that $\sum_j P_{ij} = 1$.

Using the total information, E, described above, actually gives the distribution of the n_k-s that was generated from these impact vectors. This also provides the *uncertainty distribution* for the parameters of the parametric model. In reality the number of data sets that can be generated by considering all possible combinations of various hypotheses about events is very large. As a result, the implementation of this procedure is difficult, and in practice one may restrict the calculation of the average impact vector for each event.

A discussion of some of the other uncertainties in parameter estimation are found in [5].

11.8 Conclusions

An impressive amount of work on the handling of CCFs has been carried out during the past few years. Considerable advances have been made, in particular, within the field of investigating and identifying defences against CCFs. However, further work will be required to facilitate the determination of plant specific defensive measures.

The main problem in the analysis of CCFs is still the lack of high quality data for carrying out estimation in the CCF models. Thus, improvements in the failure event reporting should be welcomed. Structured methods for arriving at plant specific CC parameter estimates are now available. The approaches can however be very time consuming, and further work is needed to implement various approaches that can be used for analyses with different levels of details. A generalization of a method like the partial beta factor method to cover dependent failures of various multiplicities would be helpful.

Handling of model and parameter uncertainty should also be improved. Subjective probabilities might be used even to describe modeling uncertainties. Further advances in the protection against CCF may be obtained by a high degree of integration of reliability modeling, engineering judgements and operating experience.

As human errors plays an important role in CCFs, the analysis of human interactions should be intensified. Further work is recommended in the field of evaluating the impact of maintenance activities on CCFs.

REFERENCES

[1] Edwards G.T. and A. Watson, *A Study of Common-Mode Failures*, SRD R-146, United Kingdom Atomic Energy Authority, Wigshaw Lane, Culcheth Warrington, WA3 4NE, 1979.

[2] Humphreys P., A.M. Games, A.M. Smith, and D.H. Worledge. *Progress Towards a better Understanding of Dependent Failures by Data Collection Classification and Improved Modelling Techniques*. In *Proceedings Reliability '87*, Vol 1, paper 2C/4, ed. A.M. Jenkins. Elsevier, London, 1987.

[3] Mosleh A., K.N. Fleming, G.W. Parry, H.M. Paula, D.H. Worledge and D.M. Rasmuson, *Procedures for Treating Common Cause Failures in Safety and Reliability Studies*, Vol. 1: Procedural Framework and Examples. NUREG/CR-4780, EPRI NP-5613, 1988.

[4] Mosleh A., K.N. Fleming, G.W. Parry, H.M. Paula, D.H. Worledge and D.M. Rasmuson, *Procedures for Treating Common Cause Failures in Safety and Reliability Studies*, Vol. 2: Analytical Background and Techniques. NUREG/CR-4780, EPRI NP-5613, 1989.

[5] Mosleh A., *Common Cause Failures: An Analysis Methodology and Examples*, Reliability Engineering and Systems Safety, Vol. 34, pp. 249-292, 1991.

[6] Pouchet A., A. Amendola and P.C. Cacciabue, *Summary of the Common Cause Failure Reliability Benchmark Exercise*. Joint Research Centre Report PER 1133/86, Ispra, Italy, April 1986.

[7] Hirschberg S. (editor), *Dependencies, Human Interactions and Uncertainties in Probabilistic Safety Assessment*, Final report of the NKA project RAS 470, Nordic Liaison Committee for Atomic Energy, April 1990. (Available on request from ABB Atom AB, Library, S-721 63 Vasterås, Sweden.)

[8] Bengtz M., S. Bjore and S. Hirschberg, *NKA Project Risk Analysis (RAS-470), Identification of Common Cause Failure Events for Motor Operated Valves in Swedish Boiling Water Reactor Plants*. Final Report RAS-470(86)4 (ABB Atom Report RPA 86-40), January 1986.

[9] Hirschberg S.(ed), *NKA Project 'Risk Analysis' (RAS-470), Summary Report on Common Cause Failure Data Benchmark Exercise*. Final Report RAS-470(86)14 (ABB Atom Report RPA 86-241), July 1987.

[10] Hirschberg S., *Experience from Dependent Failure Analysis in Nordic Countries*, Reliability Engineering and Systems Safety, Vol. 34, pp. 355-388, 1991.

[11] Humphreys P. and A.M. Jenkins, *Dependent Failures Development*, Reliability Engineering and Systems Safety, Vol. 34, pp. 417-427, 1991.

[12] Chaganty N. R., and K, Joag-dev, *Dependence Notions in Reliability Theory*, in P.R. Krishnaiah and C.R. Rao (eds), Handbook of Statistics, Vol. 7, pp 99-111, Elsevier Science Publishers, 1988.

[13] Apostolakis G. and P. Moieni, *The Foundation of Models of Dependence in Probabilistic Safety Assessment*, Reliability Engineering, Vol. 18, No.3, pp. 177-195, 1987.

[14] Watson I.A., *Analysis of Dependent Events and Multiple Unavailabilities with particular Reference to Common Cause Failures*, Nuclear Engineering and Design 93, pp. 227-244, 1986.

[15] U.S. Nuclear Regulatory Commission, *PRA Procedures Guide*. Report NUREG/CR-2300, January 1983.

[16] Paula H.M., D.J. Campbell and D.M. Rasmuson, *Quality Cause-Defense Matrices: Engineering Tools to Support the Analysis and Prevention of Common Cause Failures*, Reliability Engineering and Systems Safety, Vol. 34, pp. 389-415, 1991.

[17] Rasmuson D.M., *Some Practical Considerations in Treating Dependencies in PRAs*, Reliability Engineering and Systems Safety, Vol. 34, pp. 327-343, 1991.

[18] Parry G.W., *Common Cause Failure Analysis: A Critique and Some Suggestions*, Reliability Engineering and Systems Safety, Vol. 34, pp. 309-326, 1991.

[19] Rasmuson D.M., J.C. Shepherd, N.M. Marshall, and L.R. Fitch, *Use of COMCAN III in System Design and Reliability Analysis*. Report EGG-2187, EG&G Idaho, October 1982.

[20] Worrel R.P., *SETS Reference Manual*, NUREG/CR-4213, SAND832675, Sandia National Laboratories, Albuquerque, New Mexico, 1985.

[21] Fleming K.N., Mosleh, A. and Deremer, R.K., *A Systematic Procedure for the Incorporation of Common Cause Events into Risk and Reliability Models*, Nucl. Engineering and Design, 93, pp. 245-279, 1985.

[22] Harris B., *Stochastic Models for Common Failures*. In "Reliability and Quality Control", (editor: A.P. Basu). Elsevier Science Publishers B.V. (North Holland), 1986.

[23] Waller R.A., *A Brief Survey and Comparison of Common Cause Failure Analysis*. NUREG/CR-4314. Los Alamos National Laboratory, 1985.

[24] Dörre P., *An Event-based Multiple Malfunction Model*, Reliability Engineering, Vol. 17, No.1, pp. 73-80, 1987.

[25] Fleming K.N. and Mosleh, A., *Common Cause Data Analysis and Implications in System Modelling*, Proceedings: International Topical Meeting on Probabilistic Safety Methods and Applications. San Francisco, 1985.

[26] WASH-1400, *Reactor Safety Study, Appendix IV: Common Mode Failures - Bounding Techniques and Special Techniques*, NUREG-75/014, US Nuclear Regulatory Commission, Washington, D.C., 1975.

[27] Fleming K.N., *A Redundant Model for Common Mode Failures in Redundant Safety Systems*, Proceedings of the Sixth Pittsburgh Annual Modeling and Simulation Conference, Instrument Society of America, Pittsburgh, PA, pp. 579-581, 1975.

[28] Evans M.G.K., G.W. Parry, and J. Wreathall, *On the Treatment of Common-Cause Failures in System Analysis*, Reliability Engineering, Vol. 9, No. 2, pp 107-115, 1984.

[29] Humphreys R.A., *Assigning a Numerical Value to the Beta Factor Common Cause Evaluation*, Reliability '87 Proc. Paper 2C. 1987.

[30] Johnston B. D., *A structured Procedure for Dependent Failure Analysis DFA*, Reliability Engineering and Systems Safety, Vol. 34, pp. 125-136, 1991.

[31] Mosleh, A. and N.O. Siu, *A Multi-parameter Common Cause Failure Model*, 9th International Conference on Structural Mechanics in Reactor Technology, Lausanne, Switzerland, pp. 147-152, August 1987.

[32] Hokstad P. and L. Bodsberg, *Reliability Model for Computerized Safety System*, 1989 Proceedings, Annual Reliability and Maintainability Symposium, pp. 435-440, Atlanta, Georgia, USA, 1989.

[33] Vesely W.E., *Estimating Common Cause Failure Probabilities in Reliability and Risk Analyses: Marshall-Olkin Specializations*. In *Nuclear Systems Reliability Engineering and Risk Assessment*, eds: J.B. Fussel and G.R. Burdick. Philadelpia: Society for Industrial and Applied Mathematics, pp. 314-341, 1977.

[34] Atwood C.L., *The Binomial Failure Rate Common Cause Model*, Technometrics, No. 28, pp. 139-148, 1986.

[35] Hokstad P., *A Shock Model for Common Cause Failures*, Reliability Engineering and Systems Safety, Vol. 23, pp. 127-145, 1988.

[36] Martz H. and R.A. Waller, *Bayesian Reliability Analysis*. Wiley, 1982.

[37] Mankamo T., *Common Load Model, a Tool for Common Cause Failure Analysis*, Sahkotekniikan laborratorio, Tech. Research Centre of Finland, Tiedonanto 31, 1977.

[38] Clayton D., *A Model for Association in Bivariate Life Tables and its Application in Epidemiological Studies of Familial Tendency in Chronic Disease Incidence*, Biomerica Vol.65, pp 141-151, 1978.

[39] Oakes D., *A Model for Association in Bivariate Survival Data*, J.R. Soc. B, Vol. 44, No.3, pp. 414-422, 1982.

[40] Lydersen S., *Clayton's Model for Dependent Components in System Reliability*, Reliability Engineering and System Safety, Vol 36, pp. 53-58, 1992

[41] EG & G Idaho, Inc., *Data Summaries of Licensee Event Reports of Diesel Generators at US Commercial Nuclear Power Plants*, January 1, 1976 to December 31, 1978. Report NUREG/CR-1362, EGG-EA-5092, prepared for the US Nuclear Regulatory Commission, March 1980.

[42] EG & G Idaho, Inc., *Data Summaries of Licensee Event Reports of Pumps at US Commercial Nuclear Power Plants January 1, 1972 to April 30, 1978.* Report NUREG/CR-1205, EGG-EA-5044, prepared for the US Nuclear Regulatory Commission, January 1982.

[43] EG & G Idaho, Inc., *Data Summaries of Licensee Event Reports of Valves at US Commercial Nuclear Power Plants, Main Report, January 1, 1976 to December 31, 1978.* Report NUREG/CR-1363, EGG-EA-5125, prepared for the US Nuclear Regulatory Commission, October 1982.

[44] Miller C.R., W.H Hubble, M. Trojovsky and S.R. Brown, *Data Summaries of Licensee Event Reports of Selected Instrumentation and Control Components at US Commercial Nuclear Power Plants.* Report NUREG/CR-1740, EGG-EA-5816, Rev. 1, prepared for the US Nuclear Regulatory Commission, EG & G Idaho, Inc., October 1982.

[45] Sams D.W. and Trojovsky, *Data Summaries of Licensee Event Reports of Primary Containment Penetrations at US Commercial Nuclear Power Plants, January 1, 1972 to December 31, 1978.* Report NUREG/CR-1730, EGG-EA-51/88, September 1980, prepared for the US Nuclear Regulatory Commission, EG & G Idaho, Inc.

[46] Hubble W.H. and C.F. Miller, *Data Summaries of Licensee Event Reports of Control Rods and Drive Mechanisms at US Commercial Nuclear Power Plants, January 1, 1972 to April 30, 1978.* Report NUREG/CR-1331, EGG-EA-5079, February, 1983 prepared for the US Nuclear Regulatory Commission, EG & G Idaho, Inc.

[47] S.M. Stoller Corporation, Nuclear Power Experience, updated monthly.

[48] Atwood C.L., *Common Cause Fault Rates for Pumps.* Report NUREG/CR-2098, EPRI-695-DOC-01, EGG-EA-5289, February 1983, prepared for the US Nuclear Regulatory Commission, EG & G Idaho, Inc.

[49] Atwood C.L. and J.A. Steverson, *Common Cause Fault Rates for Valves: Estimates Based on Licensee Event Reports at US Commercial Nuclear Power Plants.* Report NUREG/CR-2770, EGG-EA-5485, prepared for the US Nuclear Regulatory Commission, EG & G Idaho, Inc., February 1983.

[50] Meachum T.R. and C.L. Atwood, *Common Cause Fault Rates for Instrumentation and Control Assemblies.* Report NUREG/CR-3289, EPRI-685-DOC-06, EGG-222258, prepared for US Nuclear Energy Commission under Department of Energy Contact No. DE-AC07:761D01570, Idaho National Engineering Laboratory, EG & G Idaho, Inc., May 1973.

[51] Fleming K.N. and A. Mosleh, *Classification and Analysis of Reactor Operating Experience Involving Dependent Events.* Report EPRI NP-3967, Electric Power Research Institute, February 1985.

[52] Atwood C.L., *Data Analysis Using the Binomial Failure Rate Common cause Model*, Report NUREG/CR-3437, September 1983.

New Trends in System Reliability Evaluation
K.B. Misra (Editor)

CHAPTER 12

SOFTWARE RELIABILITY MODELLING

Tom Downs
Intelligent Machines Laboratory
Department of Electrical and Computer Engineering
The University of Queensland, Australia, 4072

ABSTRACT

This chapter commences with an explanation of the potential usefulness of software reliability models and then describes three of the better known models that were developed during the 1970s. It is argued that these and similar models have had only modest success because they simply hypothesize a failure time distribution and take no account of the manner in which software is tested. Several models that do take testing into account are then discussed and it is shown that these models offer distinct advantages (both in terms of performance and applications) over the earlier models. The chapter also discusses current issues relating to the assessment of the performance of software reliability models.

12.1 Introduction

As computer systems find ever wider applications, many of them in safety-critical situations, the problem of producing reliable software assumes increasing importance. There can be no doubt that the discipline of software engineering has made considerable progress over the years, but in spite of this, failures in today's computer systems are much more likely to be caused by faults in software rather than faults in hardware. The faults in software that cause system malfunctions are not simply coding errors; there are faults due to inadequacies in system specifications and additional faults created during the software system design process. Many of these faults are discovered (and eliminated) by software testing but, as is explained in [1], it is not usually possible to discover all faults by this means, and it is common for software systems (particularly the large and complex ones) to be released with every expectation that they still contain many faults.

Software development is usually subject to the forces of the market and, as a consequence, is governed by quite severe time and cost constraints. Within these constraints, a project manager attempts to produce as reliable a product as possible, but reliability is often a secondary consideration after timeliness and cost. Most software is released without any warranty and, in those cases where warranties are provided, they are generally quite limited in scope. But as software development techniques become more mature, we can expect that the setting of reliability goals will become more the rule than the exception. And once this more mature position has been attained by the industry, the issuing of software warranties should become more commonplace. The formulation of warranty policies for products in other industries has always relied upon the availability of product reliability models and so there can be little doubt that software reliability models have a role to play here.

This is by no means the only potential application for software reliability models. For instance, a good model could be of assistance during software testing where it could be used to track reliability improvements and to provide an aid for making decisions on when testing can be terminated. Tracking of reliability during testing is very important. A good model could indicate whether or not reliability targets are likely to be met in an acceptable time. If the model indicated that progress was not sufficiently speedy, its predictions could be used to help make decisions on whether it would be worthwhile to deploy greater resources in testing or whether, alternatively, further testing would not be cost effective and the project should be written off as a business loss. If progress *were* being made at an adequate rate, the model could be used as a basis for estimating the optimum time to terminate testing (in terms of a suitable cost criterion).

There are several other useful applications for software reliability models. One relates to the fact that many large software systems have sections of code that are executed much more frequently than others. For instance, in computer-controlled telephone exchanges, the code relating to call handling is much more heavily used than the code associated with maintenance activities. In the testing of software of this type, it is usual for the code that is destined to be heavily used, and any other supporting software that is perceived as critical to system performance, to be tested much more thoroughly than the remainder of the code. In cases such as this, a decision has to be made regarding the amount of test effort that will be applied to different sections of the code. For manpower planning purposes, an initial decision has to be taken prior to the commencement of testing. Section 12.4.1 of this chapter describes a method of dealing with this problem which is based upon a model developed in Section 12.2.2.

Another application of software reliability models is in estimating the "jump" in failure rate that very often occurs when a software system is transferred from the test phase to the operational phase. For large and expensive software systems, this phenomenon can be a source of considerable customer dissatisfaction; an estimation technique that can assist with this problem was described in [2]. A good software reliability model could also be used in order to provide an accurate measure of the operational reliability achieved by a software system; this would allow assessment of the effectiveness of the software development methodology employed. Later in the operational phase, a good model could be used to estimate safe times for system updates (if they are introduced too frequently, they can cause reliability problems) and, ultimately, to estimate the best time to invest in a new system.

This chapter commences with a description of three of the early software reliability models that have proved to be quite influential. These three models are all based upon simple assumptions which imply a particular distributional form for software failure times. They pay no attention to the manner in which software is tested or used and, in Section 12.2.2, it is argued that this is the reason these and similar models generally exhibit only modest performance when applied to software failure data. A variety of models that take account of the testing process are then described and, in Section 12.4, it is shown that these models offer distinct advantages (both in terms of performance and applications) over the earlier models. The chapter also discusses current issues relating to the assessment of the performance of software reliability models.

12.2 Software Reliability Models

The early history of software reliability modelling has frequently been reviewed (see for instance [3], [4]). Three of the early models are often used as benchmarks against which to compare more recent models, and this section commences with a brief description of these three models.

12.2.1 Three Early Models

(i) The Jelinski-Moranda Model

The first model of software reliability that attracted wide attention was described in [5] by Jelinski and Moranda. In this model, which I shall refer to as the JM model, two major simplifying assumptions are made : (i) that faults are distributed uniformly (but randomly) over the software, and (ii) that the software is executed in a uniform (but random) fashion. These assumptions imply that the software failure rate at any given time is proportional to the number of faults in the software at that time. Thus, under this model, if the number of faults in a software system prior to any testing is N and the proportionality constant is ϕ, the initial failure rate is $N\phi$. And after the first fault is detected and eliminated, the failure rate drops to $(N-1)\phi$. Each time a fault is removed, the failure rate drops by an amount ϕ, as shown in Fig. 12.1. An attractive feature of this model is that it allows the parameter N to be estimated quite simply by the method of maximum likelihood. This is achieved as follows.

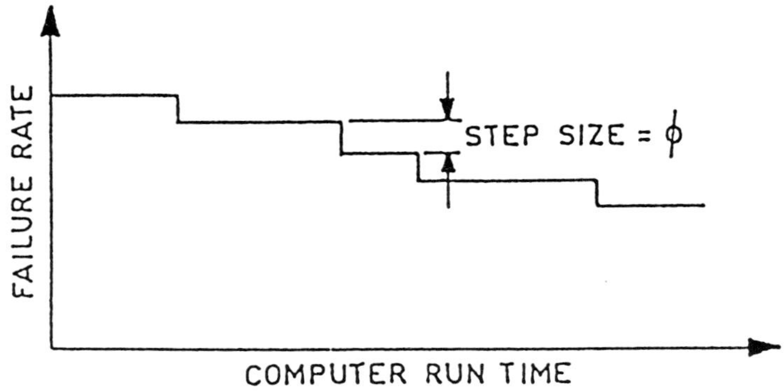

Fig. 12.1 Software Failure Rate Assumptions of JM Model

The fact that the failure rate is constant between failures implies that the times between failures are exponentially distributed. The failure rate during the ith interval in the failure process is equal to $(N-i+1)\phi$ so that the ith interval has density function

$$f_i(t) = (N-i+1)\phi\exp[-(N-i+1)\phi]t \tag{12.1}$$

Suppose that during software execution, the time intervals between successive failures are $\tau_1, \tau_2, ..., \tau_n$. The likelihood function is simply the product of the values of the density function $f_i(t)$ at the failure time intervals τ_i :

$$L(\tau_1,\tau_2,...,\tau_n) = \prod_{i=1}^{n} (N-i+1)\phi\exp[-(N-i+1)\phi\tau_i] \tag{12.2}$$

The value of N which maximises this function is the maximum likelihood estimate, $\hat{N}$, of the number of faults initially present in the software. An equation for $\hat{N}$ is easily derived from (12.2) using elementary calculus (see [5]). The equation is nonlinear in $\hat{N}$, but is readily solved using numerical techniques such as Muller's method [6].

Unfortunately, the JM model is very optimistic in its predictions; that is, it generally predicts much smaller numbers of faults in the software than actually exist. The optimism in the predictions is due to the implicit assumption that the failure rate of a software system is proportional to the number of faults it contains. Littlewood [7] criticised this assumption by pointing out that most programs have some portions that are heavily used and others that are lightly used. And it is quite clear that faults in the heavily-used portions will contribute more (individually) to the failure rate than faults in the lightly-used portions. Another reason why different faults can have different effects on failure rate is that in most software systems, some faults can be revealed by a large number of possible inputs while others can only be revealed by a small number; an example of this can be seen by comparing a fault that affects many paths of control flow with one that affects only a few.

A simple way of observing why the assumptions underlying the JM model lead to optimistic predictions is by means of a graphical plot of software failure rate after j faults have been removed (call it λ_j) against j (the number of faults removed). In the case of the JM model, because each fault contributes the same amount to the failure rate, a straight line should be obtained, as depicted in Fig. 12.2. I refer to plots of this type as plots of the discrete failure rate function (DFRF). If the straight line obtained by such a DFRF plot is extended, it crosses the abscissa at a point whose value is equal to N, the number of faults initially present in the software. But in a practical software system, the DFRF plot should be convex in shape, as is shown in Fig. 12.3. Convexity reflects the fact that faults which contribute most to the failure rate are more likely to be discovered (and eliminated) earlier than those which contribute least. Thus, the reductions in failure rate that are obtained as faults are removed early in the test phase (or the operational phase) should be greater (on the average) than those obtained later. And in practice, a DFRF plot decreases asymptotically towards the abscissa, but for large and complex software systems it rarely actually reaches the axis.

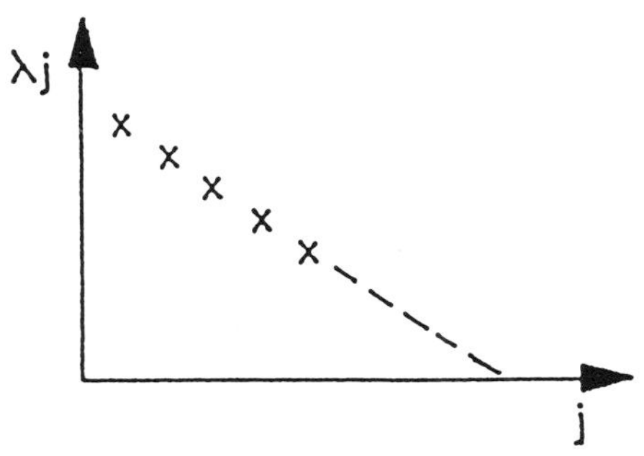

Fig. 12.2 Linear DFRF Plot

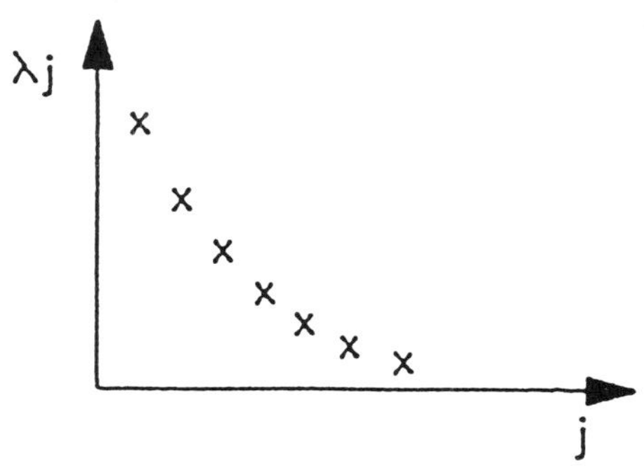

Fig. 12.3 Convex DFRF Plot

(ii) The Littlewood-Verrall model

The second model that will be considered in this section was developed by Littlewood and Verrall [8] and will be referred to as the LV model. This model is rather more realistic than the JM model in that it aims to take account of the fact that attempts at removing faults from software are not always successful. (Indeed, it is not uncommon for such attempts to lead to a net increase in the number of faults in a software system.) In the LV model, the uncertainty surrounding the effects of corrective action after a software fault has been detected are modelled by treating the failure rate as a random variable. Littlewood and Verrall did, of course, assume that debugging is generally effective so the assumption is made in the LV model that each time an attempt is made to correct a fault, it is *probable* that the failure rate decreases. Mathematically, if λ_i is the failure rate after i failures, and assuming corrective action is taken after each failure,

$$P\{\lambda_i < \mu\} \geq P\{\lambda_{i-1} < \mu\} \tag{12.3}$$

where μ is any positive constant.

Let $f_i(.)$ be the density function of the time to the ith failure and let Λ_i be the random variable representing the failure rate in the interval prior to the ith failure. Then, if $g_i(.)$ is the density function of Λ_i, we have

$$f_i(t) = \int_o^\infty f_i(t|\Lambda_i=\lambda_i)g_i(\lambda_i)d\lambda_i \tag{12.4}$$

This integral is only analytically tractable for a restricted class of functions $f_i(.)$ and $g_i(.)$. Thus, Littlewood and Verrall were significantly constrained in their choice of distributions for $f_i(.)$ and $g_i(.)$ and in fact restricted their attention to the exponential distribution for $f_i(.)$ and the gamma distribution for $g_i(.)$. Thus, in the LV model

$$f_i(t|\Lambda_i=\lambda_i) = \lambda_i e^{-\lambda_i t} \tag{12.5}$$

$$g_i(\lambda_i) = \frac{\Psi_i\{\Psi_i\lambda_i\}^{\alpha-1}\exp\{-\Psi_i\lambda_i\}}{\Gamma(\alpha)} \tag{12.6}$$

Substitution of these expressions into (12.4) and evaluating the integral gives

$$f_i(t) = \frac{\alpha\Psi_i^\alpha}{\{t+\Psi_i\}^{\alpha+1}} \tag{12.7}$$

which is the density function of a Pareto distribution. The Pareto density function differs from that of the exponential in having a longer tail, implying a greater chance of large times between failures.

Note that (12.3) places an ordering restriction on the cdfs of the λ_i and, in the gamma distribution above, this requirement can be satisfied by choosing Ψ_i as a monotonically increasing function of i. In essence, Ψ_i represents a programmer's ability in fixing faults and hence gives a measure of reliability growth. Apparently Littlewood and Verrall expected a "natural" parametric representation to emerge for Ψ_i when their model was applied to a range of software failure data. This did not occur so that the best fit has to be obtained by trial and error. Two possible representations of Ψ_i are

$$\Psi_i = \beta_0 + \beta_1 i \tag{12.8}$$

and

$$\Psi_i = \beta_0 + \beta_1 i + \beta_2 i^2 \tag{12.9}$$

Thus, to use the LV model, one must estimate, from a set of failure data, the parameter α and the β_i.

The LV model generally gives superior predictions of software reliability than are given by the JM model (see, for instance, the comparisons in [9] and [10]).

(iii) Nonhomogeneous Poisson process models

A nonhomogeneous Poisson process (NHPP) is a Poisson process whose parameter is a function of time. In the software reliability context, the NHPP is used to describe the counting process *N(t)* which represents the number of failures observed in the interval *(0,t)*, i.e.

$$Pr\{N(t)=k\} = \frac{(m(t))^k}{k!}\, e^{-m(t)} \qquad k = 0,1,2,... \tag{12.10}$$

where the time-varying parameter *m(t)* is the expected number of failures up to time *t*.

The first NHPP model of software reliability was described by Goel and Okumoto in [11] and is based upon the assumption that the average number of software failures in the small time interval $(t,t+\Delta t)$ is proportional to the number of faults in the software system at time *t*. And if *a* represents the number of faults initially in the software, this assumption can be written

$$m(t+\Delta t) - m(t) = b(a-m(t))\Delta t \tag{12.11}$$

where *b* is the constant of proportionality, equal to the fault detection rate per fault.

Dividing each side of (12.11) by Δt and letting $\Delta t \rightarrow 0$ gives the differential equation

$$\frac{d}{dt}\left[\, m(t) \,\right] = b(a-m(t)) \tag{12.12}$$

Use of the boundary conditions

$$m(t) = \begin{matrix} 0 & \quad t=0 \\ a & \quad t\rightarrow\infty \end{matrix} \tag{12.13}$$

gives the following solution to (12.12) :

$$m(t) = a(1-e^{-bt}) \tag{12.14}$$

The failure rate at time t is simply the rate of change of m(t), i.e.

$$\lambda(t) = \frac{d}{dt}\,(m(t)) = abe^{-bt} \tag{12.15}$$

In [10], Goel and Okumoto, who used maximum likelihood to estimate the parameters a and b, compared the predictions of their model with those obtained from the JM model for a real-time system used in the U.S. Navy. The NHPP model gave significantly better predictions than the JM model with the latter giving more optimistic results.

12.2.2 Models Based Upon the Testing Process

Most of the software reliability models that have appeared in the literature have been developed with little regard for the manner in which software testing is normally carried out. An important feature of the testing of large and complex software systems (and these are the kinds of systems for which software reliability models are particularly relevant) is that a number of shifts in testing strategy normally take place as the test phase unfolds. These shifts occur because there is usually a need to carry out a variety of testing activities, each with different objectives. For instance, in functional testing the execution profile normally changes each time a different function is tested.

Changes in execution profile can also occur during stress testing and performance testing. Models of the kind that have been discussed so far have been developed without any regard to these facts and yet they have frequently been applied to failure data collected during the test phase in order to predict reliability growth. It is not surprising that their performance in this task has been generally poor (see, for instance [12], [13]).

Even if the use of such models was restricted to predicting future reliability from data collected during the operational phase only (here shifts in execution profile are much less likely) the quality of the predictions obtained could still depend considerably upon the manner in which the software was tested prior to release. As a simple example, consider the case of a modular software system which has had some modules much more thoroughly tested than the others. A user of that system, whose data environment causes the thoroughly tested modules to be executed almost exclusively, should observe a very reliable system. Another user, whose data environment leads to regular execution of some of the less thoroughly tested modules, can be expected to observe a much less reliable system. The failure characteristics experienced in the two cases would surely be completely different. And yet, models of the type discussed in the previous section would attempt to represent the failure times in each case by the same distributional form, relying upon the parameters of the distribution to accommodate any differences. In my view, this is another reason why models of this type have failed to find much favour in practical applications.

It therefore appears that improved models of software reliability are likely to be developed if greater attention is paid to the manner in which software is tested. This implies a need for models of the software testing process.

Modeling Software Testing

An approach to the modelling of software testing was introduced in [2] and extended in [14] and [15]. This approach is based upon the following assumptions :

- Software testing takes the form of execution of a sequence of logic paths through the software.

- The input data supplied to the software is governed by an execution profile which defines the probabilities with which different paths are selected for execution.

- The execution profile is held invariant in the intervals between software failures.

Given these assumptions, the software testing process can be regarded as a sequence of Bernoulli trials [16] in which the two possible outcomes are "software failure" (when the selected path contains a fault) and "non-failure" (when the selected path is fault-free).

Now suppose that r paths are executed in a unit time interval and let the probability of not locating a fault in unit time be p. Then, if we make the simplifying assumption that all paths are of equal length[1], the probability of not locating a fault in a particular path is $p^{1/r}$.

Let the number of paths that have to be executed in order to locate a fault be Z. Then, since the above assumptions allow testing to be regarded as a sequence of Bernoulli trials, and given that the probability of a trial resulting in a software failure is $1-p^{1/r}$, we have, for the expected value of Z (see [16]) :

$$E(Z) = \frac{1}{1-p^{1/r}} \tag{12.16}$$

and the expected time to locate a fault is

$$\bar{t} = \frac{1}{r\left(1-p^{1/r}\right)} \tag{12.17}$$

In order to derive the fundamental result of this section, one further assumption is required. It is assumed that faults are randomly distributed over paths in such a way that, under a given execution profile, the probability of locating a fault in any small time interval Δt is constant. Then, since we have assumed that the execution profile is held invariant in the intervals between software failures, it follows that the failure rate is constant in the intervals between failures and hence that failure times are exponentially distributed. An expression for the parameter of the exponential distribution can now be obtained by extending the above argument as follows.

Let each path be split into k portions so that rk portions of paths are tested in unit time. Then, with p defined as before, the probability of not locating a fault in a given portion of a path is $p^{1/rk}$. The expected time to locate a fault can then be written

$$\bar{t} = \frac{1}{rk\left(1-p^{1/rk}\right)} \tag{12.18}$$

The parameter of the exponential distribution is then given by

[1] This assumption was relaxed somewhat in [14], but nothing of real advantage ensued and so the assumption will be maintained here.

$$\frac{1}{\lambda} = \lim_{k \to \infty} \left[\frac{1}{rk\left(1 - p^{1/rk}\right)} \right] \tag{12.19}$$

and, using L'Hôpital's rule, we obtain

$$\begin{aligned} \lambda &= -\log p = -r\log(p^{1/r}) \\ &= -r\log[\Pr\{\text{a path selected for testing is fault-free}\}] \end{aligned} \tag{12.20}$$

Note that this result has been derived with only a very mild assumption being made about the manner in which faults are distributed over paths. In addition, the only assumption made concerning the strategy under which paths are selected for testing is that the execution profile remains constant between software failures. These two factors (i.e. the distribution of faults over a software system and the test strategy employed) have considerable influence on the probability that a path selected for testing is fault-free and hence (12.20) allows account to be taken of different fault distributions and different testing strategies.

In [2,14,15] only one type of distribution of faults was considered, but the manner in which the distribution changes under different testing strategies was the subject of considerable investigation. The effects of these changes upon software failure rate were also studied. The remainder of this section gives a brief review of these investigations.

In order to model the distribution of faults over paths, it is important to take account of the fact that different faults generally affect different numbers of paths. The actual number of paths affected by a particular fault is normally unknown without a detailed examination of the code and so, for arbitrary software, it is reasonable to treat this number as a random variable. We can then proceed as follows.

Suppose we wish to model the distribution of N faults over a software system that contains M logic paths. Let each fault affect k paths with probability P_k $(k=1,2,\ldots,M)$. The distribution we require can be modelled by commencing with a situation in which the M paths are initially fault-free and then adding the N faults to the paths. Each fault is allocated to X randomly selected paths, where X is a random variable with $Pr\{X=k\}=P_k$.

We could, if we wished, obtain a full description of the distribution of faults obtained in this way, but in order to use (12.20), a full description of the distribution of faults over paths is not necessary. All we need is an expression for the probability that an arbitrary path is fault-free. This is very easily obtained :

When the first fault is allocated to X randomly selected paths, an arbitrary path remains fault-free with probability $1-E(X)/M$. And since each fault allocation is independent of the others we have, at the end of the fault allocation process :

$$Pr\{\text{an arbitrary path is fault-free}\} = [1-E(X)/M]^N \quad (12.21)$$

$E(X)$ is usually very small in comparison to M, so (12.21) can be replaced by the first-order approximation :

$$Pr\{\text{an arbitrary path is fault-free}\} \approx 1 - \frac{NE(X)}{M} \quad (12.22)$$

In [14] and [15] this result was used as a basis for the development of a variety of models of the software testing process. The simplest of these models assumed uniform (i.e. random) testing and performed less well than the other models (which assumed nonuniform testing). In these latter models it was assumed that a software system can be partitioned into two sections, with one section being heavily tested and the other lightly tested. The two sections are referred to as section 1 and section 2 respectively. The models developed for nonuniform testing can be readily extended to more than two sections (in order to reflect more complex execution profiles). The basic theory can be developed as follows.

Let N_j represent the number of faults initially in section $j(j=1,2)$ and let M_j be the number of paths through section $j(j=1,2)$. Also, let $X_{o,j}$ be the random variable representing the number of paths affected by an arbitrary fault in section $j(j=1,2)$ prior to any debugging. (The zero subscript on $X_{o,j}$ indicates that the random variable represents the situation where zero faults have been removed from the software.)

Now define the execution profile such that on any test run a path through section 1 is executed with probability p and a path through section 2 is executed with probability $1-p$. The software failure rate prior to the removal of any faults from the system is then given by use of (12.22) in a simple extension to (12.20) :

$$\lambda_o = -r\log\left\{p\left[1 - \frac{N_1E(X_{0,1})}{M_1}\right] + (1-p)\left[1 - \frac{N_2E(X_{0,2})}{M_2}\right]\right\} \quad (12.23)$$

The fact that M_1 and M_2 are large numbers allows (12.23) to be simplified to give, to a small approximation

$$\lambda_0 = K[N_1E(X_{0,1}) + CN_2E(X_{0,2})] \quad (12.24)$$

where

$$K = \frac{rp}{M_1} \quad , \quad C = \frac{(1-p)}{p} \frac{M_1}{M_2}$$

By determining the probabilities with which faults are located in the two sections of code under the given execution profile, an expression can be derived for the failure rate after k faults have been removed from the software :

$$\lambda_k = K \left[N_1^{(k)} E(X_{k,1}) + C N_2^{(k)} E(X_{k,2}) \right] \qquad (12.25)$$

In this expression, $N_j^{(k)}$ represents the number of faults in section $j (j=1,2)$ after k faults have been removed from the software and, similarly, $X_{k,j}$ is the random variable representing the number of paths affected by an arbitrary fault in section $j (j=1,2)$ after k faults have been removed. The terms $N_j^{(k)}$ and $E(X_{k,j})$ are calculated recursively; for details, see [15].

The model described by (12.25) is known as the NURP model (non-uniform testing, each fault affecting a random number of paths). The performance of this model is compared with the performance of three other models in Section 12.4.2.

Other Models that Incorporate Testing

A significant feature of models that incorporate testing in some way is that they can be used as a basis for allocating testing resources. This is very important when one considers that as much as 50% of the total development effort in a large software project can be spent on testing. Unfortunately, very few models that make a serious attempt to incorporate testing have so far appeared in the literature. Apart from the approach described in the previous section, I am aware of only three other techniques, and I will describe them here. The application of such models to resource allocation problems will be considered in Section 12.5.

(i) A model that includes test effort

Yamada and his co-workers at Hiroshima University have developed a software reliability model that incorporates test effort[2] [17]. In their approach, they take note of the fact that the Rayleigh distribution has

[2] This can represent such things as the amount of manpower involved, the amount of CPU time consumed and the number of test cases executed.

been widely used to represent the resources consumed in software development and in [17] they represent test effort by means of a Weibull distribution (for which the Rayleigh distribution is a special case).

Specifically, they assume that the test effort allocated at time t takes the form

$$w(t) = \alpha\beta m t^{m-1}\exp[-\beta t^{m}] \qquad \alpha>0,\beta>0,m>0 \qquad (12.26)$$

which, for $m=2$, reduces to the Rayleigh curve and, for $m=1$, is a simple exponential function.

Estimates of the parameters α, β and m can be obtained from observations of the amount of test effort that is applied at different times. The determination of analytic expressions for least-squares estimates of these parameters is outlined in [18].

Yamada and his colleagues incorporate the test effort function into the NHPP model (described in Section 12.2.1) in a quite straightforward manner. Note that in (12.11) - (12.15) the parameter b is a constant, representing the fault detection rate. In [17] and [18] the constant b is directly replaced by the test effort function (defined in (12.26)). This gives the following expression for the expected number of errors found by time t :

$$H(t) = a(1-\exp[-rW(t)]) \qquad (12.27)$$

where

$$W(t) = \int_{o}^{t} w(x)dx = \alpha(1-\exp[-\beta t^{m}]) \qquad (12.28)$$

(12.27) and (12.28) indicate a property that is quite realistic, at least for large and complex software systems. Substituting (12.28) in (12.27) and letting $t\rightarrow\infty$ gives

$$H(\infty) = a(1-e^{-r\alpha}) \qquad (12.29)$$

and, since a represents the number of faults originally in the software and $H(t)$ is the expected number of faults found by time t, we see that this model indicates that testing for an infinite amount of time will not result in all faults being detected.

Note that this model has five parameters. Three of these, α, β and m are estimated from test effort as described above. The estimates of these three parameters can then be used to provide maximum likelihood estimates of the reliability growth parameters a and r (see [18]).

(ii) A model based upon the hypergeometric distribution

In [19], Tohma *et al* develop a model in which the testing of software is considered to take place in a sequence of *test instances*. Each test instance consists of input data and the observed outcome. A test instance may consist of many test runs (eg. it might represent a full day's testing) so that several faults might be observed during the execution of a single test instance.

Following the notation in [19], suppose there are n test instances and let $t(i)$ represent the ith test instance, $i=1, 2, ..., n$.

For this model it is assumed that faults detected by a test instance are perfectly removed with no new faults being inserted. It is also assumed that test instance $t(i)$ is able to reveal $w(i)$ of the m initial faults with the $w(i)$ being randomly selected from the initial m. Note, however, that, in general, some faults can be revealed by more than one test instance so that, by the time we come to the ith test instance, some of the $w(i)$ might already have been detected and removed. This fact is the source of the hypergeometric distribution that underpins this model.

Let $N(i)$ represent the number of faults that are actually revealed by $t(i)$ and let $C(i)$ represent the total number of faults revealed by test instances $t(1)$ to $t(i)$; i.e. define

$$C(i) = \sum_{j=1}^{i} N(j) \tag{12.30}$$

Then, just prior to implementation of test instance $t(i)$, there are $m\text{-}C(i\text{-}1)$ faults remaining in the software. We wish to determine an expression for the probability that x faults are revealed by $t(i)$. Note that the x faults can be selected from the $m\text{-}C(i\text{-}1)$ remaining faults in $\binom{m-C(i-1)}{x}$ ways. Note also that $w(i)\text{-}x$ is the number of initial faults that $t(i)$ is capable of revealing that have already been revealed by earlier test instances and observe that there are $\binom{C(i-1)}{w(i)-x}$ ways in which this can occur. Finally note that there are $\binom{m}{w(i)}$ ways in which the set of faults $w(i)$ can be selected from the initial set m. Putting all these facts together, we have

$$Pr\{x|m, C(i-1), w(i)\} = \frac{\binom{m-C(i-1)}{x}\binom{C(i-1)}{w(i)-x}}{\binom{m}{w(i)}} \qquad (12.31)$$

x is therefore governed by a hypergeometric distribution. In order to use this model, the parameters m and $w(i)$ have to be estimated from software failure data. In [19] this was achieved by noting that from (12.31), the expected value of the number of faults $N(i)$ revealed by $t(i)$ is given by [16]

$$\overline{N(i)} = \frac{1}{m} w(i)[m-C(i-1)] \qquad (12.32)$$

and, from this, an estimate of $C(i)$ can be written :

$$\hat{C}(i) = \sum_{j=1}^{i} \overline{N(j)} \qquad (12.33)$$

Estimates of m and $w(i)$ can then be found by determining the values of these parameters that minimize the difference between the estimate $\hat{C}(i)$ and its actual value $C(i)$.

(iii) A piecewise linear hazard rate model (PLHR)

The PLHR model was described in [15] and has a failure-rate characteristic that takes the form shown in Fig. 12.4. This characteristic is intended to reflect the fact that errors in software are often located close together, in badly designed sections of code. And so, when a test reveals an error in a particular section of code, additional tests are carried out to exercise that section of code further and this often reveals several more errors. When this occurs, there is a good number of short time intervals between failures, and this is accounted for by the initially high value of the hazard rate $\lambda(t)$ in Fig. 12.4. As the errors are cleared from a poorly designed section of code, $\lambda(t)$ will steadily decrease and ultimately level off at some low level. The tester will eventually become confident that all errors have been removed from that section of code and begin testing some other section that is suspected of containing faults. The failure rate will therefore rise again, but it will not rise to the same level attained immediately after a failure (i.e. for small t) because the tester is now searching for a new section of bad code and not actively debugging one.

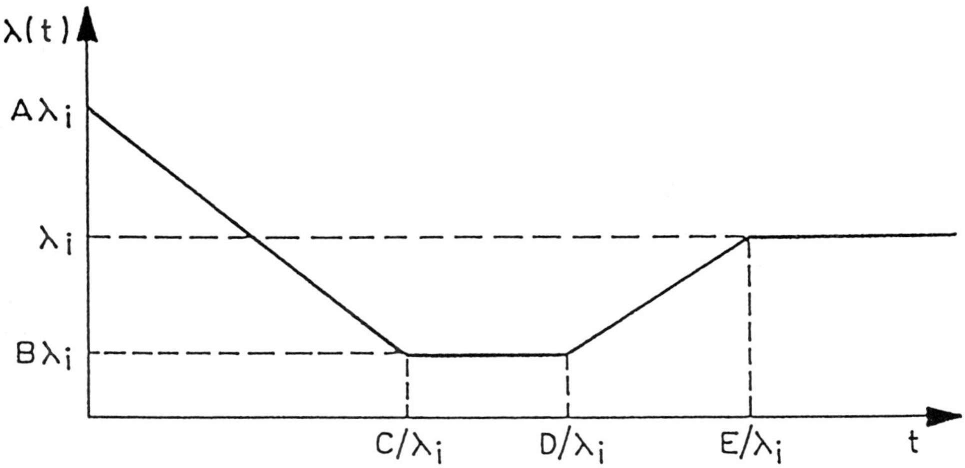

Fig. 12.4 A Piecewise-Linear Failure-Rate Function

This model assumes that the hazard rate maintains the basic shape of Fig. 12.4 throughout the test phase. But the model also takes account of the fact that there is, on average, a downward trend in failure rate as debugging proceeds, and this is achieved as follows. In Fig. 12.4 the hazard rate ultimately levels off to a value λ_i and this value is used to track the overall trend in failure rate. λ_i is estimated separately using a model such as the NURP model described in Section 12.2. This ensures that λ_i gradually decreases as faults are removed.

Having estimated λ_i, the remaining parameters A, B, C, D and E are then estimated. These parameters have been chosen in such a way that they are dimensionless quantities and Fig. 12.4 indicates that they are subject to the following constraints : $A>1$, $0<B<1$, $0<C<D<E$. The five parameters are estimated quite easily using maximum likelihood. Define

$$I(t) = \int_0^t \lambda(s)ds \qquad (12.34)$$

from which the density function of the time to next failure can be written

$$f(t) = \lambda(t)e^{-I(t)} \qquad (12.35)$$

This density function is used to form the likelihood function in the usual way. The integral I(t) is obtained by simply evaluating the area under the piecewise linear curve in Fig. 12.4.

12.3 Assessing the Performance of Software Reliability Models

In this section, techniques for assessing the performance of software reliability models are considered, with predictive validity being taken as the essential performance criterion. A method of assessing predictive validity was proposed by Littlewood in [7] and this was widely used during the 1980s. The method, which is known as the u-plot, has certain weaknesses that have recently come under scrutiny. This section commences with a description of the u-plot; the weaknesses of the method are then pointed out and some alternatives to the u-plot are then discussed. The possibility of determining the kind of performance that would be exhibited by an ideal software reliability model is also considered.

12.3.1 The u-Plot

The u-plot is an adaptation to software failure data of the Kolmogorov-Smirnov (KS) test for goodness-of-fit [20]. In its most conventional application, the KS test is used to test a *simple* hypothesis. In such a case, there are n independent observations $x_1, x_2, ..., x_n$ from an unknown cdf $F(x)$ and it is required to test the hypothesis $H_o : F(x) = F_o(x)$ where the parameters of $F_o(x)$ are specified. In order to apply the KS test, the x_i are subjected to the *probability integral transform* :

$$s_i = F_o(x_i) \qquad i = 1,2,.....,n \tag{12.36}$$

and, if the hypothesis is true, the s_i are samples from a uniform distribution on [0,1].

The s_i are ordered according to magnitude, so if the ordered variables are represented by s_i' we have $s_1' \leq s_2' \leq ... \leq s_n'$. The sample distribution function is then formed:

$$F_n(s) = \begin{cases} 0 & s < s_1' \\ i/n & s_i' \leq s < s_{i+1}' \\ 1 & s_n' \leq s \end{cases} \tag{12.37}$$

A plot of $F_n(s)$ against s gives the u-plot and, if the hypothesized distribution is correct, the u-plot should be close to a straight line drawn at 45° from the origin. The maximum vertical distance between the u-plot and this straight line is known as the Kolmogorov distance, and is given the symbol D_n. An illustrative example is shown in Fig. 12.5. For clarity, only the values of $F_n(s)$ corresponding to each of the s_i' are shown.

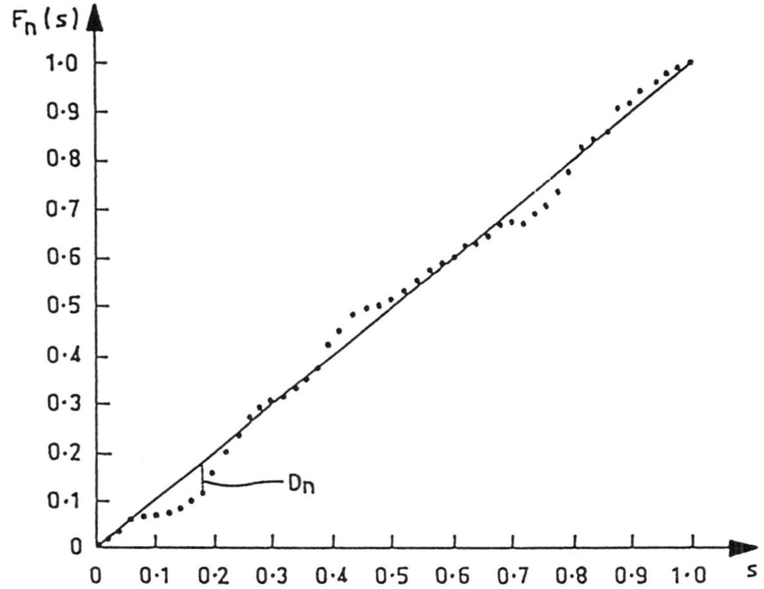

Fig. 12.5 An Example of a U-Plot

Given D_n, one assesses the goodness-of-fit of the given distribution $F_o(.)$ to the data set simply by checking D_n against a table of critical values. A single table (see [20]) is applicable to any distribution $F_o(.)$ because the KS test is distribution-free.

Littlewood's method of assessing the predictive validity of a software reliability model involves a rather less conventional use of the KS test. Let the parameters of the model be represented by the vector θ and suppose we have a set of software failure data (e.g. one of Musa's data sets [21]) consisting of n data values (failure times) $x_1, x_2, ..., x_n$. The first r of these data values are used to obtain an initial estimate of the parameters of the model. These parameters provide an estimate $\hat{\theta}$ of the failure time distribution following the rth failure, and data value x_{r+1} provides a single sample from this distribution. In the most common application of Littlewood's approach, data value x_{r+1} is now used to improve the estimates of the parameters of the model, which are then used to provide an estimate of the failure time distribution following failure $r+1$; the data value x_{r+2} is a single sample from this distribution. The probability integral transform is then implemented in the form

$$\hat{s}_i = F_o^{(i-1)}(x_i, \hat{\theta}(x_1, x_2, ..., x_{i-1})) \qquad i = r+1, r+2, ..., n \qquad (12.38)$$

where $F_o^{(i-1)}$ is the failure time distribution (according to the software reliability model) after *i-1* failures.

Clearly there are significant differences between the transformations in (12.38) and (12.36) and yet, in [7], Littlewood proposed that conventional tables of critical values for the KS test could be used for data transformed as in (12.38). Attention was drawn in [14] to the differences involved in the transformations in the two cases and doubts were expressed about the validity of Littlewood's method. The differences in the two transformations can be summarised as follows :

In (12.36), the full set of data values is transformed using the hypothesized cdf $F_o(.)$ whose parameters have been specified.

In (12.38), only the last *n-r* values are transformed, and each is transformed using a different cdf (x_i is transformed using $F_o^{(i-1)}(.)$). In addition, the different cdfs used have their parameters estimated from different numbers of data values (the parameters of $F_o^{(i-1)}(.)$ are estimated from *i-1* data values).

In [14] it was pointed out that when the KS test is applied to the *composite* hypothesis (i.e. when distribution parameters are estimated from the full set of data values) conventional tables of KS statistics are not applicable (see, for instance, [22]). And so, given the very significant differences between (12.36) and (12.38), it is not surprising that the validity of Littlewood's method has been questioned. However, a very recent study has shown conclusively that conventional tables of KS statistics are, in spite of these differences, applicable when Littlewood's method is used. This is true so long as sufficient initial data points are used to gain a good *a priori* estimate of the parameters of the model (see [23]).

12.3.2 U-Plot Performance of an Ideal Model of Software Reliability

It is not uncommon in the literature for a user of the u-plot to claim that a relatively small value of Kolmogorov distance is indicative of a relatively good fit of a set of data to the model under test. This is an incorrect interpretation of tables of the KS statistic because lower values of Kolmogorov distance are actually indicative of a higher confidence in the validity of the hypothesis under test. Thus, when data sets are generated from a particular distribution, the Kolmogorov distance can vary substantially from data set to data set, even when the hypothesized distribution is the correct one. This implies that if a software reliability model were ever developed that represented the failure behaviour of software exactly, the Kolmogorov distance obtained from a u-plot would generally differ from data set to data set. By investigating the statistical properties of the Kolmogorov distance it is possible to establish some useful facts about the u-plot performance to be expected of an ideal software reliability model. One such investigation is described in [23].

In [23], a scaled version of the Kolmogorov distance was considered, namely

$$K_n = \sqrt{n}\,.D_n \tag{12.39}$$

and use was made of the fact that, as n increases, the distribution of K_n tends to a fixed function, independent of n. That is

$$\lim_{n \to \infty} Pr\{K_n < x\} = \Phi(x) \tag{12.40}$$

where $\Phi(.)$ is independent of n.

For $n \geq 30$, the difference between $\Phi(.)$ and the cdf of K_n is negligible and so, in [23], the behaviour of $\Phi(.)$ was investigated by studying the statistical behaviour of K_n with $n = 30$. The pdf of K_n was generated by Monte Carlo simulation for the case of a simple hypothesis with $n = 30$. This led to the curve shown in Fig. 12.6, which was obtained by generating 50,000 sets of 30 samples from a known distribution $G(x)$ and determining the Kolmogorov distance for each set when testing the hypothesis $H_o : F(x) = G(x)$.

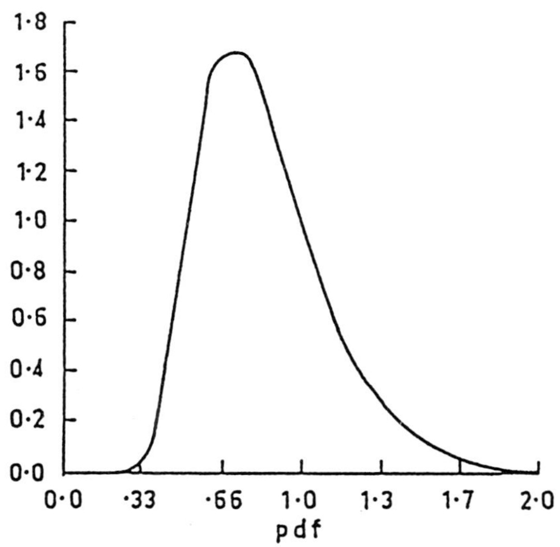

Fig. 12.6 The Density Function for K_{30}

Now note that if we had available a software reliability model that correctly represented software failure behaviour and if we used the u-plot to assess the predictive ability of this ideal model on many sets of software failure data, the values of K_n obtained for the various data sets should be distributed as in Fig. 12.6.

It is obvious that we cannot improve on a model that represents the software failure process exactly so that any attempt to develop models that give values of K_n that are distributed "better" than K_n in Fig. 12.6 would appear to be pointless. Stated another way, if a model is ever developed that gives $0.5 < K_n < 1.2$ on 90% of an adequately large range of software failure data, there would seem to be nothing to be gained by searching for an improved model. But nothing is ever exactly as it seems, and a weakness in the u-plot approach prevents us from using Fig. 12.6 as the ultimate benchmark.

12.3.3 A Weakness in the U-Plot, and Some Alternatives

The major weakness inherent in the use of the u-plot concerns the fact that it is possible for a model to make poor predictions over a whole set of data and yet be ascribed a small Kolmogorov distance (indicating good predictions) for overall performance. A very simple illustration of this is a case where a model provides a series of very pessimistic predictions in the first half of the period under study and then a series of equally optimistic predictions in the second half. An excellent value of Kolmogorov distance would be obtained in such a case because temporal information is discarded in the formation of the u-plot.

Littlewood and his colleagues addressed this problem in [24] and considered a number of possible methods of overcoming the difficulty. One method, called the y-plot, involves the transformation of the data into a form where it can be considered as representing the inter-event times of a Poisson process. Any shifts in predictive performance then appear as changes in the rate of occurrence of events in the Poisson process. The data are further transformed so that, in the absence of any shifts in predictive performance, the data are uniformly distributed on [0,1]. The y-plot is a plot of the sample distribution function of this transformed data so that the Kolmogorov distance can be used as a performance measure. The y-plot maintains the temporal ordering of the data, but interpretation of the plot is confused by the fact that increases in the Kolmogorov distance can occur due either to poor predictive performance or to shifts in predictive performance.

The other significant method considered in [24] is called the *prequential* method. The method employs the prequential likelihood function which is defined as follows.

Given a set of n observed failure times $x_1, x_2, ..., x_n$, we as before use the first r of these to estimate the failure time distribution. Each subsequent failure time is then used to improve this estimate, giving a sequence of distributions whose density functions we write $f^{(i)}(.)$, $i=r, r+1, ..., n$. The prequential likelihood function is then defined as below

$$PL = \prod_{i=r}^{n-1} f^{(i)}(x_{i+1}) \tag{12.41}$$

This function is used in the performance comparisons given in Section 12.4.2. It is worth noting here that although the prequential method is of some value in comparing the performance of models, it does not give any guide to the accuracy of an individual model's predictions.

One possible way of producing improved methods of dealing with this problem can be found by looking at how the Kolmogorov distance changes as predictions are made over a set of software failure data. In order to illustrate this, we will consider application of Littlewood's method in a slightly different way. As before, we will use the first r data points to estimate parameters and apply the probability integral transform as in (12.38).

Now recall that in the normal application of Littlewood's method, one evaluates $\hat{s}_i$ for $i=r+1$, $r+2$, etc. until all the data points have been used up. But here, rather than waiting for all data points to be used up, we will plot $\hat{s}_i$ for $i=r+1$ only, then for $i=r+1$ and $r+2$ only, and so on. For each plot, a value of the Kolmogorov distance can be determined. This allows determination of the way in which the Kolmogorov distance changes with increasing i. An illustrative example is shown in Fig. 12.7.

The plots in Fig. 12.7 are typical of the manner in which the Kolmogorov distance D_n can vary on the application of two different models to a given data set. Now note that in the normal application of Littlewood's method, we would estimate predictive performance from the value of D_n obtained after all the data have been used up (i.e. we would be guided by the value of D_n at the right-hand end of Fig. 12.7). And at the right-hand end of the figure, model A has a lower value of D_n than has model B. Thus, using Littlewood's method, we would conclude that model A exhibits superior predictive performance on this set of data than does model B. And yet we can see quite clearly from Fig. 12.7 that model B has a lower value of D_n for a much greater proportion of the data points, indicating better overall performance by model B. This makes it clear that determination of Kolmogorov distance at a single point can lead to erroneous conclusions being drawn.

From Fig. 12.7 it is fairly obvious that some form of averaging of predictive performance over the whole range of data points is required in order to obtain a more reliable assessment of overall performance. Two very simple averaging procedures were described in [25]. These measures were the average prediction error and the average bias (optimistic or pessimistic predictions). Each of these measures was computed from predictions of the number of faults still to be found in a software system. Although these measures give useful qualitative results, it is difficult to see how they could be used as a basis for determining the performance to be expected of an ideal software reliability model.

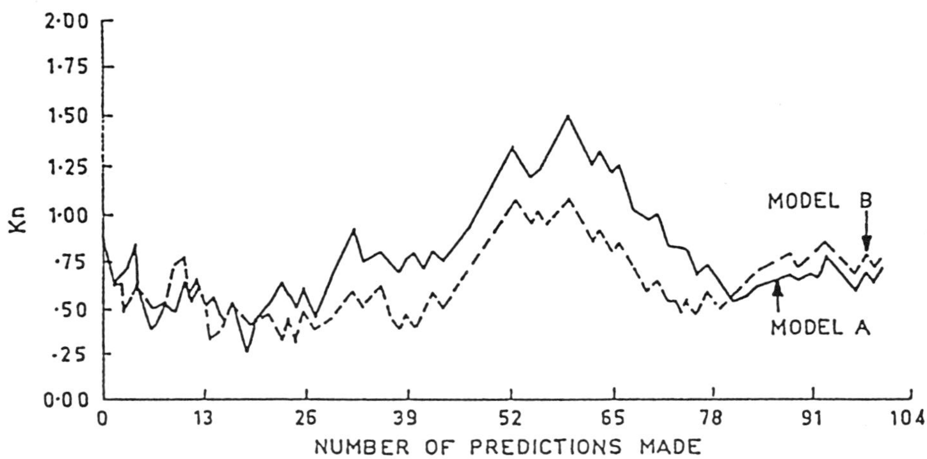

Fig. 12.7 K_n Statistics for Two Models

In [23], a method was proposed based upon the averaging of K_n (defined in (12.39)) over a set of predictions. The new statistic has the potential to provide a benchmark for assessing how close existing models are to the ideal, but the statistic was derived using essentially qualitative arguments and a thorough study of its properties is required before any definitive statements can be made.

12.4 Applications and Performance of Models that Incorporate Testing

Software reliability models that incorporate the testing process in a meaningful way can be expected to perform better (as predictors of reliability) during the test phase than models that make no attempt to account for the manner in which testing is carried out. In this section we will compare the performances of the two kinds of models on failure data collected during the test phase on a range of software systems.

Note also that models that include the testing process have at least one additional advantage over other models. When appropriately formulated, they can be used to assist with the problem of allocating testing resources. We commence this section with two examples of the application of models to the resource allocation problem and then give some comparisons of predictive performance.

12.4.1 Two Applications

Two of the models that were described in Section 12.2.2 have been applied by their authors to the problem of allocating testing resources. One method allows *a priori* estimation of the best allocation of test effort while the other requires some initial testing to be carried out in order to estimate model parameters before attempting resource allocation. Both methods are described here.

A Priori Allocation of Test Effort

It frequently occurs in practice that some parts of a software system have a much more critical effect on system performance than others, and hence need to be more thoroughly tested prior to release. In such situations, the following important question must be addressed by testing personnel : "How much testing effort should be applied to the critical software, and how much to the remainder?" The non-uniform testing model described in Section 12.2 can be used, before the commencement of testing, to assist in determining a solution to this allocation problem.

Model parameters cannot, of course, be estimated prior to the commencement of testing, and the method to be described here relies upon the user making an informed guess at the unknown parameter values. Past experience with similar software systems would obviously be of assistance here but, as will become clear, "ball-park" estimates (within an order of magnitude, or so) are quite adequate.

The NURP testing model requires the estimation of the 3 parameters[3] N_1, N_2 and C. To simplify the discussion here, we will assume that errors are evenly distributed throughout the software system so that N_1 and N_2 are given by

$$N_i = N \cdot \frac{\textit{size of section } i \textit{ of software system}}{\textit{total size of software system}} \qquad i=1,2 \quad (12.42)$$

[3] Parameter K can be eliminated from the likelihood equations.

where N is the total number of faults in the system. The assumption of a uniform distribution of faults helps clarify the following discussion because it reduces the number of parameters that have to be estimated from 3 to 2. In practice, no such assumption needs to be made, as will become clear.

Now suppose that we guess that for a software system under consideration, the total number of faults is somewhere in the range $[N_{min}, N_{max}]$. In principle, this range can be made as large as we like in order to ensure that the system we are dealing with is definitely accommodated.

The other parameter, C, is proportional to $(1\text{-}p)/p$ and, for what follows, we must keep explicit the testing profile p. So, we guess that parameter C lies somewhere in the range $\left[\frac{(1-p)}{p}A_{min}, \frac{(1-p)}{p}A_{max}\right]$, where A_{min} and A_{max} are positive constants.

The upper and lower bounds for the two unknown parameters define a boundary within which we expect system failure characteristics to lie as testing and debugging proceed. For a given boundary value and a given value of the testing profile p (recall that p represents the proportion of test effort applied to the more critical section of code) we can compute the expected time required to attain a desired reliability goal. This expected time is simply found by plugging values into (12.25) and noting that the expected time to find fault i is $1/\lambda_i$. These expected times are added until the required goal is attained.

For each boundary value, the expected time to attain the desired reliability goal can be plotted against parameter p. If this is done for each boundary value, a set of curves, typically of the form shown in Fig. 12.8, is obtained.

Now note that the bottom curve in Fig. 12.8 indicates a situation where the expected time to attain the reliability goal is minimum for $p=1.0$, i.e. when all the test effort is applied to the critical software. But note also that this curve represents a boundary of the region within which we expect the software properties to lie and that the curves for the other boundaries indicate that the expected time to attain the reliability goal is infinite with $p=1.0$. These latter curves point to the fact that if the properties of the software lie close to these boundaries, the reliability goal will not be attained even if all the faults are removed from the critical software.

However, the curves do indicate that if a value of p somewhere in the region (0.6, 0.7) is chosen, the expected time to attain the reliability goal will not be far from the minimum for any of the boundary curves. Thus, a sensible choice in this case would be to invest around 60 or 70 per cent of testing effort in the critical software.

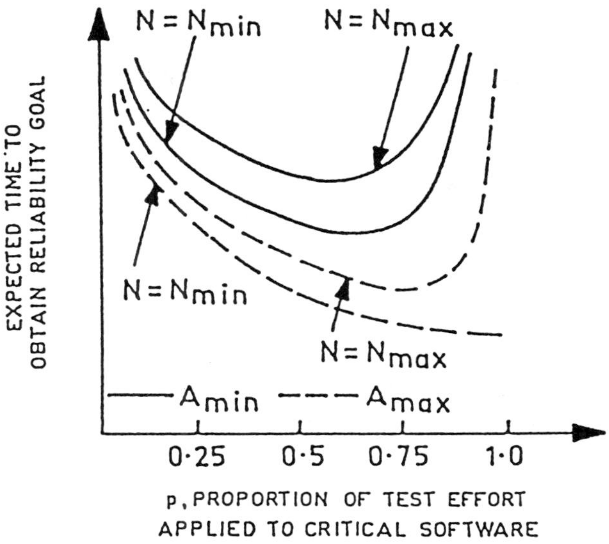

Fig. 12.8 Plots of Expected Testing Time Against p

Note that this procedure for allocating test effort would be essentially the same if 3 parameters had to be guessed (as would occur if we had not made the assumption underlying (12.42)). In such a case there would be 8 curves to consider. Note also that the procedure is unaffected by the breadth of the ball-park estimates that are made for the various parameters, although obviously, firmer conclusions can be drawn from tighter estimates. More discussion of this procedure appears in [2].

Allocation of Test Effort Based Upon Early Test Results

In [17] the question of optimally allocating test effort at the module testing stage is addressed. A method is developed which employs the model of Yamada *et al.* discussed in Section 12.2.2 and which is directed at the solution of two distinct allocation problems. The model is used to represent each module in a system with M modules and it is assumed that sufficient initial testing has been carried out to give satisfactory estimates of model parameters. For module i, (12.27) becomes

$$H_i(t) = a_i(1-exp[-r_iW_i(t)]) \qquad (12.43)$$

and parameters a_i and r_i are estimated from the initial test results.

(i) Minimization of remaining errors

In this case it is assumed that the total amount of testing resource available for module testing has been specified and the manager has to allocate this resource in such a way that it will minimize the total number of software errors remaining.

From (12.43) the number of software errors in module i (after further testing) can be written

$$z_i = a_i exp[-r_i q_i] \qquad i=1,2,\dots,M \tag{12.44}$$

where q_i is the amount of testing resource applied to module i.

The resource allocation problem can then be formulated as a relatively simple optimization problem :

$$Minimize \sum_{i=1}^{M} v_i a_i \; exp[-r_i q_i]$$

such that (12.45)

$$\sum_{i=1}^{M} q_i = Q$$

where Q is the total amount of resource available. The parameter v_i is included to allow emphasis to be placed on particular modules.

The solution to (12.45) can be found explicitly using the method of Lagrange multipliers (see [17]).

(ii) Minimization of testing resource expenditure

Here it is assumed that a target has been set for the total number of errors remaining in the system at the completion of testing and the manager is seeking to minimize the amount of testing resource expended. If Z represents the target figure for software errors, this problem can be formulated as follows :

$$Minimize \sum_{i=1}^{M} q_i$$

such that (12.46)

$$\sum_{i=1}^{M} a_i \; exp[-r_i q_i] = Z$$

And again, an explicit solution to this optimization problem can be found using Lagrange multipliers [17].

These ideas can be extended to provide the manager with additional control over the testing process (see [17]).

12.4.2 *Performance Evaluation*

In this section, the performances of some of the models described earlier are compared. This comparison is carried out using the sixteen major data sets compiled by Musa in [21]. Each of these data sets consists of a list of failure time intervals collected during the test phase of a software system. They have been used previously as a basis for comparing predictive validity, and one recent study based on these data compared the Littlewood-Verrall model with a model of the Jelinski-Moranda type [10]. The study concluded that the Littlewood-Verrall model was generally superior and, in what follows, we will compare the Littlewood-Verrall model with two of the models that have been developed to reflect aspects of the testing process. Following this comparison, a brief discussion will be given of the possibility of improving model performance by taking account of the shifts of emphasis, and changes in the line of attack, that frequently occur during the test phase.

Performance Comparisons

The discussion here is based upon the comparison reported in [15]. This comparison was carried out before the work described in Section 12.3.3 had been undertaken and so, in common with other comparisons reported in the literature, it employs the u-plot as one of its comparison techniques. Section 12.3.3 tells us that any conclusions drawn from the u-plot should be treated with caution. But in addition to the u-plot, the comparison undertaken in [15] employed the prequential method, which, as mentioned earlier, was proposed in [24] as a means of overcoming the weaknesses in the u-plot.

In [15], the predictive performance of eight different models was compared. One was the PLHR model described in Section 12.2.2, two were variants of the Littlewood-Verrall model and the remaining five were all variants of the models of software testing described in Section 12.2.2. Here we will restrict the comparison to the PLHR model, the non-uniform testing (NURP) model described in Section 12.2.2, and the two variants of the Littlewood-Verrall model. These two variants differ in that one employs the Ψ_i function defined by (12.8) and the other employs the Ψ_i function in (12.9); we refer to these two models as the LV2 and the LV3 model respectively.

Comparative statistics for the four models and the sixteen sets of data are shown in Tables 12.1 and 12.2. In each table, and for each data set, the smallest value indicates best performance (note that we employ $-\log_e(PL)$ in Table 12.2). For each of these tables, "Win Matrices" have been constructed which indicate, for each model, the number of data sets for which it out-performed each of the other models. These win matrices appear as Tables 12.3 and 12.4 and, as an example, the first row in Table 12.3 indicates that the NURP model outperformed the PLHR model on 3 data sets, and so on. The final column in Tables 12.3 and 12.4 indicates the total number of "wins" for each model.

Clearly, the PLHR model performs outstandingly well by either measure. There is little to choose between the other three models, although the LV3 model performs quite badly according to the prequential measure. This appears to be due to the fact that the prequential method is quite sensitive to noisy (i.e. highly variable) predictions and an examination of the predictions made by the LV3 model for these data sets indicates a much greater tendency to noise than is observed in the other models.

Table 12.1 Kolmogorov Distance

Data set	NURP	PLHR	LV2	LV3
1	0.076	0.060	0.129	0.178
2	0.103	0.153	0.238	0.162
3	0.146	0.286	0.334	0.339
4	0.098	0.106	0.117	0.238
5	0.227	0.090	0.097	0.083
6	0.257	0.096	0.123	0.136
14	0.222	0.176	0.334	0.247
17	0.189	0.107	0.182	0.134
27	0.227	0.134	0.301	0.207
40	0.141	0.069	0.300	0.217
ss1a	0.158	0.079	0.155	0.114
ss1b	0.239	0.122	0.129	0.118
ss1c	0.202	0.132	0.130	0.152
ss2	0.243	0.100	0.184	0.143
ss3	0.286	0.078	0.237	0.187
ss4	0.250	0.069	0.213	0.216

Table 12.2 $-\log_e(\mathrm{PL})$

Data set	NURP	PLHR	LV2	LV3
1	766	743	766	782
2	369	416	372	371
3	259	283	265	297
4	328	333	315	371
5	7057	6772	6975	6975
6	314	307	292	315
14	394	393	429	432
17	300	317	300	298
27	424	415	437	435
40	994	988	996	1023
ss1a	1116	1076	1108	1116
ss1b	3672	3538	3629	3664
ss1c	2659	2571	2632	2701
ss2	1968	1894	1992	1994
ss3	2781	2618	2801	2816
ss4	1995	1894	2003	2053

Table 12.3 Win Matrix for Kolmogorov-Smirnov Statistic

	NURP	PLHR	LV2	LV3	wins
NURP	0	3	7	7	17
PLHR	13	0	14	14	41
LV2	9	2	0	6	17
LV3	9	2	10	0	21

Table 12.4 Win Matrix for Prequential Statistic

	NURP	PLHR	LV2	LV3	wins
NURP	0	4	8	11	23
PLHR	12	0	11	14	37
LV2	6	5	0	12	23
LV3	4	2	3	0	9

The important fact here, of course, is that the PLHR model has performed so well under both measures. This is very encouraging, especially considering the *ad hoc* nature of the PLHR model. Various ways of generalizing this model are currently under investigation.

The Effects of Variations in Execution Profile

For large and complex software systems, testing is rarely, if ever, carried out under the same execution profile throughout the test phase. Usually the various system modules are first tested individually, and this is followed by gradual integration of the system and testing of the interfaces between modules. When the system has been fully integrated, a series of function and performance tests is carried out. Except where otherwise stated we have, in this chapter, been assuming that software reliability models are applied to a fully integrated system.

In their normal mode of application, the assumption is generally made (often implicitly) that the data to which a model is fitted are generated by a constant execution profile. And yet, during function testing, it is clear that the execution profile will be different for each function to be tested and that the execution profile will change yet again when performance testing is undertaken. The fact that changes in the execution profile are generally ignored by software reliability models is surely a reason for their relatively poor performance in applications and for the fact that they have so far proved to be of only minor assistance to software developers.

I am only aware of two instances where the fact that the execution profile undergoes significant change has been addressed by software reliability modellers. In both [2] and [19] it was advocated that the failure data be partitioned into separate sets, each corresponding to a period when the execution profile is constant. In [2] the predictions obtained from each partition were used as a basis for estimating the step increase in failure rate that usually occurs when a large software system is transferred from the test phase to the operational phase. In [19] it was shown that partitioning of the data in the manner advocated can lead to a radical improvement in the fit of the model to the data.

In the test phase of most software systems, it will be clear when a change in execution profile takes place, and if it is intended that predictions of future reliability be obtained from a statistical model, then the times of such changes should be recorded. No such information is available for Musa's data sets, although it is clear that some shifts in execution profile have taken place. Without such information it is impossible to get the best out of a software reliability model. An attempt was made in [26] to develop a statistical test to identify the points in a set of data where changes in execution profile have taken place, but this is a difficult problem.

12.5 Conclusions

This chapter commenced with a description of three of the better known software reliability models that emerged in the 1970s. These, like most of the models of software reliability that have appeared in the literature, are black box models in the sense that they simply hypothesize a failure time distribution and estimate the parameters of this distribution from software failure data. Models of this kind have not been widely applied by the software industry because their predictions tend not to be sufficiently accurate to use as a basis for decision making.

It was argued in Section 12.2.2 that better performing models will be obtained if the modelling process takes account of the manner in which software is tested. Several models that make some attempt to do this were then described and it was shown that at least one of these models (the PLHR model) exhibits predictive performance that is superior to that available from earlier models. It was also shown how the fact that these newer models take more account of practical issues allows for a much broader range of model applications.

The problem of assessing the performance of software reliability models was also discussed and several issues raised. Among these, a weakness in the most widely used technique of performance assessment (the u-plot) was pointed out and a new method, which appears to overcome this difficulty, was described. In addition, the possibility of determining the performance that would be exhibited by an ideal model of software reliability was discussed.

In summary, the chapter has shown that improved models of software reliability can be obtained by taking some account of the manner in which software is tested. This raises the possibility that greater improvements could be obtained by taking account of other aspects of the software lifecycle. If this were done, models with a substantially greater number of parameters are likely to be obtained; different parameters could be estimated at those stages of the lifecycle where they are significant. How far one needs to go in this direction in order to obtain models that are sufficiently accurate for decision making purposes is unclear, but this question would be easier to answer if the performance of an ideal software reliability model were available as a benchmark. Work on this latter issue, which was first discussed in [23], is continuing.

REFERENCES

[1] Downs T., *Reliability Problems in Software Engineering - A Review*, Computer Systems Science and Engineering, Vol. 2, pp. 131-147, 1987.

[2] Downs T., *An Approach to the Modeling of Software Testing with Some Applications*, IEEE Trans. on Software Engineering, Vol. SE-11, pp. 375-386, 1985.

[3] Musa J.D., A. Iannino and K. Okumoto, *Software Reliability : Measurement, Prediction and Application*, McGraw-Hill, New York, 1987.

[4] Goel A.L., *Software Reliability Models : Assumptions, Limitations and Applicability*, IEEE Trans. on Software Engineering, Vol. SE-11, pp. 1411-1423, 1985.

[5] Jelinski Z. and P.B. Moranda, *Software Reliability Research*, in Statistical Computer Performance Evaluation, W. Freiberger (ed.), Academic Press, pp. 465-484, 1972.

[6] Young D.M. and R.T. Gregory, *A Survey of Numerical Mathematics, Addison-Wesley, 1973.*

[7] Littlewood B., *Theories of Software Reliability : How Good are They and How Can They be Improved?*, IEEE Trans. on Software Engineering, Vol. SE-6, pp. 489-500, 1980.

[8] Littlewood B. and J.L. Verrall, *A Bayesian Reliability Model with a Stochastically Monotone Failure Rate*, IEEE Trans. on Rel., Vol. R-23, No. 2, pp. 108-114, 1974.

[9] Littlewood B., *The Littlewood-Verrall Model for Software Reliability Compared with Some of its Rivals*, Journal of Systems and Software, Vol. 1, pp. 251-258, 1980.

[10] Troy R. and R. Moawad, *Assessment of Software Reliability Models*, IEEE Trans. on Software Engineering, Vol. SE-11, pp. 839-849, 1985.

[11] Goel A.L. and K. Okumoto, *Time-Dependent Error-Detection Rate Model for Software Reliability and Other Performance Measures*, IEEE Trans. on Rel., Vol. R-28, No. 3, pp. 206-211, 1979.

[12] Angus J.E., *The Application of Software Reliability Models to a Major C^3I System*, Proc. of Ann. Rel. and Maint. Symposium, pp. 268-274, 1984.

[13] Ehrlich W.K. and T.J. Emerson, *Modeling Software Failures and Reliability Growth During System Testing*, Proc. of IEEE International Conference on Software Engineering, pp. 72- 77, 1987.

[14] Downs T., *Extensions to an Approach to the Modeling of Software Testing with Some Performance Comparisons*, IEEE Trans. on Software Engineering, Vol. SE-12, pp. 979-987, 1986.

[15] Downs T. and P. Garrone, *Some New Models of Software Testing with Performance Comparisons*, IEEE Trans. on Rel., Vol. R-40, No. 3, pp. 322-328, 1991.

[16] Feller W., *An Introduction to Probability Theory and its Applications*, Third Edition, Wiley, New York, 1968.

[17] Ohtera H. and S. Yamada, *Optimal Allocation and Control Problems for Software Testing Resources*, IEEE Trans. on Rel., Vol. R-39, No. 2, pp. 171-176, 1990.

[18] Yamada S., J. Hishitani and S. Osaki, *Test-Effort Dependent Software Reliability Measurement*, International Journal of System Science, Vol. 22, pp. 73-83, 1991.

[19] Tohma Y., K. Tokunaga, S. Nagase and Y. Murata, *Structural Approach to the Estimation of the Number of Residual Software Faults Based on the Hypergeometric Distribution*, IEEE Trans. on Software Engineering, Vol. SE-15, pp. 345-355, 1989.

[20] Mann, N.R., R.E. Schafer and N.D. Singpurwalla, *Methods for Statistical Analysis of Reliability and Life Data*, Wiley, New York, 1974.

[21] Musa J.D., *Software Reliability Data*, Technical Report, Data and Analysis Center for Software, Rome Air Development Center, Rome, N.Y., 1979.

[22] Lilliefors H.W., *On the Kolmogorov-Smirnov Test for Normality with Mean and Variance Unknown*, Journal of the American Statistical Association, Vol. 62, pp. 399-402, 1967.

[23] Downs T. and A. Scott, *Evaluating the Performance of Software Reliability Models*, to appear in IEEE Trans. on Rel., Dec. 1992.

[24] Abdel-Ghaly A.A., P.Y. Chan and B. Littlewood, *Evaluation of Competing Software Reliability Predictions*, IEEE Trans. on Software Engineering, Vol. SE-12, pp. 950-967, 1986.

[25] Malaiya Y.K. and N. Karunanithi, *Predictability Measures for Software Reliability Models*, in Software Models, Y.K. Malaiya and P.K. Srimani, eds. IEEE Computer Society Press, Los Alamitos, pp. 60-65, 1990.

[26] Scott A., *The Mathematical Modeling of Software Reliability*, M Eng Sc Thesis, The University of Queensland, 1991.

New Trends in System Reliability Evaluation
K.B. Misra (Editor)
1993 Elsevier Science Publishers B.V.

CHAPTER 13

MECHANICAL SYSTEMS RELIABILITY MODELING

S. S. Rao
School of Mechanical Engineering
Purdue University
West Lafayette, IN 47907-1288, U.S.A.

ABSTRACT

The ever increasing demand for better and reliable mechanical systems necessitates the explicit consideration of reliability in the modeling and design of mechanical systems. The inadequacy of the traditional factor of safety in ensuring proper degree of safety of mechanical systems is explained. The differences between the reliability modeling of electronic/electrical and mechanical systems are outlined. The role of modeling strength, load, and geometry in the reliability analysis of mechanical and structural systems is described. The methods of analyzing the probability of failure of weakest-link and fail-safe systems are presented. The modeling and analysis methods described in this chapter are illustrated through the reliability analysis of a compound gear train and a Geneva mechanism. The techniques presented in this work are quite general and are applicable for the modeling, analysis, and design of any mechanical system.

13.1. Introduction

The consideration of reliability plays an important role in virtually all mechanical systems. As the demands for high-performance and reliable products increase, the minimization of the probability of failure at the lowest possible cost becomes important in order to be able to compete in the world market. Traditionally, the concept of *factor of safety* is used to maintain a proper degree of safety and reliability in the design of machine elements and systems. This approach has several limitations when material properties, applied loads, geometric parameters, assembly conditions and other parameters have random characteristics.

It has been shown that conventional design procedures based on factor of safety are not rational in the sense that a uniform factor of safety does not lead to a uniform probability of failure in all situations. The factor of safety is taken as the ratio of mean strength and mean load. Although the strength of a component and the load acting on it are assumed to be constants in deterministic design procedures, they are treated as random variables in reliability based design. Figures 13.1(a) and (b) illustrate that it is possible for a component to have different reliabilities even though the factor of safety remains unaltered [1, 2]. In fact, Fig. 13.1(b) is obtained by multiplying the mean values of resistance or strength, $\bar{R}$, and load, $\bar{L}$, by a constant $k > 1$, such that the

$$\text{Factor of Safety} = \frac{k\bar{R}}{k\bar{L}} = \frac{\bar{R}}{\bar{L}} \tag{13.1}$$

where a bar over a symbol denotes its mean value. Thus the overlapping area between the probability distributions of strength and load, and hence the probability of failure, will be smaller in the case of Fig. 13.1(b). The reliability

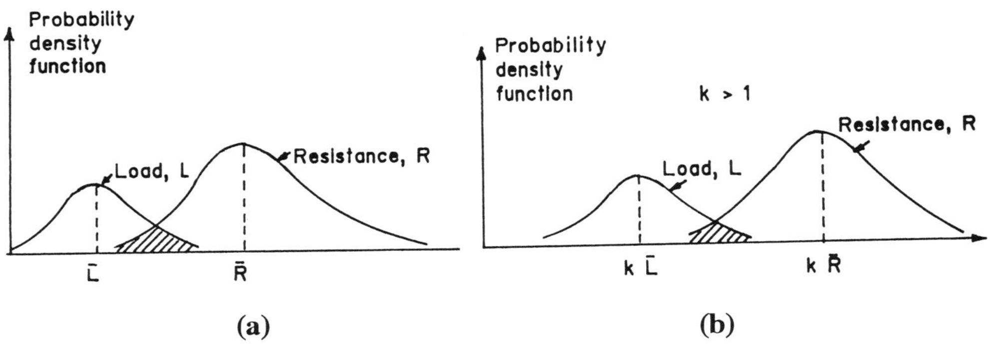

Fig. 13.1 Change in Reliability for the Same Factor of Safety

can be different, for the same factor of safety, when the variabilities of the random variables are different. For example, in Figs. 13.2(a) and (b), the mean values of strength and load are the same. However, the standard deviations of both strength and load are smaller and the overlapping area between their probability distributions, and hence the probability of failure of the component, will be smaller in the case of Fig. 13.2(b) compared to that of Fig. 13.2(a). Mischke [3] presented

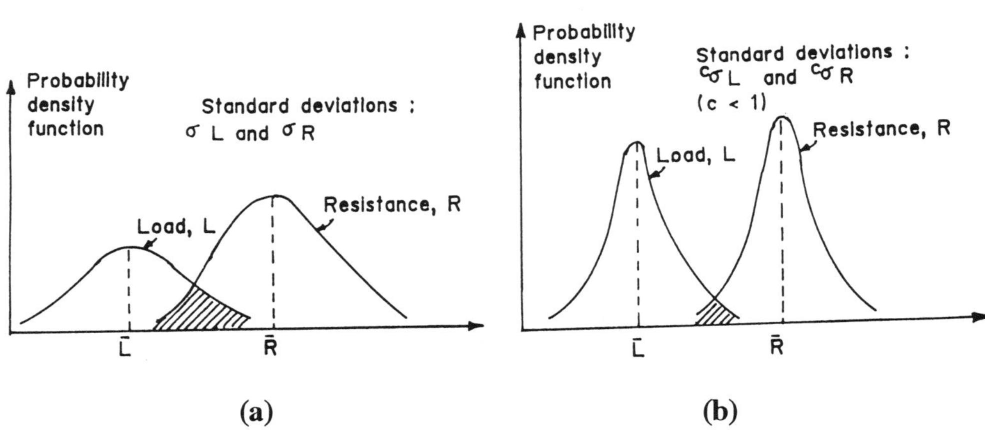

Fig. 13.2 Change in Reliability for the Same Factor of Safety

a formal relationship between the *reliability* and the *factor of safety* of a mechanical component. By treating the factor of safety as a random variable, he used Bienayme-Chabyshev and Camp-Meidell theorems to derive expressions for the mean and the variance of the factor of safety for a specified value of reliability.

13.2 Role of Reliability in Mechanical Systems Design

Reliability is important in every phase of a mechanical component/system. In fact, failure to perform an intended function at any stage during its lifetime constitutes the failure of a system. Fig. 13.3 shows the relationship of reliability with the various phases of the life cycle of a component/system. The development

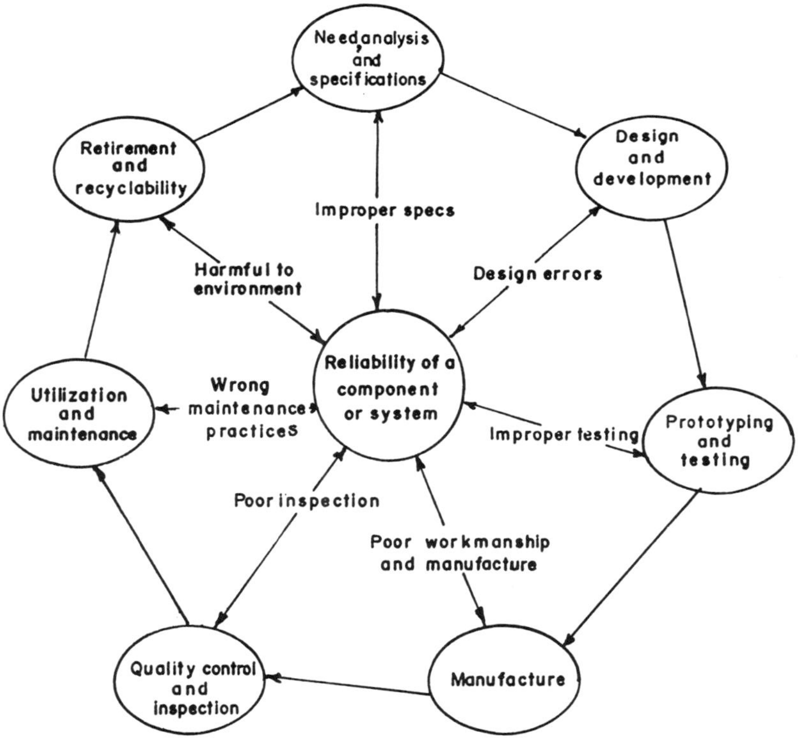

Fig. 13.3 Relationship of Reliability with Various Phases of the Life Cycle of a System

of a system starts with the identification of a need and the formulation of specifications. The formulation of proper specifications is the key to developing a reliable system. In fact, the specifications dictate the manufacturability, performance, maintenance and recyclability of the system. The design stage translates the specifications into detailed geometry and drawings of the system components. Once the detailed design is completed, the building of a prototype and its testing is undertaken. This step is aimed at revealing the weaknesses and shortcomings of the design in satisfying the specified needs. When the design is qualified properly, the system is manufactured using suitable machines and processes. The quality control and inspection techniques are used to ensure that the component/system is manufactured according to the design specifications. During the practical utilization of the component/system, proper maintenance schemes are used to ensure the continuous operation and availability of the system. At the end of the useful life of the system, it is to be retired/recycled to avoid any adverse environmental impact. The experience gained through the design, production and operation of the system forms a new technological base for the development of improved systems to satisfy additional needs of the society.

13.3 Differences Between Reliability Modeling of Electronic and Mechanical Systems

For most electrical/electronic components, the failure rates can be assumed to be constants. However, for mechanical components and subsystems, the operating environments vary considerably thereby producing a wide dispersion of failure rates even for apparently similar components and subsystems. The reliability prediction of electronic equipment is supported by standardized documentation in the form of military standards, specifications and handbooks [4,5]. These documents were developed due to the standardization and mass production characteristics of electronic parts that permitted the creation of valid failure rate data banks. Unfortunately, there is no commonly accepted method for predicting the reliability of mechanical components and equipment at the present time. Although documents such as the nonelectronic reliability notebook were developed from existing failure rate data banks, the accuracy of the resulting reliability predictions is not good due to the wide dispersion of failure rates and inconsistencies in failure patterns observed for mechanical components and subsystems.

The main characteristics of mechanical components that lead to inconsistent failure rates are the following:

- Many components such as valves and gearboxes often perform more than one function and failure data for specified applications are not available. For example, a hydraulic valve may contain a manual shut-off feature as well as an automatic control mechanism, and either one may be the cause of failure in a specific application.
- Mechanical and structural components have several modes of failure and a failure may be associated with more than one failure mode.
- The failure rates of mechanical components are not constants due to wear, fatigue and other stress-related problems. Once the failure rate is not

constant, the data gathering becomes complex since the operating time is also to be recorded.

- The mechanical component reliabilities are more sensitive to loading, geometry, materials, operating mode and utilization rate compared to electronic component reliabilities.
- The definition of failure for a mechanical component often depends upon its application. For example, failure due to excessive noise or leakage cannot be universally established. Lack of such information in the available failure rate data banks limits their usefulness.

13.4 Failure Modes of Mechanical Systems

Any mechanical or structural component or system has a finite life during which it performs satisfactorily. One of the primary causes for the failure of a component could depend on the nature, type, and magnitude of the loads acting on it. All the loads can be classified into three categories - static, dynamic, and cyclic. The failures that result from different loads include yielding, buckling, creep, fracture, crushing, tearing, spalling, abrading, and wear. Another cause of failure of mechanical systems is the corrosion or chemical attack by the environment. The nature of damage caused by corrosion includes general thinning, pitting, cavitation, hydrogen embrittlement, and intergranular corrosion. In many cases, several modes of failure occur simultaneously. Some of the important failure modes and their characteristics can be summarized as follows [6]:

Static failure

When a load, larger than the ultimate strength of the material, is applied, most materials fail by fracture leading to static failure.

Fatigue failure

When a load is varied in a cyclic manner, most materials fail by fracture even when the magnitude of the maximum cyclic stress is far below the yield strength of the material. In such a case, the failure starts with a crack at a point of stress concentration and ends with the propagation of the crack which ultimately leads to the final fracture.

Creep failure

When a load is applied over an extended period of time, some materials sustain strain and deformation. This is known as creep failure since the creep deformation either exceeds a limiting tolerable distortion or causes the material to rupture.

Corrosion failure

When a metal surface is exposed to a corrosive environment, direct chemical or electrochemical reactions take place with the environment causing damage to the structure. The corrosion damage is usually accelerated by the application of stress. In hydrogen embrittlement, the ductility of the metal increases due to the absorption of hydrogen. This results in either brittle failure under static loads at low strain rates or fracture under impact loads at high strain rates.

Wear failure

When two surfaces in contact have relative motion, gradual mechanical destruction of metal surfaces takes place. This type of failure is known as wear and can occur in all types of contact - sliding, rolling, or impact.

Instability failure

When a small change in the work done by the external forces exceeds the strain energy stored in the component or system, it becomes unstable as in the case of the buckling of columns, plates, and shells.

In practice, machine and structural components experience several other types of failure as well. For example, manufacturing tolerances can produce higher stresses than the ideal structure. When two machine parts rub together, fretting fatigue will occur, which can reduce the strength by as much as 50 percent. Casting and welding defects, poor surface finishes, overtightening or faulty bolting or riveting may induce larger stresses. Improper grinding may embrittle a surface. Similarly, hammering a part to straighten it can introduce residual stresses. Each type of failure requires a different type of analysis, inspection, and testing procedure. A reliability model must be able to consider all possible failure modes of a component or system. In general, a mechanical system consists of several components and is subjected to a variety of loads. Hence a proper reliability model must be able to consider multiple members and multiple loads along with their probability characteristics. Two types of models commonly used in reliability analysis of mechanical components are - the weakest link or series model and the fail safe or a parallel model. These have been presented in this chapter.

13.5 Modeling of Strength, Load and Geometry

The reliability of a mechanical or structural system is influenced by the probability characteristics of strength of materials, loads, and geometry. These aspects are discussed in the following paragraphs.

13.5.1 Strength

Many of the failure modes of mechanical components are associated with the variability of material properties. Most engineering materials, due to the nature of the production processes used in producing them, are inhomogeneous and anisotropic, and contain several imperfections and defects. These cause variability in the properties of materials [7]. For example, tests conducted for the static strength such as yield or ultimate strength, of a material leads to a probability distribution as indicated in Fig. 13.4.

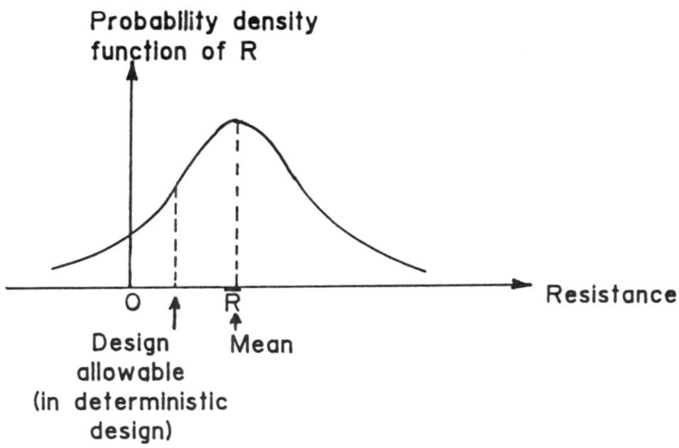

Fig. 13.4 Probabilistic Variation of Resistance or Strength of Material

If dynamic strength tests are conducted, most structural materials exhibit fatigue characteristics. The time-or cycle- dependent material properties, which lead to the fatigue behavior, can be represented as shown in Fig. 13.5. This figure

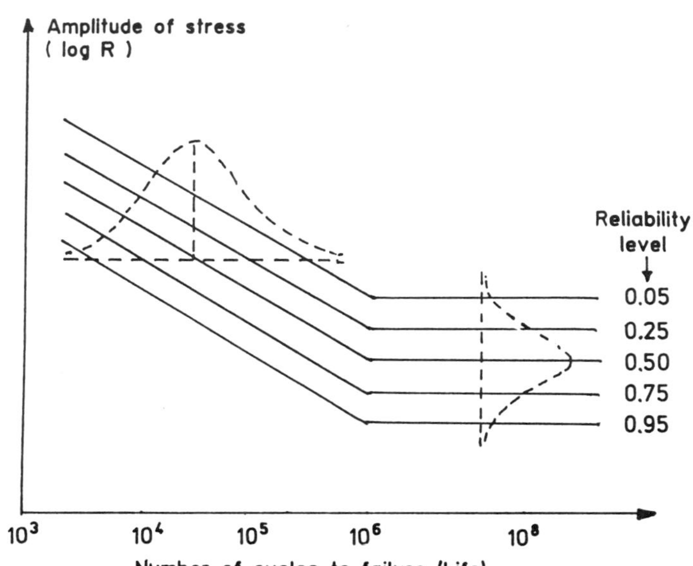

Fig. 13.5 Fatigue Characteristics of Materials

also indicates the statistical behavior of fatigue properties. For a given lifetime, the variability can be represented by a distribution in the same manner as for the static strength. Over any specified time period (or a specified number of load cycles), the variability of the fatigue properties are usually represented by constant-probability-of-survival (reliability) curves [8]. It can be seen from Fig. 13.5 that the variability in times to failure at constant stress amplitude is much larger than the variability in stress at constant life, particularly at larger values of life.

13.5.2 Loads

The loads acting on a machine or structure can be static or dynamic. The dead or gravity loads caused by the weight of the machine or structure and the live loads caused by the weights of the moving parts are the static loads. The variations in these loads are mostly caused by fluctuations in the specific weights of the materials. The static loads can be described by suitable probability distributions. The dynamic loads are caused by rotating unbalances present in machines, wind loads, and earthquake loads. The probabilistic variations in the dynamic loads are characterized as random processes [9].

13.5.3 Geometry

The geometry of machine components such as thicknesses of plates, diameters of shafts, dimensions of rectangular parts, diameters of holes, profiles of cams and gear teeth, and distances between components are all subjected to variability due to manufacturing tolerances and limitations of operators and machines. Although nominal values of geometric parameters are used in design, the physical realization of these parameters is subject to variability due to the unavoidable manufacturing errors. The statistical analysis of the various types of dimensions produced in different manufacturing processes permits a statistical description through suitable probability distributions [10].

13.6 Probability of Failure of Weakest-Link Systems

A weakest-link system is a series system in which the failure of any component or link constitutes the failure of the whole system. Figure 13.6 indicates the various types of weakest-link models that can be used in practice. For example, in Fig. 13.6(a), a system known as the fundamental case consisting of a single link subjected to a single load, is shown. For the purpose of reliability analysis, let the resistance or strength of the member be denoted by R and the load by L. If the probability density functions of R and L are known to be $f_R(r)$ and $f_L(l)$, respectively, the probability of failure of the system, P_f, is given by

$$P_f = P(R < L) = \int_{-\infty}^{\infty} F_R(r)\, f_L(l)\, dl = 1 - \int_{-\infty}^{\infty} F_L(r)\, f_R(r)\, dr \tag{13.2}$$

The probability of failure, given by (13.2), is bounded as [1]:

$$p_1\, p_2 < P_f < p_1 + p_2 - p_1 p_2 \tag{13.3}$$

where $p_1 = P(R < R_{min})$ (13.4)

and $p_2 = P(L > L_{max})$ (13.5)

with R_{min} denoting the minimum expected resistance of the component and L_{max} indicating the maximum expected load. Figure 13.6(b) shows the case where a single member is subjected to several load conditions. In this case, the failure probability can be expressed as

$$P_f = 1 - \int_{-\infty}^{\infty} \left[\prod_{j=1}^{n_l} F_{L_j}(r)\right] f_R(r)\, dr \tag{13.6}$$

where $F_{L_j}(r)$ denotes the probability distribution function of the *jth* load acting on the member(L_j) and n_l indicates the number of load conditions. Figure 13.6(c) indicates a system in which a single load acts on a chain consisting of n links. In this case, the probability of failure can be determined from the relation

$$P_f = 1 - \int_{-\infty}^{\infty} \left\{\prod_{i=1}^{n} \left[1 - F_{R_i}(l_i)\right]\right\} f_L(l)\, dl \tag{13.7}$$

where l_i is the force or stress induced in the *ith* member or link due to the load l and $F_{R_i}(r)$ is the probability distribution function of the resistance of the *ith* member. Since this is a weakest-link system, the term in braches in (13.7) represents the probability of survival of the chain and is based on all links surviving

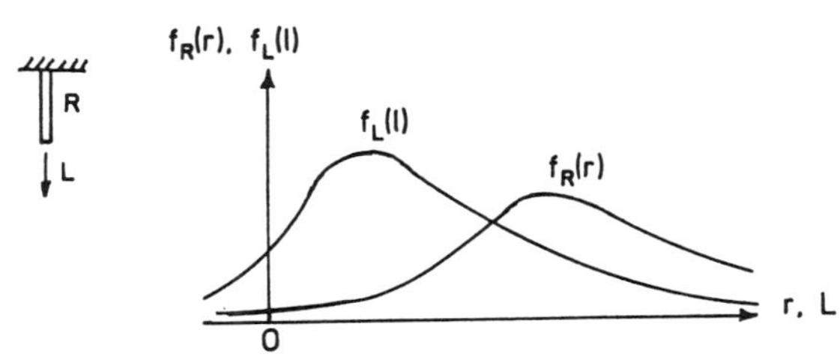

(a) Fundamental Case (one Member, One Load)
Fig. 13.6 Weakest-Link (Series Systems)

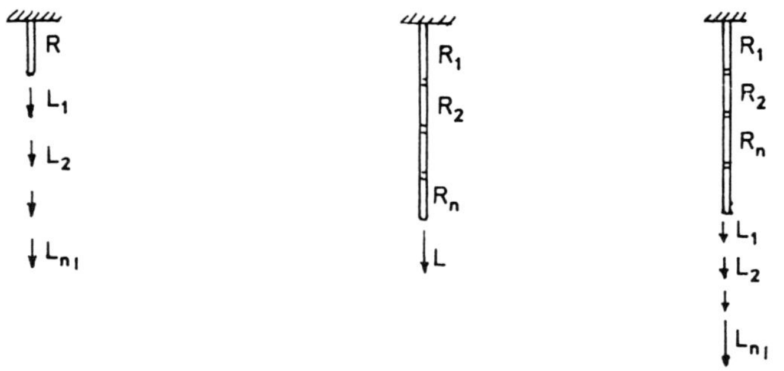

(b) One Member, Several Loads **(c) Several Members, One Load** **(d) Several Members, Several Loads**

Fig. 13.6 Weakest-Link (Series Systems)

under the load L = l. This term is evaluated from the products of probabilities of individual links surviving under the load L = l. Equation (13.7) is often approximated as

$$P_f = 1 - \prod_{i=1}^{n} [1 - \bar{P}_{f_i}] \tag{13.8}$$

where $\bar{P}_{f_i}$ denotes the probability of failure of the *ith* link under the given load. Finally, Fig. 13.6(d) shows the case of a system consisting of several members acted on by multiple loads. In this case, the probability of failure is given by

$$P_f = 1 - \int_{L_1}\int_{L_2}\cdots\int_{L_{nl}} \prod_{i=1}^{n} [1 - F_{R_i}(l_{im})]\; f_{L_1}(l_1)\cdots f_{L_{nl}}(l_{im})\, dl_1 \cdots dl_{n_l} \tag{13.9}$$

where l_{im} is the total force or stress induced in the *ith* link due to the loads l_1, l_2, ..., l_{nl} . This equation is often approximated as

$$P_f = 1 - \prod_{i=1}^{n} \prod_{j=1}^{n_l} [1 - \bar{P}_{f_{ij}}] \tag{13.10}$$

where $\bar{P}_{f_{ij}}$ is the probability of failure of member i under load j.

$$\bar{P}_{f_{ij}} = P(R_i < a_{ij} L_j) = \int_{-\infty}^{\infty} F_{R_i}(a_{ij} l_j)\, f_{L_j}(l_j)\, dl_j \tag{13.11}$$

where $a_{ij}\, l_j$ denotes the magnitude of load or stress induced in member i when a load l_j is applied to the system.

13.6.1 Examples of Weakest-link Systems

Figures 13.7, 13.8 and 13.9 indicate some weakest-link systems. Figure 13.7 denotes a statically determinate structure for which the failure of any of the 13 members of the truss constitutes the failure of the whole truss. Fig. 13.8 represents a typical multispeed compound gear train. In this gear train, the failure of any gear under the loads induced during power transmission constitutes the failure of the system. Finally, Figure 13.9 shows a valve train which consists of a shaft, cam, roller follower, push rod, rocker arm, valve spring, valve, a key, three pins, and a pivot. The failure of any of these components leads to the failure of the system.

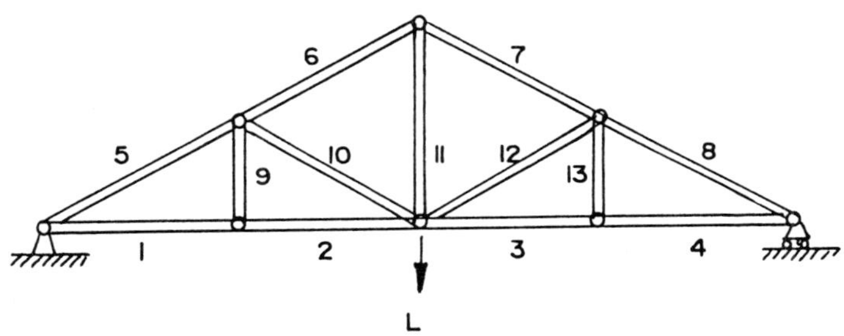

Fig. 13.7 A 13-Bar Truss

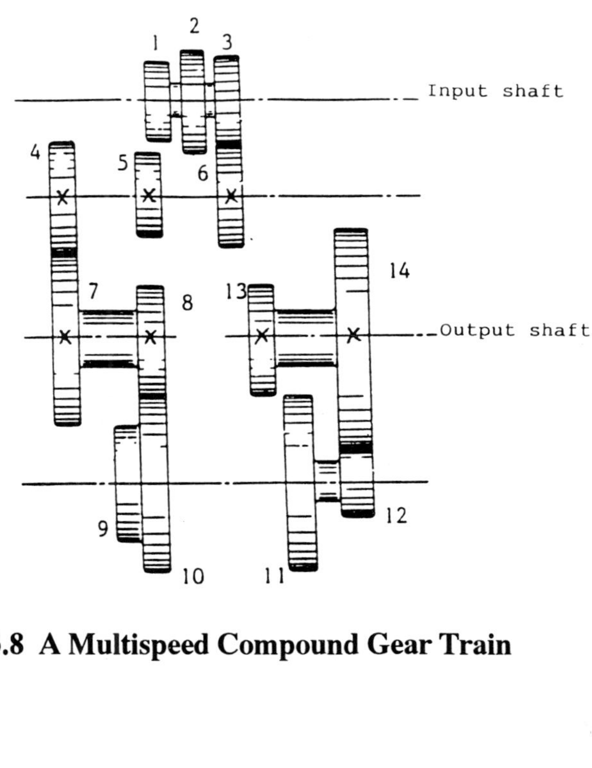

Fig. 13.8 A Multispeed Compound Gear Train

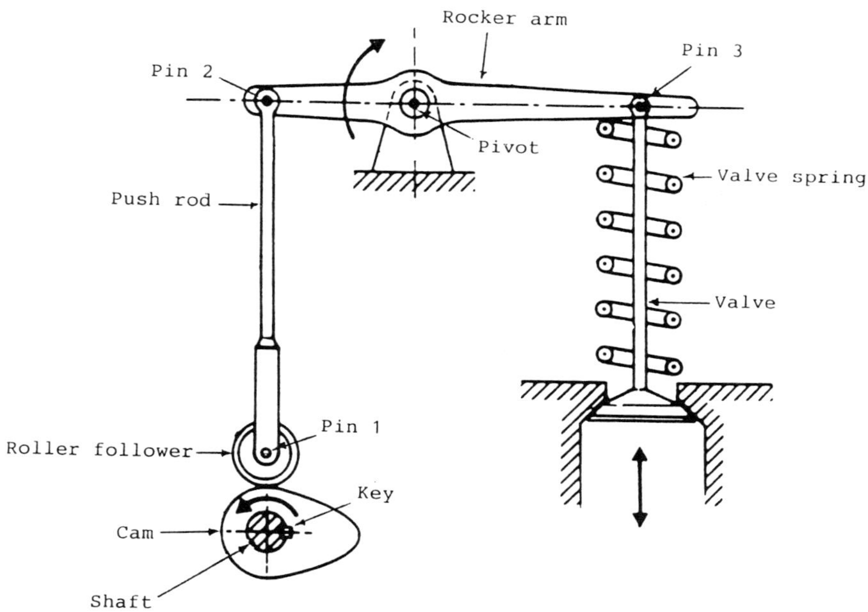

Fig. 13.9 A Cam-Follower System

13.7 Probability of Failure of Fail-Safe Systems

A fail-safe or parallel system is one in which the failure of few members may not necessarily constitute the failure of the system. In general, there will be more than one way in which a parallel system can fail under a given load. For example, a parallel system may fail due to the failure of different elements or due to the failure of different number of elements or the element failures may occur in different orders. All these considerations are important in the reliability analysis of a parallel system. For a system with n parallel members, the total resistance, S, is given by

$$S = R_1 + R_2 + \ldots + R_n \tag{13.12}$$

where R_j is the resistance of member j, and the probability of failure of the system under a load, L, can be expressed as

$$P_f = P\,(S < L) \tag{13.13}$$

By introducing the new random variables, S_i, as

$$S_i = R_1 + R_2 + \ldots + R_i\,, \tag{13.14}$$

the probability distribution functions of S_i can be determined as follows:

$$F_{S_1}(s) = P(S_1 \leq s) = P(R_1 \leq s) = F_{R_1}(s) \tag{13.15}$$

$$F_{S_2}(s) = P(S_2 \leq s) = P(R_1 + R_2 \leq s) = \int_{R_2=t=0}^{\infty} F_{s_1}(s-t)\, f_{R_2}(t)\,.\,dt \tag{13.16}$$

$$F_{S_3}(s) = P(S_3 \leq s) = P(R_1 + R_2 + R_3 \leq s) = \int_{R_3=t=0}^{\infty} F_{s_2}(s-t)\, f_{R_3}(t)\,.\,dt \tag{13.17}$$

and

$$F_{S_n}(s) \equiv F_S(s) = P(R_1 + R_2 + \ldots + R_n \leq s) = \int_{R_n=t=0}^{\infty} F_{S_{n-1}}(s-t)\, f_{R_n}(t)\, dt \tag{13.18}$$

Equation (13.18) implies that the distribution function of the total resistance can be found if the probability density and distribution functions of the individual member resistances $R_1, R_2, \ldots, R_n$ are known. Finally, the reliability of the parallel system can be determined using (13.13), as

$$R_0 = 1 - P_f = 1 - \int_{l=0}^{\infty} F_s(l)\,.\,f_L(l)\,.dl \tag{13.19}$$

13.7.1 Examples of Fail-Safe Systems

Figures 13.10, 13.11 and 13.12 represent some examples of fail- safe systems. The limit analysis of the beam shown in Figure 13.10 is based on the formation of plastic hinges at the three critical locations indicated. Then the failure is associated with a collapse mode of the system. Similarly, for the portal frame shown in Fig. 13.11, different collapse modes can be associated with the hinges

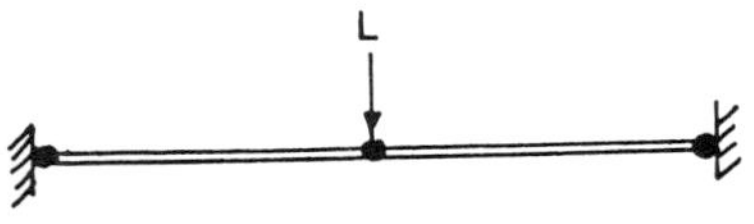

Fig. 13.10 A Beam with Plastic Hinges

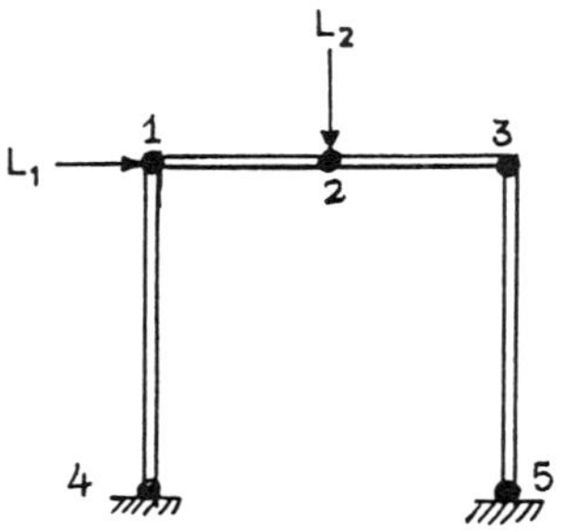

Fig. 13.11 A Portal Frame with Plastic Hinges

formed at locations 1, 2, ..., 5 [11]. Finally, for the truss shown in Figure 13.12, the failure of certain members constitutes the failure of the total system. All these systems can be modeled as parallel or fail-safe systems.

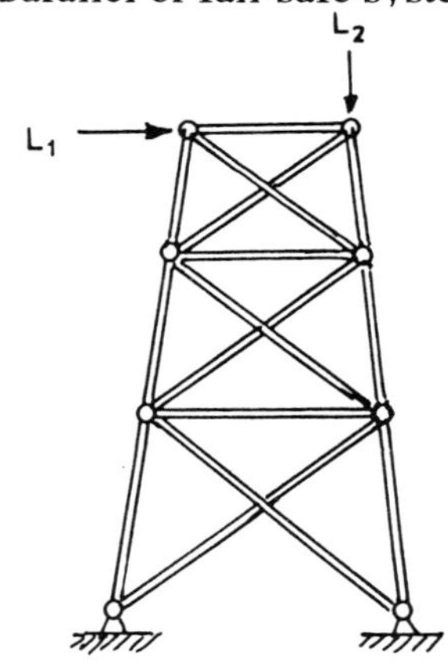

Fig. 13.12 A Statically Indeterminate Truss

13.8 Illustrative Examples

13.8.1 Reliability Modeling and Analysis of a Gear Train Mechanism

It is assumed that the structural and the kinematic designs of the gear train have been already completed. The structural design includes the schematic layout of the various gear pairs, both sliding and fixed, from the power source to the output shaft. The kinematic design gives the speeds and the interconnection of different gears. Fig. 13.13 shows the schematic diagram of a 9- speed gear train

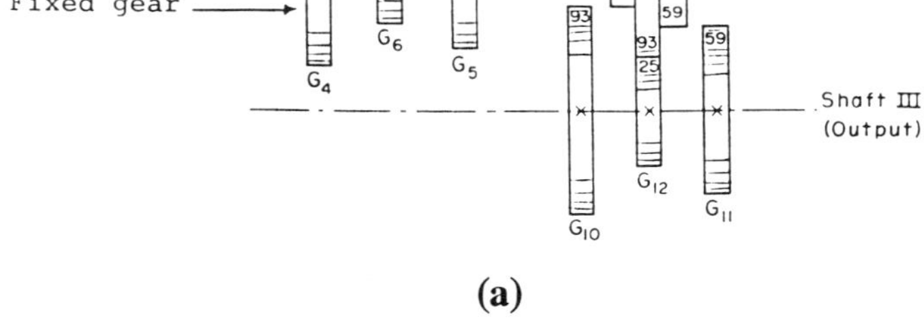

(a)

Fig. 13.13 A 9-Speed Compound Gear Train

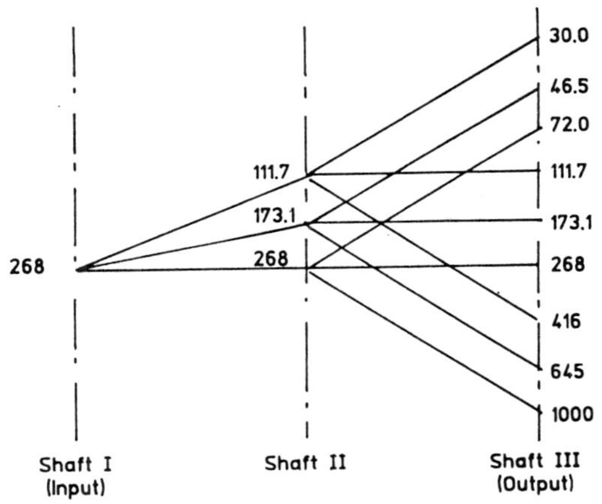

(b)

Fig. 13.13 A 9-Speed Compound Gear Train

along with its speed diagram. This section presents the dynamic design phase of the gear train and finds a design such that the probabilities in bending and wear, P_{f_b} and P_{f_w}, of any transmission chain formed to realize a particular output speed is equal to the allowable values. The mean face widths of different gear pairs are treated as design variables. For an N speed gear drive, there will be N transmission chains. If some gear pair is active in more than one transmission chain, then the face width which is the maximum out of all possible transmission chains, is taken as the feasible face width of that gear pair. When the gear train gives the j*th* speed as output speed, let n gear pairs be in mesh; thus it can be idealized as a chain consisting on n links in *series*. Since the failure of any gear pair with respect to either bending strength or surface wear resistance constitutes the failure of the whole chain, it is modeled as a weakest-link structure acted upon by two loads separately. If $P_{f_{bij}}$ represents the probability of failure in bending of the i*th* gear pair at j*th* speed, then the probability of failure of the chain in bending in j*th* speed is given by

$$P_{f_{bj}} = 1 - \prod_{i=1}^{n} (1 - P_{f_{bij}}) . \tag{13.20}$$

Similarly, the probability of failure of the chain in surface wear in j*th* speed is given by

$$P_{f_{wj}} = 1 - \prod_{i=1}^{n} (1 - P_{f_{wij}}) . \tag{13.21}$$

where $P_{f_{wij}}$ denotes the probability of failure in wear of the *ith* gear pair at *jth* speed. All the random variables are assumed to follow normal distributions for simplicity. In the absence of standard data, this assumption of normal distribution can be justified from the central limit theorem [12].

Probability of Failure of a Gear Pair in Bending and Surface Wear

By considering the *ith* gear pair in *jth* speed as a single link being acted upon by the bending stress, the probability of failure of the gear pair in bending can be found as [13,14]

$$P_{f_{bj}} = 1 - \int_{-\infty}^{\infty} F_{S_b}(r) \cdot f_{s_b}(r) \, .dr \tag{13.22}$$

where F_{S_b} and f_{s_b} denote, respectively, the probability distribution and density functions of allowable bending strength (S_b) and induced bending stress (s_b). Both s_b and S_b follow normal distribution with their density functions given by

$$f_{s_b}(r) = \frac{1}{\sqrt{2\pi}\,\sigma_{s_b}} \cdot \exp\left[-\frac{1}{2}\left(\frac{r - \bar{s}_b}{\sigma_{s_b}}\right)^2\right] \tag{13.23}$$

and

$$f_{S_b}(r) = \frac{1}{\sqrt{2\pi}\,\sigma_{S_b}} \cdot \exp\left[-\frac{1}{2}\left(\frac{r - \bar{S}_b}{\sigma_{S_b}}\right)^2\right] \tag{13.24}$$

where $\bar{s}_b$ and $\bar{S}_b$ indicate the mean values and σ_{s_b} and σ_{S_b} represent the standard derivations of s_b and S_b, respectively. If a new random variable is defined as $\xi = S_b - s_b$, the density function of ξ is given by

$$f_{\xi}(r) = \frac{1}{\sqrt{2\pi}\,\sigma_{\xi}} \cdot \exp\left[-\frac{1}{2}\left(\frac{r - \bar{\xi}}{\sigma_{\xi}}\right)^2\right] \tag{13.25}$$

where the mean and standard deviations of ξ are given by (assuming no correlation between the induced and allowable stresses)

$$\bar{\xi} = \bar{S}_b - \bar{s}_b \tag{13.26}$$

and

$$\sigma_\xi = (\sigma_{S_b}^2 + \sigma_{s_b}^2)^{1/2} \tag{13.27}$$

Then the probability of failure is given by

$$P_{f_b} = P(\xi < 0) = \int_{-\infty}^{0} f_\xi(r)\,.dr\,. \tag{13.28}$$

In order to evaluate the integral in (13.28), the relationship between the distribution of ξ and the standardized normal distribution can be utilized. The standard normal variable is given by

$$z = \frac{\xi - \bar{\xi}}{\sigma_\xi} \tag{13.29}$$

and (13.28) can be rewritten as

$$P_{f_b} = P_{f_{bij}} = \int_{-\infty}^{z_{ib}} \exp\,[-(1/2)\,z^2]\,.dz \tag{13.30}$$

where the upper limit of integration is given by

$$z_{1b} = -\frac{\bar{\xi}}{\sigma_\xi} = -\left[\frac{\bar{S}_b - \bar{s}_b}{(\sigma_{S_b}^2 + \sigma_{s_b}^2)^{1/2}}\right] \tag{13.31}$$

The value of z_{1b} can be found from the tables of normal distribution for any specified probability of failure $P_{f_{bij}}$. Similarly the upper limit of integration corresponding to the probability of failure in surface wear $P_{f_{wij}}$ is given by

$$z_{1b} = -\left[\frac{\bar{S}_w - \bar{s}_w}{(\sigma_{S_w}^2 + \sigma_{s_w}^2)^{1/2}}\right] \tag{13.32}$$

Mean and Standard Deviations of the Induced Stresses

For a gear pair transmitting a power p at a speed n_w, the torque is given by

$$M_t = 97500 \ p/n_w \tag{13.33}$$

The induced bending stress s_b, assuming the tooth to be a cantilever beam with end load, can be calculated as [15]

$$s_b = \frac{\beta M_t}{At} \tag{13.34}$$

where $$\beta = \frac{K_c K_d (i+1)}{i \,.\, m. \, y \,.\, \cos\alpha} \tag{13.35}$$

y is the Lewis form factor given by

$$y = 0.52 \left[1 + \frac{20}{T_w} \right] \tag{13.36}$$

A = center distance between the mating gears, t = face width of the gear pair, K_c = stress concentration factor, K_d = dynamic load factor, i = transmission ratio = $\frac{T_w}{T_p}$, T_w = number of teeth on the wheel, T_p = number of teeth on the pinion, m = module, and α = pressure angle. The mean and standard deviations of s_b are given by

$$\bar{s}_b = 97500\beta \, \bar{p} / [\bar{n}_w \bar{A} \bar{t}] \tag{13.37}$$

and

$$\sigma_{s_b} = \frac{1}{\bar{A}\bar{t}} \left[\frac{\beta^2 \bar{M}_t^2 (\bar{t}^2 \sigma_A^2 + \bar{A}^2 \sigma_t^2) + \beta^2 \bar{A}^2 \bar{t}^2 \sigma_{M_t}^2}{\bar{A}^2 \bar{t}^2 + \bar{t}^2 \sigma_A^2 + \bar{A}^2 \sigma_t^2} \right]^{1/2} \tag{13.38}$$

where

$$\sigma_{M_t} = \frac{97500}{\bar{n}_w} \left[\{\bar{p}^2 \sigma_{n_w}^2 + \bar{n}_w^2 \sigma_p^2\} / \{\bar{n}_w^2 + \sigma_{n_w}^2\} \right]^{1/2} \tag{13.39}$$

and the partial derivative rule has been used in deriving (13.38). The induced wear stress s_w is related to the face width t as [15]

$$s_w = \frac{\gamma}{A}\sqrt{\frac{M_t}{t}} \tag{13.40}$$

where
$$\gamma = 0 \cdot 59 \left(\frac{i+1}{i}\right)\left[\frac{(i+1)\,EK_cK_d}{\sin 2\alpha}\right]^{1/2} \tag{13.41}$$

and E is the Young's modulus of the material of the gear pair. The mean and standard deviations of s_w are given by

$$\bar{s}_w = \frac{\gamma}{\bar{A}}\sqrt{\frac{\bar{M}_t}{\bar{t}}} \tag{13.42}$$

and

$$\sigma_{s_w} = \frac{1}{\bar{A}\sqrt{\bar{t}}}\left[\frac{4\bar{M}_t^2\gamma^2\bar{t}^2\sigma_A^2 + \gamma^2\bar{A}^2\bar{M}_t^2\sigma_t^2 + \bar{A}^2\bar{t}^2\gamma^2\sigma_M^2}{4\bar{M}_t\bar{A}^2\bar{t}^2 + 4\bar{M}_t\bar{t}\sigma_A^2 + \bar{M}_t\bar{A}^2\sigma_t^2}\right]^{1/2} \tag{13.43}$$

With little manipulation, Eqs. (13.38) and (13.43) can be stated in terms of the coefficients of variation of the different parameters.

13.8.2 Reliability Modeling of Geneva Mechanisms

When a Geneva mechanism is fabricated and assembled, its performance differs from the desired one due to tolerances in link lengths, clearances in joints, variations in the input speed and power, and errors due to misalignments in the assembly. Besides affecting the kinematic accuracy, these errors affect the dynamic performance, the noise level and finally the life of the mechanism. The prediction of the reliability of the mechanism in meeting the specified kinematic and dynamic response requirements becomes important in many applications [16]. Fig. 13.14 shows an external Geneva mechanism [17]. Fig. 13.14(b) denotes a probabilistic model of the mechanism while the pin of the driving member is in engagement with one of the slots of the driven member with a consideration of tolerances and clearances. Here r_{12} denotes the radial clearance between the output

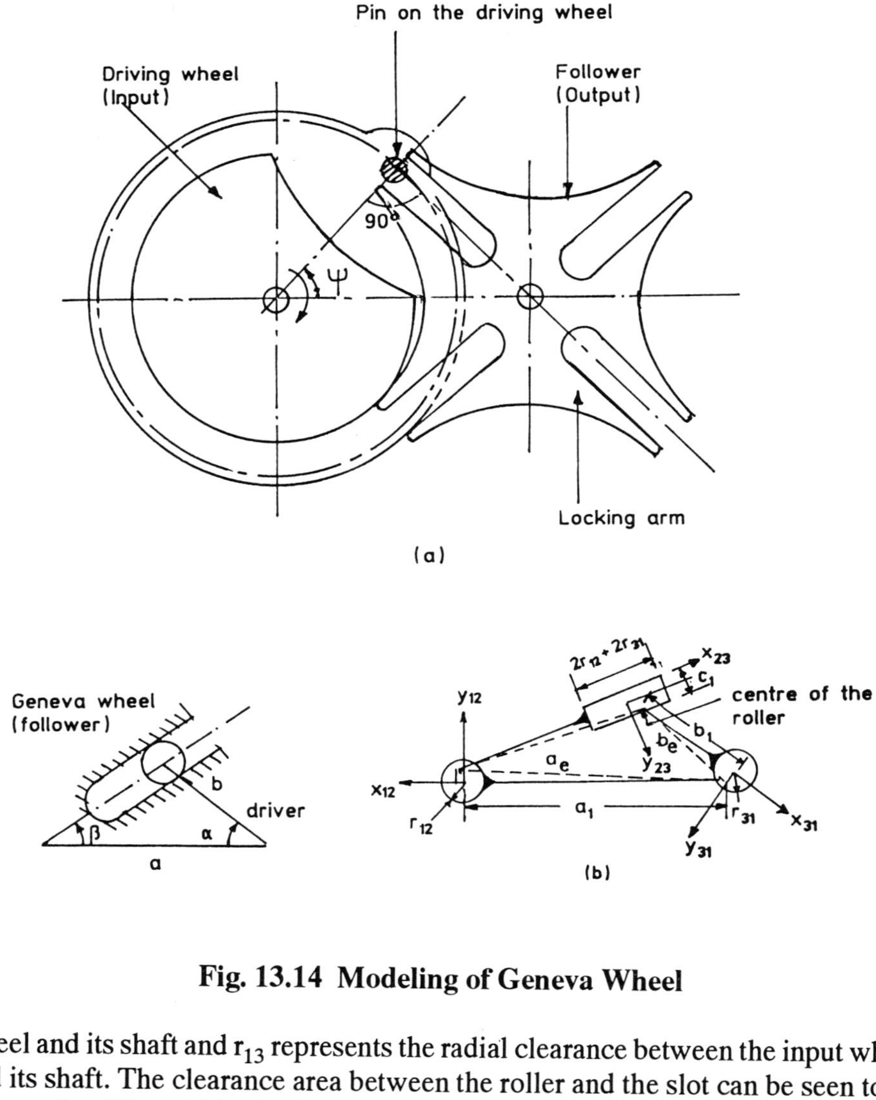

Fig. 13.14 Modeling of Geneva Wheel

wheel and its shaft and r_{13} represents the radial clearance between the input wheel and its shaft. The clearance area between the roller and the slot can be seen to be rectangular. The width of this rectangular clearance zone is denoted as r_{23}.

For the probabilistic analysis of the mechanism, a rectangular coordinate system (x_{ij}, y_{ij}) is chosen such that the x_{ij} axis coincides with the center line of the link ending in the race. For most practical problems it can be safely assumed that the probability of the pin axis lying at any point is same for all the points inside the clearance zones. The equivalent linkage is shown by the dotted lines in Fig. 13.14(b). From the geometry of Fig. 13.14(b), it can be seen that

$$a_e^2 = (a_1 + x_{12})^2 + y_{12}^2 \tag{13.44}$$

$$b_e^2 = (b_1 + x_{31})^2 + y_{31}^2 \tag{13.45}$$

$$c_e = y_{23} \tag{13.46}$$

Since $(a_1 + x_{12}) >> y_{12}$ and $(b_1 + x_{31}) >> y_{31}$, $a_e \approx a_1 + x_{12}$ and $b_e \approx b_1 + x_{31}$. The geometry leads to the relationship

$$\sin\beta_1 = \frac{b_e \sin\alpha}{(a_e^2 + b_e^2 - 2a_e b_e \cos\alpha)^{1/2}} + \frac{c_e (a_e - b_e \cos\alpha)}{(a_e^2 + b_e^2 - 2a_e b_e \cos\alpha)} \tag{13.47}$$

where a_1 = distance between the pin and the follower with tolerance = $a+\Delta a$, b_1 = distance between the pin and the driving wheel with tolerance = $b+\Delta b$, α = input angle, and β_1 = output angle. The acceleration and jerk can be determined as

$$\text{Acceleration} = \frac{d^2\beta_1}{d\alpha^2} \; ; \; \text{Jerk} = \frac{d^3\beta_1}{d\alpha^3} \tag{13.48}$$

The expressions for the acceleration and jerk can be written as

$$\text{Acceleration} = \text{Acceleration}\,(a + \Delta a + x_{12}, b + \Delta b + x_{31}, y_{23}, \alpha) \tag{13.49}$$

and $$\text{Jerk} = \text{Jerk}\,(a + \Delta a + x_{12}, b + \Delta b + x_{31}, y_{23}, \alpha) \tag{13.50}$$

By denoting the vector of random variables, $\vec{x}$, as

$$\vec{x} = \begin{Bmatrix} x_1 \\ x_2 \\ x_3 \\ x_4 \\ x_5 \end{Bmatrix} = \begin{Bmatrix} \Delta a \\ \Delta b \\ x_{12} \\ y_{23} \\ x_{31} \end{Bmatrix} \tag{13.51}$$

the mean and standard deviations of the acceleration (A) and jerk (J) can be determined as

$$\overline{A}_j = A(a + \overline{x}_1 + \overline{x}_3, b + \overline{x}_2 + \overline{x}_5, \overline{x}_4, \alpha_j) \tag{13.52}$$

$$\sigma_{A_j} = \left\{ \sum_{i=1}^{5} \left[\frac{\partial A}{\partial x_i} \Big|_{\vec{\bar{x}}, \alpha_j} \right]^2 \sigma^2_{x_i} \right\}^{1/2} \tag{13.53}$$

$$\bar{J}_j = J(a + \bar{x}_1 + \bar{x}_3, b + \bar{x}_2 + \bar{x}_5, \bar{x}_4, \alpha_j) \tag{13.54}$$

$$\sigma_{J_j} = \left\{ \sum \left[\frac{\partial J}{\partial x_i} \Big|_{\vec{\bar{x}}, \alpha_j} \right]^2 \sigma^2_{x_i} \right\}^{1/2} \tag{13.55}$$

where $\vec{\bar{x}}$ is the vector of mean values of x_i, $\bar{x}_i$, and σ_{x_i} is the standard deviation of x_i. If the permissible (mean) values and the acceptable deviations about the permissible values at the input angle α_j are specified as $\bar{A}_{oj}$ and σ_{Aoj} for acceleration and $\bar{J}_{oj}$ and $\sigma_{J_{oj}}$ for jerk, the probability of failure of the mechanism or the probability of the acceleration or jerk of the actual mechanism falling outside the specified band width is given by

$$P_{f_{Aj}} = \int_{-\infty}^{-Z_{1Aj}} e^{-\frac{1}{2} z^2} \, . \, dz \tag{13.56}$$

$$P_{f_{Jj}} = \int_{-\infty}^{-Z_{1Jj}} e^{-\frac{1}{2} z^2} \, . \, dz \tag{13.57}$$

where

$$z_{1Aj} = -\frac{\xi}{\sigma_\xi} = -\left\{ \frac{\bar{A}_{oj} - \bar{A}_j}{\sqrt{\sigma_{A_{oj}}^{\ 2} + \sigma_{A_j}^{\ 2}}} \right\} \tag{13.58}$$

$$z_{1Jj} = -\frac{\bar{\eta}}{\sigma_\eta} = -\left\{ \frac{\bar{J}_{oj} - \bar{J}_j}{\sqrt{\sigma_{J_{oj}}^{\ 2} + \sigma_{J_j}^{\ 2}}} \right\} \tag{13.59}$$

It is to be noted that the Eqs. (13.56) and (13.57) assume that the generated and the permissible accelerations and jerks follow the normal distribution.

Once the probability characteristics of the tolerances and clearances are known, the reliability of the mechanism can be estimated.

13.9 Conclusions

The inadequacy of the factor of safety and the importance of reliability in maintaining a proper degree of safety in mechanical systems has been discussed. The features that make the reliability modeling of mechanical systems different from electronic/electrical systems have been outlined. The role of variabilities in strength, geometry and loads in the reliability analysis of mechanical systems has been explained. The modeling of mechanical systems as weakest-link (series) and fail-safe (parallel) systems and the resulting reliability analysis has been explained.

The reliability analyses of a nine-speed gear train and an external Geneva mechanism have been presented to illustrate the models presented. The following aspects need to be addressed in future for a complete and more accurate modeling and analysis of mechanical systems:

- Incorporation of manufacturing limitations, in the form of component tolerances and assembly errors, in a probabilistic setting.
- Incorporation and use of practical failure data through time-varying failure-rate models.
- Development of probabilistic models with distributions other than normal distribution for the random parameters [18].
- Modelling of complex mechanical systems using a combination of weakest-link and fail-safe models.
- More studies on the relationship of factor of safety to the reliability of mechanical systems.

REFERENCES

[1] Rao S.S., *Reliability-Based Design*, McGraw-Hill, New York, 1992.

[2] Kececioglu D. and D. Cormier, *Designing a Specified Reliability Directly into a Component*, Proceedings of the Third Annual Aerospace Reliability and Maintainability Conference, Society of Automotive Engineers, Inc., New York, pp. 546-565, June 1964.

[3] Mischke C., *A New Method of Relating Factor of Safety and Reliability*, ASME Journal of Engineering for Industry, Vol. 92, pp. 537-542, August 1970.

[4] MIL-STD-790, *Reliability Assurance Program for Electronic Parts Specification*, Version 790D, May 1986.

[5] MIL-HDBK-217, "Reliability Prediction of Electronic Equipment,"Version 217E, October 1986.

[6] Tiner N.A., *Failure Analysis with the Electron Microscope*, Fox-Mathis Publications, Los Angeles, CA, 1973.

[7] Tall L. and G. A. Alpsten, *On the Scatter of Yield Strength and Residual Stresses in Steel Members*, in Symposium on Concepts of Safety of Structures and Methods of Design, IABSE, London, pp. 151-163, 1969.

[8] Bogdanoff J.L. and F. Kozin, *Probabilistic Models of Cumulative Damage*, John Wiley, New York, 1985.

[9] Gould P.L. and S. H. Abu-Sitta, Dynamic Response of Structures to Wind and Earthquake Loading, John Wiley, New York, 1980.

[10] Vorlicek M. and M. Holicky, *Analysis of Dimensional Accuracy of Building Structures*, Elsevier Science Publishing Co. Inc., Amsterdam, The Netherlands, 1989.

[11] Melchers R.E., *Reliability of Parallel Structural Systems*, Journal of the Structural Division, Proceedings of the ASCE, Vol. 109, pp. 2651-2665, 1983.

[12] Wadsworth G.P. and J. G. Bryan, *Introduction to Probability and Random Variables*, McGraw-Hill, New York, 1960.

[13] Rao S.S., *A Probabilistic Approach to the Design of Gear Trains*, International Journal of Machine Tool Design and Research, Vol. 14, pp. 267-278, 1974.

[14] Rao S.S. and Gautem Das, *Reliability Based Optimum Design of Gear Trains*, ASME Journal of Mechanisms, Transmissions, and Automation in Design, Vol. 106, pp. 17-22, 1984.

[15] Dobrovolsky V., K. Zablonsky, and S. Mak, *Machine Elements*, Foreign Languages Publishing House, Moscow, 1961.

[16] Rao S.S. and S. S. Gavane, *Analysis and Synthesis of Mechanical Error in Geneva Mechanisms - A Stochastic Approach: Part I - External Geneva Mechanism*, ASME Journal of Mechanical Design, Vol. 104, pp. 63-71, 1982.

[17] Shigley J.E. and J. J. Uicker, Jr., *Theory of Machines and Mechanisms*, McGraw-Hill, New York, 1980.

[18] Misra K.B., *Reliability Analysis and Prediction*, Elsevier Science Publishers, Amsterdam, The Netherlands, 1992.

New Trends in System Reliability Evaluation
K.B. Misra (Editor)

CHAPTER 14

USE OF FUZZY SETS THEORY

(PART - 1 : THEORY)

Krishna B. Misra
Director-grade-Scientist
Disaster Prevention & Impact Minimization Division
Nehru Marg, Nagpur 440 020, INDIA

ABSTRACT

Fuzzy sets being relatively a new concepts, a brief summary of important definitions and relevant mathematical background are essential for understanding the fuzzy set theory (FST) and its applications. Our objective, in this chapter, is to restrict our discussion to such topics in FST which will have direct bearing on their applications to system reliability or performance evaluation and other related areas like fault tree analysis and risk assessment. The applications of FST are being presented in Part-II of this chapter, separately.

14.1 INTRODUCTION

There has been a tremendous explosion of literature on the fuzzy set theory and its applications since the first paper by Zadeh [1] was published in 1965. In fact there have been several excellent texts written on the subject. However, it is not our intention to cover all the aspects of FST and we will discuss FST only in a limited sense as would be necessary for a reader to understand the applications of FST presented this chapter. This has been done for two reasons. Firstly, for the sake of completeness and secondly for better comprehension of the potentials of the FST. This was felt necessary, so as not to allow a beginner to get lost in the maize of thousands of papers and several books that have been published during past 28 years since the publication of Zadeh's paper [1]. However, for further information on the subject, references cited in the texts [2-12] could be referred to.

14.1.1 Fuzzy Sets

In conventional set theory (CST), an object (usually denoted by lower-case alphabets, viz., a,b,...,x,y,z) is considered to be *belonging* to a set (denoted by upper-case alphabets, viz., A,B,...,X,Y,Z) or *not belonging* to it at all. That is, *x is an element of A* ($x \in A$), or *x is not an element of A* ($x \notin A$). Correspondingly, an object x has a degree of belonging to a set A as unity or zero. Thus in the CST, a set has a very well-defined boundary and accordingly is called a *crisp set*.

Let X be a field of reference often called as *universe of discourse*, covering a definite range of objects, whose generic elements are denoted by x. Membership in a classical subset A of X is often characterized by a membership function (or grade of membership or degree of compatibility etc.) μ_A from X to {0,1} such that

$$\mu_A(x) = \begin{cases} 1 & iff\ x \in A \\ 0 & iff\ x \notin A \end{cases} \quad (14.1)$$

where, {0,1} is called a *valuation set*.

However, *fuzziness* is the lack of well-defined boundaries of a set of objects and refers to a class of sets with a continuous *grade of membership*, involving a gradual transition (rather than an abrupt one, as is implied in (14.1)) from membership to nonmembership.

If the membership function, $\mu_A(x)$ is allowed to be in a real interval [0,1] and is continuous as given in Fig.14.1, **A** is called a *fuzzy set*. Fig.14.1 defines a fuzzy set of real numbers that are *approximately*, equal to 5.

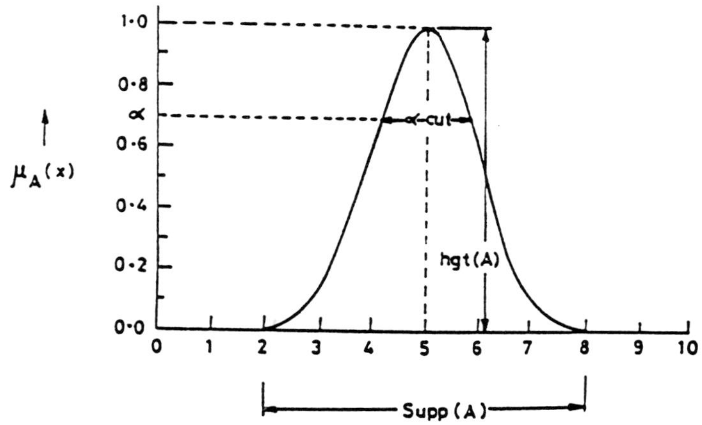

Fig.14.1 The Membership Function of a Fuzzy Set **A** (**A** represents real numbers around 5)

A is completely characterized by a set of ordered pairs, viz.,

$$\mathbf{A} \equiv \{(x,\ \mu_A(x)),\ x \in X\} \quad (14.2)$$

Generally, it is found to be convenient to write a fuzzy set **A** as follows :

$$\mathbf{A} = \mu_A(x_1)/x_1 + \mu_A(x_2)/x_2 + \ldots + \mu_A(x_n)/x_n$$

$$= \sum_{i=1}^{n} \mu_A(x_i)/x_i;\ \text{if x is finite set } (x_1, x_2, \ldots, x_n) \quad (14.3)$$

where, the summation or the operation of + stands for logical sum (i.e., OR).

Illustration

Let us consider a set of natural numbers and we define a subset of these natural numbers, which are *small natural numbers*. Obviously, the word *small* is not crisply defined (or vaguely defined), it will represent a fuzzy set. Let this set be denoted by **A**.

Now, we can make the following statements about a few numbers, which can be designated as *small* numbers, viz.,

(i) 0 definitely belongs to a set of small natural numbers, i.e.,

$$0 \in \mathbf{A} \text{ with } \mu_{\mathbf{A}}(0)=1 \quad \text{or} \quad 0 \underset{1}{\in} \mathbf{A}$$

(ii) 1 can be called to belong to a subset of small numbers with a membership grade of 0.9, i.e.,

$$1 \in \mathbf{A} \text{ with } \mu_{\mathbf{A}}(1)=0.9 \quad \text{or} \quad 1 \underset{0.9}{\in} \mathbf{A}$$

(iii) Number 2 can be called as belonging to **A** with a membership grade of 0.8, i.e.,

$$2 \in \mathbf{A} \text{ with } \mu_{\mathbf{A}}(2)=0.8 \quad \text{or} \quad 2 \underset{0.8}{\in} \mathbf{A}$$

(iv) Similarly, one can write the following about 3,

$$3 \in \mathbf{A} \text{ with } \mu_{\mathbf{A}}(3)=0.6 \quad \text{or} \quad 3 \underset{0.6}{\in} \mathbf{A}$$

In the same way, we can go on qualifying natural numbers and define a subset of *small natural numbers*, denoted by a fuzzy set **A**. Thus, we can write **A** as

$$\mathbf{A} = \frac{1}{0} + \frac{0.9}{1} + \frac{0.8}{2} + \frac{0.6}{3} + \frac{0.2}{4} + \frac{0}{5} + \frac{0}{6}$$

$$\mathbf{A} = \frac{1}{0} + \frac{0.9}{1} + \frac{0.8}{2} + \frac{0.6}{3} + \frac{0.2}{4}$$

Obviously in above representation, elements with null membership have been omitted and thus **A** has been defined over a *finite support set*. If X is not finite, we shall write

$$\mathbf{A} = \int_X \mu_{\mathbf{A}}(x)/x \tag{14.4}$$

Let us define a fuzzy set **A** such that

$$\mu_{\mathbf{A}}(x) = \begin{cases} 0 & \text{for } x \le 1 \\ 1 - \frac{1}{x} & \text{for } x > 1 \end{cases}$$

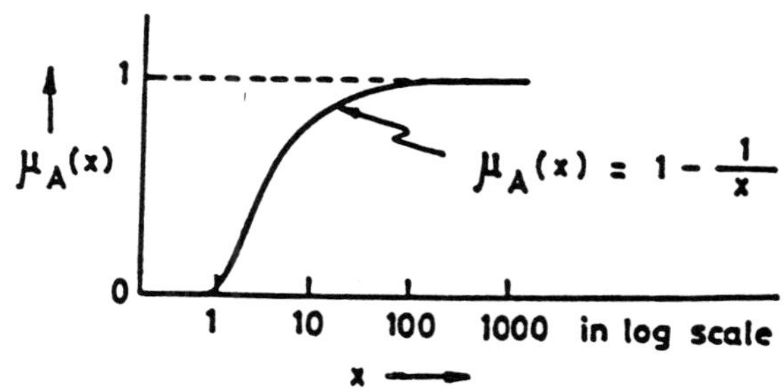

Fig. 14.2 Membership Function Defined over **x** (when **x** is not finite)

This membership function is shown in Fig.14.2 and is of real continuous type defined over X.

Therefore, $$\mathbf{A} \equiv \int_X (1-\frac{1}{x})/x, \qquad \text{for } x>1$$

14.1.2 Definitions

Support

The support of a fuzzy set **A** is the ordinary subset of X such that

$$\mathbf{Supp(A)} = \{x \in X,\ \mu_A(x) > 0\} \tag{14.5}$$

This has been indicated in Fig.14.1.

Height

The height of **A** is denoted by **hgt(A)** and is given by

$$\mathbf{hgt(A)} = \mathbf{Sup}_{x \in X}\ \mu_A(x) \tag{14.6}$$

i.e., the least upper bound of $\mu_A(x)$. This is also indicated in Fig.14.1.

Normalized FS

The fuzzy set **A** is said to be normalized *iff*

$$\exists_x \in X,\ \mu_A(x) = 1 \text{ or } \mathbf{hgt(A)} = 1 \tag{14.7}$$

For example, the fuzzy set **A** shown in Fig.14.1 is a *normalized FS*.
In case, $\forall x$, **hgt(A)** < 1, the fuzzy set **A** is known as *non-normalized* fuzzy set. For example, in Fig.14.3(a), a *non-normalized* fuzzy set has been shown.

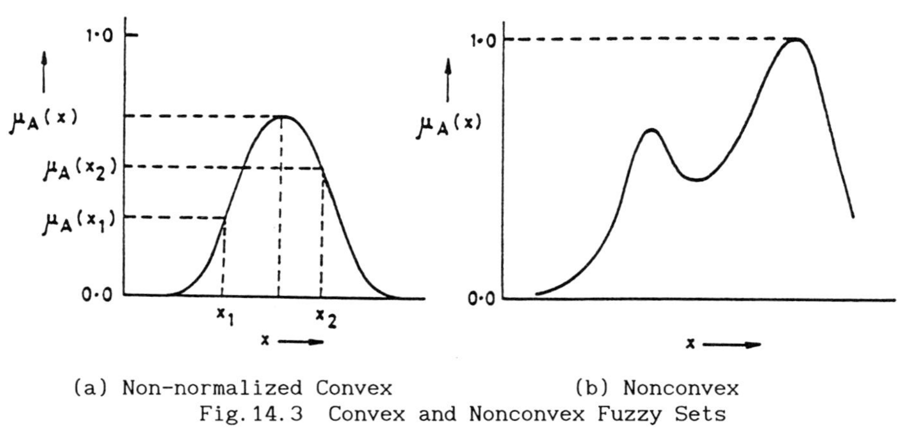

(a) Non-normalized Convex (b) Nonconvex

Fig.14.3 Convex and Nonconvex Fuzzy Sets

Convex Fuzzy Sets

A fuzzy set **A** is convex *iff*
$\forall x_1 \in X,\ \forall x_2 \in X,\ \forall \lambda \in [0,1]$,

$$\mu_A(\lambda x_1+(1-\lambda)x_2) \geq \mathbf{min}\big(\mu_A(x_1),\ \mu_A(x_2)\big) \tag{14.8}$$

alternatively, a fuzzy set is convex if its α-cuts are convex. It is important to note that the above definition does not imply that μ_A is a convex function of x. Convex and nonconvex fuzzy sets are shown in Fig. 14.3(a) and (b), respectively.

α-cut

An ordinary (crisp) set of all the elements of the universal set X, which have membership value greater than or equal to α, and belongs to a fuzzy set **A**, is called *α-cut of* **A** *or* $\mathbf{A}_\alpha$. This is shown in Fig. 14.1.

$$\mathbf{A}_\alpha = \{\alpha \in X \mid \mu_A(x) \geq \alpha\} \tag{14.9}$$

One can also define *strong α-cut,* $\mathbf{A}_{\bar{\alpha}} \equiv \{x \in X,\ \mu_A(x) > \alpha\}$

The membership of a fuzzy set can also be expressed in terms of the characteristic functions of its α-cuts, viz.,

$$\mu_A(x) = \underset{\alpha\in]0,1]}{\mathbf{Sup}}\ \mathbf{min}\big(\alpha,\ \mu_{A_\alpha}(x)\big), \tag{14.10}$$

where, $$\mu_{A_\alpha}(x) = \begin{cases} 1, & iff\ x \in \mathbf{A}_\alpha \\ 0, & \text{otherwise.} \end{cases}$$

Cross-over Points of A

The elements of X such that $\mu_A(x) = 1/2$ are known as crossover points of **A**.

Empty Set

An empty set ϕ is defined as

$$\forall x \in X,\ \mu_\phi(x) = 0 \tag{14.11}$$

Complement Set

A complement $\bar{\mathbf{A}}$ of a fuzzy set **A** is defined by membership function as (see Fig. 14.4).

$$\forall x \in X,\ \mu_{\bar{A}}(x) = 1 - \mu_A(x)$$

or simply $$\bar{\mathbf{A}} \Leftrightarrow \mu_{\bar{A}} = 1-\mu_A \tag{14.12}$$

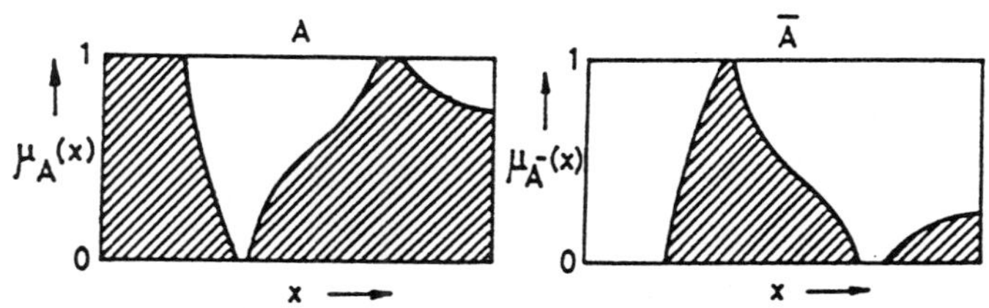

Fig. 14.4 A Complement of a Fuzzy Set **A**

It may be noted here that the complement has been defined by the membership function alone and therefore $\mu_{\bar{A}}(x)$ depends only on $\mu_{A}(x)$. If we let the complementation function be defined by $\mu_{\bar{A}}(x) \equiv f(\mu_{A}(x))$, then $f(0)=1$ and $f(1)=0$ denote the usual complementation when **A** is a crisp set. Also, when f is continuous it strictly monotonically decreases with $\mu_{A}(x)$ and is involutive, i.e., $f(f(\mu_{\bar{A}}(x)))=\mu_{A}(x)$.

Power of a Fuzzy Set

The *mth* power of a fuzzy set is defined as

$$\left.\begin{aligned} \mathbf{A}^{m} &\triangleq \int_{X} (\mu_{A}(x))^{m}/x \qquad \text{for an infinite FS} \\ &\triangleq \sum_{X} (\mu_{A}(x))^{m}/x \qquad \text{for a finite FS} \end{aligned}\right\} \tag{14.13}$$

where, m is any positive number.

If m is any non-negative real number such that $m \sup_{X} \mu_{A}(x)=1$, then

$$_{m}\mathbf{A} \triangleq \int_{X} m\mu_{A}(x)/x \tag{14.14}$$

Concentration

As a special case of (14.13), the operation of *concentration* can be defined as

$$\mathbf{CON}(\mathbf{A}) \triangleq \mathbf{A}^{2} = \int_{X} (\mu_{A}(x))^{2}/x \tag{14.15}$$

Dilation

However, *dilation* is written as

$$\left.\begin{aligned} \mathbf{DIL}(\mathbf{A}) \triangleq {}_{0.5}\mathbf{A} &= \int_{X} 0.5(\mu_{A}(x))/x, \quad \text{for infinite FS} \\ \text{or} \qquad &= \sum_{X} (0.5\mu_{A}(x))/x, \quad \text{for finite FS} \end{aligned}\right\} \tag{14.16}$$

The operations implied in (14.15) and (14.16) are extensively used to model *linguistic hedges*.

Cardinality of a Fuzzy Set

Scalar

If X is a finite set, the cardinality of **A**, $|\mathbf{A}|$ is defined as

$$|\mathbf{A}| = \sum_{x \in X} \mu_{A}(x) \tag{14.17}$$

$|\mathbf{A}|$ is also sometimes called as *power of* **A**. $||\mathbf{A}|| = |\mathbf{A}|/|X|$, is called as *relative cardinality* which can be viewed as the proportion of elements of X that are in **A**.

When X is not finite $|\mathbf{A}|$ does not always exist but if **A** has a finite support,

$$|\mathbf{A}| = \sum_{x \in \mathbf{Supp}(\mathbf{A})} \mu_A(x) \tag{14.18}$$

Otherwise, in case X is a measurable set and P is a measure on X ($\int_X dP(x)=1$), $||\mathbf{A}||$ can be equated to the weighted sum $\int_X \mu_A(x)dP(x)$, which seems to provide *fuzzification* of the universe X.

Fuzzy Cardinality

When **A** has a finite support, its fuzzy cardinality is

$$|\mathbf{A}|_f = \sum \alpha / |\mathbf{A}_\alpha| = \{(n, \alpha),\ n \in \mathbb{N} \text{ and } \alpha = \mathbf{Sup}\{\lambda,\ |\mathbf{A}_\lambda| = n\}\} \tag{14.19}$$

where, $\mathbf{A}_\alpha$ denotes the α-cut of **A** and $\mathbb{N}$ is a set of natural integers.

14.1.3 Remarks

(i) The universe is *never* fuzzy and $\mu_X(x) = 1,\ \forall x$.
(ii) A fuzzy set is actually a generalized subset of a classical set. Therefore, it should be rightly called a *fuzzy subset*. But for convenience, we would simply call it as *fuzzy set*.
(iii) Assignment of membership function of a fuzzy set is subjectively done but *not arbitrarily*.

14.2 SET-THEORETIC OPERATIONS ON FUZZY SETS

Before we provide set theoretical operations on fuzzy sets, some additional definitions are necessary for understanding of these operations. However, in all further explanations or statements, μ_A(i.e., without x) shall be used for the state of convenience.

Equality

Two fuzzy sets **A** and **B** are said to be *equal* (i.e., **A** = **B**)

$$iff\ \forall x \in X,\ \mu_A = \mu_B \tag{14.20}$$

Containment

A FS **B** is said to be *contained in* **A** (i.e., **B** ⊆ **A**),

$$iff\ \forall x \in X,\ \mu_B \le \mu_A. \tag{14.21}$$

This is shown in Fig. 14.5.

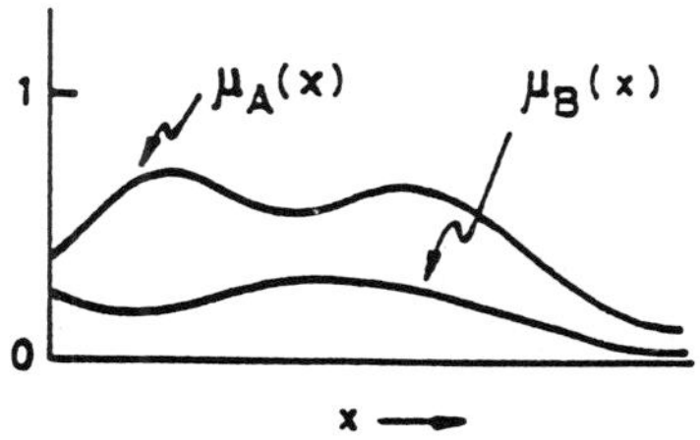

Fig. 14.5 Set Containment (**B** ⊆ **A**)

Product

A product set **AB** of fuzzy sets **A** and **B** is defined as

$$\left.\begin{aligned}\mathbf{AB} &\triangleq \int_X \mu_A\mu_B/x\ , \quad \text{for infinite FS.}\\ &= \sum_X \mu_A\mu_B/x\ , \quad \text{for finite FS.}\end{aligned}\right\} \qquad (14.22)$$

This operation is shown in Fig. 14.6.

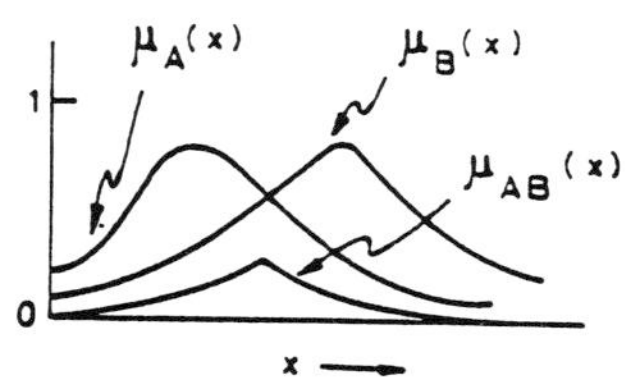

Fig. 14.6 Product of Two Fuzzy Sets

Now, we can define the following set theoretic operations on two fuzzy sets **A** and **B** :

Intersection

$$\forall x \in X,\ \mathbf{A} \cap \mathbf{B} \Leftrightarrow \mu_{\mathbf{A} \cap \mathbf{B}} = \mu_A \wedge \mu_B \triangleq \mathbf{min}(\mu_A, \mu_B) \qquad (14..23)$$

This operation is shown in Fig. 14.7.

Union

$$\forall x \in X,\ \mathbf{A} \cup \mathbf{B} \Leftrightarrow \mu_{\mathbf{A} \cup \mathbf{B}} = \mu_A \vee \mu_B \triangleq \mathbf{max}(\mu_A, \mu_B) \qquad (14.24)$$

This is also shown in Fig. 14.7.

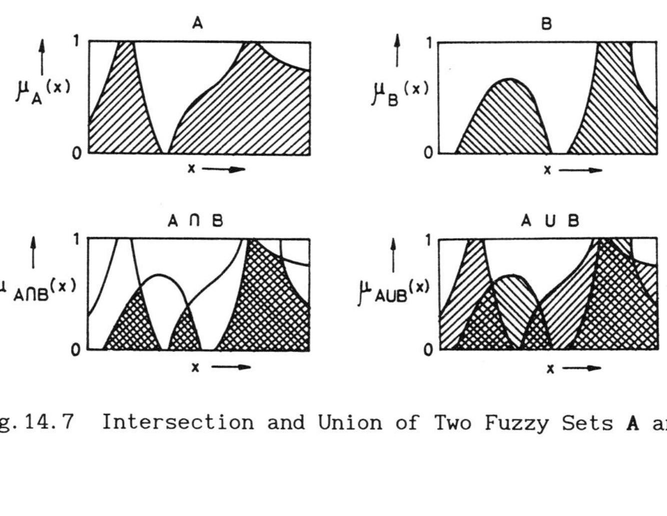

Fig. 14.7 Intersection and Union of Two Fuzzy Sets **A** and **B**

Algebraic Product

$$\forall x \in X,\ \mathbf{A.B} \Leftrightarrow \mu_{\mathbf{A.B}} = \mu_A \cdot \mu_B \qquad \text{(see Fig. 14.6)} \qquad (14.25)$$

This is called a *probabilistic intersection.*

Algebraic Sum

$$\forall x \in X,\ \mathbf{A\dot{+}B} \Leftrightarrow \mu_{\mathbf{A\dot{+}B}} \equiv \mu_A + \mu_B - \mu_A\mu_B$$
$$= 1 - (1-\mu_A)(1-\mu_B) \qquad (14.26)$$

This is also called as *probabilistic union* (see Fig. 14.8).

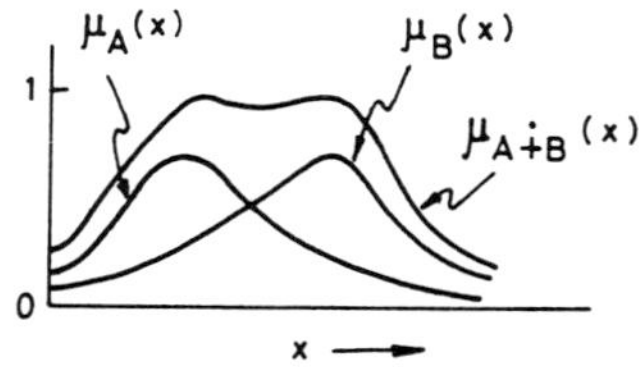

Fig. 14.8 Probabilistic Union of Two FS **A** and **B**

Bounded Product

$$\forall x \in X,\ \mathbf{A\odot B} \Leftrightarrow \mu_{\mathbf{A\odot B}} \triangleq 0 \vee (\mu_A+\mu_B-1) = \mathbf{max}(0, \mu_A+\mu_B-1) \qquad (14.27)$$

This is also called as *bold intersection.*

Bounded Sum

$$\forall x \in X,\ \mathbf{A\oplus B} \Leftrightarrow \mu_{\mathbf{A\oplus B}} \triangleq 1 \wedge (\mu_A+\mu_B) \triangleq \mathbf{min}\big(1, (\mu_A+\mu_B)\big) \qquad (14.28)$$

This is also called as *bold union.*

Bounded Difference :

$$\forall x \in X,\ \mathbf{A\ominus B} \Leftrightarrow \mu_{\mathbf{A\ominus B}} \triangleq 0 \vee (\mu_A-\mu_B) \triangleq \mathbf{max}\big(0, (\mu_A,\ \mu_B)\big) \qquad (14.29)$$

In all the above definitions (14.23 to 14.29), + and - stand for arithmetic sum and difference, respectively. Further operaters + and ., are interactive, whereas the **min** and **max** operators are non-interactive, since any change in **A** or **B** induces a change in **A.B** and **A+B**. The choice of the pair of operators is a matter of context.

14.3 FUZZY SETS AND MULTI-VALUED LOGIC

The FS theory may be embedded in a MVL by interpreting grade of membership $\mu_A(x)$ as representing the *truth value* of the statement **x is the FS A**. In MVL (or continuous logic) and FS logic, the truth value v(P) takes value in the unit interval [0,1], in which V(P)=1 indicates that the statement P is completely *true* and v(P)=0, indicates P is completely *false.*

In this section, we show that the operations of *bounded-sum* ⊕ and *bounded product* ⊙ for FS are corresponding to the operations of *sum* and *product*, respectively in Lukasiewicz's many-valued logic [13,14,15]. Further representation of implication (⟶) and equivalence (⟷) is provided by means of ⊕, ⊙ and ⊖.

The operations involved in MVL and FSL are listed in the following Fig. 14.9.

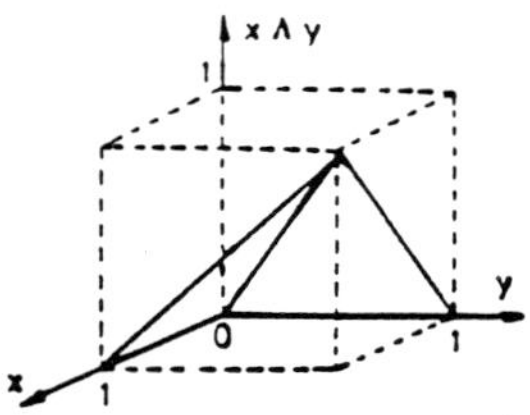

Logical Product $x \wedge y = \mathbf{min}\{x, y\}$ (14.30)

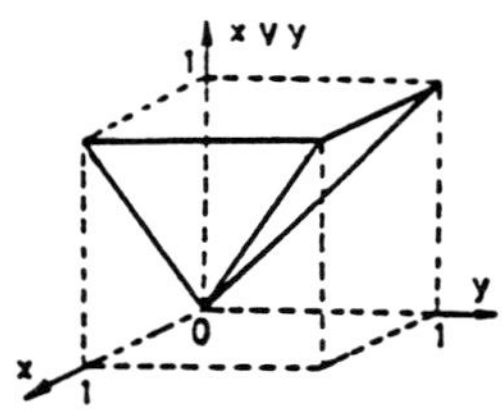

Logical Sum $x \vee y = \mathbf{max}\{x, y\}$ (14.31)

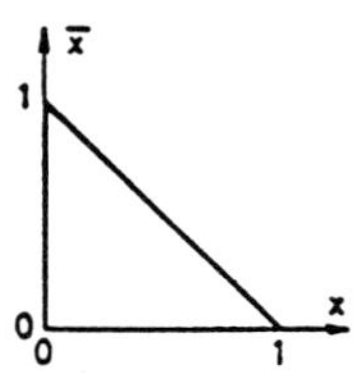

Negation $\bar{x} = 1-x$ (14.32)

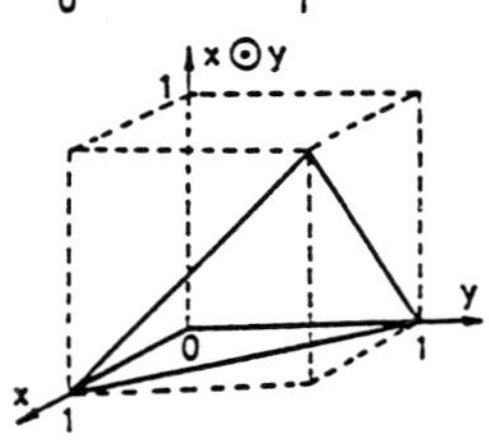

Product $x \odot y = 0 \vee (x+y-1)$ (14.33)

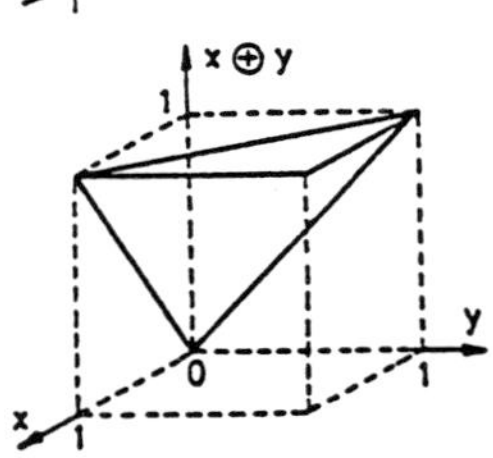

Sum $x \oplus y = 1 \wedge (x+y)$ (14.34)

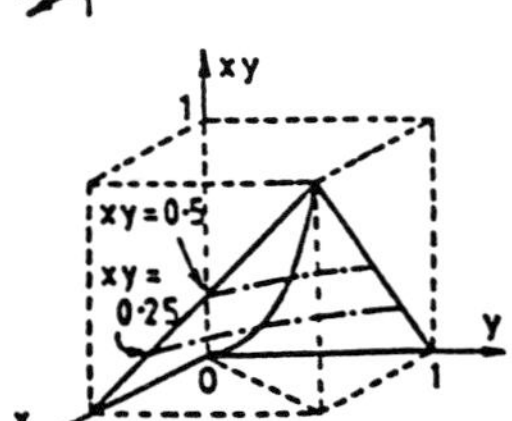

Algebraic Product $xy = x.y$ (14.35)

Fig. 14.9 Multivalued Logic vs. Fuzzy Set Logic (Contd.)

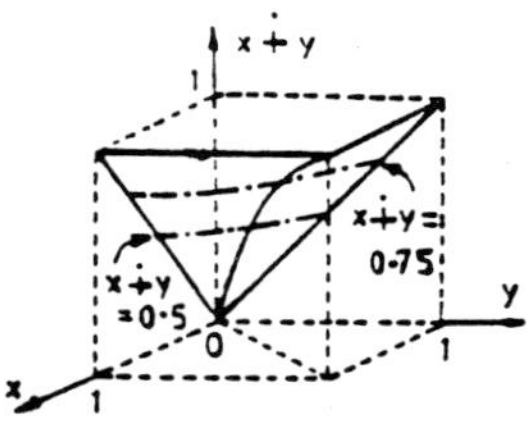

Algebraic Sum $x \dotplus y = x+y-xy$ (14.36)

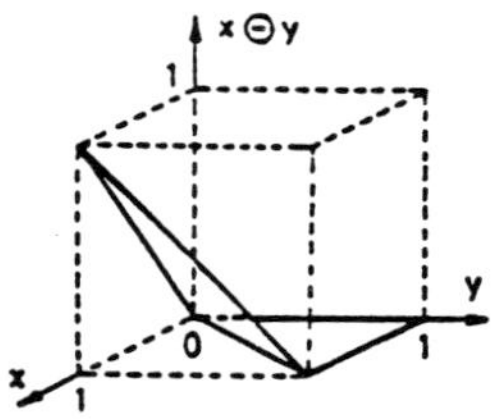

Difference $x \ominus y = 0 \wedge (x-y)$ (14.37)

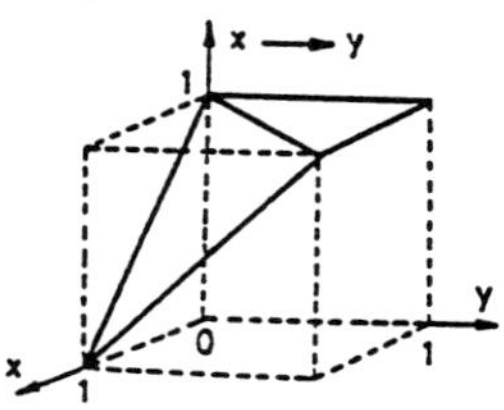

Implication $x \longrightarrow y = 1 \wedge (1-x+y)$ (14.38)

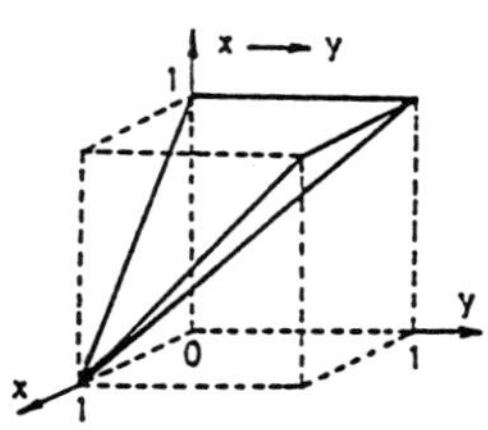

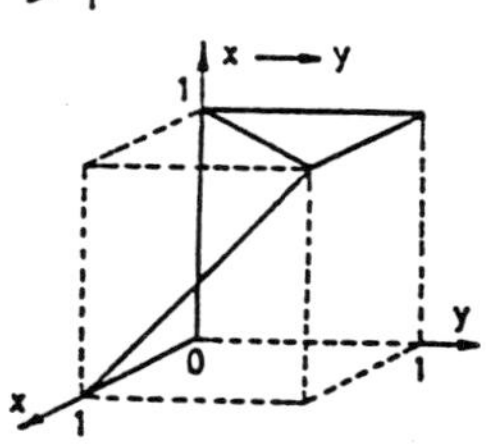

Implications

$$x \longrightarrow y = \bar{x} \vee y \quad (14.39)$$

$$x \longrightarrow y = 1 \; \ldots \; x \leq y$$

$$= \frac{y}{x} \; \ldots x > y \quad (14.40)$$

Fig. 14.9 Multivalued Logic vs. Fuzzy Set Logic

The implication of (14.38) can be defined using the logical operations of product ⊙ (14.33), sum ⊕ (14.34) and difference ⊖ (14.37).

viz.,

$$x \longrightarrow y = \bar{x} \oplus y \quad (14.41)$$

$$= \overline{x \odot \bar{y}} \quad (14.42)$$

$$= \overline{x \ominus y} \quad (14.43)$$

Conversely, ⊕, ⊙, and ⊖ are also defined by this implication.

$$x \oplus y = \bar{x} \longrightarrow y \tag{14.44}$$

$$x \odot y = \overline{x \longrightarrow \bar{y}} \tag{14.45}$$

$$x \ominus y = \overline{x \longrightarrow y} \tag{14.46}$$

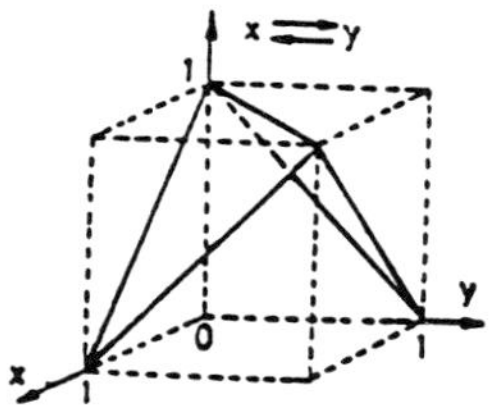

Equivalence $x \leftrightarrow y = (1-x+y) \wedge (1-y+x)$ (14.47)

Finally, the equivalence of (14.47) can be defined as

$$\begin{aligned} x \longleftrightarrow y &= (x \rightarrow y) \wedge (y \rightarrow x) \\ &= (\overline{x \ominus y}) \wedge (\overline{y \ominus x}) \\ &= \overline{(x \ominus y) \vee (y \ominus x)} \\ &= |\overline{x - y}| = 1 - |x - y| \end{aligned} \tag{14.48}$$

where $|x - y|$ stands for the absolute value of a real number $(x-y)$.

14.4 ALGEBRAIC PROPERTIES OF FUZZY SETS

In this section, some of the algebraic properties of fuzzy sets under various connectives (or operators) will be discussed.

14.4.1 Under $\cap$, $\cup$ and $^{-}$

The membership grade for fuzzy sets satisfy the following properties under **min**, **max** and **complement** operations :

$\mu_A \leq \mu_A$ **Reflexive law** (14.49)

$\mu_A \leq \mu_B,\ \mu_B \leq \mu_A \Rightarrow \mu_A = \mu_B$ **Antisymmetric law** (14.50)

$\mu_A \leq \mu_B,\ \mu_B \leq \mu_C \Rightarrow \mu_A \leq \mu_C$ **Transitive law** (14.51)

$$\left.\begin{aligned} \mu_A \vee \mu_A &= \mu_A \\ \mu_A \wedge \mu_A &= \mu_A \end{aligned}\right\} \quad \textbf{Idempotence laws} \tag{14.52}$$

$$\left.\begin{aligned} \mu_A \vee \mu_B &= \mu_B \vee \mu_A \\ \mu_A \wedge \mu_B &= \mu_B \wedge \mu_A \end{aligned}\right\} \quad \textbf{Commutative laws} \tag{14.53}$$

$$\left.\begin{aligned} (\mu_A \vee \mu_B) \vee \mu_C &= \mu_A \vee (\mu_B \vee \mu_C) \\ (\mu_A \wedge \mu_B) \wedge \mu_C &= \mu_A \wedge (\mu_B \wedge \mu_C) \end{aligned}\right\} \quad \textbf{Associative law} \tag{14.54}$$

$$\left.\begin{aligned}\mu_A \wedge (\mu_A \vee \mu_B) &= \mu_A \\ \mu_A \vee (\mu_A \wedge \mu_B) &= \mu_A\end{aligned}\right\} \quad \textbf{Absorption laws} \qquad (14.55)$$

$$\left.\begin{aligned}\mu_A \wedge (\mu_B \vee \mu_C) &= (\mu_A \wedge \mu_B) \vee (\mu_A \wedge \mu_C) \\ \mu_A \vee (\mu_B \wedge \mu_C) &= (\mu_A \vee \mu_B) \wedge (\mu_A \vee \mu_C)\end{aligned}\right\} \quad \textbf{Distributive laws} \qquad (14.56)$$

$$\mu_{\bar{\bar{A}}} = \mu_A \qquad \textbf{Involution law} \qquad (14.57)$$

$$\left.\begin{aligned}\overline{\mu_A \vee \mu_B} &= \mu_{\bar{A}} \wedge \mu_{\bar{B}} \\ \overline{\mu_A \wedge \mu_B} &= \mu_{\bar{A}} \vee \mu_{\bar{B}}\end{aligned}\right\} \quad \textbf{DeMorgan's laws} \qquad (14.58)$$

$$\left.\begin{aligned}\mu_A \vee 0 = \mu_A, \ \mu_A \wedge 1 = \mu_A \\ \mu_A \vee 1 = 1, \ \mu_A \wedge 0 = 0\end{aligned}\right\} \quad \textbf{Identity laws} \qquad (14.59)$$

However, there is a failure of complement laws, viz.,

$$\left.\begin{aligned}\mu_A \vee \mu_{\bar{A}} &= 1 \\ \mu_A \wedge \mu_{\bar{A}} &= 0\end{aligned}\right\} \qquad (14.60)$$

which otherwise are also satisfied in addition to (14.49 to 14.59) in case of ordinary sets (crisp sets). In case of fuzzy sets, we have

$$\left.\begin{aligned}\mu_A \vee \mu_{\bar{A}} &\neq 1 \\ \mu_A \wedge \mu_{\bar{A}} &\neq 0\end{aligned}\right\} \qquad (14.61)$$

From the above properties of grades of fuzzy sets, it can be said that grades constitute a distributive lattice under $\wedge$, $\vee$ and $\bar{\ }$, but do not form a Boolean lattice because of failure of complementary laws (14.61). The same holds for the fuzzy sets i.e., if **A**, **B** and **C** are fuzzy sets in a universe of discourse X, then the fuzzy sets will have the following properties satisfied.

$$\left.\begin{aligned}\mathbf{A} \cup \mathbf{A} &= \mathbf{A} \\ \mathbf{A} \cap \mathbf{A} &= \mathbf{A}\end{aligned}\right\} \quad \textbf{Idempotence} \qquad (14.62)$$

$$\left.\begin{aligned}\mathbf{A} \cup \mathbf{B} &= \mathbf{B} \cup \mathbf{A} \\ \mathbf{A} \cap \mathbf{B} &= \mathbf{B} \cap \mathbf{A}\end{aligned}\right\} \quad \textbf{Commutativity} \qquad (14.63)$$

$$\left.\begin{aligned}(\mathbf{A} \cup \mathbf{B}) \cup \mathbf{C} &= \mathbf{A} \cup (\mathbf{B} \cup \mathbf{C}) \\ (\mathbf{A} \cap \mathbf{B}) \cap \mathbf{C} &= \mathbf{A} \cap (\mathbf{B} \cap \mathbf{C})\end{aligned}\right\} \quad \textbf{Associativity} \qquad (14.64)$$

$$\left.\begin{aligned}\mathbf{A} \cup (\mathbf{A} \cap \mathbf{B}) &= \mathbf{A} \\ \mathbf{A} \cap (\mathbf{A} \cup \mathbf{B}) &= \mathbf{A}\end{aligned}\right\} \quad \textbf{Absorption} \qquad (14.65)$$

$$\left.\begin{aligned}\mathbf{A} \cup (\mathbf{B} \cap \mathbf{C}) &= (\mathbf{A} \cup \mathbf{B}) \cap (\mathbf{A} \cup \mathbf{C}) \\ \mathbf{A} \cap (\mathbf{B} \cup \mathbf{C}) &= (\mathbf{A} \cap \mathbf{B}) \cup (\mathbf{A} \cap \mathbf{C})\end{aligned}\right\} \quad \textbf{Distributivity} \qquad (14.66)$$

$$\bar{\bar{\mathbf{A}}} = \mathbf{A} \qquad \text{Involution} \qquad (14.67)$$

$$\left.\begin{array}{l} \overline{\mathbf{A} \cup \mathbf{B}} = \bar{\mathbf{A}} \cup \bar{\mathbf{B}} \\ \overline{\mathbf{A} \cap \mathbf{B}} = \bar{\mathbf{A}} \cap \bar{\mathbf{B}} \end{array}\right\} \qquad \text{DeMorgan's} \qquad (14.68)$$

$$\left.\begin{array}{l} \mathbf{A} \cup \phi = \mathbf{A},\ \mathbf{A} \cup \mathrm{U} = \mathrm{U} \\ \mathbf{A} \cap \phi = \phi,\ \mathbf{A} \cap \mathrm{U} = \mathbf{A} \end{array}\right\} \qquad \text{Identity} \qquad (14.69)$$

$$\left.\begin{array}{l} \mathbf{A} \cup \bar{\mathbf{A}} \neq \mathrm{U} \\ \mathbf{A} \cap \bar{\mathbf{A}} \neq \phi \end{array}\right\} \qquad \text{Complementary} \qquad (14.70)$$

or (14.70) can be more precisely written as

$$0.5\mathrm{U} \subseteq \mathbf{A} \cup \bar{\mathbf{A}} \subseteq \mathrm{U}$$

$$\phi \subseteq \mathbf{A} \cap \bar{\mathbf{A}} \subseteq 0.5\ \mathrm{U} \qquad (14.71)$$

where, $\mu_{0.5U} = 0.5\ \mu_U = 0.5 \times 1 = 0.5$

14.4.2 Under Algebraic Product (.) and Sum ($\dot{+}$)

$$\left.\begin{array}{l} \mathbf{A} \dot{+} \mathbf{A} \geq \mathbf{A} \\ \mathbf{A}\ .\ \mathbf{A} \subseteq \mathbf{A} \end{array}\right\} \qquad \text{Idempotency} \qquad (14.72)$$

$$\left.\begin{array}{l} \mathbf{A} \dot{+} \mathbf{B} = \mathbf{B} \dot{+} \mathbf{A} \\ \mathbf{A}.\mathrm{B} = \mathrm{B}.\mathbf{A} \end{array}\right\} \qquad \text{Commutativity} \qquad (14.73)$$

$$\left.\begin{array}{l} (\mathbf{A} \dot{+} \mathbf{B}) \dot{+} \mathbf{C} = \mathbf{A} \dot{+} (\mathbf{B} \dot{+} \mathbf{C}) \\ (\mathbf{A}.\mathbf{B}).\mathrm{C} = \mathbf{A}.(\mathrm{B}.\mathrm{C}) \end{array}\right\} \qquad \text{Associativity} \qquad (14.74)$$

$$\left.\begin{array}{l} \mathbf{A} \dot{+} (\mathbf{A}.\mathbf{B}) \geq \mathbf{A} \\ \mathbf{A}.(\mathbf{A} \dot{+} \mathbf{B}) \leq \mathbf{A} \end{array}\right\} \qquad \text{Absorption} \qquad (14.75)$$

$$\left.\begin{array}{l} \mathbf{A} \dot{+} (\mathbf{B}.\mathbf{C}) \geq (\mathbf{A} \dot{+} \mathbf{B}).(\mathbf{A} \dot{+} \mathbf{C}) \\ \mathbf{A}.(\mathbf{B} \dot{+} \mathbf{C}) \leq (\mathbf{A}.\mathbf{B}) \dot{+} (\mathbf{A}.\mathbf{C}) \end{array}\right\} \qquad \text{Distributivity} \qquad (14.76)$$

$$\left.\begin{array}{l} \overline{\mathbf{A} \dot{+} \mathbf{B}} = \bar{\mathbf{A}}.\bar{\mathbf{B}} \\ \overline{\mathbf{A}.\mathbf{B}} = \bar{\mathbf{A}} \dot{+} \bar{\mathbf{B}} \end{array}\right\} \qquad \text{DeMorgan's laws} \qquad (14.77)$$

$$\left.\begin{array}{l} \mathbf{A} \dot{+} \phi = \mathbf{A},\ \mathbf{A} \dot{+} \mathrm{U} = \mathrm{U} \\ \mathbf{A}.\phi = \phi,\ \mathbf{A}.\mathrm{U} = \mathbf{A} \end{array}\right\} \qquad \text{Identity} \qquad (14.78)$$

$$\left.\begin{array}{l}\mathbf{A} \dot{+} \mathbf{A} \neq U \\ \mathbf{A}.\bar{\mathbf{A}} \neq \phi\end{array}\right\} \quad \textbf{Complementary} \qquad (14.79)$$

Property (14.79) can be more precisely wtitten as

$$\left.\begin{array}{l}0.75\ U \subseteq \mathbf{A} \dot{+} \bar{\mathbf{A}} \subseteq U \\ \phi \subseteq \mathbf{A}.\bar{\mathbf{A}} \subseteq 0.25U\end{array}\right\} \qquad (14.80)$$

Thus fuzzy sets under algebraic product (·) and sum ($\dot{+}$) do not constitute such algebraic structures as a *lattice* and a *semiring*. Fuzzy sets, however, form a commutative monoid under $\dot{+}$ (or ·).

Mizumoto and Tanaka [16] considered various combinations of operators to analyze the type of algebraic structure, fuzzy sets could offer under the operators. Some of their findings can be summarized as follows :

(i) Fuzzy sets form a unitary (=U) commutative semiring with zero (=ϕ) under ∪(as addition) and algebraic product (as multiplication). The duality holds good for intersection ∩ (as addition) and algebraic sum $\dot{+}$ (as multiplication). Fuzzy sets also form a *lattice ordered semigroup* with zero ϕ and unity U under ∩, ∪ and ·. The duality holds for ∪, ∩ and $\dot{+}$.

(ii) Fuzzy sets under ⊕ and ⊙ do not satisfy absorption and distributive laws and hence do not form such algebraic structures as a *lattice* and *semiring*. The same is true of ⊕ and ⊖, and of ⊙ and ⊖. Fuzzy sets, however, form a *commutative monoid* under ⊕ (or ⊙), but do not form such structure under ⊖.

(iii) Fuzzy sets satisfy associative laws, commutative laws and distributive law under the operation of bounded product ⊙ and union ∪, and thus they form a unitary (=U) *commutative semiring* with zero (=ϕ) under ⊙ (as multiplication) and ∪(as addition). Dually, fuzzy sets under the operations of bounded-sum ⊕ (as multiplication) and intersection ∩ (as addition) form a unitary (=ϕ) *commutative semiring* with zero (=U). Moreover, fuzzy sets under ⊕ (as multiplication) and ∪ (as addition) form a unitary (=ϕ) commutative semiring. The same holds good for ⊙ (as multiplication) and ∩ (as addition), where U becomes a unit element for ∩. Furthermore, fuzzy sets also form a *lattice ordered semigroup* with unity U and zero ϕ under ∪, ∩ and ⊙, where ⊙ is a semigroup operation. Dually, they form a *lattice ordered semigroup* with unity ϕ and zero U under ∩, ∪ and ⊕.

14.5 EXTENDED OPERATIONS

In this section we define some very important and widely used operation on fuzzy sets but before we do that let us define a cartesian product of fuzzy sets.

14.5.1 Cartesian Product

Let X be the cartesian product of universes $X = X_1 \times X_2 \times \ldots \times X_k$ and if $\mathbf{A}_1, \mathbf{A}_2, \ldots, \mathbf{A}_k$ are fuzzy sets of $X_1, X_2, \ldots, X_k$ respectively, the cartesian product of $\mathbf{A}_1, \mathbf{A}_2, \ldots, \mathbf{A}_k$ is denoted by $\mathbf{A}_1 \times \mathbf{A}_2 \times \ldots \times \mathbf{A}_k$

and is defined as a fuzzy sets of $X = X_1 \times X_2 \times \ldots \times X_k$, whose membership function is expressed by

$$\mu_{A_1} \times \mu_{A_2} \times \ldots \times \mu_{A_k}(x_1,x_2,\ldots,x_k) = \mu_{A_1} \wedge \mu_{A_2} \wedge \ldots \wedge \mu_{A_k}$$

Equivalently,

$$A_1 \times A_2 \times \ldots A_k = \int_{X_1 \times X_2 \times \ldots \times X_k} (\mu_{A_1} \wedge \mu_{A_2} \wedge \ldots \wedge \mu_{A_k})/(x_1,x_2,\ldots,x_k)$$

$$= \int_X \min\ (\mu_{A_1} \wedge \mu_{A_2} \wedge \ldots \wedge \mu_{A_k})/(x_1,x_2,\ldots,x_k) \qquad (14.81)$$

where $(x_1,x_2,\ldots,x_k)$ is an ordered k-tuple.

The cartesian product rule helps us to extend operations on fuzzy sets. For example, let $\mathbf{A} = \sum_i \mu_A(x_i)/x_i$ and $\mathbf{B}=\sum_j \mu_B(x_j)/x_j$ be the two fuzzy sets in X and let * be a binary operation defined in X. The operation * can be extended to fuzzy sets **A** and **B** by defining a relation

$$\mathbf{A}*\mathbf{B} = \left(\sum_i \mu_A(x_i)/x_i * \sum_j \mu_B(x_j)/x_j \right)$$

$$= \sum_{i,j} \left(\mu_A(x_i) \wedge \mu_B(x_j)\right)/(x_i * x_j) \qquad (14.82)$$

where $\wedge$ stands for **min**.

14.5.2 Extension Principle

It provides a general method for extending non fuzzy mathematical concepts in order to deal with fuzzy quantities. Extension principle [17] is shown to be useful for defining set theoretic operations for higher order fuzzy sets also.

This is a basic identity that allows the domain of the definition of a mapping or a relation to be extended from points in X to fuzzy sets of X. Let f be a mapping from X to Y and **A** be a fuzzy set of X expressed as

$\mathbf{A} = \mu_1/x_1 + \mu_2/x_2 + \ldots + \mu_n/x_n$, then extension principle asserts that

$$f(\mathbf{A}) = f(\mu_1/x_1 + \mu_2/x_2 + \ldots + \mu_n/x_n) \equiv \mu_1/f(x_1)+\mu_2/f(x_2)+\ldots+\mu_n/f(x_n) \qquad (14.83)$$

Thus the image of **A** under f can be deduced from the knowledge of the images of $x_1,x_2,\ldots,x_n$ under f. Now if we have function f, which is a mapping from the cartesian product $X_1 \times X_2 \times \ldots \times X_k$ to a universe Y such that $y=f(x_1,x_2,\ldots,x_k)$, and $\mathbf{A}_1,\mathbf{A}_2,\ldots,\mathbf{A}_k$ are k fuzzy sets in $X_1,X_2,\ldots,X_k$, which are characterised by a set of membership functions $\{\mu_{A_i}(x_i)\}_{i=1}^{k}$, then the extension principle allows us to induce from k fuzzy sets $\{\mathbf{A}_i\}_{i=1}^{k}$, a fuzzy set **B** on Y through f such that

$$\mu_B(y) = \sup_{x_1,x_2,\ldots,x_k} \min\left(\mu_{A_1}(x_1),\ \ldots\ \mu_{A_k}(x_k)\right) \qquad (14.84)$$

$$\mu_B(y) = 0, \quad \text{if } f^{-1}(y) = \{x_1,x_2,\ldots,x_k \mid y = f(x_1,x_2,\ldots,x_k)\}$$

where $f^{-1}(y)$ is the inverse image of y and $\mu_B(y)$ is the greatest amongst the membership values $\mu_{A_1 \times A_2 \times \ldots \times A_k}(x_1, x_2, \ldots, x_k)$ of the realizations of y using k-tuples $(x_1, x_2, \ldots, x_k)$. The special case when k=1, was earlier provided by Zadeh [2.1]. When f is *one-to-one*, (14.84) becomes

$\mu_B(y) = \mu_A(f^{-1}(y))$, if $f^{-1}(y) \neq \phi$. Zadeh writes (14.84) as

$$\mathbf{B} = f(\mathbf{A}_1, \mathbf{A}_2, \ldots, \mathbf{A}_k) = \int_{X_1 \times X_2 \times \ldots \times X_k} \mathbf{min}\left(\mu_{A_1}(x_1), \ldots, \mu_{A_k}(x_k)\right) / f(x_1, x_2, \ldots, x_k)$$

where the **Sup** operation is implicit. The *consistency degree* of two fuzzy sets **A** and **B** can be expressed as

$$\mathbf{Sup}_{x=y}\ \mathbf{min}\left(\mu_A(x), \mu_B(x)\right) = \mathbf{hgt}(\mathbf{A} \cap \mathbf{B})$$

It is the application of the extension principle to the *equality*.

14.6 FUZZY RELATIONS

If X is the cartesian product of n universes of discourse $X_1, X_2, \ldots, X_n$, then an n-ary fuzzy relation R in X is a fuzzy set of X. Then R may be expressed as the union of its consistent fuzzy singletons $\mu_R(x_1, x_2, \ldots, x_n)/(x_1, x_2, \ldots, x_n)$; that is

$$R = \int_{X_1 \times X_2 \times \ldots \times X_n} \mu_R(x_1, x_2, \ldots, x_n)/(x_1, x_2, \ldots, x_n) \tag{14.85}$$

where μ_R is the membership function of R. When n=2, the relation is called *binary*.

Let us provide some definitions which are relevant to our discussion.

(i) The composition of two fuzzy relations of **A** and **B** denoted by **BoA**, is defined as a fuzzy relation in X whose membership function is related to those of **A** and **B** by

$$\mu_{BoA}(x, y) = \mathbf{Sup}_{v}\left[\mathbf{min}\left(\mu_A(x, v), \mu_B(v, y)\right)\right]; \quad v, x, y \in X \tag{14.86}$$

(ii) Let X and Y be two spaces of objects, and h be the mapping from X to Y. Let **B** be a fuzzy set in Y with membership function $\mu_B(y)$. The fuzzy set **A** in X induced by the inverse mapping h^{-1} is defined by

$$\mathbf{A} = \left\{\left(x, \mu_A(x)\right) \mid x = h^{-1}(y) \text{ and } \mu_A(x) = \mu_B(y), \ y \in \mathbf{B}\right\} \tag{14.87}$$

Now consider a converse problem. Suppose **A** is a fuzzy set in X and h is the mapping from X to Y. What is the membership function for fuzzy set **B** in Y which is induced by this mapping? The following two cases arise :

Case I

The mapping h is *one-to-one*. Then

$$\mathbf{B} = \left\{\left(y, \mu_B(y)\right) \mid y = h(x) \text{ and } \mu_B(y) = \mu_A(x), \ x \in \mathbf{A}\right\} \tag{14.88}$$

since h is *one-to-one* for any $y \in Y$, where y is an image of some x in **A**, i.e., y=h(x), there does not exist x′ other than x such that y=h(x′).

Case II

The mapping is *many-to-one*. In this case, we assign the largest of the grade of memberships to y. Thus we have a general definition for both cases, i.e.,

Let **A** be a fuzzy set in X with membership function $\mu_A(x)$. The fuzzy set **B** in Y induced by mapping h is defined by

$$\mathbf{B} = \{(y, \mu_B(y)) \mid y = h(x) \text{ and } \mu_B(y) = \max_{x \in T^{-1}(y)} \mu_A(x),\ x \in \mathbf{A}\} \tag{14.89}$$

where $T^{-1}(y)$ is the set of points in X that are mapped into y by h.

(iii) If R is a relation from U to V (or equivalently, a relation in $U \times V$), and S is a relation from V to W, then the composition of R and S is a fuzzy relation from U to W denoted by $R \circ S$ and defined by

$$R \circ S = \int_{U \times W} \operatorname*{Sup}_{V} \left[\left(\mu_R(u,v) \wedge \mu_S(v,w) \right) / (u,w) \right] \tag{14.90}$$

If U, V and W are finite sets, then relation matrix for $R \circ S$ is the **max-min** product of the relation matrices for R and S.

(iv) If R is an n-any fuzzy relation in $X_1 \times X_2 \times \ldots \times X_n$, then its projection (shadow) on $X_{i1} \times X_{i2} \times \ldots \times X_{ik}$ is k-ary fuzzy relation R_q in X, which is defined by

$$R_q \triangleq \mathbf{proj}\ R \text{ on } X_{i1} \times X_{i2} \times \ldots \times X_{ik} \triangleq P_q R$$

$$\triangleq \int_{X_1 \times X_2 \times \ldots \times X_k} \left(V_{x(\bar{q})} \mu_R(x_1, x_2, \ldots, x_n) \right) / (x_{i1}, x_{i2}, \ldots, x_{ik}) \tag{14.91}$$

where q is the index sequence $(i1, \ldots, ik)$;

$x(q) \triangleq (x_{i1}, x_{i2}, \ldots, x_{ik})$; $\bar{q}$ is the complement of q and $Vx(\bar{q})$ is the supremum of the $\mu_R(x_1, x_2, \ldots, x_n)$ over the x's that are in $x(\bar{q})$.

(v) The domain, range and height, respectively of a fuzzy relation R is defined as follows :

$$\mu_{\text{dom } R}(x) = \operatorname*{Sup}_{y \in Y} \left(\mu_R(x,y) \right) \qquad x \in X \tag{14.92}$$

$$\mu_{\text{Ran } R}(y) = \operatorname*{Sup}_{x \in X} \left(\mu_R(x,y) \right) \qquad y \in Y \tag{14.93}$$

$$\mathbf{hgt}(R) = \operatorname*{Sup}_{x} \left\{ \operatorname*{Sup}_{y} \left(\mu_R(x,y) \right) \right\} \tag{14.94}$$

(vi) A fuzzy relation is *subnormal*, if $\mathbf{hgt}(R)<1$ and *normal*, if $\mathbf{hgt}(R)=1$. Similarly, the support of R is denoted by $\mathbf{Supp}(R)$ and is defined to be the exact subset of $X \times Y$ over which $\mu_R(x,y)>0$.

(vii) If $Y=(Y_1, Y_2, \ldots, Y_n)$ is an n-ary variable taking values in the cartesian product space $X = X_1 \times X_2 \times \ldots \times X_n$, then an n-ary fuzzy relation $R(Y_1, Y_2, \ldots, Y_n)$ in X is a fuzzy restriction if it acts as an elastic constraint on the values that may be assigned to Y. An n-ary variable that is associated with fuzzy restriction on the values that may be assigned to it is said to be an n-ary fuzzy variable.

14.7 FUZZY SETS OF TYPE 2

A fuzzy set of type 2, **A**, in a universe of discourse X, is characterized by a fuzzy membership function μ_A as

$$\mu_A : X \longrightarrow [o,1]^{[0,1]} \text{ or else } \mu_A : X \longrightarrow [0,1]^J \tag{14.95}$$

with the value $\mu_A(x)$ being called a *fuzzy grade* and is itself a fuzzy set in [0,1] (or in the subset J of [0,1]), represented by

$$\mu_A(x) = \int f(u)/u, \ \mu \in [0,1] \text{ or } J$$

where f is a membership function for the fuzzy grade $\mu_A(x)$ and is defined as

$$f : [0,1] \longrightarrow [0,1]$$

J is assumed to be a *finite set*. However, the algebraic properties of fuzzy grades in J should also be satisfied, in case J is continuous.

Since the grade of fuzzy set of type 2 is a fuzzy set in J $\subseteq$ [0,1], by analogy, fuzzy sets of the type n, (n=1,2,...) can also be defined as

$$\mu_A : X \longrightarrow [0,1]^{J_1^{\cdot^{\cdot^{\cdot^{J_{n-1}}}}}} \tag{14.96}$$

where $J_1, J_2, \ldots, J_{n-1}$ are subsets of [0,1]

14.7.1 Operations on Fuzzy Sets of Type 2

The operations on fuzzy sets of type 2 [18] can be defined using the extension principle [17].

Let $\mu_A(x)$ and $\mu_B(x)$ be two fuzzy grades (i.e., fuzzy sets in J $\subseteq$ [0,1]) of fuzzy sets of type 2, **A** and **B**, respectively, represented by

$$\mu_A(x)=f(u_1)/u_1+f(u_2)/u_2+ \ldots +f(u_n)/u_n = \sum_i f(u_i)/u_i; \quad u_i \in J \tag{14.97}$$

$$\mu_B(x)=g(w_1)/w_1+g(w_2)/w_2+ \ldots +g(w_n)/w_n = \sum_j g(w_j)/w_j; \quad w_j \in J \tag{14.98}$$

where the functions f and g are membership functions of fuzzy grades (fuzzy sets in J $\subseteq$ [0,1]) $\mu_A(x)$ and $\mu_B(x)$, respectively and the values $f(u_i)$ and $g(w_j)$ in [0,1] represent the grades for u_i and w_j in J, respectively. Then, we can define the following operations on fuzzy sets of order 2:

Intersection

$$\begin{aligned} \mathbf{A} \cap \mathbf{B} \Leftrightarrow \mu_{\mathbf{A} \cap \mathbf{B}}(x) &\triangleq \mu_A(x) \sqcap \mu_B(x) \\ &= \Big(\sum_i f(u_i)/u_i\Big) \sqcap \Big(\sum_j g(w_j)/w_j\Big) \\ &= \sum_{i,j} \big(f(u_i) \wedge g(w_j)\big)/\big(u_i \wedge w_j\big) \end{aligned} \tag{14.99}$$

Union

$$A \cup B \Leftrightarrow \mu_{A \cup B}(x) \triangleq \mu_A(x) \sqcup \mu_B(x)$$
$$= \Big(\sum_i f(u_i)/u_i\Big) \sqcup \Big(\sum_j g(w_j)/w_j\Big)$$
$$= \sum_{i,j} \big(f(u_i) \wedge g(w_j)\big)/\big(u_i \vee w_j\big) \tag{14.100}$$

Complement

$$\bar{A} \Leftrightarrow \mu_{\bar{A}}(x) \triangleq \neg\, \mu_A(x)$$
$$= \sum_i f(u_i)/(1-u_i) \tag{14.101}$$

where V and Λ represent the **max** and **min** operations, respectively. We name these operations on fuzzy grades as **meet** ($\sqcap$), **join** ($\sqcup$) and **negation** ($\neg$), respectively.

It is important to note here that the operations of $\cap$, $\cup$ and $^{-}$ as defined in (14.23), (14.24) and (14.12) are *not* the same as defined in (14.99-14.101) for $\sqcap$, $\sqcup$ and $\neg$, respectively.

Illustration

For example, let J = {0, 0.1, 0.2, ..., 0.9, 1} and let

$$\mu_A(x) = 0.5/0 + 0.7/0.1 + 0.3/0.2$$
$$\text{and } \mu_B(x) = 0.9/0 + 0.6/0.1 + 0.2/0.2 \tag{14.102}$$

Then, we have

$$\left.\begin{aligned} \mu_A(x) \sqcap \mu_B(x) &= 0.7/0 + 0.6/0.1 + 0.2/0.2 \\ \mu_A(x) \sqcup \mu_B(x) &= 0.5/0 + 0.7/0.1 + 0.3/0.2 \\ \text{and} \qquad \neg\, \mu_A(x) &= 0.5/1 + 0.7/0.9 + 0.3/0.8 \end{aligned}\right\} \tag{14.103}$$

whereas for $\cap$, $\cup$ and $^{-}$ operations, we would have

$$\left.\begin{aligned} \mu_A(x) \cap \mu_B(x) &= 0.5/0 + 0.6/0.1 + 0.2/0.2 \\ \mu_A(x) \cup \mu_B(x) &= 0.9/0 + 0.7/0.1 + 0.3/0.2 \\ \text{and} \quad \overline{\mu_A(x)} &= 0.5/0+0.3/0.1+0.7/0.2+(1/0.3 + 1/0.4 +...+1/1) \end{aligned}\right\} \tag{14.104}$$

14.7.2 Algebraic Structures of Fuzzy Grades Under $\sqcap$, $\sqcup$, and $\neg$

As mentioned in section 2.4, that the grades of ordinary fuzzy sets form a distributive lattice, however the *fuzzy grades* of fuzzy sets of type 2 form a distributive lattice only under $\cap$, $\cup$ and $^{-}$ of (14.23), (14.24) and (14.12). Whereas under the operations $\sqcap$, $\sqcup$ and $\neg$ of (14.99), (14.100) and (14.101), the fuzzy grades in J satisfy such laws as idempotence (14.52), commutative (14.53), associative (14.54), involution (14.57), DeMorgan's laws (14.58) but do not satisfy absorption (14.55), distributive (14.56), identity (14.59) and complementary laws (14.60). The part of identity laws (14.59) is however satisfied i.e.,

$$\mu_A \sqcup 0 = \mu_A, \ \mu_A \sqcap 1 = \mu_A \tag{14.105}$$

Some of the properties of fuzzy grades which Mizumuto and Tanaka [18] studied under the operations of $\sqcap$, $\sqcup$ and $\neg$ and $\cap$, $\cup$ and $^{-}$ can be summarized as follows :

(i) If an order relation $\underset{\cdot}{\sqsubseteq}$ over arbitrary fuzzy grades in J is defined as

$$\mu_A \underset{\cdot}{\sqsubseteq} \mu_B \Leftrightarrow \mu_A \sqcap \mu_B = \mu_A \tag{14.106}$$

then the set of arbitrary fuzzy grades forms a *partially ordered set* (*poset*) under $\underset{\cdot}{\sqsubseteq}$. Similarly let $\overset{\cdot}{\sqsubseteq}$ be an order relation given as

$$\mu_A \overset{\cdot}{\sqsubseteq} \mu_B \Leftrightarrow \mu_A \sqcup \mu_B = \mu_B \tag{14.107}$$

then the arbitrary fuzzy grades also form a *partially ordered set* (*posets*) under $\overset{\cdot}{\sqsubseteq}$. In general, we have $\underset{\cdot}{\sqsubseteq} \neq \overset{\cdot}{\sqsubseteq}$.

(ii) If μ_A and μ_B are convex fuzzy grades in J, then $\mu_A \sqcap \mu_B$, $\mu_A \sqcup \mu_B$, and $\neg\mu_A$ are also convex.

(iii) For convex fuzzy grades in J, the distributive laws are satisfied, viz.,

$$\mu_A \sqcap (\mu_B \sqcup \mu_C) = (\mu_A \sqcap \mu_B) \sqcup (\mu_A \sqcap \mu_C) \tag{14.108}$$

$$\mu_A \sqcup (\mu_B \sqcap \mu_C) = (\mu_A \sqcup \mu_B) \sqcap (\mu_A \sqcup \mu_C) \tag{14.109}$$

(iv) Convex fuzzy grades in J form a semiring (more precisely, a commutative semiring with identities) under $\sqcap$ and $\sqcup$.

(v) If μ_A and μ_B are normal fuzzy grades in J, then $\mu_A \sqcap \mu_B$, $\mu_A \sqcup \mu_B$ and $\neg\mu_A$ are also normal.

(vi) For normal fuzzy grades, part of the identity (14.59) i.e.,

$$\mu_A \sqcup 1 = 1, \ \mu_A \sqcap 0 = 0 \tag{14.110}$$

is satisfied.

(vii) If μ_A and μ_B are normal convex fuzzy grades in J, then $\mu_A \sqcap \mu_B$, $\mu_A \sqcup \mu_B$ and $\neg\, \mu_A$ are also normal convex from conditions (v) and (vi) given above.

(viii)For normal convex fuzzy grades in J, the absorption laws, i.e.,

$$\mu_A \sqcap (\mu_A \sqcup \mu_B) = \mu_A, \ \mu_A \sqcup (\mu_A \sqcap \mu_B) = \mu_A \tag{14.111}$$

are satisfied.

(ix) Normal convex fuzzy grades in J form a distributive lattice under $\sqcap$, $\sqcup$, where the greatest element is 1/1 and the least element is 1/0.

(x) Fuzzy sets of type 2 in a set X form a distributive lattice under operation of $\cap$ and $\cup$ defined in (14.99) and (14.100), respectively, where the grades characterizing these fuzzy sets of type 2 are normal convex fuzzy grades in $J \subseteq [0,1]$.

If new meet and join operators (viz., $\lceil\cdot\rceil$ and $\lfloor\cdot\rfloor$) are defined, which obviously are different from $\sqcap$ and $\sqcup$ of (14.99) and (14.100), respectively, these would offer another set of properties as follows :

Let fuzzy grades μ_A and μ_B be represented by

$$\mu_A = \sum_{u \in J} f(u)/u \quad \text{and} \quad \mu_B = \sum_{w \in J} g(w)/w$$

then

$$\mu_A \lceil\cdot\rceil \mu_B = \sum_{u,w \in J} f(u)g(w)/u \wedge w \tag{14.112}$$

and

$$\mu_A \lfloor\cdot\rfloor \mu_B = \sum_{u,w \in J} f(u)g(w)/u \vee w \tag{14.113}$$

where, $f(u)g(w)$ stands for an algebraic product of $f(u)$ and $g(w)$.

Under (14.112), (14.113) and (14.101) operations, Mizumuto and Tanata [18] demonstrated that arbitrary fuzzy grades satisfy such laws as commutative, associative, involution and DeMorgan's laws. But normal fuzzy grades (needless to say, any fuzzy grades, normal fuzzy grades or convex fuzzy grades) do not satisfy idempotence, absorption, distributive and complement laws. A part of identity laws i.e.,

$$\mu_A \lfloor\cdot\rfloor 1 = \mu_A, \ \mu_A \lceil\cdot\rceil 0 = \mu_A$$

are satisfied for any fuzzy grades. Another part of identity laws i.e.,

$$\mu_A \lfloor\cdot\rfloor 1 = 1, \quad \mu_A \lceil\cdot\rceil 0 = 0$$

can be satisfied for normal fuzzy grades and normal convex fuzzy grades only.

14.7.3 Algebraic Properties of Fuzzy Grades under $\cdot$, $\dot{+}$ and $\neg$

Mizumoto and Tanaka [19] outlined the following algebraic properties of fuzzy grades under product ($\cdot$), algebraic sum ($\dot{+}$) and negation ($\neg$) :

(i) If μ_A and μ_B are convex fuzzy grades, then $\mu_A \cdot \mu_B$, $\mu_A \dot{+} \mu_B$ and $\neg\mu_A$ are also convex fuzzy grades. However if the fuzzy grades are discrete, the convexity under $\cdot$ and $\dot{+}$ does not hold good. This is for this reason that A_4 in Fig. 14.11 (although appearing to be convex) is actually a nonconvex set.

(ii) If μ_A and μ_B are normal fuzzy grades, then $\neg\mu_A$, $\mu_A \cdot \mu_B$ and $\mu_A \dot{+} \mu_B$ are also normal fuzzy grades. Further, if μ_A and μ_B are normal convex fuzzy grades, then $\mu_A \cdot \mu_A$, $\mu_A \dot{+} \mu_B$ and $\neg\mu_A$ are also normal convex fuzzy grades.

(iii) For any arbitrary fuzzy grades which include discrete fuzzy grades, the commutative, associative, involution, DeMorgan's laws and a part of identity laws, viz.,

$$\mu_A \cdot 1 = \mu_A; \quad \mu_A \dot{+} 0 = \mu_A$$

hold good.

(iv) Normal convex fuzzy grades or for that matter any fuzzy grades, normal fuzzy grades and convex fuzzy grades do not satisfy idempotence, absorption, distributive, complement and identity laws, respectively, viz.,

$$\mu_A \cdot \mu_A \neq \mu_A; \quad \mu_A \dot{+} \mu_A \neq \mu_A \tag{14.114}$$

$$\left.\begin{aligned} &\mu_A \cdot (\mu_A \dot{+} \mu_B) \neq \mu_A \\ &\mu_A \dot{+} (\mu_A \cdot \mu_B) \neq \mu_A \end{aligned}\right\} \tag{14.115}$$

$$\left.\begin{aligned} &\mu_A \cdot (\mu_B \dot{+} \mu_C) \neq (\mu_A \cdot \mu_B) \dot{+} (\mu_A \cdot \mu_C) \\ &\mu_A \dot{+} (\mu_B \cdot \mu_C) \neq (\mu_A \dot{+} \mu_B) \cdot (\mu_A \dot{+} \mu_C) \end{aligned}\right\} \tag{14.116}$$

$$\mu_A \cdot (\neg\mu_A) \neq 0; \quad \mu_A \dot{+} (\neg\mu_A) \neq 1 \tag{14.117}$$

$$\mu_A \cdot 0 \neq 0; \quad \mu_A \dot{+} 1 \neq 1 \tag{14.118}$$

However, identity laws $\mu_A \cdot 0=0$ and $\mu_A \dot{+} 1=1$ can be satisfied by normal fuzzy grades and normal convex fuzzy grades.

(v) Arbitrary fuzzy grades under algebraic product $(\cdot)$ form a commutative semigroup with identity 1. The duality holds for algebraic sum $(\dot{+})$, where 0 is an identity. The same can be said about normal fuzzy grades, convex fuzzy grades and normal convex fuzzy grades. Since normal convex fuzzy grades satisfy (14.114) to (14.118), they do not form such algebraic structures as a *lattice* and a *semiring*.

14.8 FUZZY NUMBERS

In this section, we will provide some definitions and properties of fuzzy numbers and subsequently apply all that has been discussed so far to illustrate how fuzzy numbers can be operated upon for basic arithmetic and other operations.

A Fuzzy Number

A fuzzy number **A** in the real line $\mathcal{R}$ is a fuzzy set characterized by a membership function μ_A as

$$\mu_A : \mathcal{R} \longrightarrow [0,1]$$

and the fuzzy number **A** is expressed as

$$\mathbf{A} = \int_{x \in \mathcal{R}} \mu_A(x)/x, \quad \text{with the understanding that} \quad \mu_A(x) \in [0,1]$$

represents the grade of membership of x in **A** and $\int$ denotes the union of several $\mu_A(x)/x$'s.

Simple examples of fuzzy numbers are fuzzy sets of the real line labelled *small, approximately 10, very close to 8, much larger than 10* etc. In fact, a fuzzy number **2** denoting a value *about 2* may be expressed as

$$\mathbf{2} = \int_1^2 (x-1)/x + \int_2^3 (3-x)/x \tag{14.119}$$

and is illustrated by a full line in Fig. 14.10. Here + stands for the union.

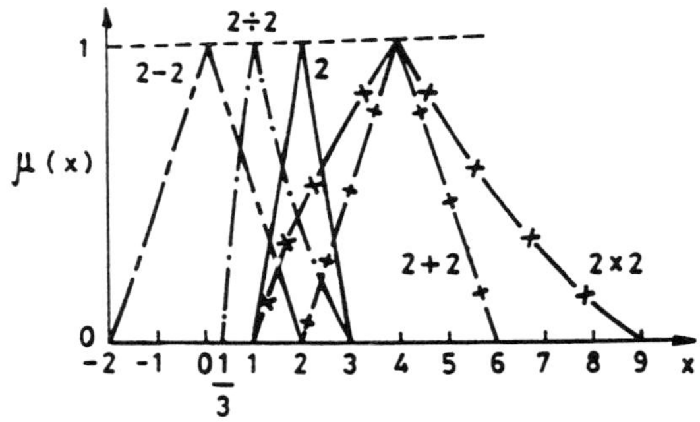

Fig. 14.10 Various Arithmetic Operations on **2** (a fuzzy number)

A fuzzy number can be viewed as a membership grade function or a possibility distribution over the possible support. Generally, a fuzzy number shall be either convex or concave fuzzy subset in real line and as a special case, it could be an interval, in which case the fuzzy arithmetic can be viewed as a generalization of interval arithmetic.

Positive Fuzzy Number

A fuzzy number **A** is said to be positive if $0<a_1 \le a_2$ holds for the support $\Gamma_{\mathbf{A}} = [a_1, a_2]$ of **A**, i.e., $\Gamma_{\mathbf{A}}$ is in positive real line, $\mathcal{R}_+$.

Negative Fuzzy Number

A fuzzy number **A** is said to be negative if $a_1 \le a_2 < 0$ for the support, $\Gamma_{\mathbf{A}}$.

Zero Fuzzy Number

A fuzzy number **A** is said to be zero if $a_1 \le 0 \le a_2$ for the support, $\Gamma_{\mathbf{A}}$. In Fig. 14.10, **2−2** is a *zero fuzzy number* while all other fuzzy numbers are positive.

Type of Fuzzy Numbers

Before we define the arithmetic operations on fuzzy numbers, it would not be out of place to talk about membership grade function. There is no explicit way of assigning membership values, which are relative grades over the real interval [0,1]. This value of relative grades is context dependent and is subjective. Very often, it is convenient to express the membership function of a fuzzy set in terms of standard forms. They can be broadly classified into three categories, viz., π, s and z type of functions.

π-Number

The possibility distribution of such a fuzzy number is bell-shaped on the real line over the interval $[a_1, a_2]$ and is characterized by two parameters, viz., peak point m at which this $\mu_{\mathbf{A}}(x)=1$ and the bandwidth b, which can be defined as the distance between the cross over points x_c on either side of the peak point at which the $\mu_{\mathbf{A}}(x)=0.5$. Thus a π-fuzzy number can be expressed as (m, b) where m is the peak point and b is the bandwidth. One can also use normalized bandwidth $b'=b/m$ in some cases and express fuzzy number as (m,b′). Thus one such example of a π-fuzzy number can be given as follows,

where,

$$\left.\begin{aligned} \mu_\pi(x) &= 0 && ;\ \text{for}\quad x\le m-b \text{ and for } x\ge m+b \\ &= \frac{2}{b^2}\,(x-m+b)^2 && ;\ \text{for}\quad m-b \le x \le m-\frac{b}{2} \\ &= 1-\frac{2}{b^2}\,(x-m)^2 && ;\ \text{for}\quad m-\frac{b}{2} \le x \le m+\frac{b}{2} \\ &= \frac{2}{b^2}\,(x-m-b)^2 && ;\ \text{for}\quad m+\frac{b}{2} \le x \le m+b \end{aligned}\right\} \qquad (14.120)$$

This number is shown in Fig. 14.11(a).

s-Number

The possibility distribution of such a number has a shape of an s and an example of equations expressing an s-number can be given as

$$\left.\begin{aligned} \mu_s(x) &= 0 && ;\ \text{for } x \le m-b \\ &= \frac{2}{b^2}\,(x-m+b)^2 && ;\ \text{for } m-b \le x \le m-\frac{b}{2} \\ &= 1-\frac{2}{b^2}\,(x-m)^2 && ;\ \text{for } m-\frac{b}{2} \le x \le m \\ &= 1 && ;\ \text{for } x \ge m \end{aligned}\right\} \qquad (14.121)$$

where, b is defined as the bandwidth or interval reckoned as a distance between the points where $\mu(x)=0$ and 1, respectively and the point m is the rightmost end point of this interval or the left peak point. This number is shown in Fig. 14.11(b).

z-Number

A z-number is a mirror image of an s-number. The defining equations can be given by :

$$\left.\begin{aligned} \mu_z(x) &= 1 && ;\ \text{for } x \le m \\ &= 1-\frac{2}{b^2}\,(x-m)^2 && ;\ \text{for } m \le x \le m+\frac{b}{2} \\ &= \frac{2}{b^2}\,(x-m-b)^2 && ;\ \text{for } m+\frac{b}{2} \le x \le m+b \\ &= 0 && ;\ \text{for } x \le m+b \end{aligned}\right\} \qquad (14.122)$$

Here, m is right peak point or the leftmost point of the interval b. This number is shown in Fig. 14.11(c).

s/z-Number

An s/z-number has a flat top possibility distribution combining distributions of a s-number and a z-number together with the left peak point of the s-number lying to the left of the right peak point of the z-number. An s/z number can be represented by an ordered pair $\{(m_1,b_1),(m_2,b_2)\}$, in which the first element consists of the parameters of an s-number and the second that of a z-number. This number is shown in Fig. 14.11(d).

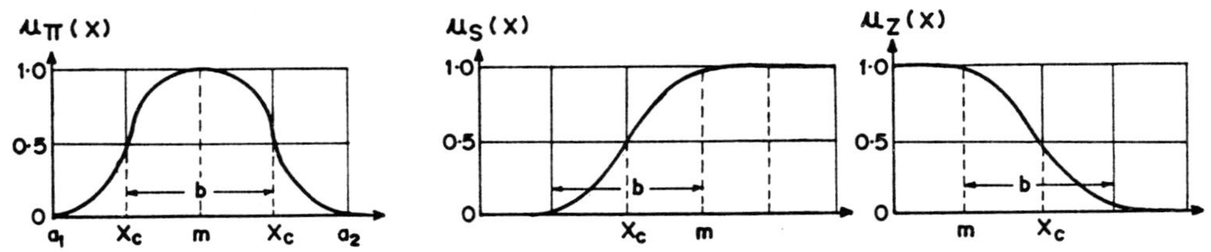

(a) A π-fuzzy Number (b) An s-fuzzy Number (c) A z-fuzzy Number

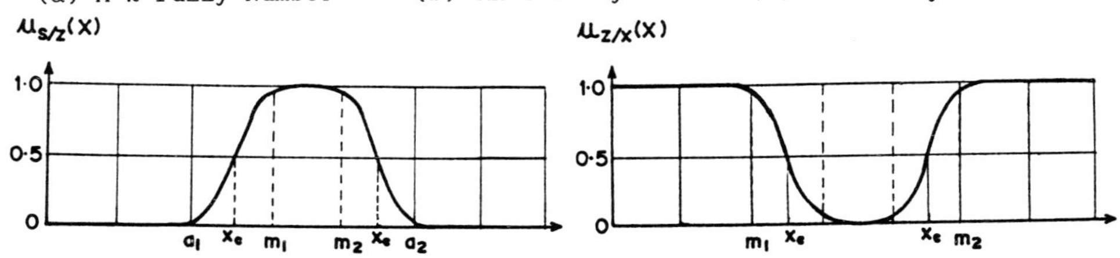

(d) An s/z-fuzzy Number (e) A z/s-fuzzy Number

Fig. 14.11 Various Types of Fuzzy Numbers

z/s-Number

The possibility distribution of a z/s number is the complement of that of an s/z number and can be regarded as the union of the possibility distributions of a z-number and an s-number. It can be represented by an ordered pair $\{(m_1, b_1), (m_2, b_2)\}$, whose first element represents the parameters of a z-number and the other that of an s-number. This number is shown in Fig. 14.11(e).

14.8.1 The Four Arithmetic Operations

The four arithmetic operations, viz., +, −, × and ÷ can be defined [20,21] for two fuzzy sets using the extension principle as given in section 14.5.2.

Let **A** and **B** be the fuzzy numbers in $\mathcal{R}$ and let * be a binary operation defined in $\mathcal{R}$. Then the operation * can be extended to fuzzy numbers **A** and **B** by defining

$$\mathbf{A} * \mathbf{B} = \int_{x,\, y \in \mathcal{R}} \big(\mu_A(x) \wedge \mu_B(y)\big) / \; x*y \tag{14.123}$$

where $\wedge$ stands for **min**.

Here, binary operation * can be replaced by the four arithmetic operations of +, −, × and of ÷. For example,

$$\left.\begin{aligned} \mathbf{A} + \mathbf{B} &= \int \big(\mu_A(x) \wedge \mu_B(y)\big)/(x+y) \\ \mathbf{A} - \mathbf{B} &= \int \big(\mu_A(x) \wedge \mu_B(y)\big)/(x-y) \\ \mathbf{A} \times \mathbf{B} &= \int \big(\mu_A(x) \wedge \mu_B(y)\big)/(x\times y) \\ \mathbf{A} \div \mathbf{B} &= \int \big(\mu_A(x) \wedge \mu_B(y)\big)/(x\div y) \end{aligned}\right\} \qquad (14.124)$$

The membership functions of these fuzzy numbers are obtained respectively as :

$$\left.\begin{aligned} \mu_{\mathbf{A+B}}(a) &= \bigvee_{x+y=a} \big(\mu_A(x) \wedge \mu_B(y)\big) \\ &= \bigvee_{x} \big(\mu_A(x) \wedge \mu_B(a-x)\big) \\ \mu_{\mathbf{A-B}}(a) &= \bigvee_{x} \big(\mu_A(x) \wedge \mu_B(x-a)\big) \\ \mu_{\mathbf{A\times B}}(a) &= \bigvee_{(x\neq 0)} \big(\mu_A(x) \wedge \mu_B(\tfrac{a}{x})\big) \\ \mu_{\mathbf{A\div B}}(a) &= \bigvee_{x} \big(\mu_A(x) \wedge \mu_B(x/a)\big) \\ &= \bigvee_{y} \big(\mu_A(ay) \wedge \mu_B(y)\big) \end{aligned}\right\} \qquad (14.125)$$

These have been illustrated in Fig.14.10 for **A** and **B**, both being equal to **2**.

It is highly convenient to use *α-level sets* of fuzzy sets to define these arithmetic operations on fuzzy numbers if the fuzzy numbers are convex. As defined in earlier sections, a fuzzy number **A** in $\mathcal{R}$ shall be convex if for any real numbers x, y, z $\in \mathcal{R}$ with $x \le y \le z$, we have

$\mu_A(y) \ge \mu_A(x) \wedge \mu_A(z)$, where $\wedge$ stands for **min** operation.

Various kinds of fuzzy numbers are shown in Fig.14.12.

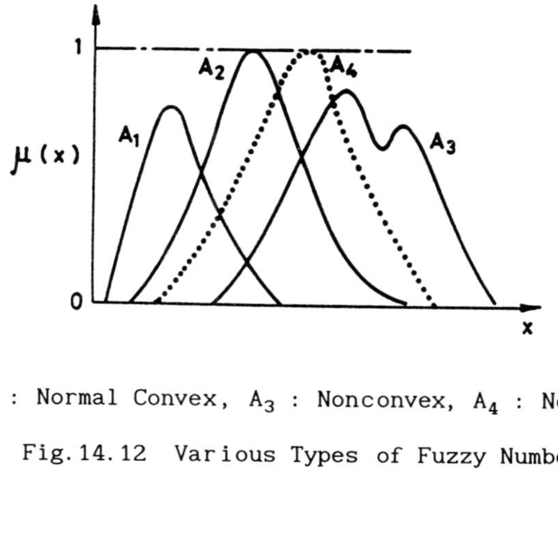

A_1 : Convex, A_2 : Normal Convex, A_3 : Nonconvex, A_4 : Normal Nonconvex

Fig.14.12 Various Types of Fuzzy Numbers

Let α-level sets of convex fuzzy numbers **A** and **B** be denoted by $\mathbf{A}_\alpha$ and $\mathbf{B}_\alpha$ and are intervals in $\mathcal{R}$, which are special convex fuzzy numbers where grades are unity at x belonging to $\mathbf{A}_\alpha$ and zero elsewhere.

Also let the α-level set for sum **A+B** and **A** and **B** be denoted by $(\mathbf{A+B})_\alpha$, then we have

$$(\mathbf{A+B})_\alpha = \mathbf{A}_\alpha + \mathbf{B}_\alpha$$

Therefore, using resolution identity i.e.,

$\mathbf{A} = \int_0^1 \alpha\, \mathbf{A}_\alpha$, we can write the four arithmetic operations as

$$\left.\begin{aligned}
\mathbf{A} + \mathbf{B} &= \int_0^1 \alpha(\mathbf{A+B})_\alpha = \int_0^1 \alpha(\mathbf{A}_\alpha + \mathbf{B}_\alpha) \\
\mathbf{A} - \mathbf{B} &= \int_0^1 \alpha\ (\mathbf{A}_\alpha - \mathbf{B}_\alpha) \\
\mathbf{A} \times \mathbf{B} &= \int_0^1 \alpha\ (\mathbf{A}_\alpha \times \mathbf{B}_\alpha) \\
\mathbf{A} \div \mathbf{B} &= \int_0^1 \alpha\ (\mathbf{A}_\alpha \div \mathbf{B}_\alpha)
\end{aligned}\right\} \tag{14.126}$$

For the fuzzy number **2** given by (14.119), we have

$$\mathbf{2} + \mathbf{2} = \int_2^4 \frac{x}{2} - 1/x + \int_4^6 3 - \frac{x}{2}\ /x,$$

$$\mathbf{2} - \mathbf{2} = \int_{-2}^0 \frac{x}{2} + 1/x + \int_0^2 1 - \frac{x}{2}\ /x,$$

$$\mathbf{2} \times \mathbf{2} = \int_1^4 \sqrt{x} - 1/x + \int_4^9 3 - \sqrt{x}\ /x,$$

$$\mathbf{2} \div \mathbf{2} = \int_{1/3}^1 3 - \frac{4}{x+1}\ /x + \int_1^3 \frac{4}{x+1} - 1/x,$$

These numbers have been distinctly illustrated in Fig.14.10.

14.8.2 Properties of Fuzzy Numbers

The following observations can be made about fuzzy numbers [21] :

(i) If **A** is a zero convex number, then 1/**A** (i.e., 1÷**A**) is *not* a convex fuzzy number.

(ii) If **A** and **B** are convex fuzzy numbers, then **A÷B** is *not*, *in general*, a convex fuzzy number.

(iii) If **A** is a convex fuzzy number and **B** is a positive (or negative) convex fuzzy number than **A÷B** is a convex fuzzy number.

(iv) If **A** and **B** are normal fuzzy numbers, then **A+B**, **A−B**, **A×B** and **A÷B** are also normal. If amongst the two fuzzy numbers **A** and **B**, one is convex and the other is nonconvex, then **A+B**, **A−B**, **A×B** and **A÷B** may be convex or nonconvex.

(v) For any fuzzy numbers **A**, **B** and **C**, the associative and commutative laws (sum and product) apply. Also the identity law holds good.

(vi) For any fuzzy number **A**, there exists no inverse fuzzy numbers **A′** and **A″** under + and ×, respectively, such that

$$\mathbf{A} + \mathbf{A'} = 0$$

$$\mathbf{A} \times \mathbf{A''} = 1$$

For $-\mathbf{A}$ and $-\frac{1}{\mathbf{A}}$of a fuzzy number **A**, we have in general

$$\mathbf{A} + (-\mathbf{A}) \neq 0$$

$$\mathbf{A} \times \left(\frac{1}{\mathbf{A}}\right) \neq 1$$

(vii) In general for any fuzzy numbers **A**, **B** and **C**, the distributive law **A**×(**B**+**C**) = (**A**×**B**)+(**A**×**C**) does not hold good. However this law holds good only if **A**, **B** and **C** are all positive convex fuzzy numbers.

(viii) The family of positive convex fuzzy numbers forms a commutative semiring with 0 and unity under + and ×.

(ix) Arbitrary fuzzy numbers satisfy idempotence, commutative and associative laws under the operation of **meet** ($\sqcap$) and **join** ($\sqcup$) as defined in (14.99) and (14.100), respectively.

(x) Convex fuzzy numbers also satisfy distributive laws and are closed under $\sqcup$ and $\sqcap$. Thus they form a commutative semiring under $\sqcup$ and $\sqcap$, but do not form a lattice because they do not satisfy absorption laws under $\sqcup$ and $\sqcap$.

(xi) Normal convex fuzzy numbers are closed and satisfy absorption laws under $\sqcup$ and $\sqcap$. Thus, they form a distributive lattice under $\dot{\sqcup}$ and $\sqcap$. Consequently, the order relation $\underset{\cdot}{\sqsubseteq}$ becomes coincident with $\sqsubseteq$ and hence the order relation for them can be defined as

$$\mathbf{A} \underset{\cdot}{\sqsubseteq} \mathbf{B} \longleftrightarrow \mathbf{A} \sqcap \mathbf{B} = \mathbf{A} \longleftrightarrow \mathbf{A} \sqcup \mathbf{B} = \mathbf{B}$$

Further properties of fuzzy numbers can be seen in [21].

14.9 REPRESENTATIONS OF A FUZZY NUMBER

Let a fuzzy number be represented by **m** such that

$\mu_{\mathbf{m}}(\mathbf{m}) = 1$; where m is the mean value of **m** and α, β are the left and right spreads of **m**. $(\alpha,\beta) \in \mathcal{R}^2_+$. Let L and R be the left and right reference functions, as shown in Fig. 14.13.

Then the membership function shall be written as

$$\mu_{\mathbf{m}}(x) = \begin{cases} L\left(\frac{m-x}{\alpha}\right); & \text{for } x \le m,\ \alpha > 0 \\ R\left(\frac{x-m}{\beta}\right); & \text{for } x \ge m,\ \beta > 0 \end{cases} \tag{14.127}$$

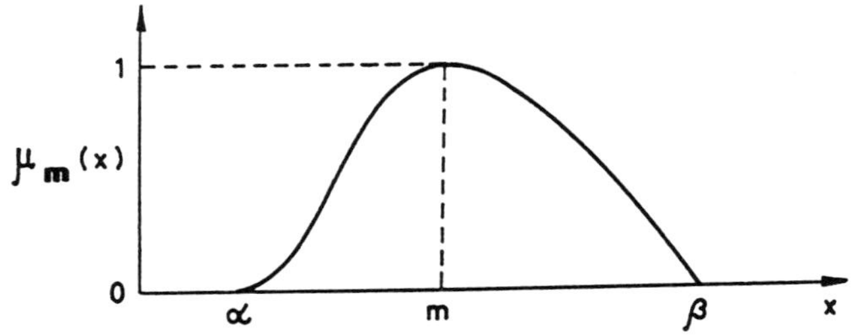

Fig. 14.13 Representation of a Fuzzy Number **m**

Symbolically, we can write such a fuzzy number as

$\mathbf{m} = (m;\ \alpha,\ \beta)_{L-R}$, where **m** is defined as a L-R fuzzy number or L-R representation of a fuzzy number **m**. Alternatively, we could have a membership function as given in Fig. 14.14.

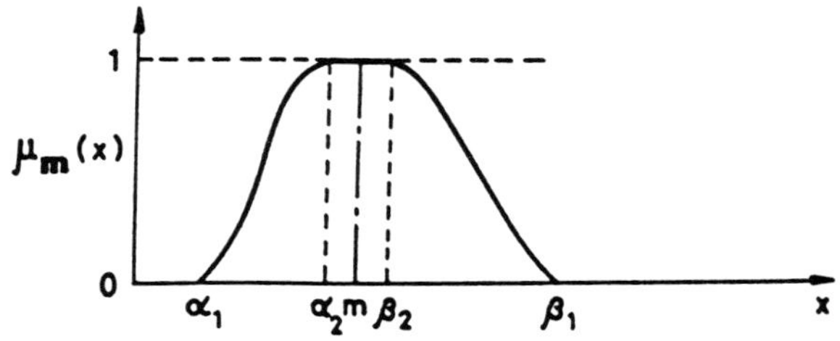

Fig. 14.14 Alternative Representation of a Fuzzy Number

where, α_1, α_2 are left hand and β_1, β_2 are right hand side points of the function such that

$$\left.\begin{aligned} \mu_{\mathbf{m}}(x) &= 0,\ \forall\ x \in (-\infty,\ \alpha_1), \\ &\text{is strictly increasing on } [\alpha_1,\ \alpha_2] \\ &= 1,\ \forall x \in [\alpha_2,\ \beta_2] \\ &\text{is strictly decreasing on } [\beta_2,\ \beta_1] \\ &= 0,\ \forall x \in [\beta_1,\ +\infty] \end{aligned}\right\} \qquad (14.128)$$

where, α_1, α_2, β_1, β_2 are real numbers.

If $\alpha_2 = \beta_2$, $\mathbf{m}$ is representation of a number, whose value is **approximately m** and it is also possible to have $m-\alpha_1=\beta_1-m$, then the number representation would look like a bell-shaped curve symmetric about m.

Further, if the number instead of being represented as in Fig. 14.14 is represented by a trapezium as shown in Fig. 14.15, we can have

$$\left.\begin{aligned} \mu_{\mathbf{m}}(x) &= \frac{x - \alpha_1}{\alpha_2-\alpha_1} &&;\ \text{for } \alpha_1 \le x \le \alpha_2 \\ &= 1 &&;\ \text{for } \alpha_2 \le x \le \beta_2 \\ &= \frac{\beta_1-x}{\beta_1-\beta_2} &&;\ \text{for } \beta_2 \le x \le \beta_1 \\ &= 0 &&;\ \text{for } x \le \alpha_1 \text{ or } x \ge \beta_1 \end{aligned}\right\} \qquad (2.129)$$

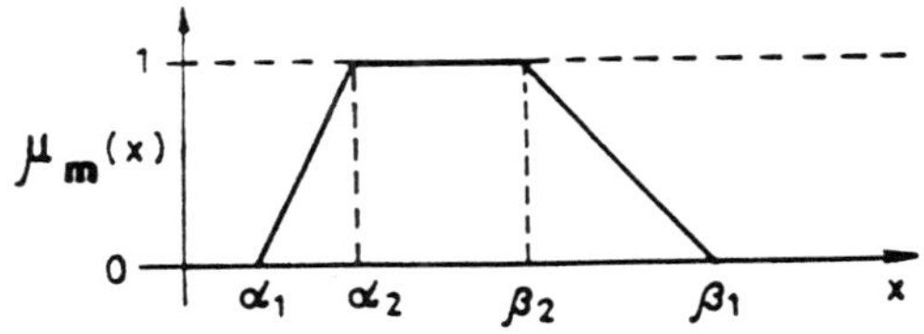

Legend

$\alpha_d \overset{\Delta}{=} \alpha_2-\alpha_1$ and $\beta_d \overset{\Delta}{=} \beta_1-\beta_2$
Here, $\alpha_d>0$, $\beta_d>0$ and $\beta_2>\alpha_2$

Fig. 14.15 Trapezoidal Representation of a Fuzzy Number

Various special cases of trapezoidal representations are shown in Fig. 14.16. Therefore any trapezoidal fuzzy number is fully determined by a quadruple of real numbers $(\alpha_1, \alpha_2, \beta_2, \beta_1)$ as is provided in (14.129). We can write a trapezoidal fuzzy number $\mathbf{m}$ as

$$\mathbf{m} = (\alpha_1,\ \alpha_2,\ \beta_2,\ \beta_1)$$

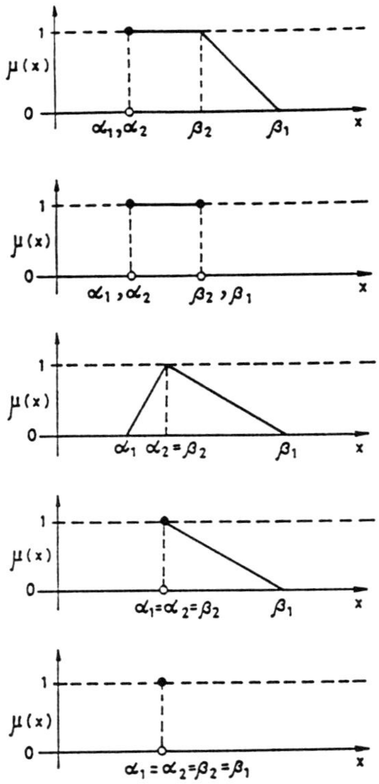

Legend

$\alpha_d \triangleq \alpha_2 - \alpha_1$ and $\beta_d \triangleq \beta_1 - \beta_2$

$\alpha_d = 0,\ \beta_d > 0,\ \beta_2 > \alpha_2$
Trapezoidal Number

$\alpha_d = \beta_d = 0,\ \beta_2 > \alpha_2$
Rectangular Number or Interval Number

$\alpha_d > 0,\ \beta_d > 0,\ \alpha_2 = \beta_2$
Triangular Number

$\alpha_d = 0,\ \beta_d > 0,\ \alpha_2 = \beta_2$
Right-triangular Number

$\alpha_d = \beta_d = 0,\ \alpha_2 = \beta_2$
Crisp Number

Fig. 14.16 Special Cases of a Trapezoidal Representation

Trapezoidal fuzzy numbers have a linearity property, viz.,

if $\mathbf{m}_k = \{\alpha_1,\ \alpha_2,\ \beta_2,\ \beta_1\}_k,\ x_k \geq 0;\ k = 1, 2, \ldots, n$

then $\mathbf{m} \triangleq \mathbf{m}_1 x_1 + \mathbf{m}_2 x_2 + \ldots + \mathbf{m}_n x_n$ is also a trapezoidal number such that

$$\left.\begin{aligned} \mathbf{m} &= (\alpha_1, \alpha_2, \beta_2, \beta_1)_{\mathbf{m}} \\ (\alpha_1)_{\mathbf{m}} &= \sum_{k=1}^{n} (\alpha_1)_k x_k \\ (\alpha_2) &= \sum_{k=1}^{n} (\alpha_2)_k x_k \\ (\beta_2)_{\mathbf{m}} &= \sum_{k=1}^{n} (\beta_2)_k x_k \\ (\beta_1)_{\mathbf{m}} &= \sum_{k=1}^{n} (\beta_1)_k x_k \end{aligned}\right\} \quad (14.130)$$

This of course follows from the extension principle as well, through linear combination of fuzzy numbers.

14.9.1 Arithmetic Operations

Various arithmetic operations on fuzzy numbers represented by L-R functions can be derived as follows :

Addition

Let $\mathbf{m} = (m;\ \alpha,\ \beta)_{L-R}$ and $\mathbf{n} = (n;\ \gamma,\ \delta)_{L-R}$ be two fuzzy numbers then their sum $\mathbf{m} + \mathbf{n}$ is an L-R fuzzy number and

$$(m;\ \alpha,\ \beta) + (n;\ \gamma,\ \delta) = (m+n;\ \alpha+\gamma,\ \beta+\delta)$$

with left side function as

$$\mathbf{m}+\mathbf{n}(z) = \left\{ \begin{array}{l} L(\frac{z-(m+n)}{\alpha+\gamma}) \\ R(\frac{z-(m+n)}{}) \end{array} \right\} \quad (14.131)$$

It may be noted that equation (14.131) does not depend upon the analytical expressions of functions L and R, provided that all fuzzy numbers are of L-R type. Choosing L and R is a subjective matter.

Multiplication by a Scalar

Let λ be a positive real number and $\mathbf{m}$ be a fuzzy number, then

$$\mu_{\lambda\mathbf{m}}(z) = \mathbf{Sup}_{z=\lambda x}\ \mu_{\mathbf{m}}(x) = \mu_{\mathbf{m}}(z/\lambda) \quad (14.132)$$

For $\lambda=-1$, we get opposite of $\mathbf{m}$ i.e., $-\mathbf{m}$
Using L-R representation, we have

$$\mu_{\lambda\mathbf{m}}(x) = \left\{ \begin{array}{l} L(\frac{\lambda m-z}{\lambda\alpha})\ ;\ \text{for } t \le \lambda m,\ \lambda>0 \\ R(\frac{z-\lambda m}{\lambda\beta});\ \text{for } t \ge \lambda m \end{array} \right\} \quad (14.133)$$

i.e.,

$$\left.\begin{aligned} \lambda\mathbf{m} &= \lambda(m;\alpha,\beta)_{LR} = (\lambda m;\ \lambda\alpha,\ \lambda\beta)_{LR} \text{ for } \lambda>0 \\ \lambda\mathbf{m} &= \lambda(m;\alpha,\beta)_{LR} = (\lambda m; -\lambda\beta, -\lambda\alpha)_{RL} \text{ for } \lambda<0 \end{aligned}\right\} \quad (14.134)$$

Product

Let **m** and **n** be two fuzzy numbers and (**m.n**) be the product of these fuzzy numbers, then its membership function $\mu_{\mathbf{mn}}(z)$ will be given by

$$\mu_{\mathbf{m}\cdot\mathbf{n}}(z) = \max_{xy=z} \min \left(\mu_{\mathbf{m}}(x),\ \mu_{\mathbf{n}}(y)\right) \tag{14.135}$$

However, if $\mathbf{m} = (m;\alpha,\beta)_{LR}$ and $\mathbf{n} = (n;\gamma,\delta)$, the fuzzy number (**m.n**) is *not* a L-R fuzzy number. If we neglect a term such as $\alpha\gamma$, provided α and γ are small compared to m and n, then an approximate formula of the type

$$(m;\alpha,\beta).(n;\gamma,\delta) \approx (mn;\ m\gamma+n\alpha,\ m\delta+n\beta) \tag{14.136}$$

follows. As a result of this one could write

$$(\mathbf{m})^2 \approx (m^2;\ 2m\alpha,\ 2m\beta) \tag{14.137}$$

and more generally, for *pth* power of **m**

$$(\mathbf{m})^p \approx (m^p;\ pm\alpha,\ pm\beta) \tag{14.138}$$

From product equation, it is also clear, that the fuzziness of a product is greater than that of each of multiplication factor.

Quotient

Before we define quotient of two fuzzy numbers of the L-R type, we would like to define the *inverse* of a fuzzy number $\mathbf{m} = (m;\alpha,\beta)$

Inverse

$$1/\mathbf{m} = 1/(m;\ \alpha,\ \beta) \approx (1/m;\ \beta/m^2,\ \alpha/m^2) \tag{14.139}$$

and *quotient*

$$\mathbf{m}/\mathbf{n} = (m;\ \alpha,\ \beta)/(n;\ \gamma,\ \delta) = \left(\frac{m}{n}\ ;\ \frac{\delta m+\alpha n}{n^2},\ \frac{\gamma m+\beta n}{n^2}\right) \tag{14.140}$$

14.9.2 Extended Min and Max Operations

Using Zadeh's extension principle of section 14.5.2, the membership function of the **extended max** operation on **m** and **n**, viz., **max(m,n)** is given by

$$\mu_{\mathbf{max}(\mathbf{m},\mathbf{n})}(z) = \sup_{z=\max(x,y)} \min \left(\mu_{\mathbf{m}}(x),\ \mu_{\mathbf{n}}(x)\right) \tag{14.141}$$

Let us assume that **m** and **n** are two fuzzy numbers as shown in Fig.14.17. Then **max(m,n)** is a fuzzy number whose membership function appears as a dotted line and those of **min(m,n)** is given by the line given in dashes as is shown in Fig.14.17.

for $z \le z_1$, the *least upper bound* is reached for x=z and $x \le z_1$

$$\therefore\ \mu_{\mathbf{max}(\mathbf{m},\mathbf{n})}(z) = \mu_{\mathbf{n}}(z) \text{ if } z \le z_1 \tag{14.142}$$

for $z_1 \le z \le z_2$, the least upper bound is reached for x = z and $z_1 \le y \le z_2$.

$$\therefore \mu_{\mathbf{max}(\mathbf{m},\mathbf{n})}(z) = \mu_{\mathbf{m}}(z) \text{ if } z_1 \le z \le z_2 \tag{14.143}$$

for $z_2 \le z \le z_3$, $\mu_{\mathbf{m}}(z) \ge \mu_{\mathbf{n}}(z)$ and $\exists(x,y)$, $\mu_{\mathbf{n}}(y) = 1$ (14.144)

and $z = \mathbf{max}(x,y)$. Here $x=z$; $y=n$

$$\mu_{\mathbf{max}(\mathbf{m}\cdot\mathbf{n})}(z) = \mu_{\mathbf{m}}(z) \text{ if } z_2 \le z \le z_3$$

for $z \ge z_3$; $\mu_{\mathbf{m}}(z) \le \mu_{\mathbf{n}}(z)$ and the least upper bound is reached for $y=z$ and $x \le z_3$

$$\therefore \mu_{\mathbf{max}(\mathbf{m},\mathbf{n})}(z) = \mu_{\mathbf{n}}(z) \text{ if } z \ge z_3 \tag{14.145}$$

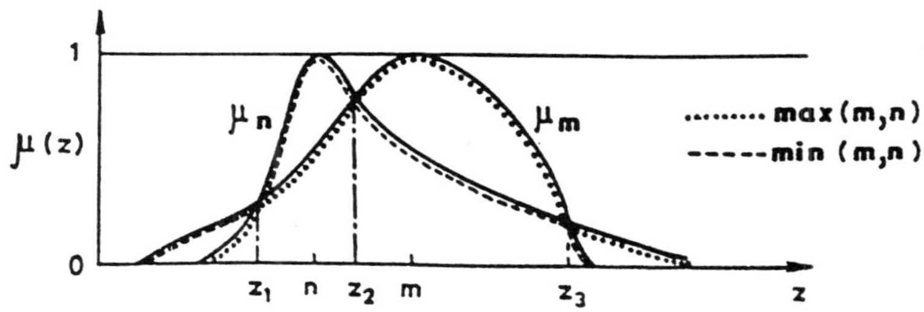

Fig. 14.17 Extended Min and Max Operations on Fuzzy Numbers **m** and **n**

Thus **max**(**m**,**n**) is a fuzzy number different from **m** and **n**. The calculation of **max**(**m**,**n**) is easy if **m**, **n** are *convex normalized fuzzy numbers*. Further **min** and **max** are idempotent, associative, commutative, mutually distributive.

14.10 FUZZY OPERATORS

In this section, we will discuss several fuzzy operators that are used to combine two fuzzy sets, whose some arbitrary elements of the universal set are given by $a = \mu_A(x)$ and $b = \mu_B(x)$. However, they must satisfy certain conditions as given in the following section.

t-Norms and t-Conorms

A function T (a two place real-valued function) whose domain is a unit square $[0,1] \times [0,1]$ i.e.,

$$T : [0,1] \times [0,1] \longrightarrow [0,1]$$

is called a *triangular norm* (or in short *t-norm*) if it satisfies the following Conditions :

$$\left.\begin{array}{lll} T_1 : T(0,0) = 0;\ T(a,1) = T(1,a) = a & & \textbf{(boundary conditions)} \\ T_2 : T(a,b) \le T(c,d) \text{ whenever } a \le c;\ b \le d & & \textbf{(monotonicity)} \\ T_3 : T(a,b) = T(b,a) & & \textbf{(symmetry)} \\ T_4 : T\big(a, T(b,c)\big) = T\big(T(a,b),c\big) & & \textbf{(associativity)} \end{array}\right\} \tag{14.146}$$

It is thus obvious that T is a *semigroup* of [0,1] with identity 1. If T is a t-norm, the function S,

$$S : [0,1] \times [0,1] \longrightarrow [0,1]$$

defined by

$$S(a,b) = 1 - T\big((1-a),(1-b)\big) \tag{14.147}$$

is called its *t-conorm* (or dual of t-norm). It can be easily seen that conditions T_2, T_3 and T_4 along with the condition that

$$S_1 : S(1,1) = 1 \text{ and } S(a,0) = S(0,a) = a \tag{14.148}$$

in place of T_1 of (14.146) are satisfied by S(a,b). Any t-conorm can be generated from t-norm using transformation defined in (14.147). Converse also holds good.

Some of the important *t-norms* and *t-conorms* are :

$$\left.\begin{aligned} T_m(a,b) &= a \wedge b = \mathbf{min}(a,b) \\ S_m(a,b) &= a \vee b = \mathbf{max}(a,b) \end{aligned}\right\} \text{Zadeh's min and max operators} \tag{14.149}$$

$$\left.\begin{aligned} T_p(a,b) &= a.b \\ S_p(a,b) &= a+b-a.b \end{aligned}\right\} \text{Probabilistic operators} \tag{14.150}$$

$$\left.\begin{aligned} T_b(a,b) &= \mathbf{max}(a+b-1,\ 0) \\ S_b(a,b) &= \mathbf{min}(a+b,\ 1) \end{aligned}\right\} \text{Bold operators} \tag{14.151}$$

$$\left.\begin{aligned} T_w(a,b) &= \begin{cases} \mathbf{min}(a,b); & \text{if } \mathbf{max}(a,b) = 1 \\ 0 \quad ; & \text{if } \mathbf{max}(a,b) < 1 \end{cases} \\ S_w(a,b) &= \begin{cases} \mathbf{max}(a,b); & \text{if } \mathbf{min}(a,b) = 0 \\ 1 \quad ; & \text{if } \mathbf{min}(a,b) > 0 \end{cases} \end{aligned}\right\} \text{Weak operators} \tag{14.152}$$

$$\left.\begin{aligned} T_{ss}(a,b) &= \mathbf{max}\big(0,\ a^{-p} + b^{-p} - 1\big)^{-1/p} \\ S_{ss}(a,b) &= 1-\mathbf{max}\Big(0,\ \big((1-a)^{-p}+(1-b)^{-p} - 1\big)\Big)^{-1/p} \end{aligned}\right\} \text{Schweizer and Sklar family of operators [22]} \tag{14.153}$$

where, $p \in (-\infty, \infty)$

Further,

$$\left.\begin{aligned} T_s(a,b) &= \mathbf{max}\big((1+\lambda)(a+b-1)-\lambda ab,\ 0\big) \\ S_s(a,b) &= \mathbf{min}(a+b+\lambda ab,\ 1)\ ;\ \lambda \ge -1 \end{aligned}\right\} \text{Sugeno's family of operators [23]} \tag{14.154}$$

$$\left.\begin{aligned} T_h(a,b) &= \frac{a.b}{\nu+(1-\nu)(a+b-a.b)}\ ;\ \nu \ge 0 \\ S_h(a,b) &= \frac{(\nu-2)(ab)+a+b}{1-(1-\nu)ab} \end{aligned}\right\} \text{Hamacher's family of operators [24]} \tag{14.155}$$

Similarly, for $\nu = 1$, we obtain $T_h(a,b) = T_p(a,b)$

$$T_F(a,b) = \log_k\left(1+\frac{(k^a-1)(k^b-1)}{k-1}\right); \text{ for } 0\le k\le\infty,\ k\ne 1$$

$$S_F(a,b) = 1-\log_k\left(1+\frac{(k^{1-a}-1)(k^{1-b}-1)}{k-1}\right)$$

Frank's Family of operators [25] (14.156)

It can be seen that for Frank's family, the following holds good.

$$T_F(a,b)_{k=0} = T_m(a,b),\quad T_F(a,b)_{k=1} = T_p(a,b) \quad\text{and}\quad T_{F\,k\to\infty} = T_b(a,b)$$

Frank [25] has also claimed that, his family is the only set of t-norms to satisfy the equation,

$$T(a,b) + S(a,b) = a+b \tag{14.157}$$

In addition to these, we have

$$T_y(a,b) = 1-\mathbf{min}\left(1, ((1-a)^p+(1-b)^p)^{1/p}\right) \;;\; p\ge 1$$

$$S_y(a,b) = \mathbf{min}\left(1, (a^p+b^p)^{1/p}\right)$$

Yager's family of operators [26] (14.158)

It can be shown that for $p=1$, $T_y(a,b) \longrightarrow T_b(a,b)$
and for $p\to\infty$, $T_y(a,b) \longrightarrow T_m(a,b)$

Further,

$$T_d(a,b) = \frac{a.b}{\mathbf{max}(a,b,\alpha)} \;;\; \alpha\in[0,1]$$

$$S_d(a,b) = \frac{a+b-ab-\mathbf{min}(a,b,1-\alpha)}{\mathbf{max}\ (1-a,1-b,\alpha)}$$

Dubois and Prade's family of operators [27] (14.159)

For this family,

for $\alpha=0$, $T_d(a,b) = T_m(a,b)$

and for $\alpha=1$, $T_d(a,b) = T_p(a,b)$

and finally,

$$T_{do}(a,b) = \frac{1}{1+\left[\left(\frac{1}{a}-1\right)^{\lambda}+\left(\frac{1}{b}-1\right)^{\lambda}\right]^{1/\lambda}}$$

$$S_{do}(a,b) = \frac{1}{1+\left[\left(\frac{1}{a}-1\right)^{-\lambda}+\left(\frac{1}{b}-1\right)^{-\lambda}\right]^{-1/\lambda}}$$

Dombi's family of operators [28] (14.160)

where $\lambda\in[0,\infty)$

The following observations can be made about the operators discussed above.

(i) The only *t-norm* which satisfies the condition

$$T(a,a) = a$$

is $T_m(a,b)$. So also is $S_m(a,b)$, which is the only *t-conorm* satisfying $S(a,a)=a$.

(ii) If we define partial ordering on t-norms assuming that we have two t-norms say T and T′ and if for all $(a,b)\in I\times I$ (I being the closed unit interval [0,1]), then

$$T(a,b) \le T'(a,b)$$

Based on this definition, we can show that

$$T_w \le T \le T_m$$

Thus T_w and T_m are the minimal and maximal elements of this partial order.

Further, $T_b \le T_p$

(iii) For Sugeno's family, for $\lambda' > \lambda''$

$$T_{s(\lambda')} \le T_{s(\lambda'')}$$

(iv) For Hamacher's family, for $\nu' > \nu''$

$$T_{h(\nu')} \le T_{h(\nu'')}$$

(v) For Frank's family, for $k' > k''$

$$T_{F(k')} \le T_{F(k'')}$$

(vi) For Yager's family, if $p' > p''$ then $T_{y(p')} \ge T_{y(p'')}$

(vii) For Dubois and Prade's family, for $\alpha' > \alpha''$

$$T_{d(\alpha')} \le T_{d(\alpha'')}$$

The importance of t-norms and t-conorms lies in the fat that they extend the set operations of *intersection* and *union* from *crisp* to *fuzzy sets*. If the intersection of two fuzzy sets **A** and **B** is defined as fuzzy set **C** and if it a *t-norm* that is

$$\mathbf{C}(x) = T\big(\mathbf{A}(x),\ \mathbf{B}(x)\big)$$

then we have an acceptable definition of interaction in fuzzy set theory. The **min** and **max** operators are the only t-norm and t-conorm, which have a property

$$\mathbf{min}(a,a) = a \quad \text{and} \quad \mathbf{max}(a,a) = a$$

besides their being *distributive* with respect to each other. For any t-norm, $T(a,b) \le \mathbf{min}(a,b)$ and thus **min** offers the *least restrictive* possible definition of *intersection*.

Whereas for any t-conorm,

$$S(a,b) \ge \mathbf{max}(a,b)$$

and thus **max** provides the *most conservative union*.

14.11 FUZZY MEASURES

In the preceding sections, we were intereseted in the grade of membership $\mu_A(x)$ of a known element $x \in X$ in a set **A** without a precise boundary, whereas in this section we are concerned with subjectively assessing or guessing the degree of belonging of an apriori non-located element in X to a subset **A** of X. **A** may or may not be fuzzy. Such an uncertainty can be expressed more conveniently using probabilities.

Non-fuzzy Subsets

Let g be a function from $\mathcal{P}(X)$ to [0,1], where $\mathcal{P}(X)$ is set of fuzzy sets on X.

Then, g is said to be a fuzzy measure iff :

$$\left.\begin{array}{lll} \text{(i)} & g(\phi) = 0;\ g(X) = 1; & \textbf{(boundary conditions)} \\ \text{(ii)} & \forall \mathbf{A}, \mathbf{B} \in \mathcal{P}(X), \text{ if } \mathbf{A} \subseteq \mathbf{B}, \text{ then } g(\mathbf{A}) \le g(\mathbf{B}) & \textbf{(monotonicity)} \\ \text{(iii)} & \text{if } \forall i \in N,\ \mathbf{A}_i \in \mathcal{P}(X) \text{ and } (\mathbf{A}_i)_i \text{ is} & \\ & \text{monotonic (i.e., } \mathbf{A}_1 \subseteq \mathbf{A}_2 \subseteq \mathbf{A}_3 \subseteq \ldots \subseteq \mathbf{A}_n \text{ or} & \\ & \mathbf{A}_1 \supseteq \mathbf{A}_2 \supseteq \mathbf{A}_3 \supseteq \ldots \supseteq \mathbf{A}_n \ldots), \text{ then} & \\ & \lim\limits_{i\to\infty} g(\mathbf{A}_i) = g(\lim\limits_{i\to\infty} \mathbf{A}_i) & \textbf{(continuity)} \end{array}\right\} \quad (4.161)$$

where g is associated with a non located element x of X and g(**A**) is called a *grade of fuzziness* of **A**. It expresses an evaluation of the statement *x belongs to* **A** in a situation in which one subjectively guesses whether x is within **A** ? The monotonicity of g means that $x \in \mathbf{A}$ is less certain then $x \in \mathbf{B}$ when $\mathbf{A} \subseteq \mathbf{B}$.

Also $\quad \forall\ \mathbf{A}, \mathbf{B} \in \mathcal{P}(X);\ g(\mathbf{A} \cup \mathbf{B}) \ge \mathbf{max}\big(g(\mathbf{A}),\ g(\mathbf{B})\big)$

and $\quad \forall\ \mathbf{A}, \mathbf{B} \in \mathcal{P}(X);\ g(\mathbf{A} \cap \mathbf{B}) \le \mathbf{min}\big(g(\mathbf{A}),\ g(\mathbf{B})\big)$

Probability Measure

P is a probability measure iff:

(i) $\forall\ \mathbf{A},\ P(\mathbf{A}) \in [0,1];\ P(X) = 1$

(ii) if $\forall\ i \in N,\ \mathbf{A}_i \in \mathcal{P}(X)\big(N \overset{\Delta}{=} \text{set of natural integers}\big)$
and $\forall i \ne j\ \mathbf{A}_i \cap \mathbf{A}_j = \phi$, then

$$P(\bigcup_{i\in N} \mathbf{A}_i) = \sum_{i=1}^{\infty} P(\mathbf{A}_i) \quad (14.162)$$

P is, therefore, a fuzzy measure.

Dirac Measure

A Dirac measure is a fuzzy measure μ defined by

$$\forall\ \mathbf{A} \in \mathcal{P}(X),\ \mu(\mathbf{A}) = \begin{cases} 1\ ; & \text{iff } x_o \in \mathbf{A} \\ 0\ ; & \text{otherwise.} \end{cases} \quad (14.163)$$

where x_o is a given element in X and μ is the membership value of x_o in any subset of X.

λ-Fuzzy Measure

If $\forall\ \mathbf{A},\ \mathbf{B} \in \mathcal{P}(X)$ such that $\mathbf{A} \cap \mathbf{B} = \phi$, then

$$g_\lambda(\mathbf{A} \cup \mathbf{B}) = g_\lambda(\mathbf{A}) + g_\lambda(\mathbf{B}) + \lambda g_\lambda(\mathbf{A}) g_\lambda(\mathbf{B}), \quad \text{for } \lambda > -1 \quad (14.164)$$

14.11.1 Belief Function

A belief function *Bel* is a measure on X finite, iff:

(i) $Bel(\phi) = 0$, $Bel(X) = 1$; $\forall\ \mathbf{A} \in \mathcal{P}(X)$, $0 \le Bel(\mathbf{A}) \le 1$;

(ii) $\forall\ \mathbf{A}_1, \mathbf{A}_2, \ldots, \mathbf{A}_n \in \mathcal{P}(X)$,

$$Bel(\mathbf{A}_1 \cup \mathbf{A}_2 \cup \ldots \cup \mathbf{A}_n) \ge \sum_{i=1}^{n} Bel(\mathbf{A}_i) - \sum_{i<j} Bel(\mathbf{A}_i \cap \mathbf{A}_j) + \ldots + (-1)^{n+1} Bel(\mathbf{A}_1 \cap \mathbf{A}_2 \cap \ldots \cap \mathbf{A}_n) \qquad (14.165)$$

$Bel(\mathbf{A})$ is known as a *grade of belief* that a given element of X belongs to **A**.

Also, $\qquad Bel(\mathbf{A}) + Bel(\bar{\mathbf{A}}) \le 1,$

which means that a *lack of belief* in $x \in \mathbf{A}$ does not imply a *strong belief* in $x \in \bar{\mathbf{A}}$. Particularly, a total ignorance is modelled by the *belief function* Bel_i such that

$$Bel_i(\mathbf{A}) = 0, \text{ if } \mathbf{A} \neq X \text{ and } Bel_i(\mathbf{A}) = 1 \text{ if } \mathbf{A} = X.$$

A probability measure is a special case of the belief function. Belief functions are fuzzy measures. Also a *λ-fuzzy measure* is a belief function iff $\lambda \ge 0$.

Consonant Belief Function

A *consonant belief function* (CBF) is a belief function whose focal elements, $\mathbf{A}_1, \mathbf{A}_2, \ldots, \mathbf{A}_n$ are nested i.e., $A_1 \subset A_2 \subset \ldots \subset A_n$. A CBF is a λ-fuzzy measure *iff* it is a Dirac measure. This definition of CBF is equivalent to

$$\left.\begin{array}{l} \text{(i)}\ \ Bel(\phi) = 0;\ Bel(X) = 1 \\ \text{(ii)}\ Bel(\mathbf{A} \cap \mathbf{B}) = \mathbf{min}\big(Bel(\mathbf{A}),\ Bel(\mathbf{B})\big) \end{array}\right\} \qquad (14.166)$$

Other properties of CBF are :

$$\left.\begin{array}{l} \forall \mathbf{A},\ \mathbf{min}\big(Bel(\mathbf{A}),\ Bel(\bar{\mathbf{A}})\big) = 0 \\ \forall\ Bel,\ \exists\ \mathbf{A},\ \mathbf{B},\ Bel\big(\mathbf{A} \cup \mathbf{B}\big) > \mathbf{max}\big(Bel(\mathbf{A}),\ Bel(\mathbf{B})\big) \\ \qquad\qquad \text{except if } Bel \text{ is a Dirac measure} \end{array}\right\} \qquad (14.167)$$

$Bel(\bar{\mathbf{A}})$ was interpreted by Shackle [29] as a *potential grade of surprise*. Particular cases of CBF are *certainty measures* such that

$$\exists\ Y_o \subseteq X;\ c(\mathbf{A}) = 1 \text{ if } \mathbf{A} \supseteq Y_o, \text{ and 0 otherwise.}$$

14.11.2 Plausibility Measure

The *plausibility* of a subset of **A** of X (finite) has been defined by Shafer [30], as

$$Pl(A) = 1 - Bel(\bar{\mathbf{A}});\ \text{where } Bel(\mathbf{A}) \text{ is a belief function.}$$

A *plausibility* measure satisfies the following axioms :

(i) $Pl(\phi) = 0$; $Pl(X) = 1$

(ii) $\forall\ \mathbf{A}_2, \mathbf{A}_2, \ldots, \mathbf{A}_n \subseteq X$

$$Pl(\mathbf{A}_1 \cap \mathbf{A}_2 \cap \ldots \cap \mathbf{A}_n) \le \sum Pl(\mathbf{A}_i) - \sum Pl(\mathbf{A}_i \cup \mathbf{A}_j) + \ldots + (-1)^{n+1} Pl(\mathbf{A}_1 \cup \mathbf{A}_2 \cup \ldots \cup \mathbf{A}_n) \quad (14.168)$$

Plausibility measures are particular cases of fuzzy measures since

$$\left.\begin{array}{l} \forall\ \mathbf{A} \subseteq X,\ \forall\ \mathbf{B} \subseteq X;\ Pl(\mathbf{A} \cup \mathbf{B}) \le Pl(\mathbf{A}) + Pl(\mathbf{B}) - Pl(\mathbf{A} \cap \mathbf{B}) \\ \text{For } \mathbf{C} \subseteq \mathbf{A} \text{ and } \mathbf{B} = \mathbf{C} \cup \bar{\mathbf{A}},\ \mathbf{A} \cup \mathbf{B} = X,\ \mathbf{A} \cap \mathbf{B} = \mathbf{C} \\ \text{and } 1 \le Pl(\mathbf{A}) + Pl(\mathbf{B}) - Pl(\mathbf{C}) \text{ and since } Pl(\mathbf{B}) = 1, \\ \therefore\ Pl(\mathbf{A}) \ge Pl(\mathbf{C}). \text{ Further } Pl(\mathbf{A}) + Pl(\bar{\mathbf{A}}) \ge 1. \end{array}\right\} \quad (14.169)$$

Actually, plausibility measures and belief functions were introduced by Dempster [31] under the names *upper* and *lower probabilities* induced from a probability measure by a multi-valued mapping.

A plausibility measure is a λ-fuzzy measure *iff* :

$$-1 < \lambda \le 0$$

14.11.3 Possibility Measure

A possibility measure [32] π, is a function from $\mathcal{P}(X)$ to $[0,1]$ such that

$$\left.\begin{array}{l} \text{(i)}\ \ \Pi(\phi) = 0,\ \Pi(X) = 1 \\ \text{(ii) for any } \mathbf{A}_i \text{ subsets of } X,\ \Pi(\cup_i \mathbf{A}_i) = \sup_i \Pi(\mathbf{A}_i) \end{array}\right\} \quad (14.170)$$

A possibility measure can be constructed from a possibility distribution, i.e., a function π from X to [0,1] such that

$$\left.\begin{array}{l} \sup_{x \in X} \pi(x) = 1 \qquad \textbf{(normalization condition)} \\ \text{More specifically,} \\ \forall\ \mathbf{A},\ \Pi(\mathbf{A}) = \operatorname{Sup}_{x \in \mathbf{A}} \pi(x) \end{array}\right\} \quad (14.171)$$

provides *possibility distribution* from the knowledge of π, for denumerable universes X, at least for $\pi(x) = \Pi(\{x\})$ A possibility measure is a belief function *iff* it is a Dirac measure. Also a possibility measure is λ-fuzzy measure *iff* it is a Dirac measure. Further it is not difficult to visualize that a possibility measure on X (finite) is a plausibility measure. Further π can be viewed as a membership function of a fuzzy set.

If g be a function from $\mathcal{P}(X)$ to [0,1] such that

$$\left.\begin{array}{l} \text{(i)}\ \ g(\phi) = 0;\ g(X) = 1 \\ \text{(ii)}\ \forall\ \mathbf{A} \subseteq X,\ \forall\ \mathbf{B} \subseteq X, \text{ if } \mathbf{A} \cap \mathbf{B} = \phi \end{array}\right\} \quad (14.172)$$

then $g(\mathbf{A} \cup \mathbf{B}) = \mathbf{max}\big(g(\mathbf{A}),\ g(\mathbf{B})\big)$

when X finite, g is a possibility measure. Further, $\prod_m(\mathbf{A}) = 1-Bel_1(\mathbf{A})$ (i.e., $\prod_m(\mathbf{A}) = 1$ *iff* $\mathbf{A} \neq \phi$) is called *maximum possibility measure.*

Further, a crisp possibility measure can also defined by

$$\exists\ Y_o \subseteq X,\ c\prod(\mathbf{A}) = 1,\ \text{if } \mathbf{A} \cap Y_o \neq \phi \tag{14.173}$$

and $$c\prod(\mathbf{A}) = 0,\ \text{otherwise,}$$

where $c\prod(\mathbf{A}) = 1-c(\bar{\mathbf{A}})$ and c is a *certainty measure.*

A constant belief function *Bel* can be built from a possibility measure by setting

$$\prod(\mathbf{A}) = 1-Bel(\bar{\mathbf{A}})$$

Hence, $$Bel(\mathbf{A}) = 1-\sup_{x\notin\mathbf{A}} \pi(x) = \inf_{x\notin\mathbf{A}}\big(1-\pi(x)\big) = \inf_{x\notin\mathbf{A}} \nu(x) \tag{14.174}$$

By analogy with modal logic, where **A** *is necessary*, is equivalent to *not-A is not possible.* Therefore CBF is a *necessity measure.* There are many other fuzzy measures. For a more detailed information, one can refer to [2 to 12].

14.12 CLASSIFICATION OF EVENTS AND PROBABILITIES

For extending the concept of events and probabilities to fuzzy set theory area, we would need to consider the following cases :

1. Event is *nonfuzzy* but the probability is *fuzzy.*
2. Event is *fuzzy* but the probability is *nonfuzzy.*
3. Event is *fuzzy* and also the probability is *fuzzy.*

14.12.1 Fuzzy Probability of a Nonfuzzy Event

In this case, it is assumed that the event is well-defined but the data pertaining to the occurrence of this event is not so reliable as to compute an exact value of the probability. Instead we define a spectrum of probability values and associated with them a membership function value (which are often called as *fuzzy grades*). These types of FS have already been introduced as *FS of type 2.* Such considerations are often encountered in reliability studies, where the failure data on components may *not* be very precisely known or the component failure data is known in a particular type of environment and the component is being used in another. Further, it is also possible that human errors have not been appropriately selected. The influence of failure of one component on the other may not have been fully understood or considered.

In all such cases, we would be right if we do provide a band of probabilities with our degree of judgement of their appropriateness through a membership grade function. Such a case is illustrated in Fig. 14.18, which shows two representations of a fuzzy probability. On abscissa, we show the likelihood of occurrence of an event and on the ordinate is shown the degree of membership of our trust in that value of probability. It may be noted that the abscissa is the probability of an event which can only have the values between zero and one. In this way, we can consider the uncertainties about the data through the concept of a fuzzy probability.

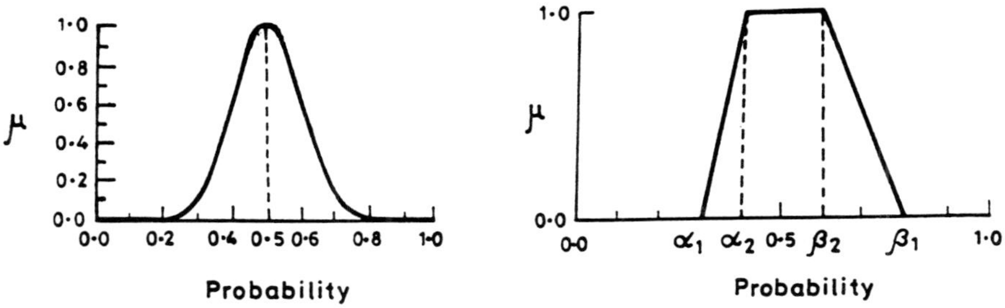

Fig. 14.18 Fuzzy Probability Representation of Events

14.12.2 Non Fuzzy Probability of a Fuzzy Event

Before we consider the other two cases individually, we would need to define a fuzzy event and the associated terms.

Fuzzy Event

Let us assume that X is the Euclidean n-space R^n. A fuzzy event in R^n is a fuzzy set **A** on R^n whose membership function is measurable.

Probability of a Fuzzy Event

Let $\mathcal{B}$ be a Borel field in R^n and P a probability measure on $\mathcal{B}$. The probability of a fuzzy event **A** is defined by the L-S integral

$$P(\mathbf{A}) = \int_{R^n} \mu_A(x)dP = E\{\mu_A(x)\} \qquad (14.175)$$

Thus the probability of a fuzzy event is the expectation of its membership function. P(**A**) evaluates the degree with which the sample set R^n has the fuzzy property **A**. The corresponding experiment is a random selection of element x_i *more or less* belonging to **A**. At each trial a membership value $\mu_A(x_i)$ is provided.

Then
$$P(\mathbf{A}) = \lim_{n_t \to \infty} \frac{\sum_{i=0}^{n_t} \mu_A(x_i)}{n_t} \qquad (14.176)$$

where n_t is the number of trials. Thus, P(**A**) can be interpreted a proportion of elements of R^n *belonging to* **A**.

Independence

Two fuzzy events **A** and **B** are independent (statistically) if

$$P(\mathbf{A}.\mathbf{B}) = P(\mathbf{A})\ P(\mathbf{B}) \qquad (14.177)$$

It may be noted that in ordinary set theory, we would have had the condition P(**A** ∩ **B**) = P(**A**) P(**B**) for independence. But in the FST, we have its equivalent as P(**A.B**). This is due to the following reasons :

Let $X_1 = R^m$, $X_2 = R^p$ and the P be the product measure $P_1 \times P_2$ where P_1 and P_2 are probability measures on X_1 and X_2, respectively. Let $\mathbf{A}_1$ and

$\mathbf{A}_2$ be the events in X_1 and X_2 characterised by their membership functions, $\mu_{A_1}(x^1,x^2) = \mu_{A_1}(x^1)$ and $\mu_{A_2}(x^1,x^2)=\mu_{A_2}(x^2)$, respectively. Then $\mathbf{A}_1$ and $\mathbf{A}_2$ are independent events. This would not be true if independence were defined in terms of $P(\mathbf{A} \cap \mathbf{B})$.

Conditional Probability

The conditional probability of a fuzzy events **A** given **B** is defined by $P(\mathbf{A}|\mathbf{B}) = P(\mathbf{A}.\mathbf{B})/P(\mathbf{B})$ provided $P(\mathbf{B})>0$. It may be noted that if **A** and **B** are independent,

then, $P(\mathbf{A}|\mathbf{B}) = P(\mathbf{A})$ and $P(\mathbf{B}|\mathbf{A}) = P(\mathbf{B})$ as in the nonfuzzy case.

Further, it can be easily seen that where $\mathcal{P}(R^n)$ is set of fuzzy set in R^n,

$$\forall\ \mathbf{A},\mathbf{B} \in \mathcal{P}(R^n) \qquad (\mu_A,\ \mu_B \text{ are measurable})$$

$$\text{if } \mathbf{A} \subseteq \mathbf{B} \text{ then } P(\mathbf{A}) \le P(\mathbf{B}); \tag{14.178}$$

$$P(\mathbf{A}\dot{+}\mathbf{B}) = P(\mathbf{A}) + P(\mathbf{B}) - P(\mathbf{A}.\mathbf{B}) \tag{14.179}$$

$$P(\mathbf{A} \cup \mathbf{B}) = P(\mathbf{A}) + P(\mathbf{B}) - P(\mathbf{A} \cap \mathbf{B}) \tag{14.180}$$

Mean

$$\text{The mean is defined as } \frac{1}{P(\mathbf{A})}\int_{R^n} x\mu_A(x)dP. \tag{14.181}$$

Probability in Multidimensional Space

Expression given in (14.175) provides the probability in one dimensional space only. However the probability in a multidimensional space can be written as follows :

A probability space is assumed to be triplet $(R^n, \mathcal{B}, P)$, whose $\mathcal{B}$ is the σ field of Borel sets in R^n and P is a probability measure over R^n. Let a point in R^n be denoted by $\vec{x}$.

Let a set $\mathbf{A} \in \mathcal{B}$. Then defining a characteristic function of the set **A** by $\mu_A(\vec{x})$: $R^n \longrightarrow [0,1]$, we have the probability of **A** given as

$$P(\mathbf{A}) = \int_{R^n} \mu_A(\vec{x})dP(\vec{x}) \tag{14.182}$$

where $\int_{R^n} dP(\vec{x}) = 1$ or $P(\mathbf{A}) = E\{\mu_A\}$ as before.

14.12.3 Fuzzy Probability of a Fuzzy Event

Let X=R and p(x) be a probability distribution and $(\mathbf{A},\mathbf{B})$ a fuzzy interval bounded by two non-overlapping convex normalized fuzzy sets on R. Then the fuzzy probability of a fuzzy event [33], *x belongs to the fuzzy interval* $(\mathbf{A},\mathbf{B})$ is $FP[(\mathbf{A},\mathbf{B})]$

$$FP[(\mathbf{A},\mathbf{B})] = \int_{\mathbf{A}}^{\mathbf{B}} p(x)dx = P(\mathbf{B}) \ominus P(\mathbf{A}) \tag{14.183}$$

when **A** and **B** are ordinary numbers, the above expression becomes the usual definition of probability if $x \in [\mathbf{A},\mathbf{B}]$. Here, the result is a fuzzy set of [0,1], which can be interpreted as *linguistic probability*.

14.13 POSSIBILITY THEORY

Possibility theory was proposed by Zadeh [32] in 1976 as a mathematical formism complementary to the probability theory, which deals with possibility measures that form a distinct subset of fuzzy measures. In many situations, the information concerning an uncertainty is not so strong enough as to make some probabilistic statements. In other cases, the information does not have a character of a probability. In both the above cases, it could then be more appropriate to take recourse to possibility theory, which has been shown to be adequate and effective in handling such situations.

It would be proper to briefly outline some of the important definitions and explanations needed for understanding the possibility theory and its applications to reliability evaluation and risk assessment problems.

14.13.1 Possibility

Let A be a nonfuzzy set of X and v be a variable on X such that when v takes its value in A, it would be equivalent to saying that any element of A can take a value of v and that any element *not* in A cannot have a value of v. The statement *v takes its value in A* induces a possibility distribution π over X associating with each element x the possibility that x is a value of v.

$$\Pi(v=x) = \pi(x) = \begin{cases} 1 & ; \text{ if } x \in A \\ 0 & ; \text{ otherwise} \end{cases} \qquad (14.184)$$

Next, we assume that **A** is a fuzzy set that acts as a fuzzy restriction on the possible value of v. In other words, **A** induces a possibility distribution that is equal to $\mu_{\mathbf{A}}$ on the value of v:

$$\Pi(v=x) = \pi(x) = \mu_{\mathbf{A}}(x) \qquad (14.185)$$

Although a fuzzy set and a possibility distribution have the same mathematical expression, a fuzzy set **A** should be viewed as a fuzzy value that we assign to a variable whereas the possibility restriction **A** is the fuzzy set of nonfuzzy values that can possibly be assigned to v. The π signifies that *it is impossible that v belongs to the complement of* **supp A** *in* X. That is, *it is necessary that v belongs to* **supp A**; because it is sure that v takes its values on X and only on X. However, v can be any element of **A** with a given possibility.

14.13.2 Possibility of a Nonfuzzy Event

If π is a possibility distribution induced by a fuzzy set F in X and A is a nonfuzzy set of X, the possibility that x belongs to A is $\Pi(A)$, where Π is the possibility measure induced by π and

$$\Pi(A) = \sup_{x \in A} \mu_F(x) = \sup_{x \in A} \Pi(x) \qquad (14.186)$$

It would be interesting to compare this with the probability of the event. If p is a probability distribution over X, the probability that x belongs to A is

$$P(A) = \int_A dP = \begin{cases} \int_A p(x)dx & ; \text{ if X is real} \\ \sum_{x \in A} p(x) & ; \text{ if X is finite} \end{cases}$$

In fact, **sup** in (14.186) acts as a Lebesgue integral. It is actually a fuzzy integral of Sugeno [33],

$$\prod(A) = \oint_X \mu_A(x) \circ \prod(\cdot) \tag{14.187}$$

In fact, possibility signifies the *degree of feasibility* or *ease of attainment*, whereas the probability is the degree of likelihood, belief, frequency or proportion.

14.13.3 Consistency Principle

This can be best understood by stating that:

(i) Whatever is *possible*, *may not be probable* and what is *improbable* need *not be impossible*.
(ii) Whatever is *impossible* is certainly *improbable*.

In otherwords, this principle states that the *degree of possibility* of an event is greater than or equal to *its degree of probability*, which itself should be greater than or equal to its *degree of necessity* (CBM as based on (14.174)). This means that $\prod(x)=1$ as soon as $P(X)=1$, which indicates that possibility distribution should be normalized, i.e.,

$$\sup_{x \in X} \prod(x) = 1 \tag{14.188}$$

14.13.4 Conditional Possibility

Let X and Y be two universes and u, v be two variables with a possibility distribution associated with them. Let $\pi_u(x)$ and $\pi_v(x)$ be the projections of $\pi_{(u,v)}(x,y)$ on X and Y, then

$$\left.\begin{aligned} \pi_u(x) &= \sup_y \pi_{(u,v)}(x,y) \\ \pi_v(x) &= \sup_x \pi_{(u,v)}(x,y) \end{aligned}\right\} \tag{14.189}$$

where $\pi_u(x)$ and $\pi_v(y)$ are said to be *marginal* possibility distributions. If $\pi_{(u,v)}(x,y)$ is separable, then we have

$$\pi_{(u,v)}(x,y) = \min\left(\pi_u(x).\pi_v(y)\right) \tag{14.190}$$

But from (14.190) and (14.189), we have

$$\pi_u(x) = \sup_y \min\left(\pi_u(x), \pi_v(y)\right)$$

which is only true if $\pi_{(u,v)}(x,y)$is separable. However, in general, we have

$$\pi_u(x) = \sup_y \min\left(\pi_{(u,v)}(x,y),\ \pi_v(y)\right) \tag{14.191}$$

$\pi_{(u,v)}(x,y)$ is interpreted as a *conditional possibility distribution.*

An analogy between conditional probability and possibility was studied by Nguyen [34], who provided the notation of *normalized possibility distribution.* Denoting such a distribution by

$$\pi(x|y) = \pi_{(u,v)}(x,y).\alpha\big(\pi_u(x).\pi_v(y)\big) \tag{14.192}$$

where $\alpha(\cdot,\cdot)$ is a normalization function, α is determined from the following two requirements :

(i) $\pi(x|y) \in [0,1]$

(ii) $\mathbf{min}\big(\pi_u(x),\ \pi_v(y)\big).\alpha\big(\pi_u(x),\pi_v(y)\big) = \pi_u(x)$

which means, if $\pi_{(u,v)}$ is separable, the normalized conditional possibility distribution equals to the projection $\pi_u(x)$. This is similar to the case of conditional probability $P(A|B)$, which equals to $P(A)$ if A and B are independent.

Parallel to the notation of independence, we have a notation of non-interactivity for possibility such as

$$\pi(x|y) = \left.\begin{array}{ll} \pi_{(u,v)}(x,y) & ;\ \text{if } \pi_u(x)\le\pi_v(y) \\ \pi_{(u,v)}(x,y)\dfrac{\pi_u(x)}{\pi_v(y)} & ;\ \text{if } \pi_u(x)>\pi_v(y) \end{array}\right\} \tag{14.193}$$

Similarly, analogous to probability expression

$$P(A) = \sum_{y\in Y} P(A|y)p(y) \qquad (Y \text{ finite})$$

we have,

$$\begin{aligned} \Pi_u(A) &= \sup_y \mathbf{min}\big(\Pi_{(u,v)}(\mathbf{A},y),\ \pi_v(y)\big) \\ &= \sup_y \mathbf{min}\big(\pi_{(u,v)}(\mathbf{A}|y),\ \pi_v(y)\big) \end{aligned} \tag{14.194}$$

where, $\quad \Pi(\mathbf{A}|y) = \sup_{x\in X} \pi(x|y)$

14.13.5 Possibility of a Fuzzy Event

The possibility of a fuzzy event $\mathbf{A}$ is a universe X with respect to the possibility measure Π and can be defined as

$$\begin{aligned} \Pi(\mathbf{A}) &= \int_{\mathbf{A}} 1 \circ \Pi(\cdot) = \int_X \mu_{\mathbf{A}}(x)\cdot\Pi(\cdot) \\ &= \sup_{x\in X} \mathbf{min}\big(\mu_{\mathbf{A}}(x),\ \pi(x)\big) \end{aligned} \tag{14.195}$$

where $\pi(x)$ is possibility distribution associated with Π. The possibility of a fuzzy event is thus the *degree of consistency* of this fuzzy event with a possibility distribution. As in nonfuzzy case, we have

$$\Pi(A \cup B) = \mathbf{max}\big(\Pi(\mathbf{A}),\ \Pi(\mathbf{B})\big) \tag{14.196}$$

and $\quad \Pi(\mathbf{A}) \le \Pi(\mathbf{B});\quad$ for $\mathbf{A} \subseteq \mathbf{B}$ $\qquad$ (14.197)

Further, two fuzzy events $\mathbf{A}$ and $\mathbf{B}$ (from universes X and Y with a possibility distribution of $\pi(x,y)$ over $X \times Y$) is said to be non-interactive *iff*

$$\Pi(A \cap B) = \min\left(\Pi(A),\ \Pi(B)\right) \tag{14.198}$$

Also Bayes' Theorem [5] holds good for possibility of fuzzy events, viz.,

$$\Pi(A,B) = \sup_{\substack{x \in X \\ y \in Y}} \min\left(\mu_A(x),\ \mu_B(y),\ \pi(x,y)\right) \tag{14.199}$$

14.13.6 Fuzzy Possibility of a Fuzzy Event

Fuzzy possibility of fuzzy event can be defined [5] through the extension principle analogous to (14.183) fuzzy probabilities.

F $\Pi[(A,B)]$ has membership function

$$\mu_{F\Pi[(A,B)]}(z) = \sup_{\substack{x,y \\ z=\max\limits_{t \in [x,y]} \Pi(t)}} \min\left(\mu_A(x),\ \mu_B(y)\right) \tag{14.200}$$

New Trends in System Reliability Evaluation
K.B. Misra (Editor)

CHAPTER 14

(Contd.)

USE OF FUZZY SETS THEORY

(PART-II : APPLICATIONS)

Takeshisa Onisawa* and Krishna B. Misra

* Institute of Engineering Mechanics
University of Tsukuba
1-1-1, Tennodai, Tsukuba 305, Japan

ABSTRACT

Since FST is necessary to deal with subjectivity and fuzziness which are inherent in the system reliability analysis, we introduce some of the important applications of FST to system reliability evaluation in this chapter. The first is fuzzy performance which is represented by a fuzzy set. The fuzzy performance is based on the consideration that the system performance can't be necessarily classified into crisp categories and that the transition from one state to the other is not sharp. The second application consists of fault tree analysis by a fuzzy probability, which is a fuzzy set defined on a probability space and by a failure possibility concept, which is a subjective unreliability measure. This approach allows the use of natural language in the system reliability analysis. Lastly, we discuss dependency considerations, environmental considerations and common cause failures in the failure possibility approach.

14.14 FST IN PERFORMANCE EVALUATION

A conventional analysis of system reliability is based on probability theory. There have been many applications of probabilistic system reliability analysis, e.g., WASH-1400[35] which is famous as risk assessment of nuclear power plants, THERP[36] which is famous as a model of human reliability. A probabilistic methodology is based on the assumption that a state is either *good* or *bad,* i.e., a *success event* or a *failure event,* and that the failure event occurs at random. A component of a system may fail at random. However an accident in a large and complex system such as a nuclear power plant, a chemical plant, does not necessarily occur at random since the system is composed of great many components and their functional relations are very complex. In practice, many accidents of large scale and complex systems have occurred frequently, e.g., accidents of nuclear power plants at Three Mile Island(1979) and Chernobyl(1986), an explosion accident of Challenger(1986), a crash accident of a JAL jumbo(1985), an accident of a chemical plant at Bhopal(1984). These accidents were evaluated to be hard to occur from the viewpoint of probability theory since their estimated probabilities were very very small. Small probability of an event does not always mean low possibility of the event.

A considerable amount of data should be collected in order to analyze system reliability probabilistically. In practice, however, it is not likely that enough amount of immediate data can be collected to estimate failure and human error probabilities.

The engineering judgement of experts, who are familiar with an analyzed system, plays an important role in the analysis of system reliability. It is said that experts' judgement holds a central position in all safety analyses of complex technical system[37]. An estimated probability of a complex system accident is usually very very small from the viewpoint of our daily life. Safety evaluation of the analyzed system is also closely related to experts' judgement. Their engineering judgement is based on their subjectivity. That is, experts' subjectivity is inherent in the area of a system reliability analysis. The probability theory does not deal with human subjectivity. Although subjective probability is considered in the probability theory, it follows probability axioms. However human subjectivity does not necessarily follow probability axioms.

As mentioned in 14.1, Zadeh[1] proposed fuzzy sets to handle human subjectivity, which cannot be represented clearly by any other formalism. Although FST has met with a great deal of criticism, it has been successfully applied to many fields, e.g., a control system, an artificial intelligence, an expert system, a decision support system, a diagnostic system[38]. In Japan, fuzzy technology has been applied even to home electronics consumer products, an automatic washing machine, a vacuum cleaner, a video camera, etc. In the field of system reliability analysis, FST finds its application to take into consideration the experts' subjectivity and fuzziness that are inherent in the analysis. Recently, there have been many applications of FST to system reliability and some related problems[39-52]. In this part of the chapter, we would introduce some of the important applications of FST to system reliability evaluation.

14.14.1 Fuzziness in the Estimate of Failure Probability and Error Probability

A failure probability and an error probability are necessary to analyze system reliability, quantitatively. Usually, these probabilities should be estimated from a large amount of data. In practice, however, it is not likely that lots of immediate data can be collected which may suffice to estimate a meaningful probability. Therefore, the failure probability and the error probability are usually estimated from reference data in other areas by experts based on their engineering and experienced judgement. Their judgement is closely related to subjectivity and fuzziness. Once hardware reliability or human reliability is expressed numerically, e.g., by the failure probability or the error probability, reliability seems to be expressed objectively and exactly. As a matter of fact, however, fuzziness is hiding behind the numerical values. Therefore, FST is necessary to deal with this kind of fuzziness.

14.14.2 Fuzziness in Consideration of Influence of Environmental Conditions

Hardware reliability and human reliability are both affected by environmental conditions, e.g., a set of conditions under which a machine is operated, or it may just be the psychological stress of the operator. That is, the estimated failure probability and error probability are not physical constants. It is, therefore, necessary to evaluate the condition under which a hardware is used, or the environmental conditions of a task, in order to modify the estimated basic probability. However, as the evaluation and the modification are performed by the experts based on their engineering judgement and experience, subjectivity and fuzziness are likely to be found in the evaluation and the modification. For example, let us consider the influence of psychological stress on human reliability. The model as shown in Table 14.1 can be used to modify the basic error probability depending on the stress level[53]. In this model the evaluation of the stress level and the determination of numerical factors to be multiplied are based on experts' engineering judgement and experience. Furthermore, in Table 14.1, the stress level is expressed by natural language. The meaning of natural language expressions cannot be expressed by a numerical value or a formula. For instance, the borderline between the moderately high level and the extremely high level is not defined clearly. Therefore, the concept of FST can be useful in this kind of consideration. The considerations outlined in 14.14.1 and 14.14.2 are shown in Fig. 14.19.

Table 14.1 Modification of Basic Error Probability Under Different Levels of Stress

Stress	Step by Step Procedures		Dynamic Interaction	
Level	Experienced	Novice	Experienced	Novice
Very Low	$p_{SM}\times 2$	$p_{SM}\times 2$	–	–
Optimum	p_{SM}	p_{SM}	p_{DM}	$p_{DM}\times 2$
Moderately High	$p_{SM}\times 2$	$p_{SM}\times 4$	$p_{DM}\times 5$	$p_{DM}\times 10$
Extremely High	0.25	0.25	0.25	0.25

* A novice is defined as a person with less than 6 months on the job in which he/she has been licensed or otherwise qualified.
* " – " means "no account".
* p_{SM} and p_{DM} are basic error probabilities of tasks.

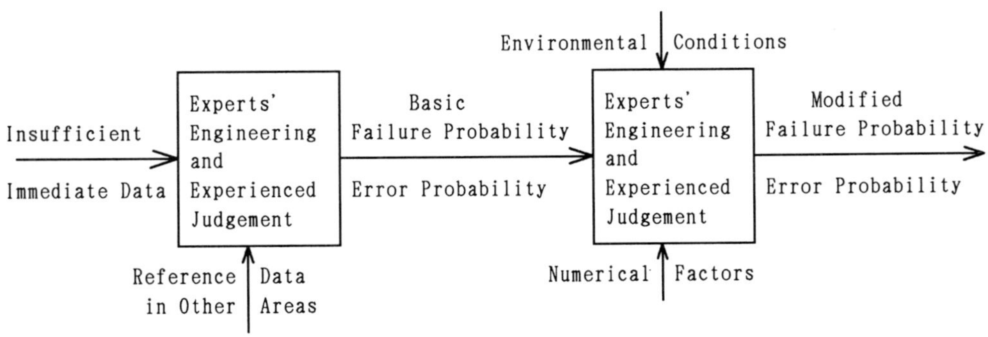

Fig. 14.19 Fuzziness in Failure Probability and Error Probability

14.14.3 Fuzziness in Consideration of Dependence

Let us consider a parallel system as shown in Fig. 14.20. If subsystems A and B are independent of each other, the parallel system works as a redundant system. The failure probability of the parallel system is estimated by $p_A\times p_B$, where p_A and p_B are failure probabilities of subsystems A and B, respectively. However if two subsystems are dependent on each other, for example, if the failure of subsystem A has an influence on subsystem B, the parallel system does not work as a redundant system. This kind of dependence is observed in a complex and large scale system, or in similar consecutive human tasks. A conditional probability is used to consider the dependence in the probabilistic method. Therefore the conditional probability should be estimated from lots of immediate data. In practice, however, we have no data to estimate the conditional probability since it is dependent on the analyzed system or human task. Information of the dependence including the conditional probability is usually obtained on the basis of engineering judgement and experience of the experts, who are familiar with the analyzed system or human task.

In human reliability, the following model is considered. Let us assume that a human operator performs tasks A and B in an order and that errors in both the tasks A and B lead to an error in the whole task. In Fig. 14.20, subsystems A and B are considered as tasks A and B, respectively. If it is difficult to obtain the conditional probability, the model shown in Table 14.2 can be employed[53]. In this

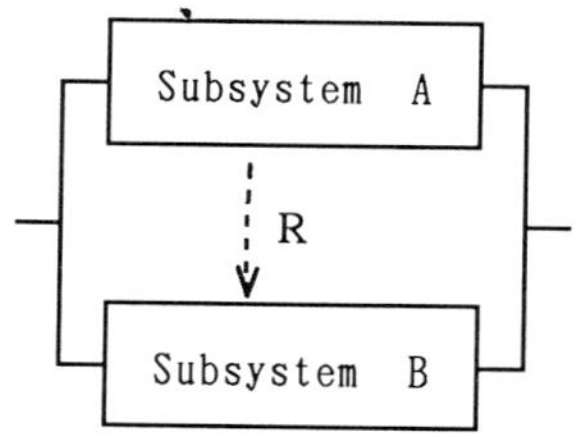

Fig. 14.20 Parallel System with Dependence

model five kinds of dependence levels are defined. Zero dependence implies that task A and task B are independent of each other. In this case, the error probability of the whole task is estimated to be $p_A \times p_B$, where p_A and p_B are the error probabilities of performing task A and task B, respectively, when they are independent of each other. On the other hand, a complete dependence implies that an error in task A leads to an error in task B completely. In this case the error probability of the whole task is not estimated to be $p_A \times p_B$ but $p_A \times 1.0$. Other three formulas shown in Table 14.2 are employed depending on the dependence levels.

In this model, the dependence level assessment and determination of formulas in the intermediate three levels are based on experts' engineering judgement. Furthermore, in Table 14.2, the dependence level is expressed by natural language. The meaning of natural language expressions cannot be expressed by a formula. For instance, the borderline between high dependence and moderate dependence cannot be determined clearly. Thus, fuzziness and subjectivity are also found in the consideration of dependence.

Table 14.2 Dependence Model in Human Reliability

Dependence Level	Error Probability of Task B
Zero Dependence	p_B
Low Dependence	$(1 + 19 \times p_B)/20$
Moderate Dependence	$(1 + 6 \times p_B)/7$
High Dependence	$(1 + p_B)/2$
Complete Dependence	1.0

14.14.4 Fuzziness in Evaluation of System Safety

Let us assume that the probability of an accident in a system is estimated to be 10^{-6} [1/year]. Can this system be evaluated as being safe ? The evaluation is dependent on an analyzed system, whether it is a nuclear power plant or an automobile plant. It is also dependent on the person evaluating it, whether he/she is concerned with the analyzed system or is a neighbor of the system.

The accident probability of 10^{-6} is very very small from the viewpoint of our daily life. This probability is usually compared with a probability of a commonly occurring event which is estimated by experts based on their engineering judgement. That is, fuzziness and subjectivity are latent in the estimation of the probability of the comonly occuring event and in the associated comparison.

What does small accident probability, e.g., 10^{-6}, mean ? Does it imply that the accident occurs once per 10^6 years ? Or does it simply mean that the accident would hardly ever occur ? Although accident probability is estimated to be very small, an accident can happen and really does happen occasionally. Small probability of an event does not necessarily imply low possibility of the event[5]. Fig. 14.21 shows a relation between probability and possibility. System reliability should be analyzed from the viewpoint of not only the probability theory but also the possibility theory, i.e., FST.

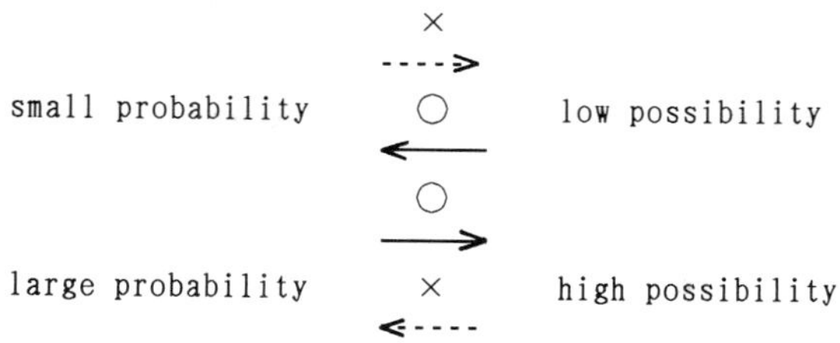

Fig. 14.21 Relation between Probability and Possibility

14.14.5 Fuzziness in Macroscopic Analysis

The following problems are considered in a reliability analysis of a complex and large scale system:

(1) To what extent should sequences from cause events to a system accident be analyzed ?

A microscopic analysis would need minute information. However, a complete information on reliability of the complex and large scale system cannot be obtained. Even experts cannot have every knowledge about the system.

(2) To what extent should a human complex task be broken down into basic tasks?

Human reliability should be considered in any system reliability analysis since it is impossible to exaggerate the fact that system reliability is dependent on human reliability in these highly technical development days. A human task must be broken down into basic tasks minutely in order to estimate the error probability of the task by the use of probabilities of basic tasks. However, even if a human task is analyzed in detail, human reliability is not estimated precisely since human behavior is complex and unpredictable.

From the above mentioned point of view, a macroscopic analysis of system reliability is necessary. The use of natural language expressions is one of the macroscopic analyses. As a matter of fact, natural language expressions are used in the construction of a fault tree or an event tree, and in the failure mode and effects analysis(FMEA). The macroscopic analysis by the use of natural language expressions is related to fuzziness since the meaning of natural language expressions is fuzzy. The macroscopic analysis may provide a hint to deal with the fuzziness described in from 14.14.1 through 14.14.4.

14.14.6 Fuzziness in System Performance Assessment

In a two-state model of a reliability analysis, a system state is considered to be either *good* or *bad*. The two states constitute the two basic events a *success event* and a *failure event*. These two events are considered to be random. The probability of the *success event* in a system is defined as the *reliability* of the system.

In practice, however, the transition from one state to the other is not sharp. That is, the system state is not necessarily divided into two states *good* and *bad* clearly. Especially, in human assessment, a system state cannot be considered to be in either of the two states precisely. The classification of the performance of a system may be *poor, satisfactory, good, very good, excellent,* etc. The state should be considered fuzzily rather than precisely since the final decision making on system performance assessment is done by the human beings, only. As mentioned before, the human judgement is related to fuzziness in nature. Therefore, FST is necessary to deal with human assessment of system performance.

A system reliability analysis is a complex problem since the system being analyzed can be often large and complex, and the analysis involves a human evaluation of the system safety. Only one methodology may not be sufficient to deal with the problem. It is necessary to use some combination of methodologies, e.g., the probability theory and FST, to handle problems in this field.

14.15 RELIABILITY ANALYSIS OF GENERAL SYSTEM

In the usual two-state models either *good* or *bad* state can constitute a system state. The probabilistic method of reliability analysis is based on this two-state model. These two states constitute the two basic or primary events. The probability of a *success event* constitutes reliability of the system based on the consideration that *good* or *bad* events are random and that probabilities associated with these random events are designated as *reliability* or *unreliability* of the system. In real system, however, the situation is not so simple. The transition from one state to the other is not sharp. There may be situations, where one could classify the performance of a system or a subsystem not in these two precise categories but many more, e.g., *poor, satisfactory, good, very good, excellent.* In this way, there could be many states possible where transition from one to the other is not very sharp or alternatively the states are fuzzy states. In the following sections, the performance index for describing the system performance is considered using FST[41].

A structure function gives the structural relation between a system and its constituent components(subsystems). A performance function gives the performance level of a system depending on the performance levels of some system components(subsystems) and the mode of components(subsystems) operation. In the following sections $\Phi(S)$, $\Phi(A)$, $\Phi(B)$, --- are employed to denote the structure functions of a system and subsystems A, B, ---, and $\Psi(S)$, $\Psi(A)$, $\Psi(B)$, --- are the performance functions of a system and subsystems A, B, ---. And let performance levels of a system or subsystems be expressed such as *excellent performance, good performance, rather good performance, poor performance, no performance,* which are defined by fuzzy sets shown in Fig. 14.22. These definitions are used merely as an illustration.

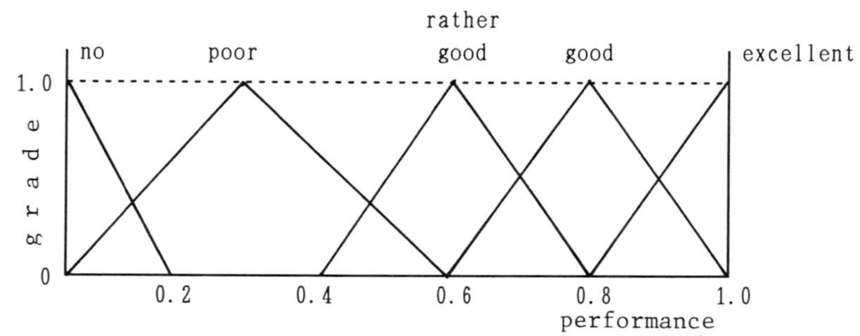

Fig. 14.22 Fuzzy Sets Representations of System Performance

14.15.1 Simple Series Parallel Configurations

Series System

In a series system, subsystems are connected in series mode of operation functionally. Let us consider a series system composed of two subsystems A and B as shown in Fig. 14.23. If two subsystems are independent of each other, the performance function of this system depends on the individual performances of the subsystems. The structure function of this system is given by

$$\Phi(S) = \Phi(A) \cap \Phi(B), \tag{14.201}$$

and the performance function is given by

$$\Psi(S) = \Psi(A) \sqcap \Psi(B). \tag{14.202}$$

Fig. 14.23 Series System

In (14.202) a **meet** operation($\sqcap$) is employed in place of a **min** operation. This implies that the performance of the system in this case depends on the subsystem having the worst performance. That is, a series system performance is not a minimal membership grade of fuzzy sets representing subsystems performances but a minimum of subsystems performances themselves. If performances of subsystems A and B are *good performance* and *rather good performance* as shown in Fig. 14.22, then the performance of the series system is *rather good performance.* If performances of subsystems A and B are given by fuzzy sets shown in Fig. 14.24, then the performance of the series system is obtained by a fuzzy set drawn by a dotted line in Fig. 14.24, which is different from fuzzy sets drawn by real lines.

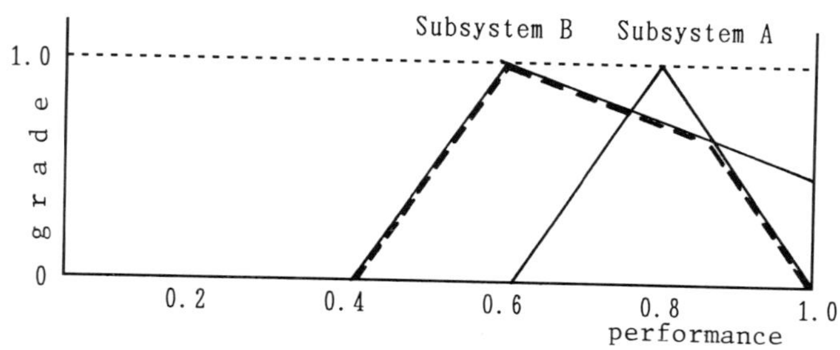

Fig. 14.24 Example of Series System Performance

Parallel System

A parallel system works if at least one subsystem functions. That is, the system fails only when all subsystems fail. Let us consider a parallel system composed of two subsystems A and B as shown in Fig. 14.25. If two subsystems are independent of each other, the performance function of this system depends on the individual performances of the subsystems. The structure function of this system is given by

$$\Phi(S) = \Phi(A) \cup \Phi(A), \tag{14.203}$$

and the performance function is given by

$$\Psi(S) = \Psi(A) \sqcup \Psi(B). \tag{14.204}$$

In (14.204) a **join** operation($\sqcup$) is employed in place of a **max** operation. This implies that the performance of the system in this case depends on the subsystem having the best performance. That is, parallel system performance is not a maximal membership grade of fuzzy sets representing subsystems performances but a maximum of subsystems performances themselves. If performance of subsystems A and B are *good performance* and *rather good performance* as shown in Fig. 14.22, then

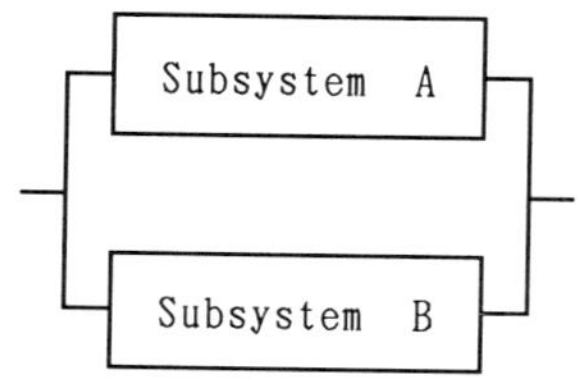

Fig. 14.25 Parallel System

the performance of the parallel system is *good performance.* If performances of subsystems A and B are given by fuzzy sets as shown in Fig. 14.26, then the performance of the parallel system is obtained by a fuzzy set drawn by a dotted line in Fig. 14.26, which is different from fuzzy sets drawn by real lines.

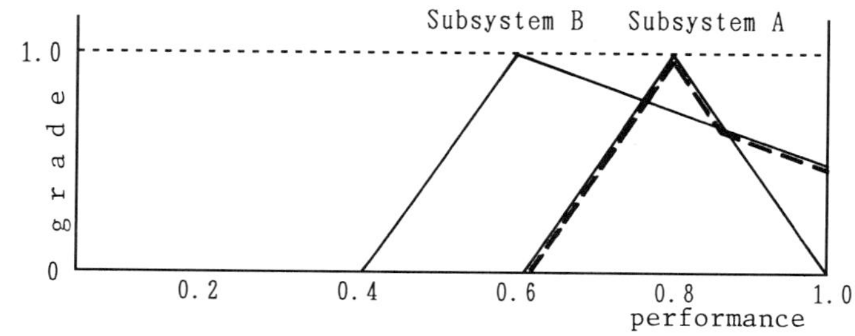

Fig. 14.26 Example of Parallel System Performance

14.15.2 Non Series Parallel Configuration

A non series parallel system is the one which does not follow the simple series or parallel system's laws of combinations. For example, let us consider a bridge network composed of five subsystems A, B, C, D, and E as shown in Fig. 14.27. If subsystems are independent of one another, the structure function is given by

$$\begin{aligned}\Phi(S) = {} & (\Phi(A) \cap \Phi(D)) \cup (\Phi(A) \cap \Phi(C) \cap \Phi(E)) \\ & \cup (\Phi(B) \cap \Phi(C) \cap \Phi(D)) \cup (\Phi(B) \cap \Phi(E)),\end{aligned} \tag{14.205}$$

and the performance function is given by

$$\begin{aligned}\Psi(S) = {} & [\Psi(A) \sqcap \Psi(D)] \sqcup [\Psi(A) \sqcap \Psi(C) \sqcap \Psi(E)] \\ & \sqcup [\Psi(B) \sqcap \Psi(C) \sqcap \Psi(D)] \sqcup [\Psi(B) \sqcap \Psi(E)].\end{aligned} \tag{14.206}$$

If performances of subsystems A, B, C, D, and E are *excellent performance, rather good performance, poor performance, good performance,* and *good performance* as shown in Fig. 14.22, then the performance of the bridge network is *good performance.* If performances of subsystems A, B, C, D, and E are given by fuzzy sets shown in Fig. 14.28, then the performance of the bridge network is obtained by a fuzzy set drawn by a dotted line in Fig. 14.28, which is different from fuzzy sets drawn by real lines.

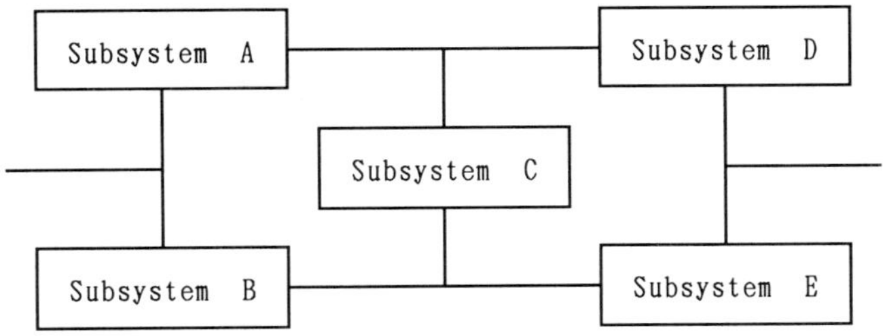

Fig. 14. 27 Bridge Network System

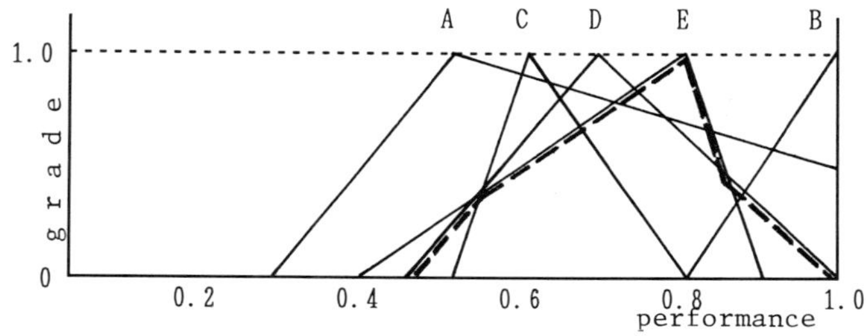

Fig. 14.28 Example of Bridge Network System Performance

14.15.3 Two-out-of-Three Configuration

In a 2-out-of-3 system three subsystems are connected in parallel mode. The system works if any two subsystems function. Let us consider a 2-out-of-3 system composed of subsystems A, B, and C. If three subsystems are independent of one another, the performance function of this system depends on the individual performances of subsystems. The structure function of this system is given by

$$\Phi(S) = (\Phi(A) \cap \Phi(B)) \cup (\Phi(A) \cap \Phi(C)) \cup (\Phi(B) \cap \Phi(C)), \tag{14.207}$$

and the performance function is given by

$$\Psi(S) = [\Psi(A) \sqcap \Psi(B)] \sqcup [\Psi(A) \sqcap \Psi(C)] \sqcup [\Psi(B) \sqcap \Psi(C)]. \tag{14.208}$$

If performances of subsystems A, B, and C are *good performance, rather good performance,* and *excellent performance* as shown in Fig. 14.22, then the performance of the 2-out-of-3 system is *good performance.* If performances of subsystems A, B, and C are given by fuzzy sets shown in Fig. 14.29, then the performance of this system is obtained by a fuzzy set drawn by a dotted line in Fig. 14.29, which is different from fuzzy sets drawn by real lines.

14.16 FUZZY PROBABILITY

A fuzzy probability has been employed in order to deal with some kinds of fuzziness such as described in section 14.14 or in [40, 42, 49, 50, 54, 55, 56]. The fuzzy probability is a fuzzy set(a fuzzy number) defined on a probability space as shown in Fig. 14. 30. In Fig. 14.30 the abscissa mean probability and the ordinate mean the grade of membership. The other types of representations have been described in sections 14.9 and 14.12.1. The fuzzy probability approach has some advantages. For example, natural language expressions such as *the failure probability of this*

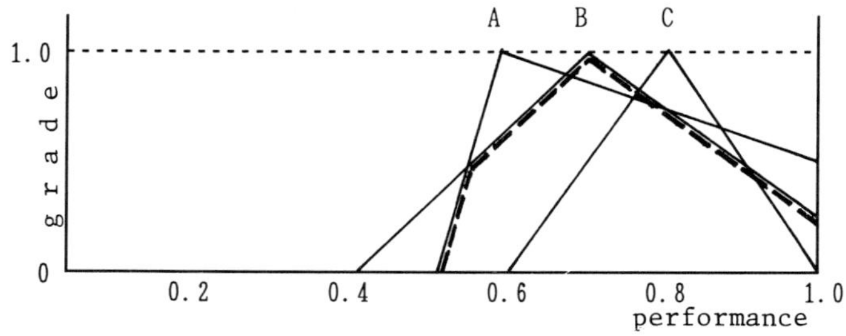

Fig. 14.29 Example of 2-out-of-3 System Performance

component is about 10^{-4} are available. And the conventional methodology is useful by the extension principle. The addition, the subtraction and the multiplication of fuzzy probabilities are performed by the extension principle. Assuming that a fuzzy set is a normalized convex set, the extension principle is expressed by another way using α-cut, which is different from the way described in section 14.5.2. Let f be a two-valued function from $X \times Y$ to Z, and A and B be fuzzy sets on X and Y, respectively. A fuzzy set C on Z through f is defined by the extension principle such as

$$\begin{aligned} f(A\ ,\ B)_\alpha &= [c_L(\alpha)\ ,\ c_U(\alpha)] \\ &= f(A_\alpha\ ,\ B_\alpha), \end{aligned} \tag{14.209}$$

where $f(A\ ,\ B)_\alpha$, A_α, and B_α, and $C_\alpha = [c_L(\alpha)\ ,\ c_U(\alpha)]$ are α-cuts of $f(A\ ,\ B)$, A, B, and C, and $\alpha \in [0,\ 1]$.

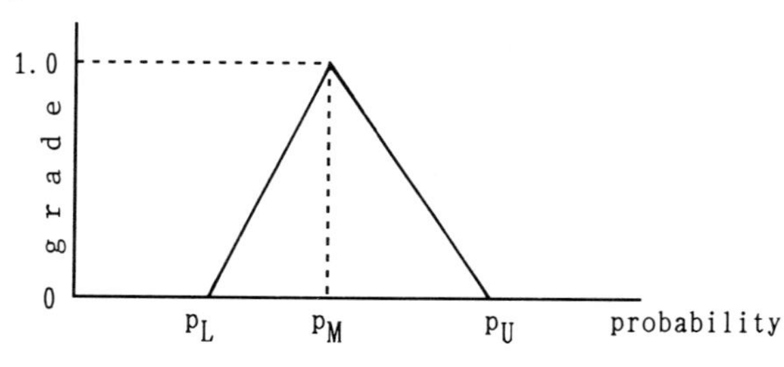

Fig. 14.30 Example of Fuzzy Probability

Let us consider the addition, the subtraction and the multiplication of two fuzzy probabilities P_A and P_B. Let α-cut of P_A and P_B be $[p_{AL}(\alpha)\ ,\ p_{AU}(\alpha)]$ and $[p_{BL}(\alpha)\ ,\ p_{BU}(\alpha)]$, where $\alpha \in [0,\ 1]$.

Addition

$$(P_A + P_B)_\alpha = [p_{AL}(\alpha)+p_{BL}(\alpha)\ ,\ p_{AU}(\alpha)+p_{BU}(\alpha)]. \tag{14.210}$$

Subtraction

$$(P_A - P_B)_\alpha = \begin{cases} [c_L(\alpha)\ ,\ c_U(\alpha)], & p_{AL}(\alpha)>p_{BU}(\alpha), \\ [0\ ,\ c_U(\alpha)], & p_{AL}(\alpha)\le p_{BU}(\alpha),\ p_{AU}(\alpha)>p_{BL}(\alpha), \\ 0, & p_{AL}(\alpha)<p_{BU}(\alpha),\ p_{AU}(\alpha)\le p_{BL}(\alpha). \end{cases} \tag{14.211}$$

where $c_L(\alpha)=p_{AL}(\alpha) - p_{BU}(\alpha)$ and $c_U(\alpha)=p_{AU}(\alpha) - p_{BL}(\alpha)$.

Multiplication

$$(p_A \times p_B)_\alpha = [p_{AL}(\alpha)\times p_{BL}(\alpha)\ ,\ p_{AU}(\alpha)\times p_{BU}(\alpha)]. \tag{14.212}$$

* "1E-2" reads as 1×10^{-2}.

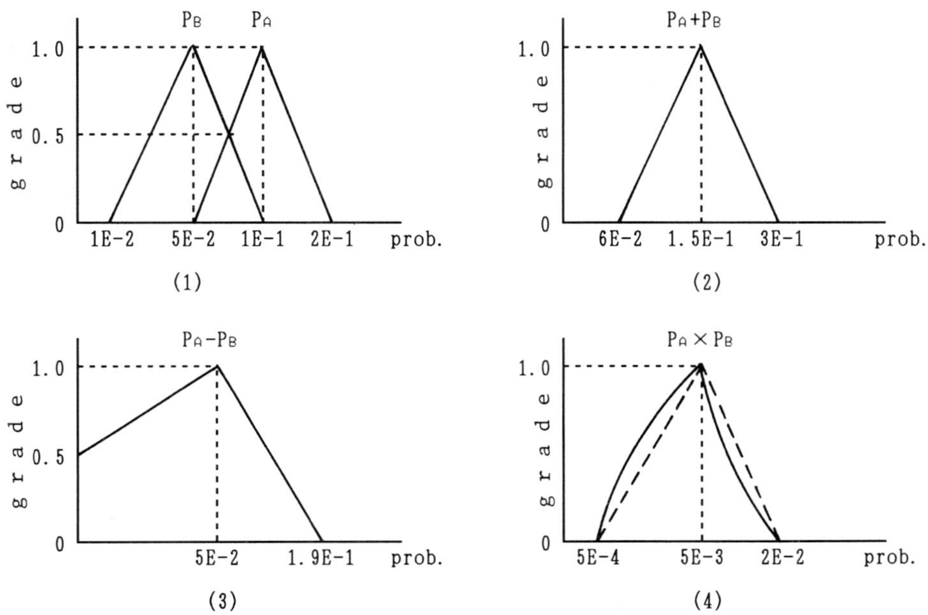

Fig. 14.31 Examples of Addition, Subtraction and Multiplication of Fuzzy Probabilities

Fig. 14.31 shows these calculations when two fuzzy probabilities are defined by triangular membership functions as shown in Fig. 14.31(1). In Fig. 14.31(4) the real line shows the multiplication result by the use of the extension principle. The membership function does not have a triangular shape any longer. A linearization can be considered[42] such as given by

$$\mu_{P_C}(p) = \begin{cases} 0, & p<p_{CL}(0), \\ (p - p_{CL}(0))\ /\ (p_C - p_{CL}(0)), & p_{CL}(0)\le p\le p_C, \\ \\ (p - p_{CU}(0))\ /\ (p_C - p_{CU}(0)), & p_C\le p\le p_{CU}(0), \\ 0, & p_{CU}(0)<p. \end{cases} \tag{14.213}$$

where, $P_C = P_A\times P_B$, $p_{CL}(0) = p_{AL}(0)\times p_{BL}(0)$, $p_{CU}(0) = p_{AU}(0)\times p_{BU}(0)$, $p_C = p_A\times p_B$. $[p_{AL}(0)\ ,\ p_{AU}(0)]$, $[p_{BL}(0)\ ,\ p_{BU}(0)]$, and $[p_{CL}(0)\ ,\ p_{CU}(0)]$ are 0-cut sets of P_A, P_B and P_C. And $\mu_{P_A}(p_A) = \mu_{P_B}(p_B) = \mu_{P_C}(p_C) = 1$.

The straight dotted line in Fig. 14.31(4) shows the linearization of the multiplication result.

14.16.1 Simple Series Parallel Configuration

The following addition, subtraction, and multiplication are extended to fuzzy sets by the extension principle.

Series System

Let us consider a series system composed of two subsytems A and B as given in Fig. 14.23. Let fuzzy failure probabilities of subsystems A and B be P_A and P_B, respectively. Assuming that two subsystems are independent of each other, the fuzzy failure probability of the series system is obtained by

$$P_S = 1 - (1 - P_A) \times (1 - P_B). \tag{14.214}$$

Parallel System

Let us consider a parallel system composed of two subsystems A and B as given in Fig. 14.25. Let fuzzy failure probabilities of subsystems A and B be P_A and P_B, respectively. Assuming that two subsystems are independent of each other, the fuzzy failure probability of the parallel system is obtained by

$$P_S = P_A \times P_A. \tag{14.215}$$

14.16.2 Non Series Parallel System

Let us consider a bridge network, as given in Fig. 14.27, comprising of five subsystems A, B, C, D, and E. Let their fuzzy failure probabilities be P_A, P_B, P_C, P_D, and P_E. Assuming that five subsystems are independent of one another, the fuzzy failure probability of this system is obtained in the following way.

The fuzzy failure probability of the path A-D is obtained by

$$P_{A-D} = 1 - (1 - P_A)\times(1 - P_D). \tag{14.216}$$

The fuzzy failure probability of the path A-C-E is obtained by

$$P_{A-C-E} = 1 - (1 - P_A)\times(1 - P_C) \times (1 - P_E). \tag{14.217}$$

Fuzzy failure probabilities of the path B-C-D and the path B-E are obtained by

$$P_{B-C-D} = 1 - (1 - P_B)\times(1 - P_C)\times(1 - P_D), \tag{14.218}$$

and

$$P_{B-E} = 1 - (1 - P_B)\times(1 - P_E), \tag{14.219}$$

respectively. The fuzzy probability of the bridge network is obtained by

$$P_S = P_{A-D}\times P_{A-C-E}\times P_{B-C-D}\times P_{B-E}, \tag{14.220}$$

where $P_i^2(i = A, B, C, D, E)$ should be P_i.

14.16.3 Two-out-of-Three System

Let us consider a 2-out-of-3 system composed of three subsystems A, B and C. Let their fuzzy failure probabilities be P_A, P_B, and P_C. Assuming that three subsystems are independent of one another, the fuzzy failure probability of this system is obtained by

$$P_S = (1 - P_A)\times P_B\times P_C+P_A\times(1 - P_B)\times P_C$$
$$+P_A\times P_B\times(1 - P_C)+P_A\times P_B\times P_C. \qquad (14.221)$$

It is found that the same consideration as the conventional methodology can be employed even if the fuzzy probability is used. The difference is only that calculations such as the addition, the subtraction, and the multiplication should be extended to a fuzzy set.

14.17 FUZZY FAULT TREE ANALYSIS

In this section, a fault tree is called a fuzzy fault tree if the estimate of reliability of each of the basic events and relations between various events are represented by fuzzy sets. Fuzzy fault tree analysis can be carried out using a fuzzy probability approach [40, 42, 54, etc.] or using a failure possibility approach[48, 58, 59, etc.]. Both the approaches have been used extensively in the safety and reliability studies.

14.17.1 By Fuzzy Failure Probability

The expressions (14.210), (14.211) and (14.212) are used for the addition , the subtraction, and the multiplication of fuzzy probabilities. And the linearization (14.213) sometimes may be used for simplicity. Fig. 14.32 shows an example of a fuzzy fault tree. Assuming that the failure probabilities of basic events are provided in Fig. 14.33, the failure probability of the top event is shown in Fig. 14.34, where the result eventually is linearized.

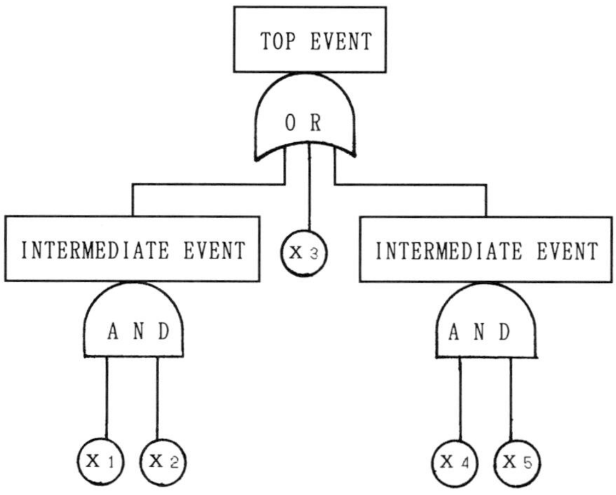

Fig. 14.32 A Fuzzy Fault Tree

The fault tree analysis by the fuzzy failure probability does not deal with all the problems mentioned in section 14.14. The fuzzy failure probability approach has the same handicap as the conventional probability approach. Both hardware reliability and human reliability should be considered in a system reliability analysis. However, the measure of hardware reliability, e.g., a failure probability, and the measure of human reliability, e.g., an error probability, should not be treated in the same way in the system reliability analysis since, the term *probability* although is used in both failure probability and error probability, their definitions are different in nature. The failure probability is defined as *the failure rate per a unit time(one hour*

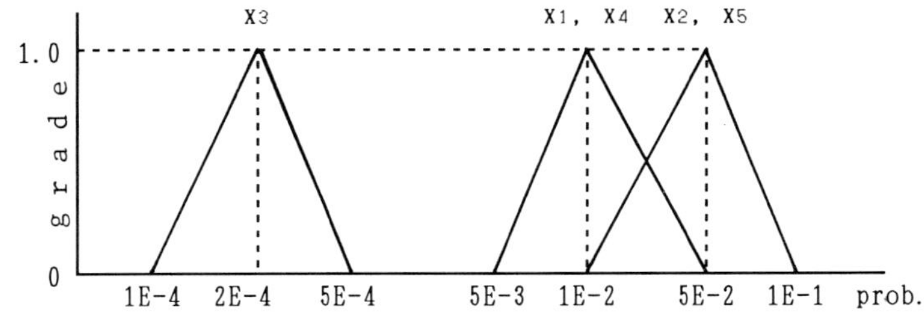

Fig. 14.33 Fuzzy Failure Probabilities of Basic Events of Fuzzy Fault Tree in Fig. 14.32

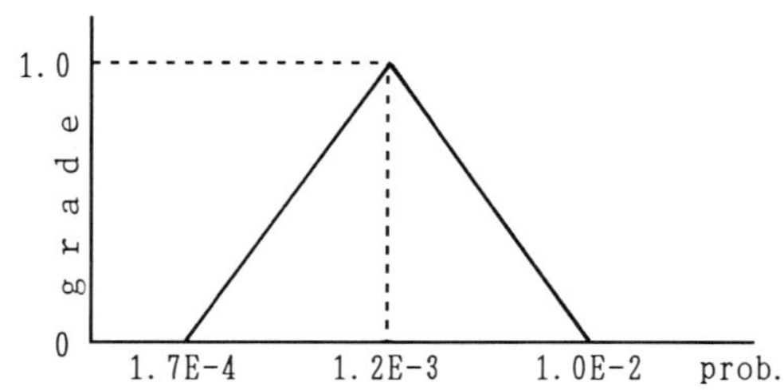

Fig. 14.34 Fuzzy Probability of Top Event for Fuzzy Fault Tree of Fig. 14.32

or one year). On the other hand the error probability is defined as *(the number of errors in a task)/(the number of chances that an error can occur in a task).* Therefore, it is hoped not only that human reliability can be expressed in the same reliability terms as equipment, but also that the measure of human reliability can be combined with the measure of hardware reliability to provide an estimate of system reliability as a whole[57]. However, in the following sections, we would describe another method by which this can be possible.

14.17.2 By Failure Possibility

Failure Possibility

A fuzzy set defined on the unit interval [0, 1] is used in this method. This fuzzy set is called a failure possibility with the membership function (14.222)[48, 58, 59].

$$F(x) = \frac{1}{1+20\times|x - x_0|^m}, \tag{14.222}$$

where x_0 and m are parameters. The parameter x_0 helps in normalizing, i.e., offers a maximal grade of 1 and the parameter m is related to fuzziness.

In (14.222), as one would notice that $F(0) \neq 0$, and $F(1) \neq 0$. Therefore, (14.222) does satisfy the condition that there exists a possibility that a system may fail certainly or a human operator may make an error in a task certainly. At the same time, it allows that there exists also a possibility that the system may not fail at all or the human operator may not make an error in the task at all. Thus the membership function (14.222) used in this method would help us in providing a more realistic model. Fig. 14.35 shows an example of the failure possibility. In this figure, the abscissa represents the subjective unreliability. The failure possibility is a fuzzy set defined on subjective unreliability. That is, the failure possibility is considered to be a subjective unreliability measure. The ordinate represents the grade of membership. In this method, the estimate of reliability is given by natural language or numerical values[60], as the case may be.

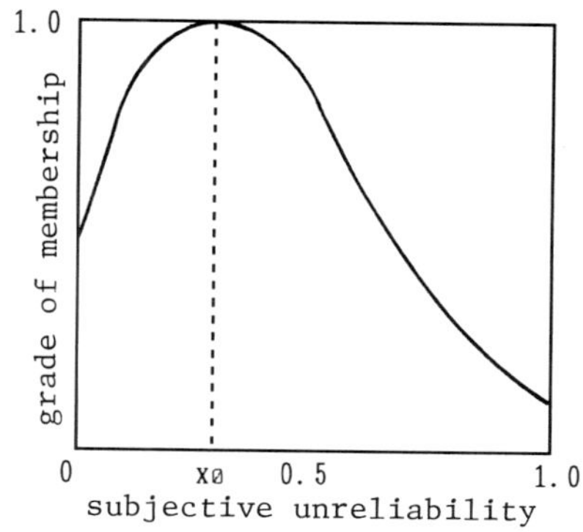

Fig. 14.35 Failure Possibility

Estimate of Reliability by Natural Language

An estimate of reliability is expressed in the form of the reliability estimate and its fuzziness. Table 14.3 shows natural language expressions about the reliability estimate and the correspondence between various terms of natural language and the representative values of the parameter x_0 described in (14.222). Table 14.4 shows natural language expressions of fuzziness of the reliability estimate and the correspondence between these expressions and the parameter m. When equipment reliability or human reliability is expressed by natural language, the parameters of the failure possibility is selected from Tables 14.3 and 14.4. The mappings between the linguistic terms and their numerical representations as provided in Tables 14.3 and 14.4 are defined subjectively. An analyst is free to define them subjectively. But once he/she defines a mapping, he/she should employ it throughout the system reliability analysis in order to keep consistency in the analysis. Failure possibilities in corresponding to the class 1 and class 11 have membership functions given by

$$F(x) = \begin{cases} 1, & x = 1, \\ \\ 0, & x \neq 1, \end{cases} \tag{14.223}$$

and

$$F(x) = \begin{cases} 0, & x \neq 0, \\ \\ 1, & x = 0, \end{cases} \tag{14.224}$$

respectively. They are not fuzzy sets but crisp sets. The term *crisp set* is often used for an ordinary set.

Table 14.3 Natural Language Expressions of Reliability Estimate and Numerical Values of Parameter x_0

Class	Natural Language Expressions of Reliability Estimate	Parameter x_0 (Representative Value)
(System, equipment or human operator has)		
1	no reliability	–
(System, equipment or human reliability is)		
2	very low	0.9 – 1.0(0.95)
3	low	0.7 – 0.9(0.8)
4	rather low	0.55 – 0.7(0.625)
5	standard	0.45 – 0.55(0.5)
6	rather high	0.3 – 0.45(0.375)
7	high	0.2 – 0.3(0.25)
8	quite high	0.1 – 0.2(0.15)
9	extremely high	0.05 – 0.1(0.075)
(Accident, failure or human error is)		
10	next to impossible	0.0 – 0.05(0.025)
11	impossible	–

Table 14.4 Natural Language Expressions of Fuzziness and Numerical Values of Parameter m

Expressions of Fuzziness	Parameter m
(Fuzziness is)	
low	2.0
medium	2.5
rather high	3.0
high	3.5

Estimate of Reliability through Numerical Values

The numerical information about the estimate of reliability is usually given by, e.g., failure probability or human error probability. Failure probability and error probability are usually given as the triplet such as $[p_L, p_M, p_U]$, where p_M is a recommended mean value of the failure or error probability, p_L is its lower bound, and p_U is its upper bound.

In this case, the parameter x_0 is derived as a function of p_M, such as

$$x_0 = f(p_M) = \frac{1}{1+(K\times log(1/p_M))^3}, \tag{14.225}$$

where $p_M \neq 0$, and $x_0 = 0$ when $p_M = 0$, and

$$K = 1/log(1/p_S), \tag{14.226}$$

where p_S is the subjective standard failure or error probability.

The subjective standard probability denotes that an analyst considers that the failure probability of equipment or the error probability of a human operator in a task should be p_S. The failure probability and the error probability are compared with the standard probability in the sense of (14.225). The parameter m is then selected from Table 14.5, where $k = p_M/p_L$, p_U/p_M. Table 14.5 implies that the larger the parameter k is, the fuzzier the evaluated unreliability could be. As mentioned before, the mapping between parameters m and k is defined subjectively.

The mapping in Table 14.5 is an example.

Table 14.5 Correspondence between Parameters m and k

Range of Parameter k	Parameter m
$k \leq 3$	2.0
$3 < k \leq 5$	2.5
$5 < k \leq 10$	3.0
$10 < k$	3.5

Fig. 14.36 indicates the process of derivation of the failure possibility. The failure possibility expresses expert's subjective estimate of hardware or human unreliability whether the estimate of reliability is given through the natural language or numerical values. Therefore, hardware unreliability and human unreliability can be treated in the same way by the failure possibility. The failure possibility has been also applied to human reliability analysis[48, 61].

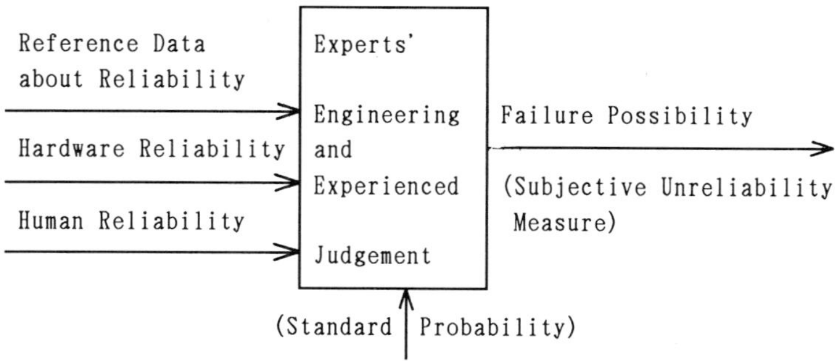

Fig. 14.36 Derivation of Failure Possibility

And and Or Operations

In this method, the failure possibility is used instead of the failure probability and the error probability for a fault tree analysis. Therefore, an **and** operation and an **or** operation of failure possibilities must be considered. From the viewpoint of reliability analysis a **min** operation is the most pessimistic one just as the operation of an **and** gate in the fault tree analysis and a **max** operation is the most optimistic operation parallel to the operation of an **or** gate in a fault tree analysis. One can also use the following functions in the method instead of the **min** and the **max** operations.

$$H(x\ ,\ y) = \frac{1}{1+(((1-x)/x)^{1/3}+((1-y)/y)^{1/3})^3}, \tag{14.227}$$

where $0 < x\ ,\ y \leq 1$, and $H(0\ ,\ y) = H(x\ ,\ 0) = 0$.

$$G(x\ ,\ y) = \frac{((x/(1-x))^3+(y/(1-y))^3)^{1/3}}{1+((x/(1-x))^3+(y/(1-y))^3)^{1/3}}, \tag{14.228}$$

where $0 \leq x\ ,\ y < 1$, and $G(1\ ,\ y) = G(x\ ,\ 1) = 1$.

Functions H and G are used with the extension principle as the **and** and the **or** operations, respectively. The function H is a special case of Dombi t-norm[28] and the function G is a special case of Dombi t-conorm[28]. These functions are compatible or approximately compatible with the algebraic product or the algebraic sum of failure or error probabilities in the following sense:

Let us consider a parallel system(task) and a series system(task) composed of two subsystems(subtasks) which are independent of each other. Let failure(error) probabilities of subsystems(subtasks) A and B be p_{MA} and p_{MB}, respectively. Now, the failure(error) probability of the parallel system(task) can be estimated by

$$p = p_{MA} \times p_{MB}. \tag{14.229}$$

We can estimate the parameter x_0 of the failure possibility of the system(task) by

$$f(p_{MA} \times p_{MB}) = H(f(p_{MA}), f(p_{MB})), \tag{14.230}$$

where p_{MA}, p_{MB} and $p_{MA} \times p_{MB}$ are evaluated by the same standard probability p_S.

On the other hand, the failure(error) probability of a series system(task) is estimated by

$$p = p_{MA} + p_{MB} - p_{MA} \times p_{MB}. \tag{14.231}$$

In this case compatibility is not complete but approximate such as

$$f(p_{MA} + p_{MB} - p_{MA} \times p_{MB}) \cong G(f(p_{MA}), f(p_{MB})), \tag{14.232}$$

where p_{MA}, p_{MB}, and $p_{MA} + p_{MB} - p_{MA} \times p_{MB}$ are evaluated by the same standard probability p_S.

Results of Analysis using Natural Language Expressions

Here, the result of an analysis is expressed by natural language in the form of reliability estimate and its fuzziness. Let F_R be a failure possibility obtained by the analysis and F_S be a failure possibility with the membership function given by (14.222). The distance between F_R and F_S is defined such as

$$d = \int_0^1 \sqrt{(x_{1R}(\alpha) - x_{1S}(\alpha))^2 + (x_{2R}(\alpha) - x_{2S}(\alpha))^2}\, d\alpha, \tag{14.233}$$

where $(F_R)_\alpha = [x_{1R}(\alpha), x_{2R}(\alpha)]$ and $(F_S)_\alpha = [x_{1S}(\alpha), x_{2S}(\alpha)]$ are α-cuts of F_R and F_S, respectively.

The parameters x_0 and m of F_S are selected in a manner so as to minimize the distance d. The result of the analysis is presented by natural language expressions in Tables 14.3 and 14.4.

Example 1

Fig. 14.37 shows a fault tree which we would like to analyze. Table 14.6 provides the data about the basic events. Table 14.6(1) gives all data in numerical values. Standard probabilities p_S are also given in Table 14.6(1).

Fig. 14.38 shows the failure possibility of the top event obtained by the use of the data given in Table 14.6(1). The failure possibility obtained by the analysis is shown by the dotted line in Fig. 14.38, whereas the real line shows the failure possibility F_S minimizing the distance (14.233). The result of the analysis is *system reliability is high and its fuzziness is low.* Assuming that a fuzzy failure probability of each of the basic events of the fuzzy fault tree given in Fig. 14.37 is defined by Fig. 14.39, the linearized fuzzy failure probability of the top event is shown in Fig. 14.40.

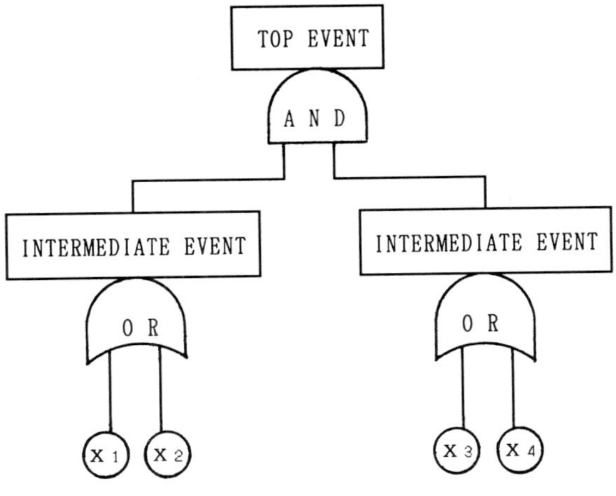

Fig. 14.37 Analyzed Fuzzy Fault Tree

Table 14.6 Data of Basic Events of the Fuzzy Fault Tree of Fig. 14.37

(1)

Basic Event	p_L	p_M	p_U	p_S
x_1	$5E-3$	$2E-2$	$5E-2$	$1E-2$
x_2	$6E-3$	$1E-2$	$3E-2$	$1E-2$
x_3	$5E-3$	$2E-2$	$5E-2$	$1E-2$
x_4	$6E-3$	$1E-2$	$3E-2$	$1E-2$

(2)

Basic Event	p_L	p_M	p_U	p_S	Reliability Estimate	Fuzziness
x_1	—	—	—	—	rather low	rather high
x_2	$1.7E-4$	$1E-3$	$8E-3$	$1E-3$	—	—
x_3	—	—	—	—	standard	rather high
x_4	$1E-3$	$7E-3$	$5E-2$	$5E-3$	—	—

* "$5E-3$" reads as 5×10^{-3}.

Fig. 14.41 shows the failure possibility of the top event obtained by the use of the data in Table 14.6(2), that is, data of basic events are given by either numerical values or natural language. The result is *system reliability is rather high and its fuzziness is rather high.*

14.18 DEPENDENCY CONSIDERATIONS

As mentioned before, the event dependence should be considered when the system being analyzed is large and complex since many subsystems may be coupled to one another functionally. The result of the analysis provides an optimistic estimate without the consideration of dependency. However, we usually don't have enough data to estimate the conditional probability. The information of the dependency is usually provided by the experts based on their engineering judgement, as they happen to be familiar with the analyzed system or task. If the conditional probability

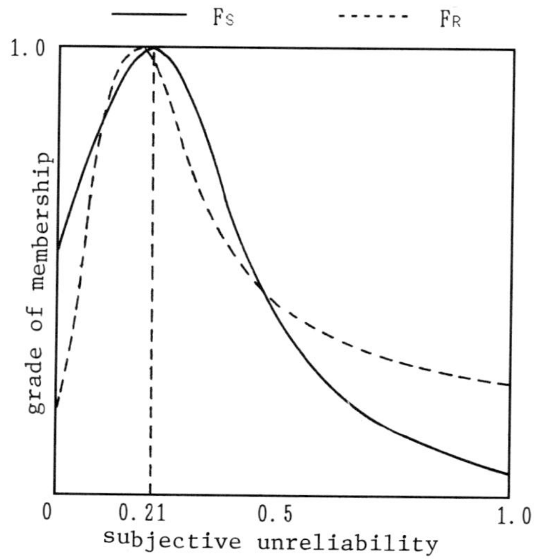

Fig. 14.38 Failure Possibility of Top Event of Fuzzy Fault Tree in Fig. 14.37 (Data of basic events are shown in Table 14.6(1).)

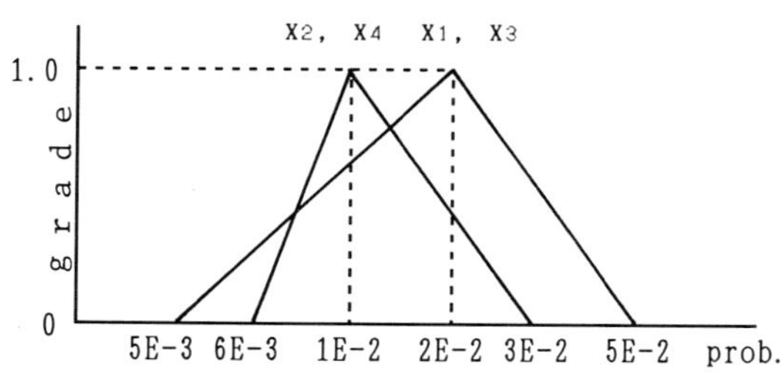

Fig. 14.39 Fuzzy Failure Probabilities of Basic Events of Fuzzy Fault Tree in Fig. 14.37

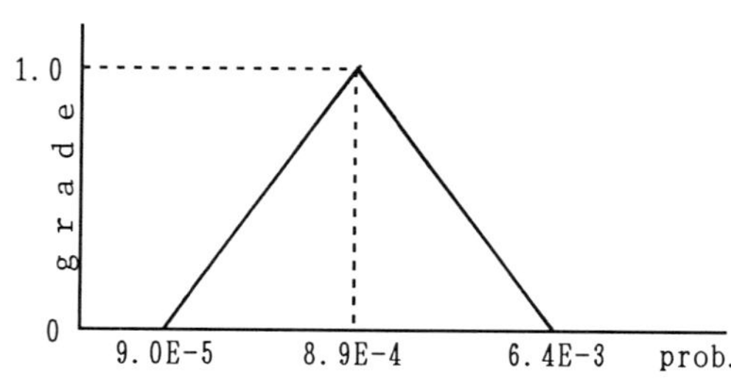

Fig. 14.40 Fuzzy Failure Probability of Top Event of Fuzzy Fault Tree in Fig.14.37

can be obtained as the fuzzy probability, the conventional analysis method can be employed by using of the extension principle.

The information of the dependency is often expressed by natural language. In the failure possibility approach the information is given by natural language in the form of dependency level estimate and its fuzziness. The natural language

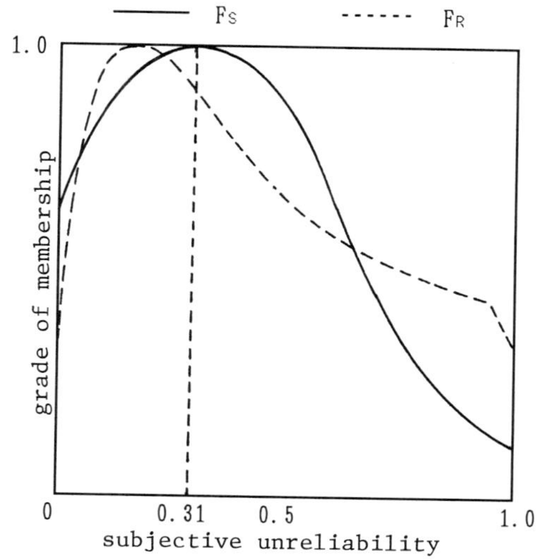

Fig. 14.41 Failure Possibility of Top Event of Fuzzy Fault Tree in Fig. 14.37 (Data of basic events are given in Table 14.6(2))

expressions can be represented by a fuzzy set, whose membership function is given by (14.234).

$$R(r) = \frac{1}{1+20\times|r - r_0|^{mr}}, \tag{14.234}$$

where r_0 and mr are parameters.

Table 14.7 provides the natural language expressions of the estimate of the dependency level and indicates the correspondence between natural language terms and the parameter r_0. Table 14.8 shows natural language expressions of fuzziness of dependency level estimate and the correspondence between natural language terms and the parameter m.

Table 14.7 Natural Language Expressions of Dependency Level Estimate and Correspondence between Expressions and Parameter x_0

Level	Expressions of Dependency Level Estimate	Parameter r_0
(Dependency level is)		
1	complete dependency	—
2	high dependency	0.95
3	moderate dependency	0.75
4	low dependency	0.5
5	zero dependency	—

As mentioned before, the mapping between linguistic terms and their numerical representatives is defined subjectively. An analyst can define this mapping subjectively. But once he/she defines it, he/she should use it throughout the analysis in order to keep consistency in the analysis. Fuzzy sets representing complete dependency and zero dependency have membership functions given by

Table 14.8 Natural Language Expressions of Fuzziness of Dependency Level Estimate and Correspondence between Expressions and Parameter m

Class	Expressions of Fuzziness	Parameter m
(Fuzziness of dependency level estimate is)		
1	low	2.0
2	medium	2.5
3	rather high	3.0
4	high	3.5

$$R(r)=\begin{cases}1, & r=1,\\ 0, & r\neq 1,\end{cases} \tag{14.235}$$

and

$$R(r)=\begin{cases}0, & r\neq 0,\\ 1, & r=0,\end{cases} \tag{14.236}$$

respectively. It is obvious that they are not fuzzy sets but crisp sets.

14.18.1 Dependency of Type I

Let us consider a parallel system composed of two subsystems A and B as shown in Fig. 14.42. A series system is not considered here since the failure of any subsystem in a series system leads to the system failure whether or not the dependency exists between the subsystems. The failure of subsystem A in Fig. 14.42 is assumed to have an influence on subsystem B. Let F_A and F_B be the failure possibilities of subsystems A and B, respectively, and R be a fuzzy set representing the information on the dependency level. Here, dependency is modelled considering two cases[48]. One is the case, in which the failure of subsystem A has an influence on subsystem B. The other is the case, in which the failure of subsystem A has no influence on subsystem B. One can obtain (14.237) through (14.240) by the use of the extension principle.

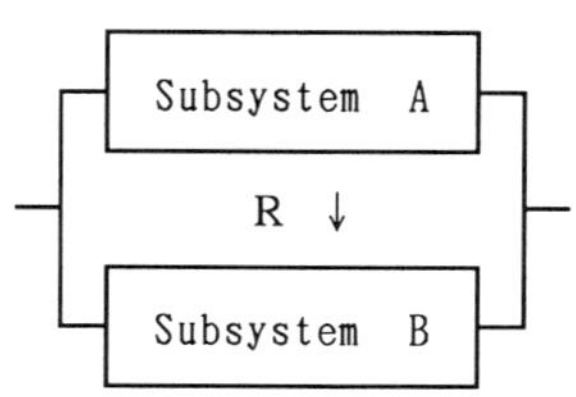

Fig. 14.42 Parallel System with Dependency of Type I

Subsystem A has an influence on subsystem B.

The failure possibility of the system given in Fig. 14.42 is obtained by F_A **and** R, that is,

$$F_B' = H(F_A , R). \tag{14.237}$$

Subsystem A has no influence on subsystem B.

If the dependency level is not complete, the failure of subsystem A has always no influence on subsystem B. The portion of the failure possibility of subsystem A is considered to be F_A', which is evaluated on the basis of the assumption that subsystem A has no influence on subsystem B.

$$F_A = G(F_A' , F_B'). \tag{14.238}$$

This implies that the failure of subsystem A has an influence on that of subsystem B(F_B') or not(F_A'). In this case the failure possibility of the system is estimated by

$$F' = H(F_A' , F_B). \tag{14.239}$$

This implies that the failure of the system occurs only when the failure of subsystem A occurs and the failure of subsystem B occurs without the influence of subsystem A.

The failure possibility considering both the above cases is obtained by

$$F = G(F_B' , F') \tag{14.240}$$

This implies that the system failure occurs when the failure of subsystem B occurs with an influence of the failure of subsystem A(F_B') or when failures of both subsystems A and B occur without dependence (F').

The above model can correspond to the model with the conditional probability in the following way. Let failure probabilities of subsystems A and B be p_A and p_B, respectively, and r(B|A) be a conditional probability, i.e., the failure probability of subsystem B subject to the condition that subsystem A has failed. The failure probability of the system can be determined as follows;

$$\begin{aligned} p &= p_A \times r(B|A) + p_A \times (1 - r(B|A)) \times p_B \\ &= p_A \times r(B|A) + (p_A - p_A \times r(B|A)) \times p_B, \end{aligned} \tag{14.241}$$

since the above two cases are considered to be exclusive to each other. The first term in (14.241) corresponds to (14.237), the inside of the parentheses of the second term in (14.241) corresponds to F_A' in (14.238), the complete second term in (14.241) corresponds to (14.239), and the entire (14.241) corresponds to (14.240). Fig. 14.43 shows failure possibilities of the system given in Fig. 14.42 for five different dependency levels, where reliability estimate of each subsystem is *standard* and its fuzziness is *medium*. These failure possibilities are F_R's. The estimate of reliability of a system with dependency of type I is represented as follows : Complete dependency ; *system reliability is standard and its fuzziness is medium,* high dependency ; *system reliability is rather high and its fuzziness is high,* moderate dependency ; *system reliability is rather high and its fuzziness is rather high,* low dependency ; *system reliability is high and its fuzziness is medium,* zero dependency ; *system reliability is quite high and its fuzziness is low.*

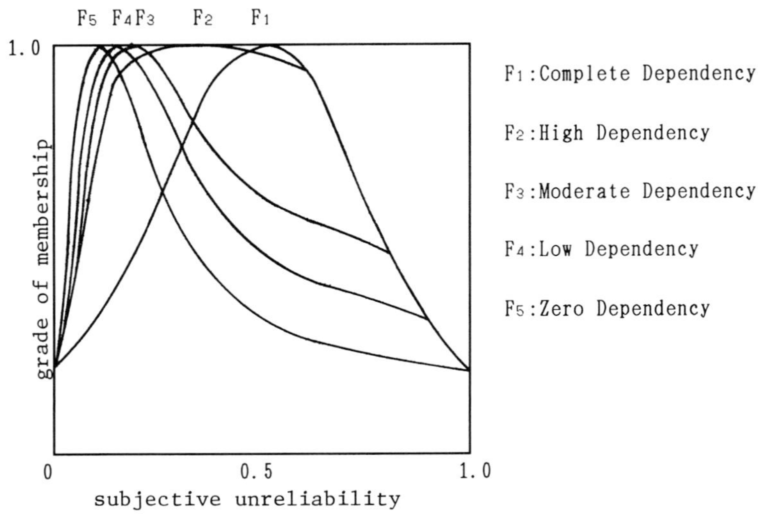

Fig. 14.43 Failure Possibilities of Parallel System for Five Dependency Levels

14.18.2 Dependency of Type II

Let us consider a system as shown in Fig. 14.44. Fig. 14.44 implies that although the failure of subsystem A does not lead to a system failure directly, it has an influence on the failure of subsystem B which then leads to a system failure consequently. There is also a possibility that the failure of subsystem B itself may occur without the influence of the failure of subsystem A. This type of dependency has been analyzed in [58]. As an example of dependency of type II, let us consider the case in which a human operator performs a task using an equipment. The failure of this equipment itself does not lead to a system failure directly. However it has an influence on the human operator error. The error leads to a system failure consequently.

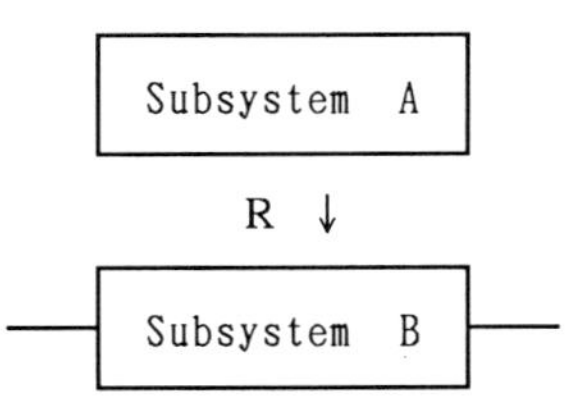

Fig. 14.44 System with Dependency of Type II

The failure possibility of the system is obtained by

$$F = G(F_B{}' \ , F_B), \tag{14.242}$$

where

$$F_B{}' = H(F_A \ , R). \tag{14.243}$$

and the extension principle is used in operations defined by (14.242) and (14.243).

This implies that a system failure occurs when the failure of subsystem B occurs as influenced by the failure of subsystem A($F_B{}'$) or not influenced(F_B), that is, F_B is added by $F_B{}'$ in the sense of the **or** operation.

Example 2

Let us analyze a fuzzy fault tree shown in Fig. 14.45. Fig. 14.46 shows the results of the analysis, where the data of Table 14.6(2) are used. Dependency of type I is considered between the intermediate events. Estimate of dependency level and its fuzziness are taken as *moderate* and *medium,* respectively. Also the dependency of type II is also considered in this tree. The basic events x_1 and x_2 have this kind of dependency. The estimate of dependency level and its fuzziness are taken as *moderate* and *high,* respectively. Dependency of type II is represented in a fuzzy fault tree as shown in Fig. 14.45 for the sake of convenience. The result is *system reliability is rather high and its fuzziness is also high.*

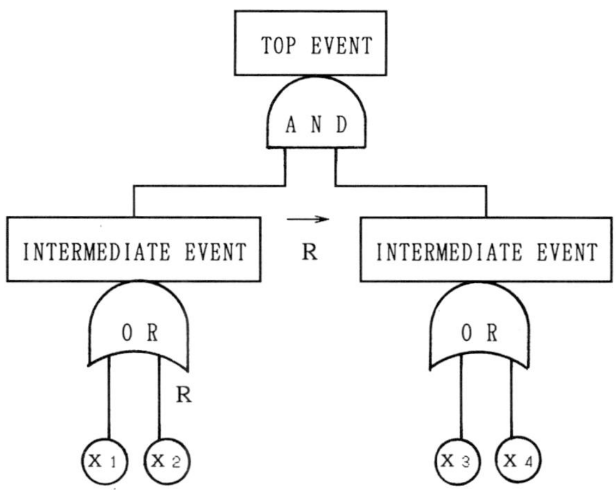

Fig. 14.45 Fuzzy Fault Tree with Dependencies of the Type I and II

14.19 ENVIRONMENTAL CONSIDERATIONS

As mentioned before, hardaware reliability and human reliability are affected by many environmental conditions[58][62]. Therefore, it is necessary to evaluate the conditions under which a hardware is operated or a task is performed in order to modify the estimated basic reliability. In the probabilistic method, the basic failure or error probability is multiplied by certain numerical factors to consider the effect of environmental conditions. Table 14.1 indicates how the consideration of environmental condition, i.e., psychological stress, in human reliability can be accounted for. Usually experts evaluate the environmental conditions and select certain numerical factors to modify the basic reliability. However the evaluation and the selection are subjective, and the relationship between the environmental conditions and reliability is not necessarily well-defined. Therefore, it is more helpful to express the relationship qualitatively than quantitatively. Let us consider the environment and maintenance quality of an equipment as the specified environmental conditions for hardware reliability, and psychological stress and human operator experience as conditions for human reliability. Now the relationship between various conditions

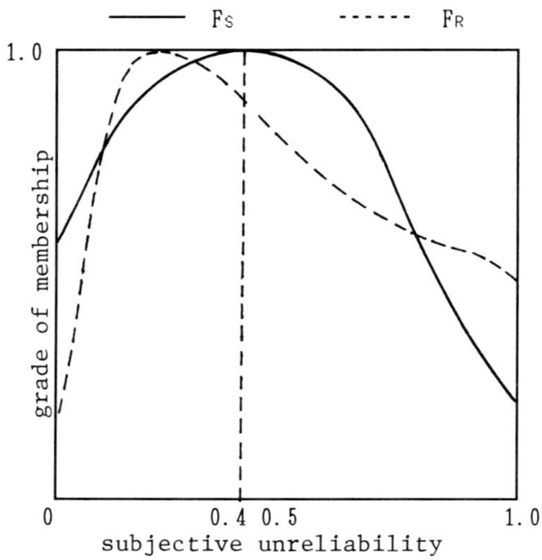

Fig. 14.46 Failure Possibility of System with Dependencies of the Type I and II

and reliability can be best expressed such as

Equipment Environment(EE)

If an equipment environment is good, the equipment reliability is high.
If an equipment environment is bad, the equipment reliability is low.

Maintenance Quality(MQ)

If maintenance quality is good, the equipment reliability is high.
If maintenance quality is bad, the equipment reliability is low.

Stress Level(SL)

If stress level is high, the operator reliability is low.
If stress level is moderate, the operator reliability is high.
If stress level is low, the operator reliability is rather low.

Operator Experience(OE)

If an operator is experienced in a task, the operator reliability is high.
If an operator is not much experienced in a task, the operator reliability is low.

The above expressions only show a tendency. Thus they imply that, for example, even if the equipment environment is good, the equipment reliability is not always high. In the above expressions *good, bad, high, low,* etc., are fuzzy terms, that is, their meanings are fuzzy and are represented well by fuzzy sets. Fig. 14.47 provides an example of fuzzy sets along with the interpretations they represent. In Fig. 14.47, the left side of the arrow shows a fuzzy set representing each of the environmental conditions. The right side of the arrow shows a corresponding fuzzy set for the reliability estimate. It may be noted here that these fuzzy sets have been defined on subjective unreliability. The membership functions of these fuzzy sets may be determined subjectively. For example, in Fig. 14.47, x^* and x' are defined as follows.

$$x^* = f(p_M), \tag{14.244}$$

where p_M is a recommended basic failure or basic error probability.

$$x' = (1+x^*)/2. \tag{14.245}$$

In the failure possibility approach when a basic failure or a basic error probability is given, or when the estimate of reliability is expressed by a natural language without considering the influence of environmental conditions, the parameter x_0 is estimated by fuzzy reasoning using the following procedure. The fuzzy reasoning has been employed in the field of fuzzy control and fuzzy expert system[63]. Here, this technique is used in the field of human performance evaluation. Let g_{ij} and h_{ij} be membership functions of fuzzy sets on the left side and on the right side of the arrow in Fig. 14.47, respectively(j = 1,2 when i = EE, MQ, and OE, j = 1,2,3 when i = SL).

(1)Each environmental condition is evaluated. Let the evaluation be denoted as e_i. In this procedure e_i is given a numerical value within the interval [0, 1]. The evaluation can also be expressed by natural language, in which case, another procedure would be necessary.
(2)The grade of membership w_{ij} is obtained such as $w_{ij} = g_{ij}(e_i)$.
(3)Subjective unreliability x_{ij}, considering the environmental condition i, is obtained such that $w_{ij} = h_{ij}(x_{ij})$.
(4)The parameter x_0 is obtained as the weighted average such as

$$x_0 = \frac{\Sigma_{i,j} w_{ij} \times x_{ij}}{\Sigma_{i,j} w_{ij}} \tag{14.246}$$

As mentioned before, the membership functions g_{ij} and h_{ij} may be defined subjectively. But they should be determined based on some considerations. For example, the following guidelines may be used for this purpose:

For Equipment:

(1)In the optimal case of each environmental condition(e_i = 1), x_0 = 0. On the other hand in the worst case(e_i = 0), x_0 = 1.
(2)In the normal case(e_i = 0.5), $x_0 = x^*$.
(3)When the evaluation of each environmental condition is about the normal(e_i = 0.5), the influence on x_0 is small even if the evaluation of each condition slightly changes. On the other hand when the evaluation is good or bad, the change in the evaluation of each condition greatly has an influence on the change in x_0.

For an Operator:

(1)In the optimal case of each environmental condition(e_i = 1 when i=OE, e_i = 0.5 when i=SL), $x_0 = x^*$, which means that no matter how good the conditions are, a human operator does not always fail in a task. Human reliability has a limit. On the other hand in the worst case(e_i = 0 when i=OE, e_i = 1 when i=SL), x_0 = 1.
(2)However, as for the stress level x' is defined by (14.245). This is based on the fact that human reliability becomes low when stress level is low, but is not so low as when stress level is high. In [53], the basic error rate is doubled as shown in Table 14.1.
(3)A human operator has flexibility. That is, even if the evaluation of each condition is not very good, human reliability does not become significantly low.

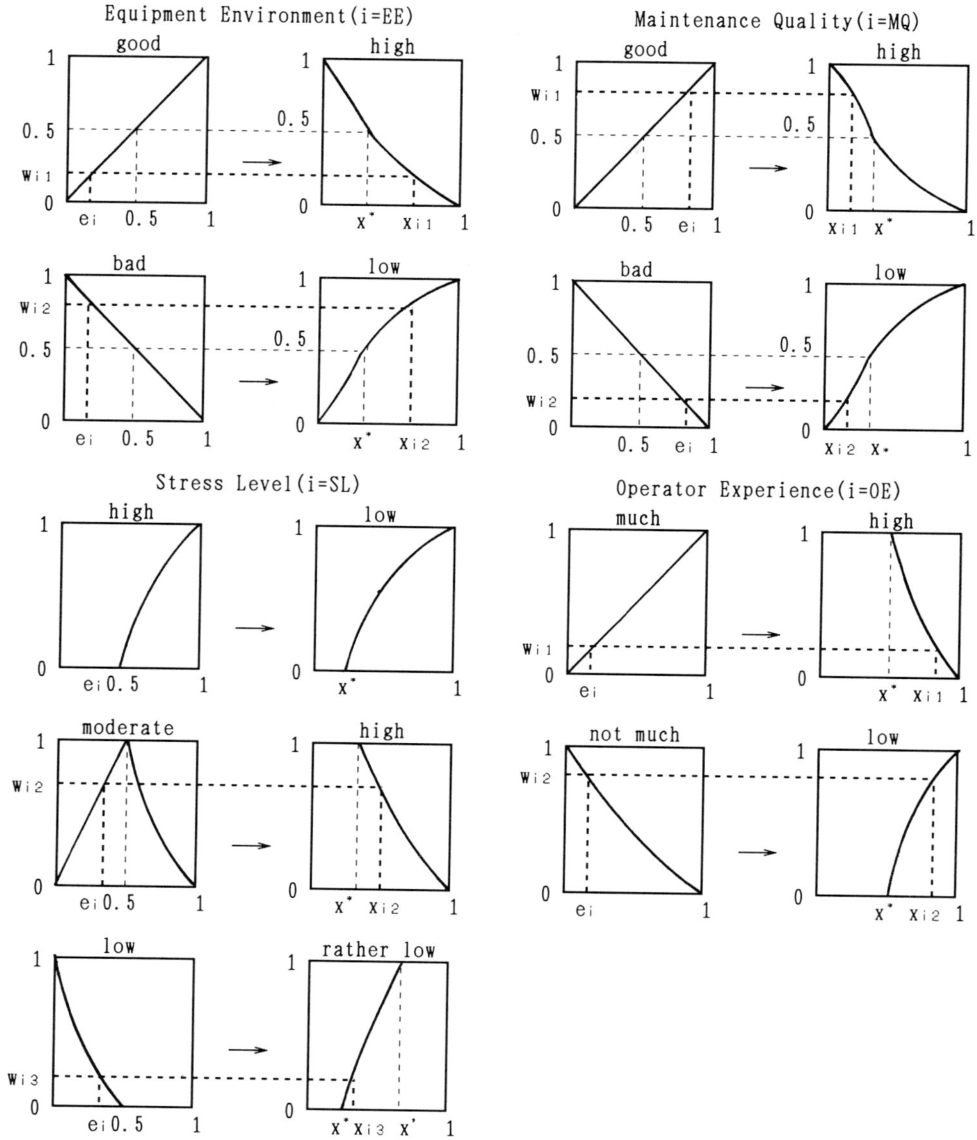

Fig. 14.47 Fuzzy Sets Representing Relations between Reliability and Environmental Conditions

If other conditions must be added, then relationships between reliability and these conditions are expressed qualitatively and the above procedure is repeated. If relationships between reliability and environmental conditions should be considered from other viewpoints, the membership functions should be redefined accordingly.

14.20 COMMON CAUSE FAILURES

A redundant system operates based on the consideration that its subsystems are functionally independent of one another. If the failure of one of subsystems has an influence on other subsystems and leads to a failure consequently, the system does not any longer work as a redundant system. The dependency situations considered in section 14.18 may be considered as one kind of common cause failures. Here a failure cause is considered from a wider point of view, that is, the cause external to the system being analyzed, such as an earthquake, a typhoon, an airplane crash etc. Whether the fuzzy sets approach or the probabilistic approach is used, it is necessary to consider the kinds of common cause failures that can occur and the situations under which they can occur in an analyzed system. For example, we should analyze whether the same or a similar component has been used in many subsystems or not, or what kinds of failure modes can be designated as common cause failures. In the construction of a fault tree, we need to analyze common cause failures and their effects.

For example, let us consider the common cause failures in the fault tree given in Fig. 14.45. Let us assume that a common cause failure should be considered in the fault tree as shown in Fig. 14.48. In this tree, the common cause failure is considered as a basic event. In the failure possibility approach, the failure possibility of the common cause failure is assumed to have a membership function given in (14.223), that is, it is a crisp set. Reliability of this system is analyzed in the same way as outlined before. The result of the analysis is shown in Fig. 14.49. The result is expressed such as *system reliability is low and its fuzziness is medium.* The failure possibility of the top event of the fuzzy fault tree of Fig. 14.45 is shown in Fig. 14.46, which implies that *system reliability is rather high and its fuzziness is high.* Comparing Fig. 14.49 with Fig. 14.46 it is found that the system structure should be improved in order to decrease the effect of the common cause failure.

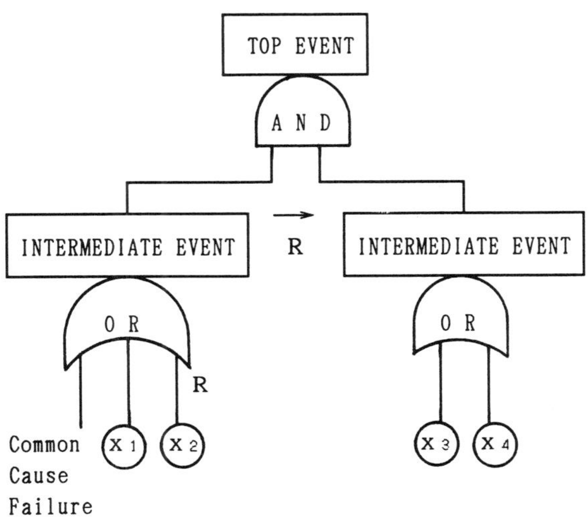

Fig. 14.48 Fuzzy Fault Tree Considering a Common Cause Failure

This analysis is related to *what if analysis*[64]. *What if analysis* implies the analysis in a hypothetical situation. In a large and complex system even if small troubles of a subsystem go undetected, a big accident sometimes does not take place until other trifling troubles occur in some other subsystems. Thus the system has a

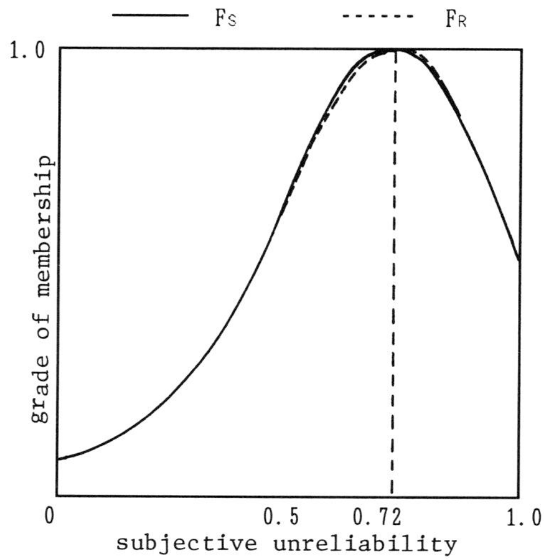

Fig. 14.49 Failure Possibility of Top Event Considering a Common Cause Failure

latent accident. In *what if analysis,* an equipment or a subsystem is assumed to be accident or trouble, or a human operator is assumed to fail in a task. Therefore its failure possibility is not a fuzzy set but a crisp set expressed by (14.223). The reliability of such a system is analyzed in the same way as mentioned earlier. Now let us consider a fuzzy fault tree shown in Fig. 14.50, where data of basic events are given in Table 14.9. Fig. 14.51 shows the analysis result which is *system reliability is rather high and its fuzziness is medium.*

Next let us assume that the basic event x_1 in Fig. 14.50 takes place certainly. In this hypothetical situation, the failure possibility of the total system is shown in Fig. 14.52. It is found that the failure possibility in Fig. 14.52 shows lower reliability than that in Fig. 14.51. However there is still a possibility that a failure does not occur since there is a possibility that other subsystems may prevent the failure from occuring on account of the parallel system with a low dependency. Next let us assume that the basic event x_7 in Fig. 14.50 also occurs certainly. It is found that the failure possibility is a crisp set as shown in Fig. 14.53, which means that the failure is bound to occur. *What if analysis* is related to the analysis of commom cause failures in the sense that failure possibilities of common cause failures events are assumed to be not fuzzy sets but crisp sets.

14.21 CONCLUSIONS

The main objective of this chapter has been to provide an insight into the problem of application of FST to various aspects of system reliability evaluation. As is now evident from the foregoing sections that FST provides an effective tool for solving many unsolved problems of system reliability evaluation, such as, uncertainty associated with data and models and of handling the problem of subjectivity.

It has been also amply brought out that the probability theory alone is not sufficient to deal with the problem of human subjectivity as it does not necessarily follow the probability axioms. The use of natural language expressions about the reliability information offers an alternative approach, to handle this problem effectively. The reliability information can be best be expressed not through probability but through a fuzzy set. Usually, the system state cannot be defined crisply as *good,* or

Table 14.9 Data Used in What If Analysis

Basic Event	Reliability Estimate	Fuzziness
x_1	standard	rather high
x_2	rather low	low
x_3	standard	rather high
x_4	standard	high
x_5	standard	low
x_6	rather high	medium
x_7	standard	low
x_8	rather high	medium

* Dependency level between intermediate events is *low* and its fuzziness is *medium*.

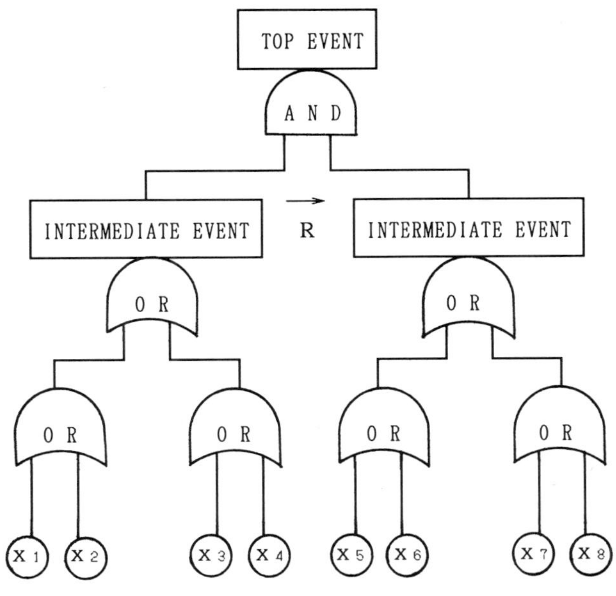

Fig. 14.50 Fuzzy Fault Tree

bad. Further, the transition from one state to the other is not discrete or sharp. Especially the classifications of the performance of a system by human being can best be done through fuzzy expressions or quantifications, and from this pont of view FST offers superior methodology than the probability theory.

As a matter of fact, the probability theory still continues to serve as a main tool in the field of system reliability evaluation. Applications of FST to safety and reliability problems have met with criticism. However, of recent, FST has been applied to a number of system reliability problems and other related areas such as fault tree analysis, event tree analysis and probabilistic risk assessment. With more and more technical articles appearing in this area, it is expected that the interest in FST would grow stronger and we would be able to overcome many of the difficulties of probabilistic approach through the use of FST.

REFERENCES

[1] Zadeh L.A., *Fuzzy Sets*, Information and Control, No. 8, pp. 338-353, 1965.

[2] Zadeh L.A., K.S. Fu, K. Tanaka and M. Shimura, *Fuzzy Sets and Their Applications to Cognitive and Decision Processes*, Academic Press, New York, 1975.

[3] Kaufmann A., *Theory of Fuzzy Subsets - Introduction to the*, Academic Press, New York, 1975.

[4] Nagoita C.V. and D.A. Ralescu, *Applications of Fuzzy Sets to System Analysis*, Birkhauser Verlag, Basel, 1975.

[5] Dubois D. and H. Prade, *Fuzzy Sets and Systems*, Academic Press, New York, 1980.

[6] Yager R.R., *Fuzzy Sets and Possibility Theory : Recent Development*, Pergamon Press Inc., New York, 1982.

[7] Gupta M.M. and E. Sanchez, *Approximate Reasoning in Decision Analysis*, North-Holland, New York, 1982.

[8] Zimmermann H.J., L.A. Zadeh and B.R. Gaines, *Fuzzy Sets and Decision Analysis*, North-Holland, New York, 1984.

[9] Zimmermann H.J., *Fuzzy Set Theory and its Applications*, Kluwer-Nijhoff Publishers, USA, 1985.

[10] Kandel A., *Fuzzy Mathematical Techniques with Applications*, Addison-Wesley Publishing Co., Masschusettes, USA, 1986.

[11] Klir G.J. and T.A. Folger, *Fuzzy Sets, Uncertainty and Information*, Prentice-Hall, Inc., Englewood Cliffs, N.J., USA, 1988.

[12] Gupta M.M. and T. Yamakawa, *Fuzzy Computing : Theory, Hardware and Applications*, North-Holland, Amsterdam, 1988.

[13] Lukasiewicz J., *Philosophische Bemerkungen zu Mehrwertigen Systemen des Aussagenkalkulus*, C.R. Seances Soc. Sci. Lettres de Varsovie, Vol. 23, pp. 51-57, 1930.

[14] Skala H.J., *On Many-Valued Logics, Fuzzy Sets, Fuzzy Logics and their Applications*, Fuzzy Sets and Systems, Vol. 1, No. 2, pp. 129-149, 1978.

[15] Mizumoto M. and K. Tanaka, *Fuzzy Sets and their Operations*, Information and Control, Vol. 48, pp. 30-48, 1981.

[16] Mizumoto M. and K. Tanaka, *Fuzzy Sets under Various Operations*, 4th International Congress of Cybernetics and Systems, Amsterdam, August 21-25, 1978.

[17] Zadeh L.A., *The Concept of a Linguistic Variable and its Application to Approximate Reasoning*, Pts. 1,2 and 3, Information Science, Vol. 8, pp. 199-249, Vol. 8, pp. 301-357 and Vol. 9, pp. 43-80, 1975.

[18] Mizumoto M. and K. Tanaka, *Some Properties of Fuzzy Sets of Type 2*, Information and Control, Vol. 31, pp. 312-340, 1976.

[19] Mizumoto M. and K. Tanaka, *Fuzzy Sets of Type 2 under Algebraic Product and Algebraic Sum*, Fuzzy Sets and Systems, Vol. 5, pp. 277-290, 1981.

[20] Dubois D. and H. Prade, *Operations on Fuzzy Numbers*, International Journal of System Sciences, Vol. 9, pp. 613-626, 1978.

[21] Mizumoto M. and K. Tanaka, *Some Properties of Fuzzy Numbers*, in Advances in Fuzzy Set Theory and Applications, edited by M.M. Gupta, R.K. Ragade and R.R. Yager, North-Holland Publishing Co., pp. 153-164, 1979.

[22] Schweizer B. and A. Sklar, *Associative Functions and Statisticaly Triangle Inequalities*, Publicationes Mathematicae Debrecen, Vol. 8, pp. 169-186, 1961.

[23] Sugeno M., *Theory of Fuzzy Integrals and their Applications*, Ph.D. Thesis, Tokyo Institute of Technology, Japan, 1974.

[24] Hamacher H., *Ueber Logische Verknupfungen Unscharfer Aussagen and deren Zugehorige Bewertungs Funktionen*, Progress in Cybernetics and Systems Research, Hemisphere Publisher, Washington D.C., pp. 276-287, 1975.

[25] Frank M.J., *On the Simultaneous Associativity of F(x,y) and x+y-F(x,y)*, Acquantiones Mathematicae, Vol. 19, pp. 194-226, 1976.

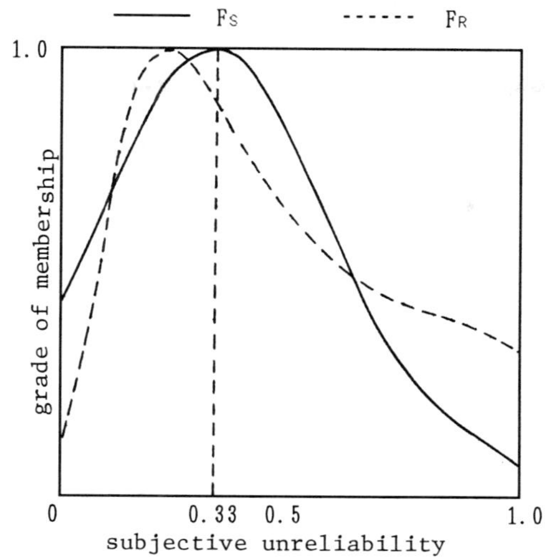

Fig. 14.51 Failure Possibility of Top Event of Fuzzy Fault Tree in Fig. 14.50

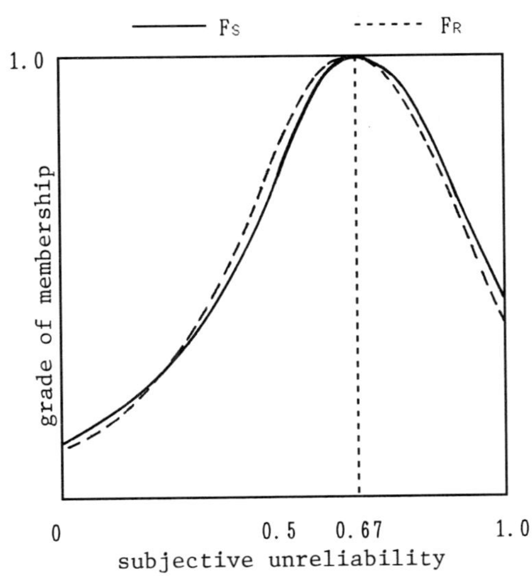

Fig. 14.52 Failure Possibility of Top Event in What If Analysis I

It may, however, be pointed out that neither FST nor probability theory stand alone can cope up to deal with complex problems arising in the area of system reliability evaluation, particulary in large and complex system. It is, therefore, possible that in near future new methodologies may be employed which may combine the best of the two approaches and therefore may involve some combinations of these two approaches to solve challenging problems of large and complex systems.

[26] Yager R., *On a Class of Fuzzy Connectives*, Fuzzy Sets and Systems, Vol. 4, pp. 235-242, 1980.

[27] Dubois D. and H. Prade, *A Class of Fuzzy Measure based on Triangular Norms*, International Journal of General Systems, Vol. 8, pp. 43-61, 1982.

[28] Dombi J., *A General Class of Fuzzy Operations, the DeMorgan Class of Fuzzy Operators and Fuzziness Measures Induced by Fuzzy Operators*, Fuzzy Sets and Systems, Vol. 8, pp. 149-163, 1982.

[29] Shackle G.L.S., *Decision, Order and Time in Human Affairs*, Cambridge University Press, London, pp. 67-103, 1961.

[30] Shafer G., *Mathematical Theory of Evidence*, Princeton University, University Press, Princeton, N.J., 1976.

[31] Dempster A.P., *Upper and Lower Probabilities Induced by Multivalued Mappings*, Annals of Math. Statistics, Vol. 38, pp. 325-339, 1967.

[32] Zadeh LA., *Fuzzy Sets as a Basis for a Theory of Possibility*, Fuzzy Sets and Systems, Vol. 1, No. 1, pp. 3-28, 1978.

[33] Dubois D. and H. Prade, *Towards Fuzzy Analysis : Integration and Derivation of Fuzzy Functions*, in Fuzzy Algebra, Analysis, Logics, Tech. Rep. TR-EE 78/13, Pt. C, Purdue University, Lafayette, 1978.

[34] Nguyen H., *On Conditional Possibility Distributions*, Fuzzy Sets and Systems, Vol. 1, No. 4, pp. 299-310, 1978.

[35] U.S. Atomic Energy Commission, *Reactor Safety Study : An Assessment of Accident Risks in U.S. Commercial Nuclear Power Plants*, WASH-1400, 1974.

[36] Swain A.D., *A Method for Performing a Human-Factors Reliability Analysis*, Report SCR-685, Sandia Corporation, 1963.

[37] Svenson O., *On Expert Judgement in Safety Analyses in the Process Industries*, Reliability Engineering and Systems Safety, Vol. 25, pp. 219-256, 1989.

[38] For example, Proc. of the 4th International Fuzzy Systems Association Congress, Brussels, Vol. Computer, Management & Systems Science, Vol. Artificial Intelligence, Vol. Engineering, 1991.

[39] Shahinpoor M. and D.J. Wells, *Applications Possibilities for Fuzzy Failure Analysis, and Diagnosis of Reactor Plant Components and Areas*, Nuclear Engineering and Design, Vol. 61, pp. 93- 100, 1980.

[40] Noma K., H. Tanaka and K. Asai, *On Fault Tree Analysis with Fuzzy Probability*, Japanese Journal of Ergonomics, Vol. 17, pp. 291-297, 1981. (in Japanese)

[41] Misra K.B. and A. Sharma, *Performance Index to Quantify Reliability Using Fuzzy Subset Theory*, Microelectronics and Reliability, Vol. 21, pp. 543-549, 1981.

[42] Tanaka H., L.T. Fan, F.S. Lai and K. Toguchi, *Fault Tree Analysis by Fuzzy Probability*, IEEE Trans. Rel., Vol. R-32, No. 3, pp. 453-457, 1983.

[43] Terano T., Y. Murayama and N. Akiyama, *Human Reliability and System Safety Evaluation of Man-Machine Systems*, Automatica, Vol. 19, pp. 719-722, 1983.

[44] Furuta H. and N. Shiraishi, *Fuzzy Importance in Fault Tree Analysis*, Fuzzy Sets and Systems, Vol. 12, pp. 205-213, 1984.

[45] Schmucker K.J., *Fuzzy Sets, Natural Language Computations, and Risk Analysis*, Computer Science Press, Rockville, MD, 1984.

[46] Karwowski W. and A. Mital, *Applications of Approximate Reasoning in Risk Analysis*, in Applications of Fuzzy Set Theory in Human Factors, eds. W. Karwowski and A. Mital, pp. 227- 243, Elsevier, 1986.

[47] Park K.S., *Fuzzy Apportionment of System Reliability*, IEEE Trans. on Rel., Vol. R-36, No.2, 129-132, 1987.

[48] Onisawa T., *An Approach to Human Reliability in Man-Machine System Using Error Possibility*, Fuzzy Sets and Systems, Vol. 27, pp. 87-103, 1988.

[49] Singer D., *A Fuzzy Set Approach to Fault and Reliability Analysis*, Fuzzy Sets and Systems, Vol. 34, pp. 145-155, 1990.

[50] Kenarangui R., *Event-tree Analysis by Fuzzy Probability*, IEEE Trans. on Rel., Vol. R-40, No. 1, pp. 120-124, 1991.

[51] Kai-Yuan C., W. Chuan-Yuan and Z. Ming-Lian, *Posbist Reliability Behaviour of Typical Systems with Two Types of Failure*, Fuzzy Sets and Systems, Vol. 43, pp. 17-32, 1991.

[52] Nishiwaki Y. and C. Preysal, *Uncertainties in Estimating Dose-Effect Relationships and Optimization of Radiation Protection Based on Fuzzy Risk Analysis*, Proc. of the 4th International Fuzzy Systems Association Congress, Brussels, Vol. Eng., pp. 151-154, 1991.

[53] Swain A.D. and H.E. Guttmann, *Handbook of Human Reliability with Emphasis on Nuclear Power Plant Applications*, NUREG/CR- 1278, 1983.

[54] Misra K.B. and G.G. Weber, *Use of Fuzzy Sets Theory for Level-1 Studies in Probabilistic Risk Assessment*, Fuzzy Sets and Systems, Vol. 37, pp. 139-160, 1990.

[55] Chowdhury S.G. and K.B. Misra, *Evaluation of Fuzzy Reliability of a Non-Series Parallel-Network*, Microelectronics and Reliability, Vol. 33, pp. 1-4, 1992.

[56] Soman K.P. and K.B. Misra, *Fuzzy Fault Tree Analysis Using Resolution Identity*, International Journal of Fuzzy Sets and Mathematics, Vol. 1, (to appear) 1993.

[57] Adams J.A., *Issues in Human Reliability*, Human Factors, Vol. 34, pp. 1-10, 1982.

[58] Onisawa T., *An Application of Fuzzy Concepts to Modeling of Reliability Analysis*, Fuzzy Sets and Systems, Vol. 37, pp. 269-286, 1990.

[59] Misra K.B. and G.G. Weber, *A New Method for Fuzzy Fault Tree Analysis*, Microelectronics and Reliability, Vol. 29, pp. 195- 216, 1989.

[60] Onisawa T., *Fuzzy Reliability Analysis Using Natural Language Information and Numerical Information*, Proc. of the 4th International Fuzzy Systems Association Congress, Brussels, Vol. Eng., pp. 163-166, 1991.

[61] Onisawa T. and Y. Nishiwaki, *Fuzzy Human Reliability Analysis on the Chernobyl Accident*, Fuzzy Sets and Systems, Vol. 28, pp. 115-127, 1988.

[62] Onisawa T., *Fuzzy Reliability Assessment Considering the Influence of Many Factors on Reliability*, Int. J. of Approximate Reasoning, Vol. 5, pp. 265-280, 1991.

[63] Sugeno M. ed., *Industrial Application of Fuzzy Control*, North-Holland, 1988.

[64] Onisawa T., *What if Analysis in Fuzzy Reliability Analysis*, Proc. of Int. Conf. on Fuzzy Logic & Neural Networks, Iizuka, Vol. 2, pp. 919-922, 1990.

OTHER RECOMMENDED READING

[1] Hirota K. and W.Pedrycz, *Concepts Formation : Representation and Processing Issues*, Special Issue on Uncertainty Management in Knowledge Based Systems (B. Bouchon-Meunier, Gu. Ed.), Int. J. Intelligent Systems, Vol.7, No.1,pp.3-13,1992.

[2] Esogbue A. O., *Fuzzy Sets Modeling and Optimization as an Aid to Disaster Control Planning*, IFES,1991,pp 863-876.

[3] Novak V.P., *Fuzzy Sets and their Application*, Published by Adam Kluger,1989.

[4] Cappelle B., E.E.Kerre, F.R.Vanmassenhove, *A Possibilistic Uncertainty Model Applied to the Barlow-Wu Class of Multistate Structures*, IFSA 1991-Math, pp 25-28.

[5] Averkin A.N.and M.Kh.Nguen, *Utilizing Fuzzy Relation in Knowledge Representation Modules*, Soviet J. Comput. Systems Sci., Vol.28, No.2, pp. 55-68,1990. Translated in Russian from Izv. Akad. Nauk. S.S.S.R. Tekhn. Kibernet, Vol.5, pp 20-33, 1989.

[6] Delgado M. and S. Morel, *On the Concept of Possibility-Probability Consistency*, Fuzzy Sets and Systems, Vol.21, pp. 311-318, 1987.

[7] Inuiguchi M., H. Ichihasi and Y. Kume, *Some Properties of Extended Fuzzy Preference Relations Using Modalities*, Inform. Sci., Vol.61, No.3, pp 187-209, 1992.

[8] Heilpern S., *The Expected Value of Fuzzy Number*, Fuzzy Sets and Systems, Vol.47, No.1, pp 81-86, Apr.10,1992.

[9] Garcia J., J. Pazos, J. Rios and J. Yague, *Methodologies of Linguistics Evaluation in Risk Situations Using Fuzzy Techniques*, Fuzzy Sets and Systems, Vol.48, No.2, pp 185-194, Jun.1992.

[10] Turksen I.B., *Fuzzy Expert Systems for IE/OR/MS*, Fuzzy Sets and Systems, Vol.51, No.1, pp 1-28, Oct.1992.

[11] Sudkamp T., *On Probability-Possibility Transformations*, Fuzzy Sets and Systems, Vol.51, No.1, pp 73-82, Oct.1992.

[12] Gupta M.M. and J. Qi, *Theory of T-norms and Fuzzy Inference Methods*, Fuzzy Sets and Systems, Vol.40, No.3, pp 431-450, Apr.1991.

[13] Dubois D., L. Lang and H. Prade, *Fuzzy Sets in Approximate Reasoning, Part1: Inference with Possibility Distributions*, Fuzzy Sets and Systems, Vol.40, No.1, pp 143-202, Mar.1991.

[14] Dubois D., J. Lang and H. Prade, *Fuzzy Sets in Approximate Reasoning, Part2: Logical Approaches*, Fuzzy Sets and Systems, Vol.40, No.1, pp 203-244, Mar.1991.

[15] Turksen I.B., *Measurement of Membership Function and their Acquisition,* Fuzzy Sets and Systems, Vol.40, No.1, pp 5-38, Mar.1992.

[16] Yuan Cai Kai-, Wen Chuan-Yuan and Zhang Ming-Lian, *Fuzzy Variables as a Basis for a Theory of Fuzzy Reliability in the Possibility Context*, Fuzzy Sets and Systems, Vol.42, No.2, pp 145-172, July 1991.

[17] Morvaj Z. and T. Kapetanovic, *Application of Fuzzy Logic in the Development of an Expert System for Security Monitoring of Electric Power System,* ASME, pp 61-70, 1990.

[18] Gutierrez Martinez S., *Fuzzy Reliability*, NAFIPS, pp 23-27, 1991.

[19] Toth H., *Probabilities and Fuzzy Events: An Operational Approach,* Fuzzy Sets and Systems, Vol.48, No.1, pp 113-127, May 1992.

[20] Zimmermann H.-J., *Description and Optimization of Fuzzy Systems,* Int. J. Gen. Syst., Vol.2, pp 209-215, 1976.

[21] Yager R.R., *Reasoning with Uncertainty for Expert Systems*, Proc. Int. Joint Confer. on Artificial Intelligence (Los Angeles, C.A.), pp 1295-1297, 1985.

[22] Higashi M. and G.J. Klir, *Resolution of Finite Fuzzy Relational Equations,* Fuzzy Sets and Systems, Vol.13, pp 65-82, 1984.

[23] Higashi M. and G.J. Klir, *Identification of Fuzzy Relational Systems,* IEEE Trans. Syst. Man and Cyber., Vol. SMC-14, pp 349-355, 1984.

[24] Tsukamoto Y. and T. Terano, *Failure Diagnosis by Using Fuzzy Logic,* IEEE Conf. on Decision Making and Control, New Orleans, pp 1390-1395, 1977.

[25] Wakabayashi J. and A. Fukumoto, *Simulation Study of a System for Diagnosis of Nuclear Power Plant Operation*, Nuclear Technology, Vol.54, pp 19-30, 1981.

[26] Dubois D. and H. Prade, *Ranking Fuzzy Numbers in the Setting of Possibility Theory*, Inform. Sci., Vol.30, pp 183-244, 1983.

[27] Bandler W. and L. Kohout, *Fuzzy Power Sets and Fuzzy Implication Operators,* Fuzzy Sets and Systems, Vol.4, pp 13-30, 1980.

New Trends in System Reliability Evaluation
K.B. Misra (Editor)

CHAPTER 15

DEMPSTER-SHAFER THEORY AND ITS APPLICATIONS

Toshiyuki Inagaki
Institute of Information Sciences and Electronics
University of Tsukuba, Tsukuba 305, Japan

ABSTRACT
This chapter gives the fundamentals of the Dempster-Shafer (DS) theory, an alternative approach for representing uncertainty where incomplete information or knowledge is a factor. Applications of the DS theory to a multiple-fault diagnosis problem and a system safety-control problem are explored.

15.1 Introduction

The Dempster-Shafer (DS) theory has been originated by Dempster [1] and developed by Shafer [2] as a tool for representing situations in which various kinds of *ignorance* exist in our knowledge or information about an object. The DS theory has been gaining popularity in various fields where incomplete knowledge is a factor.

This chapter gives first an introduction to the DS theory, where the fundamentals are explained with a variety of examples. We then explore a couple of applications of the DS theory in the area of system reliability and safety. The applications discussed are:

(1) *Multiple-fault diagnosis problem*, where faulty components must be identified based on vague information regarding occurrence and/or non-occurrence of symptoms.

(2) *Safety-control problem*, in which we must decide whether to operate or shut down a system based on available information which is possibly incomplete and imprecise.

One of the central issues in the DS theory is how we combine imperfect information given by independent *knowledge sources*, where knowledge sources include intelligent sensors, human operators. In various applications, Dempster's *rule of combination* has been used for fusing a given set of information. However Dempster's rule can give unintuitive results and exhibit numerical instability in some cases which are not very uncommon [3-4]. Yager [5] has proposed an alternative rule which is free from disadvantages of Dempster's rule. It is proven in this chapter that some other rules exist. We prove this by giving a unified rule of combination which includes Dempster's rule and Yager's rule as special cases. It is also shown that a rule of combination cannot be arbitrarily chosen in some decision making problems, especially in the area of system reliability and safety.

15.2 Dempster-Shafer Theory

In this section, we give the fundamentals of the Dempster-Shafer (DS) theory. We will see how the DS theory can represent ignorance in quite a natural manner, compared with probability theory. Methods are given for combining a set of information which are provided by independent knowledge sources.

15.2.1 Lack of Evidence

In dealing with a large-complex system such as a nuclear reactor or a spacecraft, it is unwise to assume that our knowledge is complete and perfect. Moreover, only small amount of sensory data might be available when we must identify the operating condition of the system. In some cases we may be able to judge, without much difficulty, whether the system is in its normal and safe operating condition, or in its abnormal and unsafe operating condition.

It is not very uncommon for us, however, to encounter situations in which we cannot derive a definite conclusion regarding the system state. In these situations we may say, for example, "we are 60% sure that the system is safe", where the proposition "the system is safe" is given with the *degree of belief* of 0.6 concerning the truth of the proposition. We will see, in the following, that if we define the degree of belief as the probability that the proposition is true, then the degree of belief does not necessarily represent our intention.

Let Ω be a sample space in probability theory. For simplicity we assume that Ω is a finite set which consists of n elements:

$$\Omega = \{\omega_1, ..., \omega_n\}$$

where each element ω_i is called a sample point. Let 2^Ω denote the power set of Ω; viz. 2^Ω is the set of all possible subsets of Ω. If Ω contains n sample points, 2^Ω includes 2^n subsets of Ω.

Probability measure P is defined as a function:

$$P: 2^\Omega \rightarrow [0, 1]$$

which assigns to each element of 2^Ω a number in the unit interval [0, 1]. Strictly speaking, probability measure is defined as a function which satisfies the following set of axioms:

(a) $P(A) \geq 0$ for any $A \in 2^\Omega$

(b) $P(\Omega) = 1$

(c) $P(\cup_i A_i) = \Sigma_i P(A_i)$ if $A_i \cap A_j = \emptyset$ whenever $i \neq j$

Axiom (c) imposes additivity on probability measure, which plays an important role in probability theory.

Probability theory has been providing us with many useful and powerful tools in various areas, including system reliability and safety engineering. However there exist cases in which a probability theoretic model is not very adequate, especially when there exists incompleteness in our knowledge or available information. We will see this in the following simple example.

Example 1

Consider a system which can take either one of two states, the safe state and the unsafe state. Let S and U denote the safe and the unsafe states of the system, respectively.

Suppose that we are requested to identify the current operating state of the system, but that we feel difficulty in judging with confidence whether the system is safe or unsafe. This kind of case can happen when the system is functionally very complex, or we are not given enough amount of information for judging the operating condition of the system.

Suppose we can only express our belief, based on available information and knowledge, by saying that "we are 60% sure that the system is in its safe operating condition." If we might want to represent this statement in the framework of probability theory, the probabilistic interpretation for the statement would be: "our system is safe - with probability 0.6".

What does this probabilistic interpretation really mean? The sample space Ω for this case is given as $\Omega = \{S, U\}$. The statement "our system is safe with probability 0.6" is represented as:

$$P(\{S\}) = 0.6$$

Then another statement is generated automatically. The statement is:

$$P(\{U\}) = 0.4$$

which says that "our system is unsafe - with probability 0.4", even though we had no intention of expressing a degree of belief concerning unsafety of the system. The above unintended statement is a consequence of axiom (c) of additivity for probability measure: viz.

$$P(\{S\}) + P(\{U\}) = 1$$

since $\{S\} \cap \{U\} = \emptyset$ and $\{S\} \cup \{U\} = \Omega$.

We must note that the proposition "we are 60% sure that our system is safe" is not necessarily equivalent to "we are 40% sure that our system is unsafe." We assigned the mass 60% of belief to the proposition "our system is safe" because we lack evidence which convinces us of system safety. The remaining 40% represents the degree to which we cannot say either that the system is safe or that the system is unsafe. In other words, there exists *ignorance* in our knowledge or information concerning the current and true state of the system. As a matter of fact, the following two extreme cases can happen when a new piece of evidence becomes available:

(1) Suppose we obtain additional evidence which supports unsafety of the system. Then the whole of 40% would be the degree of belief in the unsafety of the system.

(2) If we find missing evidence which supports system safety, then the remaining 40% would be added to mass 60% so that our degree of belief in system safety would become 100%.♦

15.2.2 Basic Probability Assignment

The DS theory has been proposed to give a tool for representing in a reasonable manner ignorance in our knowledge or information.

Let X be a finite set which is called the *frame of discernment* or the *universe of discourse*:

$$X = \{x_1, ..., x_n\}$$

where each element $x_i \in X$ is called an *elementary proposition* which can be true or false. It is assumed that the frame of discernment is an exhaustive set of elementary propositions of our interest, and that elementary propositions are mutually exclusive. In other words, one and only one of elementary propositions is true.

Various propositions can be expressed by applying connectives for negation, conjunction, and disjunction to elementary propositions. If we let 2^X denote the power set of X, then 2^X includes all possible propositions of our interest. If X consists of n elements, 2^X represents 2^n distinct propositions. Among them, the frame of discernment $X \in 2^X$ represents a proposition which is always true, and the empty set $\emptyset \in 2^X$ corresponds to a proposition which is always false.

Example 2

Consider a three-component system depicted in Figure 15.1. Assume that components fail statistically independently, and that no common-mode failure can happen.

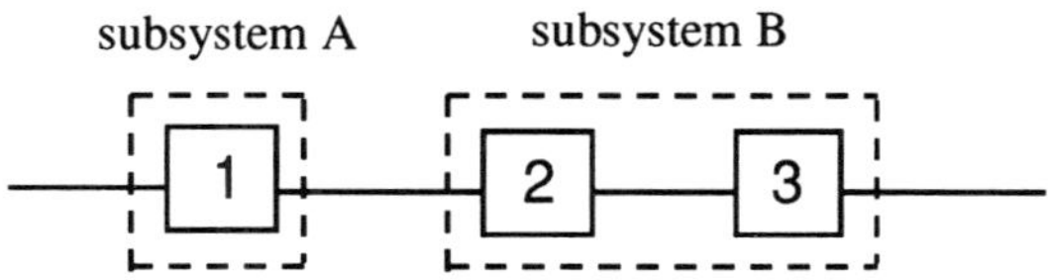

Fig.15.1 Three-Component System

Suppose the system malfunctions and we want to identify a faulty component. The frame of discernment is defined as:

$$X = \{x_1, x_2, x_3\}$$

where x_k denotes the proposition "component k is in its failed state." The power set 2^X of X includes every possible proposition of our interest. The power set 2^X is given explicitly as:

$$2^X = \{\emptyset, \{x_1\}, \{x_2\}, \{x_3\}, \{x_1, x_2\}, \{x_2, x_3\}, \{x_3, x_1\}, X\}$$

where subset $\{x_j, x_k\}$ corresponds to the proposition "either one (and only one) of components j and k is in its failed operating condition, but we do not know which component is failed".

The frame of discernment X is viewed as the proposition that "one of three components is in its failed operating condition," which is always true, because the system does not function properly. Note that proposition X itself does not convey any useful information in the sense that it gives no help in identifying a faulty component. We may be able to say that proposition X represents our complete *ignorance.*

Let us see how the connectives for negation ($\neg$), disjunction ($\cup$), and conjunction ($\cap$) work on propositions. Let $\neg\{x_1\}$ denote the negation of proposition $\{x_1\}$. Then $\neg\{x_1\}$ represent the proposition "component 1 is not failed", or its equivalent proposition "component 2 or 3 is faulty". The latter statement can be expressed as:

$$\{x_2\} \cup \{x_3\}$$

or, simply as $\{x_2, x_3\}$, which represents the proposition "subsystem B malfunctions."

We have assumed that only one component can fail, which is expressed as follows:

$$\{x_j\} \cap \{x_k\} = \emptyset$$

for any j and k such that $j \neq k$.♦

In order to express our degree how we believe in a proposition, we need the following function:

$$m : 2^X \rightarrow [0, 1]$$

which assigns to each element of 2^X a number in the unit interval [0, 1]. The number m(A) represents the *portion* of total belief assigned exactly to proposition A. We will see in Section 15.2.3 why m(A) does not necessarily represent the *total belief* in proposition A.

The function m is defined on the power set 2^X of the frame of discernment X as a measure which satisfies the following set of axioms:

I. $m(A) \geq 0$ for any $A \in 2^X$

II. $m(\emptyset) = 0$

III. $\sum_{A \in 2^X} m(A) = 1$

where measure m is called the *basic probability assignment function.*

Axiom I states that our degree of belief that a proposition is warranted is represented as some non-negative number. Axiom 2 states that our degree of belief about a proposition takes the value zero when the proposition can never be true. Axiom 3 states that degrees of belief assigned to possible propositions sum up to unity.

From Axioms I and III, we have:

$$0 \leq m(A) \leq 1$$

for any $A \in 2^X$. If $m(A) > 0$, then the subset A of X is called a *focal proposition.*

The above Axioms I-III may look similar to Axioms (a)-(c) for probability measure which are given in the previous section. However the axioms for the basic probability assignment functions are looser than those for probability measure. We will see this in the following examples.

Example 3

Let X be the frame of discernment consisting of two mutually exclusive propositions x and y: $X = \{x, y\}$. Let:

$$m(\{x\}) = 0.2, \quad m(\{y\}) = 0.5, \quad m(X) = 0.3, \quad m(\emptyset) = 0 \tag{15.1}$$

The function m defined by (15.1) satisfies Axioms I-III. Thus m is a basic probability assignment function.

For this basic probability assignment function, it is easy to see the following properties:

1. Additivity does not necessarily hold: viz.

$$m(\{x\}) + m(\{y\}) \neq m(\{x, y\}) = m(X)$$

   where $\{x\} \cap \{y\} = \emptyset$. Recall that additivity was assumed for the probability measure by Axiom (c).

2. Monotonicity does not necessarily hold: viz.

$$m(\{y\}) \geq m(X) \text{ even though } \{y\} \subset X$$

   while probability measure always satisfies monotonicity:

$$P(E) \leq P(F) \text{ for } E \subset F$$

   which is an immediate consequence of Axioms (a)-(c).

3. It is not required that $m(X) = 1$, but $m(X) \leq 1$. Actually, in the above case, we have $m(X) < 1$. Thus $m(\emptyset) = 0$ does not imply $m(X) = 1$ in the DS theory. In probability theory $P(X) = 1$ implies $P(\emptyset) = 0$, and vice versa. ♦

Example 4

Let the frame of discernment X consist of three mutually exclusive propositions: $X = \{x_1, x_2, x_3\}$. The power set 2^X has eight propositions including $\emptyset$, as shown in Example 2. Since $\emptyset$ represents a proposition which is always false, we have $m(\emptyset) = 0$ by Axiom II. Thus the total mass of unity is allocated, by Axiom III, to the remaining seven possible propositions:

$$m(\{x_1\}) + m(\{x_2\}) + m(\{x_3\}) + m(\{x_1, x_2\})$$

$$+ m(\{x_2, x_3\}) + m(\{x_3, x_1\}) + m(X) = 1$$

It would be useful to note here difference between the basic probability assignment function and a *probability mass function* (or, a *probability distribution function* for a *discrete random variable*). Let Ω be the sample space and let Y be a discrete random variable on Ω. Suppose Y can take one of values y_1, y_2, and y_3. The probability distribution for the discrete random variable Y is defined by:

$$P(\{\omega: Y(\omega) = y_1\}) + P(\{\omega: Y(\omega) = y_2\}) + P(\{\omega: Y(\omega) = y_3\}) = 1$$

If we let p denote a probability mass function, the above equation is represented as:

$$p(y_1) + p(y_2) + p(y_3) = 1$$

where:

$$p(y_k) = P(\{\omega: Y(\omega) = y_k\})$$

Note that the probability mass function p is defined only for a singleton $\{y_k\}$, k = 1, 2, 3.

The basic probability assignment function m is allowed to assign some number in the unit interval [0, 1] not only to a singleton $\{x_k\}$, but also to any subset of X such as $\{x_j, x_k\}$. When every focal proposition is given as a singleton (viz., any A for which $m(A) > 0$ is represented as $A = \{x_k\}$ for some k), the basic probability assignment function can be regarded as a probability mass function. ♦

Remark

Example 4 suggests that a basic probability assignment function is seen as a generalization of a probability mass function in probability theory. It is not wise to say from this fact, however, that probability theory is a proper subset of the DS theory. Probability theory can be constructed on an infinite sample space, while the DS theory has not yet been well-established on a frame of discernment consisting of a countably or uncountably infinite number of elementary propositions, at least at the present time. Moreover, probability theory has some useful tools which the DS theory does not have. An important point is that we should recognize differences between the DS theory and probability theory so that we can appreciate properly distinctive features of each of their theoretical frameworks. ♦

We now come to the point to see one of significant motivations for developing the DS theory. The DS theory can distinguish *lack of belief* from *disbelief* in a natural manner, which is not impossible but not very easy in conventional probability theory. Consider the following example.

Example 5

Suppose Mr. Z is a murder suspect, and we are to express our degree of belief regarding whether he is a real murderer or innocent. Let A denotes the proposition "Mr. Z is a murderer," and ¬A denotes the proposition "Mr. Z is not a murderer." Then the frame of discernment X is given as: $X = \{A, \neg A\}$.

Suppose we have some evidence, such as somebody's testimony, which makes us think that Mr. Z can be a murderer. Assume that the evidence is not perfect enough to judge with confidence that Mr. Z is definitely committed a murder.

If we express our degree of belief for the proposition A in the framework of probability theory, say as $P(A) = 0.6$, then $P(\neg A) = 0.4$ follows immediately. The latter represents the degree of belief in proposition ¬A which claims Mr. Z's innocence. It is not appropriate for us to say that Mr. Z is innocent without having any evidence for proving Mr. Z's innocence. The only evidence we have now is one which is to prove that Mr. Z is a murderer. (Recall Example 1.)

In the DS theory, on the contrary, we can express the present situation as follows:

$$m(A) = 0.6, \quad m(X) = 0.4 \tag{15.2}$$

In the above, m(X) represents the degree to which we really do not know whether Mr. Z is a murderer or not. In other words, m(X) represents our *lack of belief* in the sense that proposition X does not imply any specific proposition but a tautology. Sometimes m(X) is thus called the *degree of ignorance*. ♦

15.2.3 Belief Structures

The collection of values m(A) for $A \in 2^X$, denoted as

$$\{m(A) : A \in 2^X\}$$

is called a *belief structure*. Some typical belief structures are given below.

Example 6

Let the frame of discernment be denoted as $X = \{x_1, ..., x_n\}$. A belief structure $\{m(A) : A \in 2^X\}$ is called a *Bayesian belief structure* if any focal proposition A is given as a singleton $\{x_k\}$ for some k. For instance, let $X = \{x_1, x_2, x_3\}$ and let:

$$m(\{x_2\}) = 0.3, \quad m(\{x_3\}) = 0.7$$

The obtained belief structure is a Bayesian belief structure. ♦

Example 7

A belief structure is called a *vacuous belief structure* when:

$$m(X) = 1$$

or, in other words, $m(A) = 0$ for any A which is distinct from X.

A vacuous belief structure represents the situation of complete ignorance in which "we have no idea which proposition is true." ♦

Example 8

A belief structure is called a *simple support function focused at proposition A* if:

$$m(A) + m(X) = 1$$

or, equivalently, $m(A) = s$, $m(X) = 1 - s$ for some number $s \in [0, 1]$, and $m(B) = 0$ for any other $B \in 2^X$ such that $B \neq A$, $B \neq X$. The basic probability assignment function which is given in Example 5 determines a simple support function.

A simple support function represents the case in which "we can believe in proposition A to some extent, but we have neither belief nor disbelief in any proposition which is distinct from proposition A." ♦

Example 9

A belief structure is called a *complementary belief structure* if:

$$m(A) + m(\neg A) = 1$$

or, more precisely, $m(A) = s$, $m(\neg A) = 1 - s$ for some $s \in [0, 1]$, and $m(B) = 0$ for any B which is distinct from A and $\neg A$.

A complementary belief structure may look like a probability mass function. There exists a difference, in general. In a complementary belief structure, proposition A or $\neg A$ can consist of two or more elementary propositions. The following example shows that a complementary belief structure is not necessarily a Bayesian belief structure. ♦

Example 10

Consider a four-component system illustrated in Figure 15.2. Assume that components fail statistically independently, and that no common-mode failure exists.

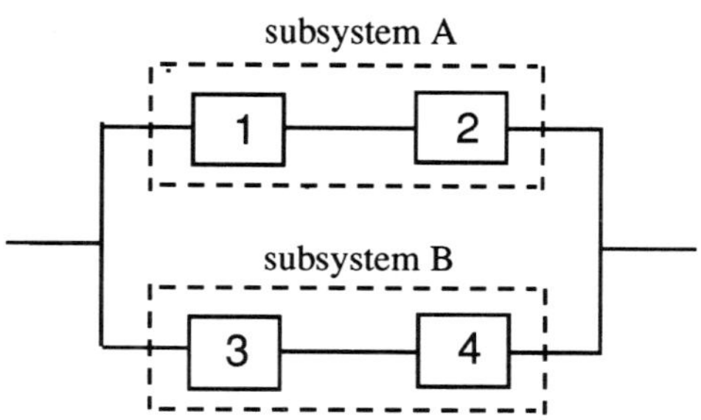

Fig.15.2 Four-Component System

Suppose that the system behaves a little bit strangely now, and that there was no problem in the system until just a few seconds ago. Some component must have become faulty during these few seconds.

Assume that we have some pieces of evidence which proves the followings:

(1) either component 1 or 2 is working properly,
(2) either component 3 or 4 is working properly, and
(3) subsystem A is more likely to be faulty than subsystem B.

Assume no further information is available for identifying which component is really faulty.

Let the frame of discernment be $X = \{x_1, x_2, x_3, x_4\}$, where x_k denotes the proposition "component k is faulty." Our situation can be represented by the following complementary belief structure:

$$m(\{x_1, x_2\}) = s, \quad m(\{x_3, x_4\}) = 1 - s \tag{15.3}$$

for some s such that $0.5 < s < 1$. Note that neither focal proposition $\{x_1, x_2\}$ nor $\{x_3, x_4\}$ is a singleton. The complementary belief structure given by (15.3) is not a Bayesian belief structure. ♦

Example 11

Suppose, in Example 10, that we do not care which component is faulty, but that we do care which subsystem is faulty, because replacement can be made only on subsystem level; we must replace a subsystem by a new one even when only a single component in the subsystem is faulty. The frame of discernment for this case would be $X = \{x_A, x_B\}$, where x_k denotes the proposition "subsystem k is faulty." The situation can then be represented by the following complementary belief structure:

$$m(\{x_A\}) = s, \quad m(\{x_B\}) = 1 - s$$

which is also a Bayesian belief structure. ♦

One of distinctive features of the basic probability assignment m(A) is seen when A is not a singleton.

Example 12

Let the frame of discernment be $X = \{x_1, x_2, x_3\}$. Suppose the following belief structure is given:

$$m(\{x_1\}) = 0.2, \quad m(\{x_1, x_2\}) = 0.3, \quad m(\{x_2, x_3\}) = 0.5 \tag{15.4}$$

The meaning of the first statement, $m(\{x_1\}) = 0.2$, would be clear. The mass 0.2 measures the amount of belief that one commits specifically to a single elementary proposition x_1.

We should be very careful in interpreting the second statement, $m(\{x_1, x_2\}) = 0.3$. The mass 0.3 measures the amount of belief that one commits to the set $\{x_1, x_2\}$ of elementary propositions x_1 and x_2, where the proposition $\{x_1, x_2\}$ means that one and only one of x_1 and x_2 is true. The point to be noted here is: *$m(\{x_1, x_2\}) = 0.3$ does not specify at all how the mass 0.3 is shared between elementary propositions x_1 and x_2.* In other words, $m(\{x_1, x_2\}) = 0.3$ can correspond to various situations, such as:

(a) mass 0.15 is allocated evenly to x_1 and x_2
(b) mass 0.1 is allocated to x_1, while 0.2 is assigned to x_2
(c) mass 0.3 is assigned to x_2, while no mass is given to x_1

Readers are recommended to give various situations which are represented by the third statement, $m(\{x_2, x_3\}) = 0.5$. Readers are also recommended to think how this approach of assigning mass is different from a conventional probability theoretic approach which utilizes a probability mass function. Suppose we want to represent the above three cases (a)-(c) which are possible under $m(\{x_1, x_2\}) = 0.3$. Then we must specify an appropriate probability mass functions for each case. The resulting probability mass function differs, of course, from case to case. ♦

Since a belief structure allows various possibilities in allocating mass to propositions, our degree of belief in a proposition cannot be represented by a single value. Instead, we need an interval lying in the unit interval [0, 1] for representing possible values which our degree of belief can take.

Let A be a proposition. Then we would say that our degree of belief for A lies in the interval [Bel(A), Pl(A)], where Bel(A) and Pl(A) are defined as follows:

$$\mathrm{Bel}(A) = \sum_{C \subset A} m(C) \tag{15.5}$$

$$\mathrm{Pl}(A) = \sum_{C \cap A \neq \emptyset} m(C) \tag{15.6}$$

Bel(A) is called a *belief function* and Pl(A) is called a *plausibility function* [6]. It is easy to see that Pl(A) can be defined alternatively by:

$$\mathrm{Pl}(A) = 1 - \mathrm{Bel}(\neg A)$$

The belief function, Bel(A), measures the *total amount* of our belief in proposition A. Suppose the set C of elementary propositions is a subset of the set A, or, $C \subset A$. Then proposition C implies proposition A, and thus the mass m(C), which is assigned precisely to C, supports also proposition A. In other words, the total degree of our belief in A is obtained by adding m(C) for every proposition C which implies proposition A. Recall that we have said in the previous section that m(A) is just the *portion* of total belief in proposition A.

The plausibility function, Pl(A), measures the *maximal amount* of our belief which can possibly be allocated to proposition A. Let us assume that $C \cap A \neq \emptyset$. Then proposition A can be true, when proposition C is true. Since the basic probability assignment function m does not specify the way how the mass m(C) is shared between elementary propositions which constitute proposition C, the total amount of m(C) can be assigned to a single elementary proposition which is included in $C \cap A$. Thus the maximal amount of our belief in proposition A is obtained by adding m(C) for C such that $C \cap A \neq \emptyset$.

Some properties of the belief function and the plausibility function are given here. From the definitions (15.5) and (15.6), it is easy to see:

$$\mathrm{Bel}(A) \leq \mathrm{Pl}(A)$$

The following two inequalities show that the belief function and the plausibility function are not additive:

$$\mathrm{Bel}(A) + \mathrm{Bel}(\neg A) \leq 1$$

$$\mathrm{Pl}(A) + \mathrm{Pl}(\neg A) \geq 1$$

Example 13

Suppose we are given the belief structure (15.4) in Example 12. Belief functions and plausibility functions for elementary propositions x_1, x_2, and x_3 are computed as:

$$\mathrm{Bel}(\{x_1\}) = m(\{x_1\}) = 0.2$$

$$\mathrm{Pl}(\{x_1\}) = m(\{x_1\}) + m(\{x_1, x_2\}) = 0.5$$

$$\mathrm{Bel}(\{x_2\}) = 0$$

$$\mathrm{Pl}(\{x_2\}) = m(\{x_1, x_2\}) + m(\{x_2, x_3\}) = 0.8$$

$$\mathrm{Bel}(\{x_3\}) = 0$$

$$\mathrm{Pl}(\{x_3\}) = m(\{x_2, x_3\}) = 0.5$$

where $\mathrm{Bel}(\{x_2\}) = 0$ and $\mathrm{Bel}(\{x_3\}) = 0$, because there exists no focal proposition A such that $A \subset \{x_2\}$, or $A \subset \{x_3\}$, respectively. ♦

15.2.4 Combining Belief Structures

One of the central issues in the DS theory is how we should combine belief structures given by independent knowledge sources, where knowledge sources include conventional or intelligent sensors, human operators, and their combinations.

Suppose we have two sets of belief structures $\{m_1(A) : A \in 2^X\}$ and $\{m_2(B) : B \in 2^X\}$ which are provided by independent knowledge sources, KS_1 and KS_2, where we assume the both belief structures are given on the same frame of discernment X. How can we fuse these two belief structures into a single belief structure? Assume knowledge source KS_1 supports proposition A with the degree $m_1(A)$, while knowledge source KS_2 allocates $m_2(B)$ to proposition B. Assume both of A and B are focal propositions. When we combine propositions A and B, we obtain a new proposition C which is defined by:

$$C = A \cap B$$

Since knowledge sources KS_1 and KS_2 are independent, it is natural to assign a number $m_1(A)m_2(B)$ to proposition C as the *portion* of total belief in C. Note that, for such C, there can exist several pairs of focal propositions A_j and B_k such that $A_j \cap B_k = C$. The degree of belief assigned precisely to proposition C is given by the following function:

$$q(C) = \sum_{A_j \cap B_k = C} m_1(A_j)\, m_2(B_k) \qquad (15.7)$$

where the sum is taken over all pairs (A_j, B_k) for which $A_j \cap B_k = C$. The function q is called a *ground assignment function* [5].

By computing q(C) for every possible $C \in 2^X$, we have:

$$\{q(C) : C \in 2^X\}$$

Can we say that $\{q(C) : C \in 2^X\}$ is a belief structure which we wanted? The answer is not affirmative, in general. Recall we have defined a belief structure, in Section 15.2.3, as $\{m(A) : A \in 2^X\}$ where m is a basic probability assignment function. Thus we have to check the obtained ground assignment function q satisfies Axioms I-III for the basic probability assignment function. The following example gives a case in which $\{q(C) : C \in 2^X\}$ is a belief structure.

Example 14

Suppose Mr. Z is a murder suspect. Let A denote a proposition "Mr. Z is a murderer," and let $\neg A$ denote "Mr. Z is not a murderer." Suppose we have two independent witnesses who express their belief structures on the frame of discernment $X = \{A, \neg A\}$.

Assume the first witness expresses his/her belief structure as:

$$m_1(A) = 0.6, \;\; m_1(X) = 0.4$$

while the second witness is rather reluctant to say that Mr. Z committed really a murder:

$$m_2(A) = 0.3, \;\; m_2(X) = 0.7$$

Then we have:

$$q(A) = m_1(A)\, m_2(A) + m_1(A)\, m_2(X) + m_1(X)\, m_2(A) = 0.72$$

$$q(X) = m_1(X)\, m_2(X) = 0.28$$

Since $q(A) + q(X) = 1$ and thus $q(\emptyset) = 0$, the ground assignment function obtained in the above satisfies Axioms I-III for the basic probability assignment function. Thus $\{q(A), q(X)\} = \{0.72, 0.28\}$ is a belief structure, which is a simple support function focused at A.♦

In general, $q(\emptyset) \geq 0$, and thus $\{q(C) : C \in 2^X\}$ does not give a belief structure, as shown in the following example.

Example 15

Let us consider a case with a belief structure which differs slightly from one discussed in Example 14. Assume the first witness gives the same belief structure as in Example 14: viz.

$$m_1(A) = 0.6, \;\; m_1(X) = 0.4$$

while the second witness expresses his/her degree of belief as:

$$m_2(A) = 0.3, \;\; m_2(\neg A) = 0.2, \;\; m_2(X) = 0.5$$

Then we have:

$$q(A) = m_1(A)\, m_2(A) + m_1(A)\, m_2(X) + m_1(X)\, m_2(A) = 0.60$$

$$q(\neg A) = m_1(X)\, m_2(\neg A) = 0.08$$

$$q(X) = m_1(X)\, m_2(X) = 0.20$$

$$q(\emptyset) = m_1(A)\, m_2(\neg A) = 0.12$$

Since $q(\emptyset) > 0$, the ground assignment function q does not satisfy Axioms I-III for the basic probability assignment function. Thus $\{q(A), q(\neg A), q(X)\} = \{0.60, 0.08, 0.20\}$ does not represent a belief structure.♦

We have seen that $\{q(C) : C \in 2^X\}$ is not a belief structure when $q(\emptyset) > 0$. If we must fuse the given two independent sets of belief structures into a single belief structure $\{m(C) : C \in 2^X\}$, what should we do? We need some *rule* which converts the ground assignment function q into a basic probability assignment function. Such a rule is called a *rule of combination.*

The oldest and commonly used rule of combination is *Dempster's rule* [1-2] which is expressed as follows:

$$m(C) = \frac{q(C)}{1 - q(\emptyset)} \quad \text{for } C \neq \emptyset \tag{15.8}$$

$$m(\emptyset) = 0 \tag{15.9}$$

By using $[1 - q(\emptyset)]^{-1}$ as a normalization constant in (15.8), Dempster's rule makes the sum of m(C) over all possible $C \in 2^X$ become unity. Mathematical meaning of the normalization would be apparent: The normalization converts the ground assignment function q, which does not necessarily satisfy Axioms I-III, into a basic probability assignment function m which satisfies Axioms I-III. The resulting $\{m(C) : C \in 2^X\}$ obtained by (15.8) and (15.9) is a belief structure.

Another interpretation of the normalization is possible. The mass $q(\emptyset)$ is *allocated proportionally* to each proposition C: viz.

$$\frac{q(C)}{1 - q(\emptyset)} q(\emptyset)$$

is added to q(C) to yield (15.8). Interpreting the normalization as *allocation of $q(\emptyset)$ into focal propositions* would make understanding Dempster's rule easy. We will see later that a scheme for allocating $q(\emptyset)$ determines a rule of combination.

How can we interpret the normalization from a practical point of view? When $q(\emptyset) > 0$, there exists some contradiction or conflict among information given by two independent knowledge sources. If such conflict occurs, we may want to evaluate credibility of information or knowledge sources so that less credible information can be discounted. We know in our daily life that information is not always true. Even when some conflict is found among information, Dempster's rule disregards every contradiction in information, and discards every contradictory portion the total mass of which is $q(\emptyset)$ completely from further consideration. Dempster's rule constructs a belief structure based on a portion of information which does not exhibit any contradiction. In other words,

Dempster's rule regards as if there did not exist any conflict among information given by independent knowledge sources.

Example 16

Let us apply Dempster's rule for combining the following belief structures given in Example 15:

$$m_1(A) = 0.6, \ m_1(X) = 0.4$$

$$m_2(A) = 0.3, \ m_2(\neg A) = 0.2, \ m_2(X) = 0.5$$

We have:

$$m(A) = q(A)\,[1- q(\emptyset)]^{-1} = 0.60 / 0.88 = 0.68$$

$$m(\neg A) = q(\neg A)\,[1- q(\emptyset)]^{-1} = 0.08 / 0.88 = 0.09$$

$$m(X) = q(X)\,[1- q(\emptyset)]^{-1} = 0.20 / 0.88 = 0.23$$

Thus we obtain a new belief structure $\{m(A), m(\neg A), m(X)\} = \{0.68, 0.09, 0.23\}$. ♦

Because of its simplicity, Dempster's rule of combination has been used in various applications, including artificial intelligence [7], medical diagnosis [8], team-oriented real-time control problems [9], and target identification [10]. It was shown recently that Dempster's rule gives counterintuitive results and exhibits numerical instability in some cases which are not uncommon [3-5]. We see these phenomena through the following two examples, which are based on the examples given in [3-5] and are adapted to system reliability settings.

Example 17

Consider a series system which consists of three components. Suppose the system malfunctions, and we must identify a faulty component. Assume that components fail independently. Then a single component has caused the system malfunction. Let the frame of discernment X consist of three mutually exclusive propositions: $X = \{x_1, x_2, x_3\}$, where x_k denotes the proposition "component k is faulty."

Assume we have two independent knowledge sources KS_k each of which gives a belief structure $\{m_k(A) : A \in 2^X\}$, $(k = 1, 2)$. Suppose we are given the following belief structures:

$$m_1(\{x_1\}) = 0.90,\ m_1(\{x_2\}) = 0.10,\ m_1(\{x_3\}) = 0.0 \tag{15.10}$$

$$m_2(\{x_1\}) = 0.0,\ m_2(\{x_2\}) = 0.10,\ m_2(\{x_3\}) = 0.90 \tag{15.11}$$

The above belief structures are contradictory. Knowledge source KS_1 suspects component 1 as being failed, while knowledge source KS_2 doubts component 3. Both knowledge sources afford little support to proposition that component 2 is failed.

By noting $\{x_j\} \cap \{x_k\} = \emptyset$ for $j \neq k$, we have:

$$q(\{x_1\}) = m_1(\{x_1\})\ m_2(\{x_1\}) = 0.0$$

$$q(\{x_2\}) = m_1(\{x_2\})\ m_2(\{x_2\}) = 0.01$$

$$q(\{x_3\}) = m_1(\{x_3\})\ m_2(\{x_3\}) = 0.0$$

$$q(\emptyset) = 1 - 0.01 = 0.99$$

Dempster's rule of combination yields the following belief structure:

$$m(\{x_1\}) = 0.0,\ m(\{x_2\}) = 1.0,\ m(\{x_3\}) = 0.0$$

which gives complete support to the proposition "component 2 is faulty" even though both knowledge sources did afford little support to the proposition. ♦

When belief structures $\{m_1(A) : A \in 2^X\}$ and $\{m_2(B) : B \in 2^X\}$ are highly contradictory, or $q(\emptyset)$ is close to unity, as in Example 17, then small perturbations in $\{m_1(A) : A \in 2^X\}$ or $\{m_2(B) : B \in 2^X\}$ can cause a drastic change in the resulting belief structure.

Example 18

Let us perturb belief structures (15.10) and (15.11) in Example 17. Assume belief structures $\{m_k(A) : A \in 2^X\}$ are given as follows:

$$m_1(\{x_1\}) = 0.89,\ m_1(\{x_2\}) = 0.10,\ m_1(\{x_3\}) = 0.01$$

$$m_2(\{x_1\}) = 0.01,\ m_2(\{x_2\}) = 0.10,\ m_2(\{x_3\}) = 0.89$$

Then we have:

$$q(\{x_1\}) = m_1(\{x_1\})\ m_2(\{x_1\}) = 0.0089$$

$$q(\{x_2\}) = m_1(\{x_2\})\ m_2(\{x_2\}) = 0.01$$

$$q(\{x_3\}) = m_1(\{x_3\})\ m_2(\{x_3\}) = 0.0089$$

$$q(\emptyset) = 1 - 0.0089 - 0.01 - 0.0089 = 0.9722$$

By applying Dempster's rule, we have:

$$m(\{x_1\}) = 0.0089 / [1 - 0.9722] = 0.32$$

$$m(\{x_2\}) = 0.01 / [1 - 0.9722] = 0.36$$

$$m(\{x_3\}) = 0.0089 / [1 - 0.9722] = 0.32$$

The combined belief structure suspects each component almost evenly as a component which caused the system malfunction. ♦

Yager has proposed an alternative rule of combination [5] which is free from the disadvantages of Dempster's rule. Yager's rule of combination is given by:

$$m(C) = q(C) \quad \text{for } C \neq \emptyset, X \tag{15.12}$$

$$m(X) = q(X) + q(\emptyset) \tag{15.13}$$

The distinctive feature of Yager's rule is seen in handling of $q(\emptyset)$ in (15.13). When any contradiction is found among information given by independent knowledge sources, Yager's rule of combination regards that contradiction as coming from ignorance, and thus allocates the conflicted portion $q(\emptyset)$ to proposition X which implies total ignorance. Yager says in his paper [5], "since we do not really know anything about the conflicted portion, we let it be distributed among all the elements (elementary propositions, according to the wording in this chapter) rather than just those in the focal sets."

Yager's rule gives the same belief structure as that obtained by applying Dempster's rule when $q(\emptyset) = 0$. For instance, in Example 14 where $q(\emptyset) = 0$, both rules give the same belief structure $\{m(A), m(X)\} = \{0.72, 0.28\}$. When $q(\emptyset) > 0$, on the other hand, belief structures derived by these two rules of combination are not identical.

Example 19

Let us combine, by using Yager's rule, the following belief structures which are given in Example 15:

$$m_1(A) = 0.6, \;\; m_1(X) = 0.4$$

$$m_2(A) = 0.3, \;\; m_2(\neg A) = 0.2, \;\; m_2(X) = 0.5$$

If we combine the belief structures by applying Yager's rule of combination, we have:

$$m(A) = q(A) = 0.60$$

$$m(\neg A) = q(\neg A) = 0.08$$

$$m(X) = q(X) + q(\emptyset) = 0.20 + 0.12 = 0.32$$

The resulting $\{m(A), m(\neg A), m(X)\} = \{0.60, 0.08, 0.32\}$ is different from the belief structure $\{m(A), m(\neg A), m(X)\} = \{0.68, 0.09, 0.23\}$ which has been obtained in Example 16 by applying Dempster's rule of combination.♦

15.3 Identification and Location of Faults

This section considers a fault diagnosis problem, in which multiple components can be in failed operating conditions. Reggia, Peng and their colleagues have developed some causal models for diagnostic expert systems [11-13]. Even though their models are formulated in the framework of probability theory, they have suggested applicability of the DS theory to some diagnostic problems [12-13]. Ishibuchi and his colleagues have developed a diagnostic model based on fuzzy set theory and the DS theory [14]. They have restricted their consideration to cases where fuzzy symptoms are expressed by belief structures which are simple support functions. Their restriction can be relaxed. This section generalizes the Ishibuchi's model a little bit so that the model can allow any belief structure.

15.3.1 Possible and Necessary Causal Relations

Let $C = \{c_1, ..., c_s\}$ be a set of disorders, and let $E = \{e_1, ..., e_t\}$ be a set of symptoms. We distinguish two types of causal relations [11-14]:

(1) *possible causal relation*, which is represented as "disorder c_k may cause symptom e_j"

(2) *necessary causal relation*, which is expressed as "disorder c_k always causes symptom e_j"

Figure 15.3 depicts causal knowledge between some disorders and symptoms, where possible causal relation is shown with a fine line, and necessary causal relation is illustrated by a heavy line. For instance, Figure 15.3 represents our knowledge "If disorder c_2 exists, then symptom e_3 is always observed. When it comes to symptom e_1, on the contrary, symptom e_1 may be observed in some cases but may not be observed in some other cases even when disorder c_2 does exist."

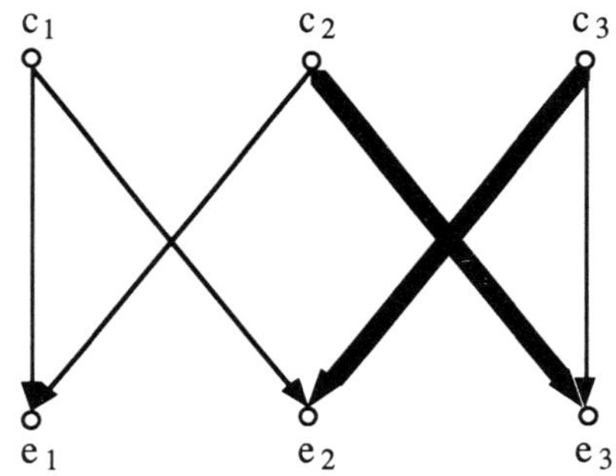

	e_1	e_2	e_3
$m^j(\{e_j\})$	0.8	0.2	0.0
$m^j(\{\neg e_j\})$	0.1	0.6	0.7
$m^j(X^j)$	0.1	0.2	0.3

Fig.15.3 Causal Relations and Belief Structures on Symptoms

A diagnostic problem which we discuss here is the following:

"We are given vague information regarding occurrence and/or non-occurrence of symptoms. Multiple disorders can be existent simultaneously. Which disorders are likely to be existent?"

We assume that *vague* information on the occurrence and/or non-occurrence of each symptom j is represented as a belief structure $\{m^j(\{e_j\}), m^j(\{\neg e_j\}), m^j(X^j)\}$, where $X^j = \{e_j, \neg e_j\}$ denotes the frame of discernment, e_j represents the proposition "symptom e_j is observed" and $\neg e_j$ denotes the proposition "symptom e_j is not observed". When symptom e_j is clearly observed, then we say that $m^j(\{e_j\}) = 1$. When symptom e_j is denied completely, then the situation is represented as $m^j(\{\neg e_j\}) = 1$. When symptom e_j is neither clearly observed nor denied definitely, such vague situation is expressed by a general belief structure $\{m^j(\{e_j\}), m^j(\{\neg e_j\}), m^j(X^j)\}$.

The belief structures $\{m^j(\{e_j\}), m^j(\{\neg e_j\}), m^j(X^j)\}$ are assumed to be given by independent knowledge source KS_j, ($j = 1, ..., t$). In other words, for each symptom, we are equipped with a specific intelligent sensor for detecting the symptom. Figure 15.3 gives an example of belief structure for each symptom.

15.3.2 Fault Diagnosis Based on Vague Information

Let us see how belief structures on symptoms are used in evaluating diagnostic hypotheses. Consider the case represented by Figure 15.3.

Let the frame of discernment X of diagnostic hypotheses be:

$$X = \{x_1, ..., x_7\}$$

where each diagnostic hypothesis is defined by:

$$x_1 = \{c_1\},\ x_2 = \{c_2\},\ x_3 = \{c_3\},\ x_4 = \{c_1, c_2\},$$

$$x_5 = \{c_1, c_3\},\ x_6 = \{c_2, c_3\},\ x_7 = \{c_1, c_2, c_3\}$$

The hypothesis $x_1 = \{c_1\}$ represents the proposition "single disorder c_1 has caused every symptom observed", which does not allow any other disorder to be existent simultaneously with disorder c_1. The hypothesis $x_4 = \{c_1, c_2\}$ represents the proposition "disorders c_1 and c_2 are causes of observed symptoms, and no other disorder exists." The diagnostic hypotheses defined in the above are exhaustive and mutually exclusive.

Some readers may be curious why the frame of discernment was not defined as $Y = \{c_1, c_2, c_3\}$. Those readers are recommended to recall that the frame of discernment must be a set of exhaustive and *mutually exclusive* elementary propositions. If we set the frame of discernment as $Y = \{c_1, c_2, c_3\}$, then it is impossible to consider cases in which multiple disorders can be existent simultaneously. The frame of discernment Y thus requires a single-disorder assumption.

Let us convert the belief structure $\{m^j(\{e_j\}), m^j(\{\neg e_j\}), m^j(X^j)\}$ in Figure 15.3 into a belief structure $\{m_j(A) : A \in 2^X\}$ on the frame of discernment X of diagnostic hypotheses.

Belief structure for symptom e_1 has been given in Figure 15.3 as:

$$\{m^1(\{e_1\}), m^1(\{\neg e_1\}), m^1(X^1)\} = \{0.8, 0.1, 0.1\}$$

Since symptom e_1 can be caused by either disorder c_1 or c_2, mass $m^1(\{e_1\}) = 0.8$ must be shared among diagnostic hypotheses which allow disorder c_1 or c_2 to occur. Thus we have:

$$m_1(\{x_1, x_2, x_4, x_5, x_6, x_7\}) = m^1(\{e_1\}) = 0.8$$

We need to be careful in interpreting $m^1(\{\neg e_1\}) = 0.1$. Since every causal relation directed to symptom e_1 is possible causal relation, proposition $\{\neg e_1\}$ does not imply non-existence of disorder c_1 or c_2. Proposition $\{\neg e_1\}$ neither supports any specific diagnostic hypothesis nor denies any specific hypothesis. Thus $m^1(\{\neg e_1\}) = 0.1$ is allocated to the frame of discernment X. It is obvious that $m^1(X^1) = 0.1$ should also be allocated to the frame of discernment X. Thus we have:

$$m_1(X) = m^1(\{\neg e_1\}) + m^1(X^1) = 0.2$$

Belief structure for symptom e_2 has been given in Figure 15.3 as:

$$\{m^2(\{e_2\}), m^2(\{\neg e_2\}), m^2(X^2)\} = \{0.2, 0.6, 0.2\}$$

It is easily seen that mass $m^2(\{e_2\}) = 0.2$ is shared among diagnostic hypotheses which allow disorder c_1 or c_3 to occur. Thus we have:

$$m_2(\{x_1, x_3, x_4, x_5, x_6, x_7\}) = m^2(\{e_2\}) = 0.2$$

How can we interpret $m^2(\{\neg e_2\}) = 0.6$? The necessary causal relation "if c_3, then e_2" implies "if $\neg e_2$, then $\neg c_3$". Thus $m^2(\{\neg e_2\}) = 0.6$ must be shared among diagnostic hypotheses which do not allow disorder c_3 to occur: viz.

$$m_2(\{x_1, x_2, x_4\}) = m^2(\{\neg e_2\}) = 0.6$$

The following equation would be obvious:

$$m_2(X) = m^2(X^2) = 0.2$$

Belief structure $\{m^3(\{e_3\}), m^3(\{\neg e_3\}), m^3(X^3)\} = \{0.0, 0.7, 0.3\}$ on symptom e_3 is converted in a similar manner into a belief structure on the diagnostic hypotheses as:

$$m_3(\{x_2, x_3, x_4, x_5, x_6, x_7\}) = m^3(\{e_3\}) = 0.0$$

$$m_3(\{x_1, x_3, x_5\}) = m^3(\{\neg e_3\}) = 0.7$$

$$m_3(X) = m^3(X^3) = 0.3$$

We thus obtain belief structures $\{m_j(A) : A \in 2^X\}$ ($j = 1, 2, 3$) on the frame of discernment X of diagnostic hypotheses.

Figure 15.4 depicts combination of belief structures $\{m_1(A) : A \in 2^X\}$ and $\{m_2(B) : B \in 2^X\}$ by applying Dempster's rule of combination.

	$m_1(\{x_1,x_2,x_4,x_5,x_6,x_7\}) = 0.8$	$m_1(X) = 0.2$
$m_2(\{x_1,x_3,x_4,x_5,x_6,x_7\}) = 0.2$	$\{x_1,x_4,x_5,x_6,x_7\}$ 0.16	$\{x_1,x_3,x_4,x_5,x_6,x_7\}$ 0.04
$m_2(\{x_1,x_2,x_4\}) = 0.6$	$\{x_1,x_2,x_4\}$ 0.48	$\{x_1,x_2,x_4\}$ 0.12
$m_2(X) = 0.2$	$\{x_1,x_2,x_4,x_5,x_6,x_7\}$ 0.16	X 0.04

Fig.15.4 Combination of Belief Structures m_1 and m_2

Since there exists no conflict between the two belief structures $\{m_1(A) : A \in 2^X\}$ and $\{m_2(B) : B \in 2^X\}$, the computed ground assignment function q_{12} can be regarded as a basic probability assignment function which is denoted here as m_{12}. For instance:

$$q_{12}(\{x_1, x_2, x_4\}) = m_{12}(\{x_1, x_2, x_4\})$$

$$= m_1(\{x_1, x_2, x_4, x_5, x_6, x_7\})\ m_2(\{x_1, x_2, x_4\})$$

$$+ m_1(X)\ m_2(\{x_1, x_2, x_4\}) = 0.48 + 0.12 = 0.60$$

Combination of $\{m_{12}(A) : A \in 2^X\}$ and $\{m_3(B) : B \in 2^X\}$ is shown in Figure 15.5, where no conflict is found between the two belief structures. Let $\{m(C) : C \in 2^X\}$ be the resulting belief structure. Then we have:

$$m(\{x_1\}) = 0.420$$

$$m(\{x_1, x_5\}) = 0.112 + 0.112 = 0.224$$

$$m(\{x_1, x_2, x_4\}) = 0.180$$

$$m(\{x_1, x_3, x_5\}) = 0.028 + 0.028$$

$$m(\{x_1, x_4, x_5, x_6, x_7\}) = 0.048$$

$$m(\{x_1, x_2, x_4, x_5, x_6, x_7\}) = 0.048$$

$$m(\{x_1, x_3, x_4, x_5, x_6, x_7\}) = 0.012$$

$$m(X) = 0.012$$

For each diagnostic hypothesis x_k, an interval $[Bel(x_k), Pl(x_k)]$ is computed as:

x_1 : [0.42, 1.0], x_2 : [0.0, 0.24], x_3 : [0.0, 0.08], x_4 : [0.0, 0.30]

x_5 : [0.0, 0.40], x_6 : [0.0, 0.12], x_7 : [0.0, 0.12]

which suggests that x_1 is the most probable diagnostic hypothesis and thus disorder c_1 must be suspected as the most probable cause.

	$m_3(\{x_1,x_3,x_5\}) = 0.7$	$m_3(X) = 0.3$
$m_{12}(\{x_1,x_2,x_4\}) = 0.60$	$\{x_1\}$ 0.420	$\{x_1,x_2,x_4\}$ 0.180
$m_{12}(\{x_1,x_4,x_5,x_6,x_7\}) = 0.16$	$\{x_1,x_5\}$ 0.112	$\{x_1,x_4,x_5,x_6,x_7\}$ 0.048
$m_{12}(\{x_1,x_2,x_4,x_5,x_6,x_7\}) = 0.16$	$\{x_1,x_5\}$ 0.112	$\{x_1,x_2,x_4,x_5,x_6,x_7\}$ 0.048
$m_{12}(\{x_1,x_3,x_4,x_5,x_6,x_7\}) = 0.04$	$\{x_1,x_3,x_5\}$ 0.028	$\{x_1,x_3,x_4,x_5,x_6,x_7\}$ 0.012
$m_{12}(X) = 0.04$	$\{x_1,x_3,x_5\}$ 0.028	X 0.012

Fig.15.5 Combination of Belief Structures m_{12} and m_3

15.4 Safety-Control and Management

This section explores an application of the DS theory to safety-control problems, in which we decide whether to start operation of the system or to shut down the system based on possibly imperfect information regarding system states.

In section 15.4.1, we give classification of safety-control policies into two categories: (1) *fault-warning* safety-control policies, and (2) *safety-preservation* safety-control policies. The classification of safety-control policies is essential for maintaining system safety [15-17].

Assume that independent knowledge sources such as intelligent sensors present belief structures regarding system states, and that the safety-control decision is made based on a combined belief structure. A rule of combination cannot be chosen in an arbitrary manner, however, because of interdependence between safety-control policy and rules of combination. To prove this, we develop first in section 15.4.2 a unified rule of combination which includes Dempster's rule and Yager's rule as its special cases. With the unified rule of combination, we show in sections 15.4.3 and 15.4.4 the interdependence between safety-control policy and rules of combination.

15.4.1 Safety-Control Policies

When the system exhibits an unusual behavior, we have to identify the operating condition of the system and must take countermeasure action, if necessary, to avoid a catastrophic accident. *Safety-control* is an integrated action of state identification and decision making whether to shut down the system or not. Identifying system state is not an easy task if the system is large and complex. Difficulty increases when human-machine interface is poorly designed. It is sometimes difficult to judge with confidence whether the system is in its safe operating condition or not.

Example 20a

Consider the case of the space shuttle *Challenger*. At the launch of *Challenger* in 1986 January, there was a discussion between one engineering group and the decisionmaker regarding the issue whether *Challenger* can be launched or the launch must be postponed. The discussion was tough because neither safety of *Challenger* nor unsafety of *Challenger* was proven explicitly. ♦

This motivates us to distinguish the following two types for safety-control policies [17]:

(1) *Fault-Warning* (FW) safety-control policy, in which the human operator shuts down the system only when he regards the system as being unsafe. The human operator lets the system run while he does not have a definite evidence which proves unsafety of the system.

(2) *Safety-Preservation* (SP) safety-control policy, where the human operator lets the system be operative only when he is definitely sure about safety of the system. If the human operator cannot be sure of system safety, he shuts down the system, even when he has no definite evidence for proving unsafety of the system.

Let S denote the proposition "the system is in its normal and safe operating condition," and let U denote the proposition "the system is in its abnormal and unsafe operating condition." Let X be the proposition "we really do not know whether the system is safe or unsafe." Figure 15.6 depicts decisions under FW and SP safety-control policies.

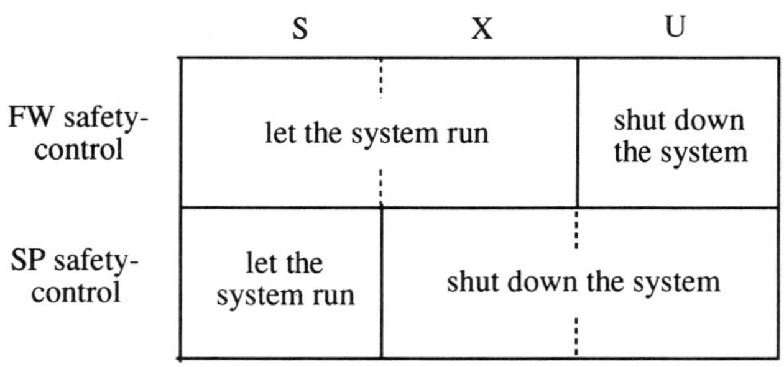

Fig.15.6 Safety-Control Policies

Example 20b

Let us come back to the case of *Challenger*. The engineering group said that *Challenger* should not be launched because they could not be sure of the safety of *Challenger* in the severe weather condition. In other words, the engineering group advised the decisionmaker to take the SP safety-control policy. The decisionmaker launched *Challenger*. He thought what the engineering group said did not prove the unsafety of *Challenger*. The decisionmaker took the FW safety-control policy. ♦

15.4.2 *Unified Rule for Combining Belief Structures*

Suppose two sets of belief structures $\{m_1(A) : A \in 2^X\}$ and $\{m_2(B) : B \in 2^X\}$ are given by independent knowledge sources, KS_1 and KS_2. We assume that the both belief structures are represented on the same frame of discernment X. As we have seen in Section 15.2.4, $\{q(C) : C \in 2^X\}$ does not give a combined belief structure when $q(\emptyset) > 0$. A *rule of combination* must be applied for converting $\{q(C) : C \in 2^X\}$ into a belief structure $\{m(C) : C \in 2^X\}$. In section 15.2.4, we have learned Dempster's rule and Yager's rule. Can any other rule of combination exist? The answer is affirmative.

Assume that $\{q(C) : C \in 2^X\}$ is given. It is essential to note that a rule of combination is fully determined when we specify a scheme for allocating mass $q(\emptyset)$ to propositions which are distinct from $\emptyset$. The idea enables us to develop a rule of combination in a systematic manner.

Any rule of combination can thus be expressed as:

$$m(C) = q(C) + f(C)\, q(\emptyset)\,, \quad C \neq \emptyset \tag{15.14}$$

$$\sum_{C \in 2^X,\, C \neq \emptyset} f(C) = 1, \quad f(C) \geq 0 \tag{15.15}$$

where f(C) denotes an allocation coefficient for proposition C.

Let us restrict our consideration to a class of combination rules for which the following property holds:

$$\frac{m(C)}{m(D)} = \frac{q(C)}{q(D)} \tag{15.16}$$

for any propositions C and D which are distinct from X, $\emptyset$. Eqn (15.16) implies that no knowledge is assumed regarding relative importance or credibility of propositions, and thus the ratio $q(C) / q(D)$ must be identical to the ratio $m(C) / m(D)$ in the belief structure $\{m(C) : C \in 2^X\}$ which is to be derived. Note that either Dempster's rule or Yager's rule satisfies (15.16). With (15.14), we can rewrite (15.16) as:

$$\frac{q(C) + f(C)\, q(\emptyset)}{q(C)} = \frac{q(D) + f(D)\, q(\emptyset)}{q(D)} \tag{15.17}$$

A necessary and sufficient condition for (15.16) is given as:

$$\frac{f(C)}{q(C)} = k \quad \text{for any } C \neq X, \emptyset \tag{15.18}$$

By substituting (15.18) into (15.14), we obtain a unified rule of combination as follows:

$$m(C) = [1 + k\, q(\emptyset)]\, q(C) \quad \text{for } C \neq X, \emptyset \tag{15.19a}$$

$$m(X) = [1 + k\, q(\emptyset)]\, q(X) + [1 + k\, q(\emptyset) - k]\, q(\emptyset) \tag{15.19b}$$

$$m(\emptyset) = 0 \tag{15.19c}$$

It would be easy to see that Dempster's rule is obtained by setting the value of parameter k in (15.19) at $k = [1 - q(\emptyset)]^{-1}$. The unified rule of combination (15.19) represents Yager's rule when k = 0. By noting:

$$f(X) = 1 - k\, [1 - q(\emptyset) - q(X)] \geq 0$$

we see that parameter k must satisfy:

$$0 \leq k \leq \frac{1}{1 - q(\emptyset) - q(X)} \tag{15.20}$$

Since k is continuous-valued, the unified rule of combination (15.19) represents infinitely many rules of combination. Figure 15.7 illustrates various rules of combination which *interpolate* or *extrapolate* the rules of combination proposed by Yager and Dempster.

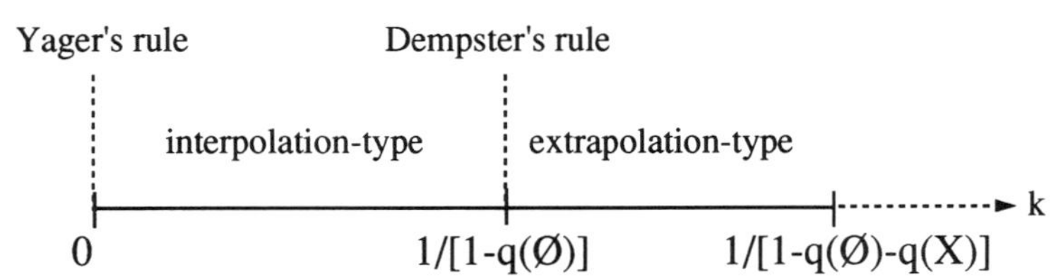

Fig.15.7 Rules of Combination Represented by (15.19)

15.4.3 *Multi-Sensor Fusion*

Let S denote the proposition "the system is safe" and U denote the proposition "the system is unsafe". Let the frame of discernment X consists of the above two propositions: X = {S, U}. The frame of discernment X represents the proposition "we do not know whether the system is safe or unsafe".

Suppose we have two independent knowledge sources, KS_1 and KS_2, such as conventional sensors or intelligent sensors. Assume that the knowledge sources represent belief structures:

$$\{m_k(S), m_k(U), m_k(X)\} \qquad k = 1, 2$$

Let us combine these belief structures. The ground assignment function $\{q(C) : C \in 2^X\}$ is computed as follows:

$$q(S) = m_1(S)\, m_2(S) + m_1(X)\, m_2(S) + m_1(S)\, m_2(X)$$

$$q(U) = m_1(U)\, m_2(U) + m_1(X)\, m_2(U) + m_1(U)\, m_2(X)$$

$$q(X) = m_1(X)\, m_2(X)$$

$$q(\emptyset) = m_1(S)\, m_2(U) + m_1(U)\, m_2(S)$$

The resulting $\{q(S), q(U), q(X), q(\emptyset)\}$ is not a belief structure because $q(\emptyset) > 0$. A belief structure $\{m(S), m(U), m(X)\}$ is obtained by using the unified rule of combination (15.19), where:

$$m(S) = [1 + k\, q(\emptyset)]\, q(S) \qquad (15.21)$$

$$m(U) = [1 + k\, q(\emptyset)]\, q(U) \qquad (15.22)$$

$$m(X) = [1 + k\, q(\emptyset)]\, q(X) + [1 + k\, q(\emptyset) - k]\, q(\emptyset) \qquad (15.23)$$

15.4.4 *Interdependence between Safety-Control Policies and Rules of Combination*

The belief structure $\{m(S), m(U), m(X)\}$ given by (15.21)-(15.23) is defined on the frame of discernment X = {S, U}. For deciding whether the system must be shut shown or not, {m(S), m(U), m(X)} should be converted into another belief structure defined on the decision space.

Let Z denote the decision space: Z = {GO, SD}, where GO denotes our decision to put a system into operation or not to shut down a system which is currently working, and SD denotes the decision to shut down the system or not to allow the system to start operation. The resulting belief structure on the decision space varies depending on the type of safety-control policy.

Suppose we take a fault-warning (FW) safety-control policy. The FW safety-control policy is represented by the map $h_{FW} : 2^X \rightarrow Z$ for which:

$$h_{FW}(\{S\}) = \{GO\},\ h_{FW}(\{X\}) = \{GO\},\ h_{FW}(\{U\}) = \{SD\}$$

Let {m*(GO|FW), m*(SD|FW)} denote a belief structure for decisions GO and SD, provided that the FW safety-control policy is taken. Then {m*(GO|FW), m*(SD|FW)} is obtained from {m(S), m(U), m(X)} immediately, where:

$$m^*(GO|FW) = m(S) + m(X)$$

$$m^*(SD|FW) = m(U)$$

More explicitly, we have:

$$m^*(GO|FW) = [1 + k\, q(\emptyset)][q(S) + q(X)] + [1 + k\, q(\emptyset) - k]\, q(\emptyset)$$

$$m^*(SD|FW) = [1 + k\, q(\emptyset)]\, q(U)$$

It is easily seen that m*(GO|FW) is monotone non-increasing with k, and m*(SD|FW) is monotone non-decreasing with k.

Recall that assigning a value to parameter k in (15.19) determines a single rule of combination. Consider the following three special cases:

(a) Yager's rule (denoted as R_Y) with k = 0
(b) Dempster's rule (denoted as R_D) with $k = [1 - q(\emptyset)]^{-1}$
(c) Extreme rule (denoted as R_E) among extrapolation-type rules of combination, where parameter k is set at $k = [1 - q(\emptyset) - q(X)]^{-1}$, the largest value that k can take; see (15.20).

Then we have:

$$m^*(GO|FW, R_Y) \geq m^*(GO|FW, R_D) \geq m^*(GO|FW, R_E) \qquad (15.24)$$

$$m^*(SD|FW, R_Y) \leq m^*(SD|FW, R_D) \leq m^*(SD|FW, R_E) \qquad (15.25)$$

These order relations (15.24) and (15.25) show that, among all rules of combination represented by (15.19), Yager's rule has the largest possibility of letting the unsafe system run. Yager's rule is the most dangerous rule under the FW safety-control policy. On the contrary, the extreme rule of combination with $k = [1 - q(\emptyset) - q(X)]^{-1}$ is the safest rule for maintaining system safety under the FW safety-control policy.

Consider next cases in which we take a safety-preservation (SP) safety-control policy. The SP safety-control policy is expressed by the map $h_{SP} : 2^X \rightarrow Z$ for which:

$$h_{SP}(\{S\}) = \{GO\}, \; h_{SP}(\{X\}) = \{SD\}, \; h_{SP}(\{U\}) = \{SD\}$$

Let $\{m^*(GO|SP), m^*(SD|SP)\}$ denote a belief structure for decisions GO and SD, provided that the SP safety-control policy is taken. We obtain $m^*(GO|SP)$ and $m^*(SD|SP)\}$ as:

$$m^*(GO|SP) = m(S) = [1 + k\,q(\emptyset)]\,q(S)$$

$$m^*(SD|SP) = m(U) + m(X)$$

$$= [1 + k\,q(\emptyset)]\,[q(U) + q(X)] + [1 + k\,q(\emptyset) - k]\,q(\emptyset)$$

Note that $m^*(GO|SP)$ is monotone non-decreasing with k, while $m^*(SD|SP)$ is monotone non-increasing with k. For the three specific rules of combination (a)-(c) which are specified in the above, we have the following order relations:

$$m^*(GO|SP, R_Y) \leq m^*(GO|SP, R_D) \leq m^*(GO|SP, R_E) \qquad (15.26)$$

$$m^*(SD|SP, R_Y) \geq m^*(SD|SP, R_D) \geq m^*(SD|SP, R_E) \qquad (15.27)$$

The order relations (15.26) and (15.27) show that, among all possible rules of combination given by (15.19), Yager's rule is the best rule of combination for maintaining system safety under the SP safety-control policy. The extreme rule of combination with $k = [1 - q(\emptyset) - q(X)]^{-1}$ has the largest possibility of letting the unsafe system be operative.

Example 21

Let $X = \{S, U\}$. Assume the following two belief structures are given:

$$m_1(S) = 0.6, \; m_1(U) = 0.3, \; m_1(X) = 0.1$$

$$m_2(S) = 0.7,\ \ m_2(U) = 0.1,\ \ m_2(X) = 0.2$$

Since q(S) = 0.61, q(U) = 0.10, q(X) = 0.02, q(Ø) = 0.27, we have:

$$m(S) = (1 + 0.27\ k)\ 0.61, \quad m(U) = (1 + 0.27\ k)\ 0.10$$

$$m(X) = (1 + 0.27\ k)\ 0.02 + (1 - 0.73\ k)\ 0.27$$

Thus we have:

$$m^*(GO|FW) = 0.90 - 0.027\ k$$

$$m^*(GO|SP) = 0.61 + 0.165\ k\ , \quad 0 \le k \le 1.41$$

where the range of parameter k in the above is computed by (15.20).

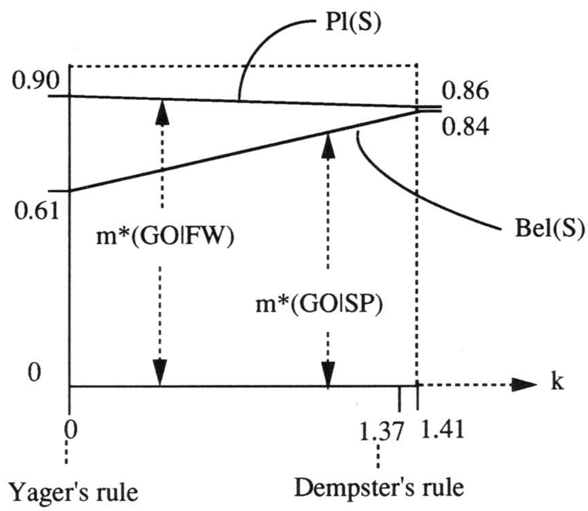

Fig.15.8 m*(GO|FW) and m*(GO|SP) in Example 21

Figure 15.8 depicts m*(GO|FW) and m*(GO|SP) for possible values of parameter k.♦

Example 22

Consider a case in which two belief structures are highly contradictory. Let X = {S, U}. Suppose we are given:

$$m_1(S) = 0.8,\ \ m_1(U) = 0.1,\ \ m_1(X) = 0.1$$

$$m_2(S) = 0.1,\ m_2(U) = 0.7,\ m_2(X) = 0.2$$

Since $q(S) = 0.25$, $q(U) = 0.16$, $q(X) = 0.02$, $q(\emptyset) = 0.57$, we have:

$$m(S) = (1 + 0.57\ k)\ 0.25,\ \ m(U) = (1 + 0.57\ k)\ 0.16$$

$$m(X) = (1 + 0.57\ k)\ 0.02 + (1 - 0.43\ k)\ 0.57$$

Thus we have:

$$m^*(GO|FW) = 0.84 - 0.091\ k$$

$$m^*(GO|SP) = 0.25 + 0.143\ k\ ,\ \ 0 \leq k \leq 2.44$$

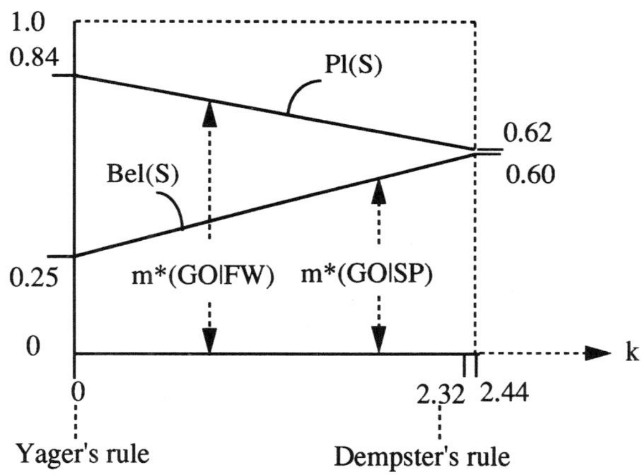

Fig.15.9 m*(GO|FW) and m*(GO|SP) for Example 22

Figure 15.9 depicts m*(GO|FW) and m*(GO|SP) for possible values of parameter k.♦

We have seen interdependence between a safety-control policy and a rule of combination. The interdependence has been discovered through analyses of the unified rule of combination (15.19). Examples 21 and 22 have shown visually the following points:

(1) No FW safety-control policy can be safer than any SP safety-control policy under any rule of combination, because:

$$m^*(GO|SP) \leq m^*(GO|FW) \quad \text{for any } k$$

which is a consequence of the relation $Bel(S) \leq Pl(S)$, where:

$$Bel(S) = m(S) = m^*(GO|SP)$$

$$Pl(S) = m(S) + m(X) = m^*(GO|FW)$$

(2) System safety varies depending not only on the type of safety-control policy but on the choice of a rule of combination. A safety-control policy and a rule of combination cannot be chosen independently. Inappropriate choice of a rule of combination degrades system safety. When the safety-control related decision must be made based on contradictory information, as in Example 22, system safety changes drastically depending on the choice of a rule of combination.

15.5 Concluding Remarks

This chapter has given the fundamentals of the DS theory, and has shown a couple of applications of the theory to system reliability and safety related problems. The multiple-fault diagnosis problem given in Section 15.3 has not yet been fully established. The model can be extended in various ways. The author and his colleagues are currently examining some more generalized models which express possible and necessary causal relations in a unified manner. Some effort would be required for developing models which can represent more realistic causal relations, such as those expressed by fault trees or event trees.

There also exist some topics for further research in safety-control related problems. Among them, seeking for an optimal value of k would be one of the most interesting problems. We are required to avoid unnecessary system shutdowns, while system safety must be maintained. Under the FW safety-control policy, we can lessen the possibility of unnecessary system shutdowns by assigning a lesser value to k in (15.19) in exchange for a higher possibility of letting the unsafe system run. Under the SP safety-control policy, on the other hand, we can lessen unnecessary system shutdowns by assigning a larger value to k in exchange for a higher possibility of accidents. An optimization problem must be formulated for determining an optimal rule of combination for each type of safety-control policy.

When an optimal rule of combination is distinct from Dempster's rule, we encounter non-associativity of a rule of combination [17]. A combination rule is said to be associative if:

$$((m_1 \oplus m_2) \oplus m_3)(C) = (m_1 \oplus (m_2 \oplus m_3))(C) \quad \text{for any } C \in 2^X$$

where $m_1 \oplus m_2$ denotes combining belief structures $\{m_1(A) : A \in 2^X\}$ and $\{m_2(B) : B \in 2^X\}$. The unified rule of combination (15.19) is associative if and only if it coincides with Dempster's rule. If a rule of combination is not associative, resulting belief structure is dependent on the ordering of information fusion when more than two belief structures must be combined. We need to analyze the effect of non-associativity of the optimal rule of combination on our safety-control decision.

ACKNOWLEDGMENT

Special thanks are due to Kenji Fujioka, graduate student at the Laboratory for Applied System Science, University of Tsukuba, for his help in preparing the example and its associated figures in Section 15.3.

REFERENCES

[1] A.P. Dempster, *Upper and lower probabilities induced by a multi-valued mapping*, Ann. Math. Statist., vol 38, pp 325-339, 1967.

[2] G. Shafer, *A Mathematical Theory of Evidence*, Princeton University Press, 1976.

[3] L.A. Zadeh, *Review of Shafer's A Mathematical Theory of Evidence*, AI Magazine, vol 5, no 3, pp 81-83, 1984.

[4] D. Dubois, H. Prade, *Combination and propagation of uncertainty with belief functions*, Proc. Ninth IJCAI, vol 1, pp 111-113, 1985.

[5] R.R. Yager, *On the Dempster-Shafer framework and new combination rules*, Info Sciences, vol 41, pp 93-137, 1987.

[6] Ph. Smets, *Belief functions*, In *Non-standard Logics for Automated Reasoning*, ed. Ph. Smets, *et al.* Academic Press, London, 1988, pp 253-286.

[7] P.R. Cohen, *Heuristic Reasoning about Uncertainty: An Artificial Intelligence Approach*, Pitman, 1985.

[8] J. Gordon, E.H. Shortliffe, *The Dempster-Shafer theory of evidence and its relevance to expert systems*, In *Rule-Based Expert Systems -- The MYCIN Experiments of the Stanford Heuristic Programming*

Project, Chapter 13, ed. B.G. Buchanan, E.H. Shortliffe. Addison-Wesley, Reading, 1984.

[9] B.G. Silverman, *Distributed inference and fusion algorithms for real-time supervisory controller positions*, IEEE Trans. Syst., Man, Cybern., vol 17, no 2, pp 230-239, 1987.

[10] P.L. Bogler, *Shafer-Dempster reasoning with applications to multisensor target identification systems*, IEEE Trans. Syst., Man, Cybern., vol 17, no 6, pp 968-977, 1987.

[11] J.A. Reggia, D.S. Nau, P.Y. Wang, *Diagnostic expert systems based on a set covering model*, Int. J. Man-Machine Studies, vol 19, pp 437-460, 1983.

[12] Y. Peng, J.A. Reggia, *Plausibility of diagnostic hypotheses: The nature of simplicity*, Proc. AAAI-86, pp 140-145, 1986.

[13] Y. Peng, J.A. Reggia, *A probabilistic causal model for diagnostic problem solving - Part I: Integrating symbolic causal inference with numeric probabilistic inference*, IEEE Trans. Syst, Man, Cybern, vol 17, no 2, pp 146-162, 1987.

[14] H. Ishibuchi, H. Ichihashi, H. Tanaka, *Set-cover theoretical diagnosis based on fuzzy symptoms*, Trans. Inst. Syst, Contr, Info Engineers, vol 2, no 2, pp 54-61, 1989 (in Japanese).

[15] T. Inagaki, Y. Ikebe, *A mathematical analysis of human-machine interface configurations for a safety monitoring system*, IEEE Trans. Reliability, vol 37, no 1, pp 35-40, 1988.

[16] T. Inagaki, Y. Ikebe, *Performance analysis of a safety-monitoring system under human-machine interface of safety-presentation type*, Microelectronics and Reliability, vol 29, no 2, pp 165-175, 1989.

[17] T. Inagaki, *Interdependence between safety-control policy and multiple-sensor schemes via Dempster-Shafer theory*, IEEE Trans. Reliability, vol 40, no 2, pp 182-188, 1991.

New Trends in System Reliability Evaluation
K.B. Misra (Editor)

CHAPTER 16

EXPERT SYSTEMS FOR RELIABILITY EVALUATION

D. J. Russomanno, R. D. Bonnell**, J. B. Bowles***

* Intergraph Corporation, Huntsville, AL 35894, U.S.A.

** Department of Electrical and Computer Engineering, University of South Carolina, Columbia, SC 29208, U.S.A.

ABSTRACT

The recent development of expert systems for system reliability evaluation is highlighted in this chapter. First, the basic expert system architecture is reviewed, then several prototype expert systems and techniques are characterized. Finally, a second generation expert system for FMEA is presented. In addition, an indication of the future trends and ongoing research in applying expert system technology to reliability analysis is covered.

16.1 Introduction

An expert system is a computer program intended for problem-solving tasks that traditionally have only been solved by human experts. Certainly, complex numerical techniques, such as matrix manipulation and other algorithmic tasks, have been solved by computers for years and would not be considered expert system applications. Rules are needed to solve algorithmic problems like ordinary differential equations; however, the rules of expert systems are intended to capture human intuition and experience, that is, the private knowledge of the expert. It is not implied that expert systems are intended to replace a human expert, but rather to play the role of an intelligent assistant or colleague. Expert systems and classical algorithmic software share many of the same underlying development methodologies; however, additional considerations must be made when designing the architecture of an expert system to be able to acquire and exploit the knowledge of a given domain.

Reliability tasks require the integrated problem solving of many human experts; therefore, expert systems in reliability applications are often very complex since the passing of information among specialists consists of semantic representations whose meanings must be preserved. Moreover, diverse knowledge representation techniques are needed to encompass the types of knowledge brought to bear by reliability experts. First-principle knowledge, that is, knowledge of what constitutes a reliable design, and reliability techniques must be captured. Knowledge about a physical system's function, behavior, and structure, as well as design heuristics (i.e., empirical associations or rules of thumb), which are acquired through experience with a system in a given domain, are needed to perform effective reliability analyses.

In this chapter, the basic expert system architecture is introduced, then recent prototype expert systems and techniques for reliability analysis are characterized, according to their knowledge and reasoning strategies, and reviewed to familiarize the reader with recent attempts in computer-aided reliability analysis. The methods reviewed emphasize expert systems for Failure Modes and Effects Analysis (FMEA), fault tree generation, and knowledge acquisition for diagnostic systems. Finally, second generation expert systems for FMEA are discussed along with a knowledge representation for relating a system's structure, behavior, and function. In addition, a problem-solving model based on the blackboard paradigm is presented which characterizes the essence of the future developments that are likely to take place in the development of expert systems for reliability evaluation.

16.2 Expert System Architecture

An expert system is an Artificial Intelligence (AI) problem-solving model for organizing the reasoning steps and domain knowledge to construct a solution to a problem. The central problem-solving issue is how and when to apply the most appropriate piece of knowledge. The classical software paradigm consists of a program (i.e., a set of procedures and a control mechanism) acting on a database. The inherent weakness of the classical software paradigm is that the problem-solving knowledge is embedded in the procedures and the control structures of the program. The expert system problem-solving model seeks to separate the domain knowledge from how and when it is manipulated to produce conclusions. Figure 16.1 depicts the basic architecture of an expert system.

The components of a simple expert system are the inference engine, knowledge base, and database. The basic architecture of an expert system may be augmented with a user interface for communication and a knowledge acquisition subsystem for inputting data, knowledge, and control strategies. The *knowledge base* holds the expertise of the system. The knowledge may be expressed in rules, frames, semantic nets, or other knowledge representation methods. The *inference engine* determines the knowledge that is applied in a problem-solving session. Strategies, such as backward and forward chaining, are used to generate consequences or decisions from the application of the knowledge base. The *database* contains the observed and derived facts about the problem domain.

The inference engine and the user interface can be viewed as one module, the expert system *shell*. The shell is typically the generic subsystem of an expert system, while the database and knowledge base are empty in the shell. A knowledge engineer tailors the knowledge base for a particular application and domain.

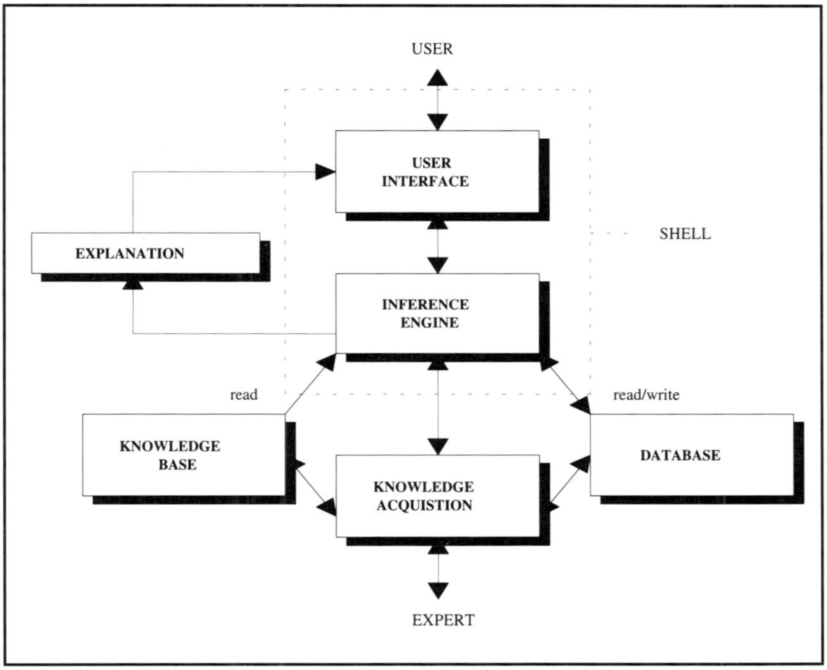

Fig. 16.1 Basic Architecture of an Expert System

16.2.1 Classical Expert System Problem-Solving Behavior

The input to the expert system and results of computations are kept in the database. The database contents are used by the inference engine, in conjunction with the knowledge base contents, to derive new conclusions and infer hypotheses that are placed in the database. The inference engine continues to access the database (the database is read/write for the inference engine) and the knowledge base (read-only for the inference engine) until a satisfactory solution is obtained or the system is halted. The inference engine controls the problem-solving activity. The most popular and successful knowledge representation formalism is *If-Then* rules, referred to as production rules. An example problem-solving behavior for a classical rule-based expert system is illustrated in Fig. 16.2 and outlined by the following steps:

1. The inference engine reads the knowledge base to determine the set of rules that are applicable based on the problem-solving state.
2. The inference engine reads the database to determine the problem-solving state.
3. A set of rules are matched with the problem-solving state to make up the conflict set.
4. A conflict resolution strategy is utilized by the inference engine to determine the appropriate rule to execute. Many conflict resolution strategies exist, for example, rule order, rule specificity ordering, and rule size ordering [1].
5. The rule selected by the conflict resolution strategy is executed.
6. The inference engine adds the conclusion of the executed rule to the database and conveys the results to the user.

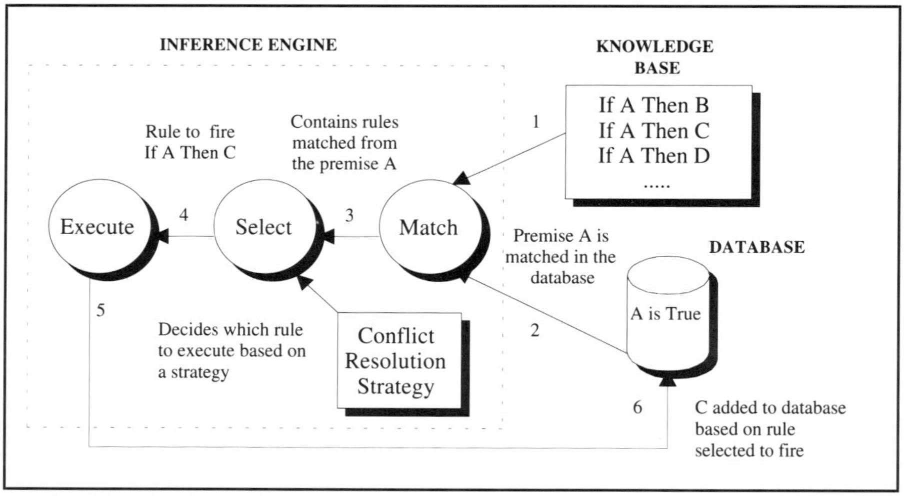

Fig. 16.2 Problem-Solving Behavior of a Forward Chaining Expert System

The preceding behavior represents a simple forward chaining control strategy that uses rules, the most elementary knowledge representation method. However, the example illustrates the more salient features of the expert system's behavior. Namely, the control of the application of the knowledge has been separated from the knowledge base. In addition, the knowledge that the system possesses, i.e., the general concepts of the domain, has been separated from the data (information about specific instances). In the classical software paradigm, the *If-Then* rules are the inferencing procedures and the implicit knowledge embedded in the program's execution. All of the *If-Then* rules in the classical software paradigm are executed. In an expert system architecture, the *If-Then* rules represent a single block of knowledge that, for a given problem, may never be executed. In the upcoming section on second generation expert systems, limitations of the classical expert system architecture will be illuminated as the development of knowledge-based systems for reliability evaluation are pursued.

16.3 Characterization of Knowledge and Reasoning in Expert Systems for Failure Modes and Effects Analysis (FMEA)

Several approaches are used to represent knowledge and reasoning in prototype computer-aided FMEA systems. The approaches to be considered include Lehtela's Computer-aided FMEA Tool (CAF) [2], FCP Analysis [3], FADES [4], FIABEX [5], and Boeing Computer Services' FMEA Assistant (BCS) [6]. The approaches can be characterized in terms of their knowledge representation and content, as well as the reasoning structures that exploit their knowledge base. Table 16.1 summarizes representative attempts to automate FMEA, and is intended to illustrate a spectrum, rather than an exact division of methods.[1] Typically, increasing knowledge results in more computer initiative and control of the problem-solving process. The approaches in Table 16.1, inherit and extend, from right to left, both reasoning and knowledge representation methods. For example, the structured knowledge paradigm may utilize frames whose slot values are derived via procedural attachment to numerical algorithms, or the deep knowledge paradigm may utilize and extend some of the structured knowledge paradigm's methods.

The most elementary approach is the database method in which the associations between an item, its failure mode, and overall system effect are stored. The problem-solving knowledge consists of associations of the form {subset of item failures} $\Rightarrow$ {subset of system failure effects}. With the database approach, the control of knowledge consists of queries to look up item failures and the user is responsible for determining the overall effect on the system's operation. Once the user infers the overall system effect, the database is updated to contain the new facts. The database approach is most useful for filing away specific facts concerning a detailed analysis; however, it is inadequate for automation of failure mode propagation and reasoning about new systems.

The second method, the rule-based paradigm, builds production rules for the consequences of item failures. In this approach, pattern matching is used on the rules to derive a chain of reasoning that results in the overall system effect (i.e., {item failure} $\Rightarrow$ {intermediate conclusion} $\Rightarrow$ {overall system effect}). Specifically, the item failure can be considered as an initial observa-

[1]Table 16.1 refers to acronyms of prototype systems and techniques characterized later in the chapter.

Table 16.1 Characterization of Computer-Aided FMEA

Approach	Deep Knowledge Paradigm	Structured Knowledge Paradigm	Rule-Based Paradigm	Database
Knowledge Represen-tation	Functional Primitives + Behavioral Primitives	Predicate Calculus Frames Semantic Nets Qualitative Equations	Production Rules	Tables
Reasoning	Deep Functional and Behavioral Reasoning	Shallow Reasoning	Forward and Backward Chaining	Queries of Tables Numerical Techniques
Prototypes		BCS FIABEX FADES FURAX	FCP	ANEKO CAF

tion and there are rules that relate this initial condition to an intermediate state, then, additional rules are used to progress from the intermediate state to a final conclusion. Backward chaining may be used to explain the reasoning steps used to infer a conclusion. The knowledge base in the rule-based approach contains direct relationships between item failures and system effects without specifying a causal path between them. The structure of the rules enables consistency to be verified. Moreover, it provides a means for constructing fault trees and developing computer-aided instruction systems for diagnostic troubleshooting and maintenance.

The third method, the structured knowledge paradigm, exploits more sophisticated reasoning and knowledge structures than the rule-based paradigm. For example, predicate calculus, frames, and semantic nets may be utilized, and depending upon the state of the solution, different control strategies can be employed to focus the reasoning process. In this approach, it is assumed that the knowledge base contains the information needed to infer the overall system effects from an item failure, and no problem-solving scenario other than human input is available for deriving new knowledge that may be needed in determining a solution. Most causal reasoning systems are implemented with this approach, that is, the causal relationships are explicitly coded in a knowledge base or library, and the reasoning system has no fundamental knowledge to support the causal relationships. These approaches begin to exploit deeper representations than rule-based approaches; however, the knowledge tends to be specialized for a particular application.

Finally, the fourth method, the deep knowledge paradigm, provides a mechanism for deriving new relationships and the knowledge necessary for producing solutions when the initial knowledge organization is inadequate. In this paradigm, the knowledge is in a more general form and is transformable into specific instances suitable for particular domains. When the problem solver is unable to determine an overall system effect because of an incomplete knowledge base, there is a mechanism for generating the needed knowledge. This approach includes underlying knowledge to support and derive new causal relationships and heuristics. Deep models are needed to *fall back* on when there is no directly applicable knowledge available for contributing to the solution state. Moreover, reasoning structures and knowledge for machine learning are required for these systems to improve their performance over time.

16.4 Prototype Computer-Aided Reliability Analysis

The following sections review noteworthy computer-aided reliability systems and automation techniques. The review begins with Lehtela's prototype which does not incorporate AI techniques but represents an advance in computer-aided FMEA methods. Next, several prototypes and techniques that advance computer-aided reliability toward an AI approach are reviewed [7].

16.4.1 Lehtela's Computer-Aided FMEA Tool (CAF)

Lehtela [2] developed a computer-aided tool for performing a FMEA of electronic circuits. The block diagram in Fig. 16.3 illustrates the key steps of the analysis. The system description is limited to electronic circuits and consists of a circuit description much like that used in SPICE [8]. Every component must be explicitly described and a library is employed that consists of component types and parameters.

The modeling of component failures consists of searching the library for an appropriate failure mode. For example, when considering a resistor that has *failed open*, a failure model consists of inserting a 100 MΩ resistor in place of the faulted resistor and running the simulation to quiescence. Steps 11-14 of Fig. 16.3 require the user to independently (i.e., no computer assistance is provided) evaluate the results of the simulation and determine if additional analyses are needed by employing heuristics. In addition, the user must be in full control of the analysis, that is, there is no mixed-initiative dialogue in which the computer shares in conducting the analysis. The program's main advantage is that the computer assistance provides a more timely and reliable analysis of electronic circuits than manual methods. Another computer-aided package utilizing the database approach is ANEKO [9].

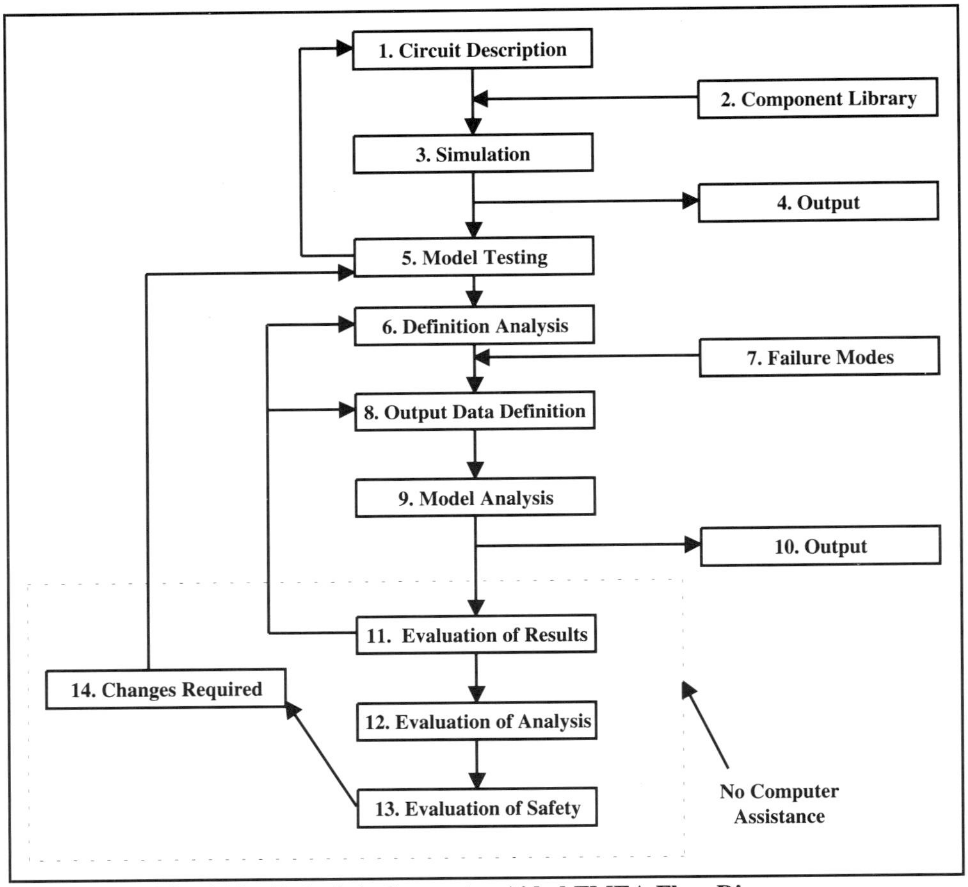

Fig. 16.3 Lehtela's Computer-Aided FMEA Flow Diagram

16.4.2 Gondran and Laleuf--FCP Analysis

Gondran and Laleuf [3] have developed a method of representing knowledge in reliability analysis referred to as FCP analysis. FCP (Functions - Components - Parameters) analysis is an attempt to formalize the representation of system failure modes to overcome the drawbacks of conventional FMEA. FCP analysis provides a formal connection between FMEA and subsequent reliability analysis, and provides a medium for building a FMEA knowledge base that can be used for automated fault tree construction, failure simulation, or diagnostic assistance. FCP analysis employs reification[2] of system functions, in addition to considering system parameters and components to provide general concepts that aid in analysis. Specifically, system functions are treated as objects just as if they were actual components. FCP analysis imposes a formalized representation on the FMEA worksheet to provide a more concise representation of the relationships among the functions-components-parameters and the failure modes. This formalized representation is used to automatically develop production rules that are used in subsequent analyses.

FCP analysis considers four of the columns in FMEA: component, failure mode, possible causes of failure, and effects on the system. The technique transforms records in the FMEA table into production rules. Table 16.2 is a traditional FMEA record of a circuit breaker that feeds the power supply of a computer.

Table 16.2 FMEA Entry for a Circuit Breaker

Component	Failure Mode	Cause of Failure	Effects
Breaker	Loss of Breaker	Primary Failure Fire	No Supply to Computer

By formalizing the entry, production rules can be generated automatically from the FCP table, Table 16.3. Notice in the following production rules that the non-component column entries are treated as predicates and the components are arguments of the predicates.

If Primary_Failure(Breaker) then Loss(Breaker)
If Fire(Equipment) then Loss(Breaker)
If Loss(Breaker) then Failure(Supply)

Table 16.3 FCP Entry for a Circuit Breaker

Component	Failure Mode	Cause of Failure	Effects
Breaker	Loss	Primary_Failure Fire(Equipment)	Failure(Supply)

The analysis goes one step further in reifying functions (i.e., treating a function as an object). The following production rule is a result of Table 16.4:

[2]Reification refers to the process of treating an abstraction as an object (e.g., treating a function as an object with a failure mode).

If Failure (Supply Computer) then
Failure (Computer)

Table 16.4 FCP Entry with Functions Treated as Objects

Component	Failure Mode	Cause of Failure	Effects
Supply Computer	Failure	Loss(Breaker)	Failure(Computer)

From the production rules, a fault tree can be generated automatically. The knowledge base can serve as the basis for an elementary simulation for answering questions, such as if component X fails in failure mode Y, what effect Z is elicited. Moreover, the knowledge base provides important background for assistance in diagnosis and control. FCP analysis assumes that a conventional FMEA has been performed and is chiefly concerned with retaining that knowledge and providing a formal connection to subsequent analyses. Expert systems are needed that derive the FMEA table from a high-level system description that is input from the designer. Although FCP analysis does not address this issue, it represents an early attempt to incorporate AI techniques in FMEA knowledge representation. In addition, FCP analysis emphasizes the importance of modeling in the development of computer-aided reliability programs.

16.4.3 FADES

FADES (Fault Analysis and Diagnosis Expert System) [4] is a prototype system developed for generating a FMEA for the fluid system domain. The tool utilizes a library of graphic symbols to generate system descriptions, while building a knowledge base that contains a behavioral description for each component. The behavior is inferred from the physical structure of the system. Specifically, while system components are described graphically via a *Picture Editor*, they are made instances of an appropriate class that contains token-type behavior knowledge about the component. The component's behavior description will inherit pertinent attributes, such as failure rate, and failure mode, etc., from the class type. The FADES model uses both path (i.e., a collection of components in the system that make up a circuit) and behavior rule modeling paradigms to form a knowledge representation of the system [4]. The program also provides a *Set Failure* analysis in which the user is in complete control of the failure selection process in the analysis. This method is useful for a novice who can utilize the system to simulate component failures to learn their effects on the system. FADES also provides a *Complete* analysis mode in which the program is in control of the failure selection process. The

Complete analysis method builds a knowledge base of the behavior of a system under fault conditions for each component in the design. Forward chaining is used to determine the system response to item failures. An explanation of the analysis is generated by tracing backward from the system response to the item failure. FADES provides both FMEA tables and fault trees and is representative of recent attempts to incorporate AI techniques in reliability programs. Similar work has been performed on FURAX [10] in the flow network domain and by Bell, et al. [11] in the electrical domain.

16.4.4 FIABEX

FIABEX was an attempt to provide an intelligent knowledge-based system for the automation of FMEA, Fault Tree Analysis (FTA) and other related reliability techniques [5]. The FIABEX model emphasizes symbolic processing and has addressed the issue of describing system behavior by using facts and rules. These facts and rules relate to system components rather than overall functionality, and their interrelationships are dependent upon the system structure, rather than a functional specification. The project has emphasized the development of physical modeling techniques for FMEA. Simple systems have been modeled as a series of components, which only propagate behavior in one direction. The structure of the FIABEX system is shown in Fig. 16.4. The *Core* is made up of the physical modeling agent and its supporting modules. The knowledge bases (KBs) make up the component libraries that contain the facts and rules describing the graphics architecture, behavior, and perception of system components from many domains. The notion of having various industries provide application-specific knowledge bases has been suggested. The *Operational Diagnosis Module* is responsible for using diagnostic trees, simulation results, fault trees, and FMEA to generate a diagnosis of a particular system during operation.

16.4.5 BCS' FMEA Assistant

Boeing Computer Services (BCS) has developed a knowledge representation of items (components and subassemblies), a FMEA intelligent interface, and tools to aid the analyst in examining FMEA information and testing of hypotheses concerning the design's reliability. The knowledge representation conceptualizes the system as a whole, utilizing frames containing information about entity attributes in a semantic network [6]. Generic component information (e.g., valve, pump, and switch, etc.) was built into the knowledge base utilizing conventional knowledge engineering methods.

In the process of adding an item to the system structure, the engineer is assisted by an intelligent editor. Upon placing an item in the system structure,

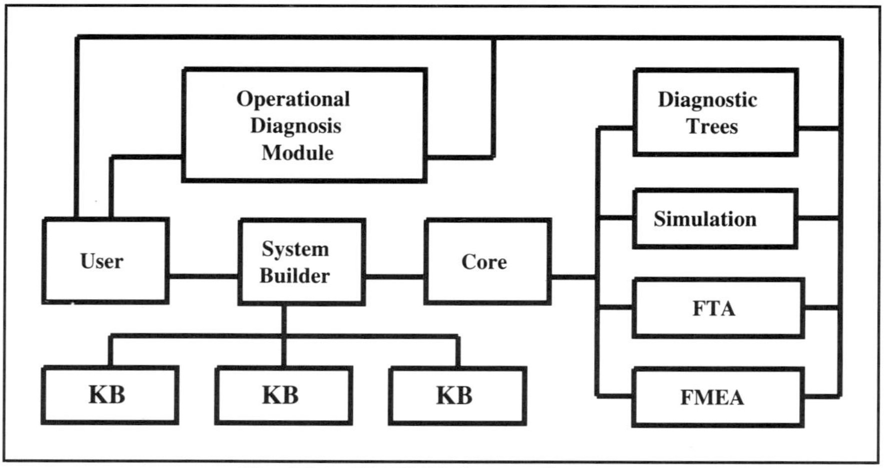

Fig. 16.4 FIABEX System Structure

the intelligent editor provides the engineer with the traditional FMEA worksheet, whose fields have been completed as much as possible for the particular item just placed. The information automatically generated in the FMEA worksheet is a result of items inheriting properties from a more general class that already exists in the knowledge base. The engineer is required to enter any information not present on the FMEA worksheet, including the result of the item failure on the overall system operation. Hence, the generation of device behavior is not entirely automated but depends upon the initial knowledge acquisition task, as well as incremental additions by the engineer in the graphics environment. Once all of the data for the FMEA has been entered, a network is depicted graphically and the engineer can examine item failure mode propagation through selected portions of the system structure. The intelligent editor provides tools for displaying a list of available failure modes for a given device, and selection of the portion of the system to graphically depict during failure mode propagation. BCS' FMEA assistant addresses three key issues in the intelligent automation of FMEA: knowledge representation, intelligent interface design, and failure propagation.

16.5 Second Generation Expert Systems for Reliability Evaluation

The goal of second generation expert systems is to reason on representations of physical systems at a deeper level than empirical associations or heuristics. In second generation expert systems for reliability evaluation, functional models, as well as causal models of a system's behavior and structure must be exploited when empirical associations or heuristics fail to provide a satisfactory analysis. A deep model of a system can be viewed as a source of knowledge to *fall back* on when the heuristics for analyzing the system are inadequate. A promising knowledge representation technique that exploits deep models is a functional representation proposed by Sembugamoorthy and Chandrasekaran [12]. The functional representation and its extensions provide a framework for organizing deep models in an expert system to simulate and predict the effects of a system's failure modes.

16.5.1 A Deep Model of Function

Second generation expert systems research[3] was initially concerned with developing a deep model of a device's structure (i.e., the static physical characteristics of the device based on its topology) and deriving numerical relationships among structural constituents to produce a simulation. The next step focused on utilizing a methodology for qualitatively simulating and deriving the system's behaviors (i.e., its dynamic characteristics). Causal reasoning, which builds upon representations of structure and behavior, addresses the issues of how systems achieve their behavior. Neither structural nor behavioral models, including causal models, explicitly describe the functionality of a device. The development of representations that focus on the function of devices is a natural evolution in model-based reasoning.

To perform a FMEA, human experts must develop an accurate mental model of how systems or subsystems work. Humans reduce the cognitive load of understanding systems by hierarchically decomposing an overall device into subsystems [13]. This requires the expert to understand, on a qualitative level, how the system is designed to perform certain functions. For example, consider the following human expert's description of the function of a hot water heater's temperature subsystem:

The temperature subsystem's function is to control the water temperature at the outlet faucet by maintaining the water temperature in the tank through the controlled addition of gas to the burner and tap water to the tank.

The description includes a function (to control), and a method by which the function is obtained (by maintaining). Such a knowledge organization, representation, and acquisition scheme is necessary to build expert systems for FMEA that use descriptions of function in device simulations of item failure effects. Moreover, the representation can be used to explain a system's

---

[3]As used here, a second generation expert system utilizes deep models in problem solving. A deep model may be a quantitative, qualitative, or causal model of a system, rather than a set of heuristics (e.g., a shallow model). Some researchers restrict the term *second generation* to imply expert systems that utilize cognitive models.

behavior in terms of the functions accomplished, which are explicitly represented in the acquired knowledge base.

The function of a system is its intended purpose (i.e., its teleology) and its behavior is how that purpose is achieved [14]. A system is made of devices and accomplishes its overall functionality via the functions of its devices. A device's function is achieved by a detailed set of causal relations that constitute the behavior of the device. Functional modeling is utilized when intelligent aids decompose a design into constituent functions that indicate purpose. A function can be accomplished by a hierarchy of behavioral segments that consist of state transitions and subcomponent functions. As a design becomes more specific, detail is added to the behavioral segments and subfunctions that make up an overall function. A behavioral segment may consist of a compiled relation such as A $\Rightarrow$ B; however, this segment may actually be made up of several behavioral segments (e.g., $A \Rightarrow A' \Rightarrow A'' \Rightarrow B$). The causal transition ($\Rightarrow$) is the result of one of the following: (1) more detailed behavioral segments (2) a component subfunction or (3) deeper domain knowledge (e.g., Kirchhoff's Law, etc.) [12]. By representing functions explicitly, systems can be simulated earlier in the design life cycle and at a more abstract level; hence, eliminating the extensive detail that results when describing a system either numerically, or qualitatively--as a long sequence of causal states.

As early as 1971, Freeman and Newell [15] identified the need and utility of a functional reasoning methodology. The functional modeling scheme described is based upon the research of how model-based reasoning is used rather than how it can be derived, and is exemplified by the work of Sembugamoorthy and Chandrasekaran [12], Keuneke [14], and Bonnet [16], among others.

The functional modeling scheme is based on a set of primitives that describe functionality at an abstract level. The scheme is top-down, that is, the functionality is explicitly represented by primitives, and if known, detailed behaviors manifested by qualitative simulations are added to the representation. This top-down approach to explicitly representing functionality is a paradigm shift from earlier work by de Kleer [17] in teleological reasoning, in which a device's function is implicit (i.e., function is derived from the system's structure and behavior). In the teleological approach, an expert recognizes structural configurations to infer purpose. For example, when analyzing an electronic circuit, the engineer could recognize a common collector configuration and infer that the circuit is being used for the purpose of increasing the bandwidth, or recognize a common emitter configuration and infer that the circuit's purpose is to increase the gain, etc. AI researchers have made little progress in this area of ongoing research.

Keuneke [14] has delineated a set of functional primitives based upon the work of Sembugamoorthy and Chandrasekaran [12]:

- *ToMake*- places an entity in a certain state,
  e.g., *ToMake* (Engine Start)

- *ToMaintain*- maintains an entity state once it has been reached,
  e.g., *ToMaintain* (Engine Temperature)

- *ToPrevent*- disallows an entity to achieve a certain state,
  e.g., *ToPrevent* (Engine Overspeed)

- *To Control*- regulates state changes via a relationship,
  e.g., *To Control*(Engine Speed)

Figure 16.5 illustrates a functional representation that has been augmented with behavior. The specification includes a functional primitive; therefore, the device is described solely by its function at the highest level of abstraction. In addition, a behavior by which the function is obtained is referenced. Sembugamoorthy and Chandrasekaran [12] indicate that causal nets should be organized by function to make them manageable for reasoning. Figure 16.5 depicts a causal network that imposes a hierarchical structure on the behaviors that achieve the overall function (e.g., behavior1 is comprised of behavior2, behavior3, and behavior4). The behavioral state sequence 1b1 ⇒ 2b1, is actually accomplished by the state transitions 1b1 ⇒ 2b2 ⇒ 3b2 ⇒ 2b1 (behavior2), and the state sequence 1b1 ⇒ 2b2 is actually accomplished by the state transitions 1b1 ⇒ 2b3 ⇒ 3b3 ⇒ 2b2 (behavior3), etc. This hierarchy indicates increasing levels of detail on how each node transition is achieved. For example, the causal relation *pressed(switch) ⇒ buzzing(buzzer)* is made up of a hierarchy of causal details that describe how applying voltage results in current flow, which produces a magnetized space; the magnetized space causes connections to separate, then reconnect, which results in the acoustic function *buzz*, etc.

A chief aspect of the functional representation is that by using primitives explicitly to represent functionality the system can be simulated at a more abstract level (e.g., simulate the failure effects of systems before the detailed design is known). Specifically, while performing a functional FMEA early in the system's conceptual development or proposal stage, the designer or analyst uses building blocks, which are tied to a functional knowledge representation, to achieve the design purpose. Then as the design progresses, the physical components that participate in the detailed behavior are configured and the causal network is built. Hence, the knowledge of the design and its failure modes and effects have been organized hierarchically from function down to

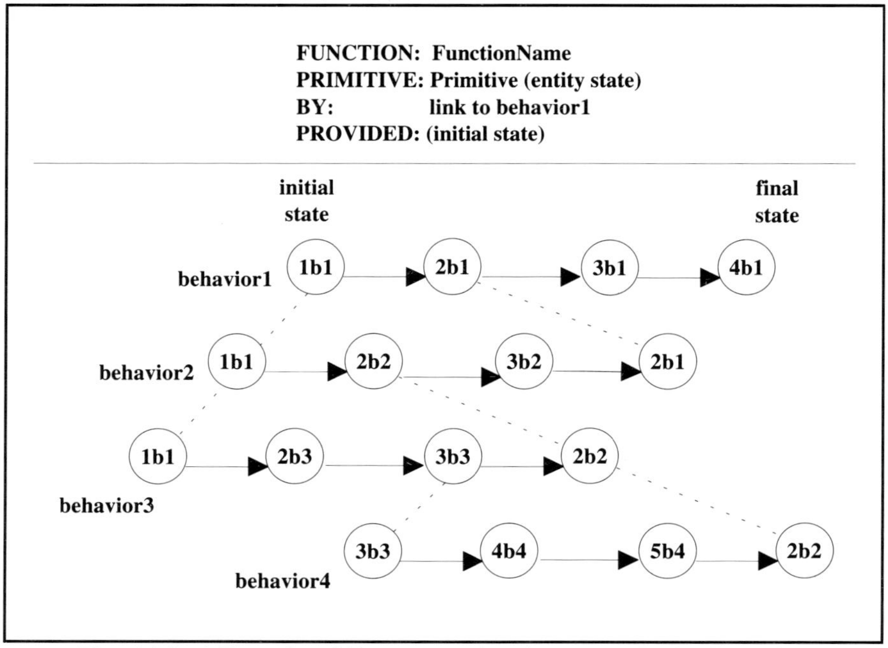

Fig. 16.5 A Functional Representation Augmented with Behavior

detailed behaviors. In the next section, a second generation expert system for FMEA, which employs a simulation subsystem based on the principles of the functional representation, will be described.

16.6 A Second Generation Expert System for FMEA

The design of an expert system is approached from a knowledge-use level perspective to provide an understanding of the problem and insight into the knowledge and expertise needed to automate the FMEA process. The blackboard conceptual model is used to provide the organizational principles required for the design of an expert system without actually specifying its realization. In the blackboard model of an Expert System for Failure Mode and Effects Analysis (XFMEA), the system is functionally decomposed into a set of knowledge sources, each containing the knowledge associated with a subfunction of the FMEA task. In addition, a learning layer is described that provides a framework for incorporating self-improving aspects into the program. The resulting conceptual model aids in developing design specifications and serves as the foundation for further research into automating FMEA and other reliability tasks in which integrated, diverse sources of knowledge are required.

16.6.1 The Blackboard Model

The blackboard model is an evolution in expert system modeling and attempts to overcome some of the inherent weaknesses of the classical expert system architecture previously discussed. The chief limitation of the classical expert system architecture is that the knowledge representation is limited by the design of the inference engine. That is, if the inference engine is a rule interpreter, then the knowledge representation utilized by the knowledge base must be rules. In the blackboard model, the knowledge of the intelligent system is functionally decomposed into modules or knowledge sources. Each module in the model can have its own knowledge representation and reasoning strategy, Fig. 16.6 [18].

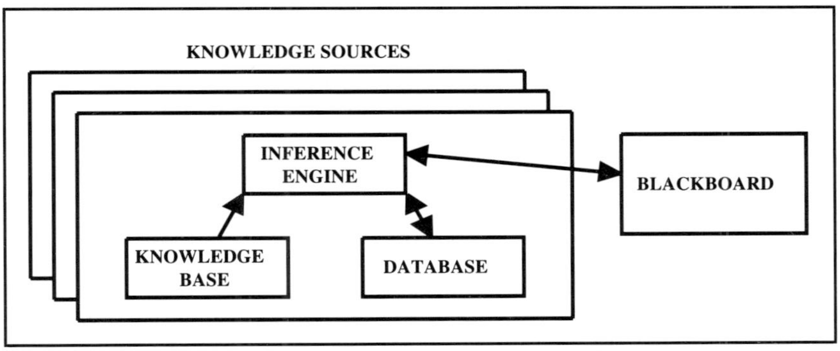

Fig. 16.6 Elementary Blackboard Model

A blackboard model consists of three components: a world model (WM), a blackboard (BB), and a set of knowledge sources (KSs), Fig. 16.7. The WM contains the fixed conceptual framework in which the intelligent system operates and the assumptions and constraints upon which its design is based. In addition, the WM includes the strategy of the user interaction format. The knowledge of the intelligent system is partitioned into a set of KSs and is independent of when, how, and where it is used. A KS can be an algorithm, a knowledge-based system, a neural network, a machine learning system, a self-improving system, or even a human. Each KS has its own inference engine; hence, the model covers the use of diverse knowledge representation techniques and inference strategies. A KS can be self-improving with its own learning system; in other words, it uses machine learning techniques to improve the performance of its task. A KS can also be a blackboard model itself with its own WM, BB, and set of KSs.

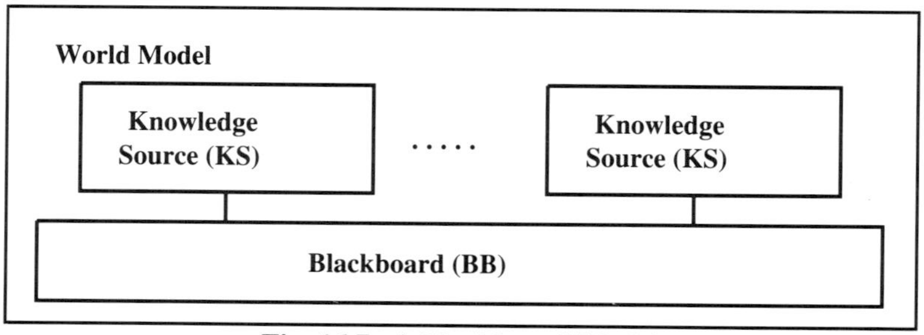

Fig. 16.7 A Blackboard Model

The BB is a structured global database that provides the only medium through which the KSs can communicate and interact. It contains data to be used by the KSs in the performance of their tasks, as well as any information relating to the current session with a user. The BB can be partitioned into several panels at different levels of abstraction. Partitioning the BB by panels imposes a structure on the BB knowledge and data.

During problem solving, the KSs read from and write to the BB, developing the solution incrementally. Each KS places information on the BB and can respond opportunistically to changes on the BB. Although no control strategy is specified in the blackboard model, opportunistic reasoning[4] as a control strategy is implicitly assumed, that is, KSs are actuated one at a time and run to completion with the choice of the KS depending upon the contribution that each can make at the current point in the problem-solving process.

16.6.2 A Blackboard Model of an Expert System for FMEA

Intelligent FMEA systems need to explore richer architectures that capture the information-processing activity of the entire analysis process. Traditionally, FMEA knowledge tends to be restricted to small groups of specialists and an analysis can only be generated by a team of engineers having a thorough knowledge of the system's design and application. Hence, a cooperative effort covering diverse areas of expertise is required. In the XFMEA blackboard model, the requisite knowledge is partitioned into KSs that embody the knowledge of a particular task or viewpoint of the domain. The top-level knowledge sources are the *User*, *Design Expert*, *Mission Expert*, *Simulation Expert*, *Analysis Generator*, and the *Discourse Manager*, as shown in Fig. 16.8.

---

[4]Opportunistic reasoning is a concept that implies that the most appropriate knowledge source will be activated based on the solution state. In a blackboard framework, the control strategy, which activates the knowledge sources, is implemented as a computational construct.

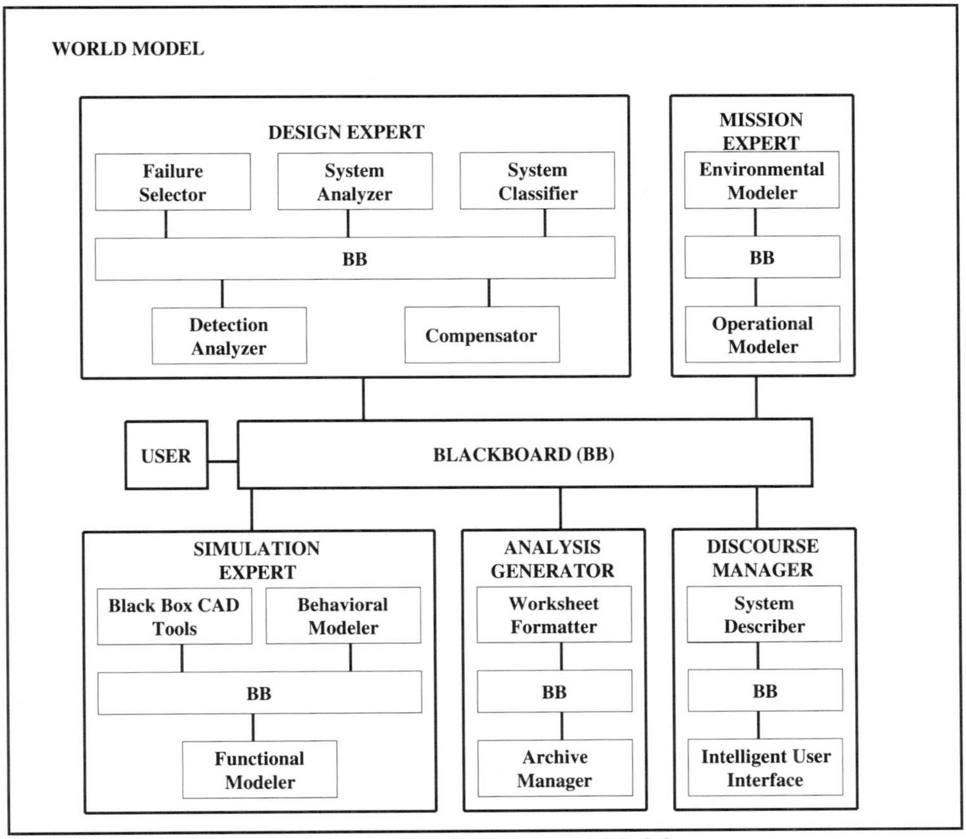

Fig. 16.8 XFMEA Blackboard Model

The fixed conceptual framework in which the XFMEA system operates is explicitly defined in the WM. The WM states the objectives, abstraction level of the analysis, system viewpoints, operating modes, and other important assumptions and constraints.

The XFMEA performs the analysis from different viewpoints, for example, system safety, mission success, availability, and failure mode detectability. The analysis viewpoints and system operating phases must be stated explicitly in the WM. As the phases change, different functional relationships and component operating modes exist. The phase includes both the operating point of the system (e.g., water heater is set on high, jet has airspeed of MACH 1, etc.) and the environment in which the system operates (e.g., water heater operates in a 120°F closet, or jet has altitude of 40,000 feet).

The information on the BB is available to all of the KSs. During the XFMEA analysis, the KSs read from and write to the BB, developing the

FMEA worksheet incrementally. The KSs in the blackboard model of XFMEA are described in the following paragraphs.

The *User* (a human KS) provides the statement of the problem, the configuration of the system, and answers questions as needed by the XFMEA system. The following is an example problem statement: *perform a detailed failure analysis on the hot water heater; provide partial solutions as completed noting any class I or II failures.*

The *Discourse Manager* provides an intelligent user interface for the *User* to interact with the XFMEA system. The interface is independent of the internal representation of the system; hence, it provides a knowledge-use level representation for user communications [19]. Moreover, this KS develops a space on the BB to match the user's conceptualization of the system under study with the internal representation of the system. The *Discourse Manager* includes the *System Describer* which provides the *User* with tools to generate a design description (e.g., schematic diagram) of the system on the BB. It is important that the *System Describer* be capable of providing system descriptions at multiple levels of abstraction to support the full design life cycle. The *Discourse Manager* is the kernel for all activity and presents information to the *User* concerning the progress of the FMEA and any partial results that may be of interest. In addition, it provides help facilities to the *User*. The *Discourse Manager* may query the *User* in the event that additional knowledge is needed; therefore, in the XFMEA blackboard model the user is viewed as another KS. The *User* is an integral constituent in the blackboard model, emphasizing the role of the analyst in intelligent FMEA. Puerta and Bonnell [20] have developed a comprehensive blackboard model of a generic, self-improving intelligent user interface.

The *Design Expert* consists of the detailed knowledge necessary to perform a FMEA. The major components of the *Design Expert* are the *Failure Selector, System Analyzer, System Classifier, Detection Analyzer* and the *Compensator*. The *Failure Selector* manages the analysis. For example, based on the level of abstraction (i.e., hardware item or functional analysis), the *Failure Selector* uses the descriptive model of the device on the BB to systematically choose an item (component or subsystem) failure for the *System Analyzer* to consider. The *System Analyzer* studies the evolving solution state to determine what knowledge is needed to solve the problem. It analyzes the effects of item failures based on its array of acquired failure-mode heuristics, or based on an appropriate system model such as a qualitative process model, a functional model, or a CAD simulation model. The *System Analyzer* must determine if an item failure mode violates an assumption, a relation, or an overall function of the system. In general, FMEA reasoning must go beyond what is obtainable from considering simulation information alone. The *System*

Analyzer must employ general strategies and knowledge gained through experience with a device in a domain; therefore, this subknowledge source must also possess strategies that determine when the analysis shifts from heuristics to deep models. The *System Classifier* analyzes each failure according to its severity and uses knowledge on the BB to take into account the environment and system operating mode impacts. The *Detection Analyzer* determines how the system will detect the failure (e.g., visual, automatic sensing device, etc.). The *Detection Analyzer* must use knowledge not only about items, but their intended functions, operating modes, environment, and relationships to other components. The *Compensator* attempts to provide a feasible solution or a compensating provision to reduce the severity of the failure effect.

The *Mission Expert* possesses the knowledge that is used to determine the action that will result when an item is in a particular operating mode and environment. Figure 16.9 depicts an elementary rule that captures the action that an item could produce in a given mode. This KS consists of the *Environmental Modeler* and *Operational Modeler* sub-knowledge sources which build environmental and operational models on the BB. The *Design Expert* depends upon the evolving operational and environmental models to classify the failure, to focus the analysis, and to select detection methods.

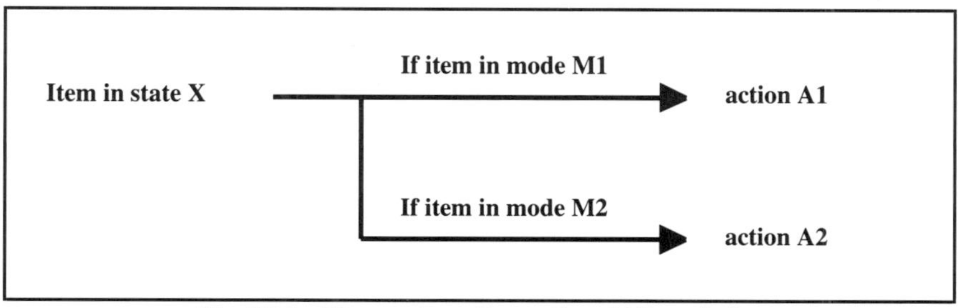

Fig. 16.9 Elementary Operational Knowledge

The *Simulation Expert* provides a simulation of the system under study. A significant challenge in the design of this KS is choosing a set of representations to describe systems and a strategy for integrating these representations at multiple levels of abstraction. The three chief components of this KS are the *Black Box CAD Tools*, the *Behavioral Modeler*, and the *Functional Modeler*. *Black Box CAD Tools* provide a detailed quantitative simulation to determine the effects of item failures. If required by the *Design Expert*, CAD tools are encapsulated to run device simulations. For example, an electronic circuit analysis program, such as SPICE [8], can simulate the effect of a given component failure. The simulation can be accomplished by replacing the faulted component with an equivalent component that provides the faulty behavior (e.g., replacing a resistor that has *failed open* by a 100 MΩ resistor). The *Design Expert* uses the derived simulation data along with information about the device's operating mode and environment to classify the failure effect. The weaknesses of CAD tools in system modeling are that they require a detailed system description to reside on the BB; they provide no explanation of how and why conclusions were reached, and at low levels of abstraction the simulation of complex devices is inefficient and tedious [21].

The second component of the *Simulation Expert* is the *Behavioral Modeler*. It provides the knowledge that underlies the ability to reason about the dynamics of the system under study based on its physical structure and behavior. Physical structural and behavioral models, exemplified by de Kleer [17], are exploited when the system can be described by assembling objects that also specify a behavioral component. The behavior of the overall system is synthesized from individual item behaviors. The advantages of this modeling scheme are that each item's structural and behavioral components can be reusable in other system descriptions; hence, a library of item structural/behavioral characteristics can be stored in a knowledge base to be available for future system synthesis. Moreover, if the system's behavioral component is accurate, the system can be simulated based on item alterations or failure modes. In general, physical structural and behavioral models progress from system descriptions (e.g., schematics) to behavior; finally, the derived behaviors can be abstracted to generate functions. Then, functionality is an implicit assumption derived from physical structure [14]. Models of a system's physical structure and behavior require that the system's behavior be synthesized from the behavior of its components.

The *Behavioral Modeler* is the principle simulation knowledge source in performing a FMEA when the device's structure is difficult to characterize in precise numerical terms. The *Behavioral Modeler* consists of the *Causal Modeler* and *Version Manager* sub-knowledge sources. The *Causal Modeler* infers the causal structure and behavior of the device based on its topology generated by the *System Describer*. The *Version Manager* stores the causal model at some level of completion so that it can be modified later, if the *Design Expert* desires a qualitative simulation of a modified or crippled system. Also, the FMEA can be performed in an iterative manner in which only partial descriptions are generated by the *System Describer*. A strength of the qualitative approach is that the reasoning process is more akin to that of an expert and explanations are easier to generate. The *Behavioral Modeler* is intended to augment the *Black Box CAD Tools*, not to replace them. Dependence upon this sub-knowledge source is determined by the abstraction level of the analysis.

The third component of the *Simulation Expert* is the *Functional Modeler* which possesses the knowledge that underlies the ability to reason about the functionality of a system. The functional representation/modeling scheme is viewed as constructing a situation-specific model on a blackboard; in addition, the blackboard model illustrates how the functional approach can be integrated with both detailed behavioral approaches and numerical techniques in modeling complex devices, Fig. 16.10. In FMEA, the functional modeling approach is most promising when system complexity requires a top-down analysis starting from a general indenture level and proceeding down to the component level.

The *Functional Modeler,* employing the functional representation, is not only useful in organizing the functionality and behavior of systems for simulation, but also for providing the formalism for expressing the FMEA worksheet in terms of primitives which have direct relationships to simulation methods. In Table 16.5, the traditional FMEA worksheet is retained; however, more precise semantics are imposed on the relationships between items, functions, and their failure effects. Moreover, the interpretation of the table is expressed in the same representation that is used in the *Functional Modeler's* knowledge base.

Table 16.5 FMEA Table Entry for a Communication System's Power Supply

ITEM	FUNCTION	FAILURE MODE	CAUSE	EFFECT ON SYSTEM
Power Supply	ToMake ($Device Energized)	Loss	Battery.Dead	¬Transmit($Signal) ∪ ¬Receive($Signal) ∪ ¬Energize($Device) ∪ ¬Buzz($Device)

The abstraction level of the analysis and the overall goals that have been stated in the WM, determine if modeling a system based on functional components is feasible. The XFMEA *Simulation Expert* must provide simulation capabilities that facilitate analyses at different levels of abstraction. These simulation capabilities aid in traversing the device hierarchy on the blackboard by allowing knowledge sources to focus their attention on the task at hand rather than manipulating representations of devices.

The *Analysis Generator* is a knowledge source that has algorithmic characteristics and it encompasses the tasks that are addressed by traditional FMEA programs. This knowledge source prepares the appropriate worksheet for the FMEA, based on the viewpoint on the BB. The *Worksheet Formatter* is a sub-knowledge source that uses the data on the BB to generate a report that is consistent with the abstraction level of the analysis. The *Analysis Generator* must search for omissions in the FMEA. For example, missing detection methods and inconsistent failure mode apportionments, etc., must be checked and additional requests may be placed on the BB. Once the FMEA has been

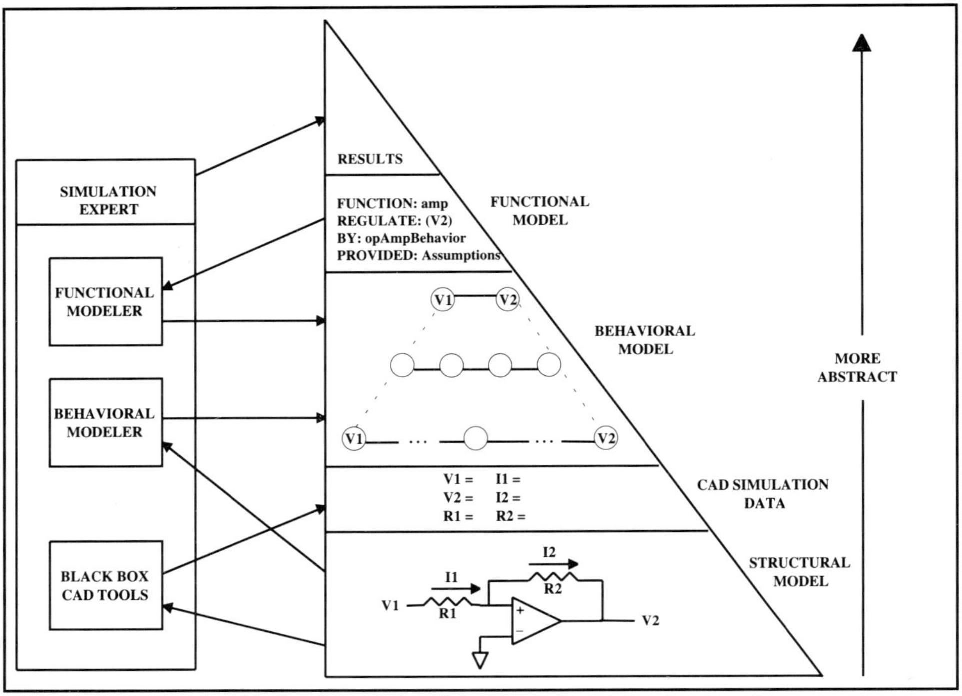

Fig. 16.10 Simulation Blackboard and Device Levels

completed for a given system, the *Analysis Generator's Archive Manager* maintains an archive of the FMEA and the system for case-based references.

Table 16.6 illustrates a typical XFMEA worksheet that would be generated by the *Analysis Generator.* The major knowledge sources are depicted under the columns that they contribute to most. The components marked with an asterisk (*) are the chief constituents of the *Design Expert.*

Table 16.6 XFMEA Worksheet

Item/Functional Identification	Function	Failure Modes & Causes	Mission Phase/ Operational Mode	Failure Effect	Failure Detection Method	Severity Class	Compensating Provisions
Failure Selector*	Functional Modeler	Failure Selector* Device Knowledge on BB	Mission Expert*	System Analyzer* Sim. Expert	Detection Analyzer*	System Classifier*	Compensator*

16.6.3 Machine Learning in XFMEA

A significant goal of intelligent FMEA automation is the ability for the system to learn. Attempts to implement an intelligent FMEA system that is capable of learning require a rigorous conceptual model as a foundation for design. After a knowledge-use level model has been developed (e.g., XFMEA blackboard model) and the requirements of the learning system have been defined, a conceptual model can be developed to encompass the learning objectives. The blackboard model's multi-layer approach to self-improvement is a principle reason for its use in XFMEA conceptual design. Ideally, a top layer will be added that can modify and extend the XFMEA blackboard model's knowledge sources, world model, and blackboard. A learning layer is necessary for an intelligent FMEA system to able to develop experiential knowledge from the application of fundamental knowledge. In other words, the XFMEA system will be capable of developing its own rules of thumb through the application of functional or physical knowledge. This paradigm is analogous to a novice, who has fundamental knowledge of how a system functions, developing into a FMEA expert.

The top layer added to the XFMEA blackboard model consists of six knowledge sources: the Performance Element (PE), the Instance Selector (IS), the Experimental Planner (EP), the Critic (C), the Learning Element (LE), and the Feedback Element (FE) [22, 23]. This top layer is useful for describing self-improvement in the XFMEA system. The Performance Element, that is, the subsystem whose performance improvement is the goal, is depicted as the *Design Expert* of the XFMEA blackboard model in Fig. 16.11. In this model, the Performance Element is the *Design Expert* and the two terms will be used interchangeably.

In the self-improving model, the Experimental Planner searches the instance space based on a strategy and recommends to the Instance Selector training instances to present to the learning system. For example, the Instance Selector provides instances that enable the *Design Expert* to develop an array of analysis heuristics through examining fundamental models developed by the *Simulation Expert*. In effect, the Experimental Planner searches for training instances from which the Instance Selector presents training that makes the *Design Expert's* assumptions concerning the underlying behavior or function of a device explicit. The Critic grades the target system's performance (i.e., the Performance Element) and suggests repair actions to correct misconceptions. For example, the *Design Expert's System Analyzer* may have learned a failure-effect heuristic that is regarded as too specific; hence, the Critic will then suggest a repair action that generalizes the rule. The Learning Element utilizes the suggestions of the Critic and reformulates the advice at an appropriate level of abstraction to be used by the Performance Element. The Feeback Element provides historical information about the Design Expert's decisions and actions to the learning system. Each knowledge source in the XFMEA blackboard model can benefit from additional research into self-improvement.

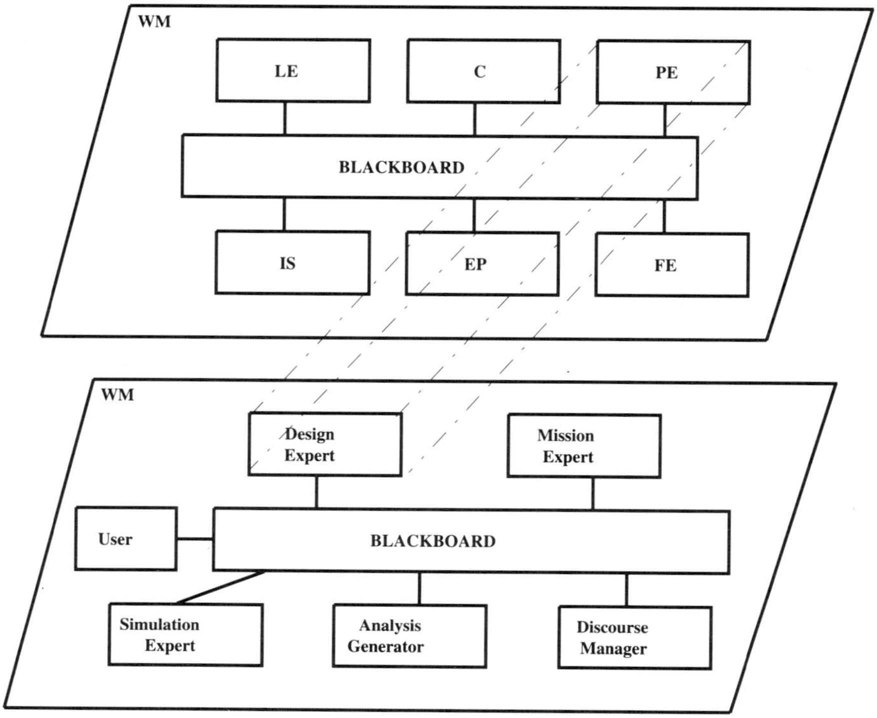

Fig. 16.11 A Self-Improving Knowledge Source for the *Design Expert*

16.7 Conclusions

The objective of this chapter has been to convey a research direction for the development of expert systems for reliability evaluation, and to present a foundation for future implementations. The XFMEA blackboard model is the result of a knowledge-use level investigation of the FMEA process and highlights the chief constituents of a second generation expert system for reliability evaluation. It identifies the major knowledge sources necessary to perform a computer-aided FMEA and the interactions between these sources of knowledge. The functional decomposition of the intelligent FMEA system reflects the source and role of the knowledge in the reasoning tasks.

The XFMEA blackboard model provides a common reference paradigm for comparing and characterizing attempts to automate FMEA. For example, in the review of prototype systems and techniques, each system can be viewed as having a particularly strong knowledge source. Lehtela's system possesses a strong *Black Box CAD Tool*; however, the program is void of any qualitative techniques. FADES employs a strong *Behavioral Modeler* and the BCS system possesses a sophisticated Discourse Manager. The XFMEA blackboard model provides a comprehensive conceptual map that emphasizes the essential information utilized by the functional components of a robust design. Moreover, the model builds a foundation upon which further research can be developed to specify the knowledge needed and the knowledge communication in an intelligent FMEA program.

REFERENCES

[1] Winston P.H., *Artificial Intelligence*, Addison-Wesley, Reading, Mass, U.S.A., 1984.

[2] Lehtela M., *Computer-Aided Failure Mode and Effect Analysis of Electronic Circuits*, Microelectronics and Reliability, Vol. 30, No. 4, pp. 761-773, 1990.

[3] Pages A. and M. Gondran, *System Reliability: Evaluation and Prediction in Engineering*, Springer-Verlag, New York, 1986.

[4] Wood C., *FADES: A Tool for Automated Fault Analysis of Complex Systems*, Proc. Artificial Intelligence in Nuclear Power Plants, IAEA/IWG NPPCI Specialists' Meeting (VT Symposium 109-110), pp. 77-89, Vol. 2, Helsinki, Finland, 1990.

[5] Grant M.M. and D.W. Harvey, *A Progress Review on the Development of FIABEX - an Intelligent Knowledge-based System for Automated Safety*, Reliability and Maintainability Analysis, Proc., Vol. 2, pp. 4C/2/1-8, UKAEA, Warrington, UK, 1989.

[6] Kamhieh C.H., D.E. Cutts and R.B. Purves, *Failure Modes and Effects Analysis Automation*, NASA, Marshall Space Flight Center, Second Conference on Artificial Intelligence for Space Applications, pp. 169-176, Huntsville, AL, 1988.

[7] Russomanno D.J., R.D. Bonnell and J.B. Bowles, *Computer-Aided FMEA: Toward an Artificial Intelligence Approach*, Fifth International Symposium on Artificial Intelligence, AAAI Press, pp. 103-112, 1992.

[8] Vladimirescu A., A.R. Newton and D.O. Pederson, *SPICE Version 2F.I User's Guide*, Dept. of Electrical Engineering and Computer Science, University of California at Berkeley, U.S.A.,1980.

[9] Stamenkovic B. and S. Holovac, *Failure Modes, Effects and Criticality Analysis: The Basic Concepts and Automatisation*, Proc. Seventh Symp. on Reliability in Electronics, Vol. 1, pp. 355-362, Budapest, Hungary, 1988.

[10] Moureau R., *FURAX: Expert System for Automatic Generation of Reliability Models for Electrical or Fluid Networks*, Proc. 7th International Conf. on Reliability and Maintainability, Brest, France, 1990.

[11] Bell D., L. Cox, S. Jackson and P. Schaefer, *Using Causal Reasoning for Automated Failure Mode and Effects Analysis (FMEA)*, Proc. Ann. Reliability and Maintainability Symp., IEEE Press, pp. 343-353, 1992.

[12] Sembugamoorthy V.and B. Chandrasekaran, *Functional Representation of Devices and Compilation of Diagnostic Problem Solving Systems, in Experience, Memory, and Reasoning*, J. Kolodner and C. Reisbeck (eds.), Lawrence Erlbaum Associates, pp. 47-73, 1986.

[13] Kurland L.C. and Y.J. Tenney, *Issues in Developing an Intelligent Tutor for a Real-World Domain: Training in Radar Mechanics*, in Intelligent Tutoring Systems Lessons Learned, J. Psotka, L.D. Massey and S.A. Mutter (eds.), Lawrence Erlbaum Associates, 1988.

[14] Keuneke A.M., *Device Representation: The Significance of Functional Knowledge*, IEEE Expert, pp. 22-25, April 1991.

[15] Freeman P. and A. Newell, *A Model for Functional Reasoning in Design*, Second International Joint Conference on Artificial Intelligence, pp. 621-640, Sept. 1971.

[16] Bonnet J., *Functional Representations: A Support for Enriched Reasoning Capabilities*, Knowledge Systems Laboratory Report No. KSL 91-58, Stanford University, Stanford, California, 1991.

[17] Kleer J.de, *How Circuits Work, in Qualitative Reasoning about Physical Systems*, D.G. Bobrow, (ed.) MIT Press, pp. 205-280, 1985.

[18] Nii H.P., *Blackboard Systems: The Blackboard Model of Problem Solving and the Evolution of the Blackboard Architecture*, AI Magazine, Part I, Vol. 7, No. 3, pp. 38-53, 1986; Part II, Vol. 7, No. 4, pp. 82-106, 1986.

[19] Newell A., *The Knowledge Level*, Artificial Intelligence, Vol. 18, pp. 87-127, 1982.

[20] Puerta A.R. and R.D. Bonnell, *A Blackboard Model for a Self-Improving Intelligent User Interface*, Proc. of the Annual Conference of the International Association of Knowledge Engineers, College Park, MD, pp. 619-635, 1989.

[21] Anderson, J.R., *The Expert Module, in Foundations of Intelligent Tutoring Systems*, M.C. Polson and J.J. Richardson (eds.), Lawrence Erlbaum Assoc. Publ., Hillsdale, NJ, 1988.

[22] Buchanan B.C., T.M. Mitchell, R.G. Smith and C.R. Johnson, *Models of Learning Systems*, in Encyclopaedia of Computer Science and Technology, J. Belzer, A.G. Holtzman and A. Kent (eds.), Vol. 2, pp. 24-51, Marcel Dekker, 1977.

[23] Whitaker E.T. and R.D. Bonnell, *Functional Modeling of Intelligent Systems using a Blackboard Model*, The Journal of Knowledge Engineering, Vol. 5, No. 1, pp. 1-12, 1992.

New Trends in System Reliability Evaluation
K.B. Misra (Editor)

CHAPTER 17

RELIABILITY ANALYSER

Krishna B. Misra

Director-grade-Scientist
Disaster Prevention & Impact Minimization Division
National Environmental Engineering Research Institute
Nagpur 440 020, India

ABSTRACT

In this chapter, we discuss an entirely new development in the area of system reliability evaluation. This development promises to provide an analyst a handy tool for carrying out various reliability analyses and make system reliability assessment a routine job, which need not be confined to highly qualified analysts only. This desk-top facility is considered preferable and economical over the sophisticated softwares on a PC, or on a mini or microcomputer keeping in view the several iterative trials that are required during the reliability design of a system. Besides, using this gadget, a reliability analyst or designer can observe the consequences of his ramifications of the designed system configuration, interactively. This gadget can also help in the learning process of system reliability designers.

17.1 Introduction

The most challenging engineering problem in recent times is probably the design and development of large scale systems particularly from the point of view of performance optimization. The reliability of such systems is a serious concern to a system designer. This is because the systems cost affects it a great deal and the consequences of a failure can be disastrous. In communication satellites, the malfunction of a single component can render the entire satellite inoperative. In nuclear plants, a small radioactive leak would be a grave health hazard. Early reliability efforts were directed only toward determining the causes of unreliability. Naturally, the field failure reports, identification and isolation of troublesome components occupied the attention of reliability analysts and designers. However, it became increasingly evident that reliability evaluation and its trade off during the early phases of system design requires an enumeration and comparison of several alternate system designs. This process, obviously, involves repeated evaluation of system reliability, which itself is not an easy task, particularly, in case of large and complex systems.

As discussed in Chapter 1, one can always prepare a graph of the system consisting of nodes and branches. As a user, one may be interested in the terminal-pair reliability of the system, whereas, as a manager of the network, one may be interested in the *global reliability* or *all-terminal reliability*.

Terminal-Pair Reliability

We know from [1] that the following methods have been widely used for the evaluation of terminal-pair reliability of systems :

(i) State Enumeration

In this method, all possible subgraphs of the main graph are formed. A graph with B branches will require formation of 2^B such subgraphs. Each subgraph corresponds to an event. The reliability of the system is determined by selecting the favourable system events. This method is useful only for very simple graphs only, as the number of states increases *very rapidly* with the number of branches.

(ii) Factoring Theorem

In general, *factoring theorem* is used to determine the reliability of a non-series parallel graph, which is *not* decomposable into the series-parallel graphs. For example, let us consider the graph of Fig. 17.1. The presence of *branch e* puts the network in the category of non-series parallel. Let R_{sc} and R_o be the reliabilities of this network, when *branch e* is short circuited and open circuited (removed), respectively. The terminal-pair reliability, R of the network is, then, given by

$$R = p_e.R_{sc} + q_e.R_o \tag{17.1}$$

where p_e and q_e are the success and failure probabilities of *branch e*. In case of non-series parallel graphs consisting of several such branches, a repeated applications of *factoring theorem* would be necessary.

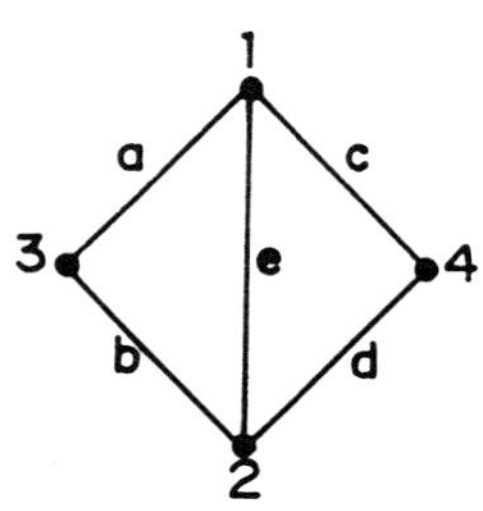

Fig. 17.1 A Non-Series Parallel System

Let n_e be the number of interconnecting branches like e, then the number of series parallel networks to be evaluated would be 2^{n_e}. Therefore, this method is useful only for smaller values of n_e. We have discussed *decomposition and transformation* procedures in detail, in Chapter 3 of this book, to overcome this difficulty.

(iii) Path sets/cut sets Enumeration

The methods based on path set and cut set enumeration can also be used with non-series parallel networks. This approach is applicable for graphs with unreliable nodes also. Once either the complete set of minimal paths or the complete set of minimal cuts have been determined, the terminal reliability can be computed using *law of union of events*. Let the path sets represent the success events, E_1 to E_n, the reliabilities of the individual path set can be computed by multiplying the success probabilities of the branches included in a path set. The terminal reliability can be determined by computing the probability of the union of events E_1 to E_n, incorporating the *law of idempotence* while resolving

$$R = P(E_1 \cup E_2 \cup E_3 \cup \ldots \cup E_n) \tag{17.2}$$

In the case of minimal cut sets, one has to identify a combination of branch failure events which would result into a system failure. The rest of the exercise is the same as described earlier, except that the final expression computes the unreliability of the system. The method based on the evaluation of minimal path set or cut set can also be used for multistate elements, where a branch has several modes of failures. For example, let us take rectifier circuits, where individual diodes can fail either in *short circuit* or *open circuit* mode.

It can be easily seen that the events E_1 to E_n in (17.2) are not *disjoint* (*mutually exclusive*). As a result, the number of terms in the direct expansion of (17.2) increases exponentially with the number of path sets or cut sets. This number can be reduced by making events E_2 to E_n mutually exclusive to event E_1. Let there be *k disjoint terms* generated through this process. Then the terminal-pair reliability is simply a sum of the probabilities of these disjointed terms, viz.,

$$R = P(A_1) + P(A_2) + \ldots P(A_k) \tag{17.3}$$

where, $A_1 = E_1$ and A_2 to A_n are (k-1) *disjointed* events with respect to E_1.

If there are n path sets, there will be 2^n-1 terms in the direct expansion of (17.2), whereas the number of disjointed terms would be $k \geq n$. There are efficient disjointing procedures available today. This has been discussed at length in Chapter 2 of this book. The choice between the methods based on finding minimal path sets or cut sets also can be made by comparing the number of minimal cut sets *vis-a-vis* the number of minimal path sets in a graph.

Global Reliability

The global reliability, for the most general case of a non-series parallel network, can be found by determining the *spanning trees* of the graph. The global reliability is the probability of the union of all spanning tree events. Here again, the terms in the direct expansion can be reduced by generating the set of mutually disjointed events.

It is clear from the foregoing discussion, that whether one uses state enumeration, or factoring theorem or even path set approach, we cannot avoid logic-based operations. For example, in state-enumeration procedure, the generation and identification of all possible success states, is a logic-based operation, which can be best executed using some kind of simulator or an analyser and not using some algorithmic steps on a computer. Similarly, the task of identification of interconnecting (or bilateral) branches in a large complex network can be efficiently performed by an analyser rather than a digital computer algorithm. The same is true for identification of path sets and cut sets of a large and complex systems. Therefore the need for development of a reliability analyzer was felt in order to carry out logic-based operations faster in analogue mode rather than using algorithmic procedure on a digital computer.

17.2 Basic Model of a Reliability Analyzer

Although, the application of graph theory in system reliability evaluation by Misra [1] in 1970, paved way for several easier and quicker algorithms for reliability evaluation of general systems, the involved mathematical analysis is still quite complicated. This is primarily due to the fact that the process basically consists of two-step methodology, viz.

(i) identification of path sets or cut sets of the system.
(ii) converting the path set or cut set information into system reliability figure.

The first part of the process is a logic-based operation, whereas the second is computationally extensive. It has been felt for long time that small desk top gadgets can be effectively used for performing these two tasks. A number of algorithms and digital computer codes are available in the literature [1], today. If the system is complex, featuring various interactions and redundancies between components, the choice of a suitable computer program and its best conditions of use for solving the problem must rely on experience, trial and error, or even sheer intuition. Sometimes, the system cannot be analyzed at all without far-reaching simplifications or by spending an enormous amount of engineer-hours in first partitioning the system into smaller subsystems, if feasible. Additionally, depending upon the required results, several programs may have to be used for the results. Besides, as was pointed out in Chapter 3, the algorithm used to deduce the path sets and cut sets of a system are computationally not attractive, particularly in a large complex system. Although a computer has many uses besides doing reliability analysis, while a Reliability Analyzer (Electronic Simulator) is a dedicated device and has a very specific task to perform.

17.2.1 Historical Background

The basic idea of a reliability analyzer was introduced by the author in [2] and developed further through [3,4 and 5], particularly, using a microprocessor platform. Later on, the idea was commercially exploited by Laviron [5] to develop a desk top aid in the form of *Electronic Simulator for Computing and Analyzing Failures* (ESCAF). Several problems [7-11] were subsequently attempted and modifications were carried out to develop a more versatile equipment-S.ESCAF. In fact, the reliability analysis of complex sequential system, which is often done by the well-known Markov technique, inherently restricts the size of a system, viz., the number of components that can be handled. However, it was demonstrated through [12] that a desk top reliability analyzer such as S.ESCAF is easy to use even by a non-specialist and offers an advantage of time and enormous convenience as it is sufficient to simulate directly the functional logic of the system under study and does not require performing fault tree analysis first. Additionally, it offers a guarantee of exhaustivity of search for all failure sequences due to its automatic search procedure. It is also easy and fast to incorporate the changes in simulation logic or reliability data during the analysis.

There is no direct experience of comparing codes in case of large complex systems, but some examples show that even for very small fault tree of less than 20 components, exact results are difficult to obtain with the computers. The studies have shown that a hardware implemented device like ESCAF is able to solve problem which are very difficult to solve (if unsolvable) by digital computer programs and does rapidly and very economically. Actually, as mentioned earlier, the task of determining path sets or cut sets can be done very fast on an electronic system simulator rather than on a digital computer using an algorithm. However, for the calculation of probabilities involving arithmetic operations, the usual capabilities of a digital computer can be appropriately exploited. Therefore, we need to combine the best of the two alternatives to devise a gadget and a procedure to determine system reliability, economically.

17.2.2 Basic Principle

Although the basic idea of a reliability analyser was formally introduced by the author in 1977 through reference 5 of [2], but a gadget based on exhaustive system state enumeration, was introduced in 1979 through [2]. To understand the basic principle involved, let us consider a series parallel network shown in Fig. 17.2. Each element of this system is represented by a 2 input AND gate. One of the inputs of an AND gate is used to simulate state of the element of the given system configuration, while the other input is fed from a *binary generator*. Each element in the system has two states, viz., *conducting* or *non-conducting.*

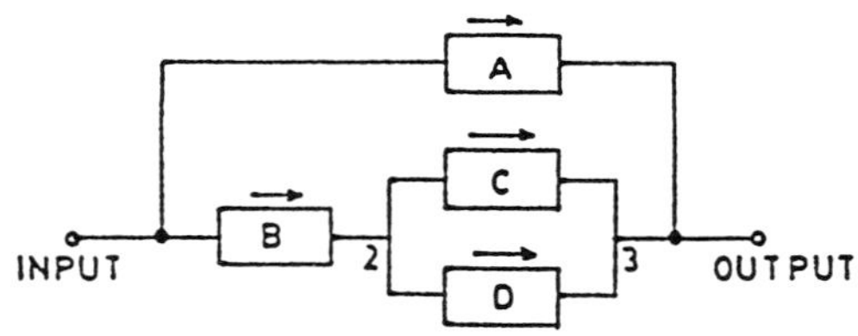

Fig. 17.2 A Series Parallel Network

Let us assume that the state 1 represents that an element is *conducting*. State 0 represents that the element is *non-conducting.* Thus we can simulate the conditions of all elements of the system by a state vector consisting of 0 and 1.

The gate i conducts, if the element i is *good* when a *pulse* is given at the input terminal of the gate marked as X_i. These pulses are generated by a *binary generator*. There are 4 elements in the Fig. 17.2, therefore, the maximum possible number of states, this system can have, is $2^4(=16)$. These 16 states will be generated sequentially by a binary generator as stated earlier.

In this way we can simulate the system shown in Fig. 17.2 by its logical representation givin in Fig. 17.3.

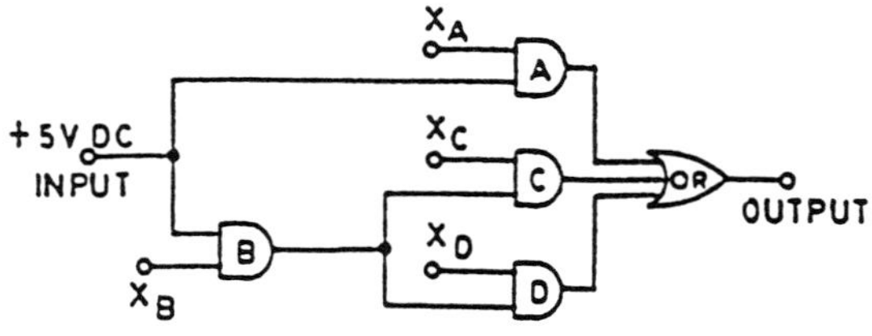

Fig. 17.3 Logical Representation of System Shown in Fig. 17.2

In general, the various path sets and cut sets of the system can be found by simulating the system by 2-input logical gates. Each state vector, which provides an output indication represents a *path set*, and each input state which does not have an output indication represents a *cut set*. It may be noted that these path sets or cut sets are not necessarily the minimal sets.

Let p_a, p_b, p_c, p_d represent the reliabilities of the system elements of Fig. 17.2, whereas q_a, q_b, q_c, q_d represent the unreliabilities of the system elements A,B,C and D, respectively. Then the system reliability given by

$$R = p_A q_B q_C q_D + p_A p_B q_C q_D + p_A q_B p_C q_D + q_A p_B p_C q_D + p_A p_B p_C q_D + p_A q_B q_C p_D + q_A p_B q_C p_D + p_A p_B q_C p_D + p_A q_B p_C p_D + q_A p_B p_C p_D + p_A p_B p_C p_D \tag{17.4}$$

The state vectors which block the passage of the input signals to the output node are known as *cut sets* (see Table 17.1) and system unreliability, Q, can be written as

$$Q = q_A q_B q_C q_D + q_A p_B q_C q_D + q_A q_B q_C p_D + q_A q_B p_C p_D + q_A q_B p_C q_D \tag{17.5}$$

from which the reliability of the system can be found by the relation R = (1-Q).

Table 17.1 Systems States

Pulse #	Binary state at the input of AND gated system X_D	X_C	X_B	X_A	State of output node
0	0	0	0	0	0 Cut
1	0	0	0	1	1 Path
2	0	0	1	0	0 Cut
3	0	0	1	1	1 Path
4	0	1	0	0	0 Cut
5	0	1	0	1	1 Path
6	0	1	1	0	1 Path
7	0	1	1	1	1 Path
8	1	0	0	0	0 Cut
9	1	0	0	1	1 Path
10	1	0	1	0	1 Path
11	1	0	1	1	1 Path
12	1	1	0	0	0 Cut
13	1	1	0	1	1 Path
14	1	1	1	0	1 Path
15	1	1	1	1	1 Path

As mentioned earlier, the path sets obtained by this procedure need not be *minimal.* If minimal paths set are required, we have to carry out an additional manipulation. Since the minimal path sets are basically the forward paths of a graph, from the input node to a specified output node. In a system consisting of r nodes, minimal path sets (forward paths) will have a maximum size of (n-1). In an example (Fig. 17.1), we have the following path sets (not necessarily the minimal path sets) of various sizes starting from order 1 to order 4 (since n=4).

Path of size 1 : A; Path of size 2 : BC, BD; Path of size 3 : ABC, ABD and BCD; Path of size 4 : ABCD. Here A, BC, BD are the forward paths and therefore the minimal paths, the remaining paths, viz., ABC, ABD, BCD, ABCD are cyclic paths (with loops). The reliability expression can be given [1] by

$$R = P[A + BC + BD] - P[ABC + ABD + BCD] + P[ABCD] \tag{17.6}$$

Out of the 16 possible pulses corresponding to the 16 possible states, the simulated network would respond corresponding to pulses # 1,6,7,10,11,14,15 only. It is possible to obtain minimal paths from the various non-minimal path sets that are generated, by ensuring that the last AND gate responds to giving an output indication the pulse corresponding to minimal path sets only.

17.2.3 Description of Modules

The schematic diagram of the elementary reliability analyzer introduced by Misra et al [2] is shown in Fig. 17.4.

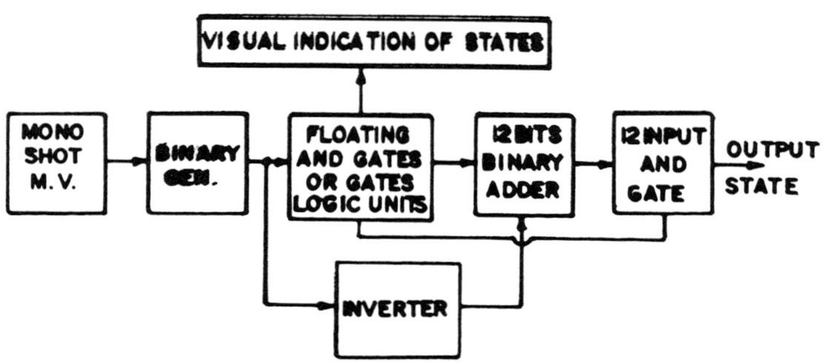

Fig. 17.4 Block Diagram of the Reliability Analyzer

The analyzer has the following basic units, namely,

- AND gates
- Monoshot Multivibrator

- 12-Bit Binary generator
- 12-Bit Binary Full adder
- 12- Input AND gate

AND Gates

One input terminal and one output terminal represent the two nodes of a branch, while the second input of an AND gate as shown in Fig. 17.5 (a) is connected to a *binary generator*. For a 12-element network, 12 AND gates are required. Therefore, three ICs of DM 7408 type were used, because each IC contained four AND gates. The output state of an AND gate is indicated by an LED, which is connected as shown in Fig. 17.5 (a).

12-Bit Binary Generator

A 12-bit binary generator was obtained by cascading three 4-bit binary counters of type DM 7493. Each monolithic counter contained four M-S Flip flops. Counter pulses were applied through *monoshot multivibrator* to terminal INPA of the first binary counter. Maximum count length of a 12-bit binary counter is 2^{12}, i.e., 4096, Thus a 12-element network will have 4096 possible combinations or states. Binary generator outputs were connected to one of the inputs of all AND gates to simulate the states of all the elements in the system. Binary generator was reset by giving logic 1 at reset pins marked as R_{01} and R_{02} as shown in Fig. 17.5 (b). The outputs were inverted by a Hex invertedr of type DM 7404. The scheme used is shown in Fig. 17.5 (b).

Monoshot Multivibrator

Clock pulses to binary generator were applied through a monoshot multivibrator, which provided a single clock pulse to the binary generator. The operation can be explained as follows : when the switch SW in Fig. 17.5 (c) is open, capacitor C charges through R and as soon as the threshold voltage is reached the output of SN 72560 goes high, which is inverted by DM 7404 to keep the output low. However, when the switch SW is closed, C discharges and output of SN 72560 goes *low*, which is inverted by DM 7404 to result into a *high* output. This provides a single clock pulse to the binary generator. This scheme is shown in Fig. 17.5 (c).

12-Bit Binary Full Adder

The output states of all AND gates and corresponding complementary outputs of the binary generator were added into a 12- bit adder, so as to obtain the vector sum. In fact, three binary adders of type DM 7483 were cascaded by tying the pin C_4 of one chip to pin C_0 of the next chip. The two sets of 12-bit inputs were applied to terminals A_1 $A_2A_3...A_{12}$ and $B_1B_2B_3...B_{12}$. The output from the binary adder was obtained from pins O_1, O_2,...,O_{12}. This scheme is shown in Fig. 17.5 (d).

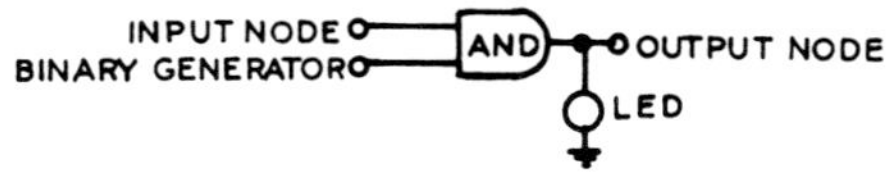

Fig. 17.5 (a) An AND Gate

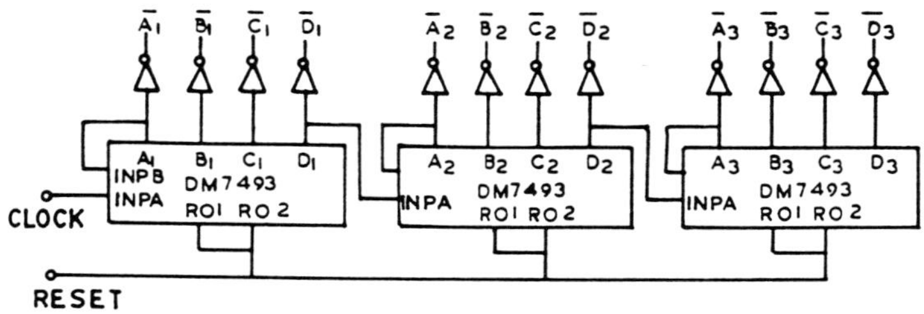

Fig. 17.5 (b) A 12-bit Binary Generator with Complementary Outputs

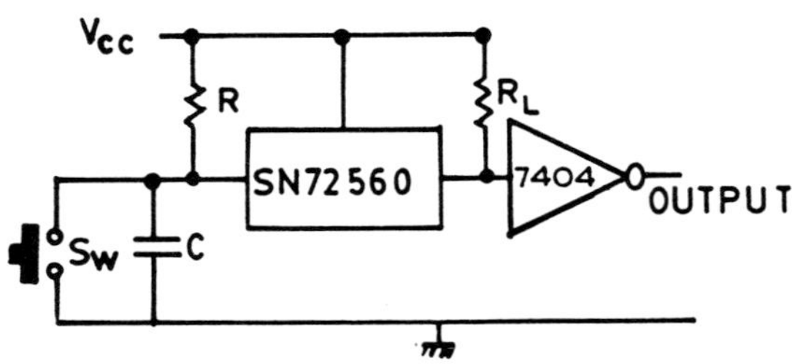

Fig. 17.5 (c) A Single Shot Multivibrator

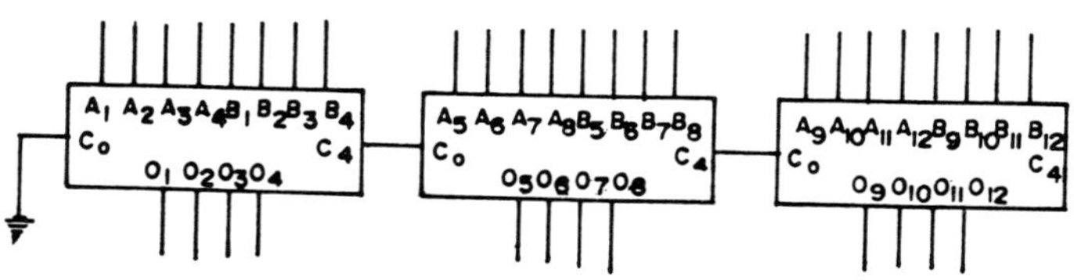

Fig. 17.5 (d) A 12-bit Binary Full Adder

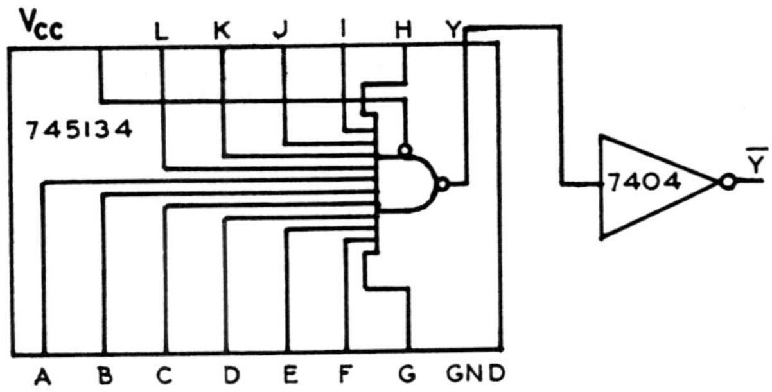

Fig. 17.5 (e) A 12-bit AND Gate

A 12-input AND Gate

A 12-bit output of a binary full adder were AND-gated using a 12- bit NAND gate of type 745134 and a Hex inverter of type DM 7404. The AND-gated output of binary full adder was further AND-gated with output node conditions of the network. Thus, if the final output went to state 1, it signified a successful system state. This is shown in Fig. 17.5 (e).

17.2.4 Operating Procedure

The following procedure was employed for using a reliability analyzer:

(1) Represent each element of system by 2-input AND gate. If two elements are terminating on a node, represent the node by an OR gate as shown in Fig. 17.3.

(2) If an element is bi-directional, represent this element by two AND gates connected *back to back.* Both the AND gates should be energized simultaneously through the same output of the binary generator.

(3) To find path sets and cut sets, observe the output node condition and all input state vector conditions as shown by LEDs. The state of input state vectors corresponding to an output node condition (corresponding to 1) represents a *path set*, whereas 0 would represent a *cut set.*

(4) To find minimal path sets, $[A_1A_2A_3...A_{12}]$ of a 12-bit full adder are connected to complementary binary generator outputs $[A_1B_1C_1D_1...A_3B_3C_3D_3]$ and other set $[B_1B_2B_3...B_{12}]$ of adder are connected to outputs of the corresponding AND gates used for simulating the system configuration.

(5) The vector sum output of adder are AND-gated in a 12-bit AND gate, which is further AND-gated with the output node condition of the system.

(6) Note the input state vector corresponding to an output node condition (*high*) indicated by the corresponding LEDs.

The above description is only to help understand the basic idea to determine the path sets or cut sets of a system and in [2], the aim was more to highlight the possibility of developing a gadget for system reliability evaluation. Later on an indication of a microprocessor-based system was provided in [3], [4] and [6].

17.3 Generation of Spanning Trees

After the basic idea [2] of a reliability analyzer was introduced it was felt necessary to develop this gadget around a microprocessor. Therefore, the efforts were directed to develop procedures to generate spanning trees [3] (for global system reliability) and the minimal path sets or cut sets [4] using a microprocessor-based gadget. A search procedure was found to be effective in this effort. Therefore, we will describe here breifly the search procedure for generating spanning trees and path sets or cut sets with the help of this gadget.

17.3.1 Search Technique

A search technique helps trace out the desired graph starting from any given node of the graph. The tracing process is subjected to a set of conditions which operate at each step to ensure the generation of the desired graph. The search consists of the following:

(a) The tracing process
(b) Constraints on tracing process, and
(c) A special process, if required.

The Tracing Process

The tracing process is organised with the help of binary counters. Each node of the graph is assigned a binary counter. Let us consider a network of Fig. 17.6. No node of the graph has more than 4 branches incident on it. Hence the node counters of 4 bits can be chosen for the graph of Fig. 17.6. Each state of a node counter represents the selection of a specific set of branches. Consider the binary counter on node C(CN_c) of the graph of Fig. 17.6. Table 17.2 shows all possibile states of the node counter CN_c and the corresponding selection of branches. It is important to note that the lower order bits of the counter address select the lower order branches incident on the node.

The complete set of states of any node counter represents all possibile combinations of branches connecting the node to the rest of the graph. The length of the counter address for a given graph depends upon the maximum number of branches incident on any node of the graph.

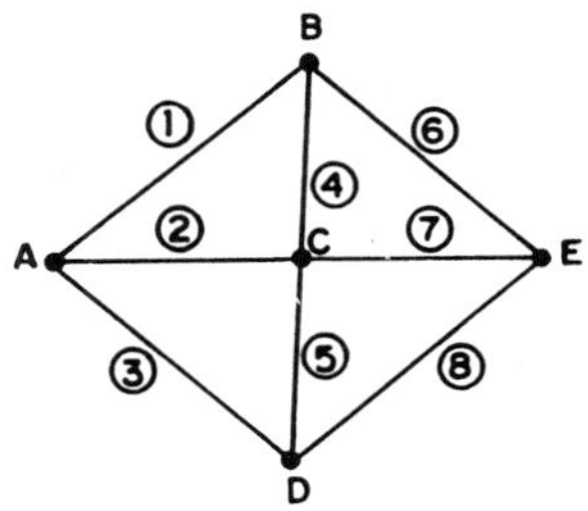

Fig. 17.6 A Sample Graph

Constraints on the Tracing Process

Any node of the graph can be chosen as the starting point. The starting node can be connected to the rest of the graph in many ways. These can be exhaustively represented by the complete set of states of the node counter assigned to the starting node. To begin with, the node counter assigned to the starting node is incremented from its *reset state*. The node at which selections are made is called a *base node*. The new selections are subjected to a set of constraints. For example,

Table 17.2 All States of the Counter CN_c

State #	State of CNc	Branch Nos selected
1	0001	2
2	0010	4
3	0011	2, 4
4	0100	5
5	0101	5, 2
6	0110	5, 4
7	0111	5, 4, 2
8	1000	7
9	1001	7, 2
10	1010	7, 4
11	1011	7, 4, 2
12	1100	7, 5
13	1101	7, 5, 2
14	1110	7, 5, 4
15	1111	7, 5, 4, 2
16	0000	Reset state, no selection

in case of determining spanning trees, the new branches and the new nodes are checked against all the nodes and branches selected in the previous steps, to avoid retracing of any branch and to avoid loop formation. A spanning tree is formed when all the nodes have been connected.

When the new selections do not conform to the complete set of constraints, the new selection is rejected. The counter of the base node, at which selections are being made, is incremented by one step and new selections are made. When a node counter, on incrementing, reaches the reset state, all possibilities of connecting a particular base node to the rest of the graph would have been exhausted.

Table 17.3 Generation of Tree 1 and Tree 2 of the Graph of Fig. 17.6 using Search Technique

#	A	B	C	D	E		Comments
	States of Node Counters						
1	1000	0000	0000	0000	0000	;	All counters reset
2	0001	0000	0000	0000	0000	;	Selection of Branch 1, Node B as new Base node
3	0001	0001	0000	0000	0000	;	Selection of Branch 1, again, Invalid
4	0001	0010	0000	0000	0000	;	Branch 4, Node C; Valid
5	0001	0010	0001	0000	0000	;	Branch 2, Loop formation; Invalid
6	0001	0010	0010	0000	0000	;	Branch 4 again, Invalid
7	0001	0010	0011	0000	0000	;	Loop formation, Invalid
8	0001	0010	0100	0000	0000	;	Branch 5, Node D; Valid
9	0001	0010	0100	0001	0000	;	Branch 3, Loop formation, Invalid
10	0001	0010	0100	0010	0000	;	Branch 5 again, Invalid
11	0001	0010	0100	0011	0000	;	Branch 3, 5; Invalid
12	0001	0010	0100	0100	0000	;	Branch 8, Node E, Valid

All the nodes have been connected, Tree 1 (Fig. 17.7) generated. The process of finding other spanning trees continues by incrementing $CN_{D'}$. However, the counter CN_D reaches its reset without yielding any valid selection. Hence the selection is Step 8, Connecting Node D to the graph (Branch 5, Node D) is set invalid.

#	A	B	C	D	E		Comments
13	0001	0010	0100	0000	0000		
14	0001	0010	0101	0000	0000	;	Branch 2, 5; Invalid
15	0001	0010	1000	0000	0000	;	Branch 7, Node E; Valid
16	0001	0010	1000	0000	0001	;	Branch 6, Node B; Invalid
17	0001	0010	1000	0000	0100	;	Branch 8, Node E; Valid

All the nodes have been joined. Tree 2 (Fig. 17.8) generated.

The tracing process in the above event shifts back to the base node of the previous step. The generation of the trees of Fig. 17.7 and Fig. 17.8 with the above technique has been detailed in Table 17.3.

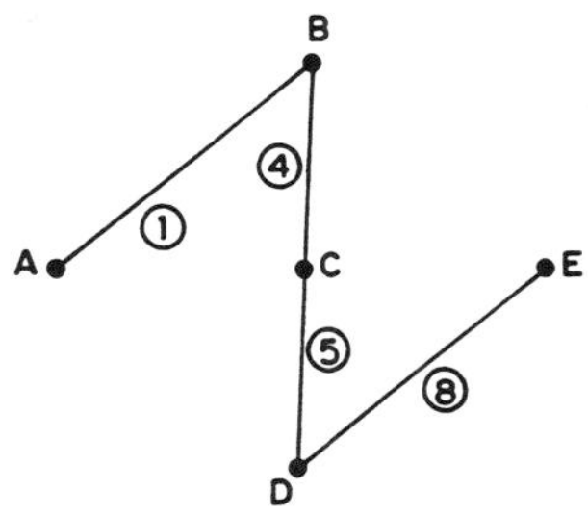

Fig. 17.7 Tree 1

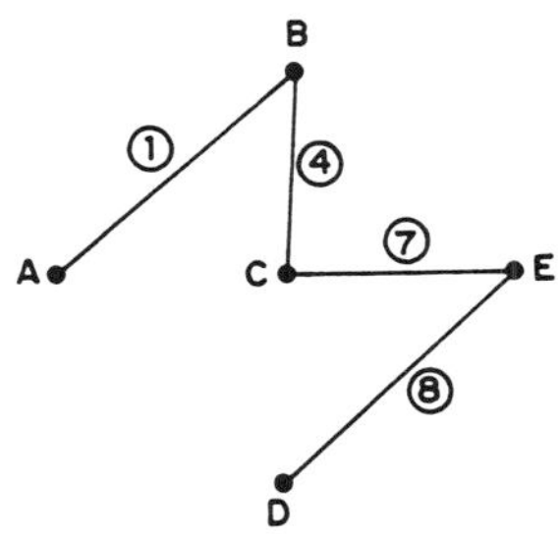

Fig. 17.8 Tree 2

The process continues till the counter of the starting node (CN_A) reaches its reset state. The above process yields an exhaustive set of desired graphs

Multiple Selection on a Base Node

When more than one branch or node is selected at a base node, the tracing process defined in an earlier section needs slight modification.

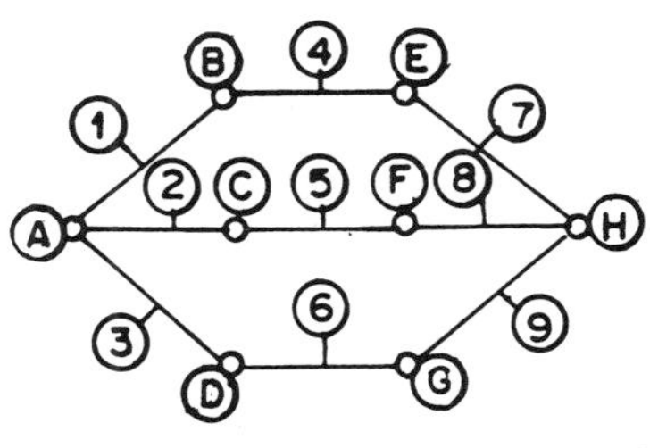

Fig. 17.9 A Sample Graph

For example, let us consider the generation of spanning trees for the graph of Fig. 17.9. When branches 1,2,3, have been selected on base node A. At this stage, 4 nodes have been already joined and another 4 nodes remain to be connected. Hence, any desired graph can be obtained exactly in 4 steps. Therefore it is possible to obtain the desired graphs by tracing the remaining graph by allotting all the 4 steps to node B, and allotting none to node C and D. It is also possible to obtain the desired graphs by allotting 2 steps to node B and one each to node C and D, respectively. Therefore, an exhaustive set of spanning trees for a multiple selection at node A with $CN_A \rightarrow 0111$ can be determined by first finding all possible methods of continuing the tracing process, and then carrying out the tracing process in all possible ways. This is done by a process of *step allotment.*

In the above example, the balance of 4 steps can be divided amongst the three new nodes B,C and D in 15 possible ways (Table 17.4).

Table 17.4 Step Allotment for a Multiple Selection $CN_A \rightarrow 0111$ for the Graph of Fig. 17.9.

#	Nodes		
	B	C	D
1	0	0	4
2	0	1	3
3	0	2	2
4	0	3	1
5	0	4	0
6	1	0	3
7	1	1	2
8	1	2	1
9	1	3	0
10	2	0	2
11	2	1	1
12	2	2	0
13	3	0	1
14	3	1	0
15	4	0	0

An exhaustive set of spanning trees with $CN_A \rightarrow 0111$ for the graph of Fig. 17.9 can be found by carrying out the tracing process as is given in Fig. 17.10. The new nodes should also be assigned a priority, in order that the tracing process always follows a set pattern. In our example, the lower order node, i.e., node B has been assigned the top priority, while the higher order node, i.e., node D has been assigned a lower priority. Therefore, for a particular step allotment, the steps allotted to node B shall have to be exhausted first, before selecting node C as the base node and so on. The exhaustive set of trees of the graph of Fig. 17.9 for $CN_A \rightarrow 0111$ have been shown in Fig. 17.10, and the details of the tracing process for the generation of tree with step allotment (1 0 3) have been shown in Table 17.5.

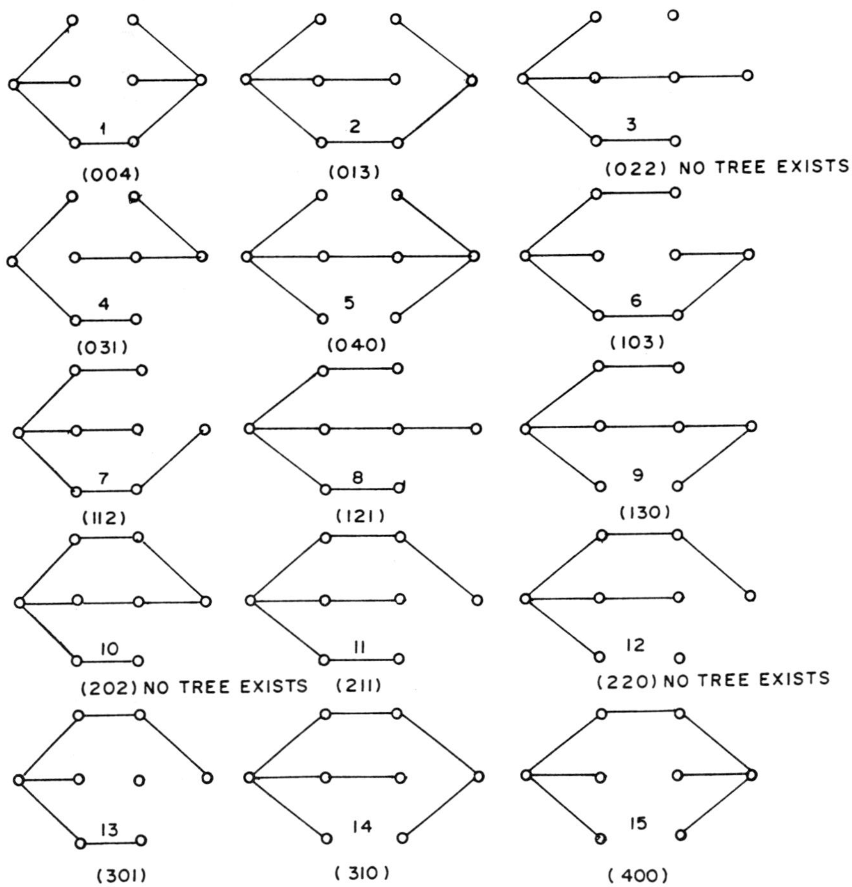

Fig. 17.10 Exhaustive Set of Trees of the Graph of 17.9

Table 17.5 Generation of Trees for the Step Allotment (1 0 3) for the Graph of Fig. 17.9.

	Node Counters of								
#	A	B	C	D	E	F	G	H	Remarks
1	0111	0010	0000	0000	0000	0000	0000	0000	; One step of B over
2	0111	0010	0000	0000	0000	0000	0000	0000	; No step to C
3	0111	0010	0000	0010	0000	0000	0000	0000	; Step 1 of D over
4	0111	0010	0000	0010	0000	0000	0010	0000	; Step 2 of D over
5	0111	0010	0000	0010	0000	0000	0010	0010	; Step 3 of D over, Tree formed
6	0111	0010	0000	0000	0000	0000	0000	0000	; CN_D, CN_G, CN_H reset
7	0111	0000	0000	0000	0000	0000	0000	0000	; CN_B also reset

17.4 Algorithms for Path Sets and Cut Sets

It has been seen from the earlier discussion that the most important and widely used methods for the evaluation of system reliability require the determination of the minimal path sets or spanning trees or the minimal cut sets . Therefore, it will be appropriate to discuss algorithms for generating these. At first we would discuss the developments described in [3], [4] and [6] and later on those due to Laviron [5,7-12].

17.4.1 Search Technique for Minimal Pathsets and Cutsets

The search technique, described in an earlier section, can also be used to determine the minimal path sets. The tracing process can be organised with the help of counters assigned to every node of the graph. The node counters, however, need not include the states with multiple selection, as only one branch is required to connect the base node to the rest of the graph. A minimal path can be declared to have been found when the new node joined is the end node.

To determine minimal cutset, first, the exhaustive set of minimal paths is found and stored. The search technique is then applied on the graph formed with the paths assumed as nodes and the branches included in a path acting as the branches incident on that assumed node. Each path is assigned a path counter. Each new branch intersects a set of paths. When all the paths have been intersected by the set of selected branches, a cut set is found. The minimal cutsets can be generated by a continuous comparison process which assures the storing of only minimal cut sets.

Minimal Path Sets

As mentioned earlier, the node counters in this case need not include the states with multiple selection. Consider the graph of Fig. 17.11. An enumeration of states of the counter corresponding to node 5 is given in Table 17.6.

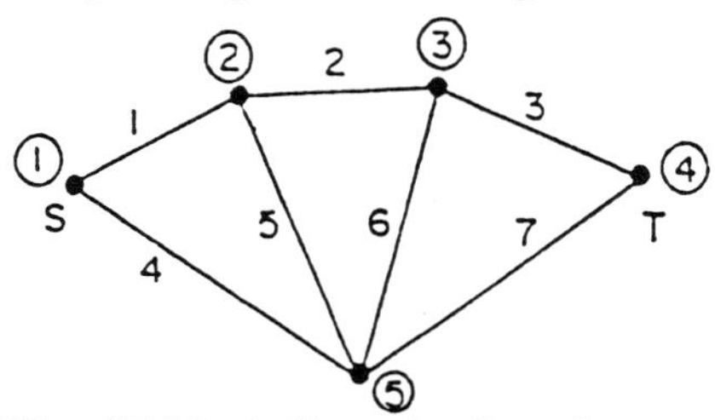

Fig. 17.11 A Sample Graph

The limiting address of any node counter depends on the number of branches incident on the node. The starting node S is the base node in step 1. The

Table 17.6 All States of the Counter of Node 5

Node address	Selected branch
00	No selection
01	04
02	05
03	06
04	07

counter states which result in the formation of a loop or retracing of a branch are rejected. When a node counter exceeds the limiting address of its node counter, the corresponding node is deleted from the traced graph and the tracing is continued on the base node of the previous step. When the new node is the same as the end node T, a minimal path is found and the tracing process is resumed on the current base node. The tracing process is over when the node counter of the starting node exceeds its limiting address.

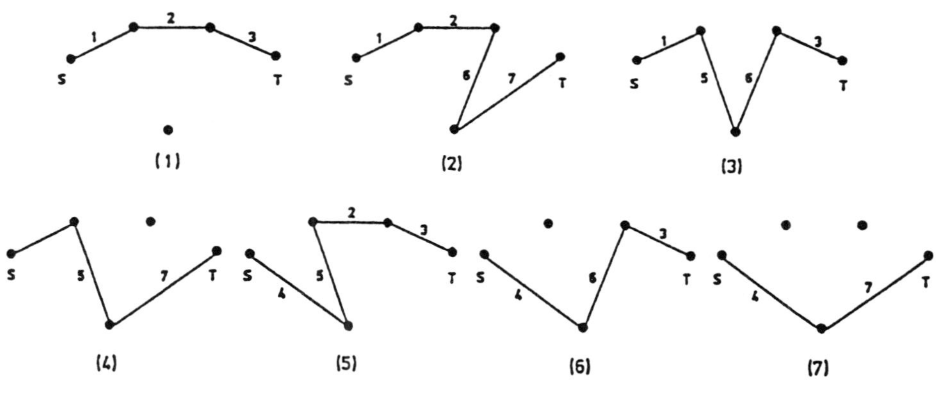

Fig. 17.12 Minimal Path Sets of the Sample Graph Given in Fig. 17.11

The complete set of minimal paths of the graph on Fig. 17.11 are shown in Fig. 17.12. The table 17.7 includes the *path vectors* and corresponding branches for the minimal paths, of the same graph. Both vectors have been formed by assigning one bit each to each branch of the graph. The bits corresponding to branches included in a path are set to 1, while other bits are set to zero. The hexadecimal word so formed is the path vector corresponding to a given path. The path vectors included in Table 17.7 have been formed by assigning higher order bits to lower order branches.

Table 17.7 Path Vectors and Corresponding Branches of the Graph of Fig. 17.11

Path #	Path vector	Branch numbers
1	EO (11100000B)	1,2,3
2	C6 (11000110B)	1,2,6,7
3	AC (10101100B)	1,3,5,6
4	8A (10001010B)	1,5,7
5	78 (01111000B)	2,3,4,5
6	34 (01100100B)	3,4,6
7	12 (00010100B)	4,7

Minimal Cutset

The minimal cut sets are determined by applying the search technique to the set of minimal paths. Each path is assigned a path counter. In this case also, the multiple selection of branches will not take place. Hence path counters will have no states with multiple selection. Consider path 3 of the sample graph with branches 1, 3, 5 and 6. The complete set of states of the path counter with corresponding branch selection are shown in Table 17.8.

The limiting address of a path counter depends on the number of branches included in a path. The tracing process is organized in a similar manner as in the case of finding minimal paths.

Table 17.8 All States of the Path Counter of Path 3

Path counter address	Selected branch
00H	No selection
01H	1
02H	3
03H	5
04H	6

The tracing process begins by selecting path 1 as the base path in step 1. Each branch intersects a set of paths. The base path in the next step is the lowest order path which has not been intersected by the branch selected in the first step. This process is continued till all the paths have been intersected.

The cuts generated in this manner may not necessarily be minimal cuts. Any new cut is compared with all the cuts obtained earlier. If the new cut contains any of the old cuts, the new cut is rejected. If any of the old cut contains the new cut, the old cut is rejected. In this manner only the minimal cuts will remain in the memory.

17.4.2 Algorithm for Determining Simple Path Sets

The method is based on the principle that if a suitable definition of sum and multiplication between set of paths is given, the set of all simple paths between all pairs of vertices can be characterized as the solution of a system of linear equations. The method of enumerating such paths corresponds to the solution of these equations. Let,

t_{ij} = Path of length one from node i to node j, if it exists, otherwise it is an empty set.

T_{ij} = Set of all simple paths between nodes i and j.

T_{ij}^{h}= Set of all simple paths between nodes i and j which pass through intermediate vertices, *if any*, of indices smaller than h.

The algorithm looks for all simple paths between vertices (n-1) and n in the following manner :

$$\text{Step 1}: T_{ij}^{1} = t_{ij} \; (i, j = 1, \ldots, n; i \neq j, n; j \neq n-1) \tag{17.6}$$

$$\text{Step 2}: T_{ij}^{h+1} = T_{ij}^{h} + T_{ih}^{h} \cdot T_{hj}^{h} \; (i, j = h+1, \ldots, n; i \neq j, n; j \neq n-1) \tag{17.7}$$

The set of all simple paths between vertices (n-1) and n are given by $T_{(n-1,n)}^{n-1}$, which can be found by an iterative procedure.

Consider the implementation of the above algorithm for the network given in Fig. 17.1. It is desired to find all the paths between vertices 3 and 4. The desired expression is T_{34}^{3}

$$T_{34}^{3} = T_{34}^{2} + T_{32}^{2} \cdot T_{24}^{2} \tag{17.8}$$

But $$T_{34}^{2} = T_{34}^{1} + T_{31}^{1} \cdot T_{14}^{1}, \tag{17.9}$$

$$T_{32}^{2} = T_{32}^{1} + T_{31}^{1} \cdot T_{12}^{1}, \tag{17.10}$$

and $$T_{24}^{2} = T_{24}^{1} + T_{21}^{1} \cdot T_{14}^{1} \tag{17.11}$$

However using (17.6), we have

$$T^1_{12} = t_{12} = e\,,\ T^1_{14} = t_{14} = c,$$
$$T^1_{21} = t_{21} = e\,,\ T^1_{24} = t_{24} = d,$$
$$T^1_{31} = t_{31} = a\,,\ T^1_{32} = t_{32} = b,\ \ T^1_{34} = t_{34} = o\,.$$

Substituting the above values, in (17.9) to (17.11) and finally in (17.8), we obtain

$$T^3_{34} = a\,c + b\,d + b\,e\,c + a\,e\,d \tag{17.12}$$

Therefore, ac, bd, bec and aed are the simple paths between nodes 3 and 4. If the above algorithm is implemented, it can be seen that in the first step approximately n^2 independent quantities have to be stored, where n is the number of vertices in the network. Hence, the memory requirement of the system will increase as the square of the number of nodes. Besides this, the implementation of the algorithm may also require renumbering of the vertices, if a new set of paths between another set of nodes has to be found.

17.4.3 Alternative Procedures

Another algorithm for finding the simple path sets and cut sets has the following special features :

(1) The minimal cut sets can be deduced for any pair of nodes (input and output) of the system, for a single description of the system topology without requiring any simplification in the system configuration.

(2) The element of the system can be either unidirectional or bidirectional.

In this method an array of predecessors is prepared for each branch and stored. The minimal path sets are traced back from the output node to the input node from this array of predecessors. After finding all the path sets, the paths with loops are eliminated. The minimal cut sets are found by first rewriting the path sets binary word of B bits where, B is the number of branches in the graph. Consider a graph with 16 components and three paths, as shown in Table 17.9.

Table 17.9 Minimal Paths in Binary Form

	Components															
Path #	1	2	3	4	5	6	7	8	9	10	11	12	13	14	15	16
1	1	1			1											
2					1			1	1		1		1		1	
3			1	1	1		1	1	1	1	1	1				

A minimal cut of first order is found by searching for a column in which every element has an entry of unity. The components corresponding to such columns are the first order minimal cuts. The logic '1' in these columns are then replaced by zeros to prevent generation of nonminimal cuts which contain first order minimal cuts. The maximum order of a minimal cut is equal to the total number of minimal paths. The second order minimal cuts are found by adding logically two columns at a time. If any of the resulting vectors has every element equal to unity, the combined column components form a second order minimal cut. This process is continued till all the cuts of order P have been found, where P is the number of paths. Each time a cut of order greater than 1 is generated, it is necessary to check whether it includes a cut of lower order. If it does, the new cut is rejected. The above procedure required 17K words memory to solve the problem of graph of 100 branches and 8 nodes.

Systems Without Feedback

The method is applicable for a system having a digraph without feedback. The parallel paths between two nodes however should be reduced to a single equivalent branch before using the technique. Consider a hypothetical system having all unidirectional branches as shown in Fig. 17.13. The network topology can be represented in a matrix form as given below :

$$\begin{array}{c} \text{i/j} \\ \\ \end{array}\ \ 1\ \ 2\ \ 3\ \ 4\ \ 5\ \ 6$$

$$X = \begin{array}{c} 1\\2\\3\\4\\5\\6 \end{array}\begin{bmatrix} 0 & A & B & C & D & E \\ 0 & 0 & F & G & H & J \\ 0 & 0 & 0 & K & L & M \\ 0 & 0 & 0 & 0 & N & P \\ 0 & 0 & 0 & 0 & 0 & Q \\ 0 & 0 & 0 & 0 & 0 & 0 \end{bmatrix} \quad ; X' = X + X^t = \begin{bmatrix} 0 & A & B & C & D & E \\ A & 0 & F & G & H & J \\ B & F & 0 & K & L & M \\ C & G & K & 0 & N & P \\ D & H & L & N & 0 & Q \\ E & J & M & P & Q & 0 \end{bmatrix}$$

All the branches in Fig. 17.13 are directed (unidirectional) and the nodes have been numbered in such a manner that no branch originating from a higher order node is incident on a lower order node. As a result of this assumption, all entries in X below the main diagonal are zero. When X is added to its transpose X^t, element by element the resultant matrix is X'. Now, consider the elements in column 2 of X'. It can be seen that the elements above the main diagonal are followed by the elements below the diagonal. From X', a set of precedence statements can be written as

$$A < F, G, H, J \quad ; \quad C,G,K \ < \ N,P$$
$$B, F < K,L,M \quad ; \quad D,H,L,N \ < Q$$

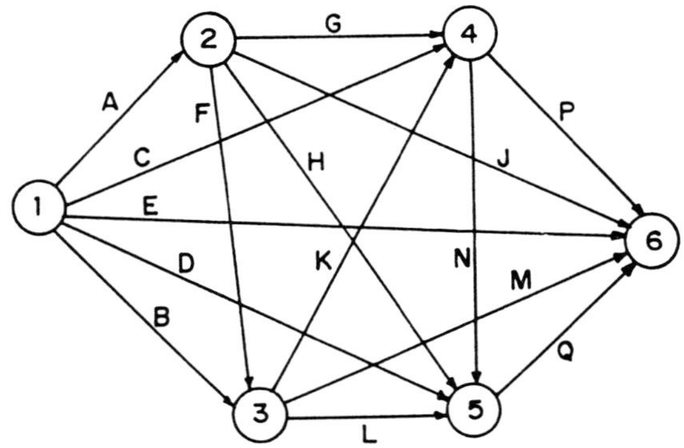

17.13 A Directed Network

These precedence statements can be used to construct the path sets as shown in Fig. 17.14. Thus the path sets of the graph of Fig. 17.13 are :

AFKNQ	AGNQ	BKNQ	CNQ
AFKP	AGP	BKP	CP
AFLQ	AHQ	BLQ	DQ
AFM	AJ	BM	E

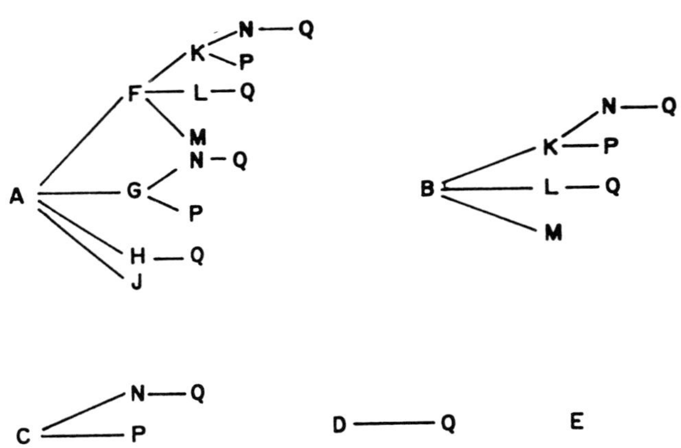

17.14 Development of Path Sets from the Matrix Representation of a System

The precedence statements as represented by Fig. 17.14 can also be used to determine the cut sets which are as given below :

A B C D E	
F G H J B C D E	(F,G,H and J Replace A)
K L M G H J C D E	(K,L,M replace B and F)
N P L M H J D E	(N and P replace K,G and C)
Q P M J E	(Q replaces N,L,H and D)

17.5 Microprocessor Based Reliability Analyser

A microprocessor-based reliability analyser was developed in 1979 (as reported in [3]) and in its basic form consisted of the following main sub-systems:

(i) The CPU
(ii) The memory system consting of PROM and RAM (Random Access Memory) to store the programme and in between results, respectively
(iii) The interface system to communicate with the outside world, such as keyboard and display unit

These are shown in Fig. 17.15. The communication between various sub-systems of a microprocessor system takes place on a bi-directional data bus with the help of address bus, which is in the control of the CPU. A demonstration unit using a CPU on Intel 8080 was developed in [3]. The system had a 8-bit wide bi-directional *data bus* and a 16-bit wide *address bus*. The CPU could access 2^{16} (approximately 65,000 bytes) separate memory locations by sending a 16-bit address on the address bus.

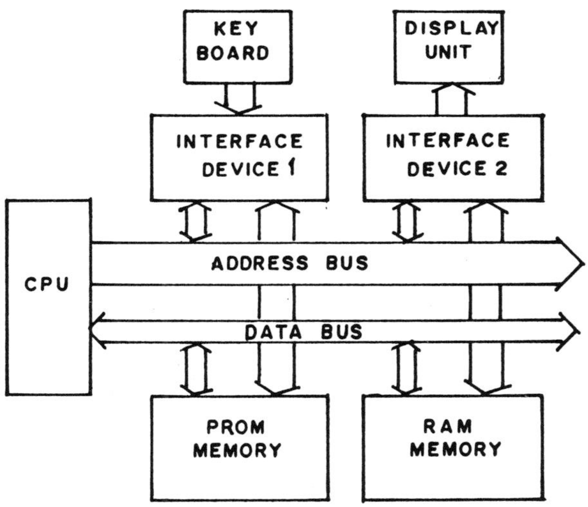

Fig. 17.15 Architecture of a Microprocessor-based System

A keyboard and display unit used with this microprocessor based-Reliability Analyser are shown in Fig. 17.16. The display of 6-digit consisted of two fields - a 4-digit *address field* and a 2-digit *data field*, which corresponded to the address and data of the memory location under reference, respectively. A user had to enter data including the *incidence matrix* of the graph in certain predetermined locations of the Read/Write (RAM) memory. The results also, available in certain known locations of memory, could be accessed by the user after executing specific programme permanently stored in the Read-only-memory (EPROM).

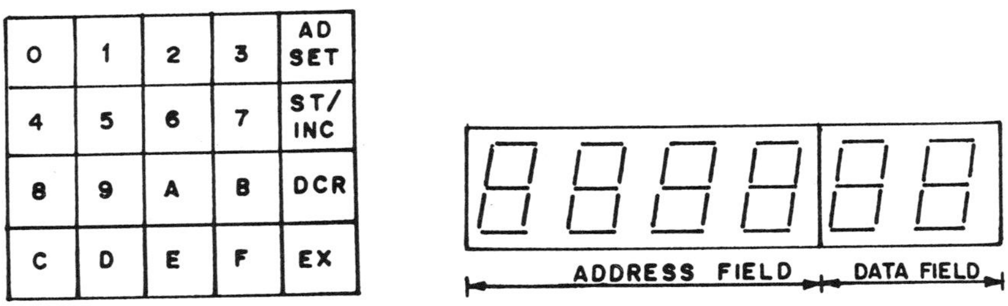

(a) Key Board **(b) Display**

Fig. 17.16 Input/Output Interfaces for a Reliability Analyzer

The analyzer had a keyboard of 20 keys, out of which 16 keys (0 to F) were meant to be *data keys*, and the remaining four were used as *function keys*. The communication between a user and the equipment was in hexadecimal numbers, where digits A to F represented decimal numbers from 11 to 15, respectively. The four function keys shown in Fig. 17.16 had the following functions :

ADSET - The user could type the hexadecimal address of the memory location to be accessed after pressing ADSET key.

ST/INC - After typing the address, typing of further two digits entered the data field. However the data contained in the above memory location was not modified unless ST/INC key was pressed.

DCR - A user could access the memory location previous to the one shown on the address field by pressing DCR key.

EX - A user could execute programmes, residing in the EPROM with their specific starting addresses, to find spanning trees, minimal path sets and minimal cut sets. A user could also type the starting address of a particular programme and press EX key to obtain the desired results.

This gadget was used to determine path sets, cut sets and spanning trees of any given network or a system. The details of the architecture, operating procedures and methodology used for achieving this are not being provided here for the reason of space but these can be found in [3], [5] and [6] and as such are not being provided here for the want of space. But the procedures described can form the basis of future developments in this area.

17.6 Development of ESCAF : A New Reliability Analyzer

The basic idea introduced by the author and his associates through [2,3,4 and 6] was commercially exploited by Laviron [5] to develop ESCAF in 1982. ESCAF stands for *Electronic Simulator to Compute and Analyze Failures*. ESCAF is a simple gadget employed typically to determine and list minimal path sets and cut sets. It has been also used to assess the importance of each component of the system and for constructing histograms or to compute system failure probability or unavailability for s-coherent and non-coherent systems. ESCAF has been also gainfully used to identify potential weak points in design and as such it serves as a powerful tool for system designers besides being useful to reliability engineers, and researchers. Its capabilities are :

(i) It simulates and analyzes complex systems with as many as 416 components, directly or through fault trees.
(ii) It provides a 2-way information flow and can be used interactively in the analysis of systems.
(iii) It can be also used to study sequential circuits and flow networks.

It will be helpful to define some of the variables that will be used in discussion to follow.

Notation

C_i component i of the system
E_i State (fail or non-fail) of component C_i or basic event i of the fault tree
l number of FCG outputs ($1 \leq$ 416)
$Q(x)$ state of simulator output T (1 for system failure, 0 for system success)
r order of simulation, i.e., the number of simultaneous events or failed components
p_i failure probability or unavailability of *ith* component or probability of event i
K_r number of cutsets of order r
$B_{k,r}$ cutset K of order r
n number of components or events of the system

17.6.1 ESCAF Architecture

ESCAF-416 system is based on two main components, viz.,

- the simulator
- the failure combination generator, to which are added a processing unit and various dialogue peripherals.

The block diagram of ESCAF is shown in Fig. 17.17.

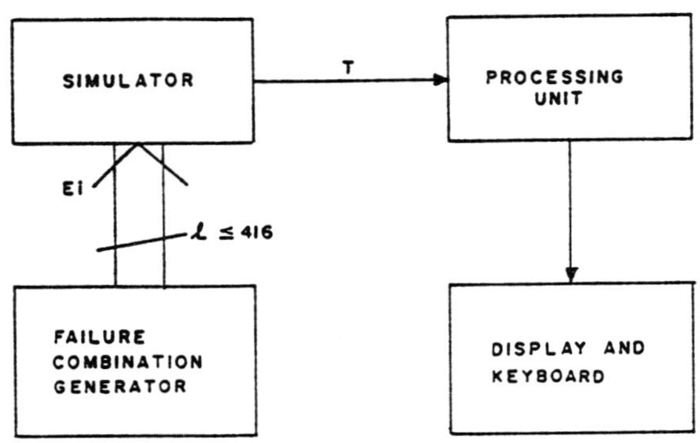

Fig. 17.17 Block Diagram of ESCAF

Failure combination generator

(a) General Structure and Assemblies

ESCAF-416 functions are organized around a *fault combination generator* that is driven by a microprocessor which manages a dialogue with the terminal and, in conjunction with a floating operator circuit. The FCG consists of several sub-assemblies, namely,

- The Micro machine which is a micro programmed computer.
- The modules for 52 outputs which, by using 8 boards, will deal with systems containing upto 416 components.

Depending on the configuration, the combination generator is capable of providing successive combinations at a rate of 2 μs to 3 μs on an average of 1.8 x 10^9 combinations per hour. The architecture of ESCAF is shown in Fig. 17.18.

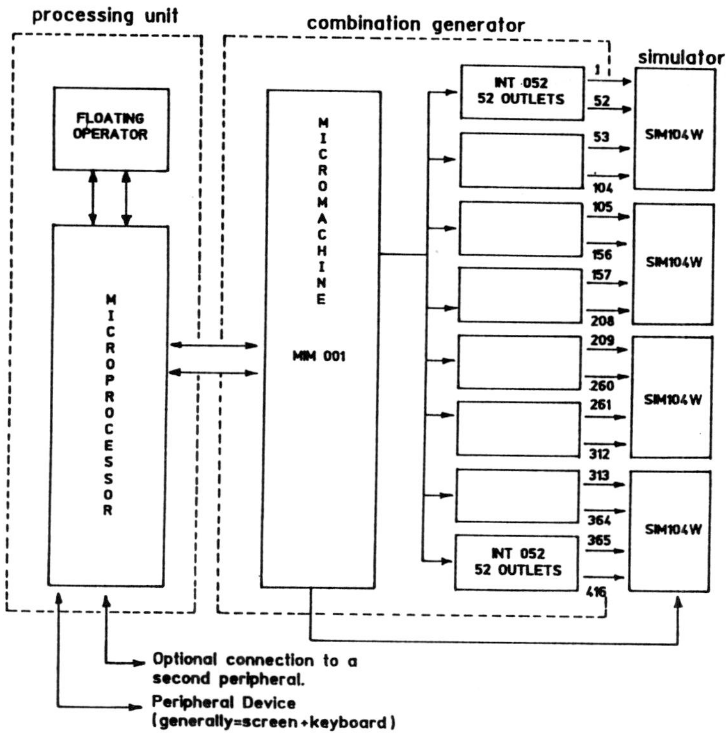

Fig. 17.18 ESCAF Architecture

(b) Circuit Boards

ESCAF-416 structure uses 5 types of board :

- a processing board
- a micro machine board
- one to eight 52 outlet boards
- one simulation board, which is wired up with logical gates by the user. Each logic gate corresponds to one component whose functioning it will simulate. For larger simulation, 1 to 4 boards can be used. Each board will accept 104 wires from the generator thus allowing up to 416 component to be simulated.

17.6.2 Physical Dimensions

The ESCAF basic unit (44 cms x 22.6 cms x 47 cms) has one to eight FCG cards, each allowing simulation of up to 52 components, giving it a maximum capacity of 416 components. Any number of cards between 1 and 8 can be provided initially. The *Basic Unit* has reserved card slots for 1 to 4 *Simulator Cards* (22 cms x 23.3 cms), each carrying the ICs simulating up to 104 components.

The *Basic Unit* has a standard RS232c connector for a serial transmission link towards an operator terminal combining a keyboard and results output device(printer or screen). A second connector is provided for a second peripheral. This is not essential for using ESCAF but does provide additional operating flexibility. For, example, a screen and keyboard terminal (VDU) can be connected at the first output, providing immediate display of results with a printer connected at second output for a permanent copy of displayed data. The second connector (output) can also be used for a data transmission link to a computer, when ESCAF results are input to more sophisticated software programs.

17.6.3 ESCAF Principle

To analyze a system with ESCAF, the standard integrated circuits mounted on one or more special cards are inserted in ESCAF basic unit (ESCAF-416), these cards are automatically connected to a *failure combination generator* (FCG) unit. The FCG generates all possible combinations of component failures (or basic events in a fault tree), first one-by-one, then in combinations of two at a time, and so on, just as the binary generator did in [2] to generate combinations of states.

The following two approaches are possible :

(i) Fault Tree Method (FTM)

In Fault tree method (FTM), the fault tree for the system is simulated by having the inputs E_i correspond to the basic events of the tree, with the electronic gates defining the tree. The output T represents the Top Event.

(ii) Direct Simulation Method (DSM)

In the Direct simulation method (DSM), a system is simulated by the electronic gates of standard integrated circuits (ICs) mounted on one or more cards which can be inserted in ESCAF. The state of a system component C_i corresponds to an input E_i in ESCAF. The Simulator output T then represents the system state S (failed or non-failed) of the entire system.

As mentioned earlier, the heart of ESCAF is the failure combination generator (FCG) with l outputs W_i (1 416). The n inputs E_i to the simulator are connected to the first n outputs W_i of the FCG. The main operation basically consists in using FCG to simulate the failure of all system components or the occurrences of all basic events, first *one by one* (Order 1 simulation), then *in combination of two* (Order 2 simulation), and so on. This, in general, provides failure or event combinations of order r. The ESCAF processing unit analyzes the system output states T for each of the combinations and determines and outputs a list of cut sets, a list of minimal cut sets and histogram showing the frequency of occurrence of each component or event in the cut set or minimal cut sets. If individual event probabilities, component failure probabilities or component availabilities are known, ESCAF can compute the overall system failure probability or unavailability.

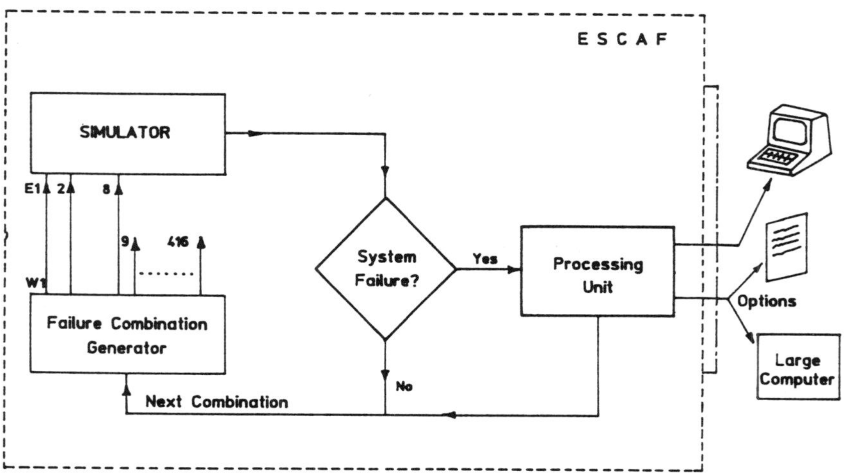

17.19 ESCAF Operations

To determine the minimal cut sets of the system, the apparatus successively simulates the simulator output, the non-failure(non occurrence) and then the failure (occurrence) of each of *r* components (events) for each cut set.

Let $\phi(x)$ represent the state of the simulator output T, which is 1 for system failure and 0 for system non- failure. A cut set is minimum, *if and only if*, the output state $\phi(x) = 0$ during each successive operation. System failure probability or unavailability is computed by ESCAF from the individual probabilities or unavailabilities P_i (entered through the keyboard) and the determined cut sets using the following formula.

$$Pr[\phi(x) = 1] = \prod_{i=1}^{n} (1 - p_i) \tag{17.13}$$

This formula assumes that the events or component failures are independent.This implies that the failure or repair of one component or element does not affect the failure or repair of other elements.

For Non-repairable systems, the p_i inputs are failure or event probabilities at a time t and above formula gives the system failure probability at the instant t. When the failure rates, λ_i are constant, each $p_i = [1\text{-}\exp(-\lambda_i t)]$. Probabilities at different times can then be computed by varying *t*.In case of time- dependent failure rates $\lambda_i(t)$, the same procedure is valid with

$$P_i = [\, 1 - \exp(-\int_0^t \lambda_i(u)\, du)\,]$$

With both repairable and non repairable systems, the p_i values are directly entered into ESCAF, whether they are probabilities or unavailabilities. If the failure rates are known, then the p_i values can be computed with hand calculator if rates are constant. Otherwise, the computation can be completed very rapidly by a quite simple program in ESCAF processing unit, for both s-coherent and non-coherent systems.

17.6.4 System Simulation and Parameter Setting

Simulation

The first step is to simulate the system on the simulator cards available with the ESCAF, using ICs to represent the various logical functions, using the logical gates such as AND,OR,NAND,NOR,NOT,XOR etc., of a fault tree. Each IC has 14 pins, 12 of the pins are connected to inputs/outputs of gates, while the other two pins are automatically connected to power supply when ICs are inserted in its sockets on a *Simulator Card.* To construct a fault tree or a network, a user interconnects the gates of the ICs using *Plug Ended Jumper Wires* (PEJW) between the corresponding auxiliary socket terminals. The gate inputs correspond to basic events and are jumpered in an identical way to a strip of 104 terminals, permanently wired to contacts as the *card's edge connector*. This is shown in Fig. 17.20 for a fault tree shown on the same figure.

When the card is inserted in the ESCAF, the (maximum) 104 basic event inputs are linked through the *Edge Connector* to an equal number of wires from the FCG. If more than 104 basic events are to be simulated, several simulator cards are used and interconnected by ribbon cables. The output of the fault tree is connected to a special terminal on one of the simulator cards, in turn wired to card's edge connector and from there to the ESCAF processing unit.

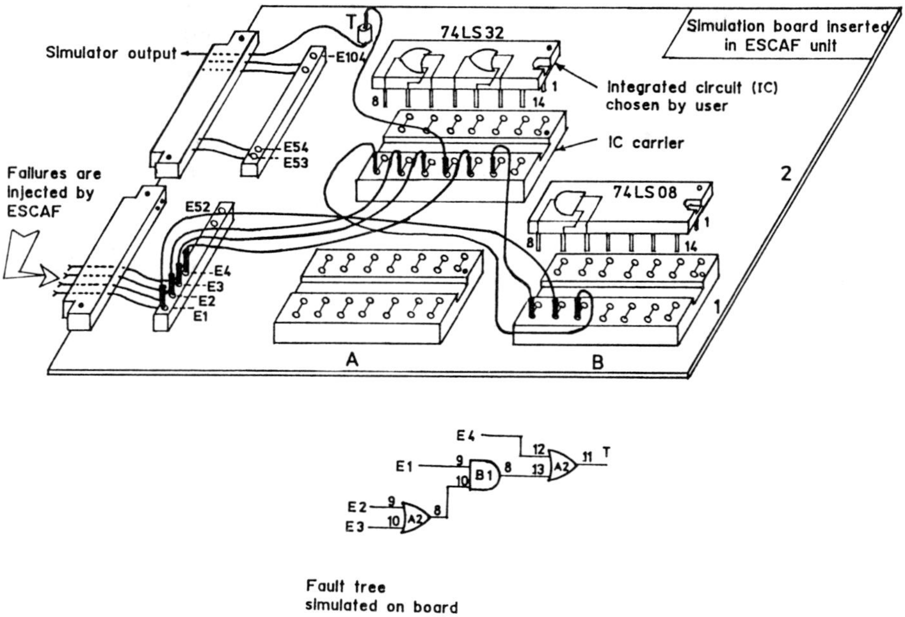

Fig. 17.20 A Fault Tree Simulation on ESCAF [10]

As is shown in Fig. 17.20, a simulator board consists of a set of IC carriers A and B. Set A provides OR-gates whereas the set B provides AND-gates. As is clear from this figure, inputs from E_2 and E_3 (coming from FCG) are connected to input leads of an OR-gate (pins # 9 and 10 of IC 74LS32) and the output of this gate is available at pin #8 of the IC, which in turn is jumpered to pin #10 of the IC 74LS08 (AND-gate IC). Pin #9 of IC 74LS08 has another input from E_1 of the edge connector. The output of this AND-gate (74LS08) is available at pin #8, which in turn is jumpered to pin #13 of another OR-gate of the earlier IC 74LS32. Pin #12 of this IC has input available directly from E_4 of the edge connector. Finally, the output of the second OR-gate of the IC 74LS32, which is available at pin #11, is jumpered to the last terminal (105 *th*) of the second edge connector, representing the simulator's output. In this way we are able to simulate on ESCAF, a fault tree, which is shown in Fig. 17.20. If it is required to simulate a network or a system, the same procedure can be used to configure its various branches or elements. It is easy to observe that the procedure of simulation in ESCAF is practically the same as described earlier by the author in [2].

Fault Tree Method

Let us further consider a complicated fault tree as is shown in Fig. 17.21. Once again, the fault tree can be simulated using standard ICs. There are 17 basic events, and the output of the fault tree is the top event, T. Since the fault tree has 19 gates, out of which 15 are OR-gates and 4 are AND-gates, we will require one IC of 74LS08 type and 4 ICs of 74LS32 type to simulate this fault tree. The first 17 outputs of the FCG are connected to the PC board simulating the fault tree. The occurence of a basic fault event is represented by logic 1. The fault tree of the example has no cut set of *order one* but has 39 minimal cut sets of *order 2*, some of which are (1,4), (4,6), (5,6), (1,7) etc. The output T=1 is available, when the basic events included in a cut set are at logic 1. The FCG output is of a 17-bit binary word, which represents the input of the simulated fault tree, and consists of various combinations of the basic events. The processing unit samples the output T for these combinations to determine cut sets or minimal cut sets. Path sets and

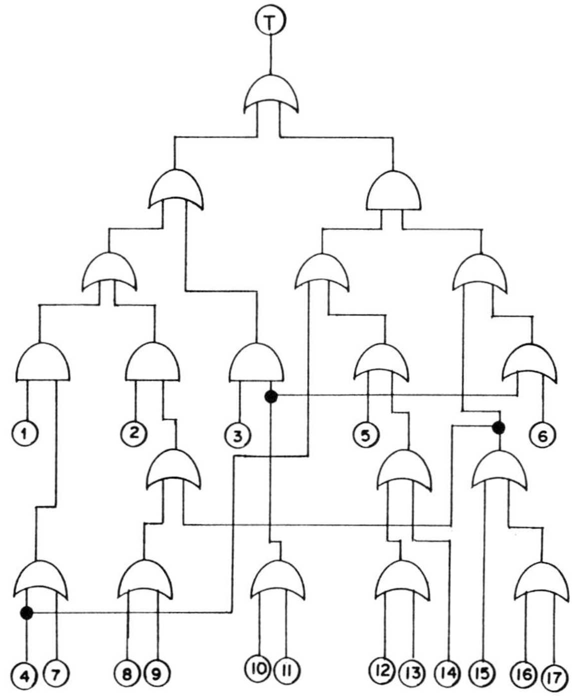

Fig. 17.21 A Sample Fault Tree

minimal path sets can be determined by inverting simultaneously the logic levels at E_i and T. There are three *eighth order* minimal path sets, viz., (1,2,3,4,5,12,13,14), (2,3,4,5,7,12,13,14) and (1,2,6,10,11,15,16,17). The output T shall be 0, when bits corresponding to the events included in a path have a level of 0.

System failure probability with limiting order C=3, can be determined by considering 39 minimal cut sets of order 2. If each basic event is assigned a probability of occurence of 10^{-3}, the system failure probability can be calculated from the expression

$$\Pr[\phi(x)=1] = \prod_{i=1}^{n}(1-p_j)\cdot\left[\delta_o + \sum_{r=1}^{c}\sum_{k=1}^{k_r}\prod_{i\varepsilon B_k,r}\frac{p_i}{(1-p_i)}\right] \qquad (17.14)$$

where, δ_o = 1 or 0, depending upon whether $\phi(x)$ is 1 or 0, when no component failure is simulated.

Substituting the data we have,

$$\Pr[\phi(x)=1] = \prod_{j=1}^{17}(1-10^{-3})\cdot\left[\delta_o + \sum_{r=1}^{2}\sum_{k=1}^{39}\prod_{i\varepsilon B_k,r}\frac{10^{-3}}{(1-10^{-3})}\right]$$

$$= (1-10^{-3})^{17}\ \text{X}\ 39\ \text{X}\left[\frac{10^{-3}}{(1-10^{-3})}\right]^2 = 3.88 \times 10^{-5}$$

It should be noted that the value of δ_o is 0, because T=0, on simulating non-occurence (logic 0) of all basic events.

Direct Simulation Method

Many systems can be directly simulated with standard ICs without requiring to go in for a fault tree approach. The DSM has a further advantage of allowing simulation of more complex systems, including the ones with 2-way information flow or bi-directional flows. Let us consider an example [8] of a network of Fig. 17.22, with 105 nodes and 127 links. The reliability criterion is that the network be simply connected. Partial simulation of the network has been shown in Fig. 17.23.

Each link of the network is represented by a 3-input NOR gate (gates L1 to L127). Two inputs of the above gate come from the terminals (N1 to N105) between which the link is connected. The third input to the above gate is the control input E_i from FCG, to simulate failure or non-failure of the specific link. Electronic gates (N1 to N105) have input coming from the links (link gates) incident on that particular node. The output of the 105 input gate is T, to be monitored by the processing unit.

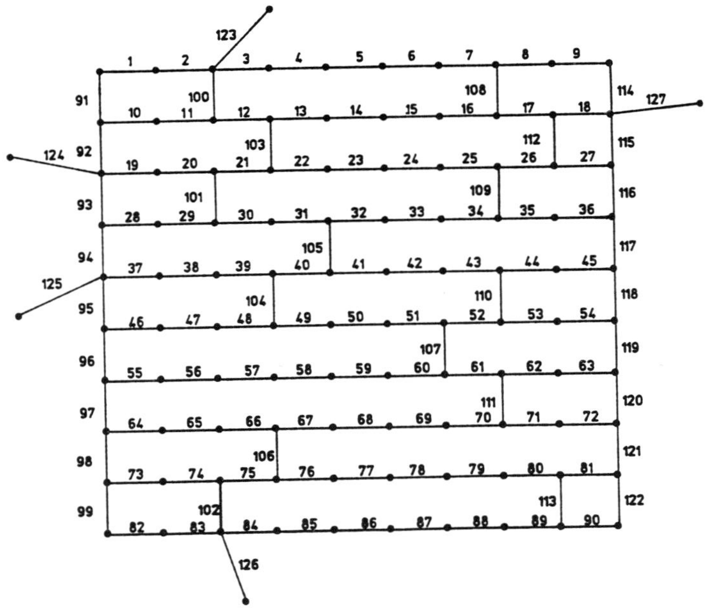

Fig. 17.22 A Sample Network for Simulation [8].

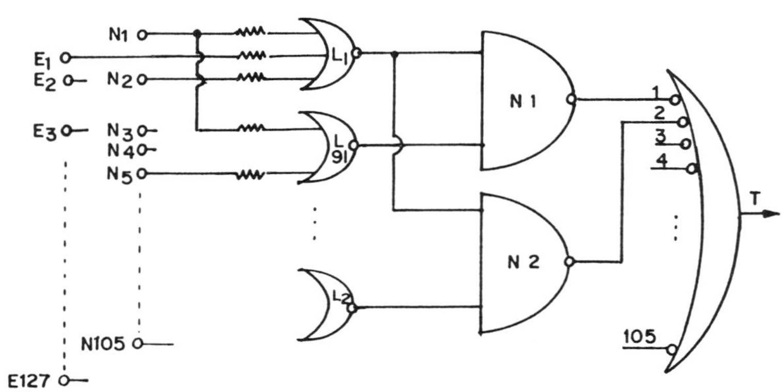

Fig. 17.23 Partial Simulation of the Network Shown in Fig. 17.22 [8].

A particular link fails, when any one of the nodes supporting it has failed (L_i output-1). Similarly, a node gets disconnected from the system only when all the links incident on it have failed. (N_i output -0). System failures takes place, if any one of the nodes has failed (T=1).

The equipment can determine all the minimal cut sets of the system shown in Fig. 17.22 and they are : 5 of order 1, 103 of order 2, 263 of order 3, 1208 of order 4, 6401 of order 5, and so on.

The probability of network failure can be calculated using (17.14)

Failure Probabilities of Links			Limiting Order	$Pr\{\phi(x)=1\}$
1-90	91-122	123-127		
$5x10^{-4}$	$1x10^{-4}$	$1x10^{-5}$	2 3	7.31×10^{-5} 7.43×10^{-5}
$5x10^{-3}$	$1x10^{-2}$	$1x10^{-3}$	2 3 4	5.36×10^{-3} 7.06×10^{-3} 7.62×10^{-3}

When individual component failure probabilities are not available, one can determine the most critical components by looking at the equipment generated histograms, which are plots of the frequency of links in the minimal cut sets.

The histogram of various order of cut sets can highlight the importance of links. For example, links 123 to 127, forming cut sets of order 1, the failure of just one of these links is sufficient to make the network disconnected. The order 2 histogram shows that the special attention should be paid to the reliability of links 55-60 and 84-89, than to links 3-7 etc.

ESCAF Performance

In the FTM, the ESCAF takes roughly 3 msec multiplied by the number of combinations, to output cut sets and minimal cut sets. This time is 7 sec. for a 50-component system at order 5, 12 sec. for 100 components at order 4 and 40 minutes for 200 components at order 5.

In the DSM, for the graph of Fig. 17.22, the equipment takes less than 1 sec to output 5 cut sets of order 1, 27 sec. to output 103 cut sets of order 2, 15 minutes to output 263 sets of order 3, and 7 hours to output 1208 cut sets of order 4.

Parameter Setting

After switching on the power, ESCAF outputs a menu to the VDU screen. This guides the user in entering run parameters through the keyboard, viz.,

- Number of basic events (or components).
- Simulation orders at which analysis to begin and end
- Type of processing and output. There are 7 basic possibilities :
  (i) determine cut sets and minimal cut sets
  (ii) output cut sets
  (iii) output minimal cut sets
  (iv) output histogram of cut sets
  (v) calculate the system's failure probability or unavailability.
- Settling time, i.e., the time between simulating an event or failure combination of the W_i terminals in the FCG to reading the resulting system state at T.

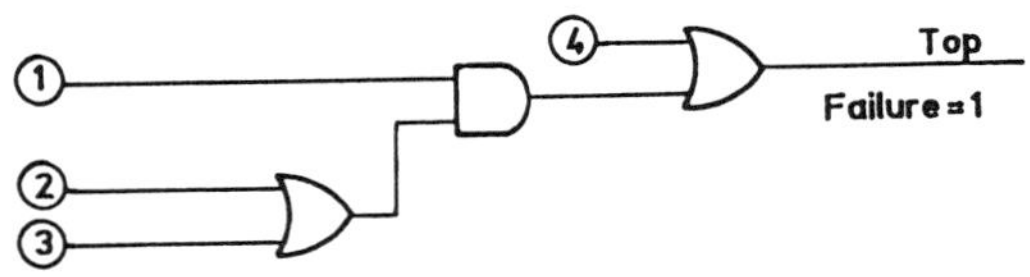

17.24 A Sample Fault Tree

After the parameters are set, the typical processing in evaluation of cut sets referring to the Fig. 17.24 can be explained as follows:

When ESCAF is started, the FCG generates a single electrical signal on the wires corresponding to basic events E_i, i=1,...,4. None of the signals applied individually at E_1, E_2 and E_3 is transmitted to the top of Tree, signifying success of the system, except when the signal is applied at E_4, which represents a cut set of order 1 and ESCAF displays number '4'. It then examines all double-event occurrences known as *order 2 combinations*, after which it simulates combinations of events taken three at a time (order 3), and so on, progressively stepping through all possible event combinations once and only once. In this way, ESCAF unit immediately identifies all the cut sets and also the minimal cut sets of the system under study, i.e., all component failures or event combinations which cause system failure.

17.7 S.ESCAF : A Modified Version

If the failures are sequential and the order in which the components fail is important, it won't be possible to directly use ESCAF for this purpose. For example, if a diesel generator (DG) is used as a standby source of power supply when there is loss of normal power supply. However, on demand the DG set may fail to start or it may start and fail later on. Therefore, the normal supply failure has to take place before the generator failure is considered. S.ESCAF provides the facility of considering such sequential failures.

S.ESCAF involves 1-4 circuit boards that are used to implement a new FCG, in which events are made to occur in a sequence rather than simultaneously as in normal ESCAF. Therefore, it would be necessary to describe the new SEQ.SIM simulator board.

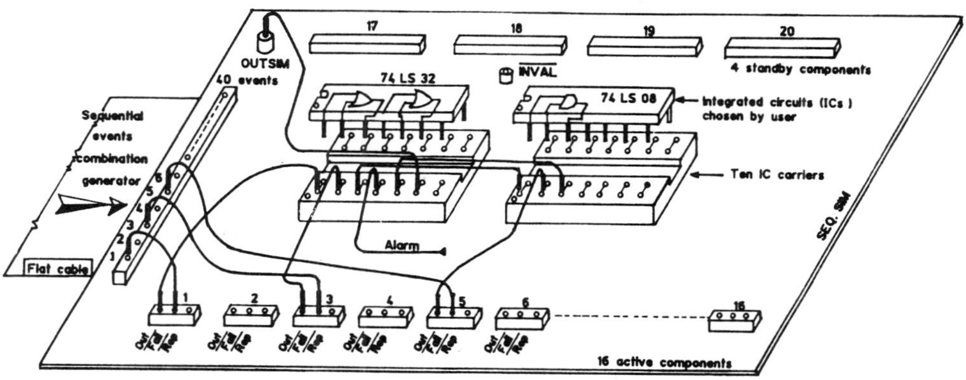

Fig. 17.25 A SEQ.SIM Simulator Board [8].

SEQ.SIM Simulator

A SEQ.SIM board (22 cms x 23 cms), shown in Fig. 17.25, can simulate a system upto 20 components, has four sets of subassemblies. The first one contains 16 terminal strips for the active components, i.e., the components that are normally operating and can fail under normal operation. The second set of subassemblies consists of terminal strips for four standby components shown as the top most strips (#17, 18, 19 and 20) in Fig. 17.25. To represent components, the simulated events can include the *correct starting* or *nonstarting, the failure during operation* or in *standby mode*. The third set of subassemblies contains carriers for ICs. These are used to simulate the configuration of the elements of the system and usually involved ICs to fabricate AND, OR or some other logic gates. The fourth subassembly includes a 40-pin connector for the flat cable from the *sequential events combination generator* corresponding to 40-event input wires. The maximum number of components that an S.ESCAF can simulate is 80.The detailed functioning of S.ESCAF for sequential failures is explained in [8] and [9].

17.8 ESCAF Capabilities

In this section, we would like to enumerate the problems that a reliability analyzer of the ESCAF type can solve. It will be observed that this gadget offers a wide list of problems that can be attempted without much difficulty thereby proving its versatility. Further improvements to extend its applicability and efficacy are always possible and may become the subject of further investigations in near future.

17.8.1 Probability Calculations

(a) Non-repairable components

Let us first consider the components that cannot be repaired and have a constant failure rate of λ_k.

The probability of failure is calculated by ESCAF using the following equation, with $\mu_k = 0$ and $\bar{a}_{ok} = 0$.

$$a_k(t) = \frac{\lambda_k}{\lambda_k + \mu_k}\left[1 - e^{(\lambda_k + \mu_k)t}\right].(1 - a_{ok}) + a_{ok} \tag{17.15}$$

where, $\bar{a}_k(t)$ = failure of probability of component at time t.

(b) Repairable components

When the components can be repaired with a constant repair rate μ_k, the ESCAF calculates probability (i.e., unavailability) of component at the given time t, using (17.15), with $\bar{a}_{ok} = 0$, which is the constant part of the probability expression and is the probability of not meeting a demand because the components may be faulty when a demand is made on them.

17.8.2 Birnbaum's Importance Factor

ESCAF can compute, (inputing data through a key board) the unavailability $\bar{A}$ of the entire system using the following equation:

Top event probability

$$A = a_k(t).\alpha_k + [1 - a_k(t)].\beta_k \tag{17.16}$$

where, α = Fraction of unavailability due to cut sets containing k*th* component.

β = Fraction of unavailability due to cut sets not containing k*th* component.

ESCAF calculates the value of co-efficient α_k for each cut set containing k*th* component. Then using (17.16), it can calculate,

$$\frac{\partial A(t)}{\partial a_k(t)} = \frac{\alpha_k - A(t)}{a_k(t)} \tag{17.17}$$

It is also called as *s-importance factor*. The ESCAF calculates the s-importance of each component's contribution to the final unavailability result.

17.8.3 Sensitivity Co-efficient Factor

This is obtained after multiplying the LHS of (17.17) by the component's unavailability (weighting factor) and dividing by the system unavailability. This is also called as *Criticality Importance Factor.*

$$\text{Criticality Factor} = \frac{\partial A(t)}{\partial a_i(t)} \cdot \frac{a_i(t)}{A(T)} = \frac{\partial \ln A(t)}{\partial \ln a_i(t)} \tag{17.18}$$

17.8.4 Time Evolution

Time evolution of each component's unavailability is obtained by computing partial derivative from (17.16). It represents the corresponding component's s-importance in the time evolution of system unavailability. The sum of these over all the components provides the parameter $d\overline{A}/da_k$ and is the time derivative of system unavailability, viz.,

$$dA/dt = \sum_{i=1}^{n} (\partial A/\partial a_i) \cdot (\partial a_i / \partial t) \tag{17.19}$$

where, n is the number of components

When calculation is performed over a sufficiently long time, the results allow a user to decide whether a steady state has been reached or not, at which $d\overline{A}/da_k$ is very close to zero plus and FAIL/UT and REP/UT intensities are equal.

17.8.5 S-importance of Minimal Cut Sets

For s-coherent systems, ESCAF supplies the list of minimal cut sets and, for each, a value representative of its probability importance. This value is computed by multiplying together the p_i values of all components in the cut set. With normalization, the values in question are all divided by N. If N is set equal to P(t), the results are *Fussell Vesely importance co-efficients.* To limit cut set listing to those sets for which the s-importance value is greater than a particular threshold K, this value can be entered.

17.8.6 Performance Indices Calculation

Let us assume that during a brief instant from t to t+dt, the probability of the system failing is w(t)dt, the probability of repair is v(t)dt, and

$$d\overline{A}(t)/dt = w(t) - v(t) \tag{17.20}$$

i.e., each observed system failure is the immediate consequence of a basic event-a component failure for an s-coherent, whereas a component repair for a non-coherent system. So the intensity w(t) can, therefore, be decomposed into two parameters representing the failure intensities due to component failure and repair respectively, viz.,

$$\text{FAIL/UT} = w(t) = w_{fail}(t) + w_{rep}(t) \tag{17.21}$$

$$\text{REP/UT} = v(t) = v_{rep}(t) + v_{fail}(t) \tag{17.22}$$

Then the *equivalent failure rate* of the system is given by,

$$\text{LAMBDA} = \frac{\text{FAIL/UT}}{A(t)} \tag{17.23}$$

Similarly, *equivalent repair rate* of the system is given by,

$$\text{MU} = \frac{\text{REP/UT}}{A(t)} \tag{17.24}$$

However, we know (for exponential failure distribution) that MTTF = (1/LAMBDA) and MTTR= 1/MU. Therefore,

$$\text{MTBF} = \frac{1}{(\text{FAIL/UT})}$$

$$\text{MTBR} = \frac{1}{(\text{REP/UT})}$$

17.8.7 Determination of Coherence Coefficients

(a) Calculation of Variables

These calculations are performed on the basis of the critical components observed in cut sets.

Let B ≡ Cut set consisting of components with index k and such that there exists a component identified by index i, for which the repair (of just one such component) leads to repair of the complete system.

j = this index implies components not belonging to the cut set.

The contribution of cut set to the variables of interest is:

$$v_{rep}(t) = Pr(B) \cdot \sum_i \mu_i = \prod_k a_k(t) \cdot \prod_j a_j(t) \sum_i \mu_i \tag{17.25}$$

$$w_{fail}(t) = \sum_i a_i(t) \cdot \lambda_i \cdot \prod_{k \neq i} a_k(t) \cdot \prod_j a_j(t) \tag{17.26}$$

Reasoning along the similar lines, (17.20) can be used for calculating the variation rate of the unavailability due to failure of components, viz.,

$$\frac{\partial A_{fail}(t)}{\partial t} = w_{fail}(t) - v_{fail}(t) \tag{17.27}$$

Variation rate due to repair of components can be written as

$$\frac{\partial A_{rep}(t)}{\partial t} = w_{rep}(t) - v_{fail}(t) \tag{17.28}$$

Using the $\bar{a}_k(t)$ value computed from (17.15), ESCAF calculates,

$$\frac{da_k(t)}{dt} = \lambda_k \cdot a_k(t) \tag{17.29}$$

$$\frac{da_k(t)}{dt} = -\mu_k \cdot (a_k(t) - a_{ok}) \tag{17.30}$$

$$\frac{dA_{fail}(t)}{dt} = \sum_{k=1}^{n} \frac{\partial A(t)}{\partial a_k(t)} \cdot \frac{da_k(t)_{fail}}{dt} \tag{17.31}$$

$$\frac{dA_{rep}(t)}{dt} = \sum_{k=1}^{n} \frac{\partial A(t)}{\partial a_k(t)} \frac{da_k(t)_{rep}}{dt} \tag{17.32}$$

The LHS of (17.32) is calculated from (17.30) and (17.17); $w_{rep}(t)$ is then determined using (17.28), (17.32) and (17.25).

The LHS of (17.31) is calculated from (17.29) and (17.17); $v_{fail}(t)$ is then determined using (17.27), (17.31) and (17.26). FAIL/UT, REP/UT are calculated using (17.21) and (17.22).

The *time derivative of Top Event*, namely, $\frac{\partial A(t)}{\partial t}$ is computed from (17.19).

17.8.8 Error Transmission

ESCAF can be used to calculate transmission of errors by the method of moments. We know that any uncertainty in the probability of component leads to an uncertainty in the calculation of probability of the system. Moreover, in many reliability studies, the estimates of probability and particularly those used to govern human reliability in system interactions, are often highly uncertain. Therefore, there exists imprecise information about a particular component failure rate, and it requires an uncertainty analysis to obtain any meaningful results.

Basically, there are two methods to account for uncertainty, viz.,

- Monte Carlo method, but it is not easy to implement and accuracy is costly.
- Method of Moments, this is based on the Taylor's series expansion of probability function (limited to first order terms) and is generally considered very efficient.

Method of Moments

Let us assume that $p_i = \overline{p}_i + dp_i$

which is probability of failure of *ith* component with the mean $\overline{p}_i$ and variation dp_i such that $E(dp_i) = 0$ and $\sigma_i^2 = Var(p_i) = E(dp_i^2)$.

Let $\overline{P} = \overline{P}(\overline{p}_1, \overline{p}_2, \ldots, \overline{p}_n)$

$$Var(P) = E[(P-\overline{P})^2] \cong \sum_{i=1}^{n} (\partial P/\partial p_i)^2 . E(dp_i)^2 + 2\sum_{i=1}^{n-1} \sum_{j=i+1}^{n} (\partial P/\partial p_i)(\partial P/\partial p_j) E(dp_i . dp_j) \tag{17.33}$$

When errors dp_i are independent, 2nd term of above equation is equal to zero. Let us assume, each component follows a *log normal distribution* with parameters μ, σ, then

$$mean = p_i = e^{\mu + (\sigma^2/2)}$$

Normalized standard deviation = $\sigma_i / p_i = \sqrt{(\sigma^2 + 1)}$.

After fault tree is simulated on board, the mean value p_i and the Normalized standard deviations $(\sigma_i/\bar{p}_i)$ are directly entered into the keyboard of ESCAF,

Top Event standard deviation= $\sigma_T = \mathrm{Var}\,[P(\bar{p}_1, \bar{p}_2, ..., \bar{p}_n, \sigma_1, \sigma_2 ..., \sigma_n)]$ (17.34)

and

$$(\partial\sigma_T / \partial\sigma_i) = (1/\sigma_T) \cdot \frac{\partial \,\mathrm{Var}\,[\,P\,(p_1, p_2, \ldots, p_n, \sigma_1, \sigma_2, \ldots \sigma_n)}{\partial\sigma_i} \tag{17.35}$$

Critical sensitivity co-efficient for standard deviation ,

$$(\partial\sigma_T / \partial\sigma_i) \cdot (\sigma_i / \sigma_T) = (\sigma_i / \sigma_T^2) \cdot \frac{\partial \,\mathrm{Var}\,[\,P\,(p_1, p_2, ..., p_n, \sigma_1, \sigma_2, ... \sigma_n)}{\partial\sigma_i} \tag{17.36}$$

ESCAF Formula for Error Calculation

Let us assume that the Top event expression, obtained through the basic events of fault tree, is

$Q = q_1 \cdot q_2 \cdot q_3$

Then, $E(Q) = E(q_1) \cdot E(q_2) \cdot E(q_3)$ (17.37)

also, $Q^2 = q_1^2 \cdot q_2^2 \cdot q_3^2$

and $E(Q^2) = E(q_1^2) \cdot E(q_2^2) \cdot E(q_3^2)$ (17.38)

We know that, V = Variance = $E(Q^2) - [\,E(Q)\,]^2$.

In same way,

$v_i = E(q_i^2) - [E(q_i)]^2$ and $E(q_i^2) = v_i - [\,E(q_i)\,]^2$ (17.39)

then $(\partial V/\partial v_i) = \partial E(Q^2) / \partial E(q_i^2)$.

Therefore, from (17.20), (17.21) and (17.22), $(\partial V / \partial v_i)$ can be found.

17.8.9 Common Cause Failures

To study the effects of common cause failures and/or permanent component degradation, a diode plug programming matrix is inserted between the FCG and the simulator, with its rows connected to the FCG outputs W_i and its columns to the simulator inputs E_i. Common cause failure studies are performed either using the direct simulation method (DSM) or the fault tree method (FTM).

The procedure is as follows :

(1) At first, the conditions underlying the dependencies between components (viz., Fire risk, Common manufacturers, maintenance, etc.) are determined.

(2) The components that share a common condition are identified and the various sub-conditions corresponding to common cause failures are described. In case of fire hazards, for instance, we must differentiate the components in room A from those in room B, and so on. This is done to take account for the fact that the influence of a fire on components depends on its location.

(3) Next, we program the CCFs at the ESCAF programming matrix. We can switch on this matrix by connecting the first n outputs W_i of the FCG to the n inputs E_i of the simulator, for simulating the individual failure of each component. If 'w' CCFs have been distinguished during steps 1 and 2, then outputs W_{n+1} to W_{n+w} of the FCG are connected to the first 'w' rows of the matrix. The first CCF, e.g., a fire in room A, is simulated by inserting diode plugs in the matrix at the intersections of the first row and the columns with the same numbers as the components that share the CCF condition. If, for example the components that may be affected by a fire in room A are components 2,4,6,8, then plugs are inserted at the intersections of row 1 and columns 2,4,6,8. When outputs W_{n+1} of the FCG is active, it then simulates simultaneous failure of all the components in room A. The other CCFs (e.g., for fire in room B, common manufacturer, etc.) are simulated with rows 2,3,....,w of the matrix.

(4) As the next step, we parameter ESCAF for new components and start the simulation run. ESCAF outputs results for both individual component failures and CCFs, i.e., the number and the list of cut sets or minimal cut sets, histograms etc. If the individual failure probabilities or unavailabilities for each of the n (s-independent) components and each of the w CCFs are known (e.g., the probability of fire in room A), then the system failure probability or unavailability can be computed by ESCAF.

When results are outputed in the form of histograms, the importance of each CCF is immediately known, i.e., how often it appears in cut sets or minimal cut sets, without the need for probability or unavailability data. If however, these data are available, ESCAF can determine the probabilistic importance of each CCF by comparison with the results obtained without CCF simulation.

The flexibility offered by the Programming Matrix is considerable. For example, in reliability analysis, the first approach to the system is usually made before progressively entering into greater detail. Each component is simulated only once at the simulator, and the analysis can be progressively refined by using the matrix to simulate the CCFs that were not initially taken into account. For instance, a circuit breaker which interrupts power to several components, simultaneously can be simulated as a CCF without modifying the simulator. The matrix also allows any component failure or CCF input from the FCG to be isolated, or placed in a permanently active condition to simulate a permanent failure. This makes it easy to study the relative importance of each component failure of the CCF, by comparing results with the input isolated and non-isolated, and to analyze the system under degraded operating conditions by simulating one or more permanent failures.

17.8.10 Analysis of Non-Coherent Systems

An s-coherent system is one which, in the failed state, cannot be returned to the non-failed state by additional component failures *and* which, in the non-failed state, cannot be made to fail by the repair of components. A system is non-coherent otherwise. It is not always possible to assess whether a system under study is s-coherent. ESCAF determines this, informs a user whether the concept of minimal cut sets is valid for the system, which in ESCAF-based analysis involves no more than a few extra commands entered from the key board. ESCAF also allows to characterize the degree of non-coherence of a system.

In fact, depending on the probabilities associated with individual components, a system can be *more or less* s-coherent. To assess the degree of non-coherence, one can define *non-coherence coefficients* NCOF(t) and NCOR(t), respectively for system failure and system repair, viz.,

$$NCOF = \frac{w_{rep}(t)}{w_{fail}(t)}, \ NCOR = \frac{v_{fail}(t)}{v_{rep}(t)} \tag{17.40}$$

As the failure probability of each component varies with time, the degree of non-coherence of a system can also vary in time. The steady-state relationships would be

$$w_{fail}(t) = v_{rep}(t), \ w_{rep}(t) = v_{fail}(t)$$

Now, for the given probability values, a non-coherence coefficient NCOH for the steady state can be defined as :

$$NCOH = NCOF(\infty) = NCOR(\infty) \tag{17.41}$$

If NCOF(t) and NCOR(t) are both zero, the system is s-coherent or behaves like an s-coherent system for the given set of probability values.

17.9 Conclusions

The concept of Reliability Analyzer, originally introduced by the author, is unique in implementing logical functions involved in system reliability evaluation process and has led to the development of a multi-purpose hardware gadget that can be used economically and efficiently even by a non-specialist. This gadget, has been successfully commercialized by Laviron of CEA, France and is being extensively used as an aid for system reliability evaluation and other studies.

Note : Incidentally, the author approached the Department of Electronics, New Delhi for funding a developmental project on Reliability Analyzer during seventies, however the idea did not find favour with them but they purchased ESCAF from CEA through hard currency in1980s.

Acknowledgement

The author would like to thank Pergamon Press plc, IEEE and Elsevier for the permission to reproduce some of the figures and tables quoted in the chapter from the papers published in their journals. The assistance rendered by Dr.V.K.Bansal is also gratefully acknowledged.

REFERENCES

[1] Misra K.B., *Reliability Analysis and Prediction (A Methodology Oriented Treatment)*, Elsevier Science Publisher BV, Amsterdam, 1992.

[2] Misra K.B. and A.K. Raja, *A Laboratory Model of System Reliability Analyzer*, Microelectronics and Reliability, Vol. 19, No. 3, pp.259-264, 1979.

[3] Bansal V.K. and K.B. Misra, *Hardware Approach for Generating Spanning Trees in Reliability Studies*, Microelectronics and Reliability, Vol. 21, No. 2, pp. 243-253, 1981.

[4] Bansal V.K., K.B. Misra and M.P. Jain, *Minimal Pathsets and Minimal Cutsets using a Search Technique*, Microelectronics and Reliability, Vol. 22, No. 6, pp. 1067-1075, 1982.

[5] Laviron A., A. Carnino and J.C. Manaranche, *ESCAF- A New and Cheap System for Complex Reliability Analysis and Computation*, IEEE Trans. Rel., Vol. R-31, No. 4, pp. 339-349, Oct. 1982.

[6] Bansal V.K., K.B. Misra and M.P. Jain, *Improved Implementation of a Search Technique to Find Spanning Trees*, Microelectronics and Reliability, Vol. 23, No. 1, pp. 141-147, 1983.

[7] Laviron A. and C.D. Heising, *Error Transmission in Large Complex Fault Trees Using the ESCAF Method*, Reliability Engineering, Vol. 12, No. 3, pp. 181-192, 1985.

[8] Laviron A., *Simulation and Reliability Analysis of Sequential Complex Systems with S. ESCAF*, Reliability Engineering, Vol. 12, No. 3, pp. 139-150, 1985.

[9] Laviron A., A. Blin, J.M. Lanore and C. Rivet, *S.ESCAF : Sequential Complex Systems are Analyzed with ESCAF through an Add-on-Option*, IEEE Trans. Rel., Vol. R-34, No. 3, pp. 229-232, Aug. 1985.

[10] Laviron A., *Analysing Critical Components in Complex System Design : A Comparison of Methods with the ESCAF Approach*, Reliability Engineering, Vol. 15, No. 3, pp. 209-233, 1986.

[11] Laviron A., *ESCAF for MTBF Time Evolution, Sensitivity Coefficients, Cutset Importance and Non-coherence of Large Systems*, IEEE Trans. Rel., Vol. R-35, No. 2, pp. 139-144, June 1986.

[12] Blot M. and A. Laviron, *Reliability Analysis with the Simulator S.ESCAF of a very Complex Sequential System : The Electrical Power Supply System of a Nuclear Reactor*, Rel. Engg. and Syst. Safety, Vol. 21, No. 2, pp. 91-106, 1988.

ABOUT THE EDITOR

Krishna B. Misra presently is a Director-grade-Scientist and Head, **Disaster Prevention and Impact Minimization Division** at National Environmental Engineering Research Institute (NEERI), Nagpur, India. Before joining NEERI on deputation for a year, he was a professor at Reliability Engineering Centre at Indian Institute of Technology, Kharagpur and the coordinator of Ministry of Human Resource Development Project on *Reliability Engineering.* He has been the founder Head of Reliability Engineering Centre and has also served as Dean, Institute Planning and Development of IIT Kharagpur.

K. B. Misra was born on January 23, 1943 in Kota (India) and received his B.E. (Electrical), M.E.(Power Systems) and Ph.D. degree (in Reliability Engineering) in 1963, 1966 and 1970, respectively. Since 1963, he has taught and researched at several reputed institutions which include University of Roorkee and IIT Kharagpur. He has worked in Federal Republic of Germany at GRS-Garching, Technical University-Munich, RWTH-Aachen and Kernforschungszentrum - Karlsruhe. He has visited several European countries and USA to deliver *invited* lectures.

He has published over 125 technical papers in IEEE Transactions on Reliability, Microelectronics and Reliability, International Journal of System Science, International Journal of Control, Reliability Engineering and System Safety, International Journal of Quality and Reliability Management, Fuzzy Sets and Systems. Dr. Misra's work is widely quoted in international research journals including at least 10 books on reliability engineering . He has also authored a 900-pages, the state-of-art book, *Reliability Analysis and Prediction*, published by Elsevier Science Publishers, Amsterdam in 1992.

Prof. Misra has served on the editorial board of several international journals like Microelectronics & Reliability, Reliability Engineering & System Safety, Quality & Reliability Management, Combinatorics Information & System Sciences, International Journal on Fuzzy Mathematics, Electrical Power Research Journal etc., and he has served as a reviewer to IEEE Transactions on Reliability, Fuzzy Sets and Systems, European Journal on Operations Research, International Journal on General Systems etc.

He is a fellow of Safety & Reliability Society (U.K.), Indian Academy of Sciences, Institution of Electronics & Telecommunications (India), Institution of Engineers (India) and has been the Vice President of System Society of India.

He is the recipient of a few awards and prizes, including the first Lal C. Verman award in 1983 *for his pioneering work in reliability engineering* in the country. He has been working in the area of reliability engineering since 1967 and has been making efforts to popularize reliability concepts in India. Due to his effort, a master's degree program in reliability engineering was conceptualised and started for the first time in India at IIT Kharagpur. He also founded *Reliability Engineering Centre* at IIT Kharagpur in 1983. This Centre is *the first* of its kind to promote research, consultancy and teaching at an advanced level in the area of reliability and safety engineering in India.

He was invited to serve as the *convener* of the *NCST Working Group on Reliability Engineering in India* set up by Govt. of India, Department of Science & Technology, New Delhi. This group submitted its report in part I & II in 1978 on the *Reliability Implementation Programme* for the country. Also served as member of the *Task Force Committee* on *Reliability Engineering* of the Department of Science and Technology, 1979. He has also served as member of *Project Assessment Committee* for National Radar Council, Department of Electronics, CSIR, UGC etc. and on Indian Standards Committee on Reliability. He is a member of *Environmental Appraisal Committee* (for nuclear power plants in India), Ministry of Environment and Forests, Govt. of India, New Delhi.

ABOUT THE CONTRIBUTORS

Beichelt, Frank received his Diplommathematiker-degree from Friedrich-Schiller-University, Jena in 1966, and the Dr. rer. nat. from Mining Academy Freiberg in 1972. From 1966-1968 he was an assistant at the Technical University, Dresden, Faculty of Forestry, and from 1969-74 head assistant at the Mining Academy Freiberg, Department of Mathematics. He received his Dr. sc. techn.-degree from the University for Transportation & Communication Friedrich List Dresden in 1978. From 1974-79 he was with the Lignite Mining Industry, as a Chief Engineer for Science and Technology. In 1979 he was appointed Associate Professor of Reliability and Maintenance at the University F. List Dresden. From 1981 to 1992 he was a full Professor of Mathematics at the Ingenieurhochschule Mittweida and Head of the Department of Mathematics. In 1990 and 1991 he was a visiting professor at the Bavarian Julius-Maximilian-University Wurzburg, Faculty of Mathematics and Informatics, and the Fachhochschule Wurzburg-Schweinfurt, respectively. He has published about 90 papers on theoretical and practical problems of Operations Research, and 7 monographs on reliability & maintenance theory. He is a member of the IEEE Reliability Society and of the New York Academy of Sciences.

Bonnell, Ronald D. received the B.S. and M.S. degrees in Mechanical Engineering from the University of Kentucky in 1957 and 1961, respectively. He completed further graduate studies in dynamics and control at the University of Tennessee from 1961 to 1965. He held an NSF Traineeship at the Georgia Institute of Technology from 1965 to 1967. He has served as Chairman of the Electrical and Computer Engineering Department and as Director of the Center for Machine Intelligence at the University of South Carolina. His research interests include expert systems, machine learning, intelligent tutoring systems, intelligent control, and logical database design. Professor Bonnell is a senior member of IEEE and a member of AAAI.

Bowles, John B. is an Associate Professor in the Electrical and Computer Engineering Department at the University of South Carolina where he teaches and conducts research in the area of reliable system design. Previously, he was employed by NCR Corporation and Bell Laboratories where he worked on several large system development projects. He holds the B.S. degree in Engineering Science from the University of Virginia, the M.S. degree in Applied Mathematics from the University of Michigan, and the Ph.D degree in Computer Science from Rutgers University. In 1987, he was a visiting scholar at the US Air Force Institute of Technology, Wright-Patterson AFB.

Collas, Gerald is an Engineering System Reliability Manager in R&D Department of Bull S.A., France. Prior to joining Bull S.A. at the end of 1984, he had over twenty years' experience in Quality Assurance, Reliability and Statistics for component industries, computer activities and aerospace and military product companies. He holds an Engineer Diploma from Ecole Superieure d'Electricite. He teaches some courses in reliability engineering at University of Angers, at Quality Master of complex systems in Grenoble and at Mouvement Francais pour la Qualite in Paris. He is a member of IEEE and SRE. He received the annual Stan Ofsthun Award of the Society of Reliability Engineer in 1991.

Deo, Narsingh is the Millican Chair Professor of Computer Science and Director of the Center for Parallel Computation at the University of Central Florida. Prior to accepting the current endowed chair in 1986, he was a Professor of Computer Science at Washington State University (1977-1986), and was the department chairman from 1980 to 1983. Before that he was a

Professor of Electrical Engineering and Computer Science at Indian Institute of Technology, Kanpur from 1971 to 1977, and a Member of Technical Staff at Jet Propulsion Laboratory from 1966 to 1971. He received his Ph.D in Electrical Engineering from Northwestern University in 1965. He was with Burroughs Corporation in Pasadena, CA from 1960 to 1963. He received his MS in 1960 from Caltech and BS from Indian Institute of Science in 1959. He was a Visiting Associate Professor at the University of Illinois, Urbana (1973), Visiting Professor at the Indian Institute of Science (1983), and a member of the Visiting Faculty at IBM's Thomas J. Watson Research Center (1984). A Fellow of IEEE, he has authored four textbooks and over 80 papers in graph theory, discrete optimization algorithms, and parallel computation. He holds a number of patents in computer hardware and is a recipient of NASA's Apollo Achievement Award. Among his other awards are Gold Medal of Patna University, Drake Scholar at Caltech, Governor's Award for Outstanding Contribution to High Tech Research in Florida (1989), and Distinguished Researcher Award-89 at UCF.

Downs, T. received the degrees B.Tech and Ph.D from the University of Bradford, England, in 1968 and 1972, respectively. From 1968 to 1973 he was employed by the Marconi Company in Chelmsford, England, in the Theoretical Sciences Laboratory where his work was concerned with CAD of electrical circuits and systems. In 1973 he was appointed Lecturer in Electrical Engineering at the University of Queensland where he is now Professor and Head of Department. His main research interests are in the testing and reliability assurance of software and in the theory and applications of artificial neural networks. He has published over 100 technical papers and is coauthor of the book Logic Design with Pascal (Van Nostrand Reinhold, 1988). Prof. Downs is a Fellow of the Institution of Engineers, Australia, a Member of the Australian Computer Society, a Member of the International Neural Network Society and a Member of the Institute of Electrical and Electronics Engineers.

Hokstad, Per is cand real in mathematical statistics from the University of Oslo, 1968. He was assistant/associate professor at the Norwegian Institute of Technology (NTH) for 15 years. He joined SINTEF (The Foundation for Scientific and Industrial Research at the Norwegian Institute of Technology) in 1985, and is now a senior research scientist at the Division of Safety and Reliability. He has been involved in assessments of reliability and production regularity for offshore production systems. Mainly he has been working with reliability modelling of computerized process safety systems, with emphasis on dependent failure modelling. Since 1991 he has also been adj. Professor in Safety/Reliability at NTH, and is currently working also with the reliability modelling of maintained systems.

Hura, Gurdeep Singh received his B.E. from GEC, Jabalpur, M.E. and Ph.D from Roorkee University in 1972, 1975 and 1985, respectively. From 1974-84, he worked at Regional Engineering College, Kurukshetra. He held a Post Doctoral Fellowship in the Department of Computer Science, Concordia University, Montreal, Canada during 1984-85 and then he joined the faculty of Computer Science and Engineering of Wright State University, Dayton, Ohio, USA in 1985-86. Currently he is with the School of Applied Sciences, Nanyang Technological University, Singapore. His research interests include Petri net modeling and their applications, computer networks, software engineering, distributed systems, real-time system design. He has authored over eighty five technical papers and guest edited a special issue on Petri Nets and Related Graph Models : Past, Present and Future for Microelectronics & Reliability, Pergamon Press, UK in 1991. He is a senior member of IEEE and a member of ACM. He was Chairman of IEEE Computer Society (Dayton Chapter) from 1988-90. He has organized and chaired numerous special sessions on Petri nets and their applications, software engineering and computer networks. He has presented a number of invited papers in various IEEE conferences.

He is currently writing a book on Computer networks. He is listed in Marquis' Who's Who in Midwest, Who's Who in America, Who's Who in Leading Technology, Who's Who in Engineering Leaders of America, and Who's Who in World. He has also been listed in International Directory of Distinguished Leadership (UK), and American Biographical Institute (USA).

Inagaki, Toshiyuki received his B.S., M.S. and Ph.D degrees in Engineering from Kyoto University in 1974, 1976, and 1979, respectively. From 1979 to 1980 he was Research Associate at the University of Houston, USA. Since 1980 he has been with the University of Tsukuba, Japan. He is currently Associate Professor at the Institute of Information Sciences and Electronics. He visited the University of Texas in 1980 and 1981 to join a research project on computer-aided curriculum analysis. From 1990 to 1991, he stayed at the University of Kassel, Germany, as an Alexander von Humboldt Foundation Research Fellow. At the University of Kassel he was engaged in research projects on man- machine systems.His research areas include system reliability and safety (probabilistic analysis, optimization), mathematical analysis of human-machine interface, reliability and safety related issues on human-computer interaction, decision making under uncertainty, approximate reasoning via Dempster-Shafer and fuzzy theoretic approaches. He is currently an Editor for the Journal of Society for Instrument and Control Engineers, Japan, and an Editor for the Journal of Japan Society for Industrial and Applied Mathematics. He is a Senior Member of IEEE, and a member of several professional societies in systems engineering and computer science.

Inoue, Koichi is a Professor of Control and Systems Engineering in the Department of Aeronautical Engineering, Faculty of Engineering, Kyoto University. He received his BS, MS and Dr.Eng. degrees all in Applied Mathematics and Physics from Kyoto University in 1963, 1965 and 1968, respectively. From 1969 to 1986, he was an Associate Professor in the Department of Precision Mechanics, Kyoto University. From 1971 to 1973, he was with the Department of Chemical Engineering, University of Houston, where he began working on systems reliability and safety. His fields of interest include systems reliability and safety, risk analysis, system optimization, human interface, knowledge engineering, and so on.

Kohda, Takehisa received his BS, MS and Ph.D degrees all in Precision Mechanics from Kyoto University in 1978, 1980, 1983, respectively. From 1983 to 1988, he was a researcher at National Mechanical Engineering Laboratory. He is an Associate Professor in the Department of Aeronautical Engineering, Faculty of Engineering, Kyoto University. During 1985-1986, he was with the Department of Chemical Engineering, University of Houston. His interests lie in systems safety and reliability, risk analysis, and systems analysis.

Koutras, Markos V. received his B.S. in Mathematics in 1975, M.Sc. in Computer Science and Operations Research in 1981 and Ph.D. in Statistics in 1983 from the University of Athens, the subject of his thesis being the Study of Classification (Discrimination) Procedures for Elliptical Multivariate Distributions. Now he is an Assistant Professor in the University of Athens (Section of Statistics and Operations Research). His research interests include Statistics, Applied Probability Theory and Combinatorics.

Medidi, Muralidhar received his M.Tech in Computer Engineering from Indian Institute of Technology, Kharagpur in 1986. He was a Lecturer in Computer Science at the Jawaharlal Nehru Technological University, Kakinada (1986-1987), and was a Project Engineer in Aeronautical Development Agency, Bangalore until 1988. Currently he is a doctoral candidate at the University of Central Florida. His research interests include algorithmic graph theory and designing data structures for parallel processing. He is a student member of the ACM and IEEE.

Onisawa, Takehisa is an Associate Professor at Institute of Engineering Mechanics in University of Tsukuba. He received his B.E. and M.E. in Control Engineering and his D.E. in Systems Science from Tokyo Institute of Technology in 1975, 1977 and 1986, respectively. He was a Research Associate at the Department of Systems Science in Tokyo Institute of Technology, 1977-1986. He was a Lecturer, 1986-1989 and an Associate Professor, 1989-1992 at the Department of Basic Engineering in Kumamoto University. Currently he is an Associate Professor in University of Tsukuba. He received the Hashimoto Award from Japan Ergonomics Association in 1987. His research interests include fuzzy sets theory and its applications. Currently his research interests are applications of fuzzy sets theory to reliability analysis, human interface, human behavior modelling and human communication modelling. He is a member of International Fuzzy Systems Association (IFSA), The Institute of Electrical and Electronics Engineers (IEEE), Japan Society for Fuzzy Theory and Systems (SOFT), Japan Ergonomics Association, The Society of Instrument and Control Engineers of Japan (SICE) and The Institute of Electronics, Information and Communication Engineers of Japan (IEICE). He has served as a reviewer to Fuzzy Sets and Systems (IFSA) and Journal of SOFT.

Papastavridis, Stavros G. received his B.S. in Mathematics from the University of Athens, and his M.A. and Ph.D. in Mathematics from Princeton University in 1974. He is Professor in the Department of Mathematics of the University of Athens. He has been Vice President of the Greek Mathematical Society, member of the National Committee of University Entrance Examination, and author of the textbooks in Mathematics used in the Greek high schools. His research interests cover areas of Algebraic Topology and Applications of Probability.

Rao, S.S. is currently a Professor in the School of Mechanical Engineering at Purdue University, West Lafayette, Indiana, USA. He received his Ph.D from Case Western Reserve University, Cleveland, Ohio. He has several years of experience as a professor and researcher at Purdue University; San Diego State University; Indian Institute of Technology, Kanpur, India and Langley Research Center, Hampton, Virginia. He served as major professor for a number of M.S. and Ph.D students. He has several years of experience as a consultant to industries. He published more than 125 technical papers in internationally reputed journals and presented numerous papers at conferences. He is the author of the following books: 1. Optimization: Theory and Applications (Wiley, New York, 2nd Edition, 1984), 2. The Finite Element Method in Engineering (Pergamon Press, Oxford, England, 2nd Edition, 1989), 3. Mechanical Vibrations (Addison-Wesley, Reading, 2nd Edition, 1990), and 4. Reliability-Based Design (McGraw-Hill, New York, 1992). Dr. Rao edited several conference proceedings and served as the Conference Chairman and Papers Review Chairman for the ASME Design Automation Committee and as an Associate Editor for the ASME Journal of Mechanisms, Transmissions, and Automation in Design. Currently he is on the editorial boards of the journals Engineering Optimization and Reliability Engineering and System Safety. He was the winner of a gold medal and several prizes for academic excellence and won awards for outstanding research, best technical paper and outstanding teaching. He is very active in conducting research on several federal and industry sponsored projects. His current research interests are in the areas of mechanical reliability, fuzzy systems, engineering optimization, design for manufacturability, and intelligent structures.

Rushdi, Ali M. received his B.Sc. degree with honours in Electrical Engineering (Electronics and Communications) from Cairo University, Cairo, Egypt in 1974 and MS and Ph.D. degrees in Electrical Engineering from the University of Illinois at Urbana-Champaign, USA, in 1977 and 1980, respectively. In 1974, he was appointed Instructor and Demonstrator at the Electrical Engineering Department of Cairo University. From 1976 to 1980, he was Research Assistant at the Electrical Engineering Department of the University of Illinois. Since 1980, he has been with King Abdul Aziz University in Jeddah, Saudi Arabia, where he is now Professor of Electrical and Computer Engineering and Head of the Computer Engineering Group. At King Abdul Aziz University, he has structured and taught a variety of graduate and undergraduate courses and supervised several theses. He has published in the areas of reliability analysis, switching theory, electromagnetics, circuit theory and engineering education. He presently serves as a reviewer for many international journals including the IEEE Transactions on Reliability, Microelectronics and Reliability and the IEEE Transactions on Instrumentation and Measurements. He is a senior member of the IEEE and six of its societies. He is also a member of the Engineers Syndicate of Egypt, the Saudi Computer Society, Eta Kappu Nu and Phi Kappa Phi.

Russomanno, David J. is working as an Applications Consultant with Intergraph Corporation. He received the B.E.E. in Electrical Engineering from Auburn University and the M.E. in Computer Engineering from the University of South Carolina in 1986 and 1989, respectively. He is currently a candidate for the Ph.D degree in Computer Engineering at the University of South Carolina. Prior to joining Intergraph, he was employed by Pratt and Whitney Aircraft and Michelin Tire Corporation as a Design Engineer. He is a member of Eta Kappu Nu, Sigma Pi Sigma, and the IEEE.

Sharit, Joseph is currently an Associate Professor in the Department of Industrial Engineering at the State University of New York at Buffalo. He joined the department in 1985 after earning his Masters and Ph.D degrees in Industrial Engineering from Purdue University, specializing in human factors. Professor Sharit teaches courses in human reliability analysis and system safety, human-machine systems, human factors research methodology, engineering statistics, and decision analysis. His research interests include human reliability analysis and human-computer interaction, with the focus in the latter area primarily on computer aiding and decision support in manufacturing industries and analysis and design of computer-based jobs for older populations. Professor Sharit is a member of the Human Factors and Ergonomics Society, the Institute of Industrial Engineers, and IEEE.

Shooman, Andrew M. received his SB in Mathematics and Computer Science from the Massachusetts Institute of Technology in 1986, his MS in Computer Science from the New York University Courant Institute of Mathematical Sciences in 1989, and his Ph.D in Computer Science from Polytechnic University in Brooklyn, New York in 1992. He has been working in industry since 1980. From 1980 to 1982, he worked for Pall Corporation in Glen Cove, New York on automated data collection and report generation for hydraulic filter testing. From 1983 to 1988, he worked for Hazeltine Corporation in Greenlawn, New York in the artificial intelligence group of the research labs. From 1989 to 1990, he worked for Grumman Data Systems in Woodbury, New York on LANDSAT multispectral image data analysis. From 1990 to 1991, he worked in the Network Design Tools group at the IBM T.J. Watson Research Center in Hawthorne, New York. He is presently a Senior Engineer at Codex Corporation, a subsidiary of Motorola Corporation in Canton, Massachusetts, working on network design and analysis. His research interests include communications networks, reliability analysis, graph theory, and artificial intelligence. His hobbies include astronomy, meteorology, music, bicycling, canoeing, and swimming.

Trivedi, Kishor S. received the B.Tech. degree from the Indian Institute of Technology, Bombay and M.S. and Ph.D degrees in computer science from the University of Illinois, Urbana- Champaign. He is the author of text entitled, Probability and Statistics with Reliability, Queuing and Computer Science Applications, published by Prentice-Hall. His research interests are in computing system reliability and performance evaluation. He has lectured and published extensively on these topics. He is a Fellow of the Institute of Electrical and Electronics Engineers. He is a Professor of Electrical Engineering at Duke University, Durham, NC. He has served as a Principal Investigator on various AFOSR, ARO, Burroughs, IBM, DEC, NASA, NIH, ONR, NSWC, Boeing, Union Switch and Signals, NSF, and SPC funded projects and as a consultant to industry and research laboratories. He was an Editor of the IEEE Transactions on Computers from 1983-1987. He is a co-designer of HARP, SAVE, SHARPE and SPNP modeling packages. These packages have been widely circulated.

Veeraraghavan, Malathi received the B.Tech. degree in Electrical Engineering from Indian Institute of Technology, Madras and M.S. and Ph.D. degrees in Electrical Engineering from Duke University, Durham. Her research interests include the design of fault-tolerant systems and networks, reliability and performance modeling, and signaling protocols in telecommunications networks. She is a member of the IEEE, IEEE Communications Society, and IEEE Reliability Society. She is currently a member of technical staff at AT&T Bell Laboratories in Holmdel, NJ. She is also an Associate Editor for IEEE Transactions on Reliability.

SUBJECT INDEX

A

abstract Petri nets 352
accident causation mechanism 266
accumulation problems 261
active errors 270
active failures 271
advanced graph reductions 129
-Δ-Y reductions 129
-polygon-chain reductions 132
ALARA 262
algebraic properties 514
-complement operation 514
-max operation 514
-min operation 514
-of fuzzy sets 514
Algorithm
-AMG 15
-CAREL 61
-for cut sets 670
-for path sets 670
-for simple path sets 673
-GKG-VT 48
-KDH 88 58
-PROB 55
alpha factor model 429
alpha-cut 507
analysis generator 647
aperture controller 290
-with commands 292
-without commands 292
appropriate graph reduction 138
-degree - 3 series 141
-k_4 contraction 138
-k_n contraction 140
- Y-Δ transformation 141
APTE 6
AR algorithm 210
attitude 277
attribute values 288
- aperture 288
- flow rate 288
- generation rate 288
automated fault tree synthesis 286

B

basic failure 269
basic graph reductions 123
-parallel 123
-pendant 124
-series 123
BCS FMEA assistant 635
behavioured modeller 645
belief function 542
belief structures 595
-Bayesian 598
-combination of 600
-complementary 596
Bell number (of u) 80
BEST 3
beta factor model 427
beta factor method 21
binomial failure rate 21
binomial failure rate model 432
Birnbaum's importance factor 692
black board model 641
black box CAD tools 645
Boolean product 43
Butterfly parallel machine, the 166

C

C-factor model, the 427
cancer fatality QDO 264
cascade failures 269
cascading failures 422
catastrophic accidents-goals for 264
causal models 646
causal relations 607
-necessary 607
-possible 607
causal tree approach 318
causal trees 18
-contribution 320
-nodes and branches 320
cause-effect diagrams 438
CC base events 421
CC component group 421
chain recursion 204
change control 283
circular lattice system 17
Clayton model 434

cognitive environment simulation 397
cognitive errors 270
coherent system 191
coloured Petri nets 348
commission errors 270
commission event trees 374
common cause failures 21,579
common cause initiating events 415
common load model 434
compensator, the 645
complement set 507
complementary cumulative profile 251
complete graphs 153
component models 318,319
composite recursion 204
computer simulation 387
concurrent engineering 3
conditional possibility 548
consecutive k-out-of-n system 17
-circular 229
-closed formulae 235
-:F system 17
-:G system 17
consequence mitigation 285
consistency principle 548
constant belief function 542
constant fatality model 265
constant likelihood model 265
convex fuzzy sets 507
cost benefit guideline 264
coupling factors 418
cross over point 507
cube 43
cube graph 156
cut set 14
cut set enumeration 120

D

data base 626
de minimis risk 262
decomposition failure 80
decomposition technique 15
defences 418
degree 2 - reduction 96
delta - star reduction 103
demand failure 270
Dempster-Shafer theory 587,588
dependability 1
dependability engineering 8
dependency 569
-considerations 569
-of type I 572
-of type II 574
dependent failure 21
dependent failure 270,412
design 271
design input 644
detection analyser 645
development rules 297
-equipment suspected to equipment failure 299
-equipment suspected to flow triple 299
-flow triple to equipment suspected 297
-flow triple to flow triple 297
digraphs 18,314
-nodes and edges 314
-system diagraphs 315
discourse manager 644
disjoint product method 197
domination 191
dual component structure 229
duality 190

E

edge disjoint subgraphs 78
edge factoring 126,127
edge set 76
elementary reduction 97
empty set 507
enabling events 270
error of commissions 20
ESCAF 679
-architecture 680
-capabilities 692
-performance 689
-physical dimensions 681
-principle 682
estimate of reliability 564
-AND or OR operations 567
-by natural language 565
-through numerical values 566
evaluation of disturbances 326
event development rules 297
event layer 273
event-likelihood model 273
events 287
-possibilities of 297
-types of 297
evidence theory 24
exact formulae 235
exhaustive state enumeration 198

expert systems 26

F

factor of safety 479,480
factoring algorithm 40,41
factoring theorem 15
factorization 99
factorization formula 78
FADES 634
FAILDIAG 7
failed subgraph 134
failure modes (of mechanical systems) 482
-corrosion failure 483
-creep failure 483
-fatigue failure 483
-instability failure 484
-static failure 483
-wear failure 484
failure prevention 283
farmer curves 251
FAST 6
fault detection 362
fault diagnosis 609
fault tree analysis 17,249
fault tree construction 316,322
-feedback loop nodes 322
-feedforward loop nodes 323
-non-loop nodes 322
FIABEX 635
flow attribute 288
flow triple 289
flows 287
FMEA 629
FMECA 3,7
form 47
form I 47
form II 48
form III 48
FT generation
-recursive algorithm 300
functional failures 6,269
functional modeller 645
fuzzy cardinality 509
fuzzy events 545
fuzzy events-probability of 545
fuzzy fault tree analysis 563
-by failure possibility 564
-by failure probability 563
fuzzy measures 540
-Dirac measure 549
-lambda measure 514
-non fuzzy subsets 540
-possibility measure 541
fuzzy numbers 525
-negative 526
-pi-number 526
-positive 526
-properties of 531
-representation of 532
-s-number 527
-s/z-number 527
-types of 526
-z-number 527
-z/s-number 527
-zero 526
fuzzy operators 537
-t-conorms 537
-t-norms 537
fuzzy probability 544,559
fuzzy relations 519
fuzzy set theoretic operations 509
-algebraic product 511
-algebraic sum 511
-bounded difference 511
-bounded product 511
-bounded sum 511
-containment 509
-equality 509
-intersection 509
-product 509
-union 510
fuzzy sets 503
-addition 535
-arithmetic operations 535
-cardinality of 508
-cartesian product of 517
-concentration operation 508
-dilation operation 508
-extended min and max operations 536
-extension principle 518
-inverse 536
-multiplication by a scalar 535
-power of 508
-product 536
-quotient 536
-theory 24
fuzzy sets of type 2 521
-complement 522
-intersection 521
-operations on 521

-union 522

G

general performance guideline 264
generalized reduction equations 107
generally accepted knowledge 325
generation rate controller 294
geometric mean model 266
geometric reliability models 33
global reliability 656
grade of membership 504
graph reduction algorithms 142
-algorithms 145
-complexity analysis 148
-heuristics 142
grids 154
GUARDBAND 3
HCR model 381
HEART 379
height 506
hidden process review 283
hierarchical goals for PRAM 259
hierarchical quorum consensus algorithm 186
high level Petri nets 348
human error probability 371
-basic HEP 37
-conditional 37
-nominal HEP 37
human performance models 404
human reliability analysis 369
hypergeometric distribution model 459

I

impact vector 438
inclusion - exclusion algorithm 40,41
inclusion - exclusion method 194
independent failures 270
individual risk 257
induced subgraphs 134
inference engine 626
initial failures 269
initiating events 270
interactive method 327
intercomponent dependencies 415
interfacing failures 269
intermediate failures 269
intersystem dependencies 415
inverted sub product 43

K

k-nodes 76
k-out-of-n : F system 191
k-out-of-n : G system 191
k-out-of-n systems 16,191
k-out-of-n voting logic 186
k-trees 156
knowledge base 626
Kolmogorov distance 463,465

L

lapse 271
latent errors 270
lattice systems 17
-circular 17
-linear 17
Lethela's CAF tool 631
likelihood layer 273
likelihood layer contributions 276
linear reliability models 22
LOCA 271
lower bound goals 259
lower bounds 90

M

m-consecutive-k-out-of-n systems 229
-circular 229
majority voting algorithm 186
malfunction simulator 6
malfunctional knowledge 6
MAPPS 388
matrix method 15
MAUD 377
mechanical failures 269
MERISE 6
MGL model 428
mini fault tree approach 18
minimal cutsets 40,323,67
minimal path sets 40,670
minproduct 43
mission expert 645
mistake 271
modeling 345
modeling-tools for 345
modularization 17
monotone 190
multi-valued logic 511
multiple dependent failure fraction method 21

multiple fault diagnosis problem 587
multiple variable inversion 15,42

N

N-modular redundancy system 187
negative feedback loops 317
negative feedforward loops 317
network 13
network decomposition 78
network nodes 39
network reduction 95
network reliability 75
neural network 18
non-series parallel configuration 558
non-shock models 428
normal activities 263
normal activities-goals for 263
normalized FS 506
NP-completeness 122
NURP model 457

O

objective likelihood 251
omission errors 270
operators 49
  -@j operator 51
  -CMB(*) operator 65
  -COM(\) operator 62
  -df operator 62
  -exclusive sharp ($) operator 52,56
  -GEN(⊕) operator 66
  -Miller cube (#) operator 56
  -percent (%) operator 49
  -pl operator 62
  -RED(/) operator 63
optimization 240

P

parallel algorithms 174
  -Dotson and Gobein 175
  -factoring 174
  -reduce and partition 176
parallel processing algorithms 16
parallel reduction 96
parallel system 557
parametric models 424
partial beta factor model 427
partition 80
partition lattice 80
Pascal's triangle 196
passive failures 271
path set 13
performance evaluation 473
Petri nets 19,339
  -classical 340
  -definition 340
  -extended 351
  -features of 342
  -generalised stochastic 19,350
  -generalized timed 19
  -Marked 341
  -stochastic 19,350
  -stochastic activity networks 351
  -timed 19
pivoted decomposition 41
pivoting 191
plant performance goal 277
plausibility measure 542
PLHR model 460
polygon-to-chain reductions 98
polygons 98
population risk 257
population size 265
possibility 547
possibility measure 542
  -of a fuzzy event 549
  -of a non-fuzzy event 547
predicate nets 347
prequential method 466
primary failures 269
proactive process control 5
probabilistic graph 13
probabilistic risk assessment 249
probability integral transform 462
progression prevention 284
prompt fatality QDO 264
propositions 47

Q

qualitative analysis 423
quality assurance 281
quality function 3
quantitative screening 423

R

r-within-consecutive k-out-of-n system 229

-circular 229
random failures 269
random graphs 76
random probability shock model 433
Rasmussen's frameworks 388
Rayleigh model 22
Reason's human error model 391
recovery failures 270
recursive formulae 234
recursive technique 200
reduction algorithm 110
reduction constants 95
reduction equations 100
reduction technique 99
regulatory cut-off level 265
relations 45
RELDIAG 7
reliability analyser 25,26,633
-basic model 656
-basic principle 658
-microprocessor based 677
reliability
-all terminal 13,40
-k terminal 13,40
-terminal pair 13
-two terminal 87,119
RELIASEP 6
replacement graphs 95
risk 250
-assessment 255
-aversion goals 264
-management 255,279
-meta uncertainty 254
-profile 250
-uncertainty 254
root cause 416
routine errors 270
run failures 270

S

s-p complex 191
s-t connectedness 41
S.ESCAF 690
SADT 6
safety goals 259
safety knowledge 277
safety margins 283
safety system unavailability 276
search technique 664
semantic networks 294
sensitivity coefficient factor 693
separating node set 78
SEQ.SIM simulator 691
sequential algorithms 167
-Dotson and Gobein algorithm 167
-Page and Perry algorithms 167
-reduce and partition algorithm 167
series reduction 96
series system 556
shell 627
shock model 21,431
Shooman's algorithms 145,146
signal flow graph 196
simple event tree 276
SIMTEST 7
simulation expert 645
siting 271
slim 377
slip 270
SOFIA 7
software reliability modelling 22
software reliability models 447
-Jelinski-Moranda model 447
-Littlewood-Verrall model 449
-nonhomogeneous Poisson process model 451
square root bounding method 425
staggered testing 420
standardization 283
state enumeration techniques 42
Stirling numbers of the second kind 80
stochastic network 97
-sp-complex 97
-sp-reducible 97
stochastic networks 75,76
STRIFE 3
structure functions 40
subcube 43
-coordinates of 44
subjective likelihood 251
sum-of-disjoint products algorithm 40,41
support 506
symmetric switching function 205
system analyser 644
system classifier 644
system knowledge 324

T

target nodes (see k-nodes)
task analysis 407
terminal nodes (see k-nodes)

terminal pair reliability 654
-cut set enumeration 6
-factoring theorem 6
-path set enumeration 6
-state enumeration 6
ternary recursive algorithm 204
TESCO 379
THERP 370
tie-set enumeration 120
time evolution 693
time reliability correlation 381
timed Petri nets 349
totally amenable system 191
transformation techniques 15
transient 271
truncation 17
two-out-of-three configuration 559
two-point separating node sets 86
two-terminal network reliability 39,40

U

u-plot 462
uniproduct 43
universe of discourse 503
untimed Petri nets 349
upper bound goals 259

V

valuation set 504
version manager 646
vertex factoring 127
vertex merging 136
voting 186

X

XFMEA 639